Springer Texts in Statistics

Series Editors:

G. Casella
S. Fienberg
I. Olkin

Springer Texts in Statistics

For other titles published in this series, go to
www.springer.com/series/417

Alan Julian Izenman

Modern Multivariate Statistical Techniques

Regression, Classification, and Manifold Learning

 Springer

Alan J. Izenman
Department of Statistics
Temple University
Speakman Hall
Philadelphia, PA 19122
USA
alan@temple.edu

ISSN: 1431-875X
ISBN: 978-0-387-78188-4   e-ISBN: 978-0-387-78189-1
DOI: 10.1007/978-0-387-78189-1

Library of Congress Control Number: 2008928720.

springer.com   **3 3001 00954 6580**

*This book is dedicated
to the memory of my parents,
Kitty and Larry,*

*and to my family,
Betty-Ann and Kayla*

Preface

Not so long ago, multivariate analysis consisted solely of linear methods illustrated on small to medium-sized data sets. Moreover, statistical computing meant primarily batch processing (often using boxes of punched cards) carried out on a mainframe computer at a remote computer facility. During the 1970s, interactive computing was just beginning to raise its head, and exploratory data analysis was a new idea. In the decades since then, we have witnessed a number of remarkable developments in local computing power and data storage. Huge quantities of data are being collected, stored, and efficiently managed, and interactive statistical software packages enable sophisticated data analyses to be carried out effortlessly. These advances enabled new disciplines called data mining and machine learning to be created and developed by researchers in computer science and statistics.

As enormous data sets become the norm rather than the exception, statistics as a scientific discipline is changing to keep up with this development. Instead of the traditional heavy reliance on hypothesis testing, attention is now being focused on information or knowledge discovery. Accordingly, some of the recent advances in multivariate analysis include techniques from computer science, artificial intelligence, and machine learning theory. Many of these new techniques are still in their infancy, waiting for statistical theory to catch up.

The origins of some of these techniques are purely algorithmic, whereas the more traditional techniques were derived through modeling, optimization, or probabilistic reasoning. As such algorithmic techniques mature, it becomes necessary to build a solid statistical framework within which to embed them. In some instances, it may not be at all obvious why a particular technique (such as a complex algorithm) works as well as it does:

> *When new ideas are being developed, the most fruitful approach is often to let rigor rest for a while, and let intuition reign — at least in the beginning. New methods may require new concepts and new approaches, in extreme cases even a new language, and it may then be impossible to describe such ideas precisely in the old language.*

> — Inge S. Helland, 2000

It is hoped that this book will be enjoyed by those who wish to understand the current state of multivariate statistical analysis in an age of high-speed computation and large data sets. This book mixes new algorithmic techniques for analyzing large multivariate data sets with some of the more classical multivariate techniques. Yet, even the classical methods are not given only standard treatments here; many of them are also derived as special cases of a common theoretical framework (multivariate reduced-rank regression) rather than separately through different approaches. Another major feature of this book is the novel data sets that are used as examples to illustrate the techniques.

I have included as much statistical theory as I believed is necessary to understand the development of ideas, plus details of certain computational algorithms; historical notes on the various topics have also been added wherever possible (usually in the Bibliographical Notes at the end of each chapter) to help the reader gain some perspective on the subject matter. References at the end of the book should be considered as extensive without being exhaustive.

Some common abbreviations used in this book should be noted: "iid" means *independently and identically distributed*; "wrt" means *with respect to*; and "lhs" and "rhs" mean *left-* and *right-hand side*, respectively.

Audience

This book is directed toward advanced undergraduate students, graduate students, and researchers in statistics, computer science, artificial intelligence, psychology, neural and cognitive sciences, business, medicine, bioinformatics, and engineering. As prerequisites, readers are expected to have had previous knowledge of probability, statistical theory and methods, multivariable calculus, and linear/matrix algebra. Because vectors and matrices play such a major role in multivariate analysis, Chapter 3 gives the matrix notation used in the book and many important advanced concepts in matrix theory. Along with a background in classical statistical theory and methods, it would also be helpful if the reader had some exposure to Bayesian ideas in statistics.

There are various types of courses for which this book can be used, including data mining, machine learning, computational statistics, and for a traditional course in multivariate analysis. Sections of this book have been used at Temple University as the basis of lectures in a one-semester course in applied multivariate analysis to statistics and graduate business students (where technical derivations are skipped and emphasis is placed on the examples and computational algorithms) and a two-semester course in advanced topics in statistics given to graduate students from statistics, computer science, and engineering. I am grateful for their feedback (including spotting typos and inconsistencies).

Although there is enough material in this book for a two-semester course, a one-semester course in traditional multivariate analysis can be drawn from the material in Sections 1.1–1.3, 2.1–2.3, 2.5, 2.6, 3.1–3.5, 5.1–5.7, 6.1–6.3, 7.1–7.3, 8.1–8.7, 12.1–12.4, 13.1–13.9, 15.4, and 17.1–17.4; additional parts of the book can be used as appropriate.

Software

Software for computing the techniques described in this book is publicly available either through routines in major computer packages or through download from Internet websites. I have used primarily the R, S-PLUS, and MATLAB packages in writing this book. In the Software Packages section at the ends of certain chapters, I have listed the relevant R/S-PLUS routines for the respective chapter as well as the appropriate toolboxes in MATLAB. I have also tried to indicate other major packages wherever relevant.

Data Sets

The many data sets that illustrate the multivariate techniques presented in this book were obtained from a wide variety of sources and disciplines and will be made available through the book's website. Disciplines from which the data were obtained include astronomy, bioinformatics, botany, chemometrics, criminology, food science, forensic science, genetics, geoscience, medicine, philately, physical anthropology, psychology, soil science, sports, and steganography. Part of the learning process for the reader is to become familiar with the classic data sets that are associated with each technique. In particular, data sets from popular data repositories are used to compare and contrast methodologies. Examples in the book involve small data sets (if a particular point or computation needs clarifying) and large data sets (to see the power of the techniques in question).

Exercises

At the end of every chapter (except Chapter 1), there is a number of exercises designed to make the reader (*a*) relate the problem to the text and fill in the technical details omitted in the development of certain techniques,

(b) illustrate the techniques described in the chapter with real data sets that can be downloaded from Internet websites, and (c) write software to carry out an algorithm described in the chapter. These exercises are an integral part of the learning experience. The exercises are not uniform in level of difficulty; some are much easier than others, and some are taken from research publications.

Book Website

The book's website is located at:

http://astro.ocis.temple.edu/~alan/MMST

where additional materials and the latest information will be available, including data sets and R and S-PLUS code for many of the examples in the book.

Acknowledgments

I would like to thank David R. Brillinger, who instilled in me a deep appreciation of the interplay between theory, data analysis, computation, and graphical techniques long before attention to their connections became fashionable.

There are a number of people who have helped in the various draft stages of this book, either through editorial suggestions, technical discussions, or computational help. They include Bruce Conrad, Adele Cutler, Gene Fiorini, Burt S. Holland, Anath Iyer, Vishwanath Iyer, Joseph Jupin, Chuck Miller, Donald Richards, Cynthia Rudin, Yan Shen, John Ulicny, Allison Watts, and Myra Wise. Special thanks go to Richard M. Heiberger for his invaluable advice and willingness to share his expertise in all matters computational. Thanks also go to Abraham "Adi" Wyner, whose conversations at Border's Bookstore kept me fueled literally and figuratively. Thanks also go to the reviewers and to all the students who read through various drafts of this book. Individuals who were kind enough to allow me to use their data or with whom I had e-mail discussions to clarify the nature of the data are acknowledged in footnotes at the place the data are first used. I would also like to thank the *Springer* editor John Kimmel, who provided help and support during the writing of this book, and the *Springer* LaTeX expert Frank Ganz for his help.

Finally, I thank my wife Betty-Ann and daughter Kayla whose patience and love these many years enabled this book to see the light of day.

Alan Julian Izenman
Philadelphia, Pennsylvania
April 2008

Contents

1
Introduction and Preview

1.1 Multivariate Analysis

This book invites the reader to learn about multivariate analysis, its modern ideas, innovative statistical techniques, and novel computational tools, as well as exciting new applications.

The need for a fresh approach to multivariate analysis derives from three recent developments. First, many of our classical methods of multivariate analysis have been found to yield poor results when faced with the types of huge, complex data sets that private companies, government agencies, and scientists are collecting today; second, the questions now being asked of such data are very different from those asked of the much-smaller data sets that statisticians were traditionally trained to analyze; and, third, the computational costs of storing and processing data have crashed over the past decade, just as we see the enormous improvements in computational power and equipment. All these rapid developments have now made the efficient analysis of more complicated data a lot more feasible than ever before.

Multivariate statistical analysis is the simultaneous statistical analysis of a collection of random variables. It is partly a straightforward extension

A.J. Izenman, *Modern Multivariate Statistical Techniques*,
doi: 10.1007/978-0-387-78189-1_1,
© Springer Science+Business Media, LLC 2008

of the analysis of a single variable, where we would calculate, for example, measures of location and variation, check violations of a particular distributional assumption, and detect possible outliers in the data. Multivariate analysis improves upon separate univariate analyses of each variable in a study because it incorporates information into the statistical analysis about the relationships between all the variables.

Much of the early developmental work in multivariate analysis was motivated by problems from the social and behavioral sciences, especially education and psychology. Thus, factor analysis was devised to provide a statistical model for explaining psychological theories of human ability and behavior, including the development of a notion of general intelligence; principal component analysis was invented to analyze student scores on a battery of different tests; canonical variate and correlation analysis had a similar origin, but in this case the relationship of interest was between student scores on two separate batteries of tests; and multidimensional scaling originated in psychometrics, where it was used to understand people's judgments of the similarity of items in a set.

Some multivariate methods were motivated by problems in other scientific areas. Thus, linear discriminant analysis was derived to solve a taxonomic (i.e., classification) problem using multiple botanical measurements; analysis of variance and its big brother, multivariate analysis of variance, derived from a need to analyze data from agricultural experiments; and the origins of regression and correlation go back to problems involving heredity and the orbits of planets.

Each of these multivariate statistical techniques was created in an era when small or medium-sized data sets were common and, judged by today's standards, computing was carried out on less-than-adequate computational platforms (desk calculators, followed by mainframe batch computing with punched cards). Even as computational facilities improved dramatically (with the introduction of the minicomputer, the hand calculator, and the personal computer), it was only recently that the floodgates opened and the amounts of data recorded and stored began to surpass anything previously available. As a result, the focus of multivariate data analysis is changing rapidly, driven by a recognition that fast and efficient computation is of paramount importance to its future.

Statisticians have always been considered as partners for joint research in all the scientific disciplines. They are now beginning to participate with researchers from some of the subdisciplines within computer science, such as pattern recognition, neural networks, symbolic machine learning, computational learning theory, and artificial intelligence, and also with those working in the new field of bioinformatics; together, new tools are being devised for handling the massive quantities of data that are routinely collected in business transactions, governmental studies, science and medical research, and for making law and public policy decisions.

We are now seeing many innovative multivariate techniques being devised to solve large-scale data problems. These techniques include nonparametric density estimation, projection pursuit, neural networks, reduced-rank regression, nonlinear manifold learning, independent component analysis, kernel methods and support vector machines, decision trees, and random forests. Some of these techniques are new, but many of them are not so new (having been introduced several decades ago but virtually ignored by the statistical community). It is because of the current focus on large data sets that these techniques are now regarded as serious alternatives to (and, in some cases, improvements over) classical multivariate techniques.

This book focuses on the areas of regression, classification, and manifold learning, topics now regarded as the core components of data mining and machine learning, which we briefly describe in this chapter. It is important to note here that these areas overlap a great deal in content and methodology: what is one person's data-mining problem may be another's machine-learning problem.

1.2 Data Mining

1.2.1 From EDA to Data Mining

Although the revolutionary concept of *exploratory data analysis (EDA)* (Tukey, 1977) changed the way many statisticians viewed their discipline, emphasis in EDA centered on quick and dirty methods (using pencil and paper) for the visualization and examination of small data sets. Enthusiasts soon introduced EDA topics into university (and high school) courses in statistics. To complete the widespread acceptance and utility of John Tukey's exploratory procedures and his idiosyncratic nomenclature, EDA techniques were included in standard statistical software packages. Nevertheless, despite the available computational power, EDA was still perceived as a collection of small-sample, data-analytic tools.

Today, measurements on a variety of related variables often produce a data set so large as to be considered unwieldy for practical purposes. Such data now often range in size from moderate (say 10^3 to 10^4 cases) to large (10^6 cases or more). For example, billions of transactions each year are carried out by international finance companies; Internet traffic data are described as "ferocious" (Cleveland and Sun, 2000); the Human Genome Project has to deal with gigabytes (2^{30} ($\sim 10^9$) bytes) of genetic information; astronomy, the space sciences, and the earth sciences have terabytes (2^{40} ($\sim 10^{12}$) bytes) and soon, petabytes (2^{50} ($\sim 10^{15}$) bytes), of data for processing; and remote-sensing satellite systems, in general, record many gigabytes of data each hour. Each of these data sets is incredibly large and complex, with millions of observations being recorded on huge numbers of variables.

Furthermore, governmental statistical agencies (e.g., the Federal Statistical Service in the United States, the National Statistical Service in the United Kingdom, and similar agencies in other countries) are accumulating greater amounts of detailed economic, labor, demographic, and census information than at any time in the past. The U.S. census file based solely on administrative records, for example, has been estimated to be of size at least 10^{12} bytes (Kirkendall, 1997). Other massive data sets (e.g., crime data, health-care data) are maintained by other governmental agencies.

The availability of massive quantities of data coupled with enormous increases in computational power for relatively low cost has led to the creation of a whole new activity called *data mining*. With massive data sets, the process of data mining is not unlike a gigantic effort at EDA for "infinite" data sets. For many companies, their data sets of interest are so large that only the simplest of statistical computations can be carried out. In such situations, data mining means little more than computing means and standard deviations of each variable; drawing some bivariate scatterplots and carrying out simple linear regressions of pairs of variables; and doing some cross-tabulations. The level of sophistication of a data mining study depends not just on the statistical software but also on the computer hardware (RAM, hard disk, etc.) and database management system for storing the data and processing the results.

Even if we are faced with a huge amount of data, if the problem is simple enough, we can sample and use standard exploratory and confirmatory methods. In some instances, especially when dealing with government-collected data, sampling may be carried out by the agency itself. Census data, for example, is too big to be useful for most users; so, the U.S. Census Bureau creates manageable public-use files by drawing a random sample of individuals from the full data set and either removes or masks identifying information (Kirkendall, 1997).

In most applications of data mining, there is no à priori reason to sample. The entire population of data values (at least, those with which we would be interested) is readily available, and the questions asked of that data set are usually exploratory in nature and do not involve inference. Because a data pattern (e.g., outliers, data errors, hidden trends, credit-card fraud) is a local phenomenon, possibly affecting only a few observations, sampling, which typically reduces the size of the data set in drastic fashion, may completely miss the specifics of whatever pattern would be of special interest.

Data mining differs from classical statistical analysis in that statistical inference in its hypothesis-testing sense may not be appropriate. Furthermore, most of the questions asked of large data sets are different from the classical inference questions asked of much smaller samples of data. This is not to say that sampling and subsequent modeling and inference have no role to play when dealing with massive data sets. Sampling, in fact, may be appropriate in certain circumstances as an accompaniment to any detailed data exploration activities.

1.2.2 What Is Data Mining?

It is usual to categorize data mining activities as either *descriptive* or *predictive*, depending upon the primary objective:

Descriptive data mining: Search massive data sets and discover the locations of unexpected structures or relationships, patterns, trends, clusters, and outliers in the data.

Predictive data mining: Build models and procedures for regression, classification, pattern recognition, or machine learning tasks, and assess the predictive accuracy of those models and procedures when applied to fresh data.

The mechanism used to search for patterns or structure in high-dimensional data might be manual or automated; searching might require interactively querying a database management system, or it might entail using visualization software to spot anomalies in the data. In machine-learning terms, descriptive data mining is known as *unsupervised learning*, whereas predictive data mining is known as *supervised learning*.

Most of the methods used in data mining are related to methods developed in statistics and machine learning. Foremost among those methods are the general topics of regression, classification, clustering, and visualization. Because of the enormous sizes of the data sets, many applications of data mining focus on dimensionality-reduction techniques (e.g., variable selection) and situations in which high-dimensional data are suspected of lying on lower-dimensional hyperplanes. Recent attention has been directed to methods of identifying high-dimensional data lying on nonlinear surfaces or manifolds.

Table 1.1 lists some of the application areas of data mining and examples of major research themes within those areas. Using the massive data sets that are routinely collected by each of these disciplines, advances in dealing with the topics depend crucially upon the availability of effective data mining techniques and software.

One of the most important issues in data mining is the computational problem of *scalability*. Algorithms developed for computing standard exploratory and confirmatory statistical methods were designed to be fast and computationally efficient when applied to small and medium-sized data sets; yet, it has been shown that most of these algorithms are not up to the challenge of handling huge data sets. As data sets grow, many existing algorithms demonstrate a tendency to slow down dramatically (or even grind to a halt).

In data mining, regardless of size or complexity of the problem (essentially, the numbers of variables and observations), we require algorithms to have good performance characteristics; that is, they have to be scalable. There is no globally accepted definition of scalability, but a general idea of what this property means is the following:

Scalability: The capability of an algorithm to remain efficient and accurate as we increase the complexity of the problem.

The best scenario is that scalability should be linear. So, one goal of data mining is to create a library of scalable algorithms for the statistical analysis of large data sets.

Another issue that has to be considered by those working in data mining is the thorny problem of *statistical inference*. The twentieth century saw Fisher, Neyman, Pearson, Wald, Savage, de Finetti, and others provide a variety of competing — yet related — mathematical frameworks (frequentist, Bayesian, fiducial, decision theoretic, etc.) from which inferential theories of statistics were built. Extrapolating to a future point in time, can we expect researchers to provide a version of statistical inference for analyzing massive data sets?

There are situations in data mining when statistical inference — in its classical sense — either has no meaning or is of dubious validity: the former occurs when we have the entire population to search for answers (e.g., gene or protein sequences, astronomical recordings), and the latter occurs when a data set is a "convenience" sample rather than being a random sample drawn from some large population. When data are collected through time (e.g., retail transactions, stock-market transactions, patient records, weather records), sampling also may not make sense; the time-ordering of the observations is crucial to understanding the phenomenon generating the data, and to treat the observations as independent when they may be highly correlated will provide biased results.

Those who now work in data mining recognize that the central components of data mining are — in addition to statistical theory and methods — computing and computational efficiency, automatic data processing, dynamic and interactive data visualization techniques, and algorithm development. There are a number of software packages whose primary purpose is to help users carry out various techniques in data mining. The leading data-mining products include the packages listed (in alphabetical order) in Table 1.2.

TABLE 1.1. *Application areas of data mining*

Marketing: Predict new purchasing trends. Identify "loyal" customers. Predict what types of customers will respond to direct mailings, telemarketing calls, advertising campaigns, or promotions. Given customers who have purchased product A, B, or C, identify those who are likely to purchase product D and, in general, which products sell together (popularly called *market basket analysis*).

Banking: Predict which customers will likely switch from one credit card company to another. Evaluate loan policies using customer characteristics. Predict behavioral use of automated teller machines (ATMs).

Financial Markets: Identify relationships between financial indicators. Track changes in an investment portfolio and predict price turning points. Analyze volatility patterns in high-frequency stock transactions using volume, price, and time of each transaction.

Insurance: Identify characteristics of buyers of new policies. Find unusual claim patterns. Identify "risky" customers.

Healthcare: Identify successful medical treatments and procedures by examining insurance claims and billing data. Identify people "at risk" for certain illnesses so that treatment can be started before the condition becomes serious. Predict doctor visits from patient characteristics. Use healthcare data to help employers choose between HMOs.

Molecular Biology: Collect, organize, and integrate the enormous quantities of data on bioinformatics, functional genomics, proteomics, gene expression monitoring, and microarrays. Analyze amino acid sequences and deoxyribonucleic acid (DNA) microarrays. Use gene expression to characterize biological function. Predict protein structure and identify related proteins.

Astronomy: Catalogue (as stars, galaxies, etc.) hundreds of millions of objects in the sky using hundreds of attributes, such as position, size, shape, age, brightness, and color. Identify patterns and relationships of objects in the sky.

Forensic Accounting: Identify fraudulent behavior in credit card usage by looking for transactions that do not fit a particular cardholder's buying habits. Identify fraud in insurance and medical claims. Identify instances of tax evasion. Detect illegal activities that can lead to suspected money laundering operations. Identify stock market behaviors that indicate possible insider-trading operations.

Sports: Identify in realtime which players and which designed plays are most effective at specific points in the game and in relation to combinations of opposing players. Identify the exact moment when intriguing play patterns occurred. Discover game patterns hidden behind summary statistics.

TABLE 1.2. *Data mining software packages.*

Company	Software Package
IBM Corp.	*Intelligent Miner*
Insightful	*Insightful Miner*
NCR Corp.	*Teradata Warehouse Miner*
Oracle	*Darwin*
SAS Institute, Inc.	*Enterprise Miner*
Silicon Graphics, Inc.	*MineSet*
SPSS, Inc.	*Clementine*

1.2.3 Knowledge Discovery

Data mining has been described (Fayyad, Piatetsky-Shapiro, and Smyth, 1996) as a step in a more general process known as *knowledge discovery in databases (KDD)*. The "knowledge" acquired by KDD has to be interesting, non-trivial, non-obvious, previously unknown, and potentially useful.

KDD is a multistep process designed to assist those who need to search huge data sets for "nuggets of useful information." In KDD, assistance is expected to be intelligent and automated, and the process itself is interactive and iterative.

KDD is composed of six primary activities:

1. selecting the target data set (which data set or which variables and cases are to be used for data mining);

2. data cleaning (removal of noise, identification of potential outliers, imputing missing data);

3. preprocessing the data (deciding upon data transformations, tracking time-dependent information);

4. deciding which data-mining tasks are appropriate (regression, classification, clustering, etc.);

5. analyzing the cleaned data using data-mining software (algorithms for data reduction, dimensionality reduction, fitting models, prediction, extracting patterns);

6. interpreting and assessing the knowledge derived from data-mining results.

In KDD, and hence in data mining, the descriptive aspect is more important than the predictive aspect, which forms the main goal of machine learning.

1.3   Machine Learning

Machine learning evolved out of the subfield of computer science known as *artificial intelligence (AI)*. Whereas the focus of AI is to make machines intelligent, able to think rationally like humans and solve problems, machine learning is concerned with creating computer systems and algorithms so that machines can "learn" from previous experience. Because intelligence cannot be attained without the ability to learn, machine learning now plays a dominant role in AI.

1.3.1   How Does a Machine Learn?

A machine learns when it is able to accumulate experience (through data, programs, etc.) and develop new knowledge so that its performance on specific tasks improves over time. This idea of learning from experience is central to the various types of problems encountered in machine learning, especially problems involving classification (e.g., handwritten digit recognition, speech recognition, face recognition, text classification). The general goal of each of these problems is to find a systematic way of classifying a future example (e.g., a handwriting sample, a spoken word, a face image, a text fragment). Classification is based upon measurements on that future example together with knowledge obtained from a *learning* (or *training*) *sample* of similar examples (where the class of each example is completely determined and known, and the number of classes is finite and known).

The need to create new methods and terminology for analyzing large and complex data sets has led to researchers from several disciplines — statistics, pattern recognition, neural networks, symbolic machine learning, computational learning theory, and, of course, AI — to work together to influence the development of machine learning.

Among the techniques that have been used to solve machine-learning problems, the topics that are of most interest to statisticians — density estimation, regression, and pattern recognition (including neural networks, discriminant analysis, tree-based classifiers, random forests, bagging and boosting, support vector machines, clustering, and dimensionality-reduction methods) — are now collectively referred to as *statistical learning* and constitute many of the topics discussed in this book. Vladimir N. Vapnik, one of the founders of statistical learning theory, relates statistics to learning theory in the following way (Vapnik, 2000, p. x):

> *The problem of learning is so general that almost any question that has been discussed in statistical science has its analog in learning theory. Furthermore, some very important general results were first found in the framework of learning theory and then formulated in the terms of statistics.*

The machine-learning community divides learning problems into various categories: the two most relevant to statistics are those of *supervised learning* and *unsupervised learning.*

Supervised learning: Problems in which the learning algorithm receives a set of continuous or categorical input variables and a correct output variable (which is observed or provided by an explicit "teacher") and tries to find a function of the input variables to approximate the known output variable: a continuous output variable yields a regression problem, whereas a categorical output variable yields a classification problem.

Unsupervised learning: Problems in which there is no information available (i.e., no explicit "teacher") to define an appropriate output variable; often referred to as "scientific discovery."

The goal in unsupervised learning differs from that of supervised learning. In supervised learning, we study relationships between the input and output variables; in unsupervised learning, we explore particular characteristics of the input variables only, such as estimating the joint probability density, searching out clusters, drawing proximity maps, locating outliers, or imputing missing data.

Sometimes there might not be a "bright-line" distinction between supervised and unsupervised learning. For example, the dimensionality-reduction technique of principal component analysis (PCA) has no explicit output variable and, thus, appears to be an unsupervised-learning method; however, as we will see, PCA can be formulated in terms of a multivariate regression model where the input variables are also used as output variables, and so PCA can also be regarded as a supervised-learning method.

1.3.2 Prediction Accuracy

One of the most important tasks in statistics is to assess the accuracy of a predictor (e.g., regression estimator or classifier). The measure of prediction accuracy typically used is that of *prediction error*, defined generically as

Prediction error: In a regression problem, the mean of the squared errors of prediction, where error is the difference between a true output value and its corresponding predicted output value; in a classification problem, the probability of misclassifying a case.

The simplest estimate of *prediction error* is the *resubstitution error*, which is computed as follows. In a regression problem, the fitted model is used to predict each of the (known) output values from the entire data set, and the resubstitution estimate is then the mean of the squared residuals, also known as the *residual mean square*. In a classification problem, the classifier predicts the (known) class of each case in the entire data set, a correct prediction is scored as a 0 and a misclassification is scored as a 1, and the resubstitution estimate is the proportion of misclassified cases.

Because the resubstitution estimate uses the same data as was used to derive the predictor, the result is an overly optimistic view of prediction accuracy. Clearly, it is important to do better.

1.3.3   Generalization

The need to improve upon the resubstitution estimator of prediction accuracy led naturally to the concept of *generalization*: we want an estimation procedure to generalize well; that is, to make good predictions when applied to a data set *independent of that used to fit the model.* Although this is not a new idea — it has existed in statistics for a long time (see, e.g., Mosteller and Tukey, 1977, pp. 37–38) — the machine-learning community embraced this particular concept (adopting the name from psychology) and made it a central issue in the theory and applications of machine learning.

Where do we find such an independent data set? One way is to gather fresh data. However, "when fresh gathering is not feasible, good results can come from going to a body of data that has been kept in a locked safe where it has rested untouched and unscanned during all the choices and optimizations" (Mosteller and Tukey, 1977, p. 38). The data in the "locked safe" can be viewed as holding back a portion of the current data from the model-fitting phase and using it instead for assessment purposes. If an independent set of data is not used, then we will overestimate the model's predictive accuracy.

In fact, it is now common practice — assuming the data set is large enough — to use a random mechanism to separate the data into three nonoverlapping and independent data sets:

a learning (or training) set $\mathcal{L}$, a data set where "anything goes ... including hunches, preliminary testing, looking for patterns, trying large numbers of different models, and eliminating outliers" (Efron, 1982, p. 49);

a validation set $\mathcal{V}$, a data set to be used for model selection and assessment of competing models (usually on the basis of predictive ability);

a test set $\mathcal{T}$, a data set to be used for assessing the performance of a completely specified final model.

The key assumption here is that the three subsets of the data are each generated by the same underlying distribution. In some instances, learning data may be taken from historical records.

As a simple guideline, the learning set should consist of about 50% of the data, whereas the validation and test sets may each consist of 25% (although these percentages are not written in stone). In some instances, we may find it convenient to merge the validation set with the test set, thus forming a larger test set. For example, we often see publicly available data sets in Internet databases divided into a learning set and a test set.

1.3.4 Generalization Error

In supervised learning problems, it is important to assess how closely a particular model (function of the inputs) fits the data (the outputs). As before, we use prediction error as our measure of prediction accuracy.

In regression problems, there are two different types of prediction error. For both types, we first fit a model to the learning set $\mathcal{L}$. Then, we use that fitted model to predict the output values of either $\mathcal{L}$ (given input values from $\mathcal{L}$) or the test set $\mathcal{T}$ (given input values from $\mathcal{T}$). Prediction error is the mean (computed only over the appropriate data set) of the squared-errors of prediction (where error = true output value – predicted output value). If we average over $\mathcal{L}$, the prediction error is called the *regression learning error* (equivalent to the resubstitution estimate computed only over $\mathcal{L}$), whereas if we average over $\mathcal{T}$, the prediction error is called the *regression test error*.

A similar strategy is used in classification problems; only the definition of prediction error is different. We first build a classifier from $\mathcal{L}$. Next, we use that classifier to predict the class of each data vector in either $\mathcal{L}$ or $\mathcal{T}$. For each prediction, we assign the value of 0 to a correct classification and 1 to a classification error. The prediction error is then defined as the average of all the 0s and 1s over the appropriate data set (i.e., the proportion of misclassified observations). If we average over $\mathcal{L}$, then prediction error is referred to as the *classification learning error* (equivalent to the resubstitution estimate computed only over $\mathcal{L}$), whereas averaging over $\mathcal{T}$ yields the *classification test error*.

If the learning set $\mathcal{L}$ is moderately sized, we may feel that using only a portion of the entire data set to fit the model is a waste of good data. Alternative data-splitting methods for estimating test error are based upon *cross-validation* (Stone, 1974) and the *bootstrap* (Efron, 1979):

V-**fold cross-validation:** Randomly divide the entire data set into, say, V nonoverlapping groups of roughly equal size; remove one of the groups and fit the model using the combined data from the other $V-1$ groups (which forms the learning set); use the omitted group as the test set, predict its output values using the fitted model, and compute the prediction error for the omitted group; repeat this procedure V times, each time removing a different group; then, average the resulting V

prediction errors to estimate the test error. The number of groups V can be any number from 2 to the sample size.

Bootstrap: Select a "bootstrap sample" from the entire data set by drawing a random sample *with replacement* having the same size as the parent data set, so that the sample may contain repeated observations; fit a model using this bootstrap sample and compute its prediction error; repeat this sampling procedure, say, 1000 times, each time computing a prediction error; then, average all the prediction errors to estimate the test error.

These are generic descriptions of the two procedures; specific descriptions are given in various sections of this book. In particular, the definition of the bootstrap is actually more complicated than that given by this description because it depends on what is assumed about the stochastic model generating the data. Although both cross-validation and the bootstrap are computationally intensive techniques, cross-validation uses the entire data set in a more efficient manner than the division into a learning set and an independent test set. We also caution that, in some applications, it may not make sense to use one of these procedures.

The expected prediction error over an independent test set is called *infinite test error* or *generalization error*. We estimate generalization error by the test error. One goal of *generalization theory* is to choose that regression model or classifier thatgives the smallest generalization error.

1.3.5 Overfitting

To minimize generalization error, it is tempting to find a model that will fit the data in the learning set as accurately as possible. This is not usually advisable because it may make the selected model too complicated. The resulting learning error will be very small (because the fitted model has been optimized for that data set), whereas the test error will be large (a consequence of *overfitting*).

Overfitting: Occurs when the model is too large or complicated, or contains too many parameters relative to the size of the learning set. It usually results in a very small learning error and a large generalization (test) error.

One can control such temptation by following the principle known as *Ockham's razor*, which encourages us to choose simple models while not losing track of the need for accuracy. Simple models are generally preferred if either the learning set is too small to derive a useful estimate of the model or fitting a more complex model would necessitate using huge amounts of computational resources.

We illustrate the idea of overfitting with a simple regression example. Using 10 equally spaced x values as the learning set, we generate corresponding y values from the function $y = 0.5 + 0.25\cos(2\pi x) + e$, where the Gaussian noise component e has mean zero and standard deviation 0.06. We try to approximate the underlying unknown function (the cosinusoid) by a polynomial in x, where the problem is to decide on the degree of the polynomial. In the top-left panel of Figure 1.1, we give the cosinusoid and the 10 generated points; in the top-right panel, a linear regression function gives a poor fit to the points and shows the result of *underfitting* by using too few parameters; in the bottom-left panel, a cubic polynomial is fitted to the data, showing an improved approximation to the cosinusoid; and in the bottom-right panel, by increasing the fit to a $9th$-degree polynomial, we ensure that the fitted curve passes through each point exactly. However, the $9th$-degree polynomial actually makes the fit much worse by introducing unwanted fluctuations and shows the result of overfitting by using too many parameters.

How would such polynomial fits affect a test set obtained by using the same x values but different noise values (hence, different y values) in the above cosinusoid model? In Figure 1.2, we plot the prediction errors for both the learning set and the test set. The learning error, as expected, decreases monotonically to zero when we fit a $9th$-degree polynomial. This behavior for the learning error is typical whenever the fitted model ranges from the very simple to the most complex. The test error decreases to a $4th$ degree polynomial and then increases, indicating that models with too many parameters will have poor generalization properties.

Researchers have suggested several methods for reducing the effects of overfitting. These include methods that employ some form of averaging of predictions made by a number of different models fit to the learning set (e.g., the "bagging" and "boosting" algorithms of Chapter 14) and regularization (where complex models are penalized in favor of simpler models). Bayesian arguments in favor of a related idea of "model averaging" have also been proposed (see Hoeting, Madigan, Raftery, and Volinsky, 1999, for an excellent review of the topic).

1.4 Overview of Chapters

This book is divided into 17 chapters. Chapter 2 describes multivariate data, database management systems, and data problems. Chapter 3 reviews basic vector and matrix notation, introduces random vectors and matrices and their distributions, and derives maximum likelihood estimates for the multivariate Gaussian mean, including the James–Stein shrinkage estimator. Chapter 4 provides the elements of nonparametric density estimation. Chapters 5 reviews topics in multiple linear regression, including

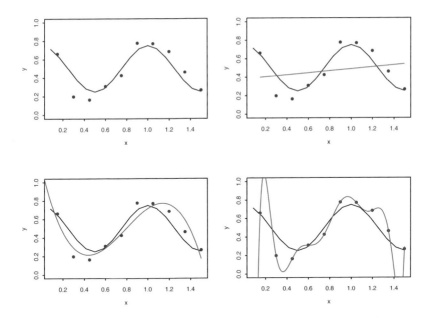

FIGURE 1.1. *Ten y-values corresponding to equally spaced x-values were generated from the cosinusoid $y = 0.5 + 0.25\cos(2\pi x) + e$, where the noise component $e \sim \mathcal{N}(0, (0.06)^2)$. Top-left panel: the true cosinusoid is shown in black with the 10 points in blue; top-right: the red line is the ordinary least-squares (OLS) linear regression fit to the points; bottom-left: the red curve is an OLS cubic polynomial fit to the points; bottom-right: the red curve is a 9th-degree polynomial that passes through every point.*

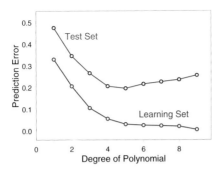

FIGURE 1.2. *Prediction error from the learning set (blue curve) and test set (red curve) based upon polynomial fits to data generated from a cosinusoid curve with noise.*

model assessment (through cross-validation and the bootstrap), biased regression, shrinkage, and model selection, concepts that will be needed in later chapters.

In Chapter 6, we discuss multivariate regression for both the fixed-X and random-X cases. We discuss multivariate analysis of variance and multivariate reduced-rank regression (RRR). RRR provides the foundation for a unified theory of multivariate analysis, which includes as special cases the classical techniques of principal component analysis, canonical variate analysis, linear discriminant analysis, factor analysis, and correspondence analysis. In Chapter 7, we introduce the idea of (linear) dimensionality reduction, which includes principal component analysis, canonical variate and correlation analysis, and projection pursuit. Chapter 8 discusses Fisher's linear discriminant analysis. Chapter 9 introduces recursive partitioning and classification and regression trees. Chapter 10 discusses artificial neural networks via analogies to neural networks in the brain, artificial intelligence, and expert systems, as well as the related statistical techniques of projection pursuit regression and generalized additive models. Chapter 11 deals with classification using support vector machines. Chapter 12 describes the many algorithms for cluster analysis and unsupervised learning.

In Chapter 13, we discuss multidimensional scaling and distance geometry, and Chapter 14 introduces committee machines and ensemble methods, such as bagging, boosting, and random forests. Chapter 15 discusses independent component analysis. Chapter 16 looks at nonlinear methods for dimensionality reduction, especially the various flavors of nonlinear principal component analysis, and nonlinear manifold learning. Chapter 17 describes correspondence analysis.

Bibliographical Notes

Books on data mining include Fayyad, Piatetsky-Shapiro, Smyth, and Uthurusamy (1996) and Hand, Mannila, and Smyth (2001). There are annual KDD workshops and conferences and a KDD journal. There is a KDD section of the ACM: www.acm.org/sigkdd. Books on machine learning include Bishop (1995), Ripley (1996), Hastie, Tibshirani, and Friedman (2001), MacKay (2003), and Bishop (2006).

2
Data and Databases

2.1 Introduction

Multivariate data consist of multiple measurements, observations, or responses obtained on a collection of selected variables. The types of variables usually encountered often depend upon those who collect the data (the *domain experts*), possibly together with some statistical colleagues; for it is these people who actively decide which variables are of interest in studying a particular phenomenon. In other circumstances, data are collected automatically and routinely without a research direction in mind, using software that records every observation or transaction made regardless of whether it may be important or not.

Data are raw facts, which can be numerical values (e.g., age, height, weight), text strings (e.g., a name), curves (e.g., a longitudinal record regarded as a single functional entity), or two-dimensional images (e.g., photograph, map). When data sets are "small" in size, we find it convenient to store them in *spreadsheets* or as *flat files* (large rectangular arrays). We can then use any statistical software package to import such data for subsequent data analysis, graphics, and inference. As mentioned in Chapter 1, massive data sets are now sprouting up everywhere. Data of such size need to be stored and manipulated in special database systems.

A.J. Izenman, *Modern Multivariate Statistical Techniques*,
doi: 10.1007/978-0-387-78189-1_2,
© Springer Science+Business Media, LLC 2008

2.2   Examples

We first describe some examples of the data sets to be encountered in this book.

2.2.1   Example: DNA Microarray Data

The DNA (deoxyribonucleic acid) microarray has been described as "one of the great unintended consequences of the Human Genome Project" (Baker, 2003). The main impact of this enormous scientific achievement is to provide us with large and highly structured microarray data sets from which we can extract valuable genetic information. In particular, we would like to know whether "gene expression" (the process by which genetic information encoded in DNA is converted, first, into mRNA (messenger ribonucleic acid), and then into protein or any of several types of RNA) is any different for cancerous tissue as opposed to healthy tissue.

Microarray technology has enabled the *expression levels* of a huge number of genes within a specific cell culture or tissue to be monitored simultaneously and efficiently. This is important because differences in gene expression determine differences in protein abundance, which, in turn, determine different cell functions. Although protein abundance is difficult to determine, molecular biologists have discovered that gene expression can be measured indirectly through microarray experiments.

Popular types of microarray technologies include cDNA microarrays (developed at Stanford University) and high-density, synthetic, oligonucleotide microarrays (developed by Affymetrix, Inc., under the GENECHIP® trademark). Both technologies use the idea of hybridizing a "target" (which is usually either a single-stranded DNA or RNA sequence, extracted from biological tissue of interest) to a DNA "probe" (all or part of a single-stranded DNA sequence printed as "spots" onto a two-way grid of dimples in a glass or plastic microarray slide, where each spot corresponds to a specific gene).

The microarray slide is then exposed to a set of targets. Two biological mRNA samples, one obtained from cancerous tissue (the *experimental sample*), the other from healthy tissue (the *reference sample*), are reverse-transcribed into cDNA (complementary DNA); then, the reference cDNA is labeled with a green fluorescent dye (e.g., Cy3) and the experimental cDNA is labeled with a red fluorescent dye (e.g., Cy5). Fluorescence measurements are taken of each dye separately at each spot on the array. High gene expression in the tissue sample yields large quantities of hybridized cDNA, which means a high intensity value. Low intensity values derive from low gene expression.

The primary goal is to compare the intensity values, R and G, of the red and green channels, respectively, at each spot on the array. The most popular statistic is the *intensity log-ratio*, $M = \log(R/G) = \log(R) - \log(G)$. Other such functions include the *probe value*, $PV = \log(R - G)$, and the *average log-intensity*, $A = \frac{1}{2}(\log R + \log G)$. The logarithm in each case is taken to base 2 because intensity values are usually integers ranging from 0 to $2^{16} - 1$.

Microarray data is a matrix whose rows are genes and whose columns are samples, although this row-column arrangement may be reversed. The genes play the role of variables, and the samples are the observations studied under different conditions. Such "conditions" include different experimental conditions (treatment vs. control samples), different tissue samples (healthy vs. cancerous tumors), and different time points (which may incorporate environmental changes).

For example, Figure 2.1 displays the heatmap for the expression levels of 92 genes obtained from a microarray study on 62 colon tissue samples, where the entries range from negative values (green) to positive values (red).[1] The tissue samples were derived from 40 different patients: 22 patients each provided both a normal tissue sample and a tumor tissue sample, whereas 18 patients each provided only a colon tumor sample. As a result, we have tumor samples from 40 patients ($T1, \ldots, T40$) and normal samples from 22 patients (Normal1, $\ldots$, Normal21), and this is the way the samples are labeled.

From the heatmap, we wish to identify expression patterns of interest in microarray data, focusing in on which genes contribute to those patterns across the various conditions. Multivariate statistical techniques applied to microarray data include supervised learning methods for classification and the unsupervised methods of cluster analysis.

2.2.2 Example: Mixtures of Polyaromatic Hydrocarbons

This example illustrates a very common problem in chemometrics. The data (Brereton, 2003, Section 5.1.2) come from a study of polyaromatic hydrocarbons (PAHs), which are described as follows:[2]

> Polyaromatic hydrocarbons (PAHs) are ubiquitous environmental contaminants, which have been linked with tumors and effects on reproduction. PAHs are formed during the burning of coal, oil, gas, wood, tobacco, rubbish, and other organic

---

[1] The data can be found in the file `alontop.txt` on the book's website. The 92 genes are a subset of a larger set of more than 6500 genes whose expression levels were measured on these 62 tissue samples (Alon et al, 1999).

[2] This quote is taken from the August 1997 issue of the *Update* newsletter of the World Wildlife Fund–UK at its website `www.wwf-uk.org/filelibrary/pdf/mu_32.pdf`.

Observed Gene Expression Matrix

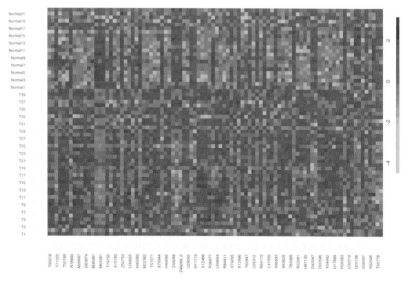

Genes = 92 # cell-lines= 62

FIGURE 2.1. *Gene expression heatmap of 92 genes (columns) and 62 tissue samples (rows) for the colon cancer data. The tissue samples are divided into 40 colon cancer samples (T1–T40) and 22 normal samples (Normal1–Normal22).*

substances. They are also present in coal tars, crude oil, and petroleum products such as creosote and asphalt. There are some natural sources, such as forest fires and volcanoes, but PAHs mainly arise from combustion-related or oil-related man-made sources. A few PAHs are used by industry in medicines and to make dyes, plastics, and pesticides.

Table 2.1 gives a list of the 10 PAHs that are used in this example.

The data were collected in the following way.[3] From the 10 PAHs listed in Table 2.1, 50 complex mixtures of certain concentrations (in mg L) of those PAHs were formed. From each such mixture, an electronic absorption

---

[3]The data, which can be found in the file PAH.txt on the book's website, can also be downloaded from the website statmaster.sdu.dk/courses/ST02/data/index.html. The fifty sample observations were originally divided into two independent sets, each of 25 observations, but were combined here so that we would have more observations than either set of data for the example.

TABLE 2.1. *Ten polyaromatic hydrocarbon (PAH) compounds.*

pyrene (`Py`), acenaphthene (`Ace`), anthracene (`Anth`), acenaphthylene (`Acy`), chrysene (`Chry`), benzanthracene (`Benz`), fluoranthene (`Fluora`), fluorene (`Fluore`), naphthalene (`Nap`), phenanthracene (`Phen`)

spectrum (EAS) was computed. The spectra were then digitized at 5 nm intervals into $r = 27$ wavelength channels from 220 nm to 350 nm. The 50 spectra are displayed in Figure 2.2. The scatterplot matrix of the 10 PAHs is displayed in Figure 2.3. Notice that most of these scatterplots appear as 5×5 arrays of 50 points, where only half the points are visible because of a replication feature in the experimental design.

Using the resulting digitized values of the spectra, we wish to predict the individual concentrations of PAHs in the mixture. In chemometrics, this type of regression problem is referred to as *multivariate inverse calibration*: although the concentrations are actually the input variables and the spectrum values are the output variables in the chemical process, the real

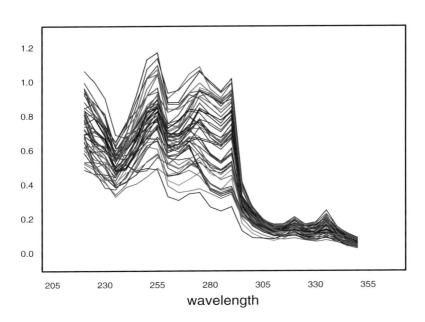

FIGURE 2.2. *Electronic absorption spectroscopy (EAS) spectra of 50 samples of polyaromatic hydrocarbons (PAH), where the spectra are measured at 25 wavelengths within the range 220–350 nm.*

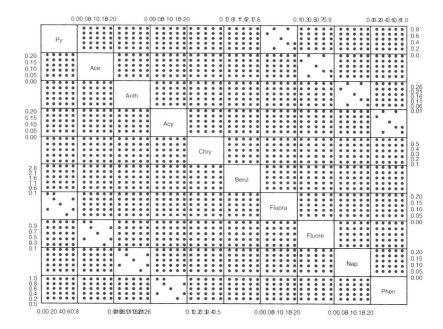

FIGURE 2.3. *Scatterplot matrix of the mixture concentrations of the 10 chemicals in Table 2.1. In each scatterplot, there are 50 points; in most scatterplots, 25 of the points appear in a 5 × 5 array, and the other 25 are replications. In the remaining four scatterplots, there are eight distinguishable points with different numbers of replications.*

goal is to predict the mixture concentrations (which are difficult to determine) from the spectra (easy to compute), and not vice versa.

2.2.3 Example: Face Recognition

Until recently, human face recognition was primarily based upon identifying individual facial features such as eyes, nose, mouth, ears, chin, head outline, glasses, and facial hair, and then putting them together computationally to construct a face. The most used approach today (and the one we describe here) is an innovative computerized system called *eigenfaces*, which operates directly on an image-based representation of faces (Turk and Pentland, 1991). Applications of such work include homeland security, video surveillance, human-computer interaction for entertainment purposes, robotics, and "smart" cards (e.g., passports, drivers' licences, voter registration).

Each face, as a picture image, might be represented by a $(c \times d)$-matrix of intensity values, which are usually quantized to 8-bit gray scale (0–255, with

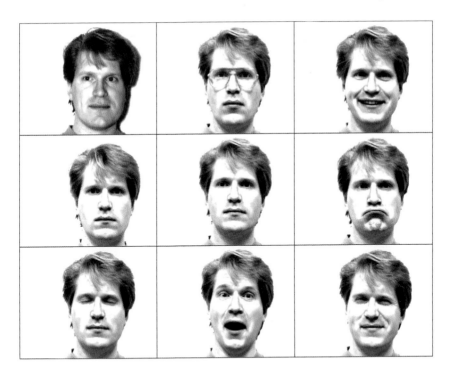

FIGURE 2.4. *Face images of the same individual under nine different conditions (1=centerlight, 2=glasses, 3=happy, 4=no glasses, 5=normal, 6=sad, 7=sleepy, 8=surprised, 9=wink). From the Yale Face Database.*

0 as black and 255 as white). These values are then scaled and converted to double precision, with values in $[0,1]$. The values of c and d depend upon the degree of resolution needed. The matrix is then "vec'ed" by stacking the columns of the matrix under one another to form a cd-vector in *image space*. For example, if an image is digitized into a (256×256)-array of pixels, that face is now a point in a 65,536-dimensional space. We can view all possible images of one particular face as a lower-dimensional manifold (*face space*) embedded within the high-dimensional image space.

There are a number of repositories of face images. The data for this example were taken from the *Yale Face Database* (Belhumeur, Hespanha, and Kriegman, 1997).[4] which contains 165 frontal-face grayscale images covering 15 individuals taken under 11 different conditions of different illumination (centerlight, leftlight, rightlight, normal), expression (happy, sad, sleepy, surprised, wink), and glasses (with and without). Each image has

---

[4]A list of the many face databases that can be accessed on the Internet, including the *Yale Face Database*, can be found at the website www.face-rec.org/databases.

size 320×243, which then gets stacked into an r-vector, where $r = 77,760$. Figure 2.4 shows the images of a single individual taken under 9 of those 11 conditions. The problem is one of *dimensionality reduction*: what is the fewest number of variables necessary to identify these types of facial images?

2.3   Databases

A *database* is a collection of persistent data, where by "persistent" we mean data that can be removed from the database only by an explicit request and not through an application's side effect. The most popular format for organizing data in a database is in the form of *tables* (also called *data arrays* or *data matrices*), each table having the form of a rectangular array arranged into rows and columns, where a row represents the values of all variables on a single multivariate *observation* (*response, case*, or *record*), and a column represents the values of a single *variable* for each observation.

In this book, a typical database table having n multivariate observations taken on r variables will be represented by an $(r \times n)$-matrix,

$$
\overset{r \times n}{\mathcal{X}} = \begin{pmatrix} x_{11} & x_{12} & \cdots & x_{1n} \\ x_{21} & x_{22} & \cdots & x_{2n} \\ \vdots & \vdots & & \vdots \\ x_{r1} & x_{r2} & \cdots & x_{rn} \end{pmatrix}, \tag{2.1}
$$

say, having r rows and n columns. In (2.1), x_{ij} represents the value in the ith row ($i = 1, 2, \ldots, r$) and jth column ($j = 1, 2, \ldots, n$) of $\mathcal{X}$. Although database tables are set up to have the form of $\mathcal{X}^\tau$, with variables as columns and observations as rows, we will find it convenient in this book to set $\mathcal{X}$ to be the transpose of the database table.

Databases exist for storing information. They are used for any of a number of different reasons, including statistical analysis, retrieving information from text-based documents (e.g., libraries, legislative records, case dockets in litigation proceedings), or obtaining administrative information (e.g., personnel, sales, financial, and customer records) needed for managing an organization. Databases can be of any size. Even small databases can be very useful if accessed often. Setting up a large and complex database typically involves a major financial committment on the part of an organization, and so the database has to remain useful over a long time period. Thus, we should be able to extend a database as additional records become available and to correct, delete, and update records as necessary.

2.3.1 Data Types

Databases usually consist of mixtures of different types of variables:

Indexing: These are usually names, tags, case numbers, or serial numbers that identify a respondent or group of respondents. Their values may indicate the location where a particular measurement was taken, or the month or day of the year that an observation was made.

There are two special types of indexing variables:

1. A *primary key* is an indexing variable (or set of indexing variables) that uniquely identifies each observation in a database (e.g., patient numbers, account numbers).

2. A *foreign key* is an indexing variable in a database where that indexing variable is a primary key of a related database.

Binary: This is the simplest type of variable, having only two possible responses, such as YES or NO, SUCCESS or FAILURE, MALE or FEMALE, WHITE or NON-WHITE, FOR or AGAINST, SMOKER or NON-SMOKER, and so on. It is usually coded 0 or 1 for the two possible responses and is often referred to as a *dummy* or *indicator* variable.

Boolean: A *Boolean* variable has the two responses TRUE or FALSE but may also have the value UNKNOWN.

Nominal: This *character-string* data type is a more general version of a binary variable and has a fixed number of possible responses that cannot be usefully ordered. These responses are typically coded alphanumerically, and they usually represent disjoint classifications or categories set up by the investigator. Examples include the geographical location where data on other variables are collected, brand preference in a consumer survey, political party affiliation, and ethnic-racial identification of respondent.

Ordinal: The possible responses for this character-string data type are linearly ordered. An example is "excellent, good, fair, poor, bad, awful" (or "strongly disagree" to "strongly agree"). Another example is bond ratings for debt issues, recorded as AA+, AA, AA-, A+, A, A-, B+, B, and B-. Such responses may be assigned scores or rankings. They are often coded on a "ranking scale" of 1–5 (or 1–10). The main problem with these ranking scales is the implicit assumption of equidistance of the assigned scores. Brand preferences can sometimes be regarded as ordered.

Integer: The response is usually a nonnegative whole number and is often a count.

Continuous: This is a measured variable in which the continuity assumption depends upon a sufficient number of digits (and decimal places) being recorded. Continuous variables are specified as *numeric* or *decimal* in database systems, depending upon the precision required.

We note an important distinction between variables that are fixed and those that are stochastic:

Fixed: The values of a fixed variable have deliberately been set in advance, as in a designed experiment, or are considered "causal" to the phenomenon in question; as a result, interest centers only on a specific group of responses. This category usually refers to indexing variables but can also include some of the above types.

Stochastic: The values of a stochastic variable can be considered as having been chosen at random from a potential list (possibly, the real line or a portion of it) in some stochastic manner. In this sense, the values obtained are representative of the entire range of possible values of the variable in question.

We also need to distinguish between input and output variables:

Input variable: Also called a *predictor* or *independent variable*, typically denoted by X, and may be considered to be fixed (or preset or controlled) through a statistically designed experiment, or stochastic if it can take on values that are observed but not controlled.

Output variable: Also called a *response* or *dependent variable*, typically denoted by Y, and which is stochastic and dependent upon the input variables.

Most of the methods described in this book are designed to elicit information concerning the extent to which the outputs depend upon the inputs.

2.3.2   Trends in Data Storage

As data collections become larger and larger, and areas of research that were once "data-poor" now become "data-rich," it is how we store those data that is of great importance.

For the individual researcher working with a relatively simple database, data are stored locally on hard disks. We know that hard-disk storage capacity is doubling annually (*Kryder's Law*), and the trend toward tiny,

TABLE 2.2. *Internet websites containing many different databases.*

www.ics.uci.edu/pub/machine-learning-databases
lib.stat.cmu.edu/datasets
www.statsci.org/datasets.html
www.amstat.org/publications/jse/jse_data_archive.html
www.physionet.org/physiobank/database
biostat.mc.vanderbilt.edu/twiki/bin/view/Main/DataSets

high-capacity hard drives has outpaced even the rate of increase in number of transistors that can be placed on an integrated circuit (*Moore's Law*). Gordon E. Moore, Intel co-founder, predicted in 1965 that the number of transistors that can be placed on an integrated circuit would continue to increase at a constant rate for at least 10 years. In 1975, Moore predicted that the rate would double every two years. So far, this assessment has proved to be accurate, although Moore stated in 2005 that his law, which may hold for another two decades, cannot be sustained indefinitely.

Because chip speeds are doubling even faster than Moore had anticipated, we are seeing rapid progress toward the manufacturing of very small, high-performance storage devices. New types of data storage devices include three-dimensional *holographic storage*, where huge quantities (e.g., a terabyte) of data can be stored into a space the size of a sugar cube.

For large institutions, such as health maintenance organizations, educational establishments, national libraries, and industrial plants, data storage is a more complicated issue, and the primary storage facility is usually a remote "data warehouse." We describe such storage facilities in Section 2.4.5.

2.3.3 Databases on the Internet

In Table 2.2, we list a few Internet websites from which databases of various sizes can be downloaded. Many of the data sets used as examples in this book were obtained through these websites.

There are also many databases available on the Internet that specialize in bioinformatics information, such as biological databases and published articles. These databases contain an amazingly rich variety of biological data, including DNA, RNA, and protein sequences, gene expression profiles, protein structures, transcription factors, and biochemical pathways. See Table 2.3 for examples of such websites.

A recent development in data-mining applications is the processing and categorization of natural-language text documents (e.g., news items, scientific publications, spam detection). With the rapid growth of the Internet and e-mail, academics, scientists, and librarians have shown enormous interest in mining the structured or unstructured knowledge present in large collections of text documents. To help those whose research interests lie in analyzing text information, large databases (having more than 10,000 features) of text documents are now available.

For example, Table 2.4 lists a number of text databases. Two of the most popular collections of documents come from Reuters, Ltd., which is the world's largest text and television news agency; the English-language collections REUTERS-21578 containing 21,578 news items and RCV1 (*Reuters Corpus Volume 1*) (Lewis, Yang, Rose, and Li, 2004) containing 806,791 news items are drawn from online databases. The 20 Newsgroups database (donated by Tom Mitchell) contains 20,000 messages taken from 20 Usenet newsgroups. The OHSUMED text database (Hersh, Buckley, Leone, and Hickam, 1994) from Ohio State University contains 348,566 references and abstracts derived from Medline, an on-line medical information database, for the period 1987–1991.

Computerized databases of scientific articles (e.g., ARXIV, see Table 2.4) are assembled to (Shiffrin and Börner, 2004):

> [I]dentify and organize research areas according to experts, institutions, grants, publications, journals, citations, text, and figures; discover interconnections among these; establish the import of research; reveal the export of research among fields; examine dynamic changes such as speed of growth and diversification; highlight economic factors in information production and dissemination; find and map scientific and social networks; and identify the impact of strategic and applied research funding by government and other agencies.

A common element of text databases is the dimensionality of the data, which can run well into the thousands. This makes visualization especially difficult. Furthermore, because text documents are typically noisy, possibly even having differing formats, some automated preprocessing may be necessary in order to arrive at high-quality, clean data. The availability of text databases in which preprocessing has already been undertaken is proving to be an important development in database research.

TABLE 2.3. *Internet websites containing microarray databases.*

```
www.broad.mit.edu/tools/data.html
sdmc.lit.org.sg/GEDatasets/Datasets.html
genome-www5.stanford.edu
www.bioconductor.org/packages/1.8/AnnotationData.html
www.ncbi.nlm.nih.gov/geo
```

TABLE 2.4. *Internet websites containing natural-language text databases.*

arXiv.org
medir.ohsu.edu/pub/ohsumed
kdd.ics.uci.edu/databases/reuters21578/reuters21578.html
kdd.ics.uci.edu/databases/20newsgroups/20newsgroups.html

2.4 Database Management

After data have been recorded and physically stored in a database, they need to be accessed by an authorized user who wishes to use the information. To access the database, the user has to interact with a database management system, which provides centralized control of all basic storage, access, and retrieval activities related to the database, while also minimizing duplications, redundancies, and inconsistencies in the database.

2.4.1 Elements of Database Systems

A *database management system* (DBMS) is a software system that manages data and provides controlled access to the database through a personal computer, an on-line workstation, or a terminal to a mainframe computer or network of computers. *Database systems* (consisting of databases, DBMS, and application programs) are typically used for managing large quantities of data. If we are working with a small data set with a simple structure, if the particular application is not complicated, and if multiple concurrent users (those who wish to access the same data at the same time) are not an issue, then there is no need to employ a DBMS.

A database system can be regarded as two entities: a *server* (or *backend*), which holds the DBMS, and a set of *clients* (or *frontend*), each of which consists of a hardware and a software component, including application programs that operate on the DBMS. Application programs typically include a query language processor, report writers, spreadsheets, natural language processors, and statistical software packages. If the server and clients communicate with each other from different machines through a *distributed processing network* (such as the Internet), we refer to the system as having a "client/server" architecture.

The major breakthrough in database systems was the introduction by 1970 of the *relational model*. We call a DBMS *relational* if the data are perceived by users only as *tables*, and if users can generate new tables from old ones. Tables in a *relational DBMS (RDBMS)* are rectangular arrays defined by their *rows* of observations (usually called *records* or *tuples*) and *columns* of variables (usually called *attributes* or *fields*); the number of tuples is called the *cardinality*, and the number of attributes is called the *degree* of the table. A RDBMS contains operators that enable users to extract specified rows (`restrict`) or specified columns (`project`) from a table and match up (`join`) information stored in different tables by checking for common entries in common columns. Also part of a DBMS is a *data dictionary*, which is a system database that stores information (*metadata*) about the database itself.

2.4.2   Structured Query Language (SQL)

Users communicate with a RDBMS through a declarative *query language* (or general interactive enquiry facility), which is typically one of the many versions of SQL (*Structured Query Language*), usually pronounced "sequel" or "ess-cue-ell." Created by IBM in the early 1970s and adopted as the industry standard in 1986, there are now many different implementations of SQL; no two are exactly the same, and each one is regarded as a *dialect*. In SQL, we can make a declarative statement that says, "From a given database, extract data that satisfy certain conditions," and the DBMS has to determine how to do it.

SQL has two main sublanguages:

- a *data definition language* (DDL) is used primarily by database administrators to define data structures by creating a database object (such as a table) and altering or destroying a database object. It does not operate on data.

- a *data manipulation language* (DML) is an interactive system that allows users to retrieve, delete, and update existing data from and add new data to the database.

There is also a *data control language* (DCL), a security system used by the database administrator, which controls the privileges granted to database users.

Before creating a database consisting of multiple tables, it is advisable to do the following: give a unique name to each table; specify which columns each table should contain and identify their data types; to each table, assign a primary key that uniquely identifies each row of the table; and have at least one common column in each table in the database.

We can then build a working data set through the DDL by using SQL `create table` statements of the following form:

```
create table <table name> (<table elements>);
```

where <table name> specifies a name for the table and <table elements> is a list separated by commas that specifies column names, their data types, and any column constraints. The set of data types depends upon the SQL dialect; they include: char(c) (a column of characters where c gives the maximum number of characters permitted in the column), integer, decimal(a, b) (where a is the total number of digits and b is the number of decimal places), date (in DBMS-approved format), and logical (True or False). The column constraints include null (that column may have empty row values) or not null (empty row values are not permitted in that column), primary keys, and any foreign keys. A semicolon ends the statement.

The DML includes such commands as select (allows users to retrieve specific database information), insert (adds new rows into an existing table), update (modifies information contained within a table), and delete (removes rows from a table). DML commands can be quite complicated and may include multiple expressions, clauses, predicates, or subqueries.

For example, the select statement (which supports restrict, project, and join operations, and is the most commonly used, but also most complicated SQL command) has the basic form

select <columns> from <table name> where <condition>;

where <columns> is a list of columns separated by commas. The select command is used to gather certain attributes from a particular RDBMS table, but where the tuples (rows) that are to be retrieved from those columns are limited to those that satisfy a given conditional Boolean search expression (i.e., True or False). One or more conditions may be joined by and or or operators as in set theory (the and always precedes the or operation). An asterisk may be used in place of the list of columns if all columns in the database are to be selected.

A primitive form of data analysis is included within the select statement through the use of five *aggregate operators*, sum, avg, max, min, and count, which provide the obvious *column statistics* over all rows that satisfy any stated conditions. For example, we can apply the command

select max(<column>) as max, min(<column>) as min from <table name> where <condition>;

to find the maximum (saved as "max") and minimum (saved as "min") of specified columns. Column statistics that are not aggregates (e.g., medians) are not available in SQL.

The smaller RDBMSs that are available include ACCESS (from Microsoft Corp.), MYSQL (open source), and MSQL (Hughes Technologies). These "lightweight" RDBMSs can support a few hundred simultaneous users and up to a gigabyte of data. All of the major statistical software packages that operate in a Windows environment can import data stored in certain of these smaller RDBMSs, especially Microsoft ACCESS.

We note that purists strongly object to SQL being thought of as a relational query language because, they argue, it sacrifices many of the fundamental principles of the relational model in order to satisfy demands of practicality and performance. RDBMSs are slow in general and, because the dialects of SQL are different enough and are often incompatible with each other, changing RDBMSs can be a nightmarish experience. Even so, SQL remains the most popular RDBMS query language.

2.4.3 OLTP Databases

A large organization is likely to maintain a DBMS that manages a *domain-specific* database for the automatic capture and storage of real-time business transactions. This type of database is essential for handling an organization's day-to-day operations. An *on-line transaction processing* (OLTP) system is a DBMS application that is specially designed for very fast tracking of millions of small, simple transactions each day by a large number of concurrent users (tellers, cashiers, and clerks, who add, update, or delete a few records at a time in the database). Examples of OLTP databases include Internet-based travel reservations and airline seat bookings, automated teller machines (ATM) network transactions and point-of-sale terminals, transfers of electronic funds, stock trading records, credit card transactions and authorizations, and records of driving license holders.

These OLTP databases are dynamic in nature, changing almost continuously as transactions are automatically recorded by the system minute-by-minute. It is not unusual for an organization to employ several different OLTP systems to carry out its various business functions (e.g., point-of-sale, inventory control, customer invoicing). Although OLTP systems are optimized for processing huge numbers of short transactions, they are not configured for carrying out complex ad hoc and data analytic queries.

2.4.4 Integrating Distributed Databases

In certain situations, data may be distributed over many geographically dispersed sites (*nodes*) connected by a *communications network* (usually some sort of local-area network or wide-area network, depending upon distances involved). This is especially true for the healthcare industry. A huge amount of information, for example, on hospital management practices may be recorded from a number of different hospitals and consist of overlapping sets of variables and cases, all of which have to be combined (or integrated) into a single database for analysis.

Distributed databases also commonly occur in multicenter clinical trials in the pharmaceutical industry, where centers include institutions, hospitals, and clinics, sometimes located in several countries. The number of total patients participating in such clinical trials rarely exceeds a few thousand, but there have been large-scale multicenter trials such as the Prostate Cancer Prevention Trial (Baker, 2001), which is a chemoprevention trial in which 18,000 men aged 55 years and older were randomized to either daily finasteride or placebo tablets for 7 years and involved 222 sites in the United States.

Data integration is the process of merging data that originate from multiple locations. When data are to be merged from different sources, several problems may arise:

- The data may be physically resident in computer files each of which was created using database software from different vendors.

- Different media formats may be used to store the information (e.g., audio or video tapes or DVDs, CDs or hard disks, hardcopy questionnaires, data downloaded over the Internet, medical images, scanned documents).

- The network of computer platforms that contain the data may be organized using different operating systems.

- The geographical locations of those platforms may be local or remote.

- Parts of the data may be duplicated when collected from different sources.

- Permission may need to be obtained from each source when dealing with sensitive data or security issues that will involve accessing personal, medical, business, or government records.

Faced with such potential inconsistencies, the information has to be integrated to become a consistent set of records for analysis.

2.4.5 Data Warehousing

An organization that needs to integrate multiple large OLTP databases will normally establish a single data warehouse for just that purpose. The term *data warehouse* was coined by W.H. Inmon to refer to a read-only, RDBMS running on a high-performance computer. The warehouse stores historical, detailed, and "scrubbed" data designed to be retrieved and queried efficiently and interactively by users through a dialect of SQL. Although data are not updated in realtime, fresh data can be added as supplements at regular intervals.

The components of a data warehouse are

DBMS: The publicly available RDBMSs that are almost mandatory for data warehousing usage include ORACLE (from Oracle Corp.), SQL SERVER (from Microsoft Corp.), SYBASE (from Sybase Inc.), POST-GRESQL (freeware), INFORMIX (from Informix Software, Inc.), and DB2 (from IBM Corp.). These "heavyweight" DBMSs can handle thousands of simultaneous users and can access up to several terabytes of data.

Hardware: It is generally accepted that large-scale data warehouse applications require either massively parallel-processing (MPP) or symmetric multiprocessing (SMP) supercomputers. Which type of hardware is installed depends upon many factors, including the complexity of the data and queries and the number of users that need to access the system.

- SMP architectures are often called "shared everything" because they share memory and resources to service more than a single CPU, they run a single copy of the operating system, and they share a single copy of each application. SMP is reputed to be better for those data warehouses whose capacity ranges between 50GB and 100GB.

- MPP architectures, on the other hand, are called "shared nothing"; they may have hundreds of CPUs in a single computer, each node of which is a self-contained computer with its own CPU, disk, and memory, and nodes are connected by a high-speed bus or switch. The larger the data warehouse (with capacity at least 200GB) and the more complex the queries, the more likely the organization will install an MPP server.

Such centralized data depositories typically contain huge quantities of information taking up hundreds of gigabytes or terabytes of disk space. Small data warehouses, which store subsets of the central warehouse for use by specialized groups or departments, are referred to as *data marts.*

More and more organizations that require a central data storage facility are setting up their own data warehouses and data marts. For example, according to Monk (2000), the Foreign Trade Division of the U.S. Census Bureau processes 5 million records each month from the U.S. Customs Service on 18,000 import commodities and 9,000 export commodities that travel between 250 countries and 50 regions within the United States. The raw import-export data are extracted, "scrubbed," and loaded into a data warehouse having one terabyte of storage. Subsets of the data that focus on specific countries and commodities, together with two years of historical data, are then sent to a number of data marts for faster and more specific querying.

It has been reported that 90 percent of all Fortune 500 companies are currently (or soon will be) engaged in some form of data warehousing activity. Corporations such as Federal Express, UPS, JC Penney, Office Depot, 3M, Ace Hardware, and Sears, Roebuck and Co. have installed data warehouses that contain multi-terabytes of disk storage, and Wal-Mart and Kmart are already at the 100 terabyte range. These retailers use their data warehouses to access comprehensive sales records (extracted from the scanners of cash registers) and inventory records from thousands of stores worldwide.

Institutions of higher education now have data warehouses for information on their personnel, students, payroll, course enrollments and revenues, libraries, finance and purchasing, financial aid, alumni development, and campus data. Healthcare facilities have data warehouses for storing uniform billing data on hospital admissions and discharges, outpatient care, long-term care, individual patient records, physician licensing, certification, background, and specialties, operating and surgical profiles, financial data, CMS (Centers for Medicare and Medicaid Services) regulations, and nursing homes, and that might soon include image data.

2.4.6   Decision Support Systems and OLAP

The failure of OLTP systems to deliver analytical support (e.g., statistical querying and data analysis) of RDBMSs caused a major crisis in the database market until the concept of data warehouses each with its own *decision support system* (DSS) emerged. In a client/server computing environment, decision support is carried out using *on-line analytical processing* (OLAP) software tools.

There are two primary architectures for OLAP systems, ROLAP (*relational* OLAP) and MOLAP (*multidimensional* OLAP); in both, multivariate data are set up using a multidimensional model rather than the standard model, which emphasizes data-as-tables. The two systems store data differently, which in turn affects their performance characteristics and the amounts of data that can be handled.

ROLAP operates on data stored in a RDBMS. Complex multipass SQL commands can create various ad hoc multidimensional views of a two-dimensional data table (which slows down response times). ROLAP users can access all types of transactional data, which are stored in 100GB to multiple-terabyte data warehouses.

MOLAP operates on data stored in a specialized multidimensional DBMS. Variables are scaled categorically to allow transactional data to be pre-aggregated by all category combinations (which speeds up response times) and the results stored in the form of a "data cube" (a large, but sparse, multidimensional contingency table). MOLAP tools can handle up to 50GB of data stored in a data mart.

OLAP users typically access multivariate databases without being aware exactly which system has been implemented. There are other OLAP systems, including a hybrid version HOLAP.

The data analysis tools provided by a multidimensional OLAP system include operators that can *roll-up* (aggregate further, producing marginals), *drill-down* (de-aggregate to search for possible irregularities in the aggregates), *slice* (condition on a single variable), and *dice* (condition on a particular category) aggregated data in a multidimensional contingency table. Summary statistics that cannot be represented as aggregates (e.g., medians, modes) and graphics that need raw data for display (e.g., scatterplots, time series plots) are generally omitted from MOLAP menus (Wilkinson, 2005).

2.4.7   Statistical Packages and DBMSs

Some statistical analysis packages (e.g., SAS, SPSS) and MATLAB can run their complete libraries of statistical routines against their OLAP database servers.

A major effort is currently under way to provide a common interface for the S language (i.e., S-PLUS and particularly R) to access the really big DBMSs so that sophisticated data analysis can be carried out in a transparent manner (i.e., DBMS and platform independent). Although a *table* in a RDBMS is very similar to the concept of *data frame* in R and S-PLUS, there are many difficulties in building such interfaces.

The R package RODBC (written by Michael Lapsley and Brian Ripley, and available from CRAN) provides an R interface to DBMSs based upon the Microsoft ODBC (Open Database Connectivity) standard. RODBC, which runs on both MS Windows and Unix/Linux, is able to copy an R data frame to a table in a database (command: `sqlSave`), read a table from a DBMS into an R data frame (`sqlFetch`), submit an SQL query to an ODBC database (`sqlQuery`), retrieve the results (`sqlGetResults`), and update the table where the rows already exist (`sqlUpdate`). RODBC works with ORACLE, MS ACCESS, SYBASE, DB2, MYSQL, POSTGRESQL, and SQL SERVER on MS Windows platforms and with MYSQL, POSTGRESQL, and ORACLE under Unix/Linux.

2.5   Data Quality Problems

Errors exist in all kinds of databases. Those that are easy to detect will most likely be found at the data "cleaning" stage, whereas those errors that can be quite resistant to detection might only be discovered during data analysis. Data cleaning usually takes place as the data are received and before they are stored in read-only format in a data warehouse. A consistent and cleaned-up version of the data can then be made available.

2.5.1 Data Inconsistencies

Errors in compiling and editing the resulting database are common and actually occur with alarming frequency, especially in cases where the data set is very large. When data from different sources are being connected, inconsistencies as to a person's name (especially in cases where a name can be spelled in several different ways) occur frequently, and matching (or "disambiguation") has to take place before such records can be merged. One popular solution is to employ Soundex (sound-indexing) techniques for name matching.

To get an idea of how poor data quality can become, consider the problem of estimating the extent of the undercount from census data collected for the 1990 U.S. census. Breiman (1994) identified a number of sources of error, including the following: *Matching errors* (incorrectly matching records from two different files of people with differing names, ages, missing gender or race identifiers, and different addresses), *fabrications* (the creation of fictitious people by dishonest interviewers), *census day address errors* (incorrectly recording the location of a person's residence on census day), *unreliable interviews* (many of the interviews were rejected as being unreliable), and *incomplete data* (a lack of specific information on certain members in the household). Most of the problems involving data fabrication, incomplete data, and unreliable interviews apparently occurred in areas that also had the highest estimated undercounts, such as the central cities and minority areas.

Massive data sets are prone to mistakes, errors, distortions, and, in general, poor data quality, just as is any data set, but such defects occur here on a far grander scale because of the size of the data set itself. When invalid product codes are entered for a product, they may easily be detected; when valid product codes, however, are entered for the wrong product, detection becomes more difficult. Customer codes may be entered inconsistently, especially those for gender identification (M and F, as opposed to 1 and 2). Duplication of records entered into the database from multiple sources can also be a problem. In these days of takeovers and buyouts, and mergers and acquisitions, what was once a code for a customer may now be a problem if the entity has since changed its description (e.g., Jenn-Air, Hoover, Norge, Magic Chef, etc., are all now part of Maytag Corp.). Any inconsistencies in historical data may also be difficult to correct if those who knew the answer are no longer with the company.

2.5.2 Outliers

Outliers are values in the data that, for one reason or another, do not appear to fit the pattern of the other data values; visually, they are located far away from the rest of the data. It is not unusual for outliers to be present in a data set.

Outliers can occur for many different reasons but should not be confused with *gross errors*. Gross errors are cases where "something went wrong" (Hampel, 2002); they include human errors (e.g., a numerical value recorded incorrectly) and mechanical errors (e.g., malfunctioning of a measuring instrument or a laboratory instrument during analysis). The density of gross errors depends upon the context and the quality of the data. In medical studies, gross error rates in excess of 10% have been quoted.

Univariate outliers are easy to detect when they indicate impossible (or "out of bounds") values. More often, an outlier will be a value that is extreme, either too large or too small. For multivariate data, outlier detection is more difficult. Low-dimensional visual displays of the data (such as histograms, boxplots, scatterplots) can encourage insight into the data and provide at the same time a method for manually detecting some of the more obvious univariate or bivariate outliers.

When we have a large data set, outliers may not be all that rare. Unlike a data set of 100 or so observations, where we may find two or three outliers, in a data set of 100,000, we should not be surprised to discover a large number (in some cases, hundreds, and maybe even thousands) of outliers. For example, Figure 2.5 shows a scatterplot of the size (in bytes) of each of 50,000 packets[5] containing roughly two minutes worth of TCP (transfer control protocol) packet traffic between Digital Equipment Corporation servers and the rest of the world on 8th March 1995 plotted against time. We see clear structure within the scatterplot: the vast majority of points occur within the 0–512 bytes range, and a number of dense horizontal bands occur inside this range; these bands show that the vast majority of packets sent consist of either 0 bytes (37% of the total packets), which are used only to acknowledge data sent by the other side, or 512 bytes (29% of the total packets). There are 952 packets each having more than 512 bytes, of which 137 points are identified as outliers (with values greater than 1.5 times IQR), including 61 points equal to the largest value, 1460 bytes.

To detect true multidimensional outliers, however, becomes a test of statistical ingenuity. A multivariate observation whose every component value may appear indistinguishable from the rest may yet be regarded as an outlier when all components are treated simultaneously. In large

---

[5]See www.amstat.org/publications/jse/datasets/packetdata.txt.

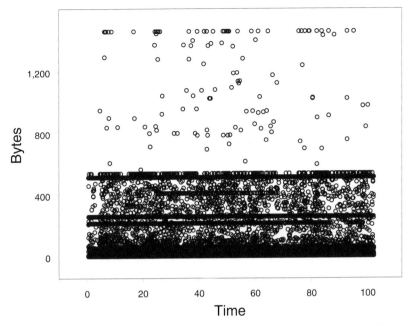

FIGURE 2.5. *Time-series plot of 50,000 packets containing roughly two minutes worth of TCP (transfer control protocol) packets traffic between Digital Equipment Corporation servers and the rest of the world on 8th March 1995.*

multivariate data sets, some combination of visual display of the data, manual outlier detection scheme, and automatic outlier detection program may be necessary: potential outliers could be "flagged" by an automatic screening device, and then an analyst would manually decide on the fate of that flagged outlier.

2.5.3 Missing Data

In the vast majority of data sets, there will be missing data values. For example, human subjects may refuse to answer certain items in a battery of questions because personal information is requested; some observations may be accidentally lost; some responses may be regarded as implausible and rejected; and in a study of financial records of a company, some records may not be available because of changes in reporting requirements and data from merged or reorganized organizations.

In R/S-PLUS, missing values are denoted by NA. In large databases, SQL incorporates the null as a *flag* or *mark* to indicate the absence of a data value, which might mean that the value is missing, unknown, nonexistent (no observation could be made for that entry), or that no value has yet been assigned. A `null` is not equivalent to a zero value or to a text string filled with spaces. Sometimes, missing values are replaced by zeroes, other times by estimates of what they should be based on the rest of the data.

One popular method deletes those observations that contain missing data and analyzes only those cases that are observed in their entirety (often called *complete-case analysis* or *listwise-deletion method*). Such a complete-case analysis may be satisfactory if the proportion of deleted observations is small relative to the size of the entire data set and if the mechanism that leads to the missing data is independent of the variables in question — an assumption referred to by Donald Rubin as *missing at random* (MAR) or *missing completely at random* (MCAR) depending upon the exact nature of the *missing-data mechanism* (Little and Rubin, 1987). Any deleted observations may be used to help justify the MCAR assumption.

If the missing data constitute a sizeable proportion of the entire data set, then complete-case methods will not work. *Single imputation* has been used to impute (or "fill in") an estimated value for each missing observation and then analyze the amended data set as if there had been no missing values in the first place. Such procedures include *hot-deck imputation*, where a missing value is imputed by substituting a value from a similar but complete record in the same data set; *mean imputation*, where the singly imputed value is just the mean of all the completely recorded values for that variable; and *regression imputation*, which uses the value predicted by a regression on the completely recorded data. Because sampling variability due to single imputation cannot be incorporated into the analysis as an additional source of variation, the standard errors of model estimates tend to be underestimated.

Since the late 1970s, Rubin and his colleagues have introduced a number of sophisticated algorithmic methods for dealing with incomplete data situations. One approach, the *EM algorithm* (Dempster, Laird, and Rubin, 1977; Little and Rubin, 1987), which alternates between an expectation (E) step and a maximization (M) step, is used to compute maximum-likelihood estimates of model parameters, where missing data are modeled as unobserved latent variables. We shall describe applications of the EM algorithm in more detail in later chapters of this book. A different approach, *multiple imputation* (Rubin, 1987), fills in the missing values $m > 1$ times, where the imputed values are generated each time from a distribution that may be different for each missing value; this creates m different data sets, which are analyzed separately, and then the m results are combined to estimate model parameters, standard errors, and confidence intervals.

2.5.4   More Variables than Observations

Many statistical computer packages do not allow the number of input variables, r, to exceed the number of observations, n, because, then, certain matrices, such as the $(r \times r)$ covariance matrix, would have less than full rank, would be singular, and, hence, uninvertible. Yet, we should not be surprised when $r > n$. In fact, this situation occurs quite routinely in certain applications, and in such instances, r can be much greater than n. Typical examples include:

Satellite images When producing maps, remotely sensed image data are gathered from many sources, including satellite and aircraft scanners, where a few observations (usually fewer than 10 spectral bands) are measured at more than 100,000 wavelengths over a grid of pixels.

Chemometrics For determining concentrations in certain chemical compounds, calibration studies often need to analyze intensity measurements on a very large number (500–1,000 or more) of different spectral wavelengths using a small number of standard chemical samples.

Gene expression data Current microarray methods for studying human malignancies, such as tumors, simultaneously monitor expression levels of very large numbers of genes (5,000–10,000 or more) on relatively small numbers (fewer than 100) of tumor samples.

When $r > n$, one way of dealing with this problem is to analyze the data on each variable separately. However, this suggestion does not take account of correlations between the variables. Researchers have recently provided new statistical techniques that are not sensitive to the $r > n$ issue. We will address this situation in various sections of this book.

2.6   The Curse of Dimensionality

The term "curse of dimensionality" (Bellman, 1961) originally described how difficult it was to perform high-dimensional numerical integration. This led to the more general use of the term to describe the difficulty of dealing with statistical problems in high dimensions. Some implications include:

1. We can never have enough data to cover every part of high-dimensional input space to learn which part of the space is important to a relationship and which is not.

To see this, divide the axis of each of r input variables into K uniform intervals (or "bins"), so that the value of an input variable is approximated by the bin into which it falls. Such a partition divides the entire r-dimensional input space into K^r "hypercubes," where K is chosen so that each hypercube contains at least one point in the input space. Given a specific hypercube in input space, an output value y_0 corresponding to a new input point in the hypercube can be approximated by computing some function (e.g., the average value) of the y values that correspond to all the input points falling in that hypercube. Increasing K reduces the sizes of the hypercubes while increasing the precision of the approximation. However, at the same time, the number of hypercubes increases exponentially. If there has to be at least one input point in each hypercube, then the number of such points needed to cover all of r-space must also increase exponentially as r increases. In practice, we have a limited number of observations, with the result that the data are very sparsely spread around high-dimensional space.

2. *As the number of dimensions grows larger, almost all the volume inside a hypercubic region of input space lies closer to the boundary or surface of the hypercube rather than near the center.*

An r-dimensional hypercube $[-A, A]^r$ with each edge of length $2A$ has volume $(2A)^r$. Consider a slightly smaller hypercube with each edge of length $2(A - \epsilon)$, where $\epsilon > 0$ is small. The difference in volume between these two hypercubes is $(2A)^r - 2^r(A - \epsilon)^r$, and, hence, the proportion of the volume that is contained between the two hypercubes is

$$\frac{(2A)^r - 2^r(A - \epsilon)^r}{(2A)^r} = 1 - \left(1 - \frac{\epsilon}{A}\right)^r \rightarrow 1 \text{ as } r \rightarrow \infty.$$

In Figure 2.6, we see a graphical display of this result for $A = 1$ and number of dimensions $r = 1, 2, 10, 20, 50$. The same phenomenon also occurs with spherical regions in high-dimensional input space (see Exercise 2.4).

Bibliographical Notes

There are many different kinds of data sets and every application field measures items in its own way. The following issues of *Statistical Science* address the problems inherent with certain types of data: consumer transaction data and e-commerce data (May 2006), Internet data (August 2004), and microarray data (February 2003).

The Human Genome Project and Celera. a private company, simultaneously published draft accounts of the human genome in *Nature* and *Science* on 15th and 16th February 2001, respectively. An excellent article on gene expression is Sebastiani, Gussoni, Kohane, and Ramoni (2003). Books on the design and analysis of DNA microarray experiments and analyzing gene expression data are Drăghici (2003), Simon, Korn, McShane, Radmacher, Wright, and Zhao (2004), and the books edited by Parmigiani, Garrett, Irizarry, and Zeger (2003), Speed (2003), and Lander and Waterman (1995).

There are a huge number of books on database management systems. We found the books by Date (2000) and Connolly and Begg (2002) most useful. The concept of a "relational" database system originates with Codd (1970),

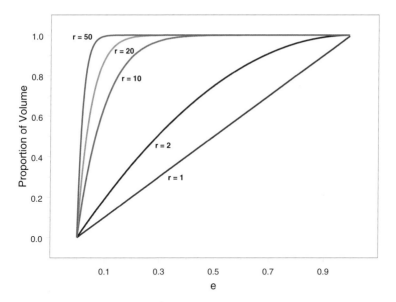

FIGURE 2.6. *Graphs of the proportion of the total volume contained between two hypercubes, one of edge length 2 and the other of edge length $2 - e$ for different numbers of dimensions r. As the number of dimensions increases, almost all the volume becomes closer to the surface of the hypercube.*

who received the 1981 ACM Turing Award for his work in the area. An excellent survey of the development and maintenance of biological databases and microarray repositories is given by Valdivia-Granda and Dwan (2006).

Books on missing data include Little and Rubin (1987) and Schafer (1997). A book on the EM algorithm is McLachlan and Krishnan (1997). For multiple imputation, see the book by Rubin (1987). Books on outlier detection include Rousseeuw and Leroy (1987) and Barnett and Lewis (1994).

Exercises

2.1 In a statistical application of your choice, what does a missing value mean? What are the traditional methods of imputing missing values in such an application?

2.2 In sample surveys, such as opinion polls, telephone surveys, and questionnaire surveys, nonresponse is a common occurrence. How would you design such a survey so as to minimize nonresponse?

2.3 Discuss the differences between single and multiple imputation for imputing missing data.

2.4 The volume of an r-dimensional sphere with radius A is given by $\mathrm{vol}_r(A) = S_r A^r / r$, where $S_r = 2\pi^{r/2}/\Gamma(r/2)$ is the surface area of the unit sphere in r dimensions, $\Gamma(x) = \int_0^\infty t^{x-1} e^{-t} dt = (x-1)!, 1x > 0$, is the gamma function, $\Gamma(x+1) = x\Gamma(x)$, and $\Gamma(1/2) = \pi^{1/2}$. Find the appropriate spherical volumes for two and three dimensions. Using a similar limiting argument as in (2) of Section 2.6, show that as the dimensionality increases, almost all the volume inside the sphere tends to be concentrated along a "thin shell" closer to the surface of the sphere than to the center.

2.5 Consider a hypercube of dimension r and sides of length $2A$ and inscribe in it an r-dimensional sphere of radius A. Find the proportion of the volume of the hypercube that is inside the hypersphere, and show that the proportion tends to 0 as the dimensionality r increases. In other words, show that all the density sits in the corners of the hypercube.

2.6 What are the advantages and disadvantages of database systems, and when would you find such a system useful for data analysis?

2.7 Find a commercial SQL product and discuss the various options that are available for the `create table` statement of that product.

2.8 Find a DBMS and investigate whether that system keeps track of database statistics. Which statistics does it maintain, how does it do that, and how does it update those statistics?

2.9 What are the advantages and disadvantages of distributed database systems?

2.10 (Fairley, Izenman, and Crunk, 2001) You are hired to carry out a survey of damage to the bricks of the walls of a residential complex consisting of five buildings, each having 5, 6, or 7 stories. The type of damage of interest is called spalling and refers to deterioration of the surface of the brick, usually caused by freeze-thaw weather conditions. Spalling appears to be high at the top stories and low at the ground. The walls consist of three-quarter million bricks. You take a photographic survey of all the walls of the complex and count the number of bricks in the photographs that are spalled. However, the photographs show that some portions of the walls are obscured by bushes, trees, pipes, vehicles, etc. So, the photographs are not a complete record of brick damage in the complex. Discuss how would you estimate the spall rate (spalls per 1,000 bricks) for the entire complex. What would you do about the missing data in your estimation procedure?

2.11 Read about MAR (missing at random) and MCAR (missing completely at random) and discuss their differences and implications for imputing missing data.

3
Random Vectors and Matrices

3.1 Introduction

This chapter builds the foundation for the statistical analysis of multivariate data. We first give the notation we use in this book, followed by a quick review of the rules for manipulating vectors and matrices. Then, we learn about random vectors and matrices, which are the fundamental building blocks for multivariate analysis. We then describe the properties of a variety of estimators of an unknown mean vector and unknown covariance matrix of a multivariate Gaussian distribution.

3.2 Vectors and Matrices

In this section, we briefly review the notation, terminology, and basic operations and results for vectors and matrices.

3.2.1 Notation

Vectors having J elements will be represented as column vectors (i.e., as $(J \times 1)$-matrices, which we will refer to as J-vectors for convenience) and will

A.J. Izenman, *Modern Multivariate Statistical Techniques*,
doi: 10.1007/978-0-387-78189-1_3,
© Springer Science+Business Media, LLC 2008

be represented by boldface letters, either uppercase (e.g., $\mathbf{X}$) or lowercase (e.g., $\mathbf{x}$, $\boldsymbol{\alpha}$) depending upon the context. Two J-vectors, $\mathbf{x} = (x_1, \cdots, x_J)^\tau$ and $\mathbf{y} = (y_1, \cdots, y_J)^\tau$, are orthogonal if $\mathbf{x}^\tau \mathbf{y} = \sum_{j=1}^{J} x_j y_j = 0$.

We denote matrices by uppercase boldface letters (e.g., $\mathbf{A}$, $\boldsymbol{\Sigma}$) or by capital script letters (e.g., $\mathcal{X}, \mathcal{Y}, \mathcal{Z}$). Thus, the $(J \times K)$ matrix $\mathbf{A} = (A_{jk})$ has J rows and K columns and jkth entry A_{jk}. If $J = K$, then $\mathbf{A}$ is said to be *square*. The $(J \times J)$ *identity matrix* $\mathbf{I}_J$ has $I_{jj} = 1$ and $I_{jk} = 0, j \neq k$, The *null matrix* $\mathbf{0}$ has all entries equal to zero.

3.2.2 Basic Matrix Operations

If $\mathbf{A} = (A_{jk})$ is a $(J \times K)$-matrix, then the *transpose* of $\mathbf{A}$ is the $(K \times J)$-matrix denoted by $\mathbf{A}^\tau = (A_{kj})$. If $\mathbf{A} = \mathbf{A}^\tau$, then $\mathbf{A}$ is said to be *symmetric*.

The *sum* of two $(J \times K)$ matrices $\mathbf{A}$ and $\mathbf{B}$ is $\mathbf{A} + \mathbf{B} = (A_{jk} + B_{jk})$, and its transpose is $(\mathbf{A} + \mathbf{B})^\tau = \mathbf{A}^\tau + \mathbf{B}^\tau = (A_{kj} + B_{kj})$. The inequality $\mathbf{A} + \mathbf{B} \geq \mathbf{A}$ holds if $\mathbf{B} \geq \mathbf{0}$ (i.e., $B_{jk} \geq 0$, all j and k).

The *product* of a $(J \times K)$-matrix $\mathbf{A}$ and a $(K \times L)$-matrix $\mathbf{B}$ is the $(J \times L)$-matrix $(C_{jl}) = \mathbf{C} = \mathbf{AB} = (\sum_{k=1}^{K} A_{jk} B_{kl})$. Note that $(\mathbf{AB})^\tau = \mathbf{B}^\tau \mathbf{A}^\tau$. Multiplication of a $(J \times K)$-matrix $\mathbf{A}$ by a scalar a is the $(J \times K)$-matrix $a\mathbf{A} = (aA_{jk})$.

A $(J \times J)$-matrix $\mathbf{A}$ is *orthogonal* if $\mathbf{AA}^\tau = \mathbf{A}^\tau \mathbf{A} = \mathbf{I}_J$ and is *idempotent* if $\mathbf{A}^2 = \mathbf{A}$. A square matrix $\mathbf{P}$ is a *projection matrix* (or a *projector*) iff $\mathbf{P}$ is idempotent. If $\mathbf{P}$ is both idempotent and orthogonal, then $\mathbf{P}$ is called an *orthogonal projector*. If $\mathbf{P}$ is idempotent, then so is $\mathbf{Q} = \mathbf{I} - \mathbf{P}$; $\mathbf{Q}$ is called the *complementary projector* to $\mathbf{P}$.

The *trace* of a square $(J \times J)$ matrix $\mathbf{A}$ is denoted by $\text{tr}(\mathbf{A}) = \sum_{j=1}^{J} A_{jj}$. Note that for square matrices $\mathbf{A}$ and $\mathbf{B}$, $\text{tr}(\mathbf{A} + \mathbf{B}) = \text{tr}(\mathbf{A}) + \text{tr}(\mathbf{B})$, and for $(J \times K)$-matrix $\mathbf{A}$ and $(K \times J)$ matrix $\mathbf{B}$, $\text{tr}(\mathbf{AB}) = \text{tr}(\mathbf{BA})$.

The *determinant* of a $(J \times J)$-matrix $\mathbf{A} = (A_{ij})$ is denoted by either $|\mathbf{A}|$ or $\det(\mathbf{A})$. The *minor* $\mathbf{M}_{ij}$ of element A_{ij} is the $(J - 1 \times J - 1)$-matrix formed by removing the ith row and jth column from $\mathbf{A}$. The *cofactor* of A_{ij} is $C_{ij} = (-1)^{i+j} |\mathbf{M}_{ij}|$. One way of defining the determinant of $\mathbf{A}$ is by using *Laplace's formula*: $|\mathbf{A}| = \sum_{j=1}^{J} A_{ij} C_{ij}$, where we expand along the ith row. Note that $|\mathbf{A}^\tau| = |\mathbf{A}|$. If a is a scalar and $\mathbf{A}$ is $(J \times J)$, then $|a\mathbf{A}| = a^J |\mathbf{A}|$. $\mathbf{A}$ is *singular* if $|\mathbf{A}| = 0$, and *nonsingular* otherwise.

Matrix decompositions include the *LR decomposition* ($\mathbf{A} = \mathbf{LR}$, where $\mathbf{L}$ is lower-triangular and $\mathbf{R}$ is upper-triangular), the *Cholesky decomposition* ($\mathbf{A} = \mathbf{LL}^\tau$, where $\mathbf{L}$ is lower-triangular and $\mathbf{A}$ is symmetric positive-definite), and the *QR decomposition* ($\mathbf{A} = \mathbf{QR}$, where $\mathbf{Q}$ is orthogonal and $\mathbf{R}$ is upper-triangular). These matrix decompositions are used as efficient methods of computing $|\mathbf{A}|$ by applying the following results: $|\mathbf{AB}| = |\mathbf{A}| \cdot |\mathbf{B}|$ if both $\mathbf{A}$ and $\mathbf{B}$ are $(J \times J)$; the determinant of a triangular matrix is the product of its diagonal entries; and for orthogonal $\mathbf{Q}$, $|\det(\mathbf{Q})| = 1$.

Let

$$\mathbf{\Sigma} = \begin{pmatrix} \mathbf{A} & \mathbf{B} \\ \mathbf{C} & \mathbf{D} \end{pmatrix} \tag{3.1}$$

be a partitioned matrix, where $\mathbf{A}$ and $\mathbf{D}$ are both square and nonsingular. Then, the determinant of $\mathbf{\Sigma}$ can be expressed in two ways:

$$|\mathbf{\Sigma}| = |\mathbf{A}| \cdot |\mathbf{D} - \mathbf{C}\mathbf{A}^{-1}\mathbf{B}| = |\mathbf{D}| \cdot |\mathbf{A} - \mathbf{B}\mathbf{D}^{-1}\mathbf{C}|. \tag{3.2}$$

The *rank* of $\mathbf{A}$, denoted $r(\mathbf{A})$, is the size of the largest submatrix of $\mathbf{A}$ that has a nonzero determinant; it is also the number of linearly independent rows or columns of $\mathbf{A}$. Note that $r(\mathbf{AB}) = r(\mathbf{A})$ if $|\mathbf{B}| \neq 0$, and, in general, $r(\mathbf{AB}) \leq \min(r(\mathbf{A}), r(\mathbf{B}))$.

If $\mathbf{A}$ is square, $(J \times J)$, and nonsingular, then a unique $(J \times J)$ *inverse matrix* $\mathbf{A}^{-1}$ exists such that $\mathbf{A}\mathbf{A}^{-1} = \mathbf{I}_J$. If $\mathbf{A}$ is orthogonal, then $\mathbf{A}^{-1} = \mathbf{A}^{\tau}$. Note that $(\mathbf{AB})^{-1} = \mathbf{B}^{-1}\mathbf{A}^{-1}$, and $|\mathbf{A}^{-1}| = |\mathbf{A}|^{-1}$. A useful result involving inverses is

$$(\mathbf{A} + \mathbf{B}\mathbf{D}^{-1}\mathbf{C})^{-1} = \mathbf{A}^{-1} - \mathbf{A}^{-1}\mathbf{B}(\mathbf{D} + \mathbf{C}\mathbf{A}^{-1}\mathbf{B})^{-1}\mathbf{C}\mathbf{A}^{-1}, \tag{3.3}$$

where $\mathbf{A}$ and $\mathbf{D}$ are $(J \times J)$ and $(K \times K)$ nonsingular matrices, respectively. If $\mathbf{A}$ is $(J \times J)$ and $\mathbf{u}$ and $\mathbf{v}$ are J-vectors, then, a special case of this result is

$$(\mathbf{A} + \mathbf{u}\mathbf{v}^{\tau})^{-1} = \mathbf{A}^{-1} - \frac{(\mathbf{A}^{-1}\mathbf{u})(\mathbf{v}^{\tau}\mathbf{A}^{-1})}{1 + \mathbf{v}^{\tau}\mathbf{A}^{-1}\mathbf{u}}, \tag{3.4}$$

which reduces the problem of inverting $\mathbf{A} + \mathbf{u}\mathbf{v}^{\tau}$ to one of just inverting $\mathbf{A}$. If $\mathbf{A}$ and $\mathbf{D}$ are symmetric matrices and $\mathbf{A}$ is nonsingular, then,

$$\begin{pmatrix} \mathbf{A} & \mathbf{B} \\ \mathbf{B}^{\tau} & \mathbf{D} \end{pmatrix}^{-1} = \begin{pmatrix} \mathbf{A}^{-1} + \mathbf{F}\mathbf{E}^{-1}\mathbf{F}^{\tau} & -\mathbf{F}\mathbf{E}^{-1} \\ -\mathbf{E}\mathbf{F}^{\tau} & \mathbf{E}^{-1} \end{pmatrix}, \tag{3.5}$$

where $\mathbf{E} = \mathbf{D} - \mathbf{B}^{\tau}\mathbf{A}^{-1}\mathbf{B}$ is nonsingular and $\mathbf{F} = \mathbf{A}^{-1}\mathbf{B}$.

If $\mathbf{A}$ is a $(J \times J)$-matrix and $\mathbf{x}$ is a J-vector, then a *quadratic form* is $\mathbf{x}^{\tau}\mathbf{A}\mathbf{x} = \sum_{j=1}^{J} \sum_{k=1}^{J} A_{jk} x_j x_k$. A $(J \times J)$-matrix $\mathbf{A}$ is *positive-definite* if, for any J-vector $\mathbf{x} \neq \mathbf{0}$, the quadratic form $\mathbf{x}^{\tau}\mathbf{A}\mathbf{x} > 0$, and is *nonnegative-definite* (or *positive-semidefinite*) if the same quadratic form is nonnegative.

3.2.3 Vectoring and Kronecker Products

The *vectoring* operation $\text{vec}(\mathbf{A})$ denotes the $(JK \times 1)$-column vector formed by placing the columns of a $(J \times K)$-matrix $\mathbf{A}$ under one another successively.

If a $(J \times K)$-matrix $\mathbf{A}$ is such that the jkth element $\mathbf{A}_{jk}$ is itself a submatrix, then $\mathbf{A}$ is termed a *block matrix*. The *Kronecker product* of a

$(J \times K)$-matrix $\mathbf{A}$ and an $(L \times M)$-matrix $\mathbf{B}$ is the $(JL \times KM)$ block matrix

$$\mathbf{A} \otimes \mathbf{B} = (\mathbf{A}B_{jk}) = \begin{pmatrix} \mathbf{A}B_{11} & \cdots & \mathbf{A}B_{1M} \\ \vdots & & \vdots \\ \mathbf{A}B_{L1} & \cdots & \mathbf{A}B_{LM} \end{pmatrix}. \qquad (3.6)$$

Strictly speaking, the definition (3.6) is commonly known as the *left Kronecker product*. There is also the *right Kronecker product* in the literature, $\mathbf{A} \otimes \mathbf{B} = (A_{ij}\mathbf{B})$, which, in our notation, is given by $\mathbf{B} \otimes \mathbf{A}$.

The following operations hold for Kronecker products as defined by (3.6):

$$(\mathbf{A} \otimes \mathbf{B}) \otimes \mathbf{C} = \mathbf{A} \otimes (\mathbf{B} \otimes \mathbf{C}) \qquad (3.7)$$
$$(\mathbf{A} \otimes \mathbf{B})(\mathbf{C} \otimes \mathbf{D}) = (\mathbf{AC}) \otimes (\mathbf{BD}) \qquad (3.8)$$
$$(\mathbf{A} + \mathbf{B}) \otimes \mathbf{C} = (\mathbf{A} \otimes \mathbf{C}) + (\mathbf{B} \otimes \mathbf{C}) \qquad (3.9)$$
$$(\mathbf{A} \otimes \mathbf{B})^{\tau} = \mathbf{A}^{\tau} \otimes \mathbf{B}^{\tau} \qquad (3.10)$$
$$\mathrm{tr}(\mathbf{A} \otimes \mathbf{B}) = (\mathrm{tr}(\mathbf{A}))(\mathrm{tr}(\mathbf{B})) \qquad (3.11)$$
$$r(\mathbf{A} \otimes \mathbf{B}) = r(\mathbf{A}) \cdot r(\mathbf{B}) \qquad (3.12)$$

If $\mathbf{A}$ is $(J \times J)$ and $\mathbf{B}$ is $(K \times K)$, then,

$$|\mathbf{A} \otimes \mathbf{B}| = |\mathbf{A}|^{K}|\mathbf{B}|^{J} \qquad (3.13)$$

If $\mathbf{A}$ is $(J \times K)$ and $\mathbf{B}$ is $(L \times M)$, then,

$$\mathbf{A} \otimes \mathbf{B} = (\mathbf{A} \otimes \mathbf{I}_{L})(\mathbf{I}_{K} \otimes \mathbf{B}) \qquad (3.14)$$

If $\mathbf{A}$ and $\mathbf{B}$ are square and nonsingular, then,

$$(\mathbf{A} \otimes \mathbf{B})^{-1} = \mathbf{A}^{-1} \otimes \mathbf{B}^{-1} \qquad (3.15)$$

One of the most useful results that combines vectoring with Kronecker products is that

$$\mathrm{vec}(\mathbf{ABC}) = (\mathbf{A} \otimes \mathbf{C}^{\tau})\mathrm{vec}(\mathbf{B}). \qquad (3.16)$$

3.2.4 Eigenanalysis for Square Matrices

If $\mathbf{A}$ is a $(J \times J)$-matrix, then $|\mathbf{A} - \lambda \mathbf{I}_{J}|$ is a polynomial of order J in λ. The equation

$$|\mathbf{A} - \lambda \mathbf{I}_{J}| = 0$$

will have J (possibly complex-valued) roots denoted by $\lambda_j = \lambda_j(\mathbf{A})$, $j = 1, 2, \ldots, J$. The root λ_j is called the *eigenvalue* (*characteristic root, latent root*) of $\mathbf{A}$, and the set $\{\lambda_j\}$ is called the *spectrum* of $\mathbf{A}$. Associated with λ_j, there is a J-vector $\mathbf{v}_j = \mathbf{v}_j(\mathbf{A})$ (not all of whose entries of zero) such that

$$(\mathbf{A} - \lambda_j \mathbf{I}_{J})\mathbf{v}_j = 0.$$

The vector $\mathbf{v}_j$ is called the *eigenvector* (*characteristic vector, latent vector*) associated with λ_j. Eigenvalues of positive-definite matrices are all positive, and eigenvalues of nonnegative-definite matrices are all nonnegative.

The following results for a real and symmetric $(J \times J)$-matrix $\mathbf{A}$ are not difficult to prove. All the eigenvalues of $\mathbf{A}$ are real and the eigenvectors can be chosen to be real. Eigenvectors $\mathbf{v}_j$ and $\mathbf{v}_k$ associated with distinct eigenvalues $(\lambda_j \neq \lambda_k)$ are orthogonal. If $\mathbf{V} = (\mathbf{v}_1, \mathbf{v}_2, \ldots, \mathbf{v}_J)$, then

$$\mathbf{A}\mathbf{V} = \mathbf{V}\Lambda, \tag{3.17}$$

where $\Lambda = \mathrm{diag}\{\lambda_1, \lambda_2, \ldots, \lambda_J\}$ is a matrix with the eigenvalues along the diagonal and zeroes elsewhere, and $\mathbf{V}^\tau \mathbf{V} = \mathbf{I}_J$.

The "outer product" of a J-vector $\mathbf{v}$ with itself is the $(J \times J)$-matrix $\mathbf{v}\mathbf{v}^\tau$, which has rank 1. The *spectral theorem* expresses the $(J \times J)$-matrix $\mathbf{A}$ as a weighted average of rank-1 matrices,

$$\mathbf{A} = \mathbf{V}\Lambda\mathbf{V}^\tau = \sum_{j=1}^{J} \lambda_j \mathbf{v}_j \mathbf{v}_j^\tau, \tag{3.18}$$

where $\mathbf{I}_J = \sum_{j=1}^{J} \mathbf{v}_j \mathbf{v}_j^\tau$, and where the weights, $\lambda_1, \ldots, \lambda_J$, are the eigenvalues of $\mathbf{A}$. The rank of $\mathbf{A}$ is the number of nonzero eigenvalues, the trace is

$$\mathrm{tr}(\mathbf{A}) = \sum_{j=1}^{J} \lambda_j(\mathbf{A}), \tag{3.19}$$

and the determinant is

$$|\mathbf{A}| = \prod_{j=1}^{J} \lambda_j(\mathbf{A}). \tag{3.20}$$

3.2.5 Functions of Matrices

If $\mathbf{A}$ is a symmetric $(J \times J)$-matrix and $\phi : \mathbf{R}^J \to \mathbf{R}^J$ is a function, then

$$\phi(\mathbf{A}) = \sum_{j=1}^{J} \phi(\lambda_j)\mathbf{v}_j \mathbf{v}_j^\tau, \tag{3.21}$$

where λ_j and $\mathbf{v}_j$ are the jth eigenvalue and corresponding eigenvector, respectively, of $\mathbf{A}$. Examples include the following:

$$\mathbf{A}^{-1} = \mathbf{V}\Lambda^{-1}\mathbf{V}^\tau = \sum_{j=1}^{J} \lambda_j^{-1}\mathbf{v}_j \mathbf{v}_j^\tau, \text{ if } \mathbf{A} \text{ is nonsingular} \tag{3.22}$$

$$\mathbf{A}^{1/2} = \mathbf{V}\Lambda^{1/2}\mathbf{V}^\tau = \sum_{j=1}^{J} \lambda_j^{1/2}\mathbf{v}_j \mathbf{v}_j^\tau \tag{3.23}$$

$$\log(\mathbf{A}) \;=\; \sum_{j=1}^{J}(\log(\lambda_j))\mathbf{v}_j\mathbf{v}_j^{\tau}, \text{ if } \lambda_j \neq 0, \text{ all } j \tag{3.24}$$

Hence, $\lambda_j(\phi(\mathbf{A})) = \phi(\lambda_j(\mathbf{A}))$ and $\mathbf{v}_j(\phi(\mathbf{A})) = \mathbf{v}_j(\mathbf{A})$. Note that $\mathbf{A}^{1/2}$ is called the *square-root* of $\mathbf{A}$.

3.2.6   Singular-Value Decomposition

If $\mathbf{A}$ is a $(J \times K)$-matrix with $J \leq K$, then

$$\lambda_j(\mathbf{A}^{\tau}\mathbf{A}) = \lambda_j(\mathbf{A}\mathbf{A}^{\tau}), \quad j = 1, 2, \ldots, J, \tag{3.25}$$

and zero for $j > J$. Furthermore, for $\lambda_j(\mathbf{A}\mathbf{A}^{\tau}) \neq 0$,

$$\mathbf{v}_j(\mathbf{A}^{\tau}\mathbf{A}) \;=\; (\lambda_j(\mathbf{A}\mathbf{A}^{\tau}))^{1/2}\mathbf{A}^{\tau}\mathbf{v}_j(\mathbf{A}\mathbf{A}^{\tau}) \tag{3.26}$$

$$\mathbf{v}_j(\mathbf{A}\mathbf{A}^{\tau}) \;=\; (\lambda_j(\mathbf{A}\mathbf{A}^{\tau}))^{-1/2}\mathbf{A}\mathbf{v}_j(\mathbf{A}^{\tau}\mathbf{A}) \tag{3.27}$$

The *singular-value decomposition* (SVD) of $\mathbf{A}$ is given by

$$\mathbf{A} = \mathbf{U}\mathbf{\Psi}\mathbf{V}^{\tau} = \sum_{j=1}^{J}\lambda_j^{1/2}\mathbf{u}_j\mathbf{v}_j^{\tau}, \tag{3.28}$$

where $\mathbf{U} = (\mathbf{u}_1, \ldots, \mathbf{u}_J)$ is a $(J \times J)$-matrix, $\mathbf{u}_j = \mathbf{v}_j(\mathbf{A}\mathbf{A}^{\tau})$, $j = 1, 2, \ldots, J$, $\mathbf{V} = (\mathbf{v}_1, \ldots, \mathbf{v}_K)$ is a $(K \times K)$-matrix, $\mathbf{v}_k = \mathbf{v}_k(\mathbf{A}^{\tau}\mathbf{A})$, $k = 1, 2, \ldots, K$, $\lambda_j = \lambda_j(\mathbf{A}\mathbf{A}^{\tau})$, $j = 1, 2, \ldots, J$,

$$\mathbf{\Psi} = \left(\mathbf{\Psi}_{\sigma} \vdots \mathbf{0} \right) \tag{3.29}$$

is a $(J \times K)$-matrix, and $\mathbf{\Psi}_{\sigma}$ is an $(J \times J)$ diagonal matrix with the non-negative *singular values*, $\sigma_1 \geq \sigma_2 \geq \ldots \geq \sigma_J \geq 0$, of $\mathbf{A}$ along the diagonal, where $\sigma_j = \lambda_j^{1/2}$ is the square-root of the jth largest eigenvalue of the $(J \times J)$-matrix $\mathbf{A}\mathbf{A}^{\tau}$, $j = 1, 2, \ldots, J$.

A corollary of the SVD is that if $r(\mathbf{A}) = t$, then there exists a $(J \times t)$-matrix $\mathbf{B}$ and a $(t \times K)$-matrix $\mathbf{C}$, both of rank t, such that $\mathbf{A} = \mathbf{B}\mathbf{C}$. To see this, take $\mathbf{B} = (\lambda_1^{1/2}\mathbf{u}_1, \ldots, \lambda_t^{1/2}\mathbf{u}_t)$ and $\mathbf{C} = (\mathbf{v}_1^{\tau}, \ldots, \mathbf{v}_t^{\tau})^{\tau}$.

3.2.7   Generalized Inverses

If $\mathbf{A}$ is either singular or nonsymmetric (or even not square), we can define a *generalized inverse* of $\mathbf{A}$. First, we need the following definition: a *g-inverse* of a $(J \times K)$-matrix $\mathbf{A}$ is any $(K \times J)$-matrix $\mathbf{A}^{-}$ such that, for any J-vector $\mathbf{y}$ for which $\mathbf{A}\mathbf{x}=\mathbf{y}$ is a consistent equation, $\mathbf{x} = \mathbf{A}^{-}\mathbf{y}$ is a solution. It can be shown that $\mathbf{A}^{-}$ exists iff

$$\mathbf{A}\mathbf{A}^{-}\mathbf{A} = \mathbf{A}; \tag{3.30}$$

we call such an $\mathbf{A}^-$ a *reflexive g-inverse*. Note that although $\mathbf{A}^-$ is not necessarily unique, it has some interesting properties. For example, a general solution of the consistent equation $\mathbf{Ax=y}$ is given by

$$\mathbf{x} = \mathbf{A}^-\mathbf{y} + (\mathbf{A}^-\mathbf{A} - \mathbf{I}_K)\mathbf{z}, \tag{3.31}$$

where $\mathbf{z}$ is an arbitrary K-vector. Furthermore, setting $\mathbf{z=0}$ shows that the $\mathbf{x}$ with minimum norm (i.e., $\| \mathbf{x} \|^2 = \mathbf{x}^\tau\mathbf{x}$) that solves $\mathbf{Ax=y}$ is given by $\mathbf{x} = \mathbf{A}^-\mathbf{y}$.

A unique g-inverse can be defined for the $(J \times K)$-matrix $\mathbf{A}$. From the SVD, $\mathbf{A} = \mathbf{U\Psi V}^\tau$, we set

$$\mathbf{A}^+ = \mathbf{V\Psi}^+\mathbf{U}^\tau, \tag{3.32}$$

where Ψ^+ is a diagonal matrix whose diagonal elements are the reciprocals of the nonzero elements of $\Psi = \Lambda^{1/2}$, and zeroes otherwise. The $(K \times J)$-matrix $\mathbf{A}^+$ is the unique *Moore–Penrose generalized inverse* of $\mathbf{A}$. It satisfies the following four conditions:

$$\mathbf{AA}^+\mathbf{A} = \mathbf{A}, \quad \mathbf{A}^+\mathbf{AA}^+ = \mathbf{A}^+, \quad (\mathbf{AA}^+)^\tau = \mathbf{AA}^+, \quad (\mathbf{A}^+\mathbf{A})^\tau = \mathbf{A}^+\mathbf{A}. \tag{3.33}$$

There are less restrictive (nonunique) types of generalized inverses than $\mathbf{A}^+$, such as the reflexive g-inverse above, involving one or two of the above four conditions.

3.2.8 Matrix Norms

Let $\mathbf{A} = (A_{jk})$ be a $(J \times K)$-matrix. It would be useful to have a measure of the size of $\mathbf{A}$, especially for comparing different matrices. The usual measure of size of a matrix $\mathbf{A}$ is the norm, $\| \mathbf{A} \|$, of that matrix. There are many definitions of a *matrix norm*, all of which satisfy the following conditions:

1. $\| \mathbf{A} \| \geq 0$

2. $\| \mathbf{A} \| = 0$ iff $\mathbf{A=0}$.

3. $\| \mathbf{A} + \mathbf{B} \| \leq \| \mathbf{A} \| + \| \mathbf{B} \|$

4. $\| \alpha\mathbf{A} \| = |\alpha| \cdot \| \mathbf{A} \|$

where $\mathbf{B}$ is a $(J \times K)$-matrix and α is a scalar. Examples of matrix norms include:

1. $\left(\sum_{j=1}^J \sum_{k=1}^K |A_{jk}|^p\right)^{1/p}$ (p-norm)

2. $\sqrt{\mathrm{tr}(\mathbf{AA}^\tau)} = \left(\sum_{j=1}^J \sum_{k=1}^K A_{jk}^2\right)^{1/2} = \left(\sum_{j=1}^J \lambda_j(\mathbf{AA}^\tau)\right)^{1/2}$ (Frobenius norm)

3. $\sqrt{\lambda_1(\mathbf{A}\mathbf{A}^\tau)}$ (spectral norm, $J = K$)

4. $\left(\sum_{j=1}^{J_0} \lambda_j(\mathbf{A}\mathbf{A}^\tau)\right)^{1/2}$, for some $J_0 < J$.

3.2.9   Condition Numbers for Matrices

The *condition number* of a square $(K \times K)$-matrix $\mathbf{A}$ is given by

$$\kappa(\mathbf{A}) = ||\mathbf{A}|| \cdot ||\mathbf{A}^{-1}|| = \frac{\sigma_1}{\sigma_K}, \tag{3.34}$$

which is the ratio of the largest to the smallest nonzero singular value. In (3.34), $|| \cdot ||$ is the spectral norm and σ_i is the square-root of the ith largest eigenvalue of the $(K \times K)$-matrix $\mathbf{A}^\tau\mathbf{A}$, $i = 1, 2, \ldots, K$. Thus, $\kappa \geq 1$. If $\mathbf{A}$ is an orthogonal matrix, all singular values are unity, and so $\kappa = 1$. $\mathbf{A}$ is said to be *ill-conditioned* if its singular values are widely spread out, so that $\kappa(\mathbf{A})$ is large, whereas $\mathbf{A}$ is said to be *well-conditioned* if $\kappa(\mathbf{A})$ is small.

3.2.10   Eigenvalue Inequalities

We shall find it useful to have the following eigenvalue inequalities.

The Eckart–Young Theorem If $\mathbf{A}$ and $\mathbf{B}$ are both $(J \times K)$-matrices, and we plan on using $\mathbf{B}$ with reduced rank $r(\mathbf{B}) = b$ to approximate $\mathbf{A}$ with full rank $r(\mathbf{A}) = \min(J, K)$, then the Eckart–Young (1936) Theorem states that

$$\lambda_j((\mathbf{A} - \mathbf{B})(\mathbf{A} - \mathbf{B})^\tau) \geq \lambda_{j+b}(\mathbf{A}\mathbf{A}^\tau), \tag{3.35}$$

with equality if

$$\mathbf{B} = \sum_{i=1}^{b} \lambda_i^{1/2} \mathbf{u}_i \mathbf{v}_i^\tau, \tag{3.36}$$

where $\lambda_i = \lambda_i(\mathbf{A}\mathbf{A}^\tau)$, $\mathbf{u}_i = \mathbf{v}_i(\mathbf{A}\mathbf{A}^\tau)$, and $\mathbf{v}_i = \mathbf{v}_i(\mathbf{A}^\tau\mathbf{A})$. Because the above choice of $\mathbf{B}$ provides a simultaneous minimization for all eigenvalues λ_j, it follows that the minimum is achieved for different functions of those eigenvalues, say, the trace or the determinant of $(\mathbf{A} - \mathbf{B})(\mathbf{A} - \mathbf{B})^\tau$.

The Courant–Fischer Min-Max Theorem A very useful result is the following expression for the jth largest eigenvalue of a $(J \times J)$ symmetric matrix $\mathbf{A}$:

$$\lambda_j(\mathbf{A}) = \inf_{\mathbf{L}} \sup_{\mathbf{x}:\mathbf{L}\mathbf{x}=0} \frac{\mathbf{x}^\tau \mathbf{A} \mathbf{x}}{\mathbf{x}^\tau \mathbf{x}}, \quad \mathbf{x} \neq \mathbf{0}, \tag{3.37}$$

where inf is an infimum over a $((j - 1) \times J)$-matrix $\mathbf{L}$ with rank at most $j-1$, and sup is a supremum over a nonzero J-vector $\mathbf{x}$ that satisfies $\mathbf{L}\mathbf{x}=\mathbf{0}$.

Equality in (3.37) is reached if $\mathbf{L} = (\mathbf{v_1}, \cdots, \mathbf{v}_{j-1})^\tau$ and $\mathbf{x} = \mathbf{v}_j = \mathbf{v}_j(\mathbf{A})$, the eigenvector associated with the jth largest eigenvalue of $\mathbf{A}$. A corollary of this result is that the jth smallest eigenvalue of $\mathbf{A}$ can be written as

$$\lambda_{J-j+1}(\mathbf{A}) = \sup_{\mathbf{L}} \inf_{\mathbf{Lx=0}} \frac{\mathbf{x}^\tau \mathbf{A} \mathbf{x}}{\mathbf{x}^\tau \mathbf{x}}, \quad \mathbf{x} \neq \mathbf{0}. \tag{3.38}$$

For a proof, see, e.g., Bellman (1970, pp. 115–117). These two results enable us to write

$$\lambda_J(\mathbf{A}) \leq \frac{\mathbf{x}^\tau \mathbf{A} \mathbf{x}}{\mathbf{x}^\tau \mathbf{x}} \leq \lambda_1(\mathbf{A}), \quad \mathbf{x} \neq \mathbf{0}, \tag{3.39}$$

where $\lambda_1(\mathbf{A})$ is the largest eigenvalue and $\lambda_J(\mathbf{A})$ is the smallest eigenvalue of $\mathbf{A}$.

The Hoffman–Wielandt Theorem   Suppose $\mathbf{A}$ and $\mathbf{B}$ are $(J \times J)$-matrices with $\mathbf{A} - \mathbf{B}$ symmetric. Suppose $\mathbf{A}$ and $\mathbf{B}$ have eigenvalues $\{\lambda_j(\mathbf{A})\}$ and $\{\lambda_j(\mathbf{B})\}$, respectively. Hoffman and Wielandt (1953) showed that

$$\sum_{j=1}^{J} (\lambda_j(\mathbf{A}) - \lambda_j(\mathbf{B}))^2 \leq \text{tr}\{(\mathbf{A} - \mathbf{B})(\mathbf{A} - \mathbf{B})^\tau\}. \tag{3.40}$$

This result is useful for studying the bias in sample eigenvalues. For a simple proof, see Exercise 3.3.

Poincaré Separation Theorem   Let $\mathbf{A}$ be a $(J \times J)$-matrix and let $\mathbf{U}$ be a $(J \times k)$-matrix, $k \leq J$, such that $\mathbf{U}^\tau \mathbf{U} = \mathbf{I}_k$. Then,

$$\lambda_j(\mathbf{U}^\tau \mathbf{A} \mathbf{U}) \leq \lambda_j(\mathbf{A}), \tag{3.41}$$

with equality if the columns of $\mathbf{U}$ are the first k eigenvectors of $\mathbf{A}$. This inequality can be proved using (3.37) from the Courant–Fischer Min-Max Theorem; see Exercise 3.4.

3.2.11   Matrix Calculus

Let $\mathbf{x} = (x_1, \cdots, x_K)^\tau$ be a K-vector and let

$$\mathbf{y} = (y_1, \cdots, y_J)^\tau = (f_1(\mathbf{x}), \cdots, f_J(\mathbf{x}))^\tau = \mathbf{f}(\mathbf{x}) \tag{3.42}$$

be a J-vector, where $\mathbf{f} : \Re^K \to \Re^J$. Then, the partial derivative of $\mathbf{y}$ wrt $\mathbf{x}$ is the JK-vector,

$$\frac{\partial \mathbf{y}}{\partial \mathbf{x}} = \left(\frac{\partial y_1}{\partial x_1}, \cdots, \frac{\partial y_J}{\partial x_1}, \cdots, \frac{\partial y_1}{\partial x_K}, \cdots, \frac{\partial y_K}{\partial x_J} \right)^\tau. \tag{3.43}$$

A more convenient form is the partial derivative of $\mathbf{y}$ wrt $\mathbf{x}^\tau$, which yields the $(J \times K)$ *Jacobian matrix*,

$$
\mathbf{J_x y} = \frac{\partial \mathbf{y}}{\partial \mathbf{x}^\tau} = \begin{pmatrix} \frac{\partial y_1}{\partial x_1} & \frac{\partial y_1}{\partial x_2} & \cdots & \frac{\partial y_1}{\partial x_K} \\ \frac{\partial y_2}{\partial x_1} & \frac{\partial y_2}{\partial x_2} & \cdots & \frac{\partial y_2}{\partial x_K} \\ \vdots & \vdots & & \vdots \\ \frac{\partial y_J}{\partial x_1} & \frac{\partial y_J}{\partial x_2} & \cdots & \frac{\partial y_J}{\partial x_K} \end{pmatrix}. \tag{3.44}
$$

The Jacobian matrix can be interpreted as the first derivative of $\mathbf{f}(\mathbf{x})$ wrt $\mathbf{x}$. It, therefore, provides a method for linearly approximating a multivariate vector-valued function: $\mathbf{f}(\mathbf{x}) \approx \mathbf{f}(\mathbf{c}) + [\mathbf{J_x f}(\mathbf{c})](\mathbf{x} - \mathbf{c})$, where $\mathbf{c} \in \Re^K$. The *Jacobian* of the transformation $\mathbf{y} = \mathbf{f}(\mathbf{x})$ is

$$
J = |\mathbf{J_x y}|. \tag{3.45}
$$

If $y = f(\mathbf{x})$ is a scalar, then the *gradient vector* is

$$
\nabla_{\mathbf{x}} y = \frac{\partial y}{\partial \mathbf{x}} = \left(\frac{\partial y}{\partial x_1}, \frac{\partial y}{\partial x_2}, \cdots, \frac{\partial y}{\partial x_K} \right)^\tau = \left(\frac{\partial y}{\partial \mathbf{x}^\tau} \right)^\tau = (\mathbf{J_x} y)^\tau, \tag{3.46}
$$

while if x is a scalar, then,

$$
\frac{\partial \mathbf{y}}{\partial x} = \left(\frac{\partial y_1}{\partial x}, \frac{\partial y_2}{\partial x}, \cdots, \frac{\partial y_J}{\partial x} \right)^\tau. \tag{3.47}
$$

For example, if $\mathbf{A}$ is a $(J \times K)$-matrix, then:

$$
\frac{\partial (\mathbf{A}\mathbf{x})}{\partial \mathbf{x}^\tau} = \mathbf{A} \tag{3.48}
$$

$$
\frac{\partial (\mathbf{x}^\tau \mathbf{x})}{\partial \mathbf{x}^\tau} = 2\mathbf{x} \tag{3.49}
$$

$$
\frac{\partial (\mathbf{x}^\tau \mathbf{A} \mathbf{x})}{\partial \mathbf{x}^\tau} = \mathbf{x}^\tau (\mathbf{A} + \mathbf{A}^\tau) \quad (J = K). \tag{3.50}
$$

The derivative of a $(J \times K)$-matrix $\mathbf{A}$ wrt an r-vector $\mathbf{x}$ is the $(Jr \times K)$-matrix of derivatives of $\mathbf{A}$ wrt each element of $\mathbf{x}$:

$$
\frac{\partial \mathbf{A}}{\partial \mathbf{x}} = \left(\frac{\partial \mathbf{A}^\tau}{\partial x_1}, \cdots, \frac{\partial \mathbf{A}^\tau}{\partial x_r} \right)^\tau. \tag{3.51}
$$

It follows that:

$$
\frac{\partial (\alpha \mathbf{A})}{\partial \mathbf{x}} = \alpha \frac{\partial \mathbf{A}}{\partial \mathbf{x}} \quad (\alpha \text{ a constant}) \tag{3.52}
$$

$$
\frac{\partial (\mathbf{A} + \mathbf{B})}{\partial \mathbf{x}} = \frac{\partial \mathbf{A}}{\partial \mathbf{x}} + \frac{\partial \mathbf{B}}{\partial \mathbf{x}} \tag{3.53}
$$

$$\frac{\partial(\mathbf{AB})}{\partial \mathbf{x}} = \left(\frac{\partial \mathbf{A}}{\partial \mathbf{x}}\right)\mathbf{B} + \mathbf{A}\left(\frac{\partial \mathbf{B}}{\partial \mathbf{x}}\right) \tag{3.54}$$

$$\frac{\partial(\mathbf{A} \otimes \mathbf{B})}{\partial \mathbf{x}} = \left(\frac{\partial \mathbf{A}}{\partial \mathbf{x}} \otimes \mathbf{B}\right) + \left(\mathbf{A} \otimes \frac{\partial \mathbf{B}}{\partial \mathbf{x}}\right) \tag{3.55}$$

$$\frac{\partial(\mathbf{A}^{-1})}{\partial \mathbf{x}} = -\mathbf{A}^{-1}\left(\frac{\partial \mathbf{A}}{\partial \mathbf{x}}\right)\mathbf{A}^{-1}, \tag{3.56}$$

where $\mathbf{A}$ and $\mathbf{B}$ are conformable matrices.

If $y = f(\mathbf{A})$ is a scalar function of the $(J \times K)$-matrix $\mathbf{A} = (A_{ij})$, define the following *gradient matrix*:

$$\frac{\partial y}{\partial \mathbf{A}} = \begin{pmatrix} \frac{\partial y}{\partial A_{11}} & \frac{\partial y}{\partial A_{12}} & \cdots & \frac{\partial y}{\partial A_{1K}} \\ \frac{\partial y}{\partial A_{21}} & \frac{\partial y}{\partial A_{22}} & \cdots & \frac{\partial y}{\partial A_{2K}} \\ \vdots & \vdots & & \vdots \\ \frac{\partial y}{\partial A_{J1}} & \frac{\partial y}{\partial A_{J2}} & \cdots & \frac{\partial y}{\partial A_{JK}} \end{pmatrix}. \tag{3.57}$$

For example, if $\mathbf{A}$ is a $(J \times J)$-matrix, then,

$$\frac{\partial(\mathrm{tr}(\mathbf{A}))}{\partial \mathbf{A}} = \mathbf{I}_J \tag{3.58}$$

$$\frac{\partial(|\mathbf{A}|)}{\partial \mathbf{A}} = |\mathbf{A}| \cdot (\mathbf{A}^\tau)^{-1}. \tag{3.59}$$

Next, we define the *Hessian matrix* as a square matrix whose elements are the second-order partial derivatives of a function. Let $y = f(\mathbf{x})$ be a scalar function of $\mathbf{x} \in \Re^K$. The $(K \times K)$-matrix,

$$\mathbf{H_x}y = \frac{\partial}{\partial \mathbf{x}}\left(\frac{\partial y}{\partial \mathbf{x}}\right)^\tau = \frac{\partial^2 y}{\partial \mathbf{x} \partial \mathbf{x}^\tau} = \begin{pmatrix} \frac{\partial^2 y}{\partial x_1^2} & \frac{\partial^2 y}{\partial x_1 \partial x_2} & \cdots & \frac{\partial^2 y}{\partial x_1 \partial x_K} \\ \frac{\partial^2 y}{\partial x_2 \partial x_1} & \frac{\partial^2 y}{\partial x_2^2} & \cdots & \frac{\partial^2 y}{\partial x_2 \partial x_K} \\ \vdots & \vdots & \ddots & \vdots \\ \frac{\partial^2 y}{\partial x_K \partial x_1} & \frac{\partial^2 y}{\partial x_K \partial x_2} & \cdots & \frac{\partial^2 y}{\partial x_K^2} \end{pmatrix}, \tag{3.60}$$

is called the Hessian of y wrt $\mathbf{x}$. Note that $\mathbf{H_x}y = \nabla_{\mathbf{x}}^2 y = \nabla_{\mathbf{x}}\nabla_{\mathbf{x}}y$, so that the Hessian is the Jacobian of the gradient of f. If the second-order partial derivatives are continuous, the Hessian is a symmetric matrix. The Hessian enables a quadratic term to be included in the Taylor-series approximation to a real-valued function:

$$f(\mathbf{x}) \approx f(\mathbf{c}) + [\mathbf{J}f(\mathbf{c})](\mathbf{x} - \mathbf{c}) + \frac{1}{2}(\mathbf{x} - \mathbf{c})^\tau[\mathbf{H}f(\mathbf{c})](\mathbf{x} - \mathbf{c}), \quad \mathbf{c} \in \Re^K. \tag{3.61}$$

3.3   Random Vectors

If we have r random variables, $X_1, X_2, \ldots, X_r$, each defined on the real line, we can write them as the r-dimensional column vector,

$$\mathbf{X} = (X_1, \cdots, X_r)^\tau. \tag{3.62}$$

which we, henceforth, call a "random r-vector." The joint distribution function F_X of the random vector $\mathbf{X}$ is given by

$$
\begin{aligned}
F_X(\mathbf{x}) &= F_X(x_1, \ldots, x_r) & (3.63) \\
&= P\{X_1 \leq x_1, \ldots, X_r \leq x_r\} & (3.64) \\
&= P\{\mathbf{X} \leq \mathbf{x}\}, & (3.65)
\end{aligned}
$$

for any vector $\mathbf{x} = (x_1, x_2, \cdots, x_r)^\tau$ of real numbers, where $P(A)$ represents the probability that the event A will occur. If F_X is absolutely continuous, then the joint density function f_X of $\mathbf{X}$, where

$$f_X(\mathbf{x}) = f_X(x_1, \ldots, x_r) = \frac{\partial^r F_X(x_1, \ldots, x_r)}{\partial x_1 \cdots \partial x_r}, \tag{3.66}$$

will exist almost everywhere. The distribution function F_X can be recovered from f_X using the relationship

$$F_X(\mathbf{x}) = \int_{-\infty}^{x_r} \cdots \int_{-\infty}^{x_1} f_X(u_1, \ldots, u_r)\, du_1 \cdots du_r. \tag{3.67}$$

Consider a subset, $X_1, X_2, \ldots, X_k$ $(k < r)$, say, of the components of $\mathbf{X}$. The marginal distribution function of that component subset is given by

$$
\begin{aligned}
F_X(x_1, \ldots, x_k) &= F_X(x_1, \ldots, x_k, \infty, \ldots, \infty) \\
&= P\{X_1 \leq x_1, \ldots, X_k \leq x_k, X_{k+1} \leq \infty, \ldots, X_r \leq \infty\},
\end{aligned} \tag{3.68}
$$

and the marginal density of that subset is

$$\int_{-\infty}^{\infty} \cdots \int_{-\infty}^{\infty} f_X(u_1, \ldots, u_r)\, du_{k+1} \cdots du_r. \tag{3.69}$$

For example, if $r = 2$, the bivariate joint density of X_1 and X_2 is given by $f_{X_1, X_2}(x_1, x_2)$, and its marginal densities are

$$f_{X_1}(x_1) = \int_{\Re} f_{X_1, X_2}(x_1, x_2)dx_2, \quad f_{X_2}(x_2) = \int_{\Re} f_{X_1, X_2}(x_1, x_2)dx_1. \tag{3.70}$$

The components of a random r-vector $\mathbf{X}$ are said to be *mutually statistically independent* if the joint distribution can be factored into the product of its r marginals,

$$F_X(\mathbf{x}) = \prod_{i=1}^{r} F_i(x_i), \tag{3.71}$$

where $F_i(x_i)$ is the marginal distribution of X_i, $i = 1, 2, \ldots, r$. This implies that a similar factorization of the joint density function holds under independence,

$$f_X(\mathbf{x}) = \prod_{i=1}^{r} f_i(x_i), \tag{3.72}$$

for any set of r real numbers $x_1, \ldots, x_r$.

3.3.1   Multivariate Moments

Let X be a continuous real-valued random variable with *probability density function* f_X; that is, $f_X(x) \geq 0$, for all $x \in \Re$, and $\int_{\Re} f_X(x)dx = 1$. The *expected value* of X is defined as

$$\mu_X = \mathrm{E}(X) = \int x f_X(x) dx, \tag{3.73}$$

and its *variance* is

$$\sigma_X^2 = \mathrm{var}(X) = \mathrm{E}\{(X - \mu_X)^2\}. \tag{3.74}$$

If $\mathbf{X}$ is a random r-vector with values in $\Re^r$, then its expected value is the r-vector

$$\boldsymbol{\mu}_X = \mathrm{E}(\mathbf{X}) = (\mathrm{E}(X_1), \cdots, \mathrm{E}(X_r))^\tau = (\mu_1, \cdots, \mu_r)^\tau, \tag{3.75}$$

and the $(r \times r)$ *covariance matrix* of $\mathbf{X}$ is given by

$$\begin{align}
\boldsymbol{\Sigma}_{XX} &= \mathrm{cov}(\mathbf{X}, \mathbf{X}) \tag{3.76}\\
&= \mathrm{E}\{(\mathbf{X} - \boldsymbol{\mu}_X)(\mathbf{X} - \boldsymbol{\mu}_X)^\tau\} \tag{3.77}\\
&= \mathrm{E}\{(X_1 - \mu_1, \cdots, X_r - \mu_r)(X_1 - \mu_1, \cdots, X_r - \mu_r)^\tau\} \tag{3.78}\\
&= \begin{pmatrix} \sigma_1^2 & \sigma_{12} & \cdots & \sigma_{1r} \\ \sigma_{21} & \sigma_2^2 & \cdots & \sigma_{2r} \\ \vdots & \vdots & \ddots & \vdots \\ \sigma_{r1} & \sigma_{r2} & \cdots & \sigma_r^2 \end{pmatrix}, \tag{3.79}
\end{align}$$

where

$$\sigma_i^2 = \mathrm{var}(X_i) = \mathrm{E}\{(X_i - \mu_i)^2\} \tag{3.80}$$

is the *variance* of X_i, $i = 1, 2, \ldots, r$, and

$$\sigma_{ij} = \mathrm{cov}(X_i, X_j) = \mathrm{E}\{(X_i - \mu_i)(X_j - \mu_j)\} \tag{3.81}$$

is the *covariance* between X_i and X_j, $i, j = 1, 2, \ldots, r$ $(i \neq j)$. It is not difficult to show that

$$\Sigma_{XX} = \mathrm{E}(\mathbf{X}\mathbf{X}^\tau) - \boldsymbol{\mu}_X \boldsymbol{\mu}_X^\tau. \tag{3.82}$$

The *correlation matrix* of $\mathbf{X}$ is obtained from the covariance matrix Σ_{XX} by dividing the ith row by σ_i and dividing the jth column by σ_j. It is given by the $(r \times r)$-matrix,

$$\mathbf{P}_{XX} = \begin{pmatrix} 1 & \rho_{12} & \cdots & \rho_{1r} \\ \rho_{21} & 1 & \cdots & \rho_{2r} \\ \vdots & \vdots & \ddots & \vdots \\ \rho_{r1} & \rho_{r2} & \cdots & 1 \end{pmatrix}, \tag{3.83}$$

where

$$\rho_{ij} = \rho_{ji} = \begin{cases} \frac{\sigma_{ij}}{\sigma_i \sigma_j} & \text{if } i \neq j \\ 1 & \text{otherwise} \end{cases} \tag{3.84}$$

is the (pairwise) *correlation coefficient* of X_i with X_j, $i, j = 1, 2, \ldots, r$. The correlation coefficient ρ_{ij} lies between -1 and $+1$ and is a measure of *association* between X_i and X_j. When $\rho_{ij} = 0$, we say that X_i and X_j are *uncorrelated*; when $\rho_{ij} > 0$, we say that X_i and X_j are *positively correlated*; and when $\rho_{ij} < 0$, we say that X_i and X_j are *negatively correlated*.

Now, suppose we have two random vectors, $\mathbf{X}$ and $\mathbf{Y}$, where $\mathbf{X}$ has r components and $\mathbf{Y}$ has s components. Let $\mathbf{Z}$ be the random $(r + s)$-vector,

$$\mathbf{Z} = \begin{pmatrix} \mathbf{X} \\ \mathbf{Y} \end{pmatrix}. \tag{3.85}$$

Then, the expected value of $\mathbf{Z}$ is the $(r + s)$-vector,

$$\boldsymbol{\mu}_Z = \mathrm{E}(\mathbf{Z}) = \begin{pmatrix} \mathrm{E}(\mathbf{X}) \\ \mathrm{E}(\mathbf{Y}) \end{pmatrix} = \begin{pmatrix} \boldsymbol{\mu}_X \\ \boldsymbol{\mu}_Y \end{pmatrix}, \tag{3.86}$$

and the covariance matrix of $\mathbf{Z}$ is the partitioned $((r + s) \times (r + s))$-matrix,

$$\begin{aligned} \Sigma_{ZZ} &= \mathrm{E}\{(\mathbf{Z} - \boldsymbol{\mu}_Z)(\mathbf{Z} - \boldsymbol{\mu}_Z)^\tau\} & (3.87) \\ &= \begin{pmatrix} \mathrm{cov}(\mathbf{X}, \mathbf{X}) & \mathrm{cov}(\mathbf{X}, \mathbf{Y}) \\ \mathrm{cov}(\mathbf{Y}, \mathbf{X}) & \mathrm{cov}(\mathbf{Y}, \mathbf{Y}) \end{pmatrix} & (3.88) \\ &= \begin{pmatrix} \Sigma_{XX} & \Sigma_{XY} \\ \Sigma_{YX} & \Sigma_{YY} \end{pmatrix}, & (3.89) \end{aligned}$$

where

$$\Sigma_{XY} = \mathrm{cov}(\mathbf{X}, \mathbf{Y}) = \mathrm{E}\{(\mathbf{X} - \boldsymbol{\mu}_X)(\mathbf{Y} - \boldsymbol{\mu}_Y)^\tau\} = \Sigma_{YX}^\tau \tag{3.90}$$

is an $(r \times s)$-matrix.

and

$$\mathbf{\Sigma}^{-1} = \frac{1}{1-\rho^2} \begin{pmatrix} \frac{1}{\sigma_1^2} & \frac{-\rho}{\sigma_1\sigma_2} \\ \frac{-\rho}{\sigma_1\sigma_2} & \frac{1}{\sigma_2^2} \end{pmatrix}. \tag{3.103}$$

The bivariate Gaussian density function of $\mathbf{X}$ is, therefore, given by

$$f(\mathbf{x}|\boldsymbol{\mu}, \mathbf{\Sigma}) = \frac{1}{2\pi\sigma_1\sigma_2\sqrt{1-\rho^2}} e^{-\frac{1}{2}Q}, \tag{3.104}$$

where

$$Q = \frac{1}{1-\rho^2} \left\{ \left(\frac{x_1-\mu_1}{\sigma_1}\right)^2 - 2\rho \left(\frac{x_1-\mu_1}{\sigma_1}\right) \left(\frac{x_2-\mu_2}{\sigma_2}\right) + \left(\frac{x_2-\mu_2}{\sigma_2}\right)^2 \right\}. \tag{3.105}$$

If X_1 and X_2 are uncorrelated, $\rho = 0$, and the middle term in the exponent (3.106) drops out. In that case, the bivariate Gaussian density function reduces to the product of two univariate Gaussian densities,

$$\begin{aligned} f(\mathbf{x}|\mu_1, \mu_2, \sigma_1^2, \sigma_2^2) &= (2\pi\sigma_1\sigma_2)^{-1} e^{-\frac{1}{2\sigma_1^2}(x_1-\mu_1)^2} e^{-\frac{1}{2\sigma_2^2}(x_2-\mu_2)^2} \\ &= f(x_1|\mu_1, \sigma_1^2) f(x_2|\mu_2, \sigma_2^2), \end{aligned} \tag{3.106}$$

implying that X_1 and X_2 are independent. (see (3.72)).

3.3.3   Conditional Gaussian Distributions

Consider the random $(r + s)$-vector $\mathbf{Z}$ in (3.85) with mean vector $\boldsymbol{\mu}_Z$ in (3.86) and partitioned covariance matrix $\mathbf{\Sigma}_{ZZ}$ in (3.89). Assume that $\mathbf{Z}$ has the multivariate Gaussian distribution. Then, the exponent in (3.95) is the quadratic form,

$$-\frac{1}{2}(\mathbf{z} - \boldsymbol{\mu}_Z)^\tau \mathbf{\Sigma}_{ZZ}^{-1}(\mathbf{z} - \boldsymbol{\mu}_Z). \tag{3.107}$$

From (3.5),

$$\mathbf{\Sigma}_{ZZ}^{-1} = \begin{pmatrix} \mathbf{A}_{11} & \mathbf{A}_{12} \\ \mathbf{A}_{21} & \mathbf{A}_{22} \end{pmatrix}, \tag{3.108}$$

where

$$\mathbf{A}_{11} = \mathbf{\Sigma}_{XX}^{-1} + \mathbf{\Sigma}_{XX}^{-1}\mathbf{\Sigma}_{XY}\mathbf{\Sigma}_{YY\cdot X}^{-1}\mathbf{\Sigma}_{YX}\mathbf{\Sigma}_{XX}^{-1}$$
$$\mathbf{A}_{12} = -\mathbf{\Sigma}_{XX}^{-1}\mathbf{\Sigma}_{XY}\mathbf{\Sigma}_{YY\cdot X}^{-1} = \mathbf{A}_{21}^\tau$$
$$\mathbf{A}_{22} = \mathbf{\Sigma}_{YY\cdot X}^{-1},$$

and $\mathbf{\Sigma}_{YY\cdot X} = \mathbf{\Sigma}_{YY} - \mathbf{\Sigma}_{YX}\mathbf{\Sigma}_{XX}^{-1}\mathbf{\Sigma}_{XY}$. As a result, we can write $\mathbf{\Sigma}_{ZZ}^{-1}$ as follows:

$$\begin{pmatrix} \mathbf{I} & -\mathbf{\Sigma}_{XX}^{-1}\mathbf{\Sigma}_{XY} \\ \mathbf{0} & \mathbf{I} \end{pmatrix} \begin{pmatrix} \mathbf{\Sigma}_{XX}^{-1} & \mathbf{0} \\ \mathbf{0} & \mathbf{\Sigma}_{YY\cdot X}^{-1} \end{pmatrix} \begin{pmatrix} \mathbf{I} & \mathbf{0} \\ -\mathbf{\Sigma}_{YX}\mathbf{\Sigma}_{XX}^{-1} & \mathbf{I} \end{pmatrix}. \tag{3.109}$$

Consider the following nonsingular transformation of the random r-vector **Z**:

$$\mathbf{U} = \begin{pmatrix} \mathbf{U}_1 \\ \mathbf{U}_2 \end{pmatrix} = \begin{pmatrix} \mathbf{I} & \mathbf{0} \\ -\boldsymbol{\Sigma}_{YX}\boldsymbol{\Sigma}_{XX}^{-1} & \mathbf{I} \end{pmatrix} \begin{pmatrix} \mathbf{X} \\ \mathbf{Y} \end{pmatrix} \tag{3.110}$$

The random vector **U** has a multivariate Gaussian distribution with mean,

$$\boldsymbol{\mu}_U = \begin{pmatrix} \mathbf{I} & \mathbf{0} \\ -\boldsymbol{\Sigma}_{XY}\boldsymbol{\Sigma}_{XX}^{-1} & \mathbf{I} \end{pmatrix} \begin{pmatrix} \boldsymbol{\mu}_X \\ \boldsymbol{\mu}_Y \end{pmatrix} \tag{3.111}$$

and covariance matrix,

$$\boldsymbol{\Sigma}_{UU} = \begin{pmatrix} \boldsymbol{\Sigma}_{XX} & \mathbf{0} \\ \mathbf{0} & \boldsymbol{\Sigma}_{YY\cdot X} \end{pmatrix}. \tag{3.112}$$

Hence, the marginal distribution of $\mathbf{U}_1 = \mathbf{X}$ is $\mathcal{N}_r(\boldsymbol{\mu}_X, \boldsymbol{\Sigma}_{XX})$, the marginal distribution of $\mathbf{U}_2 = \mathbf{Y} - \boldsymbol{\Sigma}_{YX}\boldsymbol{\Sigma}_{XX}^{-1}\mathbf{X}$ is $\mathcal{N}_s(\boldsymbol{\mu}_Y - \boldsymbol{\Sigma}_{YX}\boldsymbol{\Sigma}_{XX}^{-1}\boldsymbol{\mu}_X, \boldsymbol{\Sigma}_{YY\cdot X})$, and $\mathbf{U}_1$ and $\mathbf{U}_2$ are independent.

Now, given $\mathbf{X} = \mathbf{x}$, $\boldsymbol{\mu}_Y + \boldsymbol{\Sigma}_{YX}\boldsymbol{\Sigma}_{XX}^{-1}(\mathbf{x} - \boldsymbol{\mu}_X)$ is a constant. So, because of independence, the conditional distribution of $(\mathbf{Y} - \boldsymbol{\mu}_Y) - \boldsymbol{\Sigma}_{YX}\boldsymbol{\Sigma}_{XX}^{-1}(\mathbf{x} - \boldsymbol{\mu}_X)$ is identical to the unconditional distribution of $(\mathbf{Y} - \boldsymbol{\mu}_Y) - \boldsymbol{\Sigma}_{YX}\boldsymbol{\Sigma}_{XX}^{-1}(\mathbf{X} - \boldsymbol{\mu}_X)$, which is $\mathcal{N}_s(\mathbf{0}, \boldsymbol{\Sigma}_{YY\cdot X})$. Hence, $(\mathbf{Y} - \boldsymbol{\mu}_Y) - \boldsymbol{\Sigma}_{YX}\boldsymbol{\Sigma}_{XX}^{-1}(\mathbf{x} - \boldsymbol{\mu}_X) \sim \mathcal{N}_s(\mathbf{0}, \boldsymbol{\Sigma}_{YY\cdot X})$.

The resulting *conditional distribution* of **Y** given **X**=**x** is an s-variate Gaussian with mean vector and covariance matrix given by

$$\boldsymbol{\mu}_{Y|X} = \boldsymbol{\mu}_Y + \boldsymbol{\Sigma}_{YX}\boldsymbol{\Sigma}_{XX}^{-1}(\mathbf{x} - \boldsymbol{\mu}_X) \tag{3.113}$$

$$\boldsymbol{\Sigma}_{Y|X} = \boldsymbol{\Sigma}_{YY} - \boldsymbol{\Sigma}_{YX}\boldsymbol{\Sigma}_{XX}^{-1}\boldsymbol{\Sigma}_{XY}, \tag{3.114}$$

respectively. Note that the mean vector is a linear function of **x**, whereas the covariance matrix does not depend upon **x** at all.

3.4 Random Matrices

The $(r \times s)$-matrix

$$\mathbf{Z} = \begin{pmatrix} Z_{11} & \cdots & Z_{1s} \\ \vdots & & \vdots \\ Z_{r1} & \cdots & Z_{rs} \end{pmatrix} \tag{3.115}$$

with r rows and s columns is a matrix-valued random variable (henceforth "random $(r \times s)$-matrix") if each component Z_{ij} is a random variable, $i = 1, 2, \ldots, r, j = 1, 2, \ldots, s$. That is, if the joint distribution,

$$F_Z(\mathbf{z}) = F_Z(z_{ij}, i = 1, 2, \ldots, r, j = 1, 2, \ldots, s) \tag{3.116}$$

$$= \mathrm{P}\{Z_{ij} \leq z_{ij}, i = 1, 2, \ldots, r, j = 1, 2, \ldots, s\} \tag{3.117}$$

$$= \mathrm{P}\{\mathbf{Z} \leq \mathbf{z}\}, \tag{3.118}$$

is defined for all $\mathbf{z} = (z_{ij})$.

The expected value of the random $(r \times s)$-matrix $\mathbf{Z}$ is given by

$$\boldsymbol{\mu}_Z = \mathrm{E}(\mathbf{Z}) = \begin{pmatrix} \mathrm{E}(Z_{11}) & \cdots & \mathrm{E}(Z_{1s}) \\ \vdots & & \vdots \\ \mathrm{E}(Z_{r1}) & \cdots & \mathrm{E}(Z_{rs}) \end{pmatrix} = \begin{pmatrix} \mu_{11} & \cdots & \mu_{1s} \\ \vdots & & \vdots \\ \mu_{r1} & \cdots & \mu_{rs} \end{pmatrix}. \quad (3.119)$$

The covariance matrix of $\mathbf{Z}$ is the matrix of all covariances of pairs of elements of $\mathbf{Z}$ and has rs rows and rs columns. It is, therefore, the covariance matrix of $\mathrm{vec}(\mathbf{Z})$,

$$\boldsymbol{\Sigma}_{ZZ} = \mathrm{cov}\{\mathrm{vec}(\mathbf{Z})\} = \mathrm{E}\{(\mathrm{vec}(\mathbf{Z} - \boldsymbol{\mu}_Z))(\mathrm{vec}(\mathbf{Z} - \boldsymbol{\mu}_Z))^\tau\}. \quad (3.120)$$

If we form a new matrix-valued random variable $\mathbf{W}$ by setting

$$\mathbf{W} = \mathbf{A}\mathbf{Z}\mathbf{B}^\tau + \mathbf{C}, \quad (3.121)$$

where $\mathbf{A}$, $\mathbf{B}$, and $\mathbf{C}$ are matrices of constants, then the mean matrix of $\mathbf{W}$ is

$$\boldsymbol{\mu}_W = \mathbf{A}\boldsymbol{\mu}_Z\mathbf{B}^\tau + \mathbf{C}, \quad (3.122)$$

and, because

$$\mathrm{vec}(\mathbf{W} - \boldsymbol{\mu}_W) = \mathrm{vec}(\mathbf{A}(\mathbf{Z} - \boldsymbol{\mu}_Z)\mathbf{B}^\tau) = (\mathbf{A} \otimes \mathbf{B})\mathrm{vec}(\mathbf{Z} - \boldsymbol{\mu}_Z), \quad (3.123)$$

the covariance matrix of $\mathrm{vec}(\mathbf{W})$ is

$$\begin{aligned} \boldsymbol{\Sigma}_{WW} &= \mathrm{E}\{(\mathrm{vec}(\mathbf{W} - \boldsymbol{\mu}_W))(\mathrm{vec}(\mathbf{W} - \boldsymbol{\mu}_W))^\tau\} \\ &= (\mathbf{A} \otimes \mathbf{B})\boldsymbol{\Sigma}_{ZZ}(\mathbf{A} \otimes \mathbf{B})^\tau. \end{aligned} \quad (3.124)$$

3.4.1   Wishart Distribution

Given n independently distributed random r-vectors,

$$\mathbf{X}_i \sim \mathcal{N}_r(\boldsymbol{\mu}_i, \boldsymbol{\Sigma}), \quad i = 1, 2, \ldots, n \ (n \geq r), \quad (3.125)$$

we say that the random positive-definite and symmetric $(r \times r)$-matrix,

$$\mathbf{W} = \sum_{i=1}^{n} \mathbf{X}_i\mathbf{X}_i^\tau, \quad (3.126)$$

has the *Wishart distribution* with n degrees of freedom and associated matrix $\boldsymbol{\Sigma}$. If $\boldsymbol{\mu}_i = \mathbf{0}$ for all i, the Wishart distribution of $\mathbf{W}$ is termed *central*; otherwise, it is *noncentral*.

It can be shown that the joint density function of the $r(r+1)/2$ distinct elements of $\mathbf{W}$ is given by

$$w_r(\mathbf{W}|n, \boldsymbol{\Sigma}) = c_{r,n}|\boldsymbol{\Sigma}|^{-1/2n}|\mathbf{W}|^{\frac{1}{2}(n-r-1)}e^{-\frac{1}{2}\mathrm{tr}(\mathbf{W}\boldsymbol{\Sigma}^{-1})}, \qquad (3.127)$$

where

$$\frac{1}{c_{r,n}} = 2^{nr/2}\pi^{r(r-1)/4}\prod_{i=1}^{r}\Gamma\left(\frac{n+1-i}{2}\right). \qquad (3.128)$$

If $\mathbf{W}$ is singular, the density is 0, in which case $\mathbf{W}$ is said to have the *singular Wishart distribution*. If $\mathbf{W}$ has a Wishart density, we find it convenient to write

$$\mathbf{W} \sim \mathcal{W}_r(n, \boldsymbol{\Sigma}). \qquad (3.129)$$

Many derivations of (3.127) have appeared in the statistical literature. See Anderson (1984) for references. When $r = 1$, $\mathcal{W}_1(n, \sigma^2)$ is identical to the $\sigma^2\chi_n^2$ distribution.

The first two moments of $\mathbf{W}$ are given by

$$E(\mathbf{W}) = n\boldsymbol{\Sigma}. \qquad (3.130)$$

$$\begin{aligned}
\mathrm{cov}\{\mathrm{vec}(\mathbf{W})\} &= E\{(\mathrm{vec}(\mathbf{W} - n\boldsymbol{\Sigma}))(\mathrm{vec}(\mathbf{W} - n\boldsymbol{\Sigma}))^\tau\} &(3.131)\\
&= n(\mathbf{I}_{r^2} + \mathbf{I}_{(r,r)})(\boldsymbol{\Sigma} \otimes \boldsymbol{\Sigma}), &(3.132)
\end{aligned}$$

where $\mathbf{I}_{(p,q)}$ is a *permuted-identity matrix* (Macrae, 1974), which is a $(pq \times pq)$-matrix partitioned into $(p \times q)$-submatrices such that the ijth submatrix has a 1 in its jith position and zeroes elsewhere. For example, when $p = q = 2$, the permuted-identity matrix is given by

$$\mathbf{I}_{(2.2)} = \begin{pmatrix} 1 & 0 & 0 & 0 \\ 0 & 0 & 1 & 0 \\ 0 & 1 & 0 & 0 \\ 0 & 0 & 0 & 1 \end{pmatrix}. \qquad (3.133)$$

The permuted identity matrix $\mathbf{I}_{(r,r)}$ can be expressed as the sum of r^2 Kronecker products,

$$\mathbf{I}_{(r,r)} = \sum_{i=1}^{r}\sum_{j=1}^{r}(\mathbf{H}_{ij} \otimes \mathbf{H}_{ij}^\tau), \qquad (3.134)$$

where $\mathbf{H}_{ij}$ is an $(r \times r)$-matrix with ijth element equal to 1 and zero otherwise. Another property of the permuted identity matrix is that

$$\mathbf{I}_{(r,r)}\mathrm{vec}(\mathbf{A}) = \mathrm{vec}(\mathbf{A}^\tau), \qquad (3.135)$$

which led to it also being called a *commutation matrix*.

Properties of the Wishart Distribution

Because of the following properties of the Wishart distribution, it is not necessary to apply the density form (3.127) to obtain explicit distributional results.

1. Let $\mathbf{W}_j \sim \mathcal{W}_r(n_j, \boldsymbol{\Sigma})$, $j = 1, 2, \ldots, m$, be independently distributed (central or not). Then, $\sum_{j=1}^{m} \mathbf{W}_j \sim \mathcal{W}_r(\sum_{j=1}^{r} n_j, \boldsymbol{\Sigma})$.

2. Suppose $\mathbf{W} \sim \mathcal{W}_r(n, \boldsymbol{\Sigma})$, and let $\mathbf{A}$ be a $(p \times r)$-matrix of fixed constants with rank p. Then, $\mathbf{A}\mathbf{W}\mathbf{A}^\tau \sim \mathcal{W}_r(n, \mathbf{A}\boldsymbol{\Sigma}\mathbf{A}^\tau)$.

3. Suppose $\mathbf{W} \sim \mathcal{W}_r(n, \boldsymbol{\Sigma})$, and let $\mathbf{a}$ be a fixed r-vector. Then, $\mathbf{a}^\tau \mathbf{W} \mathbf{a} \sim \sigma_a^2 \chi_n^2$, where $\sigma_a^2 = \mathbf{a}^\tau \boldsymbol{\Sigma} \mathbf{a}$. The chi-squared distribution is central if the Wishart distribution is central.

4. Let $\mathcal{X} = (\mathbf{X}_1, \cdots, \mathbf{X}_n)^\tau$, where $\mathbf{X}_i \sim \mathcal{N}_r(\mathbf{0}, \boldsymbol{\Sigma})$, $i = 1, 2, \ldots, n$, are *independently and identically distributed (iid)*. Let $\mathbf{A}$ be a symmetric $(n \times n)$-matrix, and let $\mathbf{a}$ be a fixed r-vector. Let $\mathbf{y} = \mathcal{X}\mathbf{a}$. Then, $\mathcal{X}^\tau \mathbf{A} \mathcal{X} \sim \mathcal{W}_r(n, \boldsymbol{\Sigma})$ iff $\mathbf{y}^\tau \mathbf{A} \mathbf{y} \sim \sigma_a^2 \chi_n^2$, where $\sigma_a^2 = \mathbf{a}^\tau \boldsymbol{\Sigma} \mathbf{a}$.

3.5 Maximum Likelihood Estimation for the Gaussian

Assume that we have n random r-vectors $\mathbf{X}_1, \mathbf{X}_2, \ldots, \mathbf{X}_n$, iid as multivariate Gaussian vectors,

$$\mathbf{X}_j \sim \mathcal{N}_r(\boldsymbol{\mu}, \boldsymbol{\Sigma}), \quad j = 1, 2, \ldots, n, \tag{3.136}$$

where the *parameters*, $\boldsymbol{\mu}$ and $\boldsymbol{\Sigma}$, of this distribution are both unknown. To estimate $\boldsymbol{\mu}$ and $\boldsymbol{\Sigma}$, we use the method of *maximum likelihood (ML)*.

By independence, the joint density of the data $\{\mathbf{X}_i, i = 1, 2, \ldots, n\}$ is the product of the individual densities; that is, $\prod_{i=1}^{n} f_{\mathbf{X}_i}(\mathbf{x}_i | \boldsymbol{\mu}, \boldsymbol{\Sigma})$. If we now consider this joint density as a function of the parameters, $\boldsymbol{\mu}$ and $\boldsymbol{\Sigma}$, then we have the *likelihood function* of the parameters given the data,

$$\mathcal{L}(\boldsymbol{\mu}, \boldsymbol{\Sigma} | \{\mathbf{X}_i\}) = (2\pi)^{-nr/2} |\boldsymbol{\Sigma}|^{-n/2} \exp\left\{ -\frac{1}{2} \sum_{i=1}^{n} (\mathbf{x}_i - \boldsymbol{\mu})^\tau \boldsymbol{\Sigma}^{-1} (\mathbf{x}_i - \boldsymbol{\mu}) \right\}.$$
$$\tag{3.137}$$

Taking logarithms of this expression, we have that the log-likelihood function is

$$\ell(\boldsymbol{\mu}, \boldsymbol{\Sigma}) = \log \mathcal{L}(\boldsymbol{\mu}, \boldsymbol{\Sigma} | \{\mathbf{X}_i\})$$

$$= -\frac{nr}{2}\log(2\pi) - \frac{n}{2}\log|\mathbf{\Sigma}| - \frac{1}{2}\sum_{i=1}^{n}(\mathbf{x}_i - \boldsymbol{\mu})^\tau \mathbf{\Sigma}^{-1}(\mathbf{x}_i - \boldsymbol{\mu}).$$

$$\text{(3.138)}$$

It will be convenient to reexpress the summation term in (3.138) as follows:

$$\sum_{i=1}^{n}(\mathbf{x}_i - \boldsymbol{\mu})^\tau \mathbf{\Sigma}^{-1}(\mathbf{x}_i - \boldsymbol{\mu}) \qquad \text{(3.139)}$$

$$= \mathrm{tr}\left\{\mathbf{\Sigma}^{-1}\sum_{i=1}^{n}(\mathbf{x}_i - \bar{\mathbf{x}})(\mathbf{x}_i - \bar{\mathbf{x}})^\tau\right\} + n(\bar{\mathbf{x}} - \boldsymbol{\mu})^\tau \mathbf{\Sigma}^{-1}(\bar{\mathbf{x}} - \boldsymbol{\mu}), \quad \text{(3.140)}$$

where $\bar{\mathbf{x}} = n^{-1}\sum_{i=1}^{n}\mathbf{x}_i$ is the *sample mean*.

The ML method estimates the parameters $\boldsymbol{\mu}$ and $\mathbf{\Sigma}$ by maximizing the log-likelihood with respect to (wrt) those parameters, given the data values, $\{\mathbf{x}_i, i = 1, 2, \ldots, n\}$. First, we maximize ℓ wrt $\boldsymbol{\mu}$:

$$\frac{\partial \ell(\boldsymbol{\mu}, \mathbf{\Sigma})}{\partial \boldsymbol{\mu}} = \mathbf{\Sigma}^{-1}(\bar{\mathbf{x}} - \boldsymbol{\mu}). \qquad \text{(3.141)}$$

Setting this derivative equal to zero, the ML estimator of $\boldsymbol{\mu}$ is the random r-vector

$$\widehat{\boldsymbol{\mu}} = \bar{\mathbf{X}}, \qquad \text{(3.142)}$$

which we call the *sample mean vector*. For a given data set, the ML estimate is $\widehat{\boldsymbol{\mu}} = \bar{\mathbf{x}}$.

Deriving the ML estimate for $\mathbf{\Sigma}$ needs a little more work. If we define $\mathbf{A} = \sum_{i=1}^{n}(\mathbf{x}_i - \bar{\mathbf{x}})(\mathbf{x}_i - \bar{\mathbf{x}})^\tau$, then (3.138) can be written as

$$\ell(\boldsymbol{\mu}, \mathbf{\Sigma}) = -\frac{nr}{2}\log(2\pi) - \frac{n}{2}\log|\mathbf{\Sigma}| - \frac{1}{2}\mathrm{tr}(\mathbf{\Sigma}^{-1}\mathbf{A}) + n(\bar{\mathbf{x}} - \boldsymbol{\mu})^\tau \mathbf{\Sigma}^{-1}(\bar{\mathbf{x}} - \boldsymbol{\mu}).$$

$$\text{(3.143)}$$

The first term on the rhs of (3.143) is a constant and, at the maximum of ℓ, the last term is zero. So, we need to find $\mathbf{\Sigma}$ to maximize $-n\log|\mathbf{\Sigma}| - \mathrm{tr}(\mathbf{\Sigma}^{-1}\mathbf{A})$.

Set $\mathbf{A} = \mathbf{EE}^\tau$ and $\mathbf{E}^\tau \mathbf{\Sigma}^{-1}\mathbf{E} = \mathbf{H}$. Then, $\mathbf{\Sigma} = \mathbf{EH}^{-1}\mathbf{E}^\tau$ and $|\mathbf{\Sigma}| = |\mathbf{A}|/|\mathbf{H}|$, whence, $\log|\mathbf{\Sigma}| = \log|\mathbf{A}| - \log|\mathbf{H}|$. Also, using properties of the trace, $\mathrm{tr}(\mathbf{\Sigma}^{-1}\mathbf{A}) = \mathrm{tr}(\mathbf{\Sigma}^{-1}\mathbf{EE}^\tau) = \mathrm{tr}(\mathbf{E}^\tau \mathbf{\Sigma}^{-1}\mathbf{E}) = \mathrm{tr}(\mathbf{H})$. Putting these results together, we now need to find $\mathbf{H}$ to maximize $-n\log|\mathbf{A}| + n\log|\mathbf{H}| - \mathrm{tr}(\mathbf{H})$.

By the Cholesky decomposition of $\mathbf{H}$, there is a unique lower-triangular matrix $\mathbf{T} = (t_{ij})$ with positive diagonal elements such that $\mathbf{H} = \mathbf{TT}^\tau$.

Hence, we need to find a lower-triangular $\mathbf{T}$ to maximize $-n \log |\mathbf{A}| + \sum_{i=1}^{r}(n \log t_{ii}^2 - t_{ii}^2) - \sum_{i>j} t_{ij}^2$, where we used the facts that $|\mathbf{T}|^2 = \prod_{i=1}^{r} t_{ii}^2$ and $\operatorname{tr}(\mathbf{TT}^{\tau}) = \sum_{i=1}^{r} t_{ii}^2$. The solution is to take $t_{ii}^2 = n$ and $t_{ij} = 0$ for $i \neq j$; that is, take $\mathbf{T} = \sqrt{n}\mathbf{I}_r$. Thus, we take $\mathbf{H} = n\mathbf{I}_r$, whence, $\boldsymbol{\Sigma} = n^{-1}\mathbf{EE}^{\tau} = n^{-1}\mathbf{A}$. So, the ML estimator of $\boldsymbol{\Sigma}$ is given by the random $(r \times r)$-matrix

$$\widehat{\boldsymbol{\Sigma}} = \frac{1}{n} \sum_{i=1}^{n} (\mathbf{X}_i - \bar{\mathbf{X}})(\mathbf{X}_i - \bar{\mathbf{X}})^{\tau} = n^{-1}\mathbf{S}, \tag{3.144}$$

which we call the *sample covariance matrix*. For a given data set, the ML estimate is $\widehat{\boldsymbol{\Sigma}} = n^{-1}\mathbf{A}$.

3.5.1   Joint Distribution of Sample Mean and Sample Covariance Matrix

The ML estimator $\bar{\mathbf{X}}$ is an unbiased estimator of the population mean vector $\boldsymbol{\mu}$; that is,

$$E\{\bar{\mathbf{X}}\} = \boldsymbol{\mu}. \tag{3.145}$$

On the other hand, because

$$E\{\widehat{\boldsymbol{\Sigma}}\} = \frac{n-1}{n}\boldsymbol{\Sigma}, \tag{3.146}$$

the ML estimator $\widehat{\boldsymbol{\Sigma}}$ in (3.144) is a biased estimator of the population covariance matrix $\boldsymbol{\Sigma}$. To remove the bias from the covariance estimator (3.144), it suffices to divide $\mathbf{S}$ by $n-1$ instead of by n.

Because $\bar{\mathbf{X}}$ is a linear combination of the $\mathbf{X}_1, \ldots, \mathbf{X}_n$, each of which are i.i.d. as $\mathcal{N}_r(\boldsymbol{\mu}, \boldsymbol{\Sigma})$, then, the ML estimator, $\bar{\mathbf{X}}$ of $\boldsymbol{\mu}$ has the distribution

$$\bar{\mathbf{X}} \sim \mathcal{N}_r(\boldsymbol{\mu}, n^{-1}\boldsymbol{\Sigma}). \tag{3.147}$$

To derive the distribution of $\widehat{\boldsymbol{\Sigma}}$, we suppose for the moment that $\boldsymbol{\mu} = \mathbf{0}$. Let $\mathbf{a}$ be a fixed r-vector and consider $y_i = \mathbf{a}^{\tau}\mathbf{X}_i$, $i = 1, 2, \ldots, n$. Then, $y_i \sim \mathcal{N}_1(0, \sigma_a^2)$, where $\sigma_a^2 = \mathbf{a}^{\tau}\boldsymbol{\Sigma}\mathbf{a}$, and $\mathbf{y} = (y_1, \cdots, y_n)^{\tau} \sim \mathcal{N}_n(\mathbf{0}, \sigma_a^2\mathbf{I}_n)$. Let $\mathbf{b} = n^{-1}\mathbf{1}_n$, whence, $\mathbf{b}^{\tau}\mathbf{b} = n^{-1}$, and let $\mathbf{A} = \mathbf{I}_n - n^{-1}\mathbf{J}_n$, where $\mathbf{J}_n = \mathbf{1}_n\mathbf{1}_n^{\tau}$ is a matrix every element of which is unity. Note that $\mathbf{A}$ is idempotent with rank n. From univariate theory, $\mathbf{b}^{\tau}\mathbf{y} = \bar{y} \sim \mathcal{N}_1(0, \sigma_a^2/n)$ and, $\mathbf{y}^{\tau}\mathbf{A}\mathbf{y} = \sum_i (y_i - \bar{y})^2 \sim \sigma_a^2\chi_{n-1}^2$ are independently distributed for any $\mathbf{a}$.

Now, let $\mathcal{X} = (\mathbf{X}_1, \cdots, \mathbf{X}_n)^{\tau}$. Then, $\mathbf{b}^{\tau}\mathcal{X} \sim \mathcal{N}_r(\mathbf{0}, n^{-1}\boldsymbol{\Sigma})$ and, from Property 4 of the Wishart distribution,

$$\mathcal{X}^{\tau}\mathbf{A}\mathcal{X} \sim \mathcal{W}_r(n, \boldsymbol{\Sigma}). \tag{3.148}$$

Because $\mathbf{y} \sim \mathcal{N}_n(\mathbf{0}, \sigma_a^2 \mathbf{I}_n)$, it follows that $\mathbf{b}^\tau \mathbf{y} \sim \mathcal{N}_1(0, \sigma_a^2 \mathbf{b}^\tau \mathbf{b})$ and

$$\mathbf{y}^\tau \mathbf{b} \mathbf{b}^\tau \mathbf{y}/\mathbf{b}^\tau \mathbf{b} \sim \sigma_a^2 \chi_1^2. \tag{3.149}$$

Furthermore, $\mathbf{A}\mathbf{b}\mathbf{b}^\tau = \mathbf{0}$; postmultiplying by $\mathbf{b}$ yields $\mathbf{A}\mathbf{b} = \mathbf{0}$, so that the columns of $\mathbf{A} = (\mathbf{a}_1, \cdots, \mathbf{a}_n)$ and $\mathbf{b}$ are mutually orthogonal. Thus, $\mathcal{X}^\tau \mathbf{a}_i = \mathbf{X}_i - \bar{\mathbf{X}}$, $i = 1, 2, \ldots, n$, and $\mathbf{b}^\tau \mathcal{X}$ are statistically independent of each other. Thus, $\mathbf{b}^\tau \mathcal{X} = \bar{\mathbf{X}}$ and $\mathcal{X}^\tau \mathbf{A} \mathcal{X} = (\mathcal{X}^\tau \mathbf{A})(\mathcal{X}^\tau \mathbf{A})^\tau = \mathbf{S}$ are independently distributed.

The case of $\boldsymbol{\mu} \neq \mathbf{0}$ is dealt with by replacing $\mathbf{X}_i$ by $\mathbf{X}_i - \boldsymbol{\mu}$, $i = 1, 2, \ldots, n$. This does not change $\mathbf{S}$, and $\bar{\mathbf{X}}$ is replaced by $\bar{\mathbf{X}} - \boldsymbol{\mu}$. Thus, $\mathbf{S}$ is independent of $\bar{\mathbf{X}} - \boldsymbol{\mu}$ (and, hence, of $\bar{\mathbf{X}}$), and

$$\widehat{\boldsymbol{\Sigma}} \sim n^{-1} \mathcal{W}_r(n - 1, \boldsymbol{\Sigma}). \tag{3.150}$$

3.5.2 Admissibility

In 1955, Charles Stein rocked the statistical world by showing that the ML estimator, $\bar{\mathbf{X}}$, of the unknown mean vector, $\boldsymbol{\mu}$, of a multivariate Gaussian distribution was "admissible" in one or two dimensions but was "inadmissible" in three or higher dimensions (Stein, 1955).

The idea of inadmissibility of an estimator $\widehat{\boldsymbol{\theta}}$ of an unknown vector-valued parameter $\boldsymbol{\theta} \in \Theta$ is part of the framework of *statistical decision theory* and relates to the quality of that estimator in terms of a given *loss function* $L(\boldsymbol{\theta}, \widehat{\boldsymbol{\theta}})$. A loss function gives a quantitative description of the loss incurred if $\boldsymbol{\theta}$ is estimated by $\widehat{\boldsymbol{\theta}}$. For example, the most popular type of loss function for assessing an estimator, $\widehat{\boldsymbol{\theta}} = (\widehat{\theta}_1, \cdots, \widehat{\theta}_r)^\tau$, of the unknown parameter vector $\boldsymbol{\theta} = (\theta_1, \cdots, \theta_r)^\tau$ is the "squared-error" loss function,

$$L(\boldsymbol{\theta}, \widehat{\boldsymbol{\theta}}) = (\widehat{\boldsymbol{\theta}} - \boldsymbol{\theta})^\tau (\widehat{\boldsymbol{\theta}} - \boldsymbol{\theta}) = \sum_{j=1}^{r} (\widehat{\theta}_j - \theta_j)^2. \tag{3.151}$$

Different types of loss functions have been proposed in different situations, and we will meet several of these throughout this book.

It is usual to compare estimators through their *risk functions*, which are the expected values of the respective loss functions; that is,

$$R(\boldsymbol{\theta}, \widehat{\boldsymbol{\theta}}) = \mathrm{E}_\theta \{ L(\boldsymbol{\theta}, \widehat{\boldsymbol{\theta}}) \}. \tag{3.152}$$

Two different estimators, $\widehat{\boldsymbol{\theta}}_a$ and $\widehat{\boldsymbol{\theta}}_b$, of $\boldsymbol{\theta}$ can be compared by viewing the graphs of $R(\boldsymbol{\theta}, \widehat{\boldsymbol{\theta}}_a)$ and $R(\boldsymbol{\theta}, \widehat{\boldsymbol{\theta}}_b)$ over a suitable range of values of some function of $\boldsymbol{\theta}$, say, $\| \boldsymbol{\theta} \|$. An estimator $\widehat{\boldsymbol{\theta}}_a$ is *inadmissible* if there exists another estimator $\widehat{\boldsymbol{\theta}}_b$ for which

$$R(\boldsymbol{\theta}, \widehat{\boldsymbol{\theta}}_b) \leq R(\boldsymbol{\theta}, \widehat{\boldsymbol{\theta}}_a) \text{ for all } \boldsymbol{\theta} \in \Theta \tag{3.153}$$

and

$$R(\boldsymbol{\theta}, \widehat{\boldsymbol{\theta}}_b) < R(\boldsymbol{\theta}, \widehat{\boldsymbol{\theta}}_a) \text{ for some } \boldsymbol{\theta} \in \Theta; \qquad (3.154)$$

the estimator $\widehat{\boldsymbol{\theta}}_a$ is *admissible* if no such estimator $\widehat{\boldsymbol{\theta}}_b$ exists. In other words, an estimator is inadmissible if we can find a better estimator that has a smaller risk function, whereas an estimator that cannot be improved upon in this way is called admissible.

3.5.3 *James–Stein Estimator of the Mean Vector*

Suppose $\mathbf{X}_i, i = 1, 2, \ldots, n$, are independently drawn from an r-variate Gaussian distribution with unknown mean vector $\boldsymbol{\mu} = (\mu_1, \cdots, \mu_r)^\tau$, such that the ML estimator $\mathbf{Y} = \bar{\mathbf{X}} = n^{-1} \sum_i \mathbf{X}_i$ has the $\mathcal{N}_r(\boldsymbol{\mu}, \mathbf{I}_r)$ distribution. Thus, the components of the unknown mean vector, $\boldsymbol{\mu}$, are different, and the components of $\mathbf{Y}$ are mutually independent with unit variances. The following development can be easily modified if the covariance matrix of $\mathbf{Y}$ were $\sigma^2 \mathbf{I}_r$, where $\sigma^2 > 0$ is known (Exercise 3.17), or a more general known covariance matrix $\mathbf{V}$ (Exercise 3.18).

The risk function of the estimator $\mathbf{Y} = (Y_1, \cdots, Y_r)^\tau$ is given by

$$R(\boldsymbol{\mu}, \mathbf{Y}) = \mathrm{E}_{\boldsymbol{\mu}}\{(\mathbf{Y} - \boldsymbol{\mu})^\tau(\mathbf{Y} - \boldsymbol{\mu})\} = \mathrm{tr}\{\mathbf{I}_r\} = r. \qquad (3.155)$$

Stein's result that the sample mean vector is inadmissible for $r \geq 3$ in the case of squared-error loss was later supplemented by James and Stein (1961), who exhibited a "better" estimator of the multivariate Gaussian mean vector $\boldsymbol{\mu}$ than the sample mean $\bar{\mathbf{X}}$. Let $\boldsymbol{\theta} = (\theta_1, \cdots, \theta_r)^\tau$ be an arbitrary fixed vector, which is chosen before we look at the data. Typically, $\boldsymbol{\theta}$ is thought to be near $\boldsymbol{\mu}$.

The *James–Stein estimator*, $\delta(\mathbf{Y}) = (\delta_1(\mathbf{Y}), \cdots, \delta_r(\mathbf{Y}))^\tau$, is given by

$$\delta(\mathbf{Y}) = \boldsymbol{\theta} + \left(1 - \frac{r - 2}{S}\right)(\mathbf{Y} - \boldsymbol{\theta}), \qquad (3.156)$$

where

$$S = \| \mathbf{Y} - \boldsymbol{\theta} \|^2 = \sum_{j=1}^{r}(Y_j - \theta_j)^2 \qquad (3.157)$$

is the sum of the squared deviations of each individual mean Y_j from the constant θ_j, and $r \geq 3$. Thus, the James–Stein estimator shrinks $\mathbf{Y}$ toward $\boldsymbol{\theta}$ by a factor $c = 1 - (r - 2)/S$. Note that for fixed $\boldsymbol{\theta}$, the shrinkage factor c is the same for all components of $\mathbf{Y}$.

The estimator $\delta(\mathbf{Y})$ has a smaller risk than that of $\mathbf{Y}$ for every $\boldsymbol{\mu}$, independent of whichever vector $\boldsymbol{\theta}$ is chosen. To see this, consider the risk of $\delta(\mathbf{Y})$:

$$R(\boldsymbol{\mu}, \delta(\mathbf{Y})) = \mathrm{E}_{\boldsymbol{\mu}}\left\{\sum_{j=1}^{r}(\delta_j(\mathbf{Y}) - \mu_j)^2\right\} = \mathrm{E}_{\boldsymbol{\mu}}\{\| \delta(\mathbf{Y}) - \boldsymbol{\mu} \|^2\}. \qquad (3.158)$$

Now,

$$
\begin{aligned}
\| \, \delta(\mathbf{Y}) - \boldsymbol{\mu} \, \|^2 \;&=\; \| \, \boldsymbol{\theta} + \left(1 - \frac{r-2}{S}\right)(\mathbf{Y} - \boldsymbol{\theta}) - \boldsymbol{\mu} \, \|^2 \\
&=\; \sum_{j=1}^{r}\left\{(Y_j - \mu_j) - \frac{r-2}{S}(Y_j - \theta_j)\right\}^2.
\end{aligned}
\tag{3.159}
$$

Expand the summand to get

$$
(Y_j - \mu_j)^2 - \frac{2(r-2)}{S}(Y_j - \mu_j)(Y_j - \theta_j) + \frac{(r-2)^2}{S^2}(Y_j - \theta_j)^2. \tag{3.160}
$$

Substituting this expression back into (3.159), rearranging terms, and then taking expectations, the risk of $\delta(\mathbf{Y})$ is

$$
R(\boldsymbol{\mu}, \delta(\mathbf{Y})) =
$$
$$
r - \mathrm{E}_{\boldsymbol{\mu}}\left\{2(r-2)\sum_{j=1}^{r}\left(\frac{Y_j - \theta_j}{S}\right)(Y_j - \mu_j) - \frac{(r-2)^2}{S}\right\}. \tag{3.161}
$$

The first term inside the expectation is evaluated using *Stein's Lemma*, which says that if $Y \sim \mathcal{N}(\theta, 1)$ and g is a differentiable function such that $\mathrm{E}_\theta\{|g'(Y)|\} < \infty$, then,

$$
\mathrm{E}_\theta\{g(Y)(Y - \theta)\} = \mathrm{E}_\theta\{g'(Y)\}. \tag{3.162}
$$

Let

$$
g(Y_j) = \frac{Y_j - \theta_j}{S}, \tag{3.163}
$$

whence,

$$
g'(Y_j) = \frac{1}{S} - \frac{2(Y_j - \theta_j)^2}{S^2}. \tag{3.164}
$$

Substituting the last result into (3.162) yields

$$
R(\boldsymbol{\mu}, \delta(\mathbf{Y})) =
$$
$$
r - \mathrm{E}_{\boldsymbol{\mu}}\left\{2(r-2)\sum_{j=1}^{r}\left\{\frac{1}{S} - \frac{2(Y_j - \theta_j)^2}{S^2}\right\} - \frac{(r-2)^2}{S}\right\}; \tag{3.165}
$$

that is,

$$
R(\boldsymbol{\mu}, \delta(\mathbf{Y})) = r - \mathrm{E}_{\boldsymbol{\mu}}\left\{\frac{1}{S}\right\} < r = R(\boldsymbol{\mu}, \mathbf{Y}). \tag{3.166}
$$

This result holds as long as the expectation exists. For $r = 1$ and $r = 2$, the expectation is infinite. For $r \geq 3$, the expectation is finite. The expectation in (3.166), which represents the difference between the two risk functions, $R(\boldsymbol{\mu}, \mathbf{Y}) - R(\boldsymbol{\mu}, \delta(\mathbf{Y}))$, is sometimes called the *Stein effect*.

Thus, instead of using just the jth component, Y_j, of $\mathbf{Y}$ to estimate the jth component, μ_j, of $\boldsymbol{\mu}$, the James–Stein estimator, $\delta(\mathbf{Y})$, combines all the mutually independent components of $\mathbf{Y}$ in estimating μ_j. This estimator appears to be intuitively unappealing: why should the estimator of μ_j depend upon the estimators of $\mu_k, k \neq j$? The reason why the James–Stein estimator dominates the usual mean estimator is because we used the squared-error loss function. This surprising result is commonly referred to as *Stein's paradox* (Efron and Morris, 1977).

The James–Stein estimator (3.156) also happens to be inadmissible for $\boldsymbol{\mu}$. This follows because, for small values of S, the shrinkage factor c becomes negative, which, in turn, drags the estimator away from $\boldsymbol{\theta}$. We can avoid such anomolies by replacing the shrinkage factor c by zero if it is negative (Efron and Morris, 1973):

$$\delta_+(\mathbf{Y}) = \boldsymbol{\theta} + \left(1 - \frac{r-2}{S}\right)_+ (\mathbf{Y} - \boldsymbol{\theta}), \tag{3.167}$$

where $(x)_+ = \max\{x, 0\}$. Unfortunately, this so-called *positive-part James–Stein estimator* is still not admissible (Brown, 1971).

The James–Stein estimator of $\boldsymbol{\mu}$ shrinks $\mathbf{Y}$ toward some chosen point $\boldsymbol{\theta}$. Shrinking to different points will produce different estimates of $\boldsymbol{\mu}$. Deciding which one is best then becomes a subjective decision. If one has no information about the location of $\boldsymbol{\mu}$, then what should we take for $\boldsymbol{\theta}$? One possibility is to use $\boldsymbol{\theta} = \mathbf{0}$, so that the James–Stein estimator shrinks $\mathbf{Y}$ toward the origin. Another possibility is to shrink each component of $\mathbf{Y}$ toward the overall mean $\bar{Y} = r^{-1} \sum_{j=1}^r Y_j$. Let $\bar{\mathbf{Y}} = (\bar{Y}, \cdots, \bar{Y})^\tau$ be an r-vector whose every entry is $\bar{Y}$. The resulting James–Stein estimator is

$$\delta'(\mathbf{Y}) = \bar{\mathbf{Y}} + \left(1 - \frac{r-3}{S'}\right)(\mathbf{Y} - \bar{\mathbf{Y}}), \tag{3.168}$$

where

$$S' = \parallel \mathbf{Y} - \bar{\mathbf{Y}} \parallel^2 = \sum_{k=1}^r (Y_k - \bar{Y})^2 \tag{3.169}$$

is the sum of the squared deviations of each individual mean Y_k from the overall mean $\bar{Y}$. Note that the constant $r - 2$ is replaced by $r - 3$ because the parameter $\boldsymbol{\theta}$ is estimated by $\bar{\mathbf{Y}}$. This estimator dominates $\mathbf{Y}$ if $r \geq 4$. Thus, μ_j is estimated by $\bar{Y} + c(Y_j - \bar{Y})$, $j = 1, 2, \ldots, r$, where the shrinkage factor is

$$c = 1 - \frac{r-3}{\sum_{k=1}^r (Y_k - \bar{Y})^2} \tag{3.170}$$

which can be motivated using an empirical Bayes approach (Efron and Morris, 1975).

Bibliographical Notes

There are many books and chapters and sections of books on matrix theory. All textbooks on multivariate analysis (e.g., Anderson, 1984; Johnson and Wichern, 1998; Mardia, Kent, and Bibby, 1980; Rao, 1965; Seber, 1984) have chapters or sections on the multivariate normal distribution and the Wishart distribution and their properties.

The chi-squared distribution (the distribution of the sample variance s^2 in the univariate case) was extended to the bivariate case by Fisher (1915) and then generalized further to the multivariate case by Wishart (1928).

Excellent discussions of decision theory, including admissibility, can be found in Lehmann (1983), Casella and Berger (1990), Berger (1985), and Anderson (1984).

Exercises

3.1 Let $\mathbf{x} = (x_1, \cdots, x_p)^\tau$ and $\mathbf{y} = (y_1, \cdots, y_p)^\tau$ be any two p-vectors on $\Re^p$. Show that $\mathbf{x}^\tau \mathbf{y} \le (\mathbf{x}^\tau \mathbf{x})(\mathbf{y}^\tau \mathbf{y})$, where the equality is achieved only if $a\mathbf{x} + b\mathbf{y} = \mathbf{0}$ for $a, b \in \Re$. (Hint: Consider $(a\mathbf{x} + b\mathbf{y})^\tau (a\mathbf{x} + b\mathbf{y})$, which is nonnegative.)

3.2 Let f and g be any real functions defined in some set A, and suppose f^2 and g^2 are integrable (wrt some measure). Show that

$$\left(\int_A f(x)g(x)dx \right)^2 \le \left(\int_A [f(x)]^2 dx \right) \left(\int_A [g(x)]^2 dx \right)^2.$$

Hence, or otherwise, show that if X and Y are random variables, then, $[\text{cov}(X, Y)]^2 \le (\text{var}(X))(\text{var}(Y))$. (Hint: Consider the nonnegative integral of $(af + bg)^2$.)

3.3 Prove the Hoffman–Wielandt Theorem. (Hint: Use the spectral decomposition theorem on $\mathbf{A}$ and on $\mathbf{B}$; express $\text{tr}\{(\mathbf{A} - \mathbf{B})(\mathbf{A} - \mathbf{B})^\tau\}$ in terms of the decomposition matrices of $\mathbf{A}$ and $\mathbf{B}$, and simplify; then, show that the result is minimized by $\sum_j (\lambda_j - \mu_j)^2$.)

3.4 If $\mathbf{X} \sim \mathcal{N}_r(\boldsymbol{\mu}, \boldsymbol{\Sigma})$, show that the marginal distribution of any subset of r^* elements of $\mathbf{X}$ is r^*-variate Gaussian.

3.5 Show that $\mathbf{X} \sim \mathcal{N}_r(\boldsymbol{\mu}, \boldsymbol{\Sigma})$ if and only if $\boldsymbol{\alpha}^\tau \mathbf{X} \sim \mathcal{N}(\boldsymbol{\alpha}^\tau \boldsymbol{\mu}, \boldsymbol{\alpha}^\tau \boldsymbol{\Sigma} \boldsymbol{\alpha})$, where $\boldsymbol{\alpha}$ is a given r-vector.

3.6 If $\mathbf{X} \sim \mathcal{N}_r(\boldsymbol{\mu}, \boldsymbol{\Sigma})$, and if $\mathbf{A}$ is a fixed $(s \times r)$-matrix and $\mathbf{b}$ is a fixed s-vector, show that the random s-vector $\mathbf{Y} = \mathbf{AX} + \mathbf{b} \sim \mathcal{N}_s(\mathbf{A}\boldsymbol{\mu} + \mathbf{b}, \mathbf{A}^\top \boldsymbol{\Sigma} \mathbf{A})$.

3.7 Suppose $\mathbf{X} \sim \mathcal{N}_r(\boldsymbol{\mu}, \boldsymbol{\Sigma})$, where $\boldsymbol{\Sigma} = \mathrm{diag}\{\sigma_i^2\}$ is a diagonal matrix. Show that the elements, $X_1, X_2, \ldots, X_r$, of $\mathbf{X}$ are independent and each X_j follows a univariate Gaussian distribution, $j = 1, 2, \ldots, r$.

3.8 If $\mathbf{Z}$ in (3.85) is distributed as an $(r + s)$-variate Gaussian with mean (3.86) and partitioned covariance matrix (3.89), show that $\mathbf{X}$ and $\mathbf{Y}$ are independently distributed if and only if $\boldsymbol{\Sigma}_{XY} = \mathbf{0}$.

3.9 If $\mathbf{Z}$ in (3.85) is distributed as an $(r + s)$-variate Gaussian with mean (3.86) and partitioned covariance matrix (3.89), and if $\boldsymbol{\Sigma}_{XX}$ is nonsingular, show that $\mathbf{Y} - \boldsymbol{\Sigma}_{YX}\boldsymbol{\Sigma}_{XX}^{-1}\mathbf{X} \sim \mathcal{N}_s(\boldsymbol{\mu}_Y - \boldsymbol{\Sigma}_{YX}\boldsymbol{\Sigma}_{XX}^{-1}\boldsymbol{\mu}_X, \boldsymbol{\Sigma}_{YY \cdot X})$, where $\boldsymbol{\Sigma}_{YY \cdot X} = \boldsymbol{\Sigma}_{YY} - \boldsymbol{\Sigma}_{YX}\boldsymbol{\Sigma}_{XX}^{-1}\boldsymbol{\Sigma}_{XY}$. The conditional distribution of $\mathbf{Y}$ given $\mathbf{X}$ is $\mathcal{N}_s(\boldsymbol{\mu}_Y + \boldsymbol{\Sigma}_{YX}\boldsymbol{\Sigma}_{XX}^{-1}(\mathbf{X} - \boldsymbol{\mu}_X), \boldsymbol{\Sigma}_{YY \cdot X})$. If $\boldsymbol{\Sigma}_{XX}$ is singular, show that the above results hold, but with $\boldsymbol{\Sigma}_{XX}^{-1}$ replaced by the reflexive g-inverse $\boldsymbol{\Sigma}_{XX}^{-}$.

3.10 The conditional distribution of $\mathbf{Y}$ given $\mathbf{X}=\mathbf{x}$ can be expressed as the ratio of the joint distribution of $(\mathbf{X}, \mathbf{Y})$ to the marginal distribution of $\mathbf{X}$: $f(\mathbf{y}|\mathbf{x}) = f_{X,Y}(\mathbf{x}, \mathbf{y})/f_X(\mathbf{x})$. Using the definition of the multivariate Gaussian distribution, find the joint and marginal distributions and compute their ratio to find the conditional distribution of $\mathbf{Y}$ given $\mathbf{X}=\mathbf{x}$. Find the conditional distribution for the special case of the bivariate Gaussian distribution. (Hint: The joint distribution of $(\mathbf{U}_1, \mathbf{U}_2)$ is given by the product of their marginals; transform the variables to $\mathbf{X}$ and $\mathbf{Y}$ by substituting $\mathbf{x}$ for $\mathbf{u}_1$ and $\mathbf{y} - \boldsymbol{\Sigma}_{YX}\boldsymbol{\Sigma}_{XX}^{-1}\mathbf{x}$ for $\mathbf{u}_2$ in that joint distribution.)

3.11 If $\mathbf{X}_j \sim \mathcal{N}(\boldsymbol{\mu}_j, \boldsymbol{\Sigma}_j)$, $j = 1, 2, \ldots, n$, are mutually independent and $c_1, c_2, \ldots, c_n$ are real numbers, show that

$$\sum_{j=1}^{n} c_j \mathbf{X}_j \sim \mathcal{N}_r \left(\sum_{j=1}^{n} c_j \boldsymbol{\mu}_j, \sum_{j=1}^{n} c_j^2 \boldsymbol{\Sigma}_j \right).$$

3.12 If the s columns of the random matrix $\mathbf{Z}$ in (3.115) are independent random r-vectors with common covariance matrix $\boldsymbol{\Sigma}$, show that $\boldsymbol{\Sigma}_{ZZ} = \mathbf{I}_s \otimes \boldsymbol{\Sigma}$.

3.13 Let $\mathbf{W}_j \sim \mathcal{W}_r(n_j, \boldsymbol{\Sigma})$, $j = 1, 2, \ldots, m$, be independently distributed. Show that $\sum_{j=1}^{m} \mathbf{W}_j \sim \mathcal{W}_r(\sum_{j=1}^{m} n_j, \boldsymbol{\Sigma})$. Show that this result holds regardless of whether the distributions are central or noncentral.

3.14 If $\mathbf{W} \sim \mathcal{W}_r(n, \boldsymbol{\Sigma})$ and $\mathbf{A}$ is a $(p \times r)$-matrix of fixed constants with rank p, show that $\mathbf{AWA}^\top \sim \mathcal{W}_p(n, \mathbf{A}\boldsymbol{\Sigma}\mathbf{A}^\top)$.

3.15 Let $\mathbf{W} \sim \mathcal{W}_r(n, \boldsymbol{\Sigma})$ and let $\mathbf{a}$ be a fixed r-vector. Show that $\mathbf{a}^\tau \mathbf{W} \mathbf{a} \sim \sigma_a^2 \chi_n^2$, where $\sigma_a^2 = \mathbf{a}^\tau \boldsymbol{\Sigma} \mathbf{a}$. The chi-squared distribution is central if the Wishart distribution is central.

3.16 (Stein's Lemma) Let $X \sim \mathcal{N}(\theta, \sigma^2)$ and let g be a differentiable function such that $\mathrm{E}\{|g'(X)|\} < \infty$. Show that $\mathrm{E}\{g(X)(X - \theta)\} = \mathrm{E}\{g'(X)\}$. (Hint: Use integration by parts with $u = g(X)$ and $dv = (X - \theta)\exp\{-(X - \theta)^2/2\sigma^2\}$.)

3.17 Show that if $\mathbf{Y} = \bar{\mathbf{X}} \sim \mathcal{N}_r(\boldsymbol{\mu}, \sigma^2 \mathbf{I}_r)$, $r \geq 3$, then $\mathbf{Y}$ is inadmissible for the loss function $L(\boldsymbol{\theta}, \mathbf{Y}) = \| \boldsymbol{\theta} - \mathbf{Y} \| / \sigma^2$, where $\sigma^2 > 0$ is known.

3.18 Show that if $\mathbf{Y} = \bar{\mathbf{X}} \sim \mathcal{N}_r(\boldsymbol{\mu}, \mathbf{V})$, where $\mathbf{V}$ is a known $(r \times r)$ covariance matrix, $r \geq 3$, then $\mathbf{Y}$ is inadmissible for the loss function $L(\boldsymbol{\theta}, \mathbf{Y}) = (\mathbf{Y} - \boldsymbol{\theta})^\tau \mathbf{V}^{-1}(\mathbf{Y} - \boldsymbol{\theta})$, where $p \geq 3$. (Hint: set $S = (\mathbf{Y} - \boldsymbol{\theta})^\tau \mathbf{V}^{-1}(\mathbf{Y} - \boldsymbol{\theta})$.)

3.19 Assume that $\mathbf{X}$ is a random r-vector with mean $\boldsymbol{\mu}$ and covariance matrix $\boldsymbol{\Sigma}$. Let $\mathbf{A}$ be an $(r \times r)$-matrix of constants. Show that (a) $\mathrm{E}\{\mathbf{X}^\tau \mathbf{A} \mathbf{X}\} = \mathrm{tr}(\mathbf{A}\boldsymbol{\Sigma}) + \boldsymbol{\mu}^\tau \mathbf{A} \boldsymbol{\mu}$. Assume now that $\mathbf{A}$ is symmetric, and let $\mathbf{X} \sim \mathcal{N}_r(\boldsymbol{\mu}, \boldsymbol{\Sigma})$. Show that (b) $\mathrm{var}\{\mathbf{X}^\tau \mathbf{A} \mathbf{X}\} = 2\mathrm{tr}(\mathbf{A}\boldsymbol{\Sigma}\mathbf{A}\boldsymbol{\Sigma}) + 4\boldsymbol{\mu}^\tau \mathbf{A}\boldsymbol{\Sigma}\mathbf{A}\boldsymbol{\mu}$. If $\mathbf{B}$ is also a symmetric $(r \times r)$-matrix, show that (c) $\mathrm{cov}\{\mathbf{X}^\tau \mathbf{A} \mathbf{X}, \mathbf{X}^\tau \mathbf{B} \mathbf{X}\} = 2\mathrm{tr}(\mathbf{A}\boldsymbol{\Sigma}\mathbf{B}\boldsymbol{\Sigma}) + 4\boldsymbol{\mu}^\tau \mathbf{A}\boldsymbol{\Sigma}\mathbf{B}\boldsymbol{\mu}$.

3.20 By expressing a correlation matrix $\mathbf{R}$ with equal correlations ρ as $\mathbf{R} = (1 - \rho)\mathbf{I} + \rho\mathbf{J}$, where $\mathbf{J}$ is a matrix of ones, find the determinant and inverse of $\mathbf{R}$.

4

Nonparametric Density Estimation

4.1 Introduction

Nonparametric techniques consist of sophisticated alternatives to traditional parametric models for studying multivariate data. What makes these alternative techniques so appealing to the data analyst is that they make no specific distributional assumptions and, thus, can be employed as an initial exploratory look at the data. In this chapter, we discuss methods for nonparametric estimation of a probability density function.

Suppose we wish to estimate a continuous probability density function p of a random r-vector variate $\mathbf{X}$, where

$$ p(\mathbf{x}) \geq 0, \quad \int_{\Re^r} p(\mathbf{x}) d\mathbf{x} = 1. \tag{4.1} $$

Any p that satisfies (4.1) is called a *bona fide* density. The nonparametric density estimation (NPDE) problem is to estimate p without specifying a formal parametric structure. In other words, p is taken to belong to a large enough family of densities so that it cannot be represented through a finite number of parameters. It is usual to assume instead that p (and its derivatives) satisfy some appropriate "smoothness" conditions. However, there are applications (e.g., X-ray transition tomography) in which

A.J. Izenman, *Modern Multivariate Statistical Techniques*,
doi: 10.1007/978-0-387-78189-1_4,
© Springer Science+Business Media, LLC 2008

discontinuities in p (in that case, tissue density) are natural (Johnstone and Silverman, 1990)

Perhaps the earliest nonparametric estimator of a univariate density p was the histogram. Further breakthroughs — initially, with the kernel, orthogonal series, and nearest neighbor methods — came from researchers working in nonparametric discrimination and time series analysis. Indeed, Parzen (1962), in his seminal work on kernel density estimators, noted the resemblance between probability density estimation and spectral density estimation for stationary time series and then went on to say that "the methods employed here are inspired by the methods used in the treatment of the latter problem."

Nonparametric density estimates can be effective in the following situations. Descriptive features of the density estimate, such as multimodality, tail behavior, and skewness, are of special interest, and a nonparametric approach may be more flexible than the traditional parametric methods; NPDE is used in decision making, such as nonparametric discrimination and classification analysis, testing for modes, and random variate testing; and statistical peculiarities of the data often can be readily explained in presentations to clients through simple graphical displays of estimated density curves.

4.1.1   Example: Coronary Heart Disease

A popular application of nonparametric density estimation is that of comparing data from two independent samples. In this example, data on a large number of variables were used to compare 117 coronary heart disease patients (the "coronary group") with 117 age-matched healthy men (the "control group") (Kasser and Bruce, 1969). These variables included heart rates recorded at rest and at their maximum after a series of exercises on a treadmill.

Figure 4.1 shows kernel density estimates of resting heart rate and maximum heart rate for both groups. The maximum heart rate density estimate (see right panel) for the coronary group appears to be bimodal, possibly a mixture of the unimodal control-group density and a contaminating density having a smaller mean. The opposite conclusions appear to be the case for resting heart rate (left panel). For each density estimate, we used a smoothing parameter (window width), which reflected sample variation. Both graphs show a considerable amount of overlap in their density estimates, making it difficult to distinguish between the groups on the basis of either of these two variables.

A statistic used to monitor activity of the heart is the change in heart rate from a resting state to that after exercise; that is, maximum heart rate minus resting heart rate. As can be seen from Figure 4.1, many of the

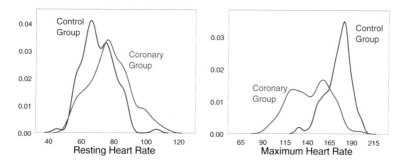

FIGURE 4.1. *Gaussian kernel density estimates for comparing a "coronary group" of 117 male heart patients (red curves) with a "control group" of 117 age-matched healthy men (blue curves) in a coronary heart disease study. Left panel: resting heart rate. Right panel: maximum heart rate after a series of exercises on a treadmill. For each density estimate, the window width was taken to reflect sample variation.*

coronary group will have very small values of this difference (one patient has a difference of 3), whereas the bulk of the control group's values will tend to be larger. Indeed, 20% of the coronary group had differences strictly smaller than the smallest of the differences of the control group, and 14% of the control group had differences lying strictly between the two largest differences of the coronary group.

4.2   Statistical Properties of Density Estimators

Like any statistical procedure, nonparametric density estimators are recommended only if they possess desirable properties. In general, research emphasis has centered upon developing large-sample properties of nonparametric density estimators.

4.2.1   Unbiasedness

An estimator $\widehat{p}$ of a probability density function p is *unbiased* for p if, for all $\mathbf{x} \in \Re^r$, $E_p\{\widehat{p}(\mathbf{x})\} = p(\mathbf{x})$. Although unbiased estimators of parametric densities, such as the Gaussian, Poisson, exponential, and geometric, do exist, no bona fide density estimator (i.e., satisfying (4.1)) based upon a finite data set can exist that is unbiased for all continuous densities (Rosenblatt, 1956). Hence, attention has focused on sequences $\{\widehat{p}_n\}$ of nonparametric density estimators that are *asymptotically unbiased* for p; that is, for all $\mathbf{x} \in \Re^r$, $E_p\{\hat{p}_n(\mathbf{x})\} \to p(\mathbf{x})$, as the sample size $n \to \infty$.

4.2.2   *Consistency*

A more important property is consistency. The simplest notion of consistency of a density estimator is where $\hat{p}$ is *weakly-pointwise consistent for p* if $\hat{p}(\mathbf{x}) \to p(\mathbf{x})$ in probability for every $\mathbf{x} \in \Re^r$, and is *strongly-pointwise consistent for p* if convergence holds almost surely. Other types of consistency depend upon the error criterion.

The L_2 Approach. This has always been the most popular approach to nonparametric density estimation. If p is assumed to be square integrable, then the performance of $\hat{p}$ at $\mathbf{x} \in \Re^r$ is measured by the *mean-squared error (MSE)*,

$$\text{MSE}(\mathbf{x}) = E_p\{\hat{p}(\mathbf{x}) - p(\mathbf{x})\}^2 = \text{var}\{\hat{p}(\mathbf{x})\} + [\text{bias}\{\hat{p}(\mathbf{x})\}]^2, \qquad (4.2)$$

where

$$\begin{align} \text{var}\{\hat{p}(\mathbf{x})\} &= E_p[\hat{p}(\mathbf{x}) - E_p\{\hat{p}(\mathbf{x})\}]^2 \qquad\qquad (4.3) \\ \text{bias}\{\hat{p}(\mathbf{x})\} &= E_p\{\hat{p}(\mathbf{x})\} - p(\mathbf{x}). \qquad\qquad (4.4) \end{align}$$

If $\text{MSE}(\mathbf{x}) \to 0$ for all $\mathbf{x} \in \Re^r$ as $n \to \infty$, then $\hat{p}$ is said to be a *pointwise consistent estimator of p in quadratic mean*.

A more important performance criterion relates to how well the entire curve $\hat{p}$ estimates p. One such measure of goodness of fit is found by integrating (4.2) over all values of $\mathbf{x}$, which yields the *integrated mean-squared error (IMSE)*,

$$\begin{align} \text{IMSE} &= \int_{\Re^r} E_p\{\hat{p}(\mathbf{x}) - p(\mathbf{x})\}^2 d\mathbf{x} \qquad\qquad (4.5) \\ &= E_p\left\{\int [\hat{p}(x)]^2 dx\right\} - 2E_p\{\hat{p}(x)\} + \int [p(x)]^2 dx. \qquad (4.6) \end{align}$$

If we let $R(g) = \int [g(x)]^2 dx$, then the last term, $R(p)$, on the rhs of (4.6) is a constant and, hence, can be removed:

$$\text{IMSE} - R(p) = E_p\{R(\hat{p}) - 2\hat{p}\}. \qquad\qquad (4.7)$$

Thus, $R(\hat{p}) - 2\hat{p}$ is an unbiased estimator for $\text{IMSE} - R(p)$.

Another popular measure is *integrated squared error (ISE, or L_2-norm)*,

$$\text{ISE} = \int_{\Re^r} [\hat{p}(\mathbf{x}) - p(\mathbf{x})]^2 d\mathbf{x}. \qquad\qquad (4.8)$$

Taking expectations over p in (4.8) gives the mean-integrated squared error; that is, $E_p(\text{ISE}) = \text{MISE} = \text{IMSE}$ (Fubini's theorem). ISE is often preferred as a performance criterion (rather than its expected value IMSE) because ISE determines how closely $\widehat{p}$ approximates p for a given data set, whereas MISE is concerned with the average over all possible data sets. For bona fide density estimates, the best possible asymptotic rate of convergence for MISE is $O(n^{-4/5})$; by dropping the restriction that p be a bona fide density, a density estimate can be constructed with MISE better than $O(n^{-1})$.

The L_1 Approach. One problem with the L_2 approach to NPDE is that the criterion pays less attention to the tail behavior of a density, possibly resulting in peculiarities in the tails of the density estimate. An alternative L_1-theory of NPDE is also available (Devroye and Gyorfi, 1985). The *integrated absolute error (IAE, or total variation or L_1-norm)* is given by

$$\text{IAE} = \int_{\Re^r} |\widehat{p}(\mathbf{x}) - p(\mathbf{x})| d\mathbf{x}. \tag{4.9}$$

IAE is always well-defined as a norm on the L_1-space, is invariant under monotone transformations of scale, and lies between 0 and 2.

If IAE $\to 0$ in probability as $n \to \infty$, then $\widehat{p}$ is said to be a *consistent estimator of p*; strong consistency of $\widehat{p}$ occurs when convergence holds almost surely. The IAE distance is related to *Kullback–Leibler relative entropy (KL),*

$$\text{KL} = \int \widehat{p}(\mathbf{x}) \log \left\{ \frac{\widehat{p}(\mathbf{x})}{p(\mathbf{x})} \right\} d\mathbf{x}, \tag{4.10}$$

and *Hellinger distance (HD),*

$$\text{HD}(m) = \left\{ \int \left([\widehat{p}(\mathbf{x})]^{1/m} - [p(\mathbf{x})]^{1/m} \right)^m \right\}^{1/m} \tag{4.11}$$

(Devroye and Gyorfi, 1985, Chapter 8). The expectation of (4.9) over all densities p yields the *mean integrated absolute error,* MIAE $= \text{E}_p\{\text{IAE}\}$. Some quite remarkable results can be proved concerning the asymptotic behavior of IAE and MIAE under little or no assumptions on p. One thing, however, is clear: The technical labor needed to get L_1 results is substantially more difficult than that needed to obtain analogous L_2 results.

4.2.3 Bona Fide Density Estimators

Some density estimation methods always yield bona fide density estimates, and others generally yield density estimates that contain negative ordinates (especially in the tails) or have an infinite integral. Negativity can occur naturally as a result of data sparseness in certain regions or it can be caused by relaxing the nonnegativity constraint in (4.1) in order to improve the rate of convergence of an estimator of p. Negativity in a density estimate can lead to an especially undesirable interpretation if a function of that estimate is needed in a practical situation. For example, Terrell and Scott (1980) remarked that "a negative hazard rate implies the spontaneous reviving of the dead." Moreover, in the quest for faster rates of convergence for density estimators, some researchers have chosen to relax the integral constraint in (4.1) rather than the nonnegativity constraint.

There are several ways of alleviating such problems. The density estimate may be truncated to its positive part and renormalized, or a transformed version of p (e.g., $\log p$ or $p^{1/2}$) may be estimated and then backtransformed to get a nonnegative estimate of p.

4.3   The Histogram

The histogram has long been used to provide a visual clue to the general shape of p. We begin with the univariate case, where $x \in \Re$. Suppose p has support $\Omega = [a, b]$, where a and b are usually taken to contain the entire collection of observed data. Create a *fixed partition* of Ω by using a grid (or mesh) of L nonoverlapping bins (or cells), $T_\ell = [t_{n,\ell}, t_{n,\ell+1})$, $\ell = 0, 1, 2, \ldots, L - 1$, where $a = t_{n,0} < t_{n,1} < t_{n,2} < \cdots < t_{n,L} = b$, and the bin edges $\{t_{n,\ell}\}$ are shown depending upon the sample size n. Let I_{T_ℓ} denote the indicator function of the ℓth bin and let $N_\ell = \sum_{i=1}^n I_{T_\ell}(x_i)$ be the number of sample values that fall into T_ℓ, $\ell = 0, 1, 2, \ldots, L - 1$, where $\sum_{\ell=0}^{L-1} N_\ell = n$.

Then, the *histogram*, defined by

$$\widehat{p}(x) = \sum_{\ell=0}^{L-1} \frac{N_\ell/n}{t_{n,\ell+1} - t_{n,\ell}} \, I_{T_\ell}(x), \tag{4.12}$$

satisfies (4.1). If we fix $h_n = t_{n,\ell+1} - t_{n,\ell}$, $\ell = 0, 1, 2, \ldots, L - 1$, to be a common bin width, and if we take $t_{n,0} = 0$, then the bins will be $T_0 = [0, h_n), T_1 = [h_n, 2h_n), \ldots, T_{L-1} = [(L - 1)h_n, Lh_n)$. Then, (4.12) reduces to

$$\widehat{p}(x) = \frac{1}{nh_n} \sum_{\ell=0}^{L-1} N_\ell I_{T_\ell}(x). \tag{4.13}$$

So, if $x \in T_\ell$, then,

$$\widehat{p}(x) = \frac{N_\ell}{nh_n}. \tag{4.14}$$

As a density estimator, the histogram leaves much to be desired, with defects that include "the fixed nature of the cell structure, the discontinuities at cell boundaries, and the fact that it is zero outside a certain range" (Hand, 1982, p. 15).

A much more serious defect relates to the sensitivity of histogram shapes to the choice of origin. Figure 4.2 displays histograms for the data set

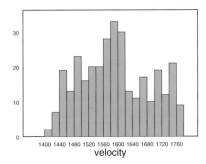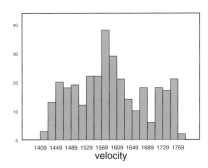

FIGURE 4.2. *Histograms of the radial velocities of 323 locations in the area of the spiral galaxy NGC7531 in the Southern Hemisphere (Buta, 1987). In both panels, the bin width is $h = 20$. In the left panel, the origin is 1,400; in the right panel, it is 1,409, the minimum data value.*

`galaxy`, which consists of the radial velocities of 323 locations in the area of the spiral galaxy NGC7531 in the Southern Hemisphere (Buta, 1987). The bin width is $h = 20$ and the origins are 1,400 (left panel) and 1,409 (right panel). We see how different the histograms look when the origin is changed.

In general, histograms tend not to have symmetric, unimodal, or Gaussian shapes. Indeed, in many large data sets, we often see histograms that are highly skewed with short left-hand tails, very long right-hand tails, several modes (some more prominent than others), and multiple outliers. In many cases, the modes can be modeled parametrically as components of a mixture of distributions.

4.3.1 The Histogram as an ML Estimator

Let $H(\Omega)$ be a specified class of real-valued functions defined on Ω. Given a random sample of observations, $X_1, X_2, \ldots, X_n$, the maximum-likelihood (ML) problem is to find a $p \in H(\Omega)$ that maximizes the likelihood function

$$L(p) = \prod_{i=1}^{n} p(X_i), \qquad (4.15)$$

or its logarithm, subject to

$$\int_{\Omega} p(t)dt = 1, \quad p(t) \geq 0 \text{ for all } t \in \Omega. \qquad (4.16)$$

If $H(\Omega)$ is finite dimensional, then a (not necessarily unique) solution to this problem exists and is called an *ML estimator of p*. The uniqueness of the solution depends upon the specification of $H(\Omega)$. If we restrict H to contain only functions of the form $p(x) = \sum_{\ell=0}^{L-1} y_\ell I_{T_\ell}(x)$, where $h \sum_{\ell=0}^{L-1} y_\ell = 1$, then the histogram (4.13) is the unique ML estimator of p based on the random sample $X_1, X_2, \ldots, X_n$; see Exercise 4.1.

4.3.2 Asymptotics

If n observations are randomly drawn from the probability density p, then the bin count N_ℓ in interval T_ℓ can be viewed as a binomial random variable; that is, $N_\ell \sim \text{Bin}(n, p_\ell)$, where $p_\ell = \int_{T_\ell} p(x)dx$. Thus, the probability that N_ℓ out of the n observations will fall into bin T_ℓ is given by

$$\text{Prob}\{N_\ell \in T_\ell\} = \binom{n}{N_\ell} p_\ell^{N_\ell} (1 - p_\ell)^{n - N_\ell}. \qquad (4.17)$$

Hence, $\text{E}\{N_\ell\} = np_\ell$ and $\text{var}\{N_\ell\} = np_\ell(1 - p_\ell)$. Under suitable continuity conditions for $p(x)$ and assuming that $p(x)$ does not vary much for $x \in T_\ell$, there exists $\xi_\ell \in T_\ell$ such that, by the mean-value theorem,

$$p_\ell = \int_{T_\ell} p(x)dx = h_n p(\xi_\ell), \qquad (4.18)$$

where h_n is the width of T_ℓ. Then, from (4.14), we have that, for $x \in T_\ell$,

$$\text{E}\{\widehat{p}(x)\} = \frac{p_\ell}{h_n} = p(\xi_\ell) \qquad (4.19)$$

and

$$\text{var}\{\widehat{p}(x)\} = \frac{\text{var}\{N_\ell\}}{n^2 h_n^2} = \frac{np_\ell(1 - p_\ell)}{n^2 h_n^2} \leq \frac{p_\ell}{nh_n^2} = \frac{p(\xi_\ell)}{nh_n}, \qquad (4.20)$$

because $p_\ell(1 - p_\ell) \leq p_\ell$.

Now, consider the bin $T_0 = [0, h_n)$. By expanding $p(y)$ around $p(x)$ using a Taylor series, we have that

$$p_0 = \int_{T_0} p(y)dy = h_n p(x) + h_n \left(\frac{h_n}{2} - x\right) p'(x) + O(h^3). \qquad (4.21)$$

The bias of $\widehat{p}(x)$ is $\text{E}_p\{\widehat{p}(x)\} - p(x)$, where, from (4.19), $\text{E}_p\{\widehat{p}(x)\} = p_0/h_n$. By the generalized mean value theorem, there exists $\xi_0 \in T_0$ such that the leading term of the integrated squared bias for bin T_0 is

$$\int_{T_0} [\text{bias}\{\widehat{p}(x)\}]^2 dx \sim p'(\xi_0) \int_{T_0} \left(\frac{h}{2} - x\right)^2 dx = \frac{h_n^3}{12} [p'(\xi_0)]^2. \qquad (4.22)$$

A similar result holds for bin T_ℓ. The *total integrated squared bias (ISB)* is obtained by multiplying this result by h_n, summing over all bins, and arguing that the sum converges to an integral. The *asymptotic integrated squared bias (AISB)*, which is defined as the leading term in ISB, is given by

$$\text{AISB} = \frac{1}{12}h_n^2 R(p'), \tag{4.23}$$

where $R(g) = \int_{\Re}\{g(u)\}^2 du$. Next, define the *integrated variance (IV)* as

$$\text{IV} = \int_{\Re} \text{var}\{\widehat{p}(x)\}dx = \sum_\ell \int_{T_\ell} \text{var}\{\widehat{p}(x)\}dx. \tag{4.24}$$

Substituting from (4.20), summing over all bins, and setting $\sum_\ell p_\ell = \int p(x)dx = 1$, we have that

$$\text{IV} = \frac{1}{nh_n} - \frac{1}{nh_n}\sum_\ell p_\ell^2. \tag{4.25}$$

Now, from (4.18), we have that $\sum_\ell p_\ell^2 = h_n \sum_\ell [p(\xi_\ell)]^2 h_n$. The summation on the rhs approximates $h_n \int [p(x)]^2 dx$. The *asymptotic integrated variance (AIV)* is defined as the leading terms in IV and is given by

$$\text{AIV} = \frac{1}{nh_n} - \frac{R(p)}{n}. \tag{4.26}$$

Combining AIV with AISB yields the *asymptotic MISE (AMISE)*,

$$\text{AMISE} = \frac{1}{nh_n} + \frac{1}{12}h_n^2 R(p'). \tag{4.27}$$

If $h_n \to 0$ and $nh_n \to \infty$ as $n \to \infty$, then IMSE $\to 0$.

Differentiating (4.27) wrt h_n, setting the result equal to zero, and solving, we have that AIMSE is minimized wrt h_n by the optimal bin width,

$$h_n^* = \left\{ \frac{6}{R(p')n} \right\}^{1/3}, \tag{4.28}$$

where $p' = p'(x) = dp(x)/dx$ is the first derivative of p wrt x, and $R(p')$ is a measure of *roughness* of the density function p (see Exercise 4.2). If $X \sim \mathcal{N}(0, \sigma^2)$, then (4.28) reduces to

$$h_n^* \approx 3.4908\sigma n^{-1/3}. \tag{4.29}$$

In Figure 4.3, we graph the histogram of 5,000 observations randomly drawn from $\mathcal{N}(0, 1)$ using bin widths 0.1, 0.2 (optimal using (4.29)), 0.3, and 0.4.

The asymptotic IMSE corresponding to the optimal choice (4.29) of bin width is given by

$$\text{AIMSE}^* = (3/4)^{2/3}[R(p')]^{1/3}n^{-2/3}, \tag{4.30}$$

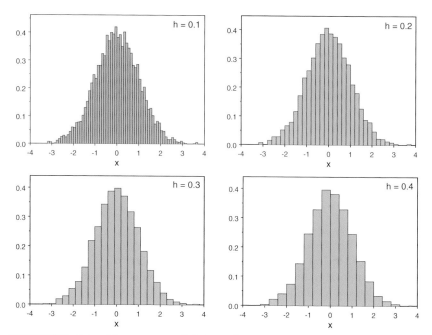

FIGURE 4.3. *Histograms of 5,000 observations randomly drawn from a standard Gaussian distribution. The optimal bin width is 0.2 (top-right panel). The other three histograms have bin widths of 0.1 (top-left panel), 0.3 (bottom-left panel), and 0.4 (bottom-right panel).*

which reduces to $\text{AIMSE}^* \approx 0.43n^{-2/3}$ in the $\mathcal{N}(0,1)$ case. This convergence rate of $O(n^{-2/3})$ is substantially slower than most other types of density estimators, which gives a more technical reason why histograms do not make good density estimators.

4.3.3 Estimating Bin Width

An important aspect of drawing histograms is choice of bin width, which operates as a smoothing parameter. The two most popular methods for choosing the most appropriate histogram bin-width for a given data set are the "plug-in" method and cross-validation.

The obvious estimate of h_n^* in the Gaussian case is given by substituting the sample standard deviation s in (4.29) in place of the unknown σ; that is, $\widehat{h}_n^* = 3.5sn^{-1/3}$ ("Scott's rule"). This "plug-in" estimator generally works well, but for non-Gaussian data, it can lead to overly smoothed histograms (via too-wide bin widths or, equivalently, too-few bins). Slightly narrower bin widths can be obtained using the more robust rule $\widehat{h}_n^* = 2(\text{IQR})n^{-1/3}$, where IQR is the interquartile range of the data. The robust rule will yield a narrower bin width than the Gaussian rule if $s/\text{IQR} > 0.57$. Although this robust rule can sometimes yield wider bin widths than the Gaussian rule, we should not see much difference between the two choices in practice.

The second method uses *leave-one-out cross-validation, CV/n*, to estimate h_n^*. From (4.8), ISE can be expanded into three terms:

$$\text{ISE} = \int [\widehat{p}(x)]^2 dx - 2 \int \widehat{p}(x)p(x)dx + \int [p(x)]^2 dx. \qquad (4.31)$$

The last term, which depends only upon the unknown p, is not affected by changes in bin-widths h, and so can be ignored. The first term only depends upon the density estimate $\widehat{p}$ and can be easily computed. Because the middle integral is the expected height of the histogram, $E_p\{\widehat{p}(X)\}$, CV/n can be used to estimate this integral. Accordingly, the *unbiased cross-validation (UCV) criterion* for a histogram is

$$
\begin{aligned}
\text{UCV}(h) &= R(\widehat{p}) - \frac{2}{n}\sum_{i=1}^{n}\widehat{p}_{-i}(x_i) \\
&= \frac{2}{(n-1)h} - \frac{n+1}{n^2(n-1)h}\sum_{\ell=1}^{L} N_\ell^2. \qquad (4.32)
\end{aligned}
$$

See Exercise 4.8. The CV/n estimate, $\widehat{h}_{UCV}$, of h is that value of h that minimizes $\text{UCV}(h)$. A *biased cross-validation (BCV) criterion* for choosing the bin width of a histogram has also been proposed and studied; for details, see Scott and Terrell (1987). The BCV bin width, $\widehat{h}_{BCV}$, is the value of h that minimizes $\text{BCV}(h)$, a similar-looking criterion to (4.32). Both UCV and BCV criteria yield consistent estimates of h, but convergence is slow in either case, the relative error being $O(n^{-1/6})$.

4.3.4   Multivariate Histograms

The univariate results on optimal bin width and asymptotically optimal IMSE can be extended to the multivariate case.

In this case, we are given a random sample, $\mathbf{X}_1, \mathbf{X}_2, \ldots, \mathbf{X}_n$, where $\mathbf{X}_i = (X_{1i}, X_{2i}, \cdots, X_{ri})^\tau$, from the multivariate density $p(\mathbf{x})$, $\mathbf{x} \in \Re^r$. Each axis is partitioned in the form of a grid of uniformly spaced bins. If the jth axis is partitioned by bins of width $h_{j,n}$, $j = 1, 2, \ldots, r$, the space $\Re^r$ is partitioned into *hyperrectangles*, each having volume $h_{1,n}h_{2,n}\cdots h_{r,n}$.

Now, suppose N_ℓ multivariate observations fall into the ℓth hyperrectangle B_ℓ, where $\sum_\ell N_\ell = n$. Then, our histogram estimate of $p(\mathbf{x})$ is

$$\widehat{p}(\mathbf{x}) = \frac{1}{nh_{1,n}h_{2,n}\cdots h_{r,n}}\sum_\ell N_\ell I_{B_\ell}(\mathbf{x}). \qquad (4.33)$$

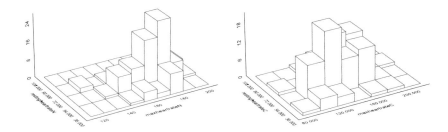

FIGURE 4.4. *Bivariate histograms for the coronary heart disease study. Variables plotted are resting heart rate and maximum heart rate. Left panel: control group. Right panel: coronary group.*

It can be shown (Scott, 1992, Theorem 3.5) that the asymptotically optimal bin width, $h_{\ell,n}^*$, for the ℓth variable is given by

$$h_{\ell,n}^* = [R(p_\ell)]^{-1/2} \left(6 \prod_{j=1}^{r} [R(p_j)]^{1/2} \right)^{1/(2+r)} n^{-1/(2+r)} \qquad (4.34)$$

and the asymptotically optimal IMSE is

$$\text{AIMSE}^* = \frac{1}{4} 6^{2/(2+r)} \left(\prod_{j=1}^{r} R(p_j) \right)^{1/(2+r)} n^{-2/(2+r)}, \qquad (4.35)$$

where $p_j = \partial p(\mathbf{x})/\partial x_j$.

In the multivariate Gaussian case, $\mathcal{N}_r(\mathbf{0}, \boldsymbol{\Sigma})$, where $\boldsymbol{\Sigma} = \text{diag}\{\sigma_1^2, \dots, \sigma_r^2\}$, (4.35) reduces to

$$h_{\ell,n}^* = 2 \cdot 3^{1/(2+r)} \pi^{r/(4+2r)} \sigma_\ell n^{-1/(2+r)}. \qquad (4.36)$$

For $r = 1$, the constant in (4.36) reduces to $2 \cdot 3^{1/3} \pi^{1/6} = 3.4908$, and as $r \to \infty$, the constant becomes $2\pi^{1/2} = 3.5449$. So, for all r, the constant lies between 3.4908 and 3.5449. A rule-of-thumb, therefore, for this particular case is to use $h_{\ell,n}^* \approx 3.5 \sigma_\ell n^{-1/(2+r)}$.

Figure 4.4 displays bivariate histograms of both the control group (left panel) and coronary group (right panel) for the coronary heart disease study (see Section 4.1.1). In particular, the control-group histogram has a unimodal and sharply skewed shape, whereas the coronary-group histogram has a bimodal and more blocky shape. Problems in visualizing important characteristics of a bivariate histogram, due to its "blocky" and discontinuous nature, often make such density estimators difficult to work with in practice.

4.4   Maximum Penalized Likelihood

The ML method of Section 4.3.3 fails miserably when the class H of densities over which the likelihood L is to be maximized is unrestricted. For that case, the likelihood is maximized by a linear combination of Dirac delta functions (or "spikes") at the n sample values, resulting in a value of $+\infty$ for the likelihood. There have been several approaches to ML density estimation in which restrictions are placed on H; these include order-restricted methods and sieve methods (see, e.g., Izenman, 1991). Here, we restrict the likelihood L by penalizing L for producing density estimates that are "too rough."

Let Φ be a given nonnegative *(roughness) penalty functional* defined on H. The Φ-*penalized likelihood* of p is defined to be

$$\widetilde{L}(p) = \prod_{i=1}^{n} p(X_i)e^{-\Phi(p)}. \tag{4.37}$$

The optimization problem calls for $\widetilde{L}(p)$, or its logarithm,

$$\mathcal{L}(p) = \log_e \widetilde{L}(p) = \sum_{i=1}^{n} \log_e p(X_i) - \Phi(p), \tag{4.38}$$

to be maximized subject to

$$p \in H(\Omega), \quad \int_{\Omega} p(u)du = 1, \quad p(u) \geq 0 \text{ for all } u \in \Omega. \tag{4.39}$$

If it exists, a solution, $\widehat{p}$, of that problem is called a *maximum penalized likelihood (MPL) estimate* of p corresponding to the penalty function Φ and class of functions H. For example, $\Phi(p) = \alpha \int_{-\infty}^{\infty} [p''(x)]^2 dx$ is used in the IMSL Fortran routine DESPL, where $\alpha > 0$ is a *smoothing parameter*. IMSL recommends $\alpha = 10$ for $\mathcal{N}(0,1)$ data and using a grid of $\alpha = 1(10)100$ for other situations.

Good and Gaskins (1971) observed that the MPL method could, for certain types of problems, be interpreted as "quasi-Bayesian" because $\widetilde{L}(p)$ in (4.37) resembles a posterior density for a parametric estimation problem. Furthermore, the MPL method is closely related to Tikhonov's *method of regularization* used for solving ill-posed inverse problems (O'Sullivan, 1986).

The existence and uniqueness of MPL density estimates have been established, and it has been shown that such estimates are intimately related to spline methods (de Montricher, Tapia, and Thompson, 1975). For example, if p has finite support Ω and if $H(\Omega)$ is a suitable class of smooth functions on Ω, then the MPL estimate $\hat{p}$ exists, is unique, and is a polynomial spline with join points (or "knots") only at the sample values.

The case when p has infinite support is more complicated. Good and Gaskins (1971) proposed penalty functionals designed to estimate the "root-density," so that $\hat{p} = \hat{\gamma}^2$ would be a nonnegative (and bona fide) estimator of p. The penalty functionals were

$$\Phi_1(p) = 4\alpha R(\gamma'), \quad \alpha > 0, \tag{4.40}$$

$$\Phi_2(p) = 4\alpha R(\gamma') + \beta R(\gamma''), \quad \alpha \geq 0, \beta \geq 0, \tag{4.41}$$

where, as before, $R(g) = \int [g(x)]^2 dx$, for any square-integrable function g, and the *hyperparameters* α and β, with $\alpha + \beta > 0$ in (4.41), control the amount of smoothing. The choice of Φ_1 or Φ_2 depends upon how best to represent the "roughness" of p. Good and Gaskins preferred Φ_2 to Φ_1, arguing that curvature as well as slope of the density estimate should be penalized.

If the optimization problem is set up correctly, and we use the penalty function Φ_1 and a given value of α, then the resulting estimator, $\hat{\gamma}_\alpha$, say, exists, is unique, and is a positive exponential spline with knots only at the sample values (de Montricher, Tapia, and Thompson, 1975). An exponential spline rather than a polynomial spline is the price to be paid for requiring nonnegativity of the density estimator. The MPL estimator is then given by $\hat{p}_\alpha = \hat{\gamma}_\alpha^2$. This density estimator is consistent over a number of norms, including L_1 and L_2. Similar statements can be made about the optimization problem where Φ_2 is the penalty function and α and β are given.

Implementation of the MPL method depends upon the quality of the numerical solutions to the restricted optimization problems. Scott, Tapia, and Thompson (1980) studied a discrete approximation to the spline solutions of the MPL problems and proved that the resulting *discrete MPL estimator* exists, is unique, converges to the spline MPL estimator, and is a strongly pointwise consistent estimator of p. Fortunately, solutions to the MPL density-estimation problem can be expressed in terms of kernel density estimates, where the kernels are weighted according to the other observations in the sample rather than with a uniform n^{-1} weight as in (4.42) below.

4.5   Kernel Density Estimation

The most popular density estimation method is the kernel density estimator. Given iid univariate observations, $X_1, X_2, \ldots, X_n \sim p$, the *kernel density estimator,*

$$\widehat{p}_h(x) = \frac{1}{nh} \sum_{i=1}^{n} K\left(\frac{x - X_i}{h}\right), \quad x \in \Re, \quad h > 0, \tag{4.42}$$

of $p(x)$, $x \in \Re$, is used to obtain a smoother density estimate than the histogram. In (4.42), K is a *kernel function*, and the *window width h* determines the smoothness of the density estimate. Choice of h is an important statistical problem: too small a value of h yields a density estimate too dependent upon the sample values, whereas too large a value of h produces the opposite effect and oversmooths the density estimate by removing interesting peculiarities. Given a kernel K and window width h, the resulting kernel density estimate is unique for a specific data set; hence, kernel density estimates do not depend upon a choice of origin as do histograms.

There are several ways to define a multivariate version of (4.42). In the following, we use the formulation provided by Scott (1992, Section 6.3.2). Given the r-vectors $\mathbf{X}_i, \mathbf{X}_2, \ldots, \mathbf{X}_n$, the *multivariate kernel density estimator* of p is defined to have the general form,

$$\widehat{p}_{\mathbf{H}}(\mathbf{x}) = \frac{1}{n|\mathbf{H}|} \sum_{i=1}^{n} K(\mathbf{H}^{-1}(\mathbf{x} - \mathbf{X}_i)), \quad \mathbf{x} \in \Re^r, \tag{4.43}$$

where $\mathbf{H}$ is an $(r \times r)$ nonsingular matrix that generalizes the window width h, and K is a multivariate function with mean $\mathbf{0}$ and integrates to 1. If, for example, we take $\mathbf{H} = h\mathbf{A}$, where $h > 0$ and $|\mathbf{A}| = 1$, the size and elliptical shape of the kernel will be determined completely by h and the matrix $\mathbf{A}\mathbf{A}^\top$, respectively. If $\mathbf{A} = \mathbf{I}_r$, then (4.43) reduces to

$$\widehat{p}_h(\mathbf{x}) = \frac{1}{nh^r} \sum_{i=1}^{n} K\left(\frac{\mathbf{x} - \mathbf{X}_i}{h}\right), \quad \mathbf{x} \in \Re^r. \tag{4.44}$$

In (4.44), the choice of kernel function K and window width h control the performance of $\widehat{p}_h$ as an estimator of p. Because $\widehat{p}_h$ inherits whatever properties the kernel K possesses, it is important that K has desirable statistical properties.

4.5.1 Choice of Kernel

The simplest class of kernels consists of multivariate probability density functions that satisfy

$$K(\mathbf{x}) \geq 0, \quad \int_{\Re^r} K(\mathbf{x})d\mathbf{x} = 1. \tag{4.45}$$

If a kernel K from this class is used in (4.44), then $\widehat{p}_h$ will always be a bona fide probability density.

TABLE 4.1. *Examples of univariate kernel functions with compact support.*

Kernel Function	$K(x)$
Rectangular	$\frac{1}{2}I_{[|x|\leq 1]}$
Triangular	$(1-|x|)I_{[|x|\leq 1]}$
Bartlett–Epanechnikov	$\frac{3}{4}(1-x^2)I_{[|x|\leq 1]}$
Biweight	$\frac{15}{16}(1-x^2)^2I_{[|x|\leq 1]}$
Triweight	$\frac{35}{32}(1-x^2)^3I_{[|x|\leq 1]}$
Cosine	$\frac{\pi}{4}\cos(\frac{\pi}{2}x)I_{[|x|\leq 1]}$

Popular choices of univariate kernels include the Gaussian kernel with unbounded support,

$$K(x) = (2\pi)^{-1/2}e^{-x^2/2}, \quad x \in \Re, \tag{4.46}$$

and the compactly supported "polynomial" kernels,

$$K(x) = \kappa_{ij}(1-|x|^i)^j I_{[|x|\leq 1]}, \quad \kappa_{ij} = \frac{i}{2\text{Beta}(j+1,1/i)}, \quad i > 0, j \geq 0. \tag{4.47}$$

Special cases of the polynomial kernel are the rectangular kernel ($j = 0$, $\kappa_{i0} = 1/2$), the triangular kernel ($i = 1, j = 1, \kappa_{11} = 1$), the Bartlett–Epanechnikov kernel ($i = 2, j = 1, \kappa_{21} = 3/4$), the biweight kernel ($i = 2, j = 2, \kappa_{22} = 15/16$), the triweight kernel ($i = 2, j = 3, \kappa_{23} = 35/32$), and, after a suitable rescaling, the Gaussian kernel ($i = 2, j = \infty$). Their specific forms are listed in Table 4.1 and graphed in Figure 4.5.

It has been known for some time that the Bartlett–Epanechnikov kernel minimizes the optimal asymptotic IMSE with respect to K. However, IMSE is, in fact, quite insensitive to the shape of the kernel, so the Gaussian or rectangular kernels are just as good in practice as the optimal kernel.

Multivariate kernels are usually radially symmetric unimodal densities, such as the Gaussian,

$$K(\mathbf{x}) = \frac{1}{(2\pi)^{r/2}}e^{-\mathbf{x}^\tau\mathbf{x}/2}, \quad \mathbf{x} \in \Re^r, \tag{4.48}$$

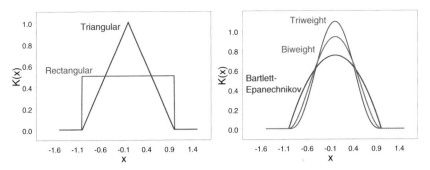

FIGURE 4.5. *Univariate kernel functions with compact support. Left panel: rectangular and triangular kernels. Right panel: Bartlett–Epanechnikov, biweight, and triweight kernels.*

and the compactly supported Bartlett–Epanechnikov,

$$K(\mathbf{x}) = \frac{r+2}{2c_r}(1 - \mathbf{x}^\tau \mathbf{x})I_{[\mathbf{x}^\tau \mathbf{x} \leq 1]}, \quad c_r = \frac{\pi^{r/2}}{\Gamma((r/2)+1)}. \tag{4.49}$$

In certain multivariate situations, it may be convenient to use *product kernels* of the form,

$$K(\mathbf{x}) = \prod_{j=1}^{r} K(x_j), \tag{4.50}$$

which is a product of univariate kernel functions, where the kernels are the same for each dimension. If we take $\mathbf{H}$ in (4.43) to be the diagonal matrix $\mathbf{H} = \text{diag}\{h_{1,n}, \cdots, h_{r,n}\} = h\mathbf{A}$ with different window widths in each dimension, where $\mathbf{A} = \text{diag}\{h_{1,n}/h, \cdots, h_{r,n}/h\}$, and let K be a product kernel, then (4.43) reduces to

$$\widehat{p}_{\mathbf{H}}(\mathbf{x}) = \frac{1}{nh^r} \sum_{i=1}^{n} \left\{ \prod_{j=1}^{r} K\left(\frac{x_j - X_{ij}}{h_{j,n}}\right) \right\}, \quad \mathbf{x} \in \Re^r, \tag{4.51}$$

where $\mathbf{x} = (x_1, \cdots, x_r)^\tau$, $\mathbf{X}_i = (X_{i1}, \cdots, X_{ir})^\tau$, and $h = (h_{1,n} \cdots h_{r,n})^{1/r}$ is the geometric mean of the r window widths.

4.5.2 Asymptotics

Early work on kernel density estimation emphasized asymptotic results, which depended upon the particular viewpoint considered.

The L_1 Approach. Among the remarkable L_1 results proved for kernel density estimates, we have that if K satisfies (4.45), then the kernel estimator (4.44) will be a strongly consistent estimator of p iff $h_n \to 0$ and

$nh_n \to \infty$, as $n \to \infty$, without any conditions on p (Devroye, 1983). Moreover, in the univariate case, MIAE is of order $O(n^{-2/5})$ (Devroye and Penrod, 1984), which is better than the corresponding L_1 rate for histograms. Explicit formulas for the minimum MIAE and the asymptotically optimal smoothing parameters for kernel estimators are available (Hall and Wand, 1988).

The L_2 Approach. Under regularity conditions on K and p, it can be shown that if $h_n \to 0$ as $n \to \infty$, then the univariate kernel density estimator is both asymptotically unbiased and asymptotically Gaussian (Parzen, 1962). In the multivariate case, the MISE is asymptotically minimized over all h satisfying the above conditions by

$$h_n^* = \alpha(K)\beta(p)n^{-1/(r+4)}, \qquad (4.52)$$

where r is the dimensionality, $\alpha(K)$ depends only upon the kernel K, and $\beta(p)$ depends only upon the unknown density p (Cacoullos, 1966). This result shows that the window width should get smaller as the sample size n gets larger; this reflects a commonsense notion that "local" smoothing information becomes more important as more data become available. Moreover, MISE $\to 0$ at the rate $O(n^{-4/(r+4)})$. These L_2 results show clearly the dimensionality effect, because these convergence rates become slower as the dimensionality r increases.

In the univariate case, the pointwise variance (4.3) and bias (4.4) of $\widehat{p}_h(x)$ are found by using Taylor-series expansions:

$$\mathrm{var}\{\widehat{p}(x)\} \approx \frac{R(K)p(x)}{nh_n} - \frac{[p(x)]^2}{n}, \qquad (4.53)$$

$$\mathrm{bias}\{\widehat{p}(x)\} \approx \frac{1}{2}\sigma_K^2 h_n^2 p''(x); \qquad (4.54)$$

where $R(g) = \int [g(x)]^2 dx$ for any square-integrable function g, and $\sigma_K^2 = \int x^2 K(x)dx$. See Exercise 4.10. Thus, we can reduce the variance by increasing the size of h_n (i.e., by oversmoothing), and bias reduction can take place if we make h_n small (i.e., by undersmoothing). This is the classical bias-variance trade-off dilemma, and so, to choose h_n, a compromise is needed.

Adding the variance term and the square of the bias term and then integrating wrt x gives us the asymptotic MISE (AMISE) for a univariate kernel density estimator:

$$\mathrm{AMISE}(h_n) = \frac{R(K)}{nh_n} + \frac{1}{4}\sigma_K^4 h_n^4 R(p''). \qquad (4.55)$$

Minimizing AMISE(h_n) wrt h_n yields the asymptotically optimal window width,

$$h_n^* = \left\{\frac{R(K)}{\sigma_K^4 R(p'')}\right\}^{1/5} n^{-1/5}, \qquad (4.56)$$

so that $\alpha(K) = \{R(K)/\sigma_K^4\}^{1/5}$ and $\beta(p) = \{R(p'')\}^{-1/5}$ in (4.52). Substituting the expression for h_n^* into AMISE shows that

$$\text{AMISE}^* = \frac{5}{4}[\sigma_K R(K)]^{4/5}[R(p'')]^{1/5}n^{-4/5}. \tag{4.57}$$

See Scott (1992, p. 131).

Consider the special case where K is a product Gaussian kernel (4.50) and the density p is multivariate Gaussian with diagonal covariance matrix, $\text{diag}\{\sigma_1^2, \ldots, \sigma_r^2\}$ (i.e., the variables are independent). Then, (4.52) reduces to

$$h_{j,n}^* = \left(\frac{4}{r+2}\right)^{1/(r+4)} \sigma_j n^{-1/(r+4)}, \quad j = 1, 2, \ldots, r. \tag{4.58}$$

In the univariate case, where K is the standard Gaussian kernel and p is a Gaussian density with variance σ^2, then

$$h_n^* = 1.06\sigma n^{-1/5} \tag{4.59}$$

is the asymptotically optimal window width. In the bivariate case, the constant in (4.58) is exactly 1. In general. $(4/(r+2))^{1/(r+4)}$ attains its minimum as a function of r when $r = 11$, where its value is 0.924. For general r, Scott (1992, p. 152) recommends the rule $h_{j,n}^* = \sigma_j n^{-1/(r+4)}$.

4.5.3 Example: 1872 Hidalgo Postage Stamps of Mexico

This example shows the effect of varying the window width h of a Gaussian kernel density estimate. The data[1] consist of 485 measurements of the thickness of the paper on which the 1872 Hidalgo Issue postage stamps of Mexico were printed (Izenman and Sommer, 1988). This example is particularly interesting because of the fact that these stamps were deliberately printed on a mixture of paper types, each having its own thickness characteristics due to poor quality control in paper manufacture.

Today, the thickness of the paper on which this particular stamp image is printed is a primary factor in determining its price. In almost all cases, a stamp printed on relatively scarce "thick" paper is worth a great deal more than the same stamp printed on "medium" or "thin" paper. It is, therefore, important for stamp dealers and collectors to know how to differentiate between thick, medium, and thin paper. Quantitative definitions of the words *thin* and *thick* do not appear in any current stamp catalogue,

---

[1]The Hidalgo stamp data can be found in the file `Hidalgo1872` on the book's website.

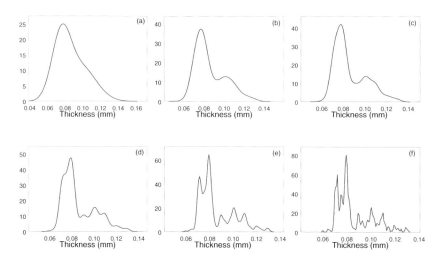

FIGURE 4.6. *Gaussian kernel density estimates of the 485 measurements on paper thickness of the 1872 Hidalgo Issue postage stamps of Mexico. The window widths are (a) $h = 0.01$; (b) $h = 0.005$; (c) $h = 0.0036$; (d) $h = 0.0025$; (e) $h = 0.0012$; and (f) $h = 0.0005$. Notice the smooth appearance of the density estimates and the emergence of more modes as h is decreased.*

and decisions as to the financial worth of such stamps are left to personal subjective judgment.

Figure 4.6 displays Gaussian kernel density estimates of the Hidalgo stamp data for six window widths: $h = 0.01, 0.005, 0.0036, 0.0025, 0.0012$, and 0.0005. As h is reduced in magnitude, more structure and detail of the underlying density become visible and more modes emerge. Clearly, the estimate in panel (a) is too smooth, and that in panel (f) is too noisy. The most reasonable density estimate is that which corresponds to a window width of $h = 0.0012$ (see panel (e)) and has seven modes. The two biggest modes occur at thicknesses of 0.072 mm and 0.080 mm; a cluster of three side modes occur at 0.090 mm, 0.100 mm, and 0.110 mm; and there are two tail modes at 0.120 mm and 0.130 mm.

Our analysis does not stop there. We have more information regarding this particular stamp issue. Every stamp from the 1972 Hidalgo Issue was overprinted with year-of-consignment information: there was an 1872 consignment (289 stamps) and an 1873–1874 consignment (196 stamps). We divided these 485 thickness measurements into two groups according to the appropriate consignment overprint.

Gaussian kernel density estimates (with common window width $h = 0.0015$) were computed for the data from each consignment. The resulting

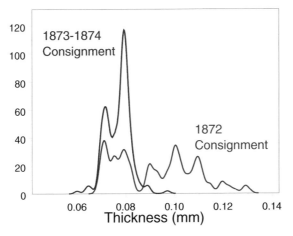

FIGURE 4.7. *Gaussian kernel density estimates from data on the 1872 consignment (n = 289) and 1873–1874 consignment (n = 196) of the 1872 Hidalgo Postage Stamp Issue of Mexico. For both density estimates, a common window width of h = 0.0015 was used.*

density estimates, which are graphed in Figure 4.7, show clearly that the paper used for printing the stamps in the two consignments had very different thickness characteristics. It appears that a large proportion of the 1872 consignment of stamps was printed on very thick paper, which was not used for the 1873–1874 consignment.

Because 1872 Hidalgo Issue stamps printed on thick paper command much higher prices, these results show that one should look at year-of-consignment as an important factor for valuation purposes.

4.5.4 *Estimating the Window Width*

For kernel density estimation, rather than trying an ad hoc sequence of different window widths until we find one with which we are satisfied, it would be much more convenient to have an automated method for determining the optimal window width for any given data set.

For the L_2 approach, we see from (4.52) that the optimal window width, h_n^*, depends explicitly on the unknown density p through the quantity $\beta(p)$, and so cannot be computed exactly. The most popular methods for estimating h_n^* are the so-called "rule-of-thumb" method, cross-validation, and the "plug-in" method.

Rule-of-Thumb Method An obvious way to estimate the window width is to insert a parametric estimate $\widehat{p}$ of p into $\beta(p)$.

In the univariate case, we can choose a "reference density" for p, find $R(p'')$, and then estimate the result using a random sample from p. If we take p to be $\mathcal{N}(0, \sigma^2)$ and K to be a standard Gaussian kernel, then the "optimal" rule-of-thumb (ROT) window width for a Gaussian reference density (see (4.61)) would be $\widehat{h}_n^{ROT} = 1.06sn^{-1/5}$, where the sample standard deviation s is the usual estimate for σ. Otherwise, a more robust estimate of σ may be used, such as $\min\{s, IQR/1.34\}$, where IQR is the interquartile range, and for Gaussian data, $IQR \approx 1.34s$ (Silverman, 1986, pp. 45–47).

For example, the Hidalgo postage stamp data has standard deviation $s = 0.015$, so that the optimal ROT window width is given by $\widehat{h}_n^{ROT} = (1.06)(0.015)(485)^{-1/5} = 0.005$; as we see from Figure 4.6(b), this value yields an overly smoothed density estimate.

Rule-of-thumb estimators for window widths are generally regarded as unsatisfactory (with some exceptions). Simulations and case studies with real data both indicate that window widths produced by this method tend to be overly large; if that happens, the density estimate will be drastically oversmoothed and the presence of an important mode may be unknowingly removed.

Cross-Validation A popular method for determining the optimal window width is *leave-one-out cross-validation (CV/n)*. In the univariate case, the basic algorithm removes a single value, say X_i, from the sample, computes the appropriate density estimate at that X_i from the remaining $n-1$ sample values,

$$\widehat{p}_{h,-i}(X_i) = \frac{1}{(n-1)h} \sum_{j \neq i} K\left(\frac{X_i - X_j}{h}\right), \qquad (4.60)$$

and then chooses h to optimize some given criterion involving all values of $\widehat{p}_{h,-i}(X_i)$, $i = 1, 2, \ldots, n$. A number of different versions of CV/n have been used for determining h in density estimation, including unbiased and biased cross-validation.

The *unbiased cross-validation* choice, h_n^{UCV}, of window width is that h that *minimizes*

$$UCV(h) = R(\widehat{p}_h) - \frac{2}{n} \sum_{i=1}^{n} \widehat{p}_{h,-i}(X_i), \qquad (4.61)$$

where $R(g) = \int_{\Re}[g(x)]^2 dx$. The criterion (4.61), which is derived in exactly the same manner as the CV-expression for the histogram given in (4.32), is referred to as an *unbiased cross-validation (UCV)* criterion because it is exactly unbiased for a shifted version of MISE; that is,

$$E_p\{UCV(h)\} = MISE(h) - R(p). \qquad (4.62)$$

Only very mild tail conditions on K and p are needed to prove that h_n^{UCV} asymptotically minimizes ISE and gives good results even for long-tailed p; it has also been shown to perform asymptotically as well as the MISE-optimal (but unattainable) window width h_n^*, and even though convergence tends to be slow, it cannot be improved upon asymptotically.

Another approach to the problem of choosing h is to minimize AMISE(h) directly. In the univariate case, AMISE depends upon the unknown $R(p'')$, which we, therefore, need to estimate. Scott and Terrell (1987) showed that $E_p\{R(\widehat{p}'')\} = R(p'') + R(K'')/nh^5 + O(h^2)$, so that $R(\widehat{p}''_h)$ asymptotically overestimates $R(p'')$. From this result, they proposed the modified estimator

$$\widehat{R}(p'') = R(\widehat{p}''_h) - \frac{R(K'')}{nh^5}, \qquad (4.63)$$

which is an asymptotically unbiased estimator of $R(p'')$. See also Hall and Marron (1987).

If we define $K_h(u) = h^{-1}K(u/h)$, then, $K''(u/h) = h^3 K''_h(u)$. Differentiating $\widehat{p}_h(x)$ (see (4.44)) twice wrt x gives

$$\widehat{p}''_h(x) = \frac{1}{n}\sum_{i=1}^n K''_h(x - X_i). \qquad (4.64)$$

Squaring (4.64), integrating the result wrt x, and then using a change of variable gives

$$
\begin{aligned}
R(\widehat{p}''_h) &= \frac{1}{n^2}\sum_{i=1}^n\sum_{j=1}^n K''_h * K''_h(X_i - X_j) \\
&= \frac{1}{n}K''_h * K''_h(0) + \frac{1}{n^2}\sum\sum_{i\neq j} K''_h * K''_h(X_i - X_j) \\
&= \frac{R(K'')}{nh^5} + \frac{1}{n^2h^5}\sum\sum_{i\neq j} K''_h * K''_h(X_i - X_j), \qquad (4.65)
\end{aligned}
$$

where the *convolution* of two functions f and g is defined by $f * g(u) = \int f(z)g(z + u)dz$. Substituting (4.65) into the expression (4.63) yields

$$\widehat{R}(p''_h) = \frac{1}{n^2h^5}\sum\sum_{i\neq j} K''_h * K''_h(X_i - X_j). \qquad (4.66)$$

Substituting (4.66) as an estimator of $R(p'')$ into AMISE (4.55) and setting $h = h_n$ yields a *biased cross-validation (BCV)* criterion,

$$\text{BCV}(h_n) = \frac{R(K)}{nh_n} + \frac{\sigma_K^4}{2n^2h_n}\sum\sum_{i<j} K''_{h_n} * K''_{h_n}(X_i - X_j). \qquad (4.67)$$

The BCV estimator of h is that value, h_n^{BCV} that (locally) minimizes the BCV(h_n) criterion.

For the Hidalgo stamp data example, the BCV choice of h is 0.0036, corresponding to Figure 4.6(c) and yielding an overly smoothed density estimate, whereas the UCV choice of h is 0.0005, corresponding to Figure 4.6(f) and yielding an undersmoothed density estimate.

Even though CV methods are popular, they have been strongly criticized. In general, we have seen that UCV tends to undersmooth, whereas BCV tends to oversmooth, especially for skewed distributions. Both methods are computationally intensive because they involve computing the differences between all pairs of data values (see (4.67) for BCV, and a similar formula can be given for UCV); thus, for large quantities of data (i.e., thousands of observations), these methods tend to becomes impractical. Furthermore, the UCV and BCV methods have been found to produce multiple local minima, and the question becomes one of which to choose (a recommended action in each case is to take the largest local minimum).

These criticisms, plus recent successful work on "plug-in" methods, have relegated the UCV and BCV methods to "first-generation" status.

Plug-in Methods The "plug-in" idea for estimating h_n^* can be traced back to Woodroofe (1970), who proposed a two-step procedure:

1. Choose a window width g_n for a "pilot" density estimate $\widehat{p}_{g_n}(x)$, and use this density estimate to compute $\widehat{R}(p'') = R(\widehat{p}_{g_n})$;

2. Plug $\widehat{R}(p'')$ into (4.59) to obtain the final window width, $\widehat{h}_n^*$.

This idea of estimating $R(p'')$ in two steps via a pilot estimate has since been modified in a number of different ways, including a fully iterated version and a version that uses (4.63) to reduce the bias. Some of these candidate ideas proved useful, others less so. For example, in certain situations, using (4.63) can produce negative values for $\widehat{R}(p'')$.

The most successful of these modifications was proposed by Sheather and Jones (1991). Estimating $R(p'')$ is different from estimating p, and so we expect the corresponding window widths, g_n and h_n, to be different, but related; that is, we expect the pilot window width $g_n = g(h_n)$. Rather than use (4.63), we estimate $R(p'')$ by $R(\widehat{p}''_{g(h_n)})$. The proposed window width, h_n^{SJ}, is that value of h_n that solves the equation,

$$h_n = \left\{ \frac{R(K)}{\sigma_K^4 R(\widehat{p}''_{g(h_n)})} \right\}^{1/5} n^{-1/5}. \tag{4.68}$$

The optimal choice for g_n is given by

$$g(h_n) = C(K) \left\{ \frac{R(p'')}{R(p''')} \right\}^{1/7} h_n^{5/7}, \tag{4.69}$$

where $C(K)$ is a constant dependent only upon the kernel K. The unknown quantities $R(p'')$ and $R(p''')$ are estimated by $R(\widehat{p}_a'')$ and $R(\widehat{p}_b''')$, respectively, where the window widths, a and b, are chosen according to the asymptotic optimality results. At this second step in the computations, $R(p'')$ and $R(p''')$ are estimated using the Gaussian reference density method, as we did for the ROT window width. The resulting convergence rate of h_n^{SJ} is $O(n^{-5/14})$.

Applying the *Sheather–Jones plug-in (SJPI)* method to the Hidalgo stamp data yields an estimated window width of 0.0012, which corresponds to the density estimate in Figure 4.6(e). Thus, the plug-in estimator clearly outperforms any of the competing window-width estimators for the Hidalgo stamp data.

Plug-in methods are currently being promoted as "second-generation" methods. This viewpoint is based upon strong evidence of superior performance from asymptotics, simulations, and experience with real data. Despite this evidence, however, there are some reservations regarding the superiority of the plug-in method. In particular, Loader (1999) makes the following points: (1) the success of plug-in methods depends crucially upon an arbitrary specification of the pilot window width, and if misspecification occurs, poor density estimates will result; (2) in difficult examples, where there are many modes in the data, the SJPI method oversmoothes and completely misses the fine structure, whereas UCV, with its tendency to undersmooth, gives a good accounting of itself; and (3) the poor performance of the UCV method may be due to an inappropriate use of a fixed window width, and that instead a more data-adaptive window width would be a better choice.

Example: Eruptions of Old Faithful Geyser

Another example of the different window-width selection methods is displayed in Figure 4.8 for the well-known Old Faithful Geyser data.[2] This data set, which has been explored at length in the density estimation literature, consists of the duration, in minutes, of 107 consecutive eruptions of Old Faithful Geyser (a hot spring that erupts hot water and steam at intervals ranging from 30 to 90 minutes, in Yellowstone National Park, Wyoming), 1–8 August 1978 (Weisberg, 1985, pp. 230–235).

We see the bimodality in the data; we also see that UCV provides a noisier density estimate than does BCV, with SJPI providing some degree of compromise between them. Compared with a histogram of the data, SJPI and BCV have substantially reduced the magnitude of the left mode

---

[2]The data can be found in the file **geyser** available on the book's website.

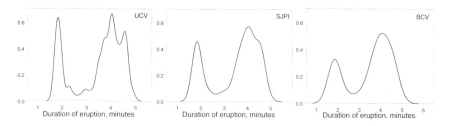

FIGURE 4.8. *Gaussian kernel density estimators of the Old Faithful Geyser data. The window widths for the estimates were selected by unbiased cross-validation (left panel), Sheather–Jones plug-in method (center panel), and biased cross-validation (right panel).*

relative to the right mode, whereas UCV retains that particular feature of the data.

4.6 Projection Pursuit Density Estimation

Multivariate kernel density estimators tend to be poor performers when it comes to high-dimensional data because extremely large sample sizes are needed to match the sort of numerical accuracy that is possible in low dimensions. In light of this, Friedman and Stuetzle (1982) and Friedman, Stuetzle, and Schroeder (1984) developed *projection pursuit density estimation (PPDE)* based upon the general projection pursuit algorithm. The PPDE method has been shown in simulations to possess excellent properties, and several striking applications of PPDE to real data have also been published.

4.6.1   The PPDE Paradigm

When dealing with small samples of high-dimensional data, the PPDE procedure may be jump-started by restricting attention to the subspace spanned by the first few significant principal components. A projection pursuit density estimator of p is then formed using the iterative procedure given in Table 4.2.

The iterative procedure is repeated as many times as necessary. At the kth iteration,

$$\widehat{p}^{(k)}(\mathbf{x}) = \widehat{p}^{(0)}(\mathbf{x}) \prod_{j=1}^{k} g_j(\mathbf{a}_j^\tau \mathbf{x}) = \widehat{p}^{(k-1)}(\mathbf{x}) g_k(\mathbf{a}_k^\tau \mathbf{x}) \qquad (4.70)$$

TABLE 4.2. *Projection pursuit density estimation algorithm.*

1. Input: $\mathcal{L} = \{\mathbf{X}_i, i = 1, 2, \ldots, n\}$. Sphere the data to have mean $\mathbf{0}$ and covariance matrix $\mathbf{I}_r$.

2. Initialize: Choose $\widehat{p}^{(0)}$ to be an initial multivariate density estimate of p, usually taken to be the standard multivariate Gaussian.

3. Do $j = 1, 2, \ldots$:

   - Find the direction $\mathbf{a}_j \in \Re^r$ for which the (model) marginal $p_{\mathbf{a}_j}$ along $\mathbf{a}_j$ differs most from the current estimated (data) marginal $\widehat{p}_{\mathbf{a}_j}$ along $\mathbf{a}_j$. Choice of direction $\mathbf{a}_j$ will not generally be unique.

   - Given $\mathbf{a}_j$, define a univariate "augmenting function"

   $$g_j(\mathbf{a}_j^\tau \mathbf{x}) = \frac{p_{\mathbf{a}_j}(\mathbf{a}_j^\tau \mathbf{x})}{\widehat{p}_{\mathbf{a}_j}(\mathbf{a}_j^\tau \mathbf{x})}.$$

   - Update the previous estimate so that

   $$\widehat{p}^{(j)}(\mathbf{x}) = \widehat{p}^{(j-1)}(\mathbf{x}) g_j(\mathbf{a}_j^\tau \mathbf{x}).$$

will be the current multivariate density estimate, where

$$g_j(\mathbf{a}_j^\tau \mathbf{x}) = \frac{p_{\mathbf{a}_j}(\mathbf{a}_j^\tau \mathbf{x})}{\widehat{p}_{\mathbf{a}_j}(\mathbf{a}_j^\tau \mathbf{x})}, \quad j = 1, 2, \ldots, k. \tag{4.71}$$

The vectors $\{\mathbf{a}_j\}$ are unit-length directions in $\Re^r$, and the augmenting (or ridge) functions $\{g_j\}$ are used to build up the structure of $\widehat{p}^{(0)}$ so that $\widehat{p}^{(k)}$ converges to p in some appropriate sense as $k \to \infty$. The number k of iterations operates as a smoothing parameter, and a stopping rule is determined by balancing bias against the variance of the estimator.

Friedman, Stuetzle, and Schroeder (1984) suggest graphical inspection of the augmenting functions (i.e., plotting $g_j(\mathbf{a}_j^\tau \mathbf{x})$ against $\mathbf{a}_j^\tau \mathbf{x}$ for $j = 1, 2, \ldots, k$) as a termination criterion for the iterative procedure. Computation of the augmenting functions $\{g_j(\mathbf{a}_j^\tau \mathbf{x})\}$ is discussed in Huber (1985, Section 15) and discussants Buja and Stuetzle (especially pp. 487–489), and Jones and Sibson (1987, Section 3). Given $\mathbf{a}_j$, estimate $p_{\mathbf{a}_j}$ by first projecting the sample data along the direction $\mathbf{a}_j$, thus obtaining $z_i = \mathbf{a}_j^\tau \mathbf{x}_i$, $i = 1, 2, \ldots, n$, and then compute a kernel density estimate from the $\{z_i\}$. Monte Carlo sampling is used to compute $\widehat{p}_{\mathbf{a}_j}$, followed by kernel density estimation. Alternatives to kernel smoothing include cubic spline functions (Friedman, Stuetzle, and Schroeder, 1984) and the average shifted histogram (Jee, 1987).

4.6.2   Projection Indexes

PPDE is driven by a projection index usually of the form

$$I(p) = \int J(p(z))p(z)dz = \mathrm{E}_p\{J(p)\}, \qquad (4.72)$$

where J is a smooth real-valued functional and z is a one-dimensional projected version of $\mathbf{x}$. As a functional of p, $I(p)$ should be absolutely continuous with easily computable first derivatives. "Interesting" projections should correspond to random or unstructured projections.

Estimates of $I(p)$ should be amenable to fast computation, unaffected by the overall covariance structure of the data or by outliers or heavy tails. A very reliable and thorough numerical optimizer is absolutely essential for finding "substantive" maxima of $I(p)$, because sampling fluctuations tend to trap ineffective optimizers within a multitude of local maxima (Friedman, 1987).

If $\{z_i\}$ are the projected data, then we can estimate (4.72) by

$$\widehat{I}(p) = \int J(\widehat{p}(z))d\widehat{F}_n(z) = \frac{1}{n}\sum_{i=1}^{n} J(\widehat{p}(z_i)). \qquad (4.73)$$

Thus, if $J(p(z)) = p(z)$, then $I(p) = \int [p(z)]^2 dz$ can be estimated by $\widehat{I}(p) = (1/n)\sum_{i=1}^{n}\widehat{p}_h(z_i)$, where $\widehat{p}_h$ is a kernel estimator with window width h. Another choice is to take $J(p(z)) = \log_e p(z)$, so that $I(p) = \int p(z)\log_e p(z)dz$, which is (negative) cross-entropy, and (4.73) can be estimated at the kth iteration by $(1/n)\sum_{i=1}^{n}\log_e \widehat{p}^{(k)}(z_i)$.

Other projection indexes that have been used for PPDE include a moment index based upon the sum of squares of the third and fourth sample cumulants of the projected data (Jones and Sibson, 1987) and the ISE criterion (Friedman, 1987; Hall, 1989a). The latter approaches, though related, differed on whether or not to transform the projected data first. Friedman used ISE between the transformed projected data density and the uniform density, and Hall's version used the ISE between the untransformed projected data density and the standard Gaussian. Both Friedman and Hall used orthogonal series density estimators (Legendre polynomials and Hermite functions, respectively) to study their projection indexes.

Each of these indexes was designed to search for deviations from "uninterestingness," whose definition depended upon the specific context. Thus, the Friedman–Tukey index searched for evidence of "clottedness" as well as departures from a parabolic density; the entropy index searched for departures of the projected data from Gaussian form because the Gaussian distribution maximizes entropy; and the moment index and ISE criteria also set up the Gaussian distribution as the least-interesting data feature.

4.7   Assessing Multimodality

As we have seen, it is not unusual for a data set, large or small, to have several modes (or local maxima) in its density estimate. Multiple modes strongly suggest that the underlying probability distribution can be modeled parametrically as a mixture of several probability distributions (each usually Gaussian), where initial values of the EM algorithm can be set by centering each mixture component at the location of a mode and setting the weight attached to that component according to the relative magnitude of the corresponding mode.

Of course, there is no guarantee that a mixture of unimodal densities will produce a multimodal density with the same number of modes as there are densities in the mixture; similarly, there is no guarantee that those individual modes will remain at the same locations in such a mixture. Indeed, the shape of the mixture distribution depends upon both the spacings of the modes and the relative shapes of the component distributions.

In many practical instances, however, the presence of more than a single mode does suggest evidence for a mixture; this has led to several tests being proposed for detecting multimodality in a distribution (see, e.g., Hartigan and Hartigan, 1985). Given a sample of data and some degree of assurance in multimodality, the modes can be evaluated in several ways. For example, Good and Gaskins (1980) used the MPL method of density estimation together with certain "bump-hunting" surgical techniques, whereas Silverman (1981, 1983) combined kernel-based density estimation with a hierarchical bootstrap testing procedure to determine the most probable number of modes in the underlying density. See Izenman and Sommer (1988) for an extensive discussion of Silverman's test and application to the 1872 Hidalgo postage stamp data. Both methods are nonparametric, data-adaptive, and computationally intensive.

Bibliographical Notes

There is a huge literature on nonparametric density estimation. Because of the amount of material published, we cannot list all pertinent articles or even books on the subject. Furthermore, due to space considerations, there are many nonparametric density estimation methods, including orthogonal series estimators and adaptive-kernel estimators, that are not described in this chapter. For descriptions of these methods, see Izenman (1991) and the references therein.

The most useful books on the subject are by Scott (1992), Silverman (1986), and Simonoff (1996). Chapters on nonparametric density estimation in books include Bishop (1995, Chapter 2), Ripley (1996, Chapter 6), and Duda, Hart, and Stork (2001, Chapter 4).

The origin of the histogram has been traced variously back to Galileo's star observations of 1632, John Graunt's mortality tables of 1662, the bar charts of William Playfair in 1786, and Karl Pearson in 1805 for the name.

There are several surveys on choices of window width, including Jones, Marron, and Sheather (1996).

Multivariate kernel density estimation was studied by Cacoullos (1966) and Epanechnikov (1969). Cacoullos (1966) appears to have been the first to call K in (4.28) a kernel function; previously, K was known as a *weight function*. He also was the first to use product kernels.

Exercises

4.1 Consider the class of functions of the form $p(x) = \sum_{\ell=0}^{L-1} y_\ell I_{T_i}(x)$, where $h \sum_{\ell=0}^{L-1} y_\ell = 1$. Given an iid sample, $X_1, X_2, \ldots, X_n$ from $p(x)$, maximize the log-likelihood function, $\mathcal{L} = \sum_{i=1}^{n} \log_e [\sum_{\ell=0}^{L-1} y_\ell I_{T_\ell}(x_i)]$, subject to the condition that $h \sum_{\ell=0}^{L-1} y_\ell = 1$. Show that the histogram (4.13) is the unique ML estimator of p. [Hint: Use Lagrangian multipliers.]

4.2 By minimizing AMISE in (4.27) wrt h_n, show that the optimal bin width, h_n^*, is given by (4.28) and that the AMISE* = AMISE(h_n^*) of the histogram with the optimal bin width is (4.30).

4.3 The *average shifted histogram (ASH)* (Scott, 1985a) is constructed by taking m histograms, $\hat{p}_1, \hat{p}_2, \ldots, \hat{p}_m$, say, each of which has the same bin width h_n, but with different bin origins, $0, h_n/m, 2h_n/m, \ldots, (m-1)h_n/m$, respectively, and then averaging those histograms,

$$\hat{p}_{\text{ASH}}(x) = m^{-1} \sum_{k=1}^{m} \hat{p}_k(x).$$

The resulting ASH is piecewise constant over intervals $[k\delta, (k+1)\delta)$ of width $\delta = h_n/m$; it has a similar block-like structure as a histogram but is defined over narrower bins. Derive the integrated variance and integrated squared-bias of the average shifted histogram. Show that the asymptotic MISE of the ASH is

$$\text{AMISE} = \frac{2}{3nh_n}\left(1 + \frac{1}{2m^2}\right) + \frac{h_n^2}{12m^2}R(p') + \frac{h_n^4}{144}\left(1 - \frac{2}{m^2} + \frac{3}{5m^2}\right)R(p'').$$

4.4 The *frequency polygon (FP)* (Scott, 1985b) connects the center of each pair of adjacent histogram bin-values with a straight line. If two adjacent bin-values are $\widehat{p}_\ell = N_\ell/nh_n$ and $\widehat{p}_{\ell+1} = N_\ell/nh_n$, then the value of the FP at $x \in [(\ell - \frac{1}{2})h_n, (\ell + \frac{1}{2})h_n)$ is

$$\widehat{p}_{FP}(x) = \left(\left(\ell + \frac{1}{2}\right) - \frac{x}{h_n}\right)\widehat{p}_\ell + \left(\frac{x}{h_n} - \left(\ell - \frac{1}{2}\right)\right)\widehat{p}_{\ell+1}.$$

Whereas the histogram is discontinuous, the FP is a continuous density estimator. Derive the integrated variance and integrated squared-bias of the frequency polygon. [Hint: For ISB, use a Taylor series expansion of $p(x)$ to the term involving p''; then, for IV, use $\text{var}(X + Y) = \text{var}(X) + \text{var}(Y) + 2\text{cov}(X, Y)$ for binomial X and Y.] Show that if p'' is absolutely continuous and $R(p''') < \infty$, then the asymptotic MISE is given by

$$\text{AMISE}(h_n) = \frac{2}{3nh_n} + \frac{49}{2880}h_n^4 R(p'').$$

Show that the h_n that minimizes $\text{AMISE}(h_n)$ is

$$h_n^* = 2\left(\frac{15}{49R(p'')}\right)^{1/5} n^{-1/5}.$$

4.5 Write a computer program to compute the FP and the ASH and try them out on a data set of your choice.

4.6 By considering m shifted histograms, let $B_k = [k\delta, (k+1)\delta)$ be the kth bin of the ASH, where $\delta = h_n/m$, and let ν_k be the bin count in B_k. Note that the ASH bin count for bin B_k is the average of the bin counts of the m shifted histograms, each of width δ, in bin B_k. Show that, for $x \in B_k$ and m large, the ASH can be expressed as a kernel density estimator with triangular kernel on $(-1, 1)$.

4.7 The ASH is not continuous but can be made continuous by linearly interpolating using the FP approach. Show that this ASH-FP density estimate can be expressed as a kernel estimator.

4.8 Rosenblatt's density estimator is

$$\widehat{p}_n(x) = h^{-1}\left[F_n\left(x + \frac{h}{2}\right) - F_n\left(x - \frac{h}{2}\right)\right],$$

where $F_n(x)$ is the empirical cumulative distribution function, $x \in \mathfrak{R}$. Show that this estimator is a kernel density estimator. Which type of kernel corresponds to Rosenblatt's estimator? Apply this kernel to estimate the density of the 1872 Hidalgo stamp data. What do you notice about the smoothness of the resulting density estimate?

4.9 Find the bias and variance of Rosenblatt's estimator (Exercise 4.8). From these expressions, find the MISE of that estimator.

4.10 Verify equation (4.32).

4.11 Verify equations (4.53) and (4.54).

4.12 Generate n observations from the *claw density*,

$$p(x) = 0.5\mathcal{N}(0,1) + (0.1)\sum_{k=0}^{4}\mathcal{N}\left(\frac{k}{2} - 1, (0.1)^2\right),$$

and estimate that density using a kernel density estimator. Take $n = 100, 200,$ and 300, and repeat 1,000 times at each sample size. Compare the performances of UCV, BCV, and SJPI window-width estimators for each simulation. Which window-width estimation method best finds the claws?

4.13 The `galaxy` velocity data consist of the radial velocities of 323 locations in the area of the spiral galaxy NGC7531 in the Southern Hemisphere; the data can be found on the book's website. Compare the kernel density estimates of the `galaxy` data using UCV, BCV, and SJPI window-width estimators. Pay special attention to the number of modes in the estimates. Use Silverman's test to determine the number of modes (see Silverman, 1981; Izenman and Sommer, 1988).

4.14 The `ushighways` data consist of the approximate length (in miles) of all 212 U.S. 3-digit interstate highways (spurs and connectors). The data were extracted by L. Winner from the *Rand McNally 1993 Business Traveler's Road Atlas and Guide to Major Cities* and can be found on the book's website. Compare the kernel density estimates for these data using UCV, BCV, and SJPI window-width estimators.

5

Model Assessment and Selection in Multiple Regression

5.1 Introduction

Regression, as a scientific method, first appeared around 1885, although the *method of least squares* was discovered 80 years earlier. Least squares owes its origins to astronomy and, specifically, to Legendre's 1805 pioneering work on the determination of the orbits of planets in which he introduced and named the method of least squares. Adrien Marie Legendre estimated the coefficients of a set of linear equations by minimizing the error sum of squares. Gauss stated in 1809 that he had been using the method since 1795, but could not prove his claim with documented evidence. Within a few years, Gauss and Pierre Simon Laplace added a probability component — a Gaussian curve to describe the error distribution — that was crucial to the success of the method. Gauss went on to devise an elimination algorithm to compute least-squares estimates. Once introduced, least squares caught on immediately in astronomy and geodetics, but it took 80 years for these ideas to be transported to other disciplines.

The ideas of regression and correlation were developed in the mid-1880s by Francis Galton in studies of heredity stature, and he applied those ideas to a comparison of the heights of parents and their children (which led to his famous phrase of "regression to mediocrity"). Galton (and also

A.J. Izenman, *Modern Multivariate Statistical Techniques*,
doi: 10.1007/978-0-387-78189-1_5,
© Springer Science+Business Media, LLC 2008

Francis Ysidro Edgeworth and Karl Pearson), however, failed to connect least squares to regression. It was George Udny Yule, in 1897, who showed that an assumption of a Gaussian error curve in regression could be replaced by an assumption that the variables were linearly related, and that, as a result, least squares could be applied to regression. Thus, the wealth of numerical algorithms already developed by astronomers and geodesists for finding least-squares solutions could be put to work solving regression equations.

Since then, regression has evolved into many different forms, including linear and nonlinear regression and parametric and nonparametric regression. Linear regression models, in particular, are referred to as simple, multiple, or multivariate depending upon the number of input and output variables considered. Simple linear regression deals with one input and one output, multiple regression deals with many inputs and one output, and multivariate regression deals with many inputs and many outputs.

5.2   The Regression Function and Least Squares

We assume that the *output* (or *dependent, response*) variable Y is linearly related to the *input* (or *independent, predictor*) variables $X_1, \ldots, X_r$ in the following way,

$$Y = \beta_0 + \sum_{j=1}^{r} \beta_j X_j + e, \tag{5.1}$$

where e is an unobservable random variable (the *error component*) with mean 0 and variance σ^2. The relationship (5.1) is known as a *linear regression model*, where $\beta_0, \beta_1, \ldots, \beta_r$ are unknown parameters and $\sigma^2 > 0$ is an unknown *error variance*. The linearity of the model (5.1) is a result of its linearity in the parameters $\beta_0, \beta_1, \ldots, \beta_r$. Thus, transformations of the input variables (such as powers X_j^d and products $X_j X_k$) can be included in (5.1) without it losing its characterization as a linear regression model.

The goal is to estimate the true values of $\beta_0, \beta_1, \ldots, \beta_r$, and σ^2, and to assess the impact of each input variable on the behavior of Y. In the likely event that some of the input variables have negligible effects on Y, we may also wish to reduce the number of input variables to a smaller number, especially if r is large. In many uses of multiple regression, we are interested in *predicting* future values of Y, given future values of the input variables, and we would like to be able to measure the accuracy of those predictions.

The way we treat the model (5.1) depends upon our assumptions about how the input variables $X_1, \ldots, X_r$ were generated. We distinguish between the case when the values of $X_1, \ldots, X_r$ are randomly selected according to some probability distribution (the "random-X" case), a situation that occurs with observational data, and the case when the values of $X_1, \ldots, X_r$ are fixed in repeated sampling (the "fixed-X" case), possibly set through a designed experiment.

5.2.1 Random-X Case

Suppose we have an input vector of random variables $\mathbf{X} = (X_1, \ldots, X_r)^\tau$ and a random output variable Y, and suppose that these $r + 1$ real-valued random variables are jointly distributed according to $P(\mathbf{X}, Y)$ with means $E(\mathbf{X}) = \boldsymbol{\mu}_X$ and $E(Y) = \mu_Y$, respectively, and covariance matrices $\boldsymbol{\Sigma}_{XX}$, $\boldsymbol{\Sigma}_{YY} = \sigma_Y^2$, and $\boldsymbol{\Sigma}_{XY}$.

Consider the problem of predicting Y by a function, $f(\mathbf{X})$, of $\mathbf{X}$. We measure prediction accuracy by a real-valued *loss function* $L(Y, f(\mathbf{X}))$, that gives the loss incurred if Y is predicted by $f(\mathbf{X})$. The expected loss is the *risk function*,

$$R(f) = E\{L(Y, f(\mathbf{X}))\}, \tag{5.2}$$

which measures the quality of f as a predictor. The *Bayes rule* is the function f^* which minimizes $R(f)$, and the *Bayes risk* is $R(f^*)$.

For squared-error loss, $R(f)$ becomes the *mean squared error* criterion by which we judge $f(\mathbf{X})$ as a predictor of Y. We have that

$$
\begin{aligned}
R(f) &= E(Y - f(\mathbf{X}))^2 &\quad (5.3)\\
&= E_\mathbf{X}[E_{Y|\mathbf{X}}\{(Y - f(\mathbf{X}))^2|\mathbf{X}\}], &\quad (5.4)
\end{aligned}
$$

where the subscripts indicate the distribution over which the expectation is taken. Hence, $R(f)$ can be minimized pointwise (at each $\mathbf{x}$). We can write

$$Y - f(\mathbf{x}) = (Y - \mu(\mathbf{x})) + (\mu(\mathbf{x}) - f(\mathbf{x})), \tag{5.5}$$

where $\mu(\mathbf{x}) = E_{Y|\mathbf{X}}\{Y|\mathbf{X} = \mathbf{x}\}$ is the mean of the conditional distribution of Y given $\mathbf{X} = \mathbf{x}$ and is called the *regression function* of Y on $\mathbf{X}$. Squaring both sides of (5.5) and taking conditional expectations, we have that

$$
\begin{aligned}
E_{Y|\mathbf{X}}\{(Y - f(\mathbf{x}))^2|\mathbf{X} = \mathbf{x}\} &= E_{Y|\mathbf{X}}\{(Y - \mu(\mathbf{x}))^2|\mathbf{X} = \mathbf{x}\}\\
&\quad + (\mu(\mathbf{x}) - f(\mathbf{x}))^2, \quad (5.6)
\end{aligned}
$$

where the cross-product term vanishes because $E_{Y|\mathbf{X}}\{Y - \mu(\mathbf{x})|\mathbf{X} = \mathbf{x}\} = 0$. Therefore, (5.6) is minimized with respect to f by taking

$$f^*(\mathbf{x}) = \mu(\mathbf{x}) = E_{Y|\mathbf{X}}\{Y|\mathbf{X} = \mathbf{x}\}, \tag{5.7}$$

so that the pointwise minimum of (5.6) is given by

$$E_{Y|\mathbf{X}}\{(Y - f^*(\mathbf{x}))^2|\mathbf{X} = \mathbf{x}\} = E_{Y|\mathbf{X}}\{(Y - \mu(\mathbf{x}))^2|\mathbf{X} = \mathbf{x}\}. \tag{5.8}$$

Taking expectations of both sides, we have that the Bayes risk is

$$R(f^*) = \min_f R(f) = \mathrm{E}\{(Y - \mu(\mathbf{X}))^2\}. \tag{5.9}$$

Thus, the best predictor of Y at $\mathbf{X}=\mathbf{x}$, using minimum mean squared error to define "best," is given by $\mu(\mathbf{x})$, the regression function of Y on $\mathbf{X}$, evaluated at $\mathbf{X}=\mathbf{x}$, which is also the unique Bayes rule.

To be more specific, suppose the relationship (5.1) holds, where we assume that e is uncorrelated with the $X_1, \ldots, X_r$. The regression function, which is linear in $\mathbf{X}$, is given by

$$\mu(\mathbf{X}) = \beta_0 + \sum_{i=1}^{r} \beta_i X_i = \beta_0 + \mathbf{X}^\tau \boldsymbol{\beta} = \mathbf{Z}^\tau \boldsymbol{\alpha}, \tag{5.10}$$

where β_0 is the *intercept*, $\boldsymbol{\beta} = (\beta_1, \ldots, \beta_r)^\tau$ is an r-vector of *regression coefficients*, $\boldsymbol{\alpha} = (\beta_0 \vdots \boldsymbol{\beta}^\tau)^\tau$ is an $(r+1)$-vector, and $\mathbf{Z} = (1 \vdots \mathbf{X}^\tau)^\tau$ is an $(r+1)$-vector. We then choose β_0 and $\boldsymbol{\beta}$ to minimize the quadratic objective function (5.8). Let

$$S(\boldsymbol{\alpha}) = \mathrm{E}\{(Y - \mathbf{Z}^\tau \boldsymbol{\alpha})^2\}, \tag{5.11}$$

and define $\boldsymbol{\alpha}^* = \arg\min_{\boldsymbol{\alpha}} S(\boldsymbol{\alpha})$. Differentiating $S(\boldsymbol{\alpha})$ with respect to $\boldsymbol{\alpha}$ yields:

$$\frac{\partial S(\boldsymbol{\alpha})}{\partial \boldsymbol{\alpha}} = -2\mathrm{E}(\mathbf{Z}Y - \mathbf{Z}\mathbf{Z}^\tau \boldsymbol{\alpha}). \tag{5.12}$$

Setting (5.12) equal to zero for a minimum, we get:

$$\boldsymbol{\alpha}^* = [\mathrm{E}(\mathbf{Z}\mathbf{Z}^\tau)]^{-1}\mathrm{E}(\mathbf{Z}Y). \tag{5.13}$$

From (5.13), and noting that $\boldsymbol{\alpha}^* = (\beta_0^* \vdots \boldsymbol{\beta}^{*\tau})^\tau$, it is not difficult to show (Exercise 5.1) that

$$\boldsymbol{\beta}^* = \boldsymbol{\Sigma}_{XX}^{-1}\boldsymbol{\Sigma}_{XY}, \tag{5.14}$$

$$\beta_0^* = \mu_Y - \boldsymbol{\mu}_X^\tau \boldsymbol{\beta}^*. \tag{5.15}$$

In practice, because $\boldsymbol{\mu}_X$, μ_Y, $\boldsymbol{\Sigma}_{XX}$ and $\boldsymbol{\Sigma}_{XY}$ will be unknown, we estimate them by ML using data generated by the joint distribution of $(\mathbf{X}, Y)$.

Suppose that

$$\mathcal{D} = \{(\mathbf{X}_i, Y_i), i = 1, 2, \ldots, n\}, \tag{5.16}$$

are iid observations from $\mathrm{P}(\mathbf{X}, Y)$, where $\mathbf{X}_i = (X_{i1}, \cdots, X_{ir})^\tau$ is the ith observed value of $\mathbf{X} = (X_1, X_2, \cdots, X_r)^\tau$ and Y_i is the ith observed value of Y, $i = 1, 2, \ldots, n$. Let $\mathcal{X} = (\mathbf{X}_1, \cdots, \mathbf{X}_n)^\tau$ be an $(n \times r)$-matrix and $\mathcal{Y} = (Y_1, \cdots, Y_n)^\tau$ be an n-vector. We estimate $\boldsymbol{\mu}_X$ and μ_Y by the r-vector $\bar{\mathbf{X}} = n^{-1}\sum_{j=1}^{n}\mathbf{X}_j$ and scalar $\bar{Y} = n^{-1}\sum_{j=1}^{n}Y_j$, respectively. Let $\bar{\mathcal{X}} = (\bar{\mathbf{X}}, \cdots, \bar{\mathbf{X}})^\tau$ be an $(n \times r)$-matrix and $\bar{\mathcal{Y}} = (\bar{Y}, \cdots, \bar{Y})^\tau$ be an n-vector.

Let $\mathcal{X}_c = \mathcal{X} - \bar{\mathcal{X}}$ and $\mathcal{Y}_c = \mathcal{Y} - \bar{\mathcal{Y}}$ be the mean-centered forms of $\mathcal{X}$ and $\mathcal{Y}$, respectively, and estimate $\boldsymbol{\Sigma}_{XX}$ by $n^{-1}\mathcal{X}_c^\tau \mathcal{X}_c$ and $\boldsymbol{\Sigma}_{XY}$ by $n^{-1}\mathcal{X}_c^\tau \mathcal{Y}_c$. The least-squares estimates of (5.14) and (5.15) are given by

$$\widehat{\boldsymbol{\beta}}^* = (\mathcal{X}_c^\tau \mathcal{X}_c)^{-1}\mathcal{X}_c^\tau \mathcal{Y}_c. \tag{5.17}$$

$$\widehat{\beta}_0^* = \bar{Y} - \bar{\mathbf{X}}^\tau \widehat{\boldsymbol{\beta}}^*, \tag{5.18}$$

respectively.

5.2.2 Fixed-X Case

In the "fixed-X" case, we view the input variables $X_1, \ldots, X_r$ as being fixed in repeated sampling. Thus, the value of Y may depend upon input variables whose values are selected by an experimentalist within the framework of a designed experiment, or Y may be observed conditional on the $X_1, \ldots, X_r$.

Suppose the n observations (5.16) satisfy (5.1), so that

$$Y_i = \beta_0 + \sum_{j=1}^{r} \beta_j X_{ij} + e_i, \quad i = 1, 2, \ldots, n, \tag{5.19}$$

where $e_1, e_2, \ldots, e_n$ are i.i.d. random variables having the same distribution as e. Equations (5.19) can be written as

$$Y_i = \mathbf{Z}_i^\tau \boldsymbol{\beta} + e_i = \mu(\mathbf{X}_i) + e_i, \quad i = 1, 2, \ldots, n, \tag{5.20}$$

where $\mu(\mathbf{X}_i) = \mathbf{Z}_i^\tau \boldsymbol{\beta}$ is the regression function, $\mathbf{Z}_i^\tau = (1, X_{i1}, \cdots, X_{ir})$, and $\boldsymbol{\beta}^\tau = (\beta_0, \beta_1, \cdots, \beta_r)$. The n equations (5.20) can be written more compactly as

$$\mathcal{Y} = \mathcal{Z}\boldsymbol{\beta} + \mathbf{e}, \tag{5.21}$$

where $\mathcal{Y} = (Y_1, \cdots, Y_n)^\tau$ is a random n-vector, $\mathcal{Z} = (\mathbf{Z}_1, \cdots, \mathbf{Z}_n)^\tau$ is an $(n \times (r+1))$-matrix with ith row $\mathbf{Z}_i^\tau$ $(i = 1, 2, \ldots, n)$, $\boldsymbol{\beta}$ is an $(r+1)$-vector, and $\mathbf{e}$ is a random n-vector of unobservable errors with $\mathrm{E}(\mathbf{e}) = \mathbf{0}$ and $\mathrm{var}(\mathbf{e}) = \sigma^2 \mathbf{I}_n$. To account for the intercept β_0, the first column of $\mathcal{Z}$ consists only of 1s.

We form the *error sum of squares (ESS)*,

$$ESS(\boldsymbol{\beta}) = \sum_{i=1}^{n} e_i^2 = \mathbf{e}^\tau \mathbf{e} = (\mathcal{Y} - \mathcal{Z}\boldsymbol{\beta})^\tau (\mathcal{Y} - \mathcal{Z}\boldsymbol{\beta}), \tag{5.22}$$

and estimate $\boldsymbol{\beta}$ by minimizing $ESS(\boldsymbol{\beta})$ with respect to $\boldsymbol{\beta}$. Differentiating $ESS(\boldsymbol{\beta})$ with respect to $\boldsymbol{\beta}$ yields

$$\frac{\partial ESS(\boldsymbol{\beta})}{\partial \boldsymbol{\beta}} = -2\mathcal{Z}^\tau (\mathcal{Y} - \mathcal{Z}\boldsymbol{\beta}), \tag{5.23}$$

$$\frac{\partial^2 ESS(\boldsymbol{\beta})}{\partial \boldsymbol{\beta} \, \partial \boldsymbol{\beta}^\tau} = -2\mathcal{Z}^\tau \mathcal{Z}, \tag{5.24}$$

and setting result (5.23) equal to 0 for a minimum yields the *normal equations*,

$$\mathcal{Z}^\tau \mathcal{Z} \widehat{\boldsymbol{\beta}} = \mathcal{Z}^\tau \mathcal{Y}. \tag{5.25}$$

Assuming that the $((r+1) \times (r+1))$-matrix $\mathcal{Z}^\tau \mathcal{Z}$ is nonsingular (and, hence, invertible), the unique *ordinary least-squares (OLS) estimator* of $\boldsymbol{\beta}$ in the model (5.21) is given by

$$\widehat{\boldsymbol{\beta}}_{\mathrm{ols}} = (\mathcal{Z}^\tau \mathcal{Z})^{-1}\mathcal{Z}^\tau \mathcal{Y}. \tag{5.26}$$

Note the resemblance of (5.26) to (5.13).

We can write $\mathcal{Z} = (\mathbf{1}_n \vdots \mathcal{X}^\tau)$, where $\mathcal{X}^\tau$ is an $(r \times n)$-matrix, with a corresponding partition of $\boldsymbol{\beta}$ as $\boldsymbol{\beta} = (\beta_0 \vdots \boldsymbol{\beta}_*^\tau)^\tau$, where $\boldsymbol{\beta}_* = (\beta_1, \cdots, \beta_r)^\tau$. Let $\bar{\mathbf{X}} = n^{-1}\mathcal{X}\mathbf{1}_n$ and $\bar{Y} = n^{-1}\mathbf{1}_n^\tau \mathcal{Y}$. As before, let $\bar{\mathcal{X}} = (\bar{\mathbf{X}}, \cdots, \bar{\mathbf{X}})$ be an $(n \times r)$-matrix, each column of which is $\bar{\mathbf{X}}$, and let $\bar{\mathcal{Y}} = (\bar{Y}, \cdots, \bar{Y})^\tau$, be an n-vector each element of which is $\bar{y}$. Then, $\mathcal{X}_c = \mathcal{X} - \bar{\mathcal{X}}$ is an $(n \times r)$-matrix and $\mathcal{Y}_c = \mathcal{Y} - \bar{\mathcal{Y}}$ is an n-vector. It is not difficult to show (Exercise 5.2) that

$$\widehat{\boldsymbol{\beta}}_* = (\mathcal{X}_c^\tau \mathcal{X}_c)^{-1}\mathcal{X}_c^\tau \mathcal{Y}_c \tag{5.27}$$

$$\widehat{\beta}_0 = \bar{Y} - \bar{\mathbf{X}}^\tau \widehat{\boldsymbol{\beta}}_* \tag{5.28}$$

Clearly, the estimates (5.17) and (5.18) are identical to the corresponding estimates (5.27) and (5.28). *Even though the descriptions differ as to how the input data are generated, the OLS estimates turn out to be the same for the random-X case and the fixed-X case.*

For fixed $\mathcal{X}$ and assuming that $\mathrm{var}(\mathbf{y}) = \sigma^2 \mathbf{I}_n$, the mean and variance of $\widehat{\boldsymbol{\beta}}_{\mathrm{ols}}$ in (5.26) are given by $\mathrm{E}(\widehat{\boldsymbol{\beta}}_{\mathrm{ols}}) = \boldsymbol{\beta}_*$ and

$$\begin{aligned}
\mathrm{var}(\widehat{\boldsymbol{\beta}}_{\mathrm{ols}}) &= (\mathcal{Z}^\tau \mathcal{Z})^{-1}\mathcal{Z}^\tau \{\mathrm{var}(\mathbf{y})\}\mathcal{Z}(\mathcal{Z}^\tau \mathcal{Z})^{-1} \\
&= \sigma^2(\mathcal{Z}^\tau \mathcal{Z})^{-1},
\end{aligned} \tag{5.29}$$

respectively.

The OLS regression estimator $\widehat{\boldsymbol{\beta}}_{\mathrm{ols}}$ has some very desirable properties that are characterized by the Gauss–Markov Theorem (Exercise 5.3). If we are looking for a linear unbiased estimator of $\boldsymbol{\beta}$ with minimum variance, the Gauss–Markov Theorem states that we need only consider $\widehat{\boldsymbol{\beta}}_{\mathrm{ols}}$.

The components of the n-vector of OLS *fitted values* are the vertical projections of the n points onto the LS regression surface (or hyperplane) $\widehat{y}_i = \widehat{\mu}(\mathbf{x}_i) = \mathbf{x}_i^\tau \widehat{\boldsymbol{\beta}}_{\mathrm{ols}}, i = 1, 2, \ldots, n$. See Figure 5.1 for a geometrical view. The variance of $\widehat{y}_i$ for fixed $\mathbf{x}_i$ is given by

$$\mathrm{var}(\widehat{y}_i \mid \mathbf{x}_i) = \mathbf{x}_i^\tau \{\mathrm{var}(\widehat{\boldsymbol{\beta}}_{\mathrm{ols}})\}\mathbf{x}_i = \sigma^2 \mathbf{x}_i^\tau (\mathcal{Z}^\tau \mathcal{Z})^{-1}\mathbf{x}_i. \tag{5.30}$$

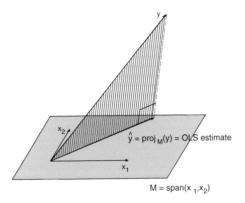

FIGURE 5.1. *A geometrical view of the ordinary least-squares method, using two input variables, X_1 and X_2. The hyperplane spanned by the input variables is denoted by M, and the OLS fitted value $\widehat{y}$ is the orthogonal projection of the output value y onto M.*

The n-vector of fitted values $\widehat{\mathcal{Y}} = (\widehat{y}_1, \ldots, \widehat{y}_n)^\tau$ is

$$\widehat{\mathcal{Y}} = \mathcal{Z}\widehat{\boldsymbol{\beta}}_{\text{ols}} = \mathcal{Z}(\mathcal{Z}^\tau \mathcal{Z})^{-1}\mathcal{Z}^\tau \mathcal{Y} = \mathbf{H}\mathcal{Y}, \tag{5.31}$$

where the $(n \times n)$-matrix $\mathbf{H} = \mathcal{Z}(\mathcal{Z}^\tau \mathcal{Z})^{-1}\mathcal{Z}^\tau$ is often called the *hat matrix* because it puts the "hat" on $\mathcal{Y}$. Note that $\mathbf{H}$ and $\mathbf{I}_n - \mathbf{H}$ are both symmetric, idempotent matrices with $\mathbf{H}(\mathbf{I}_n - \mathbf{H}) = \mathbf{0}$. Furthermore, $\mathbf{H}\mathcal{Z} = \mathcal{Z}$ and $(\mathbf{I}_n - \mathbf{H})\mathcal{Z} = \mathbf{0}$. The variance of $\widehat{\mathcal{Y}}$ is given by

$$\text{var}(\widehat{\mathcal{Y}}|\mathbf{X}) = \mathbf{H}\{\text{var}(\mathcal{Y})\}\mathbf{H}^\tau = \sigma^2 \mathbf{H}. \tag{5.32}$$

The ijth component h_{ij} of $\mathbf{H}$ is the amount of *leverage* (or impact) that the observed value of y_j exerts on the fitted value $\widehat{y}_i$. The hat matrix $\mathbf{H}$ is, therefore, used to identify *high-leverage points*. In particular, the diagonal components h_{ii} satisfy $0 \leq h_{ii} \leq 1$, their sum is the number, r, of input variables, and the average leverage magnitude is r/n. From this, high-leverage points have been defined as those points having $h_{ii} > 2r/n$.

The *residuals*, $\widehat{\mathbf{e}} = \mathcal{Y} - \widehat{\mathcal{Y}} = (\mathbf{I}_n - \mathbf{H})\mathcal{Y}$ are the OLS estimates of the unobservable errors $\mathbf{e}$. The residual vector can also be written as

$$\widehat{\mathbf{e}} = \mathcal{Y} - \mathcal{Z}\widehat{\boldsymbol{\beta}}_{\text{ols}} = (\mathcal{Z}\boldsymbol{\beta} + \mathbf{e}) - \mathcal{Z}(\boldsymbol{\beta} + (\mathcal{Z}^\tau \mathcal{Z})^{-1}\mathcal{Z}^\tau \mathbf{e}) = (\mathbf{I}_n - \mathbf{H})\mathbf{e}, \tag{5.33}$$

whence, assuming again that $\mathcal{Z}$ is fixed, it follows that $\text{E}(\widehat{\mathbf{e}}) = \mathbf{0}$ and $\text{var}(\widehat{\mathbf{e}}) = \sigma^2(\mathbf{I}_n - \mathbf{H})$. Hence, $\text{var}(\widehat{e}_i) = \sigma^2(1 - h_{ii})$, where h_{ii} is the ith diagonal element of $\mathbf{H}$, $i = 1, 2, \ldots, n$. The *residual sum of squares (RSS)* is given by

$$RSS = \sum_{i=1}^{n} \widehat{e}_i^2 = \widehat{\mathbf{e}}^\tau \widehat{\mathbf{e}} = ESS(\widehat{\boldsymbol{\beta}}_{\text{ols}}). \tag{5.34}$$

Note that

$$RSS = ESS(\boldsymbol{\beta}) + (\boldsymbol{\beta} - \widehat{\boldsymbol{\beta}}_{\text{ols}})^\tau \mathcal{Z}^\tau \mathcal{Z} (\boldsymbol{\beta} - \widehat{\boldsymbol{\beta}}_{\text{ols}}). \qquad (5.35)$$

Dividing RSS by its number of degrees of freedom, $n - r - 1$, gives us an unbiased estimate of the error variance σ^2,

$$\widehat{\sigma}^2 = \frac{RSS}{n - r - 1}, \qquad (5.36)$$

which is known as the *residual variance*. Hence, the OLS estimate of the $\text{var}(\widehat{\boldsymbol{\beta}}_{\text{ols}})$ is given by

$$\widehat{\text{var}}(\widehat{\boldsymbol{\beta}}_{\text{ols}}) = \widehat{\sigma}^2 (\mathcal{Z}^\tau \mathcal{Z})^{-1}. \qquad (5.37)$$

Residuals are often rescaled into *internally Studentized residuals* (which are more usually called *standardized residuals*) by dividing them by an estimate of their standard error,

$$\widehat{e}_i^S = \frac{\widehat{e}_i}{\widehat{\sigma}(1 - h_{ii})^{1/2}}, \quad i = 1, 2, \ldots, n. \qquad (5.38)$$

An *externally Studentized residual* can also be defined by omitting the ith case from the regression.

Because the n fitted values $\widehat{\mathcal{Y}} = \mathbf{H}\mathcal{Y}$ and the n residuals $\widehat{\mathbf{e}} = (\mathbf{I}_n - \mathbf{H})\mathcal{Y}$ have zero covariance and, hence, are uncorrelated, it follows that the regression of $\widehat{\mathcal{Y}}$ on $\widehat{\mathbf{e}}$ has zero slope. If the multiple regression model is correct, then a scatterplot of residuals (or Studentized residuals) against fitted values should show no discernible pattern (i.e., a slope of approximately zero). Anomolous patterns to look out for include nonlinearity, nonconstant variance, and possible outliers.

Now, consider the identity $y_i - \bar{y} = (y_i - \widehat{y}_i) + (\widehat{y}_i - \bar{y})$. Squaring both sides, summing over all n observations, and noting that the cross-product term disappears, we have that the *total sum of squares*,

$$S_{YY} = \sum_{i=1}^{n} (y_i - \bar{y})^2 = (\mathcal{Y} - \bar{\mathcal{Y}})^\tau (\mathcal{Y} - \bar{\mathcal{Y}}), \qquad (5.39)$$

can be written as $S_{YY} = SS_{reg} + RSS$, where the *regression sum of squares*,

$$SS_{reg} = \sum_{i=1}^{n} (\widehat{y}_i - \bar{y}_i)^2 = \widehat{\boldsymbol{\beta}}_{\text{ols}}^\tau (\mathcal{Z}^\tau \mathcal{Z}) \widehat{\boldsymbol{\beta}}_{\text{ols}}, \qquad (5.40)$$

and the *residual sum of squares*,

$$RSS = \sum_{i-1}^{n} (y_i - \widehat{y}_i)^2 = (\mathcal{Y} - \mathcal{Z}\widehat{\boldsymbol{\beta}}_{\text{ols}})^\tau (\mathcal{Y} - \mathcal{Z}\widehat{\boldsymbol{\beta}}_{\text{ols}}), \qquad (5.41)$$

TABLE 5.1. *ANOVA table for a multiple regression model.*

Source of Variation	df	Sum of Squares
Regression on $X_1, \ldots, X_r$	r	$SS_{\text{reg}} = \widehat{\boldsymbol{\beta}}_{\text{ols}}^{\tau} (\mathcal{Z}^{\tau} \mathcal{Z}) \widehat{\boldsymbol{\beta}}_{\text{ols}}$
Residual	$n - r - 1$	$RSS = (\mathcal{Y} - \mathcal{Z}\widehat{\boldsymbol{\beta}}_{\text{ols}})^{\tau} (\mathcal{Y} - \mathcal{Z}\widehat{\boldsymbol{\beta}}_{\text{ols}})$
Total	$n - 1$	$S_{YY} = (\mathcal{Y} - \bar{\mathcal{Y}})^{\tau}(\mathcal{Y} - \bar{\mathcal{Y}})$

form an orthogonal decomposition, which can be summarized by an *analysis of variance (ANOVA) table*; see Table 5.1. The *squared multiple correlation coefficient*, $R^2 = SS_{reg}/S_{YY}$, lies between 0 and 1 and is used to measure the proportion of the total variation in Y that can be explained by a linear regression on the r Xs.

So far, no assumptions have been made about the probability distribution of the errors. If $e_i \sim \mathcal{N}(0, \sigma^2)$, $i = 1, 2, \ldots, n$, it follows that

$$\widehat{\boldsymbol{\beta}}_{\text{ols}} \sim \mathcal{N}_{r+1} \left(\boldsymbol{\beta}, \sigma^2 (\mathcal{Z}^{\tau} \mathcal{Z})^{-1} \right), \tag{5.42}$$

$$RSS = (n - r - 1)\widehat{\sigma}^2 \sim \sigma^2 \chi^2_{n-r-1}, \tag{5.43}$$

and $\widehat{\boldsymbol{\beta}}_{\text{ols}}$ and $\widehat{\sigma}^2$ are independently distributed. From the ANOVA table, we can determine whether there is a linear relationship between Y and the Xs. We compute the *F-statistic*,

$$F = \frac{SS_{reg}/r}{RSS/(n - r - 1)}, \tag{5.44}$$

and compare the resulting F-value with an appropriate percentage point of the $\mathcal{F}_{r,n-r-1}$ distribution. A small value for F implies that the data did not provide sufficient evidence to reject $\boldsymbol{\beta} = \mathbf{0}$, whereas a large value indicates that at least one β_j is not zero. Under normality, if $\beta_j = 0$, the statistic

$$t_j = \frac{\widehat{\beta}_j}{\widehat{\sigma}\sqrt{v_{jj}}}, \tag{5.45}$$

where v_{jj} is the jth diagonal entry of $(\mathcal{Z}^{\tau}\mathcal{Z})^{-1}$, follows the Student's t distribution with $n - r - 1$ degrees of freedom, $j = 1, 2, \ldots, r$. A large value of $|t_j|$ is evidence that $\beta_j \neq 0$, whereas a small, near-zero value of $|t_j|$ is evidence that $\beta_j = 0$. For large n, t_j reduces to a Gaussian-distributed random variable, and the cutoff value for $|t_j|$ is usually taken to be 2.0. For $0 < \alpha < 1$, it follows that a $(1 - \alpha) \times 100\%$ confidence region for $\boldsymbol{\beta}$ is given by the set of $\boldsymbol{\beta}$-vectors such that

$$(r + 1)^{-1}(\widehat{\boldsymbol{\beta}}_{\mathrm{ols}} - \boldsymbol{\beta})^{\tau}(\mathcal{Z}^{\tau}\mathcal{Z})(\widehat{\boldsymbol{\beta}}_{\mathrm{ols}} - \boldsymbol{\beta}) \leq \widehat{\sigma}^2 F_{r+1,n-r-1}^{\alpha} . \qquad (5.46)$$

Geometrically, the confidence region (5.46) is an $(r + 1)$-dimensional ellipsoid with center $\boldsymbol{\beta}$ and orientation controlled by the matrix $\mathcal{Z}^{\tau}\mathcal{Z}$.

5.2.3 Example: Bodyfat Data

These data were used to produce predictive equations for lean body weight, a measure of health.[1] Measurements were made on $n = 252$ men in order to relate the percentage of bodyfat determined by underwater weighing (bodyfat), which is inconvenient and costly to obtain, to a number of body circumference measurements, recorded using only a scale and measuring tape.

The $r = 13$ input variables are age in years (age), weight in lb (weight), height in inches (height), neck circumference in cm (neck), chest circumference in cm (chest), abdomen 2 circumference in cm (abdomen), hip circumference in cm (hip), thigh circumference in cm (thigh), knee circumference in cm (knee), ankle circumference in cm (ankle), extended biceps circumference in cm (biceps), forearm circumference in cm (forearm), and wrist circumference in cm (wrist).

The pairwise correlations of the input variables are given in Table 5.2. We see 13 correlations greater than 0.8 and two greater than 0.9. One observation (#39) appears to be an outlier in all variables except age, height, forearm, and wrist. Using these 13 body measurements, we wish to derive accurate predictive measurements of bodyfat.

To study the relationship between bodyfat and the 13 input variables, we formulate the regression equation as follows:

$$\begin{aligned}
\mathtt{bodyfat} \; = \; & \beta_0 + \beta_1(\mathtt{age}) + \beta_2(\mathtt{weight}) + \beta_3(\mathtt{height}) + \beta_4(\mathtt{neck}) \\
& + \beta_5(\mathtt{chest}) + \beta_6(\mathtt{abdomen}) + \beta_7(\mathtt{hip}) + \beta_8(\mathtt{thigh}) \\
& + \beta_9(\mathtt{knee}) + \beta_{10}(\mathtt{ankle}) + \beta_{11}(\mathtt{biceps}) \\
& + \beta_{12}(\mathtt{forearm}) + \beta_{13}(\mathtt{wrist}) + e, \qquad (5.47)
\end{aligned}$$

where e is a random variable with mean zero and constant variance σ^2. The results of the multiple regression are given in Table 5.3 and summarized in Figure 5.2 by the ordered absolute values of the t-ratios of the 13 estimated

---

[1]The data and literature references can be downloaded from the StatLib–Datasets Archive, lib.stat.cmu.edu/datasets/, under the filename bodyfat.

TABLE 5.2. *Correlations between all pairs of input variables for the body-fat data. For these data, $r = 13$, $n = 252$.*

	age	weight	height	neck	chest	abdomen
weight	−0.013					
height	−0.245	0.487				
neck	0.114	0.831	0.321			
chest	0.176	0.894	0.227	0.785		
abdomen	0.230	0.888	0.190	0.754	0.916	
hip	−0.050	0.941	0.372	0.735	0.829	0.874
thigh	−0.200	0.869	0.339	0.696	0.730	0.767
knee	0.018	0.853	0.501	0.672	0.719	0.737
ankle	−0.105	0.614	0.393	0.478	0.483	0.453
biceps	−0.041	0.800	0.319	0.731	0.728	0.685
forearm	−0.085	0.630	0.322	0.624	0.580	0.503
wrist	0.214	0.730	0.398	0.745	0.660	0.620
	hip	thigh	knee	ankle	biceps	forearm
thigh	0.896					
knee	0.823	0.799				
ankle	0.558	0.540	0.612			
biceps	0.739	0.761	0.679	0.485		
forearm	0.545	0.567	0.556	0.419	0.678	
wrist	0.630	0.559	0.665	0.566	0.632	0.586

regression coefficients. We see a few large values in the residual analysis: 12 standardized residuals have absolute values greater than 2.0, and two of them (observations 39 and 224) have absolute values greater than 2.6. We estimate the error variance σ^2 by the residual variance, $\widehat{\sigma}^2 = 18.572$ on 238 degrees of freedom. If the errors are Gaussian distributed (an assumption that is supported by the residual analysis), the t statistics for abdomen, wrist, forearm, neck, and age are significant.

5.3 Prediction Accuracy and Model Assessment

Prediction is the art of making accurate guesses about new response values that are independent of the current data. Good predictive ability is often recognized as the most useful way of assessing the fit of a model to data. Thus, the two aims of prediction and model assessment (or validation) are closely related to each other.

For prediction in regression, we use the learning data,

$$\mathcal{L} = \{(\mathbf{X}_i, Y_i), i = 1, 2, \ldots, n\}, \tag{5.48}$$

to regress Y on $\mathbf{X}$, and then predict a new Y-value, Y^{new}, by applying the fitted model to a brand-new $\mathbf{X}$-value, $\mathbf{X}^{\text{new}}$, from the test set $\mathcal{T}$. The resulting prediction is compared with the actual response value. The predictive ability of the regression model is assessed by its *prediction* (or *generalization*) *error*, an overall measure of the quality of the prediction, usually taken to be mean squared error. The definition of prediction error depends upon whether we consider $\mathbf{X}$ as fixed or as random.

TABLE 5.3. *OLS estimation of coefficients for the regression model using the bodyfat data with* $r = 13$, $n = 252$. *The multiple* R^2 *is 0.749, the residual sum of squares is 4420.1, and the F-statistic is 54.5 on 13 and 238 degrees of freedom. A multiple regression using only those variables having* $|t| > 2$ *(i.e.,* abdomen, wrist, forearm, neck, *and* age*) has residual sum of squares 4724.9,* $R^2 = 0.731$, *and an F-statistic of 133.85 on 5 and 246 degrees of freedom.*

Coefficient	Estimate	Std.Error	t-value
(Intercept)	-21.3532	22.1862	-0.9625
age	0.0646	0.0322	2.0058
weight	-0.0964	0.0618	-1.5584
height	-0.0439	0.1787	-0.2459
neck	-0.4755	0.2356	-2.0184
chest	-0.0172	0.1032	-0.1665
abdomen	0.9550	0.0902	10.5917
hip	-0.1886	0.1448	-1.3025
thigh	0.2483	0.1462	1.6991
knee	0.0139	0.2477	0.0563
ankle	0.1779	0.2226	0.7991
biceps	0.1823	0.1725	1.0568
forearm	0.4557	0.1993	2.2867
wrist	-1.6545	0.5332	-3.1032

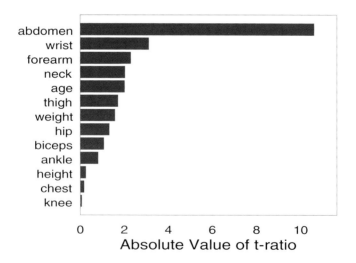

FIGURE 5.2. *Multiple regression results for the bodyfat data. The variable names are given on the vertical axis (listed in descending order of their absolute t-ratios) and the absolute value of the t-ratio for each variable on the horizontal axis.*

5.3.1 Random-X Case

In the random-X case, the learning data $\mathcal{L}$ are iid observations from the joint distribution of $(\mathbf{X}, Y)$. The observed responses Y_i, $i = 1, 2, \ldots, n$, are assumed to have been generated by the regression model,

$$Y = \beta_0 + \mathbf{X}^\tau \boldsymbol{\beta} + e = \mu(\mathbf{X}) + e, \tag{5.49}$$

where $\mu(\mathbf{X}) = \mathrm{E}(Y|\mathbf{X}) = \beta_0 + \mathbf{X}^\tau \boldsymbol{\beta}$, $\mathrm{E}(e|\mathbf{X}) = 0$, and $\mathrm{var}(e|\mathbf{X}) = \sigma^2$. From $\mathcal{T}$, we draw a new observation, $(\mathbf{X}^{\mathrm{new}}, Y^{\mathrm{new}})$, where we assume Y^{new} is unknown, from the same distribution as $(\mathbf{X}, Y)$, but independent of the learning set $\mathcal{L}$. We assess the fitted model by predicting Y^{new} from $\mathbf{X}^{\mathrm{new}}$.

If the estimated OLS regression function at $\mathbf{X}$ is

$$\widehat{\mu}(\mathbf{X}) = \widehat{\beta}_0 + \mathbf{X}^\tau \widehat{\boldsymbol{\beta}}_{\mathrm{ols}}, \tag{5.50}$$

then the predicted value of Y at $\mathbf{X}^{\mathrm{new}}$ is given by $\widehat{\mu}(\mathbf{X}^{\mathrm{new}})$. The prediction error (PE_R) in this case is defined as the mean squared error in predicting Y^{new} using $\widehat{\mu}(\mathbf{X}^{\mathrm{new}})$,

$$PE_R = \mathrm{E}\left\{Y^{\mathrm{new}} - \widehat{\mu}(\mathbf{X}^{\mathrm{new}})\right\}^2 = \sigma^2 + ME_R, \tag{5.51}$$

where the expectation is taken over $(\mathbf{X}^{\mathrm{new}}, Y^{\mathrm{new}})$, and

$$
\begin{aligned}
ME_R &= \mathrm{E}\{\mu(\mathbf{X}^{\mathrm{new}}) - \widehat{\mu}(\mathbf{X}^{\mathrm{new}})\}^2 & (5.52) \\
&= (\boldsymbol{\beta} - \widehat{\boldsymbol{\beta}}_{\mathrm{ols}})^\tau \boldsymbol{\Sigma}_{XX} (\boldsymbol{\beta} - \widehat{\boldsymbol{\beta}}_{\mathrm{ols}}), & (5.53)
\end{aligned}
$$

is the *model error* (i.e., the mean squared error of $\widehat{\mu}(\mathbf{x}^{\mathrm{new}})$ as a predictor of $\mu(\mathbf{X}^{\mathrm{new}})$, a quantity also called the "expected bias-squared"), and $\boldsymbol{\Sigma}_{XX}$ is the covariance matrix of $\mathbf{X}$.

5.3.2 Fixed-X Case

In the fixed-X case, the r-vectors $\{\mathbf{X}_i\}$, whose transposes are the rows of the design matrix $\mathcal{X}$, are fixed by the experimental conditions, so that only Y is random. We assume that the true model generating the observations $\{y_i\}$ on Y is

$$Y_i = \beta_0 + \mathbf{X}_i^\tau \boldsymbol{\beta} + e_i = \mu(\mathbf{X}_i) + e_i, \tag{5.54}$$

where $\mu(\mathbf{X}_i) = \beta_0 + \mathbf{X}_i^\tau \boldsymbol{\beta}$ is the regression function evaluated at $\mathbf{X}_i$, and the errors $e_i, i = 1, 2, \ldots, n$, are iid with mean 0 and variance σ^2 and are uncorrelated with the $\{\mathbf{X}_i\}$. We assume that the test data in $\mathcal{T}$ are generated by using "future-fixed" $\{\mathbf{X}^{\mathrm{new}}\}$ points (Breiman, 1992), which may either be the same fixed design points $\{\mathbf{X}_i\}$ as in the learning data $\mathcal{L}$ or they may be future values of $\mathbf{X}$ that are considered by the experimenter to be known and fixed (i.e., new design points). For convenience in this discussion, we assume the former situation holds. Thus, we assume that $\mathcal{T} = \{(\mathbf{X}_i, Y_i^{\mathrm{new}}), i = 1, 2, \ldots, m\}$, where

$$Y_i^{\mathrm{new}} = \mu(\mathbf{X}_i) + e_i^{\mathrm{new}}, \tag{5.55}$$

and the $\{e_i^{\mathrm{new}}\}$ are independent of the $\{e_i\}$ but have the same distribution. We further assume that the $\mathcal{X}^\tau \mathcal{X}$ matrix for the $\{\mathbf{X}_i\}$ is known.

The *predicted value* of Y^{new} at a future-fixed $\mathbf{X}$ is given by

$$\widehat{\mu}(\mathbf{X}) = \widehat{\beta}_0 + \mathbf{X}^\tau \widehat{\beta}_{\mathrm{ols}}, \tag{5.56}$$

where $\widehat{\beta}_{\mathrm{ols}}$ is the OLS estimate of the regression coefficients. The *prediction error* in the fixed-X case is defined as

$$PE_F = \mathrm{E}\left(m^{-1} \sum_{i=1}^{m} (Y_i^{\mathrm{new}} - \widehat{\mu}(\mathbf{X}_i))^2\right) = \sigma^2 + ME_F, \tag{5.57}$$

where the expectation is taken only over the $\{Y_i^{\mathrm{new}}\}$, and

$$ME_F = m^{-1} \sum_{i=1}^{n} (\mu(\mathbf{X}_i) - \widehat{\mu}(\mathbf{X}_i))^2 \tag{5.58}$$

$$= (\beta - \widehat{\beta}_{\mathrm{ols}})^\tau (m^{-1} \mathcal{X}^\tau \mathcal{X})(\beta - \widehat{\beta}_{\mathrm{ols}}) \tag{5.59}$$

is the *model error* due to the lack of fit to the true model. Compare (5.65) with (5.59).

5.4   Estimating Prediction Error

In the random-X case, when the entire data set $\mathcal{D}$ is large enough, we can use the partition into learning, validation, and test sets to do a thorough job of estimating the regression function, predicting future outcomes, and validating the model. However, in cases where such a division may not be practical, we have to use alternative methods.

5.4.1   Apparent Error Rate

As before, let $\widehat{\mu}(\mathbf{X}^{\mathrm{new}})$ be the predicted value of Y at $\mathbf{X} = \mathbf{X}^{\mathrm{new}}$, and let $L(Y, \mu(\mathbf{X})) = (Y - \mu(\mathbf{X}))^2$ be the loss incurred by predicting Y by $\mu(\mathbf{X})$. The prediction error PE for $\widehat{\mu}(\mathbf{X}^{\mathrm{new}})$ is given by (5.57). We can estimate PE by

$$\widehat{PE}(\widehat{\mu}, \mathcal{D}) = \frac{1}{n} \sum_{i=1}^{n} (Y_i - \widehat{\mu}(\mathbf{X}_i))^2 = \frac{RSS}{n}, \tag{5.60}$$

which we call the *apparent error rate* (or *resubstitution error rate*) *for* $\mathcal{D}$. This estimate of PE is computed by fitting the OLS regression function to the idiosyncrasies of the original sample $\mathcal{D}$ and then applying that function to see how well it predicts those same members of $\mathcal{D}$. The apparent error rate is a misleadingly optimistic value because it estimates the predictive ability of the fitted model from the same data that was used to fit that model. Consequently, we expect that RSS/n will be too optimistic an estimate of PE with $\widehat{PE}(\hat{\mu}, \mathcal{D}) < PE$.

Rather than using the apparent error rate for estimating prediction error, we use resampling methods (cross-validation and the bootstrap). Which resampling methodology we use depends upon whether the fixed-X or the random-X model is more appropriate. For the random-X case, we can use cross-validation or the "unconditional bootstrap," and in the fixed-X case, we can use the "conditional bootstrap." Cross-validation is not appropriate for estimating prediction error in the fixed-X case.

5.4.2 Cross-Validation

Among the methods available for estimating prediction error (and model error) for the random-X case, the most popular is *cross-validation* (Stone, 1974), of which there are several versions.

Suppose $\mathcal{D}$ is a random sample drawn from the joint probability distribution of $(\mathbf{X}, Y)$ in $(r + 1)$-dimensional space. If $n = 2m$, we can randomly split $\mathcal{D}$ into two equal subsets, treating one subset as the *learning set* $\mathcal{L}$ and the other as the *test set* $\mathcal{T}$, where $\mathcal{D} = \mathcal{L} \cup \mathcal{T}$ and $\mathcal{L} \cap \mathcal{T} = \emptyset$. Let $\mathcal{T} = \{(\mathbf{X}'_i, Y'_i), \ i = 1, 2, \ldots, m\}$. An estimate of PE_R obtained from the test set is

$$\widehat{PE} = \frac{1}{m} \sum_{i=1}^{m} (Y'_i - \hat{\mu}(\mathbf{X}'_i))^2, \tag{5.61}$$

where $\hat{\mu}(\mathbf{X}'_i) = \hat{\beta}_0 + \mathbf{X}'^{\tau}_i \hat{\boldsymbol{\beta}}_{\text{ols}}$. The learning set and the test set are then switched and the resulting two estimates of PE_R are averaged to yield a final estimate.

To generalize the above precedure, assume that $n = Vm$, where $V \geq 2$ is a small integer, such as 5 or 10. We split the data set $\mathcal{D}$ randomly into V disjoint subsets $\mathcal{T}_v, v = 1, 2, \ldots, V$, of equal size, where $\mathcal{D} = \bigcup_{v=1}^{V} \mathcal{T}_v, \mathcal{T}_v \cap \mathcal{T}_{v'} = \emptyset, v \neq v'$. We next create V different versions of the data set, each version of which has a learning set consisting of $V - 1$ of the subsets (i.e., $(V-1)m$ observations) and a test set of the one remaining subset (of m observations). In other words, we drop the $\mathcal{T}_v$ cases and consider the remaining learning set of $\mathcal{L}_v = \mathcal{D} - \mathcal{T}_v$ cases. Using only the $\mathcal{L}_v$ cases, we obtain the OLS regression function $\hat{\mu}_{-v}(\mathbf{X})$. We then evaluate this regression function at the $\mathcal{T}_v$ test-set cases, yielding the values $\hat{\mu}_{-v}(\mathbf{X}_i), \mathbf{X}_i \in \mathcal{T}_v$. We compute the prediction error from the vth test set $\mathcal{T}_v$, repeating the procedure V

times, while cycling through each of the test sets, $\mathcal{T}_1, \mathcal{T}_2, \ldots, \mathcal{T}_V$. This procedure is called *V-fold cross-validation* (*CV/V*). Combining these results gives us a CV/V-estimate of *PE*,

$$\widehat{PE}_{\text{CV/V}} = \frac{1}{V} \sum_{v=1}^{V} \sum_{(\mathbf{X}_i, Y_i) \in \mathcal{T}_v} (Y_i - \widehat{\mu}_{-v}(\mathbf{X}_i))^2. \tag{5.62}$$

Then, subtract $\widehat{\sigma}^2$ from $\widehat{PE}$ to get $\widehat{ME}$, where $\widehat{\sigma}^2$ is the residual variance obtained from the full data set.

The most computationally intensive version of cross-validation occurs when $m = 1$ (so that $V = n$). In this case, each learning set $\mathcal{L}_v$ has size $n-1$, and the test set $\mathcal{T}_v$ has size one. At the ith stage, the ith case $(\mathbf{x}_i, y_i)$ is omitted from the ith learning set, and the OLS regression function $\widehat{\mu}_{-i}(\mathbf{x})$ is computed from that learning set and evaluated at $\mathbf{x}_i$. This type of balanced split is referred to as the *leave-one-out rule* (*CV/n* or LOO). The prediction error is then estimated by

$$\widehat{PE}_{\text{CV/n}} = \frac{1}{n} \sum_{i=1}^{n} (Y_i - \widehat{\mu}_{-i}(\mathbf{X}_i))^2. \tag{5.63}$$

As before, we obtain $\widehat{ME}$ by subtracting $\widehat{\sigma}^2$ from $\widehat{PE}$.

As well as issues of computational complexity, the difference between taking $V = 5$ or 10 and taking $V = n$ is one of "bias *versus* variance." The leave-one-out rule yields an estimate of PE_R that has low bias but high variance (arising from the high degree of similarity between the leave-one-out learning sets), whereas the 5–fold or 10–fold rule yields an estimate of PE_R with higher bias but lower mean squared error (and also lower variance). Furthermore, 10–fold (and even 5-fold) cross-validation appears to be better at model assessment than is leave-one-out cross-validation.

5.4.3 Bootstrap

For estimating prediction error in regression models, we can also use the bootstrap technique (Efron, 1979). In general, the specific version of the bootstrap to be applied has to depend upon what we actually assume about the stochastic model that may have generated the data. In regression models, it again boils down to whether we are in the random-X case (using the "unconditional" bootstrap) or the fixed-X case ("conditional" bootstrap).

Unconditional Bootstrap

The unconditional bootstrap is used for the random-X case. We first sample n times *with replacement* from the original sample, $\mathcal{D}$, to get a *random-X bootstrap sample*, which we denote by

$$\mathcal{D}_R^{*b} = \{(\mathbf{X}_i^{*b}, Y_i^{*b}), i = 1, 2, \ldots, n\}. \tag{5.64}$$

Next, we regress Y_i^{*b} on $\mathbf{X}_i^{*b}$, $i = 1, 2, \ldots, n$, and obtain an OLS regression function $\widehat{\mu}_R^{*b}(\mathbf{X})$. If we then apply $\widehat{\mu}_R^{*b}$ to the *original sample*, $\mathcal{D}$, the resulting estimate of PE is given by

$$\widehat{PE}(\widehat{\mu}_R^{*b}, \mathcal{D}) = \frac{1}{n} \sum_{i=1}^{n} (Y_i - \widehat{\mu}_R^{*b}(\mathbf{X}_i))^2. \tag{5.65}$$

Averaging $\widehat{PE}(\widehat{\mu}_R^{*b}, \mathcal{D})$ over all B bootstrap samples yields the *simple bootstrap estimator* of PE,

$$\widehat{PE}_R(\mathcal{D}) = \frac{1}{B} \sum_{b=1}^{B} \widehat{PE}(\widehat{\mu}_R^{*b}, \mathcal{D}) = \frac{1}{Bn} \sum_{b=1}^{B} \sum_{i=1}^{n} (Y_i - \widehat{\mu}_R^{*b}(\mathbf{X}_i))^2, \tag{5.66}$$

which is not a particularly good estimate of PE because there are observations common to the bootstrap samples $\{\mathcal{D}_R^{*b}\}$ (that determined $\{\widehat{\mu}_R^{*b}\}$) and the original sample $\mathcal{D}$, and so an estimate of PE such as (5.66) will also be overly optimistic.

As another estimator of PE, an *apparent error rate for $\mathcal{D}_R^{*b}$* is computed by applying $\widehat{\mu}_R^{*b}$ to $\mathcal{D}_R^{*b}$:

$$\widehat{PE}(\widehat{\mu}_R^{*b}, \mathcal{D}_R^{*b}) = \frac{1}{n} \sum_{i=1}^{n} (Y_i^{*b} - \widehat{\mu}_R^{*b}(\mathbf{X}_i^{*b}))^2. \tag{5.67}$$

Averaging (5.67) over all B bootstrap samples yields

$$\widehat{PE}(\mathcal{D}_R^*) = \frac{1}{B} \sum_{b=1}^{B} \widehat{PE}(\widehat{\mu}_R^{*b}, \mathcal{D}_R^{*b}) = \frac{1}{Bn} \sum_{b=1}^{B} \sum_{i=1}^{n} (Y_i^{*b} - \widehat{\mu}_R^{*b}(\mathbf{X}_i^{*b}))^2. \tag{5.68}$$

This estimate of PE has the same disadvantages as the apparent error rate for $\mathcal{D}$.

We can improve on these estimates of PE by estimating the bias in using RSS/n (the apparent error rate for $\mathcal{D}$) as an estimate of PE and then correcting RSS/n by subtracting its estimated bias. An estimate of that bias for $\mathcal{D}_R^{*b}$ is the bth *optimism*,

$$\widehat{\text{opt}}_R^b = \widehat{PE}(\widehat{\mu}_R^{*b}, \mathcal{D}) - \widehat{PE}(\widehat{\mu}_R^{*b}, \mathcal{D}_R^{*b}). \tag{5.69}$$

Averaging $\widehat{\text{opt}}_R^b$ over all B bootstrap samples yields an overall estimate,

$$\widehat{\text{opt}}_R = \frac{1}{B} \sum_{b=1}^{B} \widehat{\text{opt}}_R^b = \widehat{PE}(\mathcal{D}) - \widehat{PE}(\mathcal{D}_R^*), \tag{5.70}$$

of the *average optimism*, opt $= E\{PE(\widehat{\mu}, \mathcal{D}) - \widehat{PE}(\widehat{\mu}, \mathcal{D})\}$, which is generally positive. The bootstrap estimator of PE is given by the sum of the apparent error rate for $\mathcal{D}$ and the bias in that apparent error,

$$\widehat{PE}_R = \frac{RSS}{n} + \widehat{\text{opt}}_R , \tag{5.71}$$

and ME is estimated by $\widehat{ME}_R = \widehat{PE}_R - \widehat{\sigma}^2$. In simulations, $\widehat{PE}_R$ (which is computationally more expensive than cross-validation) appears to have low bias and is slightly better for model assessment than is 10-fold cross-validation.

Recall that $\widehat{PE}_R(\mathcal{D})$ in (5.66) underestimates PE_R because there are observations common to the bootstrap samples $\{\mathcal{D}_R^{*b}\}$ (operating as learning sets) and to the original data set $\mathcal{D}$ (operating as the test set). In fact, the chance that the ith observation $(\mathbf{X}_i, Y_i)$ from $\mathcal{D}$ is selected at least once to be in the bth bootstrap sample $\mathcal{D}_R^{*b}$ is

$$\text{Prob}((\mathbf{X}_i, Y_i) \in \mathcal{D}_R^{*b}) = 1 - \left(1 - \frac{1}{n}\right)^n$$
$$\rightarrow 1 - e^{-1} \approx 0.632, \tag{5.72}$$

as $n \rightarrow \infty$. Thus, on average, about 37% of the observations in $\mathcal{D}$ are left out of each bootstrap sample, which contains about $0.632n$ distinct observations. One unfortunate consequence of this result is that if n is close to r, this will lead to numerical difficulties in computing $\widehat{\mu}_R^{*b}$, because in such cases it is likely that $\mathcal{X}^\tau \mathcal{X}$ will be singular or nearly singular when computed from a bootstrap sample.

We now use (5.72) to improve upon $\widehat{\text{opt}}_R$ (and also $\widehat{PE}_R$) by including in the computation the prediction errors for the ith observation $(\mathbf{X}_i, Y_i)$ only from those bootstrap samples that do *not* contain that observation, $i = 1, 2, \ldots, n$.

Let $PE_R^{(1)}$ be the expected bootstrap prediction error at those points $(\mathbf{X}_i, Y_i) \in \mathcal{D}$ that are *not* included in the B bootstrap samples. We estimate $PE_R^{(1)}$ as follows. Define n_{ib} to be the number of times that the ith observation $(\mathbf{X}_i, Y_i)$ appears in the bth bootstrap sample, and set $I_{ib} = 1$ if $n_{ib} = 0$ and zero otherwise. Then, we estimate $PE_R^{(1)}$ by

$$\widehat{PE}_R^{(1)} = \frac{1}{n} \sum_{i=1}^n \widehat{PE}_i, \tag{5.73}$$

where

$$\widehat{PE}_i = \frac{\sum_b I_{ib}(Y_i - \widehat{\mu}_b(\mathbf{X}_i))^2}{\sum_b I_{ib}} . \tag{5.74}$$

Efron and Tibshirani (1997) called $\widehat{PE}_R^{(1)}$ the *leave-one-out bootstrap estimator* because of its similarity to the leave-one-out cross-validation estimator. Another way of writing (5.74) is

$$\widehat{PE}_i = \frac{1}{B_i} \sum_{b \in C_i} (Y_i - \hat{\mu}_b(\mathbf{X}_i))^2, \tag{5.75}$$

where C_i is the set of indices of the bootstrap samples that do not contain $(\mathbf{X}_i, Y_i)$, and $B_i = |C_i|$ is the number of such bootstrap samples. These observations are often referred to as *out-of-bootstrap (OOB)* observations. Efron (1983) showed that $\widehat{PE}_R^{(1)}$ is biased upwards compared to $\widehat{PE}_{CV/n}$, which is nearly unbiased.

Based upon (5.72), the 0.632 *bootstrap estimator of optimism* is given by

$$\widehat{opt}_R^{(0.632)} = 0.632(\widehat{PE}_R^{(1)} - \widehat{PE}(\hat{\mu}, \mathcal{D})). \tag{5.76}$$

Replacing $\widehat{opt}_R$ in (5.71) by $\widehat{opt}_R^{(0.632)}$ in (5.76) yields the 0.632 *bootstrap estimator of prediction error*,

$$\begin{aligned}\widehat{PE}_R^{(0.632)} &= \widehat{PE}(\hat{\mu}, \mathcal{D}) + \widehat{opt}_R^{(0.632)} \\ &= 0.368 \cdot \frac{RSS}{n} + 0.632 \cdot \widehat{PE}_R^{(1)}. \end{aligned} \tag{5.77}$$

Although the 0.632 bootstrap estimator is an improvement over the apparent error rate, it still underestimates PE_R (Efron, 1983).

Example: Bodyfat Data (Continued)

Cross-validation and the unconditional bootstrap were used to estimate the prediction error for the bodyfat data. The results are summarized in Tables 5.4 and 5.5.

From Table 5.4, we see that the estimates obtained from $CV/5$, $CV/10$, CV/n, and the bootstrap (with $B = 500$) are reasonably close to each other. The apparent error rate, $RSS/n = 4420.064/252 = 17.5399$, underestimates the leave-one-out cross-validation estimate of the prediction error by more than 12%. Dividing RSS by its degrees of freedom to give an unbiased estimate of σ^2 yields $RSS/238 = 18.5717$, still well below the other estimates.

$B{=}10$ For a simple bootstrap illustration, let $B = 10$. The bootstrap computations are detailed in Table 5.5. The simple bootstrap estimate, $\widehat{PE}_R(\mathcal{D}) = 18.4692$, is the average of the first column and is much too small. The average of the third column, $\widehat{opt}_R = 18.4692 - 15.9535 = 2.5157$, is the difference between the average of the first column and the average of the second column and yields a measure of how optimistic the apparent error

TABLE 5.4. *Estimated prediction errors for the bodyfat data when the multiple regression model is fit. Listed are the apparent error rate (RSS/n) and the error rates from using 5-fold (CV/5), 10-fold (CV/10), leave-one-out cross-validation (CV/n), and the unconditional bootstrap and 0.632 bootstrap using $B = 500$. The subscript "R" indicates that the bootstrap computations are made for the random-X case. These results show the very optimistic value of the apparent error rate.*

RSS/n	$\widehat{PE}_{CV/5}$	$\widehat{PE}_{CV/10}$	$\widehat{PE}_{CV/n}$	$\widehat{PE}_R$	$\widehat{PE}_R^{(0.632)}$
17.5399	20.2578	20.7327	20.2948	19.6891	19.9637

rate is in estimating the prediction error. Finally, $\widehat{PE}_R = RSS/n + \widehat{\text{opt}}_R = 17.5399 + 2.5815 = 20.1214$.

B=500 When we use $B = 500$ bootstrap samples, we obtain $\widehat{PE}_R(\mathcal{D}) = 18.7683$ and $\widehat{PE}(\mathcal{D}_R^*) = 16.6191$, so that $\widehat{\text{opt}}_R = 18.7683 - 16.6191 = 2.1492$, whence, $\widehat{PE}_R = 17.5399 + 2.1492 = 19.6891$. We see a small difference between the bootstrap estimates of PE using $B = 10$ and $B = 500$ bootstrap samples.

Conditional Bootstrap

The conditional bootstrap for the fixed-X case operates by sampling *with replacement* from the *residuals* obtained from fitting the regression model to the non-stochastic inputs $\mathbf{X}_1, \mathbf{X}_2, \ldots, \mathbf{X}_n$ (Efron, 1979).

We first fit the model (5.21) and obtain the OLS regression coefficients $\widehat{\boldsymbol{\beta}}_{\text{ols}} = (\mathcal{Z}^\tau \mathcal{Z})^{-1} \mathcal{Z}^\tau \mathcal{Y}$, the estimated regression function $\widehat{\mu}(\mathbf{X}) = \mathbf{X}^\tau \widehat{\boldsymbol{\beta}}_{\text{ols}}$, the residuals $\widehat{e}_1, \widehat{e}_2, \ldots, \widehat{e}_n$, and the residual variance $\widehat{\sigma}^2$. When applying the conditional bootstrap, we assume that the errors of the model are iid and homoscedastic. For an extensive discussion of the effect of error variance heteroscedasticity on the conditional bootstrap, see Wu (1986).

Because $\text{E}(RSS/n) = (1 - p/n)\sigma^2$, where $p = r + 1$ is the number of parameters, RSS/n is biased downwards as an estimator of σ^2, and the residuals tend to be smaller than the errors of the model. Some statisticians advocate rescaling the residuals upwards by multiplying each of them by the factor $(n/(n-p))^{1/2}$; Efron and Tibshirani (1993, p. 112) feel that the scaling issue becomes important only when $p > n/4$.

Suppose we consider $\widehat{\boldsymbol{\beta}}_{\text{ols}}$ to be the true value of the regression parameter. For the bth bootstrap sample, we sample with replacement from the *residuals* to get the bootstrapped residuals, $\widehat{e}_1^{*b}, \widehat{e}_2^{*b}, \ldots, \widehat{e}_n^{*b}$, and then compute a new set of responses

$$Y_i^{*b} = \widehat{\mu}(\mathbf{X}_i) + \widehat{e}_i^{*b} , \quad i = 1, 2, \ldots, n. \tag{5.78}$$

TABLE 5.5. *Unconditional bootstrap estimates of prediction error for the bodyfat data, where $B = 10$ bootstrap samples are taken. Each row of the table represents a bootstrap sample b, and the multiple regression model is fit to that sample. For each b, the first column is the simple bootstrap estimate of prediction error, the second column is the bootstrap apparent error rate, and the third column is the difference between the first two columns. The average optimism, in this case 2.4806, is the difference between the average of the first column and the average of the second column.*

b	$\widehat{PE}(\widehat{\mu}_R^{*b}, \mathcal{D})$	$\widehat{PE}(\widehat{\mu}_R^{*b}, \mathcal{D}_R^{*b})$	$\widehat{\mathrm{opt}}_R^b$
1	18.5198	15.8261	2.6937
2	18.2555	13.5946	4.6609
3	17.9683	18.2385	-0.2702
4	18.9317	14.5406	4.3911
5	18.6249	15.7998	2.8251
6	18.0191	15.1146	2.9045
7	18.5381	17.7595	0.7786
8	18.9265	13.8298	5.0967
9	18.6881	18.8233	-0.1352
10	18.2201	16.0080	2.2121
ave	18.4692	15.9535	2.5157

The bth *fixed-X bootstrap sample* is now given by

$$\mathcal{D}_F^{*b} = \{(\mathbf{X}_i, Y_i^{*b}), \quad i = 1, 2, \ldots, n\}. \tag{5.79}$$

We regress Y^{*b} on $\mathbf{X}$ to get a bootstrapped estimator,

$$\widehat{\boldsymbol{\beta}}^{*b} = (\mathcal{Z}^\tau \mathcal{Z})^{-1} \mathcal{Z}^\tau \mathcal{Y}^{*b}, \tag{5.80}$$

of the regression coefficients, where $\mathcal{Y}^{*b} = (Y_1^{*b}, \ldots, Y_n^{*b})^\tau$. Under this bootstrap sampling scheme, $\sqrt{n}(\widehat{\boldsymbol{\beta}}^{*b} - \widehat{\boldsymbol{\beta}}_{\mathrm{ols}})$ is approximately distributed as $\sqrt{n}(\widehat{\boldsymbol{\beta}}_{\mathrm{ols}} - \boldsymbol{\beta})$ (Freedman, 1981). The bootstrap regression function is $\widehat{\mu}_F^{*b}(\mathbf{x}) = \widehat{\beta}_0 + \mathbf{x}^\tau \widehat{\boldsymbol{\beta}}^{*b}$. Straightforward analogues of the estimates for the fixed-X case, similar to those for the unconditional case, can now be computed.

5.5   Instability of LS Estimates

If $\mathcal{X}_c$ has less than full rank, then $\mathcal{X}_c^\tau \mathcal{X}_c$ will be singular, and the OLS estimate of $\boldsymbol{\beta}$ will not be unique. Singularity occurs when the matrix $\mathcal{X}_c$ is ill-conditioned, or the columns of $\mathcal{X}_c$ are collinear, or when there are more variables than observations (i.e., $r > n$). If the assumptions for the regression model do not hold (e.g., due to ill-conditioned data, collinearity, correlated errors), then we have to look for alternative solutions.

Data are *ill-conditioned* for a given problem whenever the quantities to be computed for that problem are sensitive to small changes in the data. When that is the case, computational results, especially those obtained using matrix inversion routines, are likely to be *numerically unstable*. As a result, major errors (due to rounding and cancellations) tend to accumulate and severely skew the calculations. In some regression situations, the matrix $\mathcal{X}$ (or its mean-centered version $\mathcal{X}_c$) may be *rank-deficient* or almost so because of too many highly correlated variables, which exhibit *collinearity*. Exact collinearity rarely occurs, but problems involving variables that are almost collinear ("near collinearity") are not unusual.

In linear regression models, ill-conditioning and collinearity problems coincide. Near collinearity in linear regression problems is of major concern to statisticians and econometricians, especially when an overly large number of input variables is included in the initial model (the so-called kitchen-sink approach to modeling). Among the effects of near collinearity are overly large (positive or negative) estimated coefficient values whose signs may be reversed if negligible changes are made to the data. The standard errors of the estimated regression coefficients may also be dramatically inflated, thereby masking the presence of what would otherwise be significant regression coefficients.

There are several measures of ill-conditioning of a square matrix $\mathbf{M}$, the most popular of which is the *condition number*, $\kappa(\mathbf{M})$; see Section 3.2.9. In regression, $\mathbf{M} = \mathcal{X}^\tau \mathcal{X}$. Each variable may be scaled to have equal length (e.g., replacing x_{ij} by x_{ij}/s_i, where s_i is the sample standard deviation of the ith variable). The condition number of $\mathcal{X}^\tau \mathcal{X}$ (or $\mathcal{X}$) reduces to the ratio of the largest to the smallest nonzero singular value, $\kappa = \sigma_1/\sigma_r$, of $\mathcal{X}$. If κ is large, $\mathcal{X}$ is said to be *ill-conditioned*. When exact collinearity occurs, $\kappa = \infty$.

As an alternative to κ, we can compute the set of *collinearity indices*,

$$\kappa_k(\mathcal{X}) = \sqrt{VIF_k}\ , \quad k = 1, 2, \ldots, r, \tag{5.81}$$

where

$$VIF_k = (1 - R_k^2)^{-1}, \tag{5.82}$$

is the kth *variance inflation factor*, and R_k^2 is the squared multiple correlation coefficient of the kth column of $\mathcal{X}$ on the other $r - 1$ columns of $\mathcal{X}$, $k = 1, 2, \ldots, r$. Large values of VIF_k (typically, $VIF_k > 10$) imply that R_k^2 is close to unity, which in turn suggests near collinearity may be present. The collinearity indices have value at least one and are invariant under scale changes of the columns of $\mathcal{X}$. For example, the bodyfat data has some very large VIF values: each of the variables `weight`, `chest`, `abdomen`, and `hip` has a VIF value in the range 10–50. The high VIF values for those particular four variables appear to reflect their high pairwise correlations.

5.6   Biased Regression Methods

Because the OLS estimates depend upon $(\mathcal{Z}^\tau \mathcal{Z})^{-1}$, we would experience numerical complications in computing $\widehat{\boldsymbol{\beta}}_{\mathrm{ols}}$ if $\mathcal{Z}^\tau \mathcal{Z}$ were singular or nearly singular. If $\mathcal{Z}$ is ill-conditioned, small changes to the elements of $\mathcal{Z}$ lead to large changes in $(\mathcal{Z}^\tau \mathcal{Z})^{-1}$, the estimator $\widehat{\boldsymbol{\beta}}_{\mathrm{ols}}$ becomes computationally unstable, and the individual component estimates may either have the wrong sign or be too large in magnitude. So, even though the regression model may be a good fit to the learning data, it will not generalize sufficiently well to the test data.

One way out of this situation is to abandon the requirement of an unbiased estimator of $\boldsymbol{\beta}$ and, instead, consider the possibility of using a biased estimator of $\boldsymbol{\beta}$. There are several such estimators that are superior (in terms of MSE) to $\widehat{\boldsymbol{\beta}}_{\mathrm{ols}}$ when $\mathcal{Z}$ is ill-conditioned or when $\mathcal{Z}^\tau \mathcal{Z}$ is singular (or nearly singular). Biased regression methods have primarily been used in chemometrics (e.g., food research, environmental pollution studies). In such applications, it is not unusual to see the number of input variables greatly exceed the number of observations, so that the OLS regression estimator does not exist.

We assume only that the Xs and the Y have been centered, so that we have no need for a constant term in the regression. Thus, $\mathcal{X}$ is an $(n \times r)$-matrix with centered columns and $\mathcal{Y}$ is a centered n-vector. Each of these biased estimators can be written in the form

$$\widehat{\boldsymbol{\beta}} = \sum_j f(\lambda_j)\lambda_j^{-1}\mathbf{v}_j\mathbf{v}_j^\tau \mathbf{s}, \qquad (5.83)$$

where $f(\lambda_j)$ is the jth "shrinkage" factor, $\mathbf{v}_j$ is the eigenvector associated with the jth largest eigenvalue λ_j of $\mathbf{S} = \mathcal{X}^\tau \mathcal{X}$, and $\mathbf{s} = \mathcal{X}^\tau \mathcal{Y}$. For a t-component PCR, the shrinkage factor is $f(\lambda_j) = 1$ if $j \leq t$, and 0 otherwise; for a t-component PLSR, $f(\lambda_j)$ is a polynomial of degree t; and for RR with ridge parameter $k > 0$, $f(\lambda_j) = f_k(\lambda_j) = \lambda_j/(\lambda_j + k)$.

5.6.1   Example: PET Yarns and NIR Spectra

These data[2] were obtained from a calibration study (Swierenga, de Weijer, van Wijk, and Buydens, 1999) of polyethylene terephthalate (PET) yarns, which are used for textile (e.g., clothing materials) and industrial purposes

---

[2]The datafile PET.txt can be downloaded from the book's website. It was originally provided by Erik Swierenga and is available as an R data set as part of *The* pls *Package*. see www.maths.lth.se/help/R/.R/library/pls/html/NIR.html.

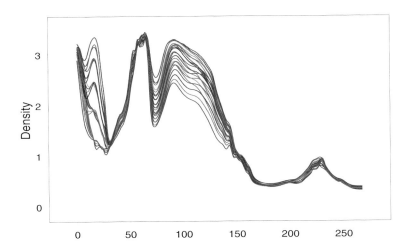

FIGURE 5.3. *Raman NIR spectra of a sample of 21 polyethylene tereph-thalate (PET) yarns. The 21 spectra are each measured at 268 frequencies. Note that the horizontal axis is variable number, not frequency.*

(e.g., tires, seat belts, and ropes). PET yarns are produced by a process of melt-spinning, whose settings largely determine the final semi-crystalline structure of the yarn (i.e., its physical structure), which, in turn, determines its thermo-mechanical properties. As a result, parameters that characterize the physical structure of PET yarns are important quality parameters for the end use of the yarn.

Raman near-infrared (NIR) spectroscopy has recently become an important tool in the pharmaceutical and semiconductor industries for investigating structural information on polymers; in particular, it is used to reveal information on the chemical nature, conformational order, state of the order, and orientation of polymers. Thus, Raman spectra are used to predict the physical structure parameters of polymers.

In this example, we study the relationship between the overall density of a PET yarn to its NIR spectrum. The data consist of a sample of $n = 21$ PET yarns having known mechanical and structural properties. For each PET yarn, the Y-variable is the density (measured in kg/m^3) of the yarn, and the $r = 268$ X-variables (measured at 268 frequencies in the range 598–1900 cm^{-1}) are selected from the NIR spectrum of that yarn. This example is quite representative of data sets in the chemometrics literature, in that $r \gg n$. The 21 NIR spectra are displayed graphically in Figure 5.3; the spectra appear to have very similar characteristics, although there are noticeable differences in some curves.

5.6.2 Principal Components Regression

An obvious way of dealing with a matrix $\mathcal{X}^\tau\mathcal{X}$ that is singular (or nearly singular) is to substitute a generalized inverse $\mathbf{G}$ in place of $(\mathcal{X}^\tau\mathcal{X})^{-1}$. Suppose $\mathcal{X}^\tau\mathcal{X}$ has known rank t ($1 \le t \le r$), so that the smallest $r - t$ eigenvalues of $\mathcal{X}^\tau\mathcal{X}$ are all zero. Then, the spectral decomposition of $\mathcal{X}^\tau\mathcal{X}$ can be written as $\mathcal{X}^\tau\mathcal{X} = \mathbf{V}\mathbf{\Lambda}\mathbf{V}^\tau$, where $\mathbf{\Lambda} = \mathrm{diag}\{\lambda_1,\ldots,\lambda_t\}$ is a diagonal matrix of the first t eigenvalues of $\mathcal{X}^\tau\mathcal{X}$ with diagonal elements ordered in magnitude from largest to smallest, and $\mathbf{V} = (\mathbf{v}_1,\ldots,\mathbf{v}_t)$ is an $(r \times t)$-matrix whose columns are the eigenvectors associated with the eigenvalues in $\mathbf{\Lambda}$. The unique rank-t Moore–Penrose inverse $\mathbf{G}$ of $\mathcal{X}^\tau\mathcal{X}$ is, therefore, given by

$$\mathbf{G} = (\mathcal{X}^\tau\mathcal{X})^+ = \mathbf{V}\mathbf{\Lambda}^{-1}\mathbf{V}^\tau = \sum_{j=1}^{t}\lambda_j^{-1}\mathbf{v}_j\mathbf{v}_j^\tau, \tag{5.84}$$

and the *generalized-inverse regression* (GIR) estimator is

$$\widehat{\boldsymbol{\beta}}_{\mathrm{gir}}^{(t)} = \mathbf{G}\mathcal{X}^\tau\mathcal{Y} = \sum_{j=1}^{t}\lambda_j^{-1}\mathbf{v}_j\mathbf{v}_j^\tau\mathbf{s}, \tag{5.85}$$

where $\mathbf{s} = \mathcal{X}^\tau\mathcal{Y}$. The GIR fitted values are then given by

$$\widehat{\mathcal{Y}}_{\mathrm{gir}}^{(t)} = \mathcal{X}\widehat{\boldsymbol{\beta}}_{\mathrm{gir}}^{(t)} = \mathcal{X}\mathbf{V}(\mathbf{\Lambda}^{-1}\mathbf{V}^\tau\mathbf{s}). \tag{5.86}$$

Marquardt (1970) showed that $\widehat{\boldsymbol{\beta}}_{\mathrm{gir}}$ minimizes the error sum of squares, $ESS(\boldsymbol{\beta})$, in (5.22) within the t-dimensional linear subspace spanned by $\mathbf{V}$. It follows that $\widehat{\boldsymbol{\beta}}_{\mathrm{gir}}$ is a constrained least-squares estimator of $\boldsymbol{\beta}$ and so is said to be *conditionally unbiased*. If $\mathcal{X}^\tau\mathcal{X}$ actually has a rank greater than t and we incorrectly use $\mathbf{G}$ in (5.85) to define the estimator of $\boldsymbol{\beta}$, then $\widehat{\boldsymbol{\beta}}_{\mathrm{gir}}^{(t)}$ is a biased estimator of $\boldsymbol{\beta}$.

The rows of the $(n \times t)$-matrix $\mathbf{Z}_t = \mathcal{X}\mathbf{V}$ are the scores of the first t *principal components* of $\mathcal{X}$ (see Chapter 7). Regressing $\mathcal{Y}$ on $\mathbf{Z}_t$ is a technique usually referred to as *principal components regression* (PCR) (Massy, 1965). This regression method is popularly used in chemometrics, where, for example, we may be interested in calibrating the fat concentration in n chemical samples to highly collinear absorbance measurements recorded at r fixed wavelength channels of an X-spectrum (Martens and Naes, 1989, sec. 3.4). In such situations, the number of variables r will likely be much greater than the number of observations n. PCR can be used to reduce the dimensionality of the regression by dropping those dimensions that contribute to the collinearity problem. PCR has also been used for mapping quantitative trait loci in statistical genetics, where Y repesents a quantitative trait value (e.g., blood pressure, yield) and $\mathbf{X}$ consists of the genotypes of a mouse or plant, etc., at each of r molecular markers (Hwang and Nettleton, 2003).

The estimated regression coefficients for the t principal components are given by the t-vector,

$$\widehat{\boldsymbol{\beta}}_{\mathrm{pcr}}^{(t)} = (\mathbf{Z}_t^\tau \mathbf{Z}_t)^{-1} \mathbf{Z}_t^\tau \mathcal{Y} = \mathbf{\Lambda}^{-1} \mathbf{V}^\tau \mathbf{s}, \tag{5.87}$$

where we have used $\mathbf{V}^\tau \mathbf{V} = \mathbf{I}_t$. Note that because of the orthogonality of the columns of $\mathbf{V}$, the elements of (5.87) do not change as t increases. Thus, (5.85) and (5.87) are related by $\widehat{\boldsymbol{\beta}}_{\mathrm{gir}}^{(t)} = \mathbf{V}\widehat{\boldsymbol{\beta}}_{\mathrm{pcr}}^{(t)}$, and the corresponding fitted values are given by

$$\widehat{\mathcal{Y}}_{\mathrm{pcr}}^{(t)} = \mathbf{Z}_t \widehat{\boldsymbol{\beta}}_{\mathrm{pcr}}^{(t)} = \mathcal{X}\mathbf{V}(\mathbf{\Lambda}^{-1}\mathbf{V}^\tau \mathbf{s}) = \mathcal{X}\widehat{\boldsymbol{\beta}}_{\mathrm{gir}}^{(t)} = \widehat{\mathcal{Y}}_{\mathrm{gir}}^{(t)}, \tag{5.88}$$

So, the fitted values obtained by GIR and PCR are identical.

It is usual to transform the PCR coefficients (5.87) into coefficients of the original input variables. Given $\widehat{\boldsymbol{\beta}}_{\mathrm{pcr}}^{(t)} = (\widehat{\beta}_{\mathrm{pcr},1}, \cdots, \widehat{\beta}_{\mathrm{pcr},t})^\tau$, we compute the r-vectors,

$$\widehat{\boldsymbol{\beta}}_{\mathrm{pcr},j}^* = \widehat{\beta}_{\mathrm{pcr},j} \mathbf{v}_j, \quad j = 1, 2, \ldots, t. \tag{5.89}$$

Then, the first k partial sums of the $\{\widehat{\boldsymbol{\beta}}_{\mathrm{pcr},j}^*\}$ give the k-component PCR coefficients of the original input variables; that is,

$$\widehat{\boldsymbol{\beta}}_{\mathrm{pcr}}^{*(k)} = \sum_{j=1}^{k} \widehat{\boldsymbol{\beta}}_{\mathrm{pcr},j}^* = \mathbf{V}\widehat{\boldsymbol{\beta}}_{\mathrm{pcr}}^{(k)}, \quad 1 \le k \le t. \tag{5.90}$$

Note that $\widehat{\boldsymbol{\beta}}_{\mathrm{pcr}}^{*(t)} = \widehat{\boldsymbol{\beta}}_{\mathrm{ols}}$.

In practice, the rank of $\mathcal{X}^\tau \mathcal{X}$ and, hence, the number of components is an unknown *metaparameter* to be determined from the data. If we extract principal components from the correlation matrix, Kaiser's rule (Kaiser, 1960) suggests we retain only those principal components whose eigenvalues are greater than one. Another way of determining t is by cross-validation (Wold, 1978).

A caveat: Although PCR aims to relate Y and the $\{X_j\}$ in the presence of severe collinearity, there is also the potential for PCR to fail dramatically. The principal components, $Z_1, \ldots, Z_t$ $(1 \le t < r)$, which are used as inputs to a multiple regression, are chosen to correspond to the t highest-variance directions of $\mathbf{X} = (X_1, \cdots, X_r)^\tau$ while dropping the remaining $r - t$ (low-variance) directions. Because the extraction of the principal components is accomplished without any reference to the output variable Y, we have no reason to expect Y to be highly correlated with any of the principal components, in particular those having the largest eigenvalues. Indeed, Y may actually have its highest correlation with one of the last few principal components (Jolliffe, 1982) or even only the last one (Hadi and Ling, 1998) which is always dropped from the regression equation.

Example: The PET Yarn Data (Continued)

Each variable (Y and all the Xs) from the PET yarn data has been centered. The (21×268)-matrix $\mathcal{X}$ yields at most $t = \min\{20, 268\} = 20$ principal components. The 20 nonzero eigenvalues from the correlation matrix in descending order of magnitude are

11.86	8.83	6.75	1.61	0.76	0.54	0.40	0.25	0.24	0.19
0.14	0.11	0.08	0.07	0.06	0.05	0.05	0.04	0.03	0.02

There are four eigenvalues larger than one. The first component accounts for 52.5% of total variance, the first two components account for 81.6% of total variance, the first three components account for 98.6% of total variance, and the first four components account for 99.5% of total variance.

Figure 5.4 displays the PCR coefficients for $t = 1, 3, 4, 20$ components. This figure shows that a single component yields regression estimates with almost no structure. By three components, the final structure is certainly visible, and the graph appears to settle down when we use four components. After four components, all that is added to the graph of the coefficient estimates is noise, which reinforces the information gained from the eigenvalues.

5.6.3 Partial Least-Squares Regression

In *partial least-squares regression (PLSR)*, the derived variables (usually referred to as *latent variables*, *components*, or *factors*) are specifically constructed to retain most of the information in the X variables that helps predict Y, while at the same time reducing the dimensionality of the regression. Whereas PCR constructs its latent variables using only data on the input variables, PLSR uses data on both the input and output variables. Chemometricians have adopted the name *PLSR1* to refer to PLSR using a single output variable and *PLSR2* to refer to PLSR using multiple output variables.

PLSR is typically obtained using an algorithm rather than as the result of an optimization procedure. The are several such algorithms. The most popular one is sequential, starting with an empty set and adding a single latent variable at each subsequent step of the process. The result is a sequence of prediction models, $\mathcal{M}_1, \ldots, \mathcal{M}_t$, where $\mathcal{M}_k$ predicts the output variable Y through a linear function of the first k latent variables. The "best" of these PLSR models is that model that minimizes a cross-validation estimate of prediction error. (How well cross-validation actually selects the best model is as yet unknown, however.)

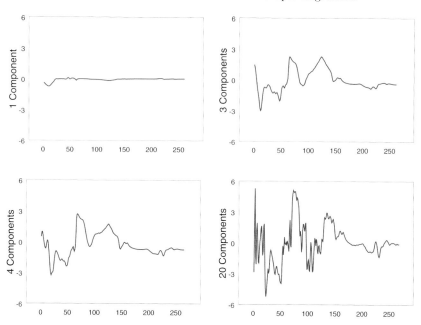

FIGURE 5.4. *Principal component regression estimates for the PET yarn data. There are 268 coefficients. The numbers of PCR components are $t = 1$ (upper-left panel), $t = 3$ (upper-right panel), $t = 4$ (lower-left panel), $t = 20$ (lower-right panel). The horizontal axis is coefficient number.*

The PLSR algorithm in Table 5.6 (Wold, Martens, and Wold, 1983) uses only a series of simple linear regression routines. We build the latent variables, $Z_1, \ldots, Z_t$, in a stepwise fashion. At the kth step, Z_k is a weighted average of the X-residuals from the previous step, where the weights are proportional to covariances of the X-residuals from the previous step with the Y-residuals from the previous step. The resulting PLSR function is a linear combination of the $Z_1, \ldots, Z_t$.

Empirical studies (Frank and Friedman, 1993) show that PLSR gives slightly better overall performance than does PCR, that fewer components are needed in PLSR than in PCR to provide a similar fit to the data, and that as the problem becomes increasingly more ill-conditioned, both biased methods yield substantial improvements in predictive ability over OLS. De Jong (1995) also showed that, in an R^2 sense and using t components, the PLSR fitted values are closer to the OLS fitted values than are the PCR fitted values.

The PLSR estimator, $\widehat{\boldsymbol{\beta}}_{\text{plsr}}^{(t)}$, where t is the number of components, is a shrinkage estimator. This is a difficult result to prove. De Jong (1995) showed that, for $1 \leq k \leq t$, $\|\widehat{\boldsymbol{\beta}}_{\text{plsr}}^{(k)}\|$ is a strictly nondecreasing function of

TABLE 5.6. *PLSR algorithm (Wold, Martens, and Wold, 1983).*

---

1. Standardize each n-vector $\mathbf{x}_j$ of data on X_j so that it has mean 0 and standard deviation 1, and set $\mathbf{x}_j^{(0)} = \mathbf{x}_j$, $j = 1, 2, \ldots, r$. Center the n-vector $\mathcal{y}$ of data on Y so that it has mean 0, and set $\mathcal{y}^{(0)} = \mathcal{y}$. Set $\widehat{\mathcal{y}}^{(0)} = \bar{y}\mathbf{1}_n$.

2. For $k = 1, 2, \ldots, t$:

   - For $j = 1, 2, \ldots, r$, regress $\mathcal{y}^{(k-1)}$ on $\mathbf{x}_j^{(k-1)}$ to get the OLS regression coefficient

     $$\widehat{\beta}_{k-1,j} = \mathrm{cov}(\mathbf{x}_j^{(k-1)}, \mathcal{y}^{(k-1)})/\mathrm{var}(\mathbf{x}_j^{(k-1)}),$$

     where, for any n-vectors $\mathbf{x}$ and $\mathbf{y}$, $\mathrm{cov}(\mathbf{x}, \mathbf{y}) = \mathbf{x}^T\mathbf{y}$ and $\mathrm{var}(\mathbf{x}) = \mathbf{x}^T\mathbf{x}$. Compute $\widehat{\beta}_{k-1,j}\mathbf{x}_j^{(k-1)}$.

   - Compute the weighted average $\mathbf{z}_k = \sum_{j=1}^{r} w_{k-1,j}\widehat{\beta}_{k-1,j}\mathbf{x}_j^{(k-1)}$ as a predictor of $\mathcal{y}$, where $w_{k-1,j} \propto \mathrm{var}(\mathbf{x}_j^{(k-1)})$. Thus,

     $$\mathbf{z}_k \propto \sum_{j=1}^{r} \mathrm{cov}(\mathbf{x}_j^{(k-1)}, \mathcal{y}^{(k-1)}) \cdot \mathbf{x}_j^{(k-1)}.$$

   - Regress $\mathcal{y}^{(k-1)}$ on $\mathbf{z}_k$ to get the OLS regression coefficient

     $$\widehat{\theta}_k = \mathrm{cov}(\mathbf{z}_k, \mathcal{y}^{(k-1)})/\mathrm{var}(\mathbf{z}_k)$$

     and the residual vector $\mathcal{y}^{(k)} = \mathcal{y}^{(k-1)} - \widehat{\theta}_k\mathbf{z}_k$.

   - Set $\widehat{\mathcal{y}}^{(k)} = \widehat{\mathcal{y}}^{(k-1)} + \widehat{\theta}_k\mathbf{z}_k$.

   - For $j = 1, 2, \ldots, r$, regress $\mathbf{x}_j^{(k-1)}$ on $\mathbf{z}_k$ to get the OLS regression coefficient

     $$\widehat{\phi}_{kj} = \mathrm{cov}(\mathbf{z}_k, \mathbf{x}_j^{(k-1)})/\mathrm{var}(\mathbf{z}_k)$$

     and residual vector $\mathbf{x}_j^{(k)} = \mathbf{x}_j^{(k-1)} - \widehat{\phi}_{kj}\mathbf{z}_k$.

   - Stop when $\sum_{j=1}^{r} \mathrm{var}(\mathbf{x}_j^{(k)}) = 0$.

3. The PLSR function fitted with t components is, therefore, given by

   $$\widehat{\mathcal{y}}_{\mathrm{plsr}}^{(t)} = \bar{y}\mathbf{1}_n + \sum_{k=1}^{t} \widehat{\theta}_k\mathbf{z}_k.$$

---

k, which implies that every PLSR iterate improves upon OLS; that is,

$$\|\widehat{\boldsymbol{\beta}}_{\mathrm{plsr}}^{(1)}\| \leq \|\widehat{\boldsymbol{\beta}}_{\mathrm{plsr}}^{(2)}\| \leq \cdots \leq \|\widehat{\boldsymbol{\beta}}_{\mathrm{plsr}}^{(t)}\| = \|\widehat{\boldsymbol{\beta}}_{\mathrm{ols}}\|. \tag{5.91}$$

Goutis (1996) used a geometric argument to give a direct proof that, for every $1 \leq k \leq t$, $\|\widehat{\boldsymbol{\beta}}_{\mathrm{plsr}}^{(k)}\| \leq \|\widehat{\boldsymbol{\beta}}_{\mathrm{ols}}\|$, and Phatak and de Hoog (2002) derived an explicit expression relating the PLSR estimator to the OLS estimator. The shrinkage behavior of individual PLSR coefficients turns out to be quite "peculiar": Frank and Friedman (1993) noted from empirical evidence and certain heuristics that whereas PLSR shrunk some OLS coefficients, it also expanded others. This shrinkage behavior was further studied by Butler and Denham (2000) and Lingjaerde and Christophersen (2000).

The *orthogonal loadings* algorithm uses a sequence of multiple regressions to arrive at the same PLSR solution as Wold's algorithm (Helland, 1988). Also, Exercise 5.11 provides the theory behind the S-PLUS PLSR algorithm given in Brown (1993, Appendix E). The PLSR algorithm in Table 5.6 is an extension of the NIPALS algorithm (Wold, 1975). See also the SIMPLS algorithm (de Jong, 1993).

Example: The PET Yarn Data (Continued)

Each variable in the PET yarn data was centered. The PLSR estimates of all 268 regression coefficients in the vector $\widehat{\boldsymbol{\beta}}_{\mathrm{plsr}}^{(t)}$ for the PET yarn data are displayed in Figure 5.5. for $t = 1, 3, 4, 20$ components. The 20-component PLSR estimate is the minimum-length LS estimator of the regression coefficient vector $\boldsymbol{\beta}$.

We see from Figure 5.5 that using only one PLSR component results in a set of regression estimates with little visible structure. Most of the variability in the regression coefficients occurs in the first 150 coefficients. The final shape of the coefficient estimates can already be discerned by 3 components, and a useful representation is given by 4 components. As additional components are added to the model, more and more high-frequency noise is added to the PLSR estimates.

5.6.4 Ridge Regression

Hoerl and Kennard (1970a) proposed that potential instability in the OLS estimator, $\widehat{\boldsymbol{\beta}}_{\mathrm{ols}} = (\mathcal{X}^\tau \mathcal{X})^{-1} \mathcal{X}^\tau \mathcal{Y}$, of $\boldsymbol{\beta}$ could be tracked by adding a small constant value k to the diagonal entries of the matrix $\mathcal{X}^\tau \mathcal{X}$ before taking its inverse. The result is the *ridge regression estimator* (or *ridge rule*),

$$\widehat{\boldsymbol{\beta}}_{\mathrm{rr}}(k) = (\mathcal{X}^\tau \mathcal{X} + k\mathbf{I}_r)^{-1} \mathcal{X}^\tau \mathcal{Y} = \mathbf{W}(k)\widehat{\boldsymbol{\beta}}_{\mathrm{ols}}, \tag{5.92}$$

where

$$\mathbf{W}(k) = (\mathcal{X}^\tau \mathcal{X} + k\mathbf{I}_r)^{-1} \mathcal{X}^\tau \mathcal{X}. \tag{5.93}$$

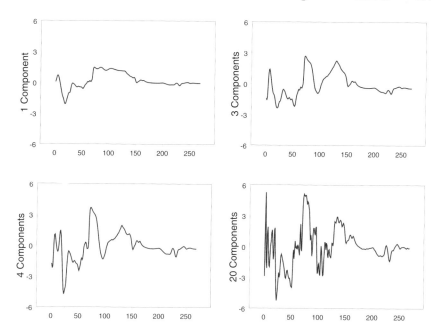

FIGURE 5.5. *Partial least-squares regression estimates for the PET yarn data. There are 268 coefficients. The numbers of PLSR components are $t = 1$ (upper-left panel), $t = 3$ (upper-right panel), $t = 4$ (lower-left panel), $t = 20$ (lower-right panel). The horizontal axis is coefficient number.*

Thus, we have a class of estimators (5.92), indexed by a parameter k. When $k > 0$, $\hat{\boldsymbol{\beta}}_{\mathrm{rr}}(k)$ is a biased estimator of $\boldsymbol{\beta}$. In the special case $\mathcal{X}^{\tau}\mathcal{X} = \mathbf{I}_r$ (the *orthonormal design* case), (5.92) reduces to $\hat{\boldsymbol{\beta}}_{\mathrm{rr}}(k) = (1 + k)^{-1}\hat{\boldsymbol{\beta}}_{\mathrm{ols}}$. When $k = 0$, (5.92) reduces to the OLS estimator.

Properties

The ridge regression estimator (5.92) can be characterized in three different ways — as an estimator with restricted length that minimizes the residual sum of squares, as a shrinkage estimator that shrinks the least-squares estimator toward the origin, and, given suitable priors, as a Bayes estimator.

1. A ridge regression estimator is the solution of a penalized least-squares problem. Specifically, it is the r-vector $\boldsymbol{\beta}$ that minimizes the error sum of squares,

$$ESS(\boldsymbol{\beta}) = (\mathcal{Y} - \mathcal{X}\boldsymbol{\beta})^{\tau}(\mathcal{Y} - \mathcal{X}\boldsymbol{\beta}), \qquad (5.94)$$

$$subject\ to\ \ \|\boldsymbol{\beta}\|^2 \leq c, \qquad (5.95)$$

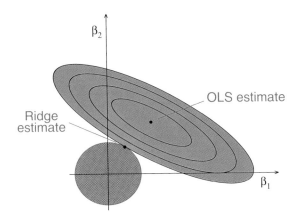

FIGURE 5.6. *The ridge regression estimator, $\widehat{\boldsymbol{\beta}}_{rr}(k)$, as the solution of a penalized least-squares problem. The ellipses show the contours of the error sum-of-squares function, and the circle shows the boundary of the penalty function, $\beta_1^2 + \beta_2^2 \leq c$, where c is the radius of the circle. The ridge estimator is the point at which the innermost elliptical contour touches the circular penalty.*

where $\|\boldsymbol{\beta}\|^2 = \boldsymbol{\beta}^\tau \boldsymbol{\beta}$ and $c > 0$ is an arbitrary constant. To see this, form the function

$$\phi(\boldsymbol{\beta}) = (\mathcal{Y} - \mathcal{X}\boldsymbol{\beta})^\tau (\mathcal{Y} - \mathcal{X}\boldsymbol{\beta}) - \lambda \boldsymbol{\beta}^\tau \boldsymbol{\beta}, \tag{5.96}$$

where $\lambda > 0$ is a Lagrangian multiplier (or *ridge parameter*) that regularizes the stability of a ridge regression estimator, and $\boldsymbol{\beta}^\tau \boldsymbol{\beta}$ is a *penalty function*. Differentiate ϕ with repect to $\boldsymbol{\beta}$, set the result equal to zero, and at the minimum, set $\boldsymbol{\beta} = \widehat{\boldsymbol{\beta}}_{rr}(\lambda)$ to get

$$(\mathcal{X}^\tau \mathcal{X} - \lambda \mathbf{I}_r)\widehat{\boldsymbol{\beta}}_{rr}(\lambda) = \mathcal{X}^\tau \mathcal{Y}. \tag{5.97}$$

The result is obtained by solving this last equation for $\widehat{\boldsymbol{\beta}}_{rr}(\lambda)$ and then setting $k = \lambda$. Note that the restriction $\boldsymbol{\beta}^\tau \boldsymbol{\beta} \leq c$ on $\boldsymbol{\beta}$ is a hypersphere centered at the origin with bounded squared radius c, where the value of c determines the value of k. Figure 5.6 shows the two-parameter case.

2. *A ridge regression estimator is a shrinkage estimator that shrinks the OLS estimator toward zero.* The singular value decomposition of the $(n \times r)$-matrix $\mathcal{X}$ is given by $\mathcal{X} = \mathbf{U}\boldsymbol{\Lambda}^{1/2}\mathbf{V}^\tau$, where $\boldsymbol{\Lambda} = \mathrm{diag}[\lambda_j]$, $\mathbf{U}\mathbf{U}^\tau = \mathbf{U}^\tau \mathbf{U} = \mathbf{I}_n$, $\mathbf{V}\mathbf{V}^\tau = \mathbf{V}^\tau \mathbf{V} = \mathbf{I}_r$, and $\mathcal{X}^\tau \mathcal{X} = \mathbf{V}\boldsymbol{\Lambda}\mathbf{V}^\tau$. The $\{\lambda_j\}$ are the ordered eigenvalues of $\mathcal{X}^\tau \mathcal{X}$. Let $\mathbf{P} = \mathcal{X}\mathbf{V} = \mathbf{U}\boldsymbol{\Lambda}^{1/2}$ so that $\mathbf{P}^\tau \mathbf{P} = \boldsymbol{\Lambda}$. Then, we can write (5.92) as follows:

$$\begin{aligned}
\widehat{\boldsymbol{\beta}}_{rr}(k) &= (\mathcal{X}^\tau \mathcal{X}^\tau + k\mathbf{I}_r)^{-1}\mathcal{X}^\tau \mathcal{Y} \\
&= (\mathbf{V}\boldsymbol{\Lambda}\mathbf{V}^\tau + k\mathbf{V}\mathbf{V}^\tau)^{-1}\mathbf{V}\boldsymbol{\Lambda}^{1/2}\mathbf{U}^\tau \mathcal{Y} \\
&= \mathbf{V}(\boldsymbol{\Lambda} + k\mathbf{I}_r)^{-1}\boldsymbol{\Lambda}^{1/2}\mathbf{U}^\tau \mathcal{Y}
\end{aligned}$$

$$= \mathbf{V}(\boldsymbol{\Lambda} + k\mathbf{I}_r)^{-1}\mathbf{P}^\tau \mathcal{Y}. \tag{5.98}$$

Now, if we let $\boldsymbol{\alpha} = \mathbf{V}^\tau \boldsymbol{\beta}$ (so that $\boldsymbol{\beta} = \mathbf{V}\boldsymbol{\alpha}$), then, the canonical form of the multiple regression model is

$$\mathcal{Y} = \mathcal{X}\boldsymbol{\beta} + \mathbf{e} = \mathbf{P}\boldsymbol{\alpha} + \mathbf{e}, \tag{5.99}$$

whence the OLS estimator of $\boldsymbol{\alpha}$ is $\widehat{\boldsymbol{\alpha}}_{\text{ols}} = (\mathbf{P}^\tau\mathbf{P})^{-1}\mathbf{P}^\tau\mathcal{Y} = \boldsymbol{\Lambda}^{-1}\mathbf{V}^\tau\mathbf{s}$, where $\mathbf{s} = \mathcal{X}^\tau\mathcal{Y}$. Set

$$\begin{aligned}
\widehat{\boldsymbol{\alpha}}_{\text{rr}}(k) &= \mathbf{V}^\tau\widehat{\boldsymbol{\beta}}_{\text{rr}}(k) \\
&= (\boldsymbol{\Lambda} + k\mathbf{I}_r)^{-1}\mathbf{P}^\tau\mathcal{Y} \\
&= (\boldsymbol{\Lambda} + k\mathbf{I}_r)^{-1}\boldsymbol{\Lambda}\widehat{\boldsymbol{\alpha}}_{\text{ols}}.
\end{aligned} \tag{5.100}$$

The jth component in the r-vector $\widehat{\boldsymbol{\alpha}}_{\text{rr}}(k)$ is, therefore, given by

$$\widehat{\alpha}_{\text{rr},j}(k) = \left(\frac{\lambda_j}{\lambda_j + k}\right)\widehat{\alpha}_{\text{ols},j} = f_k(\lambda_j)\widehat{\alpha}_{\text{ols},j}, \tag{5.101}$$

say, where $0 < f_k(\lambda_j) \leq 1$, $j = 1, 2, \ldots, r$. For $k > 0$, $\widehat{\alpha}_{\text{rr},j}(k) < \widehat{\alpha}_{\text{ols},j}$, so that $\widehat{\alpha}_{\text{rr},j}(k)$ shrinks $\widehat{\alpha}_{\text{ols},j}$ toward zero. Also, $\widehat{\alpha}_{\text{rr},j}(k)$ can be written as $\widehat{\alpha}_{\text{rr},j}(k) = w_j \cdot 0 + (1 - w_j)\widehat{\alpha}_{\text{ols},j}$, with weight $0 < w_j = k/(\lambda_j + k) < 1$, whence it follows that the smaller the value of λ_j (for a given $k > 0$), the larger the value of w_j, and, hence, the greater is the shrinkage toward zero. Thus, ridge regression shrinks low-variance directions (small λ_j) more than it does high-variance directions (large λ_j).

Note that these conclusions hold for the canonical form of the regression model with $\boldsymbol{\alpha}$ as the coefficient vector. We can transform back by setting $\widehat{\boldsymbol{\beta}}_{\text{rr}}(k) = \mathbf{V}\widehat{\boldsymbol{\alpha}}_{\text{rr}}(k)$. However, $\widehat{\boldsymbol{\beta}}_{\text{rr}}(k)$ may not shrink every component of $\widehat{\boldsymbol{\beta}}_{\text{ols}}$. Indeed, for some j, the jth component, $\widehat{\beta}_{rr,j}(k)$, of $\widehat{\boldsymbol{\beta}}_{\text{rr}}(k)$ may actually have the opposite sign from the corresponding component, $\widehat{\beta}_{\text{ols},j}$, of $\widehat{\boldsymbol{\beta}}_{\text{ols}}$, or that $|\widehat{\beta}_{\text{rr},j}(k)| > |\widehat{\beta}_{\text{ols},j}|$. What we can say, however, is that

$$\|\widehat{\boldsymbol{\beta}}_{\text{rr}}(k)\|^2 = \|\widehat{\boldsymbol{\alpha}}_{\text{rr}}(k)\|^2 = \sum_{j=1}^{r}\left(\frac{\lambda_j}{\lambda_j + k}\right)^2\widehat{\alpha}_{\text{ols},j}^2, \tag{5.102}$$

which is monotonically decreasing function of k. Thus, $\|\widehat{\boldsymbol{\beta}}_{\text{rr}}(k)\| < \|\widehat{\boldsymbol{\beta}}_{\text{ols}}\|$, so that $\widehat{\boldsymbol{\beta}}_{\text{rr}}(k)$ is a shrinkage estimator.

3. *A ridge regression estimator is a Bayes estimator when $\boldsymbol{\beta}$ is given a suitable multivariate Gaussian prior.* Suppose $\mathcal{Y} = \mathcal{X}\boldsymbol{\beta} + \mathbf{e}$, where now $\mathbf{e} \sim \mathcal{N}_n(\mathbf{0}, \sigma^2\mathbf{I}_n)$ and σ^2 is known. In other words, $\mathcal{Y} \sim \mathcal{N}_n(\mathcal{X}\boldsymbol{\beta}, \sigma^2\mathbf{I}_n)$. The likelihood is

$$\begin{aligned}
L(\mathcal{Y}|\boldsymbol{\beta},\sigma) &\propto \exp\left\{-\frac{1}{2\sigma^2}(\mathcal{Y} - \mathcal{X}\boldsymbol{\beta})^\tau(\mathcal{Y} - \mathcal{X}\boldsymbol{\beta})\right\} \\
&\propto \exp\left\{-\frac{1}{2\sigma^2}(\boldsymbol{\beta} - \widehat{\boldsymbol{\beta}})^\tau\mathcal{X}^\tau\mathcal{X}(\boldsymbol{\beta} - \widehat{\boldsymbol{\beta}})\right\},
\end{aligned} \tag{5.103}$$

which has the form $\mathcal{N}_r(\widehat{\boldsymbol{\beta}}, \sigma^2(\mathcal{X}^\tau\mathcal{X})^{-1})$. Next, assume that the components of $\boldsymbol{\beta}$ are each independently distributed as Gaussian with mean 0 and known variance σ_β^2, so that $\boldsymbol{\beta} \sim \mathcal{N}_r(\mathbf{0}, \sigma_\beta^2 \mathbf{I}_r)$ with prior density

$$\pi(\boldsymbol{\beta}) \propto \exp\left\{ -\frac{\boldsymbol{\beta}^\tau \boldsymbol{\beta}}{2\sigma_\beta^2} \right\}. \tag{5.104}$$

The posterior density of $\boldsymbol{\beta}$ is proportional to the likelihood times the prior, that is,

$$p(\boldsymbol{\beta}|\mathcal{Y},\sigma) = L(\mathcal{Y}|\boldsymbol{\beta},\sigma)\pi(\boldsymbol{\beta}) \tag{5.105}$$

$$\propto \exp\left\{ -\frac{1}{2\sigma^2}\left[(\boldsymbol{\beta} - \widehat{\boldsymbol{\beta}})^\tau \mathcal{X}^\tau\mathcal{X}(\boldsymbol{\beta} - \widehat{\boldsymbol{\beta}}) + k\boldsymbol{\beta}^\tau\boldsymbol{\beta} \right] \right\}, \tag{5.106}$$

where $k = \sigma^2/\sigma_\beta^2$. Now, for the first term in the exponent, set $\boldsymbol{\beta} - \widehat{\boldsymbol{\beta}} = (\boldsymbol{\beta} - \widehat{\boldsymbol{\beta}}(k)) + (\widehat{\boldsymbol{\beta}}(k) - \widehat{\boldsymbol{\beta}})$, and, for the second term, $\boldsymbol{\beta} = (\boldsymbol{\beta} - \widehat{\boldsymbol{\beta}}(k)) + \widehat{\boldsymbol{\beta}}(k)$. Multiplying out both expressions and gathering like terms, we find that the posterior density of $\boldsymbol{\beta}$ is given by

$$p(\boldsymbol{\beta}|\mathcal{Y},\sigma) \propto \exp\left\{ -\frac{1}{2\sigma^2}\left[(\boldsymbol{\beta} - \widehat{\boldsymbol{\beta}}(k))^\tau(\mathcal{X}^\tau\mathcal{X} + k\mathbf{I}_r)(\boldsymbol{\beta} - \widehat{\boldsymbol{\beta}}(k)) \right] \right\}. \tag{5.107}$$

In other words, the posterior density of $\boldsymbol{\beta}$ is multivariate Gaussian with mean vector (and posterior mode) $\widehat{\boldsymbol{\beta}}(k)$ and covariance matrix $\sigma^2(\mathcal{X}^\tau\mathcal{X} + k\mathbf{I}_r)^{-1}$, where $k = \sigma^2/\sigma_\beta^2$. Note that if σ_β^2 is very large, the prior density becomes vague, and a ridge regression estimator approaches the OLS estimator.

The Bias-Variance Trade-off

Consider the mean squared error of the ridge regression estimator,

$$MSE(k) = \mathrm{E}\{(\widehat{\boldsymbol{\beta}}_{\mathrm{rr}}(k) - \boldsymbol{\beta})^\tau(\widehat{\boldsymbol{\beta}}_{\mathrm{rr}}(k) - \boldsymbol{\beta})\} \tag{5.108}$$

$$= \mathrm{VAR}(k) + \mathrm{BIAS}^2(k), \tag{5.109}$$

where the first term on the right-hand side is the variance and the second term is the bias-squared. The variance term is

$$\mathrm{VAR}(k) = \mathrm{tr}\{\sigma^2(\mathcal{X}^\tau\mathcal{X} + k\mathbf{I}_r)^{-1}\mathcal{X}^\tau\mathcal{X}(\mathcal{X}^\tau\mathcal{X} + k\mathbf{I}_r)^{-1}\}$$

$$= \sigma^2\mathrm{tr}\{(\boldsymbol{\Lambda} + k\mathbf{I}_r)^{-1}\boldsymbol{\Lambda}(\boldsymbol{\Lambda} + k\mathbf{I}_r)^{-1}\}$$

$$= \sigma^2\sum_{j=1}^{r} \frac{\lambda_j}{(\lambda_j + k)^2}. \tag{5.110}$$

The bias is

$$\mathrm{E}(\widehat{\boldsymbol{\beta}}_{\mathrm{rr}}(k) - \boldsymbol{\beta}) = \mathrm{E}\{(\mathcal{X}^\tau\mathcal{X} + k\mathbf{I}_r)^{-1}\mathcal{X}^\tau\mathcal{Y} - \boldsymbol{\beta}\}$$

$$\begin{aligned}
&= \{(\mathcal{X}^\tau \mathcal{X} + k\mathbf{I}_r)^{-1}\mathcal{X}^\tau \mathcal{X} - \mathbf{I}_r\}\boldsymbol{\beta} \\
&= \{(\mathbf{V}\boldsymbol{\Lambda}\mathbf{V}^\tau + k\mathbf{I}_r)^{-1}\mathbf{V}\boldsymbol{\Lambda}\mathbf{V}^\tau - \mathbf{I}_r\}\mathbf{V}\boldsymbol{\alpha} \\
&= \mathbf{V}\{(\boldsymbol{\Lambda} + k\mathbf{I}_r)^{-1}\boldsymbol{\Lambda} - \mathbf{I}_r\}\boldsymbol{\alpha}, \qquad (5.111)
\end{aligned}$$

whence the bias-squared term is

$$\begin{aligned}
BIAS^2(k) &= (\mathrm{E}(\hat{\boldsymbol{\beta}}_{\mathrm{rr}}(k) - \boldsymbol{\beta}))^\tau (\mathrm{E}(\hat{\boldsymbol{\beta}}_{\mathrm{rr}}(k) - \boldsymbol{\beta})) \\
&= \boldsymbol{\alpha}^\tau \{\boldsymbol{\Lambda}(\boldsymbol{\Lambda} + k\mathbf{I}_r)^{-1} - \mathbf{I}_r\}\{(\boldsymbol{\Lambda} + k\mathbf{I}_r)^{-1}\boldsymbol{\Lambda} - \mathbf{I}_r\}^\tau \boldsymbol{\alpha} \\
&= k^2 \sum_{j=1}^r \frac{\alpha_j^2}{(\lambda_j + k)^2} . \qquad (5.112)
\end{aligned}$$

Thus, the mean squared error for a ridge estimator (5.92) is given by

$$MSE(k) = \sum_{j=1}^r \frac{\sigma^2 \lambda_j + k^2 \alpha_j^2}{(\lambda_j + k)^2} , \qquad (5.113)$$

where λ_j is the jth largest eigenvalue of $\mathcal{X}^\tau \mathcal{X}$, α_j is the jth element of $\boldsymbol{\alpha}$ (the orthogonally transformed $\boldsymbol{\beta}$), and σ^2 is the error variance, $j = 1, 2, \ldots, r$.

When $k = 0$, the squared-bias term is zero. The variance term decreases monotonically as k increases from zero, whereas the squared-bias term increases. For large values of k, the squared-bias term dominates the mean squared error. For these reasons, k has often been called the *bias parameter*.

Estimating the Ridge Parameter

We can use very small values of k to study how the OLS estimates would behave if the input data were mildly perturbed. If we observe large fluctuations in ridge estimates for very small k, such instability would reflect the presence of collinearity in the input variables. The main problem of ridge regression is to decide upon the best value of k. Choice of k is supposed to balance the "variance *vs.* bias" components of the mean squared error when estimating $\boldsymbol{\beta}$ by (5.92); the larger the value of k, the larger the bias, but the smaller the variance. In applications, k is determined from the data in $\mathcal{X}$.

Hoerl and Kennard recommend use of the *ridge trace*, a graphical display of all components of the vector $\hat{\boldsymbol{\beta}}_{\mathrm{rr}}(k)$ plotted on the same scatterplot against a range of values of k. The ridge trace is often touted as a diagnostic tool that exhibits the degree of stability of the regression coefficients. Because k controls the amount of bias in the ridge estimate, the value of k is estimated (albeit subjectively) by the *smallest* value at which the trace stabilizes for all coefficients. Thisted (1976, 1980) argues that choosing an estimate of k to reflect stability of the ridge trace does not necessarily yield a meaningful reduction in mean squared error.

The ridge trace is also used as a variable selection procedure. If an estimated regression coefficient changes sign in the graph of its ridge trace, this is taken to mean that the OLS estimator of that coefficient has an incorrect sign, so that that variable should not be included in the regression model. Such a variable selection rule has been criticized as being "dangerous" (Thisted, 1976) because it eliminates variables without taking into account their virtues as predictors. Thisted argues that it is possible for a variable to be a poor predictor but have a small stable ridge trace, and, vice versa, to have a very unstable ridge trace but be an important variable for the regression model.

spaceskip3pt plus2pt minus2pt In an alternative version of the ridge trace, Hastie, Tibshirani, and Friedman (2001, Section 3.4.3) choose instead to plot the components of $\widehat{\boldsymbol{\beta}}_{rr}(k)$ against what they call the *effective degrees of freedom*,

$$\mathrm{df}(k) = \mathrm{tr}(\mathbf{W}(k)) = \sum_{j=1}^{r} \lambda_j/(\lambda_j + k), \tag{5.114}$$

where the matrix $\mathbf{W}(k)$ in (5.93) shrinks the OLS estimator.

The ridge parameter k can also be estimated using cross-validation techniques. A prescription for determining a V-fold cross-validatory choice of the ridge parameter k is given in Table 5.7.

Example: The PET Yarn Data (Continued)

As before, all variables in the PET yarn data are centered. The ridge trace for the first 60 RR coefficients is displayed in Figure 5.7. We see that several of the coefficient estimates change sign as k increases. The ridge trace (not shown here) for all 268 curves indicates that the ridge parameter k stabilizes for the centered PET yarn data at about the value 0.9.

Figure 5.8 shows the 268 ridge regression coefficient estimates for selected values of the ridge parameter k. The values of k are, from the top panel, $k = 0.00001, 0.01, 0.1,$ and 1.0. We see that the smaller the value of k, the more noisy the estimates, whereas the larger the value of k, the less noisy the estimates. If $k = 0$ (which is not possible in this application, where $r \gg n$), then we would have the minimum-length LS estimate. The computations for this example were carried out using the data augmentation algorithm (see Exercise 5.8).

5.7   Variable Selection

It's very easy to include too many input variables in a regression equation. When that happens, too many parameters will be estimated, the

TABLE 5.7. *V-fold cross-validatory choice of ridge parameter k.*

---

1. Standardize each x_j so that it has mean 0 and standard deviation 1, $j = 1, 2, \ldots, r$.

2. Partition the data into V learning and test sets corresponding to one of the versions of cross-validation ($V = 5, 10$, or n).

3. Choose $k_1, k_2, \ldots, k_N$ to be N (possibly equally spaced) values of k.

4. For $i = 1, 2, \ldots, N$, and for $v = 1, 2, \ldots, V$,

   - Use the vth learning set to compute the ridge regression coefficients $\widehat{\boldsymbol{\beta}}_{-v}(k_i)$, say.

   - Obtain an estimate of prediction error, $\widehat{PE}_v(k_i)$, say, by applying $\widehat{\boldsymbol{\beta}}_{-v}(k_i)$ to the corresponding vth test set.

5. For $i = 1, 2, \ldots, N$,

   - Average the V prediction error estimates to get an overall estimate of prediction error, $\widehat{PE}_{CV/V}(k_i) = V^{-1} \sum_v \widehat{PE}_v(k_i)$, say.

   - Plot the value of $\widehat{PE}_{CV/V}(k_i)$ against k_i.

6. Choose that value of k that minimizes prediction error. In other words, the V-fold cross-validatory choice of k is given by

$$\widehat{k}_{CV/V} = \arg \min_{k_i} \widehat{PE}_{CV/V}(k_i).$$

---

regression function will have an inflated variance, and *overfitting* will take place. At the other extreme, if too few variables are included, the variance will be reduced, but the regression function will have increased bias, it will give a poor explanation of the data, and *underfitting* will occur. Some compromise between these extremes is, therefore, desirable. The notion of what makes a variable "important" is still not well understood, but one interpretation (Breiman, 2001b) is that a variable is important if dropping it seriously affects prediction accuracy.

The driving force behind variable selection is a desire for a parsimonious regression model (one that is simpler and more easily interpretable than is the model with the entire set of variables) combined with a need for greater accuracy in prediction. Selecting variables in regression models is a complicated problem, and there are many conflicting views on which type of variable selection procedure is best. In this section, we discuss several of these procedures.

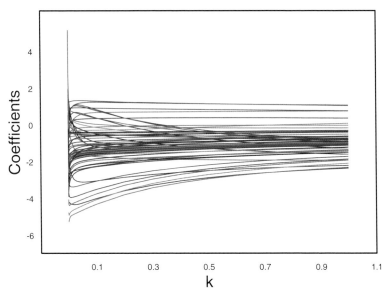

FIGURE 5.7. *Ridge trace of the first 60 ridge estimates of the 268 regression coefficients for the centered PET yarn data. Each curve represents a ridge regression coefficient estimate for varying values of k.*

5.7.1   Stepwise Methods

There are two main types of stepwise procedures in regression: backwards elimination, forwards selection, and a hybrid version that incorporates ideas from both main types.

Backwards elimination (BE) begins with the full set of variables. At each step, we drop that variable whose F-ratio,

$$F = \frac{(RSS_0 - RSS_1)/(df_0 - df_1)}{RSS_1/df_1}, \qquad (5.115)$$

is smallest, where RSS_0 is the residual sum of squares (with df_0 degrees of freedom) for the reduced model, and RSS_1 is the residual sum of squares (with df_1 degrees of freedom) for the larger model, where the "reduced" model is a *submodel* of the "larger" model. Then, we refit the reduced model and iterate again. Here, $df_0 - df_1 = 1$ and $df_1 = n - k - 1$, where k is the number of variables in the larger model.

Because of the relationship between the t and F distribution ($t_\nu^2 = F_{1,\nu}$), this procedure is equivalent to dropping that variable with the smallest ratio of the least-squares regression coefficient estimate to its respective estimated standard error. For large samples, this ratio behaves like a standard Gaussian deviate Z. A regression coefficient is, therefore, declared

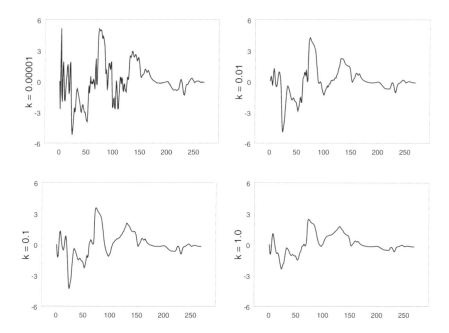

FIGURE 5.8. *Ridge regression estimates of the 268 regression coefficients for the centered PET yarn data. The values of the ridge parameter k are k=0.00001 (top-left panel), 0.01 (top-right panel), 0.1 (lower-left panel), 1.0 (lower-right panel). The horizontal axis is coefficient number.*

significant at the 5% level if the absolute value of its Z-ratio is larger than 2.0, and nonsignificant otherwise. Those variables having nonsignificant coefficients (using either the F or Z definition) are dropped from the model. We stop when all variables retained in the model are larger than some predetermined value F_{delete}, usually taken as the 10% point of the $\mathcal{F}_{1,n-k-1}$ distribution.

Forwards selection (FS) begins with an empty set of variables. At each step, we select from the variable list that variable with the largest F value (5.115) with $df_0 - df_1 = 1$ and $df_1 = n - k - 2$, where k is the number of variables in the smaller model, add that variable to the regression model, and then refit the enlarged model. We stop selecting variables for the model when the F value for each variable not currently in the model is smaller than some predetermined value F_{enter}, which is typically taken to be equal to 2 or 4 or the 25% point of the $\mathcal{F}_{1,n-k-2}$ distribution.

A hybrid *stepwise procedure* alternates backwards and forwards in its model selection and stops when all variables have either been retained for inclusion or removed.

For the bodyfat data, when we use $F_{enter} = F_{delete} = 4.0$, only four input variables (abdomen, weight, wrist, and forearm) appear in the final model using any of the above stepwise procedures. If we set $F_{enter} = F_{delete} = 2.0$, three further variables, neck, age, and thigh, are retained for the equation, although neck and thigh each have t-values smaller than 2.0.

Criticisms of Stepwise Methods. Stepwise procedures have been severely criticized for the following reasons: (1) When the input variables are highly correlated, stepwise methods can yield confusing conclusions. (2) The maximum (or minimum) of a set of correlated F statistics is not an F statistic. Hence, the decision rules used in stepwise regression to add or drop an input variable can be misleading, We should be very cautious in evaluating the significance (or not) of a regression coefficient when the associated variable is a candidate for inclusion or exclusion in a stepwise regression procedure. (3) There is no guarantee that the subsets obtained from either forwards selection or backwards elimination stepwise procedures will contain the same variables or even be the "best" subset. (4) When there are more variables than observations ($r > n$), backwards elimination is typically not a feasible procedure. (5) A stepwise procedure produces a single answer (a very specific subset) to the variable selection problem, although several different subsets may be equally good for regression purposes.

5.7.2   All Possible Subsets

An alternative method of variable selection involves examining *all possible subsets* of a given size and evaluating their powers of prediction. Thus, if we start out with r variables, each variable can be in or out of the subset; this implies that there are $2^r - 1$ different possible subsets that have to be examined (ignoring the empty subset). This number of candidate subsets quickly becomes very large even for moderate r (e.g., with 20 variables, there are more than a million subsets). Branch-and-bound algorithms (e.g., Furnival and Wilson, 1974) reduce this number to a more manageable size by eliminating large numbers of candidate models from consideration.

Let $k \in \{0, 1, 2, \ldots, r\}$ be the number of variables in a given regression *submodel P* with $|P| = p = k+1$ parameters (k variables and an intercept). There are $\binom{r}{k}$ different subsets each having k variables. Using a variable selection criterion, each of those subsets may be compared and ranked.

Most subset selection procedures choose the best submodel by minimizing a selection criterion of the form,

$$\frac{RSS_P}{n} + \lambda \cdot p \cdot \frac{\widehat{\sigma}^2}{n} , \tag{5.116}$$

where λ is a *penalty coefficient*, $\widehat{\sigma}^2$ is the residual variance from the full model R^+, and RSS_P is the residual sum of squares for submodel P. In the neural networks literature, RSS_P/n is called the *learning* (or *training*) *error*; we saw it before as the apparent error rate or resubstitution error rate. The term $\lambda p \widehat{\sigma}^2/n$ is called the *complexity term*. Special cases of (5.116) are *Akaike Information Criterion (AIC)* (Akaike, 1973) and *Mallows C_P* (Mallows, 1973, 1995), both of which have $\lambda = 2$, and the *Bayesian Information Criterion (BIC)* (Akaike, 1978; Schwarz, 1978) with $\lambda = \log n$. The best submodel found using minimum-BIC will have fewer variables than by using minimum-C_P. Asymptotically, AIC and C_P are equivalent but have different properties than BIC.

The most popular of these criteria is $C_P = RSS_P/\widehat{\sigma}^2 - (n - 2p)$. To compare submodels, we draw a scatterplot of C_P values against p. (Usually, we only plot the smallest few C_P values for each p.) Certain regions of the C_P-plot deserve special mention. For the full model,

$$C_{R^+} = |R^+| = r + 1, \tag{5.117}$$

"good" subsets (those with small bias) will have $C_P \approx p$, and those subsets with large bias will have C_P values greater than p. Furthermore, any subset with $C_P \leq r + 1$ also has $F \leq 2$ (a criterion used in stepwise regression for adding or eliminating a variable) and so is a candidate for a good subset. Analytical and empirical results suggest that C_P (and related criteria) tend to overfit when the full model has very high dimensionality.

The C_P plot for the bodyfat data is given in Figure 5.9, where we have plotted those subsets with the five smallest C_P values for each value of p. There are 27 subsets with $C_P < p$. The overall lowest $C_P = 5.9$ is obtained from a 7-variable subset with variables age, weight, neck, abdomen, thigh, forearm, and wrist.

5.7.3 Criticisms of Variable Selection Methods

There have been many criticisms leveled at variable selection methods in general. These include (1) inferential methods applied to a regression model assume that the variables are selected à priori. Subset selection procedures, however, use the data to add or delete variables and, hence, change the model. As such, they violate the inferential model and should be considered only as "heuristic data analysis tools" (Breiman, Friedman, Olshen, and Stone, 1984, p. 227). (2) When subset selection is data-driven, then the OLS estimates of the regression coefficients based upon the same data will be biased (even for large sample sizes) on the order 1–2 standard errors (Miller, 2002). (3) If the (learning) data are changed a small amount, this may drastically change the variables chosen for the optimal regression subset, rendering subset selection procedures very "unstable" (Breiman, 1996).

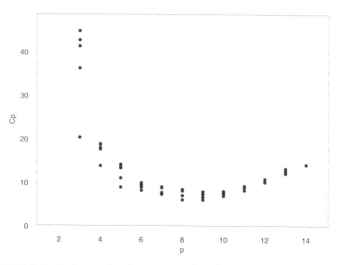

FIGURE 5.9. *Subset selection for the bodyfat data. The smallest five values of C_P are plotted against the number of parameters p in the subset model P.*

5.8   Regularized Regression

Both ridge regression and variable selection have their advantages and disadvantages. It would, therefore, be useful if we could construct a hybrid of these two ideas that would combine the best properties of each method — subset selection, shrinkage to improve prediction accuracy, and stability in the face of data perturbations.

Consider the general form of the penalized least-squares criterion, which can be written as

$$\phi(\boldsymbol{\beta}) = (\mathcal{Y} - \mathcal{X}\boldsymbol{\beta})^{\tau}(\mathcal{Y} - \mathcal{X}\boldsymbol{\beta}) + \lambda p(\boldsymbol{\beta}), \tag{5.118}$$

for a given penalty function $p(\cdot)$ and *regularization parameter* λ. We can define a family (indexed by $q > 0$) of penalized least-squares estimators in which the penalty function,

$$p_q(\boldsymbol{\beta}) = \sum_{j=1}^{r} |\beta_j|^q, \tag{5.119}$$

bounds the L_q-norm of the parameters in the model as

$$\sum_j |\beta_j|^q \le c \tag{5.120}$$

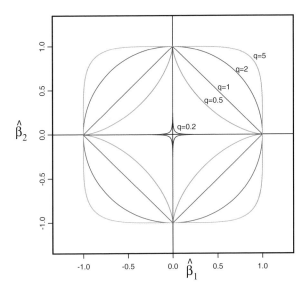

FIGURE 5.10. *Two-dimensional contours of the symmetric penalty function* $p_q(\boldsymbol{\beta}) = |\beta_1|^q + |\beta_2|^q = 1$ *for* $q = 0.2, 0.5, 1, 2, 5$. *The case* $q = 1$ *(blue diamond) yields the lasso and* $q = 2$ *(red circle) yields ridge regression.*

(Frank and Friedman, 1993). The two-dimensional contours of this symmetric penalty function for different values of q are given in Figure 5.10.

If we substitute the penalty function $p_q(\boldsymbol{\beta})$ in (5.119) in place of $p(\boldsymbol{\beta})$ in (5.118), we can write the criterion as $\phi_q(\boldsymbol{\beta})$, $q > 0$. Then, $\phi_q(\boldsymbol{\beta})$ is a smooth, convex function when $q > 1$, and is convex for $q = 1$, so that we can use classical optimization methods to minimize $\phi_q(\boldsymbol{\beta})$. By contrast, $\phi_q(\boldsymbol{\beta})$ is not convex when $q < 1$, and so its minimization is more complicated, especially when r is large.

Ridge regression corresponds to $q = 2$, and its corresponding penalty function is a circular disk ($r = 2$) or sphere ($r = 3$), or, for general r, a rotationally invariant hypersphere centered at the origin. The ridge regression estimator is that point on the elliptical contours of $ESS(\boldsymbol{\beta})$, centered at $\widehat{\boldsymbol{\beta}}$, which first touches the hypersphere $\sum_j \beta_j^2 \leq c$. The *tuning parameter* c controls the size of the hypersphere and, hence, how much we shrink $\widehat{\boldsymbol{\beta}}$ toward the origin.

If $q \neq 2$, the penalty is no longer rotationally invariant. The most interesting case is $q < 2$, where the penalty function collapses toward the coordinate axes, so that not only does it shrink the coefficients toward zero, but it also sets some of them to be exactly zero, thus combining elements of ridge regression and variable selection. When q is set very close to 0, the penalty function places all its mass along the coordinate axes, and the contours of the elliptical region of $ESS(\boldsymbol{\beta})$ touch an undetermined number of axes (so that the resulting regression function has an unknown number of zero coefficients); the result is variable selection. The case $q = 1$ produces the lasso method having a diamond-shaped penalty function with the corners of the diamond on the coordinate axes.

A hybrid penalized LS regression method called *the elastic net* (Zou and Hastie, 2005) uses as $p(\beta)$ a linear combination of the ridge regression L_2 penalty function and the Lasso L_1 penalty function.

The Lasso

The *Lasso* (least absolute shrinkage and selection operator) is a constrained OLS minimization problem in which

$$ESS(\boldsymbol{\beta}) = (\mathcal{Y} - \mathcal{X}\boldsymbol{\beta})^{\tau}(\mathcal{Y} - \mathcal{X}\boldsymbol{\beta}) \tag{5.121}$$

is minimized for $\boldsymbol{\beta} = (\beta_j)$ subject to the diamond-shaped condition that $\sum_{j=1}^{r} |\beta_j| \leq c$ (Tibshirani, 1996b). The regularization form of the problem is to find $\boldsymbol{\beta}$ to minimize

$$\phi(\boldsymbol{\beta}) = (\mathcal{Y} - \mathcal{X}\boldsymbol{\beta})^{\tau}(\mathcal{Y} - \mathcal{X}\boldsymbol{\beta}) + \lambda \sum_{j=1}^{r} |\beta_j|. \tag{5.122}$$

This problem can be solved using complicated quadratic programming methods subject to linear inequality constraints.

The Lasso has a number of desirable features that have made it a popular regression algorithm. Just like ridge regression, the Lasso is a shrinkage estimator of $\boldsymbol{\beta}$, where the OLS regression coefficients are shrunk toward the origin, the value of c controlling the amount of shrinkage. At the same time, it also behaves as a variable-selection technique: for a given value of c, only a subset of the coefficient estimates, $\widehat{\beta}_j$, will have nonzero values, and reducing the value of c reduces the size of that subset. The coefficient values will be exactly zero when one of the elliptical contours of the function

$$ESS(\boldsymbol{\beta}) = RSS + (\boldsymbol{\beta} - \widehat{\boldsymbol{\beta}}_{\mathrm{ols}})^{\tau}\mathcal{X}^{\tau}\mathcal{X}(\boldsymbol{\beta} - \widehat{\boldsymbol{\beta}}_{\mathrm{ols}}), \tag{5.123}$$

where $RSS = ESS(\widehat{\boldsymbol{\beta}})$ is a constant, touches a corner of the diamond-shaped penalty function.

In Figure 5.11, we display all 13 Lasso paths for the bodyfat data, both for the coefficients (left panel) and for the standardized coefficients (right panel). Variables are added to the regression model in the following order: 6 (abdomen), 3 (height), 1 (age), 13 (wrist), 4 (neck), 12 (forearm), 7 (hip), 11 (biceps), 8 (thigh), 2 (weight), 10 (ankle), 5 (chest), and 9 (knee). None of the coefficient paths cross zero and so no variables are dropped from the regression model at any stage of the Lasso process.

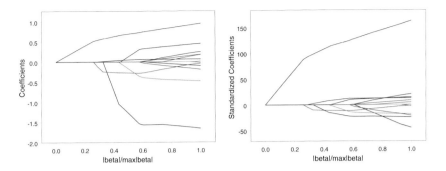

FIGURE 5.11. *Lasso paths for the bodyfat data. The paths are plots of the coefficients $\{\widehat{\beta}_j\}$ (left panel) and the standardized coefficients, $\{\widehat{\beta}_j \| \mathcal{X}_j \|_2\}$ (right panel) plotted against $\sum_j |\widehat{\beta}_j| / \max \sum_j |\widehat{\beta}_j|$. The variables are added to the regression model in the order: 6, 3, 1, 13, 4, 12, 7, 11, 8, 2, 10, 5, 9.*

The Garotte

A different type of penalized least-squares estimator is due to Breiman (1995). Let $\widehat{\boldsymbol{\beta}}_{\mathrm{ols}}$ be the OLS estimator and let $\mathbf{W} = \mathrm{diag}\{\mathbf{w}\}$ be a diagonal matrix with nonnegative weights $\mathbf{w} = (w_j)$ along the diagonal. The problem is to find the weights $\mathbf{w}$ that minimize

$$\phi(\mathbf{w}) = (\mathcal{Y} - \mathcal{X} \mathbf{W} \widehat{\boldsymbol{\beta}}_{\mathrm{ols}})^{\tau} (\mathcal{Y} - \mathcal{X} \mathbf{W} \widehat{\boldsymbol{\beta}}_{\mathrm{ols}}) \qquad (5.124)$$

subject to one of the following two constraints,

1. $\mathbf{w} \geq \mathbf{0}$, $\mathbf{1}_r^{\tau} \mathbf{w} = \sum_{j=1}^{r} w_j \leq c$ *(nonnegative garotte)*

2. $\mathbf{w}^{\tau} \mathbf{w} = \sum_{j=1}^{r} w_i^2 \leq c$ *(garotte)*.

Either version of the garotte seeks to find some desirable scaling of the regression coefficients. As c is decreased, more of the w_j become 0 (thus eliminating those particular variables from the regression function), while the nonzero $\widehat{\beta}_{\mathrm{ols},j}$ shrink toward 0. Note that both versions of the garotte, which depend upon the existence of the OLS estimator, $\widehat{\boldsymbol{\beta}}_{\mathrm{ols}}$, fail in situations where $r > n$.

The regularization parameter λ effects a compromise between how well the regression function fits the data and a size constraint on the coefficient vector. A large value of λ means that the size constraint dominates, whereas a small value of λ allows the OLS estimator to dominate. The value of λ can be determined in an objective fashion by V-fold cross-validation (see, e.g., Table 5.7).

Comparisons Extensive simulations comparing prediction accuracy under a wide variety of conditions and models (see, e.g., Breiman, 1995, 1996; Tibshirani, 1996b; Öjelund, Brown, Madsen, and Thyregod, 2002) show that ridge regression is very stable and is more accurate when there are many small coefficients, but does not do well when faced with a mixture of large and small coefficients; the nonnegative garotte is relatively stable and is more accurate when there are a few nonzero coefficients; the lasso performs well when there are a small-to-medium number of moderate-sized coefficients (while its estimates tend to have large biases); and subset selection, although very unstable, performs well only when there are a few nonzero coefficients.

5.9 Least-Angle Regression

The *least-angle regression (LAR)* algorithm (Efron, Hastie, Johnstone, and Tibshirani, 2004) is an automatic variable-selection method that improves upon Forwards Selection in multiple regression. It can also be used for situations in which $r \gg n$. Simple modifications of LAR enable the Lasso and Forwards-Stagewise algorithms to be computed efficiently. The three algorithms are referred to jointly as LARS.

In this section, we describe the LARS and Forwards-Stagewise algorithms and relate them to the Lasso. For these algorithms, $\mathcal{X} = (X_{ij})$ is an $(n \times r)$-matrix and $\mathcal{Y} = (Y_1, \cdots, Y_n)^\tau$. We assume that the input variables have been standardized to have mean zero, $\sum_{i=1}^n X_{ij} = 0$, and length one, $\sum_{i=1}^n X_{ij}^2 = 1$, $j = 1, 2, \ldots, r$, and that the output variable has mean zero, $\sum_{i=1}^n Y_i = 0$. The "current" estimate of the regression function $\boldsymbol{\mu} = \mathcal{X}\boldsymbol{\beta}$ is given by $\widehat{\boldsymbol{\mu}} = \mathcal{X}\widehat{\boldsymbol{\beta}}$, where the jth column, $\mathcal{X}_j = (X_{1j}, \cdots, X_{nj})^\tau$, of $\mathcal{X} = (\mathcal{X}_1, \cdots, \mathcal{X}_r)$ represents n observations on the jth covariate X_j. The vector of "current" correlations of $\mathcal{X}$ with the "current" residual vector $\mathbf{r} = \mathcal{Y} - \widehat{\boldsymbol{\mu}}$ is given by $\widehat{\mathbf{c}} = (\widehat{c}_1, \cdots, \widehat{c}_r)^\tau = \mathcal{X}^\tau \mathbf{r}$. The LARS algorithm builds up $\widehat{\boldsymbol{\mu}}$ sequentially by piecewise-linear steps, where Forwards-Stagewise steps are much smaller than LARS steps.

5.9.1 The Forwards-Stagewise Algorithm

1. Initialize $\widehat{\boldsymbol{\beta}} = \mathbf{0}$, so that $\widehat{\boldsymbol{\mu}} = \mathbf{0}$ and $\mathbf{r} = \mathcal{Y}$.

2. Find the covariate vector, $\mathcal{X}_{j_1}$, say, most highly correlated with $\mathbf{r}$, where $j_1 = \arg\max_j |\widehat{c}_j|$.

3. Update $\widehat{\beta}_{j_1}$ $\widehat{\beta}_{j_1} \leftarrow \widehat{\beta}_{j_1} + \delta_{j_1}$, where $\delta_{j_1} = \epsilon \cdot \text{sign}(\widehat{c}_{j_1})$ and ϵ is a small constant that controls the step-length.

4. Update $\widehat{\boldsymbol{\mu}} \leftarrow \widehat{\boldsymbol{\mu}} + \delta_{j_1} \mathcal{X}_{j_1}$ and $\mathbf{r} \leftarrow \mathbf{r} - \delta_{j_1} \mathcal{X}_{j_1}$.

5. Repeat steps 2 and 3 many times until $\hat{\mathbf{c}} = \mathbf{0}$. This is the OLS solution.

5.9.2   The LARS Algorithm

1. Initialize $\hat{\boldsymbol{\beta}} = \mathbf{0}$, so that $\hat{\boldsymbol{\mu}} = \mathbf{0}$ and $\mathbf{r} = \mathcal{Y}$. Start with the "active" set $\mathcal{A}$ an empty subset of indices of the set $\{1, 2, \ldots, r\}$.

2. Find the covariate vector, $\mathcal{X}_{j_1}$, say, most highly correlated with $\mathbf{r}$, where $j_1 = \arg\max_j |\hat{c}_j|$; the new active set is $\mathcal{A} \leftarrow \mathcal{A} \cup \{j_1\}$, and X_{j_1} is added to the regression model.

3. Move $\hat{\beta}_{j_1}$ toward $\text{sign}(\hat{c}_{j_1})$ (see Step 3 of Forwards-Stagewise algorithm) until some other covariate vector, $\mathcal{X}_{j_2}$, say, has the same correlation with $\mathbf{r}$ as does $\mathcal{X}_{j_1}$; the new active set is $\mathcal{A} \leftarrow \mathcal{A} \cup \{j_2\}$, and X_{j_2} is added to the regression model.

4. Update $\mathbf{r}$ and move $(\hat{\beta}_{j_1}, \hat{\beta}_{j_2})$ toward the joint OLS direction for the regression of $\mathbf{r}$ on $(\mathcal{X}_{j_1}, \mathcal{X}_{j_2})$ (i.e., equiangular between $\mathcal{X}_{j_1}$ and $\mathcal{X}_{j_2}$), until a third covariate vector, $\mathcal{X}_{j_3}$, say, is as correlated with $\mathbf{r}$ as are the first two variables; the new active set is $\mathcal{A} \leftarrow \mathcal{A} \cup \{j_3\}$, and X_{j_3} is added to the regression model.

5. After k LARS steps, $\mathcal{A} = \{j_1, j_2, \ldots, j_k\}$, $\hat{\boldsymbol{\mu}}_{\mathcal{A}}$ is the current LARS estimate (where exactly k estimated coefficients, $\hat{\beta}_{j_1}, \hat{\beta}_{j_2}, \ldots, \hat{\beta}_{j_k}$, are nonzero and $X_{j_1}, X_{j_2}, \ldots, X_{j_k}$ define the linear regression model), and the current vector of correlations is $\hat{\mathbf{c}} = \mathcal{X}^\tau (\mathcal{Y} - \hat{\boldsymbol{\mu}}_{\mathcal{A}})$.

6. Continue until all r covariates have been added to the regression model and $\hat{\mathbf{c}} = \mathbf{0}$. This is the OLS solution.

Modifications for LARS

LARS-Lasso   The entire Lasso sequence of paths can be generated by a slight modification of the LARS algorithm. We start with the LARS algorithm; then, if a nonzero estimated coefficient becomes 0 (e.g., changes its sign), stop and remove that variable from $\mathcal{A}$ and from the calculation of the next equiangular direction. The LARS algorithm recomputes the best direction and continues on its way. All additions and subtractions of variables are made "one-at-a-time," so that the number of steps for the LARS-Lasso algorithm can be larger than that of the LARS algorithm.

The LARS algorithm is efficient, involving of the order $O(r^3 + nr^2)$ computations, equivalent to carrying out OLS on the r input variables. The LARS-Lasso algorithm, in which we may need to drop a variable (costing at most an additional $O(r^2)$ computational operations for each variable dropped), generates the Lasso solution without difficulty.

Figure 5.11 was computed by the LARS-Lasso algorithm applied to the bodyfat data. The LARS algorithm yielded the same paths.

LARS-Stagewise A modified LARS algorithm in which $\mathcal{A}$ can drop one or more indices yields the Forwards-Stagewise algorithm, so that more steps than the LARS algorithm are needed to arrive at the OLS solution.

For the bodyfat data, the Forwards-Stagewise algorithm took the following sequence of steps: variables 6, 3, 1, 13, 4, 12, and 7 were added successively to the model; variables 3 and 1 were dropped; then variable 3 was added back, but in the next step was dropped again. Then, variables 11, 8, and 2 were added, but variable 13 was dropped. Variables 1, 10, 3, 13, 5, and 9 were next added. Then, variable 4 was dropped, then added back, then dropped again, and added back again; and variable 1 was dropped, added, dropped again, and then finally added back in. Thus, 29 modified LARS steps were needed to reach the OLS solution.

The R package `lars` includes a C_P-type statistic as a stopping rule to choose between possible LARS models. Because of its propensity to overfit in high-dimensional problems, however, there is some doubt as to how reliable C_P can be in selecting a parsimonious model.

Bibliographical Notes

There is a huge literature on multiple linear regression, and it is the area of statistics about which most is known. See, for example, Weisberg (1985) and Draper and Smith (1981, 1998).

The material on prediction error (Sections 5.4 and 5.5) is based upon the work of Efron (1983, 1986). The use of cross-validation for model selection purposes was introduced by Stone (1974) and Geisser (1975). (It is amusing to read that one discussant of Stone's article likened cross-validation to witchcraft!) Based upon a conviction that "prediction is generally more relevant for inference than parameter estimation," Geisser (1974, 1975) called the cross-validation technique the *predictive sample-reuse method*.

Book-length accounts of the bootstrap include Efron (1982), Hall (1992), Efron and Tibshirani (1993), and Chernick (1999). The names "unconditional" and "conditional" bootstrap were taken from Breiman (1992). Freedman (1981) distinguishes the two regression models for bootstrapping by calling the fixed-X case the "regression model" and the random-X case the "correlation model." An account of regression problems with collinear data from an econometric point of view is given by Belsley, Kuh, and Welsch (1980).

The ridge regression estimator first appeared in 1962 in an article in a chemical engineering journal by A.E. Hoerl. This was followed by Hoerl and Kennard (1970a,b). For the Bayesian characterization of the ridge estimator, see Lindley and Smith (1972), Chipman (1964), and Goldstein and Smith (1974).

In many texts, it is common to recommend standardizing (centering and scaling) the input variables prior to carrying out ridge regression. Such recommendations are not accepted by everyone, however. Thisted (1976), for example, states that "no argument has ever been advanced, nor does a single theorem in the ridge literature require, that $\mathcal{X}^\tau \mathcal{X}$ be in 'correlation form'." He goes on to argue that "because ridge rules are *not* invariant with respect to changes in origin of the predictor variables, it is important to recognize that origins are *not* arbitrary and that centering, taken as a rule of thumb always to be followed, can lead to misleading results and poor mean square error behavior."

Some notes on terminology and notation origins ... The penalized least-squares regression with penalty function (5.125) is widely referred to as *bridge regression* with the origin of the name ascribed to Frank and Friedman (1993). Although this name never appears in that reference, it apparently was first used by Friedman in a talk (Tibshirani, personal communication). ... Mallows (1973) states that the use of the letter C in C_P was specifically chosen to honor Cuthbert Daniel, who helped Mallows develop the idea behind C_P at the end of 1963. ... In an interview (Findley and Parzen, 1995), Akaike explains how AIC was named. Akaike had previously used the notation IC (for information criterion) in a 1974 article, and for another article had asked his assistant to compute some values of the IC. His assistant knew that if she called the quantity "IC," Fortran would assume that it was integer-valued, which it was not. So, she put an A in front of IC to turn it into a noninteger-valued quantity. Akaike apparently thought that calling it AIC was a "good idea" because it could then be used as the first of a sequence of information criteria, AIC, BIC, etc.

Exercises

5.1 From the solution (5.12) to the least-squares problem in the random-X case, use the formula for inverting a partitioned matrix to show that (5.13) and (5.14) follow.

5.2 From the solution (5.26) to the least-squares problem in the fixed-X case, use the same matrix-inversion formula to show that (5.27) and (5.28) follow.

5.3 Show that $\mathrm{cov}((\mathbf{a}^\tau - \mathbf{d}^\tau \mathcal{Z}^\tau)\mathcal{Y}, \mathbf{d}^\tau \mathcal{Z}^\tau \mathcal{Y}) = 0$ for the multiple regression model, where $\mathbf{a}$ is an n-vector and $\mathbf{d}$ is an $(r+1)$-vector.

5.4 (*Gauss–Markov Theorem*) Assume that $\widehat{\boldsymbol{\beta}}_{\text{ols}}$ is any solution of the normal equations (5.25) and that $\mathcal{Z}$ is a matrix of fixed constants. Make no assumption that $\mathcal{Z}^{\tau}\mathcal{Z}$ has full rank. Call $\mathbf{c}^{\tau}\boldsymbol{\beta}$ *estimable* if we can find a vector $\mathbf{a}$ such that $\mathrm{E}(\mathbf{a}^{\tau}\mathcal{Y}) = \mathbf{c}^{\tau}\boldsymbol{\beta}$. If $\mathbf{c}^{\tau}\boldsymbol{\beta}$ is estimable, show that $\mathbf{c}^{\tau}\widehat{\boldsymbol{\beta}}_{\text{ols}}$ is linear in $\mathcal{Y}$ and is unbiased for $\mathbf{c}^{\tau}\boldsymbol{\beta}$. Using Exercise 5.3 or otherwise, show also that $\mathbf{c}^{\tau}\widehat{\boldsymbol{\beta}}_{\text{ols}}$ has minimum variance among all linear (in $\mathcal{Y}$) unbiased estimators of $\mathbf{c}^{\tau}\boldsymbol{\beta}$.

5.5 Suppose $\mathcal{Z}^{\tau}\mathcal{Z}$ is nonsingular and that the solution of the normal equations is $\widehat{\boldsymbol{\beta}}_{\text{ols}} = (\mathcal{Z}^{\tau}\mathcal{Z})^{-1}\mathcal{Z}^{\tau}\mathcal{Y}$. Show that the Gauss–Markov Theorem holds.

5.6 Let $\mathbf{G}$ be a generalized inverse of $\mathcal{Z}^{\tau}\mathcal{Z}$ and let a solution of the normal equations be given by the *generalized-inverse regression estimator*, $\widehat{\boldsymbol{\beta}}^{*} = \mathbf{G}\mathcal{Z}^{\tau}\mathcal{Y}$. Show that the Gauss–Markov Theorem holds.

5.7 Show that a *generalized ridge regression estimator*,

$$\widehat{\boldsymbol{\beta}}_{\text{rr}}(k) = (\mathcal{X}^{\tau}\mathcal{X} + k\boldsymbol{\Omega})^{-1}\mathcal{X}^{\tau}\mathbf{y},$$

can be obtained as a solution of minimizing $ESS(\boldsymbol{\beta})$ subject to the elliptical restriction that $\boldsymbol{\beta}^{\tau}\boldsymbol{\Omega}\boldsymbol{\beta} \leq c$.

5.8 (Marquardt, 1970) Consider the following operation of *data augmentation*. Center and scale all input and output variables. Augment the $(n \times r)$-matrix $\mathcal{X}$ with r additional rows of the form $\mathbf{H}_k = \sqrt{k}\mathbf{I}_r$, where k is given, and denote the resulting $((n+r) \times r)$-matrix by $\mathcal{X}^{*}$. Augment the n-vector $\mathcal{Y}$ using r 0s, and denote the resulting $(n+r)$-vector by $\mathcal{Y}^{*}$. Show that the ridge estimator can be obtained by applying OLS to the regression of $\mathcal{Y}^{*}$ on $\mathcal{X}^{*}$. Thus, one can carry out ridge regression using standard OLS regression software and obtain the correct ridge estimator. However, much of the rest of the regression output will be inappropriate for the original data $(\mathcal{X}, \mathcal{Y})$.

5.9 In the PET yarn example, the variables were all centered, but not scaled. Standardize the input variables (the spectrum values) by centering and dividing each input variable by its standard deviation, and center the output variable (density). For the standardized data, recompute: (1) the PCR coefficient estimates, (2) the PLSR coefficient estimates, and (3) the RR coefficient estimates for various values of k (including $k > 1$), and redraw the ridge trace. What effect does standardizing have on the results that is not provided by centering alone? How would the results be affected by neither centering nor standardizing the variables?

5.10 Consider data on the composition of a liquid detergent. The datafile detergent can be downloaded from the book's website. There are five Y output variables, representing four compounds in an aqueous solution (the fifth Y variable is the amount of water in the solution), and they sum to unity. The X input variables consist of mid-infrared spectrum values recorded as the absorbances at $r = 1168$ equally spaced frequencies in the range 3100–759 cm^{-1}. The data consist of $n = 12$ sample preparations of the detergent. Graph the 12 absorbance spectra and apply PCR, PLSR, and RR to the data using each of the first four Y variables in separate regressions.

5.11 (Mallows, 1973) Consider the C_P statistic. Let P^* be a subset with $p+1$ parameters that contains P. Show that $C_{P^*} - C_P$ is distributed as $2 - t_1^2$, where t_1 is the Student's t variable having 1 degree of freedom. Show also that if the additional variable is unimportant, then the difference $C_{P^*} - C_P$ has mean and variance approximately equal to 1 and 2, respectively.

5.12 What is the relationship between R^2 and C_P?

5.13 If the regression model is correct, show that C_P can be used as an estimate of $|P|$, the number of parameters in the model.

5.14 For the OLS estimator $\widehat{\boldsymbol{\beta}}$ in the linear regrssion model $\mathcal{Y} = \mathcal{X}\boldsymbol{\beta} + e$, where e has mean zero, show that $ESS(\boldsymbol{\beta}) = RSS + (\boldsymbol{\beta} - \widehat{\boldsymbol{\beta}}_{\mathrm{ols}})^\tau \mathcal{X}^\tau \mathcal{X}(\boldsymbol{\beta} - \widehat{\boldsymbol{\beta}}_{\mathrm{ols}})$, where $RSS = ESS(\widehat{\boldsymbol{\beta}})$.

5.15 Consider the matrix $\mathcal{X}$. Center and scale each column of $\mathcal{X}$ so that $\mathcal{X}^\tau \mathcal{X}$ is the correlation matrix. Regress the kth column of $\mathcal{X}$ on the other $r - 1$ columns of $\mathcal{X}$ in a multiple regression. Compute the residual sum of squares, RSS_k, $k = 1, 2, \ldots, r$, for each column. Near collinearity exhibits irself when at least one of the $RSS_1, RSS_2, \ldots, RSS_r$ is small. Show that RSS_k is the square-root of the kth diagonal element of $(\mathcal{X}^\tau \mathcal{X})^{-1}$, which is referred to as the reciprocal square-root of VIF_k. Show that $VIF_k = (1 - R_k^2)^{-1}$, where R_k^2 is the squared multiple correlation coefficient of the kth column of $\mathcal{X}$ regressed on the other $r - 1$ columns of $\mathcal{X}$, $k = 1, 2, \ldots, r$.

5.16 Suppose the error component $\mathbf{e}$ of the linear regression model has mean $\mathbf{0}$, but now has $\mathrm{var}(\mathbf{e}) = \sigma^2 \mathbf{V}$, where $\mathbf{V}$ is a known $(n \times n)$ positive-definite symmetric matrix and $\sigma^2 > 0$ may not be necessarily known. Let $\widehat{\boldsymbol{\beta}}_{\mathrm{gls}}$ denote the *generalized least-squares (GLS)* estimator:

$$\widehat{\boldsymbol{\beta}}_{\mathrm{gls}} = \arg \min_{\boldsymbol{\beta}} \ (\mathcal{Y} - \mathcal{Z}\boldsymbol{\beta})^\tau \mathbf{V}^{-1} (\mathcal{Y} - \mathcal{Z}\boldsymbol{\beta}).$$

Show that

$$\widehat{\boldsymbol{\beta}}_{\mathrm{gls}} = (\mathcal{Z}^\tau \mathbf{V}^{-1} \mathcal{Z})^{-1} \mathcal{Z}^\tau \mathbf{V}^{-1} \mathcal{Y}$$

has expectation $\boldsymbol{\beta}$ and covariance matrix

$$\mathrm{var}(\widehat{\boldsymbol{\beta}}_{\mathrm{gls}}) = \sigma^2 (\mathcal{Z}^\tau \mathbf{V}^{-1} \mathcal{Z})^{-1}.$$

5.17 What would be the consequences of incorrectly using the ordinary least-squares estimator $\widehat{\boldsymbol{\beta}}_{\mathrm{ols}} = (\mathcal{Z}^{\tau}\mathcal{Z})^{-1}\mathcal{Z}^{\tau}\mathcal{Y}$, of $\boldsymbol{\beta}$ when $\mathrm{var}(\mathbf{e}) = \sigma^2\mathbf{V}$?

5.18 The Boston housing data can be downloaded from the STATLIB website `lib.stat.cmu.edu/datasets/boston_corrected.txt`. There are 506 observations on census tracts in the Boston Standard Metropolitan Statistical Area (SMSA) in 1970. The response variable is the logarithm of the median value of owner-occupied homes in thousands of dollars; there are 13 input variables (plus information on location of each observation). Compute the OLS estimates and compare them with those obtained from the following variable-selection algorithms: Forwards Selection (stepwise), C_p, the Lasso, LARS, and Forwards Stagewise.

5.19 Repeat comparisons between variable-selection algorithms in Exercise 5.18 for The Insurance Company Benchmark data set. The data gives information on customers of an insurance company and contains 86 variables on product-usage data and socio-demographic data derived from zip area codes. There are 5,822 customers in the learning set and another 4,000 in the test set. The data were collected to answer the following question: Can you predict who would be interested in buying a caravan insurance policy and give an explanation why? The data can be downloaded from `kdd.ics.uci.edu/databases/tic/tic.html`.

6
Multivariate Regression

6.1 Introduction

Multivariate linear regression is a natural extension of multiple linear regression in that both techniques try to interpret possible linear relationships between certain input and output variables. Multiple regression is concerned with studying to what extent the behavior of a single output variable Y is influenced by a set of r input variables $\mathbf{X} = (X_1, \cdots, X_r)^\tau$. Multivariate regression has s output variables $\mathbf{Y} = (Y_1, \cdots, Y_s)^\tau$, each of whose behavior may be influenced by exactly the same set of inputs $\mathbf{X} = (X_1, \cdots, X_r)^\tau$.

So, not only are the components of $\mathbf{X}$ correlated with each other, but in multivariate regression, the components of $\mathbf{Y}$ are also correlated with each other (and with the components of $\mathbf{X}$). In this chapter, we are interested in estimating the regression relationship between $\mathbf{Y}$ and $\mathbf{X}$, taking into account the various dependencies between the r-vector $\mathbf{X}$ and the s-vector $\mathbf{Y}$ and the dependencies within $\mathbf{X}$ and within $\mathbf{Y}$.

We describe two different multivariate regression scenarios, analogous to the fixed-X and random-X scenarios of multiple regression. In particular, we consider restricted versions of the multivariate regression problem based upon constraining the relationship between $\mathbf{Y}$ and $\mathbf{X}$ in some way. Such

A.J. Izenman, *Modern Multivariate Statistical Techniques*,
doi: 10.1007/978-0-387-78189-1_6,
© Springer Science+Business Media, LLC 2008

constraints may be linear or nonlinear in form, and they may be known or unknown to the researcher prior to statistical analysis. Our approach is guided by the well-known principle that major theoretical, computational, and practical advantages may result if one is able to express a wide variety of statistics problems in terms of a common focus, especially where that focus is regression analysis.

With this in mind, we describe the *multivariate reduced-rank regression model (RRR)* (Izenman, 1975), which is an enhancement of the classical multivariate regression model and has recently received research attention in the statistics and econometrics literature. The following reasons explain the popularity of this model: RRR provides a unified approach to many of the diverse classical multivariate statistical techniques; it lends itself quite naturally to analyzing a wide variety of statistical problems involving reduction of dimensionality and the search for structure in multivariate data; and it is relatively simple to program because the regression estimates depend only upon the sample covariance matrices of $\mathbf{X}$ and $\mathbf{Y}$ and the eigendecomposition of a certain symmetric matrix that generalizes the multiple squared correlation coefficient R^2 from multiple regression.

6.2   The Fixed-X Case

Let $\mathbf{Y} = (Y_1, \cdots, Y_s)^\tau$ be a random s-vector-valued output variate with mean vector $\boldsymbol{\mu}_Y$ and covariance matrix $\boldsymbol{\Sigma}_{YY}$, and let $\mathbf{X} = (X_1, \cdots, X_r)^\tau$ be a fixed (nonstochastic) r-vector-valued input variate. The components of the output vector $\mathbf{Y}$ will typically be continuous responses, and the components of the input vector $\mathbf{X}$ may be indicator or "dummy" variables that are set up by the researcher to identify known groupings of the data associated with distinct subpopulations or experimental conditions.

Suppose we observe n replications,

$$(\mathbf{X}_j^\tau, \mathbf{Y}_j^\tau)^\tau, \quad j = 1, 2, \ldots, n, \tag{6.1}$$

on the $(r + s)$-vector $(\mathbf{X}^\tau, \mathbf{Y}^\tau)^\tau$. We define an $(r \times n)$-matrix $\mathcal{X}$ and an $(s \times n)$-matrix $\mathcal{Y}$ by

$$\overset{r \times n}{\mathcal{X}} = (\mathbf{X}_1, \cdots, \mathbf{X}_n), \quad \overset{s \times n}{\mathcal{Y}} = (\mathbf{Y}_1, \cdots, \mathbf{Y}_n). \tag{6.2}$$

Form the mean vectors,

$$\overset{r \times 1}{\bar{\mathbf{X}}} = n^{-1} \sum_{j=1}^{n} \mathbf{X}_j, \quad \overset{s \times 1}{\bar{\mathbf{Y}}} = n^{-1} \sum_{j=1}^{n} \mathbf{Y}_j, \tag{6.3}$$

and let

$$\overset{r \times n}{\bar{\mathcal{X}}} = (\bar{\mathbf{X}}, \cdots, \bar{\mathbf{X}}), \quad \overset{s \times n}{\bar{\mathcal{Y}}} = (\bar{\mathbf{Y}}, \cdots, \bar{\mathbf{Y}}) \tag{6.4}$$

be an $(r \times n)$-matrix and an $(s \times n)$-matrix, respectively. The centered versions of $\mathcal{X}$ and $\mathcal{Y}$ are defined by

$$\overset{r \times n}{\mathcal{X}_c} = \mathcal{X} - \bar{\mathcal{X}} = (\mathbf{X}_1 - \bar{\mathbf{X}}, \cdots, \mathbf{X}_n - \bar{\mathbf{X}}), \tag{6.5}$$

$$\overset{s \times n}{\mathcal{Y}_c} = \mathcal{Y} - \bar{\mathcal{Y}} = (\mathbf{Y}_1 - \bar{\mathbf{Y}}, \cdots, \mathbf{Y}_n - \bar{\mathbf{Y}}), \tag{6.6}$$

respectively.

6.2.1 Classical Multivariate Regression Model

Consider the multivariate linear regression model

$$\overset{s \times n}{\mathcal{Y}} = \overset{s \times n}{\boldsymbol{\mu}} + \overset{s \times r}{\boldsymbol{\Theta}} \overset{r \times n}{\mathcal{X}} + \overset{s \times n}{\mathcal{E}}, \tag{6.7}$$

where $\boldsymbol{\mu}$ is an $(s \times n)$-matrix of unknown constants, $\boldsymbol{\Theta} = (\theta_{jk})$ is an $(s \times r)$-matrix of unknown regression coefficients, and $\mathcal{E} = (\mathcal{E}_1, \mathcal{E}_2, \cdots, \mathcal{E}_n)$ is the $(s \times n)$ error matrix whose columns are each random s-vectors with mean $\mathbf{0}$ and the same unknown nonsingular $(s \times s)$ *error covariance matrix* $\boldsymbol{\Sigma}_{\mathcal{E}\mathcal{E}}$, and pairs of column vectors, $(\mathcal{E}_j, \mathcal{E}_k)$, $j \neq k$, are uncorrelated with each other.

When the Xs are considered to be fixed in repeated sampling (e.g., in designed experiments), the so-called *design matrix* $\mathcal{X}$ consists of known constants and possibly also observed values of covariates, $\boldsymbol{\Theta}$ is a full-rank matrix of unknown *fixed effects*, and $\boldsymbol{\mu} = \boldsymbol{\mu}_0 \mathbf{1}_n^\tau$, where $\boldsymbol{\mu}_0$ is an unknown s-vector of constants.

Consider the problem of estimating arbitrary linear combinations of the $\{\theta_{jk}\}$,

$$\text{tr}(\mathbf{A}\boldsymbol{\Theta}) = \sum_j \sum_k A_{jk} \theta_{jk}, \tag{6.8}$$

where $\mathbf{A} = (A_{jk})$ is an arbitrary matrix of constants. There are two equivalent ways to proceed. On the one hand, we can write

$$\boldsymbol{\mu} + \boldsymbol{\Theta}\mathcal{X} = \boldsymbol{\Theta}^* \mathcal{X}^*, \tag{6.9}$$

where $\boldsymbol{\Theta}^* = (\boldsymbol{\mu}_0 \vdots \boldsymbol{\Theta})$ and $\mathcal{X}^* = (\mathbf{1}_n \vdots \mathbf{X}^\tau)^\tau$, and then estimate $\boldsymbol{\Theta}^*$. The other way is to remove $\boldsymbol{\mu}$ from the equation by centering $\mathcal{X}$ and $\mathcal{Y}$ and then estimate $\boldsymbol{\Theta}$ directly. It is the latter procedure we give here. The reader should verify that both procedures lead to the same results (see Exercise 6.7).

LS Estimation

If we set $\boldsymbol{\mu} = \bar{\mathcal{Y}} - \boldsymbol{\Theta}\bar{\mathcal{X}}$, the model (6.7) reduces to

$$\overset{s \times n}{\mathcal{Y}_c} = \overset{r \times n}{\boldsymbol{\Theta}} \overset{r \times n}{\mathcal{X}_c} + \overset{s \times n}{\mathcal{E}}. \tag{6.10}$$

Applying the "vec" operation to equation (6.10), we get

$$\overset{sn \times 1}{\text{vec}(\mathcal{Y}_c)} = \overset{sn \times sr}{(\mathbf{I}_s \otimes \mathcal{X}_c^\tau)}\overset{sr \times 1}{\text{vec}(\mathbf{\Theta})} + \overset{sn \times 1}{\text{vec}(\mathcal{E})} . \tag{6.11}$$

We see that the relationship (6.11) is just a multiple linear regression. The error variate $\text{vec}(\mathcal{E})$ has mean vector $\mathbf{0}$ and $(sn \times sn)$ block-diagonal covariance matrix,

$$\text{cov}(\text{vec}(\mathcal{E})) = \text{E}\{(\text{vec}(\mathcal{E}))(\text{vec}(\mathcal{E}))^\tau\} = \mathbf{\Sigma}_{\mathcal{E}\mathcal{E}} \otimes \mathbf{I}_n. \tag{6.12}$$

Assuming that $\mathcal{X}_c\mathcal{X}_c^\tau$ is nonsingular and using Exercise 5.16, the generalized least-squares estimator of $\text{vec}(\mathbf{\Theta})$ is given by

$$\text{vec}(\widehat{\mathbf{\Theta}}) = \tag{6.13}$$
$$((\mathbf{I}_s \otimes \mathcal{X}_c)(\mathbf{\Sigma}_{\mathcal{E}\mathcal{E}} \otimes \mathbf{I}_n)^{-1}(\mathbf{I}_s \otimes \mathcal{X}_c^\tau))^{-1}(\mathbf{I}_s \otimes \mathcal{X}_c)(\mathbf{\Sigma}_{\mathcal{E}\mathcal{E}} \otimes \mathbf{I}_n)^{-1}\text{vec}(\mathcal{Y}_c)$$
$$= (\mathbf{I}_s \otimes (\mathcal{X}_c\mathcal{X}_c^\tau)^{-1}\mathcal{X}_c)\text{vec}(\mathcal{Y}_c), \tag{6.14}$$

using results on Kronecker products of matrices. By "un-vec'ing" (6.14), it follows that

$$\widehat{\mathbf{\Theta}} = \mathcal{Y}_c\mathcal{X}_c^\tau(\mathcal{X}_c\mathcal{X}_c^\tau)^{-1}, \tag{6.15}$$

$$\widehat{\boldsymbol{\mu}} = \bar{\mathcal{y}} - \widehat{\mathbf{\Theta}}\bar{\mathcal{x}}, \tag{6.16}$$

so that $\widehat{\boldsymbol{\mu}}_0 = \bar{\mathbf{Y}} - \widehat{\mathbf{\Theta}}\bar{\mathbf{X}}$.

Thus, *under the above conditions and if $\mathcal{X}_c\mathcal{X}_c^\tau$ is nonsingular, then the minimum-variance linear unbiased estimator of* $\text{tr}(\mathbf{A}\mathbf{\Theta})$ *is given by* $\text{tr}(\mathbf{A}\widehat{\mathbf{\Theta}})$. This is the multivariate form of the Gauss–Markov theorem.

We can interpret the estimator $\widehat{\mathbf{\Theta}}$ in an important way. Suppose we transpose the regression equation (6.10) so that

$$\overset{n \times s}{\mathbf{Z}} = \overset{n \times r}{\mathbf{W}}\overset{r \times s}{\boldsymbol{\beta}} + \overset{n \times s}{\mathbf{E}} , \tag{6.17}$$

where $\mathbf{Z} = \mathcal{Y}_c^\tau$, $\mathbf{W} = \mathcal{X}_c^\tau$, $\boldsymbol{\beta} = \mathbf{\Theta}^\tau$, and $\mathbf{E} = \mathcal{E}^\tau$. The ith row vector, $\mathcal{Y}_{c(i)}$, of $\mathcal{Y}_c$ corresponds to the ith column vector, $\mathbf{z}_i$, of $\mathbf{Z}$ and represents all the n (mean-centered) observations on the ith output variable $\mathcal{Y}_{cij} = \mathcal{Y}_{ij} - \bar{\mathcal{Y}}_i$, $j = 1, 2, \ldots, n$. Thus, the n-vector $\mathbf{z}_i$ can be modeled by the multiple regression equation,

$$\overset{n \times 1}{\mathbf{z}_i} = \overset{n \times r}{\mathbf{W}}\overset{r \times 1}{\boldsymbol{\beta}_i} + \overset{n \times 1}{\mathbf{e}_i}, \tag{6.18}$$

where $\boldsymbol{\beta}_i$ is the ith column of $\boldsymbol{\beta}$, and $\mathbf{e}_i$ is the ith column of $\mathbf{E}$. The OLS estimate of $\boldsymbol{\beta}_i$ is

$$\widehat{\boldsymbol{\beta}}_i = (\mathbf{W}^\tau\mathbf{W})^{-1}\mathbf{W}^\tau\mathbf{z}_i. \tag{6.19}$$

Transforming back, we get that the least-squares estimator of $\boldsymbol{\theta}_{(i)}$ (i.e., the ith row of $\mathbf{\Theta}$) is

$$\widehat{\boldsymbol{\theta}}_{(i)} = \mathcal{Y}_{c(i)}\mathcal{X}_c^\tau(\mathcal{X}_c\mathcal{X}_c^\tau)^{-1}, \tag{6.20}$$

which is the ith row of $\widehat{\boldsymbol{\Theta}}$.

Thus, simultaneous (unrestricted) least-squares estimation applied to all the s equations of the multivariate regression model yields the same results as does equation-by-equation least-squares. As a result, nothing is gained by estimating the equations jointly, even though the output variables $\mathbf{Y}$ *may be correlated.*

In other words, even though the variables in $\mathbf{Y}$ may be correlated, perhaps even heavily correlated, the LS estimator, $\widehat{\boldsymbol{\Theta}}$, of $\boldsymbol{\Theta}$ does not contain any reference to that correlation. Indeed, the result says that in order to estimate the matrix of regression coefficients $\boldsymbol{\Theta}$ in a multivariate regression, all we need to do is (1) run s multiple regressions, each using a different Y variable, on all the X variables, (2) compute the vector of regression coefficient estimates, $\widehat{\boldsymbol{\theta}}_{(i)}$, $i = 1, 2, \ldots, s$, from each multiple regression, and then (3) arrange those estimates together into a matrix, which will be $\widehat{\boldsymbol{\Theta}}$. To those who encounter this result for the first time, it can be quite surprising!

In its basic classical formulation, therefore, we see that *multivariate regression is a procedure that has no true multivariate content*. That is, there is no reason to create specialized software to carry out a multivariate regression of $\mathbf{Y}$ on $\mathbf{X}$ when the same result can more easily be obtained by using existing multiple regression routines. This is one reason why many books on multivariate analysis do not contain a separate chapter on multivariate regression and also why the topics of multiple regression and multivariate regression are so often confused with each other.

Covariance Matrix of $\widehat{\boldsymbol{\Theta}}$

Using the "vec" operation and Kronecker products, it is not difficult to obtain the covariance matrix for $\widehat{\boldsymbol{\Theta}}$. Substituting (6.10) for $\mathcal{Y}_c$ into (6.15), we have that

$$\widehat{\boldsymbol{\Theta}} = (\boldsymbol{\Theta}\mathcal{X}_c + \mathcal{E})\mathcal{X}_c^\tau(\mathcal{X}_c\mathcal{X}_c^\tau)^{-1} = \boldsymbol{\Theta} + \mathcal{E}\mathcal{X}_c^\tau(\mathcal{X}_c\mathcal{X}^\tau)^{-1}. \tag{6.21}$$

Using the fact that $\mathcal{X}_c$ is a fixed matrix and that $\mathcal{E}$ has mean zero, we have that $\mathrm{vec}(\widehat{\boldsymbol{\Theta}})$ has mean $\mathrm{vec}(\boldsymbol{\Theta})$. Now, from (6.21),

$$\mathrm{vec}(\widehat{\boldsymbol{\Theta}} - \boldsymbol{\Theta}) = \mathrm{vec}(\mathcal{E}\mathcal{X}_c^\tau(\mathcal{X}_c\mathcal{X}_c^\tau)^{-1}) = (\mathbf{I}_s \otimes (\mathcal{X}_c\mathcal{X}^\tau)^{-1}\mathcal{X}_c)\mathrm{vec}(\mathcal{E}),$$

whence,

$$
\begin{aligned}
\mathrm{cov}(\mathrm{vec}(\widehat{\boldsymbol{\Theta}})) &= \mathrm{E}\{(\mathrm{vec}(\widehat{\boldsymbol{\Theta}} - \boldsymbol{\Theta}))(\mathrm{vec}(\widehat{\boldsymbol{\Theta}} - \boldsymbol{\Theta}))^\tau\} \\
&= (\mathbf{I}_s \otimes (\mathcal{X}_c\mathcal{X}_c^\tau)^{-1}\mathcal{X}_c)(\boldsymbol{\Sigma}_{\mathcal{E}\mathcal{E}} \otimes \mathbf{I}_n)(\mathbf{I}_s \otimes \mathcal{X}_c^\tau(\mathcal{X}_c\mathcal{X}_c^\tau)^{-1}) \\
&= \boldsymbol{\Sigma}_{\mathcal{E}\mathcal{E}} \otimes (\mathcal{X}_c\mathcal{X}_c^\tau)^{-1}, \tag{6.22}
\end{aligned}
$$

by using the multiplicative properties of Kronecker products.

So far, we have obtained the LS estimators of the multivariate linear regression model without imposing any distributional assumptions on the errors. If we now assume that the errors in the model are distributed as iid Gaussian random vectors,

$$\mathcal{E}_j \overset{iid}{\sim} \mathcal{N}_s(\mathbf{0}, \Sigma_{\mathcal{E}\mathcal{E}}), \ j = 1, 2, \ldots, n, \tag{6.23}$$

then,

$$\text{vec}(\widehat{\Theta}) \sim \mathcal{N}_{rs}(\text{vec}(\Theta), \Sigma_{\mathcal{E}\mathcal{E}} \otimes (\mathcal{X}_c \mathcal{X}_c^\tau)^{-1}). \tag{6.24}$$

Furthermore, the distribution of the least-squares estimator (6.20) is

$$\widehat{\boldsymbol{\theta}}_{(i)} \sim \mathcal{N}_r(\boldsymbol{\theta}_{(i)}, \sigma_i^2 (\mathcal{X}_c \mathcal{X}_c^\tau)^{-1}), \tag{6.25}$$

where σ_i^2 is the ith diagonal entry of $\Sigma_{\mathcal{E}\mathcal{E}}$, $i = 1, 2, \ldots, s$. Compare with (5.42).

If $\mathcal{X}_c$ has less than full rank, then the $(r \times r)$-matrix $\mathcal{X}_c \mathcal{X}_c^\tau$ will be singular. In this case, we can replace the $(\mathcal{X}_c \mathcal{X}_c^\tau)^{-1}$ term either by a generalized inverse $(\mathcal{X}_c \mathcal{X}_c^\tau)^-$ or by a ridge-regression-like term such as $(\mathcal{X}_c \mathcal{X}_c^\tau + k\mathbf{I}_r)^{-1}$, where k is a positive constant; see Section 5.6.4.

Fitted Values and Multivariate Residuals

The $(s \times n)$ matrix $\widehat{\mathcal{Y}}$ of *fitted values* is given by

$$\widehat{\mathcal{Y}} = \widehat{\boldsymbol{\mu}} + \widehat{\Theta}\mathcal{X} = \bar{\mathcal{Y}} + \widehat{\Theta}(\mathcal{X} - \bar{\mathcal{X}}), \tag{6.26}$$

or

$$\widehat{\mathcal{Y}}_c = \widehat{\Theta}\mathcal{X}_c = \mathcal{Y}_c \mathcal{X}_c^\tau (\mathcal{X}_c \mathcal{X}_c^\tau)^{-1} \mathcal{X}_c = \mathcal{Y}_c \mathbf{H}, \tag{6.27}$$

where the $(n \times n)$ matrix $\mathbf{H} = \mathcal{X}_c^\tau (\mathcal{X}_c \mathcal{X}_c^\tau)^{-1} \mathcal{X}_c$ is the *hat-matrix*.

The $(s \times n)$ *residual matrix* $\widehat{\mathcal{E}}$ is the difference between the observed and fitted values of $\mathcal{Y}$, namely,

$$\widehat{\mathcal{E}} = \mathcal{Y} - \widehat{\mathcal{Y}} = \mathcal{Y}_c - \widehat{\Theta}\mathcal{X}_c = \mathcal{Y}_c - \widehat{\mathcal{Y}}_c = \mathcal{Y}_c(\mathbf{I}_n - \mathbf{H}), \tag{6.28}$$

and, using (6.27), can also be written as

$$\begin{aligned}
\widehat{\mathcal{E}} &= \mathcal{Y}_c - \widehat{\Theta}\mathcal{X}_c \\
&= (\Theta\mathcal{X}_c + \mathcal{E}) - (\Theta + \mathcal{E}\mathcal{X}_c^\tau(\mathcal{X}_c\mathcal{X}_c^\tau)^{-1})\mathcal{X}_c \\
&= \mathcal{E}(\mathbf{I}_n - \mathbf{H}).
\end{aligned} \tag{6.29}$$

It follows immediately that $\text{E}(\text{vec}(\widehat{\mathcal{E}})) = \mathbf{0}$. A straightforward calculation shows that

$$\text{cov}(\text{vec}(\widehat{\mathcal{E}})) = \Sigma_{\mathcal{E}\mathcal{E}} \otimes (\mathbf{I}_n - \mathbf{H}). \tag{6.30}$$

The $(s \times s)$ matrix version of the residual sum of squares is

$$\mathbf{S}_e = \widehat{\mathcal{E}}\widehat{\mathcal{E}}^\tau = (\mathcal{Y}_c - \widehat{\Theta}\mathcal{X}_c)(\mathcal{Y}_c - \widehat{\Theta}\mathcal{X}_c)^\tau = \mathcal{Y}_c(\mathbf{I}_n - \mathbf{H})\mathcal{Y}_c^\tau. \qquad (6.31)$$

It is not difficult to show that $\mathbf{S}_e = \mathcal{E}(\mathbf{I}_n - \mathbf{H})\mathcal{E}^\tau$. Let $\mathcal{E}_{(j)}$ be the jth row of $\mathcal{E}$. Then, the jkth element of $\mathbf{S}_e$ can be written as

$$(\mathbf{S}_e)_{jk} = \mathcal{E}_{(j)}(\mathbf{I}_n - \mathbf{H})\mathcal{E}_{(k)}^\tau,$$

whence,

$$
\begin{aligned}
\mathrm{E}\{(\mathbf{S}_e)_{jk}\} &= \mathrm{E}\{\mathrm{tr}((\mathbf{I}_n - \mathbf{H})\mathcal{E}_{(k)}^\tau\mathcal{E}_{(j)})\} \\
&= \mathrm{tr}(\mathbf{I}_n - \mathbf{H}) \cdot (\boldsymbol{\Sigma}_{\mathcal{E}\mathcal{E}})_{jk} \\
&= (n - r)(\boldsymbol{\Sigma}_{\mathcal{E}\mathcal{E}})_{jk}.
\end{aligned}
$$

We can now state the statistical properties of an estimate of the error covariance matrix. *The residual covariance matrix,*

$$\widehat{\boldsymbol{\Sigma}}_{\mathcal{E}\mathcal{E}} = \frac{1}{n - r} \mathbf{S}_e, \qquad (6.32)$$

is statistically independent of $\widehat{\Theta}$ and has a Wishart distribution with $n - r$ degrees of freedom and expectation $\boldsymbol{\Sigma}_{\mathcal{E}\mathcal{E}}$. We see that the residual covariance matrix $\widehat{\boldsymbol{\Sigma}}_{\mathcal{E}\mathcal{E}}$ is an unbiased estimator for the error covariance matrix $\boldsymbol{\Sigma}_{\mathcal{E}\mathcal{E}}$.

The covariance matrix of $\widehat{\Theta}$ can, therefore, be estimated by

$$\widehat{\mathrm{cov}}(\mathrm{vec}(\widehat{\Theta})) = \widehat{\boldsymbol{\Sigma}}_{\mathcal{E}\mathcal{E}} \otimes (\mathcal{X}_c\mathcal{X}_c^\tau)^{-1}, \qquad (6.33)$$

where $\widehat{\boldsymbol{\Sigma}}_{\mathcal{E}\mathcal{E}}$ is given by (6.32).

Confidence Intervals

We can now construct confidence intervals for arbitrary linear combinations of $\mathrm{vec}(\Theta)$. Let $\boldsymbol{\gamma}$ be an arbitrary sr-vector and consider $\boldsymbol{\gamma}^\tau\mathrm{vec}(\widehat{\Theta})$. Assuming the error vectors are s-variate Gaussian as in (6.23), the independence of (6.15) and (6.32) means that the pivotal quantity

$$t = \frac{\boldsymbol{\gamma}^\tau(\mathrm{vec}(\widehat{\Theta} - \Theta))}{\{\boldsymbol{\gamma}^\tau(\widehat{\boldsymbol{\Sigma}}_{\mathcal{E}\mathcal{E}} \otimes (\mathcal{X}_c\mathcal{X}_c^\tau)^{-1})\boldsymbol{\gamma}\}^{1/2}} \qquad (6.34)$$

has the Student's t-distribution with $n - r$ degrees of freedom. Thus, a $(1 - \alpha) \times 100\%$ confidence interval for $\boldsymbol{\gamma}^\tau\mathrm{vec}(\Theta)$ can be given by

$$\boldsymbol{\gamma}^\tau\mathrm{vec}(\widehat{\Theta}) \pm t_{n-r}^{\alpha/2} \{\boldsymbol{\gamma}^\tau(\widehat{\boldsymbol{\Sigma}}_{\mathcal{E}\mathcal{E}} \otimes (\mathcal{X}_c\mathcal{X}_c^\tau)^{-1})\boldsymbol{\gamma}\}^{1/2}, \qquad (6.35)$$

where $t_{n-r}^{\alpha/2}$ is the $(1 - \alpha/2) \times 100\%$-point of the t_{n-r}-distribution.

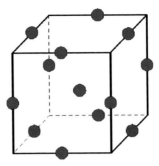

FIGURE 6.1. *Three-variable Box–Behnken design for the Norwegian Paper Quality experiment. The three variables, X_1, X_2, and X_3, each have values $-1, 0$, or 1. There are 13 design points consisting of the midpoints of each of the 12 edges of a three-dimensional cube and a point at the center of the cube. Source: NIST/SEMATECH e-Handbook of Statistical Methods,* www.itl.nist.gov/div898/handbook/pri/section3/pri3362.htm.

6.2.2   Example: Norwegian Paper Quality

These data[1] were obtained from a controlled experiment carried out in the paper-making factory of Norske Skog located in Skogn, Norway (Aldrin, 2000), which is the world's second-largest producer of publication paper. There are $s = 13$ response variables, $Y_1, \ldots, Y_{13}$, which measure different characteristics of paper.

The purpose of the experiment was to uncover how these response variables were influenced by three predictor variables, X_1, X_2, X_3, each of which is controlled exactly with values $-1, 0$, or 1 according to a 3-variable Box–Behnken design (Box and Behnken, 1960). See Figure 6.1. The 13-point design can be represented as the midpoints of each of the 12 edges of a three-dimensional cube and a point $(0, 0, 0)$ at the center of the cube.

At each of 11 design points, the response variables were measured twice; at the design point $(0, 1, 1)$, the response variables were measured only once; at the center point, the response variables were measured six times. To allow for interactions and nonlinear effects, the standard model for such designs includes an additional six predictor variables defined as $X_4 = X_1^2, X_5 = X_2^2, X_6 = X_3^2, X_7 = X_1 X_2, X_8 = X_1 X_3, X_9 = X_2 X_3$, so that $r = 9$. The data set, therefore, consists of 29 observations measured on each of $r + s = 9 + 13 = 22$ variables.

---

[1]The data, which originally appeared in Aldrin (1996), can be found in the file norwaypaper1.txt on the book's website or can be downloaded from the STATLIB website lib.stat.cmu.edu/datasets.

TABLE 6.1. *Norwegian paper quality data. This is the* (13×9)*-matrix of estimated regression coefficients,* $\widehat{\Theta}$*. The number of X-variables is $r = 9$, the number of Y-variables is $s = 13$, and the number of observations is $n = 29$.*

0.752	-0.449	-0.365	0.105	-0.291	0.545	0.111	0.390	0.217
-0.844	0.350	0.369	-0.039	0.226	-0.567	-0.141	-0.537	-0.324
0.286	-0.670	-0.572	0.044	-0.283	0.534	0.065	0.408	-0.163
0.497	-0.491	-0.666	0.142	-0.391	0.450	0.068	0.195	0.020
0.515	0.143	-0.570	-0.182	-0.372	0.420	-0.158	0.792	0.602
-0.717	0.039	-0.215	0.346	-0.362	0.055	0.139	0.462	0.125
0.878	0.051	-0.269	-0.324	-0.015	0.228	-0.243	0.126	0.255
-0.564	0.194	-0.357	-0.002	-0.427	0.046	0.236	-0.446	0.257
0.287	0.497	-0.600	-0.382	-0.011	0.837	0.143	0.380	-0.121
-0.654	-0.145	0.111	0.221	-0.354	-0.524	0.057	-0.682	0.336
0.174	-0.714	0.329	0.146	0.143	-0.144	0.086	-0.826	-0.731
-0.526	0.283	0.541	-0.832	0.428	0.339	0.214	0.125	0.173
0.505	0.052	0.428	-0.704	0.561	0.557	-0.231	-0.245	-0.181

Regressing $\mathbf{Y} = (Y_1, \cdots, Y_{13})^\tau$ on $\mathbf{X} = (X_1, \cdots, X_9)^\tau$, using formulas (6.15) and (6.16), yields the estimated mean vector $\widehat{\boldsymbol{\mu}}$,

$$\widehat{\boldsymbol{\mu}} = (32.393, 31.678, 7.034, 7.826, 14.734, 12.455, 9.996, 18.502,$$
$$22.414, 17.817, 21.405, 90.166, 23.547)^\tau, \tag{6.36}$$

and the (13×9)-matrix of estimated regression coefficients $\widehat{\Theta}$, which is given in Table 6.1. Each row of Table 6.1 can also be obtained by regressing the Y variable corresponding to that row on all nine X variables; see Ex. 6.8.

6.2.3 Separate and Multivariate Ridge Regressions

As we have seen, multivariate OLS regression reduces to a collection of s separate multiple OLS regressions. We can improve substantially upon OLS while still pursuing an equation-by-equation regression strategy by applying a biased regression procedure, such as ridge regression, separately to each output variable.

Using the penalized least-squares formulation of uniresponse ridge regression (see Section 5.8.3), let

$$\phi_j(\boldsymbol{\beta}) = (\mathbf{y}_j - \mathbf{X}\boldsymbol{\beta})^\tau (\mathbf{y}_j - \mathbf{X}\boldsymbol{\beta}) + \lambda_j \boldsymbol{\beta}^\tau \boldsymbol{\beta}, \quad j = 1, 2, \ldots, s, \tag{6.37}$$

where we allow the possibility for different ridge parameters, $\{\lambda_j\}$, for each equation. Separate ridge-regression estimators are the solutions to

$$\widehat{\boldsymbol{\beta}}(\lambda_j) = \arg\min_{\boldsymbol{\beta}} \phi_j(\boldsymbol{\beta}), \quad j = 1, 2, \ldots, s, \tag{6.38}$$

and the separate ridge parameters can be estimated using leave-one-out cross-validation,

$$\widehat{\lambda}_j = \arg\min_\lambda \left\{ \sum_{i=1}^n (y_{j,i} - \widehat{y}_{j,-i}(\lambda))^2 \right\}, \quad j = 1, 2, \ldots, s, \qquad (6.39)$$

where $\widehat{y}_{j,-i}(\lambda)$ is the predicted value (using ridge regression with ridge parameter λ) of the ith case of the jth response variable when the entire ith case is deleted from the learning set (Breiman and Friedman, 1997). Variations on this idea have been used to predict the outcome on election night in every British general election (and British elections to the European parliament) since 1974 (Brown, Firth, and Payne, 1999).

Although ridge regression can be predictively more accurate than is OLS in the case of a single output variable, this equation-by-equation strategy is unsatisfactory because it circumvents the issue that the output variables are correlated and that the combined ridge estimators do not yield a proper Bayes procedure.

Several extensions of (5.99) for the multivariate case have since been proposed that recognize the true multivariate nature of the problem. From (6.15), we have that

$$\text{vec}(\widehat{\Theta}) = (\mathbf{I}_s \otimes \mathcal{X}_c \mathcal{X}_c^\tau)^{-1}(\mathbf{I}_s \otimes \mathcal{X}_c)\text{vec}(\mathcal{Y}_c). \qquad (6.40)$$

A multivariate analogue of (5.99) can be based upon (6.40) by introducing a positive-definite $(s \times s)$ *ridge matrix* $\mathbf{K}$ so that

$$\text{vec}(\widehat{\Theta}(\mathbf{K})) = ((\mathbf{I}_s \otimes \mathcal{X}_c \mathcal{X}_c^\tau) + (\mathbf{K} \otimes \mathbf{I}_r))^{-1}(\mathbf{I}_s \otimes \mathcal{X}_c)\text{vec}(\mathcal{Y}_c) \qquad (6.41)$$

is a *multivariate ridge regression estimator* of $\text{vec}(\Theta)$ (Brown and Zidek, 1980, 1982). The application of (6.41) to predicting British elections uses a diagonal $\mathbf{K}$. Even if $\mathcal{X}_c \mathcal{X}_c^\tau$ is almost singular, (6.41) is still computable. Note that (6.41) reduces to (6.40) if $\mathbf{K} = \mathbf{0}$. If $\mathbf{K}$ is chosen from the data, then the multivariate ridge estimator (6.41) becomes adaptive. A more complicated version of (6.41) was proposed by Haitovsky (1987).

6.2.4 Linear Constraints on the Regression Coefficients

It is sometimes necessary to consider a more restricted model than the classical multivariate regression model. In certain practical situations, we might need the elements of the regression coefficient matrix Θ in the classical model $\mathcal{Y}_c = \Theta\mathcal{X}_c + \mathcal{E}$ to satisfy a set of known linear constraints.

A variety of applications can be based upon the general set of linear constraints,

$$\underset{m \times s}{\mathbf{K}} \; \underset{s \times r}{\Theta} \; \underset{r \times u}{\mathbf{L}} = \underset{m \times u}{\Gamma}, \qquad (6.42)$$

where the matrix $\mathbf{K}$ $(m \leq s)$ and the matrix $\mathbf{L}$ $(u \leq r)$ are full-rank matrices of known constants, and $\mathbf{\Gamma}$ is a matrix of parameters (known or unknown). We often take $\mathbf{\Gamma} = \mathbf{0}$.

In (6.42), the matrix $\mathbf{K}$ is used to set up relationships between the different columns of $\mathbf{\Theta}$ (e.g., treatments), whereas $\mathbf{L}$ generates possible relationships between the different responses. In many problems of this kind, it is common to take $\mathbf{L} = (\mathbf{I}_u \,\vdots\, \mathbf{0})^\tau$, where $\mathbf{0}$ is a $(u \times (r - u))$-matrix of zeroes. There are also situations in which $\mathbf{L}$ can be made more specific; in fact, $\mathbf{L}$ is peculiar to the multiresponse problem and does not have any analogue in the uniresponse situation.

Variable Selection

For example, suppose we wish to study whether a specific subset of the r input variables has little or no effect on the behavior of the output variables. Suppose we arrange the rows of $\mathcal{X}_c$ so that

$$\overset{r \times n}{\mathcal{X}_c} = (\ \overset{n \times r_1}{\mathcal{X}_{c1}^\tau} \ \vdots\ \overset{n \times r_2}{\mathcal{X}_{c2}^\tau}\ )^\tau, \tag{6.43}$$

where $\mathcal{X}_{c1}$ has r_1 rows and $\mathcal{X}_{c2}$ has $r_2 = r - r_1$ rows. Suppose we believe that the variables included in $\mathcal{X}_{c2}$ do not belong in the regression. Corresponding to the partition of $\mathcal{X}_c$, we set $\mathbf{\Theta} = (\mathbf{\Theta}_1 \,\vdots\, \mathbf{\Theta}_2)$, so that

$$\overset{s \times n}{\mathcal{Y}_c} = \overset{s \times r_1}{\mathbf{\Theta}_1} \overset{r_1 \times n}{\mathcal{X}_{c1}} + \overset{s \times r_2}{\mathbf{\Theta}_2} \overset{r_2 \times n}{\mathcal{X}_{c2}} + \overset{s \times n}{\mathcal{E}}\ . \tag{6.44}$$

To study whether the input variables included in $\mathcal{X}_{c2}$ can be eliminated from the model, we set $\mathbf{K} = \mathbf{I}_s$ and $\mathbf{L} = (\mathbf{0} \,\vdots\, \mathbf{I}_{r_2 \times u}^\tau)^\tau$, where $\mathbf{0}$ is a $(u \times r_1)$-matrix of zeroes and $\mathbf{I}_{r_2 \times u}$ is an $(r_2 \times u)$-matrix of ones along the "diagonal" and zeroes elsewhere, so that $\mathbf{K\Theta L} = \mathbf{\Theta}_2 = \mathbf{0}$.

Profile Analysis

The constraints (6.42) can be used to handle a variety of experimental design problems. Such problems include *profile analysis*, where scores on a battery of tests (e.g., different treatments) are recorded on several independent groups of subjects and compared with each other. Typically, profile analysis is carried out on multivariate data obtained from longitudinal studies or clinical trials, where the components of each data vector are ordered by time.

The simplest form of profile analysis deals with a *one-way layout* in which there are r groups of subjects, where the jth group consists of n_j subjects selected randomly to receive one of r treatments, and $n_1 + n_2 + \cdots + n_r = n$.

The scores, which are assumed to be expressed in comparable units, on the s tests by the ith subject are given by the ith column in the $(s \times n)$-matrix $\mathcal{Y} = (\mathbf{Y}_1, \cdots, \mathbf{Y}_n)$. We assume the model,

$$\mathbf{Y}_i = \boldsymbol{\mu} + \boldsymbol{\mu}_i + \mathcal{E}_i, \quad i = 1, 2, \ldots, n, \tag{6.45}$$

where $\mathbf{Y}_i$ is a random s-vector, $\boldsymbol{\mu}$ is an s-vector of constants that represents an overall mean vector, $(\boldsymbol{\mu}_1, \cdots, \boldsymbol{\mu}_n) = \boldsymbol{\Theta}\mathcal{X}$ is an $(s \times n)$-matrix of fixed constants, and $\mathcal{E}_i$ is a random s-vector with mean $\mathbf{0}$ and covariance matrix $\boldsymbol{\Sigma}_{\mathcal{E}\mathcal{E}}$, $i = 1, 2, \ldots, n$. For convenience, we assume $\boldsymbol{\mu} = \mathbf{0}$.

The design matrix $\mathcal{X}$ is constructed using n dummy variables as columns, where the jth row value of the ith column equals 1 if the ith subject is in the jth group, and 0 otherwise:

$$\overset{r \times n}{\mathcal{X}} = \begin{pmatrix} 1 & \cdots & 1 & 0 & \cdots & 0 & \cdots & 0 & \cdots & 0 \\ 0 & \cdots & 0 & 1 & \cdots & 1 & \cdots & 0 & \cdots & 0 \\ \vdots & & \vdots & \vdots & & \vdots & & \vdots & & \vdots \\ 0 & \cdots & 0 & 0 & \cdots & 0 & \cdots & 1 & \cdots & 1 \end{pmatrix}. \tag{6.46}$$

The matrix of regression coefficients $\boldsymbol{\Theta}$ is given by:

$$\overset{s \times r}{\boldsymbol{\Theta}} = \begin{pmatrix} \theta_{11} & \cdots & \theta_{1r} \\ \vdots & & \vdots \\ \theta_{s1} & \cdots & \theta_{sr} \end{pmatrix}. \tag{6.47}$$

The *treatment-mean profile* for the jth group is defined as the s-vector

$$\overset{s \times 1}{\boldsymbol{\theta}_j} = (\theta_{1j}, \cdots, \theta_{sj})^\top, \quad j = 1, 2, \ldots, r. \tag{6.48}$$

The profile of the jth group is displayed as a graph of the points (k, θ_{kj}), $k = 1, 2, \ldots, s$; we connect successive points, (k, θ_{kj}) and $(k + 1, \theta_{k+1,j})$, $k = 1, 2, \ldots, s - 1$, by straight lines. All group profiles are plotted on the same graph for visual comparison.

The population profiles of the r groups are said to be *similar* if the line segments joining successive points of each group's profile are parallel to the corresponding line segments of the profiles of all the other groups. In other words, the population profiles of the different groups are identical but with a constant difference between each pair of profiles. Figure 6.2 displays an example of parallel treatment-mean profiles of three groups $(r = 3)$ at five different timepoints $(s = 5)$. Restricting the profiles to be similar is equivalent to asserting that there is no interaction between treatments and groups.

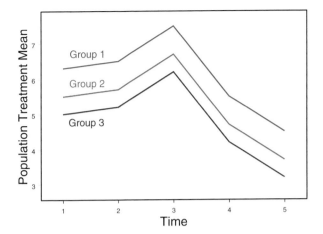

FIGURE 6.2. *Profile plots of population treatment means at five time-points (s = 5) on each of three hypothetical groups (r = 3), where the group profiles are parallel to each other.*

This similarity of the r profiles can be expressed as a set of linear constraints on Θ. To do this, we set the matrix $\mathbf{K}$ to be

$$
\overset{(s-1)\times s}{\mathbf{K}} = \begin{pmatrix} 1 & -1 & 0 & \cdots & 0 \\ 0 & 1 & -1 & \cdots & 0 \\ \vdots & \vdots & \vdots & & \vdots \\ 0 & 0 & 0 & \cdots & -1 \end{pmatrix} \tag{6.49}
$$

and the matrix $\mathbf{L}$ to be

$$
\overset{r\times(r-1)}{\mathbf{L}} = \begin{pmatrix} 1 & 0 & 0 & \cdots & 0 \\ -1 & 1 & 0 & \cdots & 0 \\ 0 & -1 & 1 & \cdots & 0 \\ \vdots & \vdots & \vdots & & \vdots \\ 0 & 0 & 0 & \cdots & -1 \end{pmatrix}, \tag{6.50}
$$

so that $\mathbf{K}\mathbf{1}_s = \mathbf{0}$ and $\mathbf{L}^\tau \mathbf{1}_r = \mathbf{0}$. Setting $\mathbf{K}\Theta\mathbf{L} = \mathbf{0}$ gives constraints on Θ that reduce to

$$
\begin{pmatrix} \theta_{11} - \theta_{12} \\ \vdots \\ \theta_{1,r-1} - \theta_{1r} \end{pmatrix} = \cdots = \begin{pmatrix} \theta_{s1} - \theta_{s2} \\ \vdots \\ \theta_{s,r-1} - \theta_{sr} \end{pmatrix}. \tag{6.51}
$$

Thus, the r treatment mean profiles are to be piecewise-parallel to each other. Alternative $\mathbf{K}$ and $\mathbf{L}$ for this problem are

$$\mathbf{K} = (\mathbf{I}_{s-1} \vdots - \mathbf{1}_s), \quad \mathbf{L} = (\mathbf{I}_{r-1} \vdots - \mathbf{1}_r)^\tau, \tag{6.52}$$

where $\mathbf{1}_s$ is an s-vector of ones.

We can constrain the population treatment mean profiles further, so that not only are they parallel, but also we could require them to be "coincidental" (i.e., identical). To do this, take $\mathbf{K} = \mathbf{1}_s^\tau$ and $\mathbf{L}$ as in (6.52), whence, $\mathbf{K\Theta L} = \mathbf{0}$ translates to $\mathbf{1}_s^\tau \boldsymbol{\theta}_1 = \mathbf{1}_s^\tau \boldsymbol{\theta}_2 = \cdots = \mathbf{1}_s^\tau \boldsymbol{\theta}_r$, which is the condition needed for coincidental profiles.

Constrained Estimation

Consider the problem of finding $\boldsymbol{\Theta}^*$ that solves the following constrained minimization problem:

$$\widehat{\boldsymbol{\Theta}}^* = \arg \min_{\substack{\boldsymbol{\Theta} \\ \mathbf{K\Theta L} = \boldsymbol{\Gamma}}} \mathrm{tr}\{(\mathcal{Y}_c - \boldsymbol{\Theta}\mathcal{X}_c)^\tau (\mathcal{Y}_c - \boldsymbol{\Theta}\mathcal{X}_c)\}. \tag{6.53}$$

Let $\boldsymbol{\Lambda} = (\lambda_{ij})$ be a matrix of Lagrangian coefficients. The normal equations are:

$$\widehat{\boldsymbol{\Theta}}^* \mathcal{X}_c \mathcal{X}_c^\tau + \mathbf{K}^\tau \boldsymbol{\Lambda} \mathbf{L}^\tau = \mathcal{Y}_c \mathcal{X}_c^\tau \tag{6.54}$$

$$\mathbf{K}\widehat{\boldsymbol{\Theta}}^* \mathbf{L} = \boldsymbol{\Gamma}. \tag{6.55}$$

From (6.54), we get

$$\widehat{\boldsymbol{\Theta}}^* = \widehat{\boldsymbol{\Theta}} - \mathbf{K}^\tau \boldsymbol{\Lambda} \mathbf{L}^\tau (\mathcal{X}_c \mathcal{X}_c^\tau)^{-1}, \tag{6.56}$$

where $\widehat{\boldsymbol{\Theta}}$ is given by (6.15). Substituting (6.56) into (6.55) gives

$$\mathbf{K}\mathbf{K}^\tau \boldsymbol{\Lambda} \mathbf{L}^\tau (\mathcal{X}_c \mathcal{X}_c^\tau)^{-1} \mathbf{L} = \mathbf{K}\widehat{\boldsymbol{\Theta}} \mathbf{L} - \boldsymbol{\Gamma}. \tag{6.57}$$

Solving this last expression for $\boldsymbol{\Lambda}$ gives

$$\boldsymbol{\Lambda} = (\mathbf{K}\mathbf{K}^\tau)^{-1} (\mathbf{K}\widehat{\boldsymbol{\Theta}} \mathbf{L} - \boldsymbol{\Gamma}) (\mathbf{L}^\tau (\mathcal{X}_c \mathcal{X}_c^\tau)^{-1} \mathbf{L})^{-1}, \tag{6.58}$$

assuming the appropriate inverses exist. Substituting (6.58) into (6.56) yields

$$\widehat{\boldsymbol{\Theta}}^* = \widehat{\boldsymbol{\Theta}} - \mathbf{K}^\tau (\mathbf{K}\mathbf{K}^\tau)^{-1} (\mathbf{K}\widehat{\boldsymbol{\Theta}} \mathbf{L} - \boldsymbol{\Gamma}) (\mathbf{L}^\tau (\mathcal{X}_c \mathcal{X}_c^\tau)^{-1} \mathbf{L})^{-1} \mathbf{L}^\tau (\mathcal{X}_c \mathcal{X}_c^\tau)^{-1}. \tag{6.59}$$

Check that premultiplying (6.59) by $\mathbf{K}$ and postmultiplying by $\mathbf{L}$ leads to $\mathbf{K}\widehat{\boldsymbol{\Theta}}^* \mathbf{L} = \boldsymbol{\Gamma}$ as required by the constraint in (6.55).

It is common practice in profile analysis to plot the points $(k, \widehat{\theta}_{kj}^*)$, $k = 1, 2, \ldots, s$, corresponding to the jth group, and connect them by straight lines. The treatment-mean profiles for all r groups are usually plotted on the same graph for easy visual comparison.

Multivariate Analysis of Variance (MANOVA)

We now set up the *multivariate analysis of variance (MANOVA)* table for the constrained model. The matrix version of the residual sum of squares, $\mathbf{S}_e^*$, under the constrained model is given by

$$
\begin{aligned}
\mathbf{S}_e^* &= (\mathcal{Y}_c - \widehat{\mathbf{\Theta}}^* \mathcal{X}_c)(\mathcal{Y}_c - \widehat{\mathbf{\Theta}}^* \mathcal{X}_c)^\tau \\
&= ((\mathcal{Y}_c - \widehat{\mathbf{\Theta}} \mathcal{X}_c) + (\widehat{\mathbf{\Theta}} - \widehat{\mathbf{\Theta}}^*) \mathcal{X}_c)((\mathcal{Y}_c - \widehat{\mathbf{\Theta}} \mathcal{X}_c) + (\widehat{\mathbf{\Theta}} - \widehat{\mathbf{\Theta}}^*) \mathcal{X}_c)^\tau \\
&= (\mathcal{Y}_c - \widehat{\mathbf{\Theta}} \mathcal{X}_c)(\mathcal{Y}_c - \widehat{\mathbf{\Theta}} \mathcal{X}_c)^\tau + (\widehat{\mathbf{\Theta}} - \widehat{\mathbf{\Theta}}^*) \mathcal{X}_c \mathcal{X}_c^\tau (\widehat{\mathbf{\Theta}} - \widehat{\mathbf{\Theta}}^*)^\tau, \quad (6.60)
\end{aligned}
$$

where the first term on the rhs of (6.60) is the matrix version of the residual sum of squares, $\mathbf{S}_e$, for the unconstrained model, and the second term is the additional source of variation, $\mathbf{S}_h = \mathbf{S}_e - \mathbf{S}_e^*$, due to dropping the constraints. The cross-product terms disappear because $(\mathcal{Y}_c - \widehat{\mathbf{\Theta}} \mathcal{X}_c) \mathcal{X}_c^\tau = \mathbf{0}$. Note that $\mathbf{S}_e$ is given by (6.31). Furthermore, the matrix version of the regression sum of squares, $\mathbf{S}_{\mathrm{reg}}$, for the unconstrained model is given by

$$
\begin{aligned}
\mathbf{S}_{\mathrm{reg}} &= \widehat{\mathbf{\Theta}} \mathcal{X}_c \mathcal{X}_c^\tau \widehat{\mathbf{\Theta}}^\tau \\
&= (\widehat{\mathbf{\Theta}}^* + (\widehat{\mathbf{\Theta}} - \widehat{\mathbf{\Theta}}^*)) \mathcal{X}_c \mathcal{X}_c^\tau (\widehat{\mathbf{\Theta}}^* + (\widehat{\mathbf{\Theta}} - \widehat{\mathbf{\Theta}}^*))^\tau \\
&= \widehat{\mathbf{\Theta}}^* \mathcal{X}_c \mathcal{X}_c^\tau \widehat{\mathbf{\Theta}}^{*\tau} + (\widehat{\mathbf{\Theta}} - \widehat{\mathbf{\Theta}}^*) \mathcal{X}_c \mathcal{X}_c^\tau (\widehat{\mathbf{\Theta}} - \widehat{\mathbf{\Theta}}^*)^\tau, \quad (6.61)
\end{aligned}
$$

where the cross-product terms disappear. The first term on the rhs of (6.61) is $\mathbf{S}_{\mathrm{reg}}^*$, the matrix version of the regression sum of squares for the constrained model, and the second term is, again, $\mathbf{S}_h$.

We can collect these results in a MANOVA table — see Table 6.2 — in which both the constrained and unconstrained regression models are set out so that their sums of squares and degrees of freedom add up appropriately.

Using (6.58), we can write $\mathbf{S}_h$ more explicitly as follows:

$$
\mathbf{S}_h = \mathbf{K}^\tau (\mathbf{K}\mathbf{K}^\tau)^{-1} (\mathbf{K}\widehat{\mathbf{\Theta}} \mathbf{L} - \mathbf{\Gamma})(\mathbf{L}^\tau (\mathcal{X}_c \mathcal{X}_c^\tau)^{-1} \mathbf{L})^{-1} (\mathbf{K}\widehat{\mathbf{\Theta}} \mathbf{L} - \mathbf{\Gamma})^\tau (\mathbf{K}\mathbf{K}^\tau)^{-1} \mathbf{K}.
$$
$$(6.62)$$

Substituting (6.15) into (6.62), expanding, and taking expectations, we get

$$
\begin{aligned}
\mathrm{E}(\mathbf{S}_h) &= \mathbf{D}(\mathbf{K}\mathbf{\Theta}\mathbf{L} - \mathbf{\Gamma})(\mathbf{L}^\tau (\mathcal{X}_c \mathcal{X}_c^\tau)^{-1} \mathbf{L})^{-1} (\mathbf{K}\mathbf{\Theta}\mathbf{L} - \mathbf{\Gamma})^\tau \mathbf{D}^\tau \\
&\quad + \mathbf{F} \cdot \mathrm{E}(\mathcal{E}\mathbf{G}\mathcal{E}^\tau) \cdot \mathbf{F}^\tau, \quad (6.63)
\end{aligned}
$$

where $\mathbf{D} = \mathbf{K}^\tau (\mathbf{K}\mathbf{K}^\tau)^{-1}$, $\mathbf{F} = \mathbf{D}\mathbf{K}$, and

$$
\mathbf{G} = \mathcal{X}_c^\tau (\mathcal{X}_c \mathcal{X}_c^\tau)^{-1} \mathbf{L} (\mathbf{L}^\tau (\mathcal{X}_c \mathcal{X}_c^\tau)^{-1} \mathbf{L})^{-1} \mathbf{L}^\tau (\mathcal{X}_c \mathcal{X}_c^\tau)^{-1} \mathcal{X}_c^\tau. \quad (6.64)
$$

Notice that $\mathbf{F}^2 = \mathbf{F} = \mathbf{F}^\tau$ and $\mathbf{G}^2 = \mathbf{G} = \mathbf{G}^\tau$, so that $\mathbf{F}$ and $\mathbf{G}$ are both projections. Now, the jkth entry in the $(s \times s)$-matrix $\mathcal{E}\mathbf{G}\mathcal{E}^\tau$ in (6.63) is the quadratic form $\mathcal{E}_{(j)} \mathbf{G} \mathcal{E}_{(k)}^\tau = \sum_u \sum_v G_{uv} \mathcal{E}_{ju} \mathcal{E}_{kv}$, where $\mathcal{E}_{(j)} = (\mathcal{E}_{ju})$

TABLE 6.2. *MANOVA table for the constrained and unconstrained multivariate regression models, where $u = \text{rank}(\mathbf{K})$.*

Source of Variation	df	Sum of Squares
Constrained model	$r - u$	$\mathbf{S}_{\text{reg}}^* = \widehat{\boldsymbol{\Theta}}^* \mathcal{X}_c \mathcal{X}_c^\tau \widehat{\boldsymbol{\Theta}}^{*\tau}$
Due to dropping constraints	u	$\mathbf{S}_h = (\widehat{\boldsymbol{\Theta}} - \widehat{\boldsymbol{\Theta}}^*) \mathcal{X}_c \mathcal{X}_c^\tau (\widehat{\boldsymbol{\Theta}} - \widehat{\boldsymbol{\Theta}}^*)^\tau$
Unconstrained model	r	$\mathbf{S}_{\text{reg}} = \widehat{\boldsymbol{\Theta}} \mathcal{X}_c \mathcal{X}_c^\tau \widehat{\boldsymbol{\Theta}}^\tau$
Residual	$n - r - 1$	$\mathbf{S}_e = (\mathcal{Y}_c - \widehat{\boldsymbol{\Theta}} \mathcal{X}_c)(\mathcal{Y}_c - \widehat{\boldsymbol{\Theta}} \mathcal{X}_c)^\tau$
Total	$n - 1$	$\mathcal{Y}_c \mathcal{Y}_c^\tau$

is the jth row of $\mathcal{E}$. So, its expected value is given by $\text{E}(\mathcal{E}^{(j)} \mathbf{G} \mathcal{E}^{(k)\tau}) = \sum_u G_{uu} (\boldsymbol{\Sigma}_{\mathcal{E}\mathcal{E}})_{jk} = (\boldsymbol{\Sigma}_{\mathcal{E}\mathcal{E}})_{jk} \cdot \text{tr}(\mathbf{G})$. Thus, $\text{E}(\mathcal{E} \mathbf{G} \mathcal{E}^\tau) = u \boldsymbol{\Sigma}_{\mathcal{E}\mathcal{E}}$, because $\text{tr}(\mathbf{G}) = \text{tr}(\mathbf{I}_u) = u$.

General Linear Hypothesis

From Table 6.2, we can test the *general linear hypothesis*,

$$\mathcal{H}_0 : \mathbf{K}\boldsymbol{\Theta}\mathbf{L} = \boldsymbol{\Gamma} \quad vs. \quad \mathcal{H}_1 : \mathbf{K}\boldsymbol{\Theta}\mathbf{L} \neq \boldsymbol{\Gamma}. \tag{6.65}$$

Under $\mathcal{H}_0$, $\text{E}\{\mathbf{S}_h/u\} = \mathbf{F}\boldsymbol{\Sigma}_{\mathcal{E}\mathcal{E}}\mathbf{F}^\tau$. Furthermore, $\text{E}\{\mathbf{S}_e/(n - r - 1)\} = \boldsymbol{\Sigma}_{\mathcal{E}\mathcal{E}}$. A formal significance test of $\mathcal{H}_0$ vs. $\mathcal{H}_1$ can, therefore, be realized through a function (e.g., determinant, trace, or largest eigenvalue) of the quantity $\mathbf{F}\mathbf{S}_h\mathbf{F}^\tau (\mathbf{F}\mathbf{S}_e\mathbf{F}^\tau)^{-1}$, where we use the fact that $\mathbf{F}$ is a projection matrix. Related test statistics have been proposed in the literature, including the following functions of $\mathbf{S}_h$ and $\mathbf{S}_e$:

1. *Hotelling–Lawley trace statistic:* $\text{tr}\{\mathbf{S}_h\mathbf{S}_e^{-1}\}$

2. *Roy's largest root:* $\lambda_{\max}\{\mathbf{S}_h\mathbf{S}_e^{-1}\}$

3. *Wilks's lambda (likelihood ratio criterion):* $|\mathbf{S}_e|/|\mathbf{S}_h + \mathbf{S}_e|$

Under $\mathcal{H}_0$ and appropriate distributional assumptions, Hotelling–Lawley's trace statistic and Roy's largest root should both be small, whereas Wilk's lambda should be large (i.e., close to 1) under $\mathcal{H}_0$. In other words, we would reject $\mathcal{H}_0$ in favor of $\mathcal{H}_1$ if the trace statistic or largest root were large and if Wilk's lambda were small (i.e., close to 0). Properties of these statistics are given in Anderson (1984, Chapter 8).

We can also compute an appropriate confidence region for $\mathbf{K\Theta L} - \boldsymbol{\Gamma}$ by using the statistic $\mathbf{K\hat{\Theta} L} - \boldsymbol{\Gamma}$. A formal significance test can be constructed from the resulting confidence region; if the confidence region does not contain $\mathbf{0}$, we say that the evidence from the data favors $\mathcal{H}_1$ rather than $\mathcal{H}_0$.

6.3 The Random-X Case

In this section, we treat the case where

$$\overset{r \times 1}{\mathbf{X}} = (X_1, \cdots, X_r)^\tau, \quad \overset{s \times 1}{\mathbf{Y}} = (Y_1, \cdots, Y_s)^\tau, \tag{6.66}$$

are jointly distributed, with $\mathbf{X}$ having mean vector $\boldsymbol{\mu}_X$ and $\mathbf{Y}$ having mean vector $\boldsymbol{\mu}_Y$, and with joint covariance matrix,

$$\begin{pmatrix} \boldsymbol{\Sigma}_{XX} & \boldsymbol{\Sigma}_{XY} \\ \boldsymbol{\Sigma}_{YX} & \boldsymbol{\Sigma}_{YY} \end{pmatrix}. \tag{6.67}$$

For convenience in exposition, we assume $s \leq r$. Although $\mathbf{X}$ is presumed to be the larger of the two sets of variates, this reflects purely a mathematical convenience, and similar expressions as appear here can be obtained in the case in which $r \leq s$. The variables $\mathbf{X}$ and $\mathbf{Y}$ are assumed to be continuous but may also include transformations (e.g., logs, square-roots, reciprocals), powers (e.g., squares, cubes), products, or ratios of the input variables. Notice that we have not assumed that the joint distribution of (6.66) is Gaussian.

6.3.1 Classical Multivariate Regression Model

Suppose $\mathbf{Y}$ is related to $\mathbf{X}$ by the following multivariate linear model:

$$\overset{s \times 1}{\mathbf{Y}} = \overset{s \times 1}{\boldsymbol{\mu}} + \overset{s \times r}{\boldsymbol{\Theta}} \overset{r \times 1}{\mathbf{X}} + \overset{s \times 1}{\boldsymbol{\mathcal{E}}}, \tag{6.68}$$

where $\boldsymbol{\mu}$ and the *regression coefficient matrix* $\boldsymbol{\Theta}$ are the unknown parameters and $\mathcal{E}$ is the unobservable error component of the model with mean $\mathrm{E}(\mathcal{E}) = \mathbf{0}$ and unknown $(s \times s)$ *error covariance matrix* $\mathrm{cov}(\mathcal{E}) = \boldsymbol{\Sigma}_{\mathcal{E}\mathcal{E}}$, and $\mathcal{E}$ is distributed independently of $\mathbf{X}$. Our first goal is to obtain suitable expressions for $\boldsymbol{\mu}$, $\boldsymbol{\Theta}$, and $\boldsymbol{\Sigma}_{\mathcal{E}\mathcal{E}}$ that are optimal in a least-squares sense.

We are interested in finding the s-vector $\boldsymbol{\mu}$ and $(s \times r)$-matrix $\boldsymbol{\Theta}$ that minimize the $(s \times s)$-matrix,

$$W(\boldsymbol{\mu}, \boldsymbol{\Theta}) = \mathrm{E}\{(\mathbf{Y} - \boldsymbol{\mu} - \boldsymbol{\Theta}\mathbf{X})(\mathbf{Y} - \boldsymbol{\mu} - \boldsymbol{\Theta}\mathbf{X})^{\tau}\}, \tag{6.69}$$

where the expectation is taken over the joint distribution of $(\mathbf{X}^{\tau}, \mathbf{Y}^{\tau})^{\tau}$. Set $\mathbf{Y}_c = \mathbf{Y} - \boldsymbol{\mu}_Y$ and $\mathbf{X}_c = \mathbf{X} - \boldsymbol{\mu}_X$, and assume that $\boldsymbol{\Sigma}_{XX}$ is nonsingular. Expanding the right-hand-side of (6.69), we get that

$$
\begin{aligned}
W(\boldsymbol{\mu}, \boldsymbol{\Theta}) &= \mathrm{E}\{\mathbf{Y}_c\mathbf{Y}_c^{\tau} - \mathbf{Y}_c\mathbf{X}_c^{\tau}\boldsymbol{\Theta}^{\tau} - \boldsymbol{\Theta}\mathbf{X}_c\mathbf{Y}_c^{\tau} + \boldsymbol{\Theta}\mathbf{X}_c\mathbf{X}_c^{\tau}\boldsymbol{\Theta}^{\tau}\} \\
&\quad + (\boldsymbol{\mu} - \boldsymbol{\mu}_Y + \boldsymbol{\Theta}\boldsymbol{\mu}_X)(\boldsymbol{\mu} - \boldsymbol{\mu}_Y + \boldsymbol{\Theta}\boldsymbol{\mu}_X)^{\tau} \\
&= (\boldsymbol{\Sigma}_{YY} - \boldsymbol{\Sigma}_{YX}\boldsymbol{\Sigma}_{XX}^{-1}\boldsymbol{\Sigma}_{XY}) \\
&\quad + (\boldsymbol{\Sigma}_{YX}\boldsymbol{\Sigma}_{XX}^{-1/2} - \boldsymbol{\Theta}\boldsymbol{\Sigma}_{XX}^{1/2})(\boldsymbol{\Sigma}_{YX}\boldsymbol{\Sigma}_{XX}^{-1/2} - \boldsymbol{\Theta}\boldsymbol{\Sigma}_{XX}^{1/2})^{\tau} \\
&\quad + (\boldsymbol{\mu} - \boldsymbol{\mu}_Y + \boldsymbol{\Theta}\boldsymbol{\mu}_X)(\boldsymbol{\mu} - \boldsymbol{\mu}_Y + \boldsymbol{\Theta}\boldsymbol{\mu}_X)^{\tau} \\
&\geq \boldsymbol{\Sigma}_{YY} - \boldsymbol{\Sigma}_{YX}\boldsymbol{\Sigma}_{XX}^{-1}\boldsymbol{\Sigma}_{XY}, \tag{6.70}
\end{aligned}
$$

with equality when

$$\boldsymbol{\mu} = \boldsymbol{\mu}_Y - \boldsymbol{\Theta}\boldsymbol{\mu}_X \tag{6.71}$$

$$\boldsymbol{\Theta} = \boldsymbol{\Sigma}_{YX}\boldsymbol{\Sigma}_{XX}^{-1}. \tag{6.72}$$

The minimum achieved is $\boldsymbol{\Sigma}_{YY} - \boldsymbol{\Sigma}_{YX}\boldsymbol{\Sigma}_{XX}^{-1}\boldsymbol{\Sigma}_{XY}$. The $\boldsymbol{\mu}$ and $\boldsymbol{\Theta}$ given by (6.71) and (6.72), respectively, minimize (6.69) and also minimize the trace, determinant, and jth largest eigenvalue of (6.69).

The $(s \times r)$-matrix $\boldsymbol{\Theta}$ is called the *(full-rank) regression coefficient matrix* of $\mathbf{Y}$ on $\mathbf{X}$, and

$$\mathbf{Y} = \boldsymbol{\mu}_Y + \boldsymbol{\Sigma}_{YX}\boldsymbol{\Sigma}_{XX}^{-1}(\mathbf{X} - \boldsymbol{\mu}_X) \tag{6.73}$$

is the *(full-rank) linear regression function* of $\mathbf{Y}$ on $\mathbf{X}$, where "full rank" refers to the rank of $\boldsymbol{\Theta}$. At the minimum, the error variate is

$$\mathcal{E} = \mathbf{Y} - \boldsymbol{\mu}_Y - \boldsymbol{\Sigma}_{YX}\boldsymbol{\Sigma}_{XX}^{-1}(\mathbf{X} - \boldsymbol{\mu}_X) = \mathbf{Y}_c - \boldsymbol{\Sigma}_{YX}\boldsymbol{\Sigma}_{XX}^{-1}\mathbf{X}_c. \tag{6.74}$$

From (6.74), we see that $\mathrm{E}(\mathcal{E}) = \mathbf{0}$, $\boldsymbol{\Sigma}_{\mathcal{E}\mathcal{E}} = \boldsymbol{\Sigma}_{YY} - \boldsymbol{\Sigma}_{YX}\boldsymbol{\Sigma}_{XX}^{-1}\boldsymbol{\Sigma}_{XY}$, and $\mathrm{E}(\mathcal{E}\mathbf{X}_c^{\tau}) = \mathbf{0}$.

6.3.2 Multivariate Reduced-Rank Regression

In Section 6.2.4, we described how to place constraints on $\boldsymbol{\Theta}$ when $\mathbf{X}$ is considered fixed. An alternative way of constraining a multivariate regression model is through a rank condition on the matrix of regression coefficients. The resulting model is called the *multivariate reduced-rank regression (RRR) model* (Izenman, 1972, 1975). In this section, we describe the RRR scenario in which $\mathbf{X}$ and $\mathbf{Y}$ are jointly distributed (i.e., the random-X case). The reader is encouraged to develop the RRR model for the fixed-X case (see Exercises 6.4, 6.5, and 6.6).

Most applications of reduced-rank regression have been directed toward problems in time series (time domain and frequency domain) and econometrics. This development has led to the introduction of the related topic of *cointegration* into the econometric literature.

The Reduced-Rank Regression Model

Consider the multivariate linear regression model given by

$$\overset{s\times 1}{\mathbf{Y}} = \overset{s\times 1}{\boldsymbol{\mu}} + \overset{s\times r}{\mathbf{C}}\ \overset{r\times 1}{\mathbf{X}} + \overset{s\times 1}{\boldsymbol{\mathcal{E}}}\ , \tag{6.75}$$

where $\boldsymbol{\mu}$ and $\mathbf{C}$ are unknown regression parameters, and the unobservable error variate, $\mathcal{E}$, of the model has mean $\mathrm{E}(\mathcal{E}) = \mathbf{0}$ and covariance matrix $\mathrm{cov}(\mathcal{E}) = \mathrm{E}\{\mathcal{E}\mathcal{E}^\tau\} = \boldsymbol{\Sigma}_{\mathcal{E}\mathcal{E}}$, and is distributed independently of $\mathbf{X}$. The difference between this model and that of (6.68) is that we allow the possibility that the rank of the regression coefficient matrix $\mathbf{C}$ is deficient; that is,

$$\mathrm{rank}(\mathbf{C}) = t\ \leq\ \min(r, s). \tag{6.76}$$

The "reduced-rank" condition (6.76) on the regression coefficient matrix $\mathbf{C}$ brings a true multivariate feature into the model. The rank condition implies that there may be a number of linear constraints on the set of regression coefficients in the model. Unlike the model studied in Section 6.2.4, however, the value of t and, hence, the number and nature of those constraints may not be known prior to statistical analysis. The name *reduced-rank regression* was introduced to distinguish the case $1 \leq t < s$ from *full-rank regression*, where $t = s$.

When $\mathbf{C}$ has reduced-rank t, then, there exist two (nonunique) full-rank matrices, an $(s \times t)$ matrix $\mathbf{A}$ and a $(t \times r)$ matrix $\mathbf{B}$, such that $\mathbf{C} = \mathbf{AB}$. The nonuniqueness occurs because we can always find a nonsingular $(t \times t)$-matrix $\mathbf{T}$ such that $\mathbf{C} = (\mathbf{AT})(\mathbf{T}^{-1}\mathbf{B}) = \mathbf{DE}$, which gives a different decomposition of $\mathbf{C}$. The model (6.75) can now be written as

$$\overset{s\times 1}{\mathbf{Y}} = \overset{s\times 1}{\boldsymbol{\mu}} + \overset{s\times t}{\mathbf{A}}\ \overset{t\times r}{\mathbf{B}}\ \overset{r\times 1}{\mathbf{X}} + \overset{s\times 1}{\boldsymbol{\mathcal{E}}}\ . \tag{6.77}$$

Given a sample, $(\mathbf{X}_1^\tau, \mathbf{Y}_1^\tau)^\tau, \ldots, (\mathbf{X}_n^\tau, \mathbf{Y}_n^\tau)^\tau$ of observations on $(\mathbf{X}^\tau, \mathbf{Y}^\tau)^\tau$, our goal is to estimate the parameters $\boldsymbol{\mu}$, $\mathbf{A}$, and $\mathbf{B}$ (and, hence, $\mathbf{C}$) in some optimal manner.

Such a setup can be motivated within a time-series context (Brillinger, 1969). Suppose we wish to send a message based upon the r components of a vector $\mathbf{X}$ so that the message received, $\mathbf{Y}$, will be composed of s components. Suppose, further, that such a message can only be transmitted using t channels ($t \leq s$). We would, therefore, first need to encode $\mathbf{X}$ into a t-vector $\boldsymbol{\xi} = \mathbf{BX}$, where $\mathbf{B}$ is a $(t \times r)$-matrix, and then on receipt of the coded message to decode it using an $(s \times t)$-matrix $\mathbf{A}$ to form the s-vector

$\mathbf{A}\boldsymbol{\xi}$, which, it would be hoped, would be as "close" as possible to the desired $\mathbf{Y}$.

One of the primary aspects of reduced-rank regression is to assess the unknown value of the metaparameter t, which we call the *effective dimensionality* of the multivariate regression (Izenman, 1980).

Minimizing a Weighted Sum-of-Squares Criterion

We, therefore, wish to find an s-vector $\boldsymbol{\mu}$, an $(s \times t)$-matrix $\mathbf{A}$, and a $(t \times r)$-matrix $\mathbf{B}$ to minimize a weighted sum-of-squares criterion,

$$W(t) = \mathrm{E}\{(\mathbf{Y} - \boldsymbol{\mu} - \mathbf{ABX})^{\tau}\boldsymbol{\Gamma}(\mathbf{Y} - \boldsymbol{\mu} - \mathbf{ABX})\}, \qquad (6.78)$$

where $\boldsymbol{\Gamma}$ is a positive-definite symmetric $(s \times s)$-matrix of weights and the expectation is taken over the joint distribution of $(\mathbf{X}^{\tau}, \mathbf{Y}^{\tau})^{\tau}$. In practice, we try out different forms of $\boldsymbol{\Gamma}$.

We minimize $W(t)$ in two steps. As before, let $\mathbf{X}_c$ and $\mathbf{Y}_c$ denote the centered versions of $\mathbf{X}$ and $\mathbf{Y}$, respectively. The first step makes no rank condition on $\mathbf{C}$. The minimizing criterion becomes:

$$\begin{aligned}
W(t) \;\geq\; & \mathrm{E}\{(\mathbf{Y}_c - \mathbf{CX}_c)^{\tau}\boldsymbol{\Gamma}(\mathbf{Y}_c - \mathbf{CX}_c)\} \\
=\; & \mathrm{E}\{\mathbf{Y}_c^{\tau}\boldsymbol{\Gamma}\mathbf{Y}_c + \mathbf{Y}_c^{\tau}\boldsymbol{\Gamma}\mathbf{CX}_c + \mathbf{X}_c^{\tau}\mathbf{C}^{\tau}\boldsymbol{\Gamma}\mathbf{Y}_c + \mathbf{X}_c^{\tau}\mathbf{C}^{\tau}\boldsymbol{\Gamma}\mathbf{CX}_c\} \\
=\; & \mathrm{tr}\{\boldsymbol{\Sigma}_{YY}^* - \mathbf{C}^*\boldsymbol{\Sigma}_{XY}^* - \boldsymbol{\Sigma}_{YX}^*\mathbf{C}^{*\tau} + \mathbf{C}^*\boldsymbol{\Sigma}_{XX}^*\mathbf{C}^{*\tau}\} \\
=\; & \mathrm{tr}\{(\boldsymbol{\Sigma}_{YY}^* - \boldsymbol{\Sigma}_{YX}^*\boldsymbol{\Sigma}_{XX}^{*-1}\boldsymbol{\Sigma}_{XY}^*) \\
& + (\mathbf{C}^*\boldsymbol{\Sigma}_{XX}^{*1/2} - \boldsymbol{\Sigma}_{YX}^*\boldsymbol{\Sigma}_{XX}^{*-1/2})(\mathbf{C}^*\boldsymbol{\Sigma}_{XX}^{*1/2} - \boldsymbol{\Sigma}_{YX}^*\boldsymbol{\Sigma}_{XX}^{*-1/2})^{\tau}\},
\end{aligned}$$
$$\tag{6.79}$$

where $\boldsymbol{\Sigma}_{XX}^* = \boldsymbol{\Sigma}_{XX}$, $\boldsymbol{\Sigma}_{YY}^* = \boldsymbol{\Gamma}^{1/2}\boldsymbol{\Sigma}_{YY}\boldsymbol{\Gamma}^{1/2}$, $\boldsymbol{\Sigma}_{XY}^* = \boldsymbol{\Sigma}_{XY}\boldsymbol{\Gamma}^{1/2}$, and $\mathbf{C}^* = \boldsymbol{\Gamma}^{1/2}\mathbf{C}$. Next, we assume that $\mathbf{C}$ has rank t. From the Eckart–Young Theorem (see Section 3.2.10), the last expression is minimized by setting

$$\mathbf{C}^*\boldsymbol{\Sigma}_{XX}^{*1/2} = \sum_{j=1}^{t} \lambda_j^{1/2}\mathbf{v}_j\mathbf{w}_j^{\tau}, \qquad (6.80)$$

where $\mathbf{v}_j$ is the eigenvector associated with the jth largest eigenvalue λ_j of the matrix

$$\boldsymbol{\Sigma}_{YX}^*\boldsymbol{\Sigma}_{XX}^{*-1}\boldsymbol{\Sigma}_{XY}^* = \boldsymbol{\Gamma}^{1/2}\boldsymbol{\Sigma}_{YX}\boldsymbol{\Sigma}_{XX}^{-1}\boldsymbol{\Sigma}_{XY}\boldsymbol{\Gamma}^{1/2} \qquad (6.81)$$

and

$$\mathbf{w}_j = \lambda_j^{-1/2}\boldsymbol{\Sigma}_{XX}^{*-1/2}\boldsymbol{\Sigma}_{XY}^*\mathbf{v}_j = \lambda_j^{-1/2}\boldsymbol{\Sigma}_{XX}^{-1/2}\boldsymbol{\Sigma}_{XY}\boldsymbol{\Gamma}^{1/2}\mathbf{v}_j. \qquad (6.82)$$

Thus, the minimizing $\mathbf{C}$ with reduced-rank t is given by

$$\mathbf{C}^{(t)} = \boldsymbol{\Gamma}^{-1/2}\left(\sum_{j=1}^{t}\mathbf{v}_j\mathbf{v}_j^{\tau}\right)\boldsymbol{\Gamma}^{1/2}\boldsymbol{\Sigma}_{YX}\boldsymbol{\Sigma}_{XX}^{-1}. \qquad (6.83)$$

The matrix $\mathbf{C}^{(t)}$ in (6.83) is called the *reduced-rank regression coefficient matrix* with rank t and weight matrix $\mathbf{\Gamma}$.

It follows that $W(t)$ in (6.78) is minimized by taking $\boldsymbol{\mu}$, $\mathbf{A}$, and $\mathbf{B}$ to be the following functions of t,

$$\boldsymbol{\mu}^{(t)} = \boldsymbol{\mu}_Y - \mathbf{A}^{(t)}\mathbf{B}^{(t)}\boldsymbol{\mu}_X, \tag{6.84}$$

$$\mathbf{A}^{(t)} = \mathbf{\Gamma}^{-1/2}\mathbf{V}_t, \tag{6.85}$$

$$\mathbf{B}^{(t)} = \mathbf{V}_t^\tau \mathbf{\Gamma}^{1/2}\boldsymbol{\Sigma}_{YX}\boldsymbol{\Sigma}_{XX}^{-1}, \tag{6.86}$$

respectively, where $\mathbf{V}_t = (\mathbf{v}_1, \ldots, \mathbf{v}_t)$ is an $(s \times t)$-matrix, where the jth column, $\mathbf{v}_j$, is the eigenvector associated with the jth largest eigenvalue λ_j of the $(s \times s)$ symmetric matrix

$$\mathbf{\Gamma}^{1/2}\boldsymbol{\Sigma}_{YX}\boldsymbol{\Sigma}_{XX}^{-1}\boldsymbol{\Sigma}_{XY}\mathbf{\Gamma}^{1/2}. \tag{6.87}$$

A stronger result (Rao, 1979) uses the Poincaré Separation Theorem (see Section 3.2.10) to show that if $\mathbf{\Gamma} = \boldsymbol{\Sigma}_{YY}^{-1}$, then all the eigenvalues of the matrix

$$\mathbf{\Gamma}^{1/2}(\mathbf{Y} - \boldsymbol{\mu} - \mathbf{ABX})(\mathbf{Y} - \boldsymbol{\mu} - \mathbf{ABX})^\tau \mathbf{\Gamma}^{1/2} \tag{6.88}$$

are simultaneously minimized by the above $\boldsymbol{\mu}^{(t)}$, $\mathbf{A}^{(t)}$, and $\mathbf{B}^{(t)}$. Hence, any function of those eigenvalues, which is increasing in each argument (e.g., trace or determinant), is also minimized by that choice.

The minimum value of the criterion $W(t)$ is given by

$$
\begin{aligned}
W_{\min}(t) &= \mathrm{E}\,\mathrm{tr}\left\{(\mathbf{Y}_c - \mathbf{C}^{(t)}\mathbf{X}_c)(\mathbf{Y}_c - \mathbf{C}^{(t)}\mathbf{X}_c)^\tau\mathbf{\Gamma}\right\} \\
&= \mathrm{tr}\left\{\boldsymbol{\Sigma}_{YY} - \mathbf{\Gamma}^{-1/2}\left(\sum_{j=1}^t \lambda_j \mathbf{v}_j \mathbf{v}_j^\tau\right)\mathbf{\Gamma}^{-1/2}\mathbf{\Gamma}\right\} \\
&= \mathrm{tr}\left\{(\boldsymbol{\Sigma}_{YY} - \boldsymbol{\Sigma}_{YX}\boldsymbol{\Sigma}_{XX}^{-1}\boldsymbol{\Sigma}_{XY})\mathbf{\Gamma} + \sum_{j=t+1}^s \lambda_j \mathbf{v}_j \mathbf{v}_j^\tau\right\} \\
&= \mathrm{tr}\left\{(\boldsymbol{\Sigma}_{YY} - \boldsymbol{\Sigma}_{YX}\boldsymbol{\Sigma}_{XX}^{-1}\boldsymbol{\Sigma}_{XY})\mathbf{\Gamma}\right\} + \sum_{j=t+1}^s \lambda_j \\
&= \mathrm{tr}\{\boldsymbol{\Sigma}_{YY}\mathbf{\Gamma}\} - \sum_{j=1}^t \lambda_j \, . \tag{6.89}
\end{aligned}
$$

When $t = s$, we have that $\sum_{j=1}^s \mathbf{v}_j \mathbf{v}_j^\tau = \mathbf{I}_s$, whence $\mathbf{C}^{(t)}$ in (6.83) reduces to the full-rank regression coefficient matrix $\boldsymbol{\Theta} = \mathbf{C}^{(s)}$. Furthermore, for any t and positive-definite matrix $\mathbf{\Gamma}$, the matrices $\mathbf{C}^{(t)}$ and $\boldsymbol{\Theta}$ are related by the expression $\mathbf{C}^{(t)} = \mathbf{P}_{\mathbf{\Gamma}}^{(t)}\boldsymbol{\Theta}$, where

$$\mathbf{P}_{\mathbf{\Gamma}}^{(t)} = \mathbf{\Gamma}^{-1/2}\left(\sum_{j=1}^t \mathbf{v}_j \mathbf{v}_j^\tau\right)\mathbf{\Gamma}^{1/2} \tag{6.90}$$

is an idempotent, but not symmetric (unless $\mathbf{\Gamma} = \mathbf{I}_s$), $(s \times s)$-matrix.

Special Cases of RRR

We have seen how the RRR model can be used to generalize the classical multivariate regression model by relaxing the implicit constraint on the rank of $\mathbf{C}$. More importantly, by carefully choosing the input vector $\mathbf{X}$, the output vector $\mathbf{Y}$, and the matrix $\mathbf{\Gamma}$ of weights, RRR can be used to play an important role as a unifying treatment of several classical multivariate procedures that were developed separately from each other.

The primary uses of RRR in the exploratory analysis of multivariate data include the following special cases:

- If we set $\mathbf{X} \equiv \mathbf{Y}$ (and $r = s$) by making the output variables identical to the input variables, and set $\mathbf{\Gamma} = \mathbf{I}_s$, then we have Harold Hotelling's *principal component analysis* (see Section 7.2) and *exploratory factor analysis* (see Section 15.4).

- If we set $\mathbf{\Gamma} = \mathbf{\Sigma}_{YY}^{-1}$, then we have Hotelling's *canonical variate and correlation analysis* (see Section 7.3).

- Using the canonical variate analysis setup for RRR, if we set $\mathbf{Y}$ to be a vector of binary variables whose component values (0 or 1) indicate the group or class to which an observation belongs, then we have R.A. Fisher's *linear discriminant analysis* (see Section 8.5).

- Using the canonical variate analysis setup for RRR, if we set $\mathbf{X}$ and $\mathbf{Y}$ each to be a vector of binary variables whose component values (0 or 1) indicate the row and column of a two-way contingency table to which an observation belongs, then we have *correspondence analysis* (see Section 18.2).

These special cases of multivariate reduced-rank regression show that the RRR model can be used as a general model for many different types of multivariate statistical analysis. Extensions of this model in other directions (e.g., to multiresponse generalized linear models, wavelets, functional data) are currently undergoing development.

Sample Estimates

The mean vectors and covariance matrix of $\mathbf{X}$ and $\mathbf{Y}$ are typically unknown and have to be estimated before we can draw any useful inferences on the regression problem. Accordingly, we assume that a random sample of n independent observations, $(\mathbf{X}_j^\tau, \mathbf{Y}_j^\tau)^\tau, j = 1, 2, \ldots, n$, is obtained on the $(r + s)$-vector $(\mathbf{X}^\tau, \mathbf{Y}^\tau)^\tau$.

First, we estimate $\boldsymbol{\mu}_X$ and $\boldsymbol{\mu}_Y$ by

$$\widehat{\boldsymbol{\mu}}_X = \bar{\mathbf{X}} = n^{-1}\sum_{j=1}^{n}\mathbf{X}_j, \quad \widehat{\boldsymbol{\mu}}_Y = \bar{\mathbf{Y}} = n^{-1}\sum_{j=1}^{n}\mathbf{Y}_j, \tag{6.91}$$

respectively. We set

$$\overset{r\times 1}{\mathbf{X}_{cj}} = \mathbf{X}_j - \bar{\mathbf{X}}, \quad \overset{s\times 1}{\mathbf{Y}_{cj}} = \mathbf{Y}_j - \bar{\mathbf{Y}}, \quad j = 1, 2, \ldots, n, \tag{6.92}$$

and let

$$\overset{r\times n}{\mathcal{X}_c} = (\mathbf{X}_{c1}, \cdots, \mathbf{X}_{cn}), \quad \overset{s\times n}{\mathcal{Y}_c} = (\mathbf{Y}_{c1}, \cdots, \mathbf{Y}_{cn}). \tag{6.93}$$

Then, we estimate the components of the covariance matrix (6.67) by

$$\widehat{\boldsymbol{\Sigma}}_{XX} = n^{-1}\mathcal{X}_c\mathcal{X}_c^{\tau} \tag{6.94}$$

$$\widehat{\boldsymbol{\Sigma}}_{YX} = n^{-1}\mathcal{Y}_c\mathcal{X}_c^{\tau} = \widehat{\boldsymbol{\Sigma}}_{XY}^{\tau} \tag{6.95}$$

$$\widehat{\boldsymbol{\Sigma}}_{YY} = n^{-1}\mathcal{Y}_c\mathcal{Y}_c^{\tau}. \tag{6.96}$$

All estimates of the unknowns in the multivariate regression models are based upon the appropriate elements of (6.94), (6.95), and (6.96).

Thus, $\mathbf{A}^{(t)}$ in (6.85) and $\mathbf{B}^{(t)}$ in (6.86) are estimated by

$$\widehat{\mathbf{A}}^{(t)} = \boldsymbol{\Gamma}^{-1/2}\widehat{\mathbf{V}}_t, \tag{6.97}$$

$$\widehat{\mathbf{B}}^{(t)} = \widehat{\mathbf{V}}_t^{\tau}\boldsymbol{\Gamma}^{1/2}\widehat{\boldsymbol{\Sigma}}_{YX}\widehat{\boldsymbol{\Sigma}}_{XX}^{-1}, \tag{6.98}$$

respectively, where

$$\widehat{\mathbf{V}}_t = (\widehat{\mathbf{v}}_1, \ldots, \widehat{\mathbf{v}}_t) \tag{6.99}$$

is an $(s \times t)$-matrix, the jth column, $\widehat{\mathbf{v}}_j$, of which is the eigenvector associated with to the jth largest eigenvalue $\widehat{\lambda}_j$ of the $(s \times s)$ symmetric matrix

$$\boldsymbol{\Gamma}^{1/2}\widehat{\boldsymbol{\Sigma}}_{YX}\widehat{\boldsymbol{\Sigma}}_{XX}^{-1}\widehat{\boldsymbol{\Sigma}}_{XY}\boldsymbol{\Gamma}^{1/2}, \tag{6.100}$$

$j = 1, 2, \ldots, s$. The reduced-rank regression coefficient matrix $\mathbf{C}^{(t)}$ in (6.83) is estimated by

$$\widehat{\mathbf{C}}^{(t)} = \boldsymbol{\Gamma}^{-1/2}\left(\sum_{j=1}^{t}\widehat{\mathbf{v}}_j\widehat{\mathbf{v}}_j^{\tau}\right)\boldsymbol{\Gamma}^{1/2}\widehat{\boldsymbol{\Sigma}}_{YX}\widehat{\boldsymbol{\Sigma}}_{XX}^{-1}, \tag{6.101}$$

and the full-rank regression coefficient matrix $\boldsymbol{\Theta}$ is estimated by

$$\widehat{\boldsymbol{\Theta}} = \widehat{\mathbf{C}}^{(s)} = \widehat{\boldsymbol{\Sigma}}_{YX}\widehat{\boldsymbol{\Sigma}}_{XX}^{-1}. \tag{6.102}$$

The sample estimators (6.97), (6.98), (6.100), (6.101), and (6.102) are identical to the estimators that appear in the reduced-rank regression solution and full-rank regression solution when $\mathbf{X}$ is fixed (Exercise 6.4). It follows that the matrix of fitted values and the matrix of residuals for the random-X case are identical to those for the fixed-X case. Although the two formulations of the regression model are different, they yield identical sample estimates.

In many applications, it is not unusual to find that the matrix $\widehat{\boldsymbol{\Sigma}}_{XX}$ and/or the matrix $\widehat{\boldsymbol{\Sigma}}_{YY}$ are singular, or at least difficult to invert. This happens, for example, when $r, s > n$. We could replace their inverses by generalized inverses, but, based upon practical experience with the methods described in Section 6.3.4, we suggest the following alternative solution.

We borrow an idea from ridge regression, where we replace $\widehat{\boldsymbol{\Sigma}}_{XX}$ and $\widehat{\boldsymbol{\Sigma}}_{YY}$ in the RRR computations by a slight perturbation of their diagonal entries,

$$\widehat{\boldsymbol{\Sigma}}_{XX}^{(k)} = \widehat{\boldsymbol{\Sigma}}_{XX} + k\mathbf{I}_r, \quad \widehat{\boldsymbol{\Sigma}}_{YY}^{(k)} = \widehat{\boldsymbol{\Sigma}}_{YY} + k\mathbf{I}_s, \tag{6.103}$$

respectively, where $k > 0$. The estimates (6.103) of $\boldsymbol{\Sigma}_{XX}$ and $\boldsymbol{\Sigma}_{YY}$ are now invertible. The matrix (6.100) is then replaced by

$$\boldsymbol{\Gamma}^{1/2}\widehat{\boldsymbol{\Sigma}}_{YX}\widehat{\boldsymbol{\Sigma}}_{XX}^{(k)-1}\widehat{\boldsymbol{\Sigma}}_{XY}\boldsymbol{\Gamma}^{1/2}, \tag{6.104}$$

where $\widehat{\boldsymbol{\Sigma}}_{XX}^{(k)-1}$ is the inverse of $\widehat{\boldsymbol{\Sigma}}_{XX}^{(k)}$, and its eigenvalues and eigenvectors are denoted by

$$(\widehat{\lambda}_j^{(k)}, \widehat{\mathbf{v}}_j^{(k)}), \quad j = 1, 2, \ldots, t. \tag{6.105}$$

The estimated reduced-rank regression coefficient matrix $\widehat{\mathbf{C}}^{(t)}$ is replaced by

$$\widehat{\mathbf{C}}^{(t)}(k) = \boldsymbol{\Gamma}^{-1/2}\left(\sum_{j=1}^{t} \widehat{\mathbf{v}}_j^{(k)}\widehat{\mathbf{v}}_j^{(k)\tau}\right)\boldsymbol{\Gamma}^{1/2}\widehat{\boldsymbol{\Sigma}}_{YX}\widehat{\boldsymbol{\Sigma}}_{XX}^{(k)-1}, \tag{6.106}$$

and the full-rank regression coefficient matrix $\widehat{\boldsymbol{\Theta}}$ is replaced by

$$\widehat{\boldsymbol{\Theta}}^{(k)} = \widehat{\mathbf{C}}^{(s)}(k) = \widehat{\boldsymbol{\Sigma}}_{YX}\widehat{\boldsymbol{\Sigma}}_{XX}^{(k)-1}. \tag{6.107}$$

How to choose k will be discussed in Section 6.3.4.

Asymptotic Distribution of Estimates

Because of the form of the LS estimates of matrices involved in the RRR solution, exact distribution results are not available. Fortunately, asymptotic results are available in some generality.

The asymptotic distribution of $\widehat{\mathbf{C}}^{(t)}$ is Gaussian with mean zero; that is,

$$\sqrt{n}\,\mathrm{vec}(\widehat{\mathbf{C}}^{(t)} - \mathbf{C}) \xrightarrow{\mathcal{D}} \mathcal{N}_{sr}(\mathbf{0}, \boldsymbol{\Psi}^{(t)}), \quad \text{as } n \to \infty, \tag{6.108}$$

where convergence is in distribution. This result has been proved by several authors for the fixed-$\mathbf{X}$ case with Gaussian assumptions on the error variate. The most general result (Anderson, 1999), which applies to both fixed-$\mathbf{X}$ and random-$\mathbf{X}$ cases without any assumption of Gaussian errors, expresses the asymptotic covariance matrix, $\mathbf{\Psi}^{(t)}$, in the form

$$\mathbf{\Psi}^{(t)} = (\mathbf{\Sigma}_{\mathcal{E}\mathcal{E}} \otimes \mathbf{\Sigma}_{XX}^{-1}) - (\mathbf{M}^{(t)} \otimes \mathbf{N}^{(t)}), \tag{6.109}$$

where

$$\begin{align}
\mathbf{M}^{(t)} &= \mathbf{\Sigma}_{\mathcal{E}\mathcal{E}} - \mathbf{A}^{(t)}(\mathbf{A}^{(t)\tau}\mathbf{\Sigma}_{\mathcal{E}\mathcal{E}}^{-1}\mathbf{A}^{(t)})^{-1}\mathbf{A}^{(t)\tau} \tag{6.110} \\
\mathbf{N}^{(t)} &= \mathbf{\Sigma}_{XX}^{-1} - \mathbf{B}^{(t)\tau}(\mathbf{B}^{(t)}\mathbf{\Sigma}_{XX}\mathbf{B}^{(t)\tau})^{-1}\mathbf{B}^{(t)}. \tag{6.111}
\end{align}$$

Thus, $\mathbf{\Psi}^{(t)}$ consists of the full-rank covariance matrix, $\mathbf{\Sigma}_{\mathcal{E}\mathcal{E}} \otimes \mathbf{\Sigma}_{XX}^{-1}$, with an adjustment by the matrix $\mathbf{M}^{(t)} \otimes \mathbf{N}^{(t)}$ for reduced-rank t. Anderson also notes that $\mathbf{\Psi}^{(t)}$ is invariant wrt any decomposition $\mathbf{C}^{(t)} = \mathbf{A}^{(t)}\mathbf{B}^{(t)} = (\mathbf{A}^{(t)}\mathbf{T})(\mathbf{T}^{-1}\mathbf{B}^{(t)})$, where $\mathbf{T}$ is an arbitrary nonsingular matrix. Such general results allow asymptotic confidence regions to be constructed in situations when the errors are non-Gaussian.

6.3.3 Example: Chemical Composition of Tobacco

This is a small worked example designed to show the computations of RRR. The data[2] are taken from a study on the chemical composition of tobacco leaf samples (Anderson and Bancroft, 1952, p. 205). There are $n = 25$ observations on $r = 6$ input variables, percent nitrogen (X_1), percent chlorine (X_2), percent potassium (X_3), percent phosphorus (X_4), percent calcium (X_5), and percent magnesium (X_6), and $s = 3$ output variables, rate of cigarette burn in inches per 1,000 seconds (Y_1), percent sugar in the leaf (Y_2), and percent nicotine in the leaf (Y_3). The covariance matrices are as follows:

$$\widehat{\mathbf{\Sigma}}_{XX} = \begin{pmatrix} 0.0763 & -0.0150 & -0.0005 & -0.0010 & 0.0682 & 0.0211 \\ -0.0150 & 0.3671 & -0.0145 & 0.0015 & 0.0330 & 0.0091 \\ -0.0005 & -0.0145 & 0.0659 & -0.0017 & -0.0595 & -0.0198 \\ -0.0010 & 0.0015 & -0.0017 & 0.0011 & 0.0002 & 0.0006 \\ 0.0682 & 0.0330 & -0.0595 & 0.0002 & 0.1552 & 0.0380 \\ 0.0211 & 0.0091 & -0.0198 & 0.0006 & 0.0380 & 0.0160 \end{pmatrix}$$

$$\widehat{\mathbf{\Sigma}}_{YY} = \begin{pmatrix} 0.0279 & -0.1098 & 0.0189 \\ -0.1098 & 4.2277 & -0.7565 \\ 0.0189 & -0.7565 & 0.2747 \end{pmatrix}$$

---

[2]These data are available in the file `tobacco.txt`, which can be downloaded from the book's website.

$$\widehat{\boldsymbol{\Sigma}}_{XY} = \begin{pmatrix} 0.0104 & -0.4004 & 0.1112 \\ -0.0631 & 0.5355 & -0.0859 \\ 0.0209 & 0.1002 & -0.0396 \\ -0.0018 & 0.0164 & -0.0008 \\ -0.0080 & -0.3904 & 0.1417 \\ -0.0066 & -0.1364 & 0.0486 \end{pmatrix} = \widehat{\boldsymbol{\Sigma}}_{YX}^{\tau}.$$

We run these data through a reduced-rank regression using the weight matrix $\boldsymbol{\Gamma} = \mathbf{I}_s$. First, we compute (6.100):

$$\widehat{\boldsymbol{\Sigma}}_{YX}\widehat{\boldsymbol{\Sigma}}_{XX}^{-1}\widehat{\boldsymbol{\Sigma}}_{XY} = \begin{pmatrix} 0.019 & -0.101 & 0.013 \\ -0.101 & 3.090 & -0.760 \\ 0.013 & -0.760 & 0.221 \end{pmatrix},$$

which has eigenvalues $\widehat{\lambda}_1 = 3.2821$, $\widehat{\lambda}_2 = 0.0378$, and $\widehat{\lambda}_3 = 0.0102$, and matrix of eigenvectors

$$\widehat{\mathbf{V}} = (\widehat{\mathbf{v}}_1, \widehat{\mathbf{v}}_2, \widehat{\mathbf{v}}_3) = \begin{pmatrix} 0.031 & -0.470 & 0.882 \\ -0.970 & 0.198 & 0.140 \\ 0.241 & 0.860 & 0.450 \end{pmatrix}.$$

For the rank-1 solution, $\widehat{\mathbf{V}}_1$ is the first column of $\widehat{\mathbf{V}}$; for the rank-2 solution, $\widehat{\mathbf{V}}_2$ is the first two columns of $\widehat{\mathbf{V}}$; and the full-rank solution is $\widehat{\mathbf{V}}_3 = \widehat{\mathbf{V}}$.

The matrices $\widehat{\mathbf{A}} = \widehat{\mathbf{A}}^{(3)} = \widehat{\mathbf{V}}$ and $\widehat{\mathbf{B}} = \widehat{\mathbf{B}}^{(3)} = \widehat{\mathbf{V}}\widehat{\boldsymbol{\Sigma}}_{YX}\widehat{\boldsymbol{\Sigma}}_{XX}^{-1}$ are given by:

$$\widehat{\mathbf{A}} = \begin{pmatrix} 0.031 & -0.470 & 0.882 \\ -0.970 & 0.198 & 0.140 \\ 0.241 & 0.860 & 0.450 \end{pmatrix}$$

$$\widehat{\mathbf{B}} = \begin{pmatrix} 4.324 & -1.359 & -1.481 & -13.729 & -0.453 & 3.867 \\ -0.411 & 0.099 & 0.365 & 2.457 & 0.306 & 1.230 \\ -0.302 & -0.081 & 0.578 & 1.048 & 0.375 & 0.034 \end{pmatrix},$$

respectively. The matrix $\widehat{\mathbf{A}}^{(1)}$ is the first column of $\widehat{\mathbf{A}}$, and $\widehat{\mathbf{A}}^{(2)}$ is the first two columns of $\widehat{\mathbf{A}}$. Similarly, the matrix $\widehat{\mathbf{B}}^{(1)}$ is the first row of $\widehat{\mathbf{B}}$, and $\widehat{\mathbf{B}}^{(2)}$ is the first two rows of $\widehat{\mathbf{B}}$. Estimates of the RRR coefficient matrices, $\widehat{\mathbf{C}}^{(t)} = \widehat{\mathbf{A}}^{(t)}\widehat{\mathbf{B}}^{(t)}$, $t = 1, 2, 3$, are given by

$$\widehat{\mathbf{C}}^{(1)} = \begin{pmatrix} 0.134 & -0.042 & -0.046 & -0.427 & -0.014 & 0.120 \\ -4.195 & 1.318 & 1.436 & 13.318 & 0.439 & -3.751 \\ 1.042 & -0.327 & -0.357 & -3.308 & -0.109 & 0.932 \end{pmatrix},$$

$$\widehat{\mathbf{C}}^{(2)} = \begin{pmatrix} 0.328 & -0.089 & -0.218 & -1.582 & -0.158 & -0.459 \\ -4.276 & 1.338 & 1.509 & 13.806 & 0.500 & -3.507 \\ 0.688 & -0.242 & -0.043 & -1.195 & 0.154 & 1.989 \end{pmatrix},$$

$$\widehat{\mathbf{C}}^{(3)} = \widehat{\boldsymbol{\Theta}} = \begin{pmatrix} 0.062 & -0.160 & 0.292 & -0.658 & 0.173 & -0.428 \\ -4.319 & 1.326 & 1.590 & 13.953 & 0.553 & -3.502 \\ 0.552 & -0.279 & 0.218 & -0.723 & 0.323 & 2.005 \end{pmatrix}.$$

and the vectors $\widehat{\boldsymbol{\mu}}^{(t)}$, $t = 1, 2, 3$, by

$$
\widehat{\boldsymbol{\mu}}^{(1)} = \begin{pmatrix} 1.750 \\ 14.688 \\ 2.640 \end{pmatrix}, \quad \widehat{\boldsymbol{\mu}}^{(2)} = \begin{pmatrix} 3.474 \\ 13.961 \\ -0.512 \end{pmatrix}, \quad \widehat{\boldsymbol{\mu}}^{(1)} = \begin{pmatrix} 1.411 \\ 13.633 \\ -1.565 \end{pmatrix}.
$$

6.3.4 Assessing the Effective Dimensionality

The most difficult part of the reduced-rank regression procedure is to assess the value of the metaparameter, t, of the multivariate regression. In order to determine t for a given multivariate sample, we recognize that such data will introduce noise into the relationship and, hence, will tend to obscure the actual structure of the matrix $\mathbf{C}$, so that rank determination for any particular problem will be made more dificult.

We, therefore, distinguish between the "true" or "mathematical" rank of $\mathbf{C}$, which will *always* be full (because it will be based upon a sample estimate of $\mathbf{C}$) and the "practical" or "statistical" rank of $\mathbf{C}$ — the one of real interest — which will typically be unknown. We refer to t as the "effective dimensionality" of the multivariate regression.

The problem of determining the value of t is a selection problem. From the integers 1 through s (assuming without loss of generality that $s \leq r$), we are to choose the smallest integer such that the reduced-rank regression of $\mathbf{Y}$ on $\mathbf{X}$ with that integer as rank will be close (in some sense) to the corresponding full-rank regression.

From (6.89), $W_{\min}(t)$ denotes the minimum value of (6.78) for a fixed value of t. The reduction in $W_{\min}(t)$ obtained by increasing the rank from $t = t_0$ to $t = t_1$, where $t_0 < t_1$, is given by

$$
W_{\min}(t_0) - W_{\min}(t_1) = \sum_{j=t_0+1}^{t_1} \lambda_j. \tag{6.112}
$$

Note that (6.112) depends upon $\boldsymbol{\Gamma}$ only through the eigenvalues, $\{\lambda_j\}$, of the matrix (6.86). As a result, the rank of $\mathbf{C}$ can be assessed through some monotone function of the sequence of ordered sample eigenvalues $\{\hat{\lambda}_j, j = 1, 2, \ldots, s\}$, in which $\hat{\lambda}_j$ is compared with suitable reference values for each j, or by using the sum of some monotone function of the smallest $s - t_0$ sample eigenvalues. For example, Bartlett's likelihood-ratio statistic for testing whether the last $s - t_0$ eigenvalues are zero is proportional to $\sum_{j=t_0+1}^{s} \log(1 + \hat{\lambda}_j)$.

An obvious disadvantage of relying solely on such formal testing procedures is that any routine application of them might fail to take into account the possible need for a preliminary screening of the data. Robustness of sample estimates of the eigenvalues and hence of the various tests

TABLE 6.3. *Algorithm for using the rank trace to assess the effective dimensionality of a multivariate regression.*

1. Define $\widehat{\mathbf{C}}^{(0)} = \mathbf{0}$ and $\widehat{\boldsymbol{\Sigma}}_{\mathcal{E}\mathcal{E}}^{(0)} = \widehat{\boldsymbol{\Sigma}}_{YY}$.

2. Carry out a sequence of s reduced-rank regressions for specific values of t. For $t = 1, 2, \ldots, s$,

   - compute $\widehat{\mathbf{C}}^{(t)}$ and $\widehat{\boldsymbol{\Sigma}}_{\mathcal{E}\mathcal{E}}^{(t)}$, and set $\widehat{\mathbf{C}}^{(s)} = \widehat{\boldsymbol{\Theta}}$ and $\widehat{\boldsymbol{\Sigma}}_{\mathcal{E}\mathcal{E}}^{(s)} = \widehat{\boldsymbol{\Sigma}}_{\mathcal{E}\mathcal{E}}$.

   - compute

   $$\Delta\widehat{\mathbf{C}}^{(t)} = \frac{\| \widehat{\boldsymbol{\Theta}} - \widehat{\mathbf{C}}^{(t)} \|}{\| \widehat{\boldsymbol{\Theta}} \|}, \quad \Delta\widehat{\boldsymbol{\Sigma}}_{\mathcal{E}\mathcal{E}}^{(t)} = \frac{\| \widehat{\boldsymbol{\Sigma}}_{\mathcal{E}\mathcal{E}} - \widehat{\boldsymbol{\Sigma}}_{\mathcal{E}\mathcal{E}}^{(t)} \|}{\| \widehat{\boldsymbol{\Sigma}}_{\mathcal{E}\mathcal{E}} - \widehat{\boldsymbol{\Sigma}}_{YY} \|},$$

   where $\| \mathbf{A} \| = (\mathrm{tr}(\mathbf{A}\mathbf{A}^{\tau}))^{1/2} = \left(\sum_i \sum_j a_{ij}^2 \right)^{1/2}$ is the classical Euclidean norm.

3. Make a scatterplot of the s points

   $$(\Delta\widehat{\mathbf{C}}^{(t)}, \Delta\widehat{\boldsymbol{\Sigma}}_{\mathcal{E}\mathcal{E}}^{(t)}), \quad t = 0, 1, 2, \ldots, s,$$

   and join up successive points on the plot. This is called the *rank trace* for the multivariate reduced-rank regression of $\mathbf{Y}$ on $\mathbf{X}$.

4. Assess the rank of $\mathbf{C}$ as the *smallest* rank for which both coordinates from step (3) are approximately zero.

---

when outliers or distributional peculiarities are present in the data can be a serious statistical obstacle to overcome.

Rank Trace

Suppose t^* is the true rank of $\mathbf{C}$. The basic idea behind the *rank trace* (Izenman, 1980) is that for $1 \leq t < t^*$, the entries in both the estimated regression coefficient matrix and the residual covariance matrix will "change" quite significantly each time we increase the rank in our sequence of reduced-rank regressions; as soon as the true rank is reached, these matrices will then cease to change significantly and will stabilize.

Let $\widehat{t}$ be an estimate of t. We expect the estimated rank-$\widehat{t}$ regression coefficient matrix, $\widehat{\mathbf{C}}^{(t)}$, to be very close to the estimated full-rank regression coefficient matrix $\widehat{\boldsymbol{\Theta}}$ when $\widehat{t} = t^*$. Similarly, we can expect the rank-$\widehat{t}$ residual covariance matrix, $\widehat{\boldsymbol{\Sigma}}_{\mathcal{E}\mathcal{E}}^{(t)}$, to be very close to the full-rank residual covariance matrix, $\widehat{\boldsymbol{\Sigma}}_{\mathcal{E}\mathcal{E}}$, when $\widehat{t} = t^*$. The steps in the computation of the rank trace and the estimation of t are detailed in Table 6.3.

Thus, the first point (corresponding to $t = 0$) is always plotted at $(1,1)$ and the last point (corresponding to $t = s$) is always plotted at $(0,0)$. The horizontal coordinate, $\Delta\widehat{\mathbf{C}}^{(t)}$, gives a quantitative representation of the difference between a reduced-rank regression coefficient matrix and its full-rank analogue, whereas the vertical coordinate, $\Delta\widehat{\boldsymbol{\Sigma}}_{\mathcal{E}\mathcal{E}}^{(t)}$, shows the proportionate reduction in the residual variance matrix in using a simple full-rank model rather than the computationally more elaborate reduced-rank model. The reason for including a special point for $t = 0$ is that without such a point, it would be impossible to assess the statistical rank of $\mathbf{C}$ at $t = 1$. In this formulation, $t = 0$ corresponds to the completely random model $\mathbf{Y} = \boldsymbol{\mu} + \mathcal{E}$.

Assessing the effective dimensionality of the multivariate regression by using step (4) in Table 6.3 involves a certain amount of subjective judgment, but from experience with many of these types of plots, the choice should not be too difficult. Because of the nature of $\widehat{\mathbf{C}}^{(t)}$, the sequence of values for the horizontal coordinate is not guaranteed to decrease monotonically from 1 to 0. It does appear, however, that in many of the applications of this method, and especially when we take $\boldsymbol{\Gamma} = \mathbf{I}_s$ as the weight matrix, the plotted points appear within the unit square, but below the $(1,1)$–$(0,0)$ diagonal line, indicating that the residual covariance matrices typically stabilize faster than do the regression coefficient matrices.

For example, the estimated RRR coefficient matrices, $\widehat{\mathbf{C}}^{(1)}$, $\widehat{\mathbf{C}}^{(2)}$, and $\widehat{\mathbf{C}}^{(3)}$, for the tobacco data (see Section 6.3.3) do not appear to have stabilized at any specific rank $t \leq 3$. In Figure 6.3, we display the rank trace for the tobacco data with weight matrix the identity. Note that dC is shorthand for $\Delta\widehat{\mathbf{C}}^{(t)}$ and dE is shorthand for $\widehat{\boldsymbol{\Sigma}}_{\mathcal{E}\mathcal{E}}^{(t)}$. The rank-trace plot shows that a RRR solution with rank 1 is best, with no discernible difference between that solution and the full-rank solution. In this simple example, this conclusion agrees with the dominant magnitude of the largest sample eigenvalue, $\widehat{\lambda}_1$, of $\widehat{\boldsymbol{\Sigma}}_{YX}\widehat{\boldsymbol{\Sigma}}_{XX}^{-1}\widehat{\boldsymbol{\Sigma}}_{XY}$, which accounts for 98.6% of the trace of that matrix.

In certain applications, and when the weight matrix $\boldsymbol{\Gamma}$ is more complicated than $\mathbf{I}_s$ (e.g., $\boldsymbol{\Gamma} = \widehat{\boldsymbol{\Sigma}}_{YY}^{-1}$), the rank trace often displays a different shape; for example, we may see points plotted outside the unit square or a non-monotonic pattern within the unit square. In such situations, we fix a positive constant k and replace the sample covariance matrices, $\widehat{\boldsymbol{\Sigma}}_{XX}$ and $\widehat{\boldsymbol{\Sigma}}_{XX}$ by $\widehat{\boldsymbol{\Sigma}}_{XX}^{(k)} = \widehat{\boldsymbol{\Sigma}}_{XX} + k\mathbf{I}_r$ and $\widehat{\boldsymbol{\Sigma}}_{YY}^{(k)} = \widehat{\boldsymbol{\Sigma}}_{YY} + k\mathbf{I}_s$, respectively, as in (6.103). Then, we compute $\widehat{\mathbf{C}}^{(t)}(k)$ as in (6.106) and $\widehat{\boldsymbol{\Sigma}}_{\mathcal{E}\mathcal{E}}^{(t)}(k)$ from the residuals. Using these adjusted estimates, we plot $\Delta\widehat{\mathbf{C}}^{(t)}(k)$ against $\Delta\widehat{\boldsymbol{\Sigma}}_{\mathcal{E}\mathcal{E}}^{(t)}(k)$. This gives us a rank trace for a specific value of k. Start with $k = 0$; if the rank trace has monotonic shape, stop, and estimate the value of t as

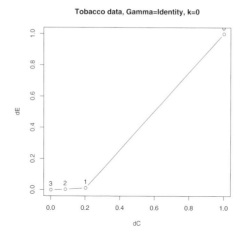

FIGURE 6.3. *Rank trace for the tobacco data.*

in Table 6.3. If the rank trace does not have monotonic shape, increase the value of k slightly and draw the resulting rank trace; if that rank trace is monotonic, stop, and estimate t. Continue increasing k until the associated rank trace is monotonic, at which point, stop and estimate t.

Cross-Validation

An alternative method for assessing the value of t is the use of cross-validation. For each rank t, compute a sequence of estimates of prediction error using any of CV/5, CV/10, or CV/n. Then, identify the smallest rank such that, for larger ranks, the prediction error has stabilized and does not decrease significantly; this is similar to saying that at $\hat{t}$, there is an elbow in the plot of prediction error vs. rank.

6.3.5 Example: Mixtures of Polyaromatic Hydrocarbons

This example refers to the data on the polyaromatic hydrocarbons (PAHs) and digitized spectra that were described in Section 2.2.2. The 50 spectra are displayed in Figure 2.2 and the scatterplot matrix of the 10 PAHs is displayed in Figure 2.3.

We use these data to carry out a reduced-rank regression of the PAH mixture concentrations (the Y variables) on the values of the digitized spectra (the X variables), where we treat the X variables as random. For this example, we take $\mathbf{\Gamma} = \mathbf{I}_s$. Because of the high correlations between neighboring spectrum values, collinearities in the X variables may make the (27×27)-matrix $\widehat{\mathbf{\Sigma}}_{XX}$ difficult to invert. So, we replace $\widehat{\mathbf{\Sigma}}_{XX}$ and $\widehat{\mathbf{\Sigma}}_{YY}$

in the RRR computations by $\widehat{\mathbf{\Sigma}}_{XX}^{(k)}$ and $\widehat{\mathbf{\Sigma}}_{YY}^{(k)}$ respectively, as in (6.102). These covariance matrix estimates and the RRR estimates now depend upon the constant $k > 0$.

The rank trace for $\mathbf{\Gamma} = \mathbf{I}_s$ and $k = 0$ is plotted in Figure 6.4 (top-left panel). We see the rank trace is monotone within the unit square and so we estimate t as $\widehat{t} = 5$. In the other panels, we show rank-trace plots for $\mathbf{\Gamma} = \widehat{\mathbf{\Sigma}}_{YY}^{-1}$, the weight matrix for canonical variate analysis (CVA). In the top-right panel, the rank-trace plot for $k = 0$ (i.e., no regularization) is not monotonic; so, we increase the value of k slightly away from $k = 0$. The bottom-left and bottom-right panels show the rank-trace plot for $k = 0.000001$ and for $k = 0.001$, respectively. At $k = 0.000001$, the rank trace is monotone but not smooth, whereas at $k = 0.001$, the rank trace is a smooth, monotone sequence of points. The most appropriate estimate for t if we apply the weight matrix $\mathbf{\Gamma} = \widehat{\mathbf{\Sigma}}_{YY}^{-1}$ is $\widehat{t} = 5$, which agrees with our estimate for $\mathbf{\Gamma} = \mathbf{I}_s$.

Applying CV to the PAH data yields the CV prediction errors (PEs) as a function of the rank t, and these are given in Table 6.4 and Figure 6.5. As a method for estimating the true rank, t, of $\mathbf{C}$, the CV PEs appear to level off at $t = 5$, which agrees with the rank assessments from the rank-trace plots.

6.4 Software Packages

A good source for SAS programs and discussion of SAS output for multivariate regression and MANOVA is Khattree and Naik (1999). It should be noted that although there is an RRR method implemented in the SAS procedure PROC PLS, it is not the same as and has no connection to the RRR method discussed in this book. The examples in this chapter were computed using the R program MULTANL+RRR (written by Charles Miller), which can be downloaded from the book's website. An S-PLUS package rrr.s (written by Magne Aldrin) for carrying out RRR can be downloaded from the STATLIB website at lib.stat/cmu.edu/S/.

Bibliographical Notes

In textbooks, multivariate regression is usually discussed within the context of the multivariate general linear model or multivariate analysis of variance (MANOVA), where the emphasis is most often placed on the fixed-X case.

The reduced-rank regression model has its origins in the work of Anderson (1951), Rao (1965), and Brillinger (1969). The deliberately alliterative

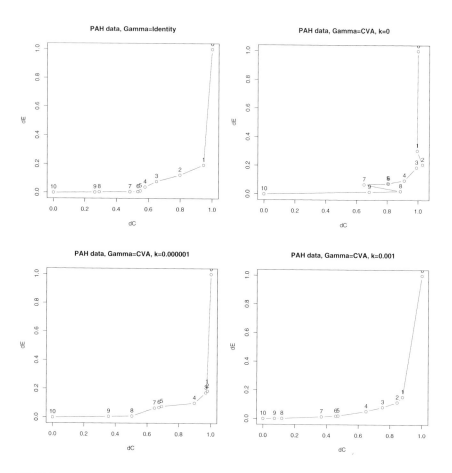

FIGURE 6.4. *Rank trace for reduced-rank regression on the PAH data. There are $r = 27$ wavelengths, $s = 10$ PAHs, and $n = 50$ mixtures. Top-left panel: $\mathbf{\Gamma} = \mathbf{I}_s$. Other panels have $\mathbf{\Gamma} = \widehat{\mathbf{\Sigma}}_{YY}^{-1}$ and $k = 0$ (top-right); $k = 0.000001$ (bottom-left); $k = 0.001$ (bottom-right).*

TABLE 6.4. *CV prediction errors for reduced-rank regression of the PAH data.*

Rank	CV/5	CV/10	CV/n
1	0.254	0.242	0.248
2	0.186	0.171	0.166
3	0.143	0.124	0.117
4	0.102	0.086	0.082
5	0.077	0.060	0.054
6	0.070	0.054	0.047
7	0.070	0.054	0.047
8	0.070	0.053	0.047
9	0.068	0.052	0.046
10	0.064	0.047	0.040

name "reduced-rank regression" was coined by Izenman (1972). Since then, the amount of research into the theory of reduced-rank regression models has steadily increased, leading to the monographs by van der Leeden (1990) and Reinsel and Velu (1998).

Because many authors mistakenly omit the hyphen in the name "reduced-rank regression," we give reasons why it should be included. The terms "reduced-rank" and "full-rank" are *compound adjectives* describing the type of regression and, therefore, must take a hyphen. Further, without hyphens the methodology is apt to be confused with the topic of "rank regression," which deals with multivariate regression of rank data (see, e.g., Davis and McKean, 1993). Of course, we could also study reduced-rank regression of rank data.

Exercises

6.1 Using the result in the fixed-$\mathbf{X}$ case that the covariance matrix of the matrix of residuals $\mathcal{E}$ is $\mathrm{cov}(\mathrm{vec}(\widehat{\mathcal{E}})) = \boldsymbol{\Sigma}_{\mathcal{E}\mathcal{E}} \otimes (\mathbf{I}_n - \mathbf{H})$, find expressions for the means, variances, and covariances of the elements of the rows and columns of the matrix $\mathcal{E}$. Simplify your results when $\boldsymbol{\Sigma}_{\mathcal{E}\mathcal{E}} = \mathrm{diag}\{\sigma_1^2, \cdots, \sigma_s^2\}$.

6.2 If $\boldsymbol{\Sigma}_{XX}$ and $\boldsymbol{\Sigma}_{YY}$ are nonsingular, show that the eigenvalues of $\mathbf{R}$ lie between 0 and 1.

6.3 Let $\mathbf{X}' = \boldsymbol{\Psi} + \boldsymbol{\Lambda}\mathbf{X}$ and $\mathbf{Y}' = \boldsymbol{\Phi} + \boldsymbol{\Delta}\mathbf{Y}$, where $\boldsymbol{\Lambda}$ and $\boldsymbol{\Delta}$ are nonsingular. Show that the minimizing criterion (6.79) with $\boldsymbol{\Gamma} = \boldsymbol{\Sigma}_{YY}^{-1}$ is invariant under these nonsingular transformations.

6.4 Develop a theory of reduced-rank regression for the "fixed-$\mathbf{X}$" case.

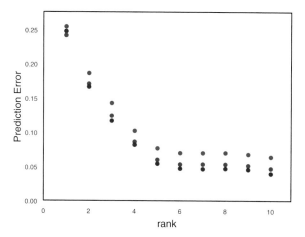

FIGURE 6.5. *Prediction errors for PAH example (n=50, r=27, s=10) plotted against rank of the regression coefficient matrix. The PEs were computed using cross-validation: CV/5 (red dots), CV/10 (blue dots), and CV/n (purple dots). The results show a leveling-off of the PE at rank $t = 5$.*

6.5 Use the results from Exercise 6.1 to develop a theory of residual diagnostics from a multivariate reduced-rank regression (RRR) for the "fixed-**X**" case. In particular, derive the distribution theory for RRR residuals and the distribution of quadratic forms in RRR residuals. How could you use this theory to detect outliers?

6.6 Consider the likelihood-ratio test statistic for the dimensionality of a multivariate regression. Let the null hypothesis be that the true rank is at most t with the alternative that the regression is full-rank. Let $Q_e^{(t)} = \widehat{e}^{(t)}\widehat{e}^{(t)\tau}$ and $Q_e = \widehat{e}\,\widehat{e}^\tau$ denote the residual sum of squares matrices for a rank-t reduced-rank regression and a full-rank regression, respectively. Let $\Lambda_{LR}^{(t)} = \det\{Q_e^{(t)}\}/\det\{Q_e\}$. Show that

$$-2\log_e \Lambda_{LR}^{(t)} = -n \sum_{j=t+1}^{s} \log_e(1 - \widehat{\lambda}_j),$$

where $\widehat{\lambda}_j$ is the jth largest eigenvalue of $\widehat{\mathbf{R}}$. (Asymptotically, under the null hypothesis, $-2\log_e \Lambda_{LR}^{(t)} \sim \chi^2_{(s-t)(r-t)}$.)

6.7 Show that the two procedures described in Section 6.2.1 lead to the same results in estimating $\text{tr}(\mathbf{A\Theta})$. The two procedures are (1) write $\boldsymbol{\mu} + \boldsymbol{\Theta}\mathcal{X} = \boldsymbol{\Theta}^*\mathcal{X}^*$, where $\boldsymbol{\Theta}^* = (\boldsymbol{\mu}_0 \vdots \boldsymbol{\Theta})$ and $\mathcal{X}^* = (\mathbf{1}_n \vdots \mathbf{X}^\tau)^\tau$, and then estimate $\boldsymbol{\Theta}^*$; (2) remove $\boldsymbol{\mu}$ by centering $\mathcal{X}$ and $\mathcal{Y}$, and then estimate $\boldsymbol{\Theta}$ directly.

6.8 Using the data from the Norwegian paper quality example (Section 6.2.2), show that Table 6.1 can also be derived by regressing each of the 13 Ys on all the 9 Xs.

6.9 In the classical multivariate regression model (Section 6.2.1), show that $S_e = \mathcal{Y}_c(\mathbf{I}_n - \mathbf{H})\mathcal{Y}_c^\tau$, where $\mathbf{H} = \mathcal{X}_c^\tau(\mathcal{X}_c\mathcal{X}_c^\tau)^{-1}\mathcal{X}_c$. Hence, or otherwise, show that $S_e = \mathcal{E}(\mathbf{I}_n - \mathbf{H})\mathcal{E}^\tau$.

6.10 Write a computer program to carry out a multivariate ridge regression, and then apply it to the Norwegian paper quality data. Compare the results with those obtained from separate univariate ridge regressions.

6.11 The data for this exercise is `Table 60.1` in Andrews and Herzberg (1985, pp. 357–360), which can be downloaded from the STATLIB website `lib.stat.cmu.edu/datasets/Andrews/`. The data consist of 8 measurements on each of 4 variates on 13 different types of root-stocks of apple trees. The 4 variates are: trunk girth in mm (Y_1) and extension growth in cm (Y_2) at 4 years after planting, and trunk girth in mm (Y_3) and weight of tree above ground in lb (Y_4) at 15 years after planting. So, there are $s = 4$ measurements on each of $n = 8 \times 13 = 104$ trees. Rescaling each variable might be appropriate. The design matrix $\mathcal{X}$ is a (13×104)-matrix of 0s and 1s depending upon which tree is derived from which root-stock. Regress the (4×104)-matrix $\mathcal{Y}$ on $\mathcal{X}$ and estimate the (4×13) regression coefficient matrix $\boldsymbol{\Theta}$. Estimate the (4×4) error covariance matrix $\boldsymbol{\Sigma}_{\mathcal{E}\mathcal{E}}$. Estimate the standard errors for these regression coefficient estimates. Compute the (unconstrained) MANOVA table for these data.

6.12 Extend the MANOVA analysis to a two-way layout of vector observations $\mathbf{Y} = (\mathbf{Y}_{ij})$, where i denotes the row and j denotes the column. The two-way model with one observation in each cell is defined by

$$\mathbf{Y}_{ij} = \boldsymbol{\mu} + \boldsymbol{\mu}_{i\cdot} + \boldsymbol{\mu}_{\cdot j} + \mathcal{E}_{ij}, \quad i = 1, 2, \ldots, I, \ j = 1, 2, \ldots, J,$$

where we assume that $\sum_i \boldsymbol{\mu}_{i\cdot} = \sum_j \boldsymbol{\mu}_{\cdot j} = \mathbf{0}$, and the $\mathcal{E}_{ij}$ are random s-vectors with mean $\mathbf{0}$. Write down the design matrix $\mathcal{X}$ and the matrix of regression coefficients $\boldsymbol{\Theta}$. Write down the partition of $\mathbf{Y}_{ij} - \bar{\mathbf{Y}}$, where $\bar{\mathbf{Y}}$ is the average of all IJ observations, in terms of the ith row effect $\bar{\mathbf{Y}}_{i\cdot} - \bar{\mathbf{Y}}$, the jth column effect $\bar{\mathbf{Y}}_{\cdot j} - \bar{\mathbf{Y}}$, and the residual effect $\mathbf{Y}_{ij} - \bar{\mathbf{Y}}_{i\cdot} - \bar{\mathbf{Y}}_{\cdot j} + \bar{\mathbf{Y}}$, where $\bar{\mathbf{Y}}_{i\cdot}$ is the average over all columns for the ith row, and $\bar{\mathbf{Y}}_{\cdot j}$ is the average over all rows for the jth column. Derive the corresponding partition in terms of sums-of-squares and determine their respective degrees of freedom. Write down the corresponding two-way MANOVA table.

6.13 Generalize Exercise 6.11 to the case of m observations $\mathbf{Y}_{ijk}$ in each cell ($k = 1, 2, \ldots, m$), where an interaction term $\boldsymbol{\mu}_{ij}$ satisfying $\sum_i \boldsymbol{\mu}_{ij} = \sum_j \boldsymbol{\mu}_{ij} = \mathbf{0}$ is added to the model. The error term now becomes $\mathcal{E}_{ijk}$. The ith row effect is $\bar{\mathbf{Y}}_{i\cdot\cdot} - \bar{\mathbf{Y}}$, the jth column effect is $\bar{\mathbf{Y}}_{\cdot j\cdot} - \bar{\mathbf{Y}}$, the interaction effect is $\bar{\mathbf{Y}}_{ij\cdot} - \bar{\mathbf{Y}}_{i\cdot\cdot} - \bar{\mathbf{Y}}_{\cdot j\cdot} + \bar{\mathbf{Y}}$, and the residual is $\mathbf{Y}_{ijk} - \bar{\mathbf{Y}}_{ij\cdot}$. Derive the two-way MANOVA table for this case.

6.14 Write a program to carry out a constrained multivariate regression including the MANOVA Table 6.2.

6.15 Run a RRR on the Norwegian paper quality data. Plot the rank trace using $\mathbf{\Gamma} = \mathbf{I}_s$ as the weight matrix. Estimate the effective dimensionality of the multivariate regression. Compare the estimate with one obtained using CV.

6.16 Using the results (6.109), (6.110), and (6.111), show that the asymptotic covariance of the regression coefficient matrix $\mathrm{vec}(\widehat{\mathbf{C}}^{(t)})$ reduces to $\mathbf{\Sigma}_{\mathcal{EE}} \otimes \mathbf{\Sigma}_{XX}^{-1}$ when $t = s$ (i.e., full rank).

7
Linear Dimensionality Reduction

7.1 Introduction

When faced with situations involving high-dimensional data, it is natural to consider the possibility of projecting those data onto a lower-dimensional subspace without losing important information regarding some characteristic of the original variables. One way of accomplishing this reduction of dimensionality is through variable selection, also called *feature selection* (see Section 5.7). Another way is by creating a reduced set of linear or nonlinear transformations of the input variables. The creation of such composite variables (or *features*) by projection methods is often referred to as *feature extraction*. Usually, we wish to find those low-dimensional projections of the input data that enjoy some sort of optimality properties.

Early examples of projection methods were *linear* methods such as principal component analysis (PCA) (Hotelling, 1933) and canonical variate and correlation analysis (CVA or CCA) (Hotelling, 1936), and these have become two of the most popular dimensionality-reducing techniques in use today. Both PCA and CVA are, at heart, eigenvalue-eigenvector problems. Furthermore, both can be viewed as special cases of multivariate reduced-rank regression. This latter connection to regression is fortuitous. Whereas PCA and CVA were once regarded as isolated statistical tools, their now

A.J. Izenman, *Modern Multivariate Statistical Techniques*,
doi: 10.1007/978-0-387-78189-1_7,
© Springer Science+Business Media, LLC 2008

being part of such a well-traveled tool as regression means that we should be able to carry out feature selection and extraction, as well as outlier detection within an integrated framework.

7.2 Principal Component Analysis

Principal component analysis (PCA) (Hotelling, 1933) was introduced as a technique for deriving a reduced set of orthogonal linear projections of a single collection of correlated variables, $\mathbf{X} = (X_1, \cdots, X_r)^\tau$, where the projections are ordered by decreasing variances. Variance is a second-order property of a random variable and is an important measurement of the amount of information in that variable. PCA has also been referred to as a method for "decorrelating" $\mathbf{X}$; as a result, the technique has been independently rediscovered by many different fields, with alternative names such as *Karhunen–Loève transform* and *empirical orthogonal functions*, which are used in communications theory and atmospheric sciences, respectively.

PCA is used primarily as a dimensionality-reduction technique. In this role, PCA is used, for example, in lossy data compression, pattern recognition, and image analysis. We have already seen in Section 5.7.2 how PCA is used in chemometrics to construct derived variables in biased regression situations, when the number of input variables is too large for useful analysis.

In addition to reducing dimensionality, PCA can be used to discover important features of the data. Discovery in PCA takes the form of graphical displays of the principal component scores. The first few principal component scores can reveal whether most of the data actually live on a linear subspace of $\Re^r$ and can be used to identify outliers, distributional peculiarities, and clusters of points. The last few principal component scores show those linear projections of $\mathbf{X}$ that have smallest variance; any principal component with zero or near-zero variance is virtually constant, and, hence, can be used to detect collinearity, as well as outliers that pop up and alter the perceived dimensionality of the data.

7.2.1 Example: The Nutritional Value of Food

Nutritional data from 961 food items are listed alphabetically in this data set.[1] The nutritional components of each food item are given by the following seven variables: fat (grams), food energy (calories), carbohydrates

---

[1]The data are given in the file food.txt, which can be downloaded from the book's website or from http://www.ntwrks.com/~mikev/chart1.html.

TABLE 7.1. *Coefficients of the six principal components of the covariance matrix of the transformed food nutrition data.*

Food Component	PC1	PC2	PC3	PC4	PC5	PC6
Fat	0.557	0.099	0.275	0.130	0.455	0.617
Food energy	0.536	0.357	-0.137	0.075	0.273	-0.697
Carbohydrates	-0.025	0.672	-0.568	-0.286	-0.157	0.344
Protein	0.235	-0.374	-0.639	0.599	-0.154	0.119
Cholesterol	0.253	-0.521	-0.326	-0.717	0.210	-0.003
Saturated fat	0.531	-0.019	0.261	-0.150	-0.791	0.022
Variance	2.649	1.330	1.020	0.680	0.267	0.055
% Total Variance	44.1	22.2	17.0	11.3	4.4	0.9

(grams), `protein` (grams), `cholesterol` (milligrams), `weight` (grams), and `saturated fat` (grams). Food items are listed according to very disparate serving sizes, which include teaspoon, tablespoon, cup, loaf, slice, cake, cracker, package, piece, pie, biscuit, muffin, spear, pat, wedge, stalk, cookie, and pastry. To equalize out the different types of servings for each food, we first divide each variable by weight of the food item (which leaves us with 6 variables), and then, because of wide variations in the different variables, each variable is standardized by subtracting its mean and dividing the result by its standard deviation. The resulting data are $\mathcal{X} = (X_{ij})$.

A PCA of the transformed data yields six principal components ordered by decreasing variances. The first three principal components, PC1, PC2, and PC3, which account for more than 83% of the total variance, have coefficients given in Table 7.1. Notice that PC1 puts little weight on `carbohydrates`, and PC2 puts little weight on `fat` and `saturated fat`.

The scatterplot of the first two principal components is given in Figure 7.1. The scatterplot appears to show a number of interesting features. Notice the almost straight-line edge to the plotted points at the upper left-hand corner. We also can identify various groups of points in this display, where the food items in each group have been ordered by magnitude of that nutritional component, starting at the largest value:

1. Cholesterol: 318 (raw egg yolk), 189 (chicken liver), 62 (beef liver), 312 (fried egg), 313 (hard-cooked egg), 314 (poached egg), 315 (scrambled egg), and 317 (raw whole egg).

2. Protein: 357 (dry gelatin), 778 (raw seaweed), 952 and 953 (yeast), and 578–580 (parmesan cheese).

3. Saturated fat: 124–129 (butter), 441 and 442 (lard), 212 (bitter chocolate), 224–226 (coconut), 326 and 327 (cooking fat), and 166–168 (cheddar cheese).

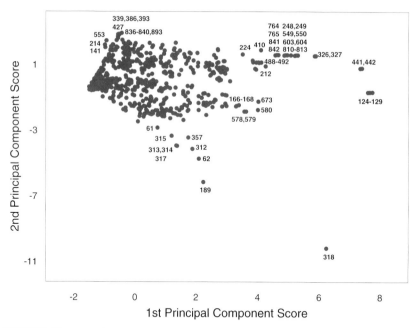

FIGURE 7.1. *Scatterplot of the first two principal components of the food nutrition data. Numbers next to certain points indicate the food item corresponding to that point. Multiple food items may be plotted at the same point.*

4. Fat and food energy: 326 and 327 (cooking fat), 441 and 442 (lard), 603 and 604 (peanut oil), 549–550 (olive oil), 248 and 249 (corn oil), 764 and 765 (safflower oil), 810–813 (soybean cottonsead oil), 841 and 842 (sunflower oil), 124–129 (salted butter), and 488–492 (margarine).

5. Carbohydrates: 837–840 (white sugar), 393 (hard candy), 836 (brown sugar), 553 (onion powder), 339 (fondant), 834 (Kellogg Sugar Frosted Flakes), 843 (sunflower seeds), 844 (Super Sugar Crisp Cereal), 427 (jelly beans), 141 (carob flour), and 221 (coca powder).

Most of these points are identified in the scatterplot, but some are covered too well to be displayed clearly. We see that food item 318 (raw egg yolk) is an outlier along an imaginary cholesterol axis and 124–129 (butter) and 441 and 442 (lard) are outliers along an imaginary saturated-fat axis. Similarly, in scatterplots of PC1 and PC3, and of PC2 and PC3 (not shown here), we see that food items 357 (dry gelatin) and 779 (raw seaweed) are outliers along an imaginary protein axis.

7.2.2   Population Principal Components

Assume that the random r-vector

$$\mathbf{X} = (X_1, \cdots, X_r)^\tau \tag{7.1}$$

has mean $\boldsymbol{\mu}_X$ and $(r \times r)$ covariance matrix $\boldsymbol{\Sigma}_{XX}$. PCA seeks to replace the set of r (unordered and correlated) input variables, $X_1, X_2, \ldots, X_r$, by a (potentially smaller) set of t (ordered and uncorrelated) linear projections, $\xi_1, \ldots, \xi_t$ $(t \leq r)$, of the input variables,

$$\xi_j = \mathbf{b}_j^\tau \mathbf{X} = b_{j1}X_1 + \cdots + b_{jr}X_r, \quad j = 1, 2, \ldots, t, \tag{7.2}$$

where we minimize the loss of information due to replacement.

In PCA, "information" is interpreted as the "total variation" of the original input variables,

$$\sum_{j=1}^{r} \text{var}(X_j) = \text{tr}(\boldsymbol{\Sigma}_{XX}). \tag{7.3}$$

From the spectral decomposition theorem (Section 3.2.4), we can write

$$\boldsymbol{\Sigma}_{XX} = \mathbf{U}\boldsymbol{\Lambda}\mathbf{U}^\tau, \quad \mathbf{U}^\tau\mathbf{U} = \mathbf{I}_r, \tag{7.4}$$

where the diagonal matrix $\boldsymbol{\Lambda}$ has diagonal elements the eigenvalues, $\{\lambda_j\}$, of $\boldsymbol{\Sigma}_{XX}$, and the columns of $\mathbf{U}$ are the eigenvectors of $\boldsymbol{\Sigma}_{XX}$. Thus, the total variation is $\text{tr}(\boldsymbol{\Sigma}_{XX}) = \text{tr}(\boldsymbol{\Lambda}) = \sum_{j=1}^{r} \lambda_j$.

The jth coefficient vector, $\mathbf{b}_j = (b_{1j}, \cdots, b_{rj})^\tau$, is chosen so that:

- The first t linear projections ξ_j, $j = 1, 2, \ldots, t$, of $\mathbf{X}$ are ranked in importance through their variances $\{\text{var}\{\xi_j\}\}$, which are listed in decreasing order of magnitude: $\text{var}\{\xi_1\} \geq \text{var}\{\xi_2\} \geq \ldots \geq \text{var}\{\xi_t\}$.

- ξ_j is uncorrelated with all ξ_k, $k < j$.

The linear projections (7.2) are then known as the *first t principal components* of $\mathbf{X}$.

There are two popular derivations of the set of principal components of $\mathbf{X}$: PCA can be derived using a least-squares optimality criterion, or it can be derived as a variance-maximizing technique. In the next two subsections, we discuss these two definitions.

7.2.3   Least-Squares Optimality of PCA

Let

$$\mathbf{B} = (\mathbf{b}_1, \cdots, \mathbf{b}_t)^\tau, \tag{7.5}$$

be a $(t \times r)$-matrix of weights $(t \leq r)$. The linear projections (7.2) can be written as a t-vector,

$$\boldsymbol{\xi} = \mathbf{BX}, \tag{7.6}$$

where $\boldsymbol{\xi} = (\xi_1, \cdots, \xi_t)^\tau$. We want to find an r-vector $\boldsymbol{\mu}$ and an $(r \times t)$-matrix $\mathbf{A}$ such that the projections $\boldsymbol{\xi}$ have the property that $\mathbf{X} \approx \boldsymbol{\mu} + \mathbf{A}\boldsymbol{\xi}$ in some least-squares sense. We use the least-squares error criterion,

$$\mathrm{E}\{(\mathbf{X} - \boldsymbol{\mu} - \mathbf{A}\boldsymbol{\xi})^\tau (\mathbf{X} - \boldsymbol{\mu} - \mathbf{A}\boldsymbol{\xi})\}, \tag{7.7}$$

as our measure of how well we can reconstruct $\mathbf{X}$ by the linear projection $\boldsymbol{\xi}$.

We can write the criterion (7.7) in a more transparent manner by substituting $\mathbf{BX}$ for $\boldsymbol{\xi}$. The criterion is now a function of an $(r \times t)$-matrix $\mathbf{A}$ and a $(t \times r)$-matrix $\mathbf{B}$ (both of full rank t), and an r-vector $\boldsymbol{\mu}$. The goal is to choose $\mathbf{A}$, $\mathbf{B}$, and $\boldsymbol{\mu}$ to minimize

$$\mathrm{E}\{(\mathbf{X} - \boldsymbol{\mu} - \mathbf{ABX})^\tau (\mathbf{X} - \boldsymbol{\mu} - \mathbf{ABX})\}. \tag{7.8}$$

For example, when $t = 1$, we can write (7.8) as the least-squares problem,

$$\min_{\boldsymbol{\mu}, \mathbf{A}, \mathbf{B}} \mathrm{E} \sum_{j=1}^{r} (X_j - \mu_j - a_{j1} \mathbf{b}_1^\tau \mathbf{X})^2, \tag{7.9}$$

where $\boldsymbol{\mu} = (\mu_1, \cdots, \mu_r)^\tau$, $\mathbf{A} = \mathbf{a}_1 = (a_{11}, \cdots, a_{r1})^\tau$, and $\mathbf{B} = \mathbf{b}_1^\tau$.

The criterion (7.8) is just (6.80) with $\mathbf{Y} \equiv \mathbf{X}$, $s = r$, and $\boldsymbol{\Gamma} = \mathbf{I}_r$. Hence, (7.8) is minimized by the reduced-rank regression solution,

$$\mathbf{A}^{(t)} = (\mathbf{v}_1, \cdots, \mathbf{v}_t) = \mathbf{B}^{(t)\tau}, \tag{7.10}$$

$$\boldsymbol{\mu}^{(t)} = (\mathbf{I}_r - \mathbf{A}^{(t)} \mathbf{B}^{(t)}) \boldsymbol{\mu}_X, \tag{7.11}$$

where $\mathbf{v}_j = \mathbf{v}_j(\boldsymbol{\Sigma}_{XX})$ is the eigenvector associated with the jth largest eigenvalue, λ_j, of $\boldsymbol{\Sigma}_{XX}$. Thus, our best rank-t approximation to the original $\mathbf{X}$ is given by

$$\widehat{\mathbf{X}}^{(t)} = \boldsymbol{\mu}^{(t)} + \mathbf{C}^{(t)} \mathbf{X} = \boldsymbol{\mu}_X + \mathbf{C}^{(t)} (\mathbf{X} - \boldsymbol{\mu}), \tag{7.12}$$

where

$$\mathbf{C}^{(t)} = \mathbf{A}^{(t)} \mathbf{B}^{(t)} = \sum_{j=1}^{t} \mathbf{v}_j \mathbf{v}_j^\tau \tag{7.13}$$

is the reduced-rank regression coefficient matrix with rank t for the principal components case. From (6.91), the minimum value of (7.8) is given by $\sum_{j=t+1}^{r} \lambda_j$, the sum of the smallest $r - t$ eigenvalues of $\boldsymbol{\Sigma}_{XX}$.

It may be helpful to think of these results in the following way. Let $\mathbf{V} = (\mathbf{v}_1, \cdots, \mathbf{v}_r)$ be the $(r \times r)$-matrix whose columns are the complete set of r ordered eigenvectors of $\boldsymbol{\Sigma}_{XX}$. We have shown that the most accurate rank-t least-squares reconstruction of $\mathbf{X}$ can be obtained by using the composition of two linear maps $L' \circ L$. The first map $L : \Re^r \rightarrow \Re^t$ takes the first

t columns of $\mathbf{V}$ to form t linear projections of $\mathbf{X}$, and then the second map $L' : \Re^t \to \Re^r$ uses those same t columns of $\mathbf{V}$ to carry out a linear reconstruction of $\mathbf{X}$ from those projections.

The *first t principal components* (also known as the *Karhunen–Loève transform*) of $\mathbf{X}$ are given by the linear projections, $\xi_1, \ldots, \xi_t$, where

$$\xi_j = \mathbf{v}_j^\tau \mathbf{X}, \quad j = 1, 2, \ldots, t. \tag{7.14}$$

The covariance between ξ_i and ξ_j is

$$\mathrm{cov}(\xi_i, \xi_j) = \mathrm{cov}(\mathbf{v}_i^\tau \mathbf{X}, \mathbf{v}_j^\tau \mathbf{X}) = \mathbf{v}_i^\tau \boldsymbol{\Sigma}_{XX} \mathbf{v}_j = \lambda_j \mathbf{v}_i^\tau \mathbf{v}_j = \delta_{ij} \lambda_j, \tag{7.15}$$

where δ_{ij} is the Kronecker delta, which equals 1 if $i = j$ and zero otherwise. Thus, λ_1, the largest eigenvalue of $\boldsymbol{\Sigma}_{XX}$, is $\mathrm{var}\{\xi_1\}$; λ_2, the second-largest eigenvalue of $\boldsymbol{\Sigma}_{XX}$, is $\mathrm{var}\{\xi_2\}$; and so on, while all pairs of derived variables are uncorrelated, $\mathrm{cov}(\xi_i, \xi_j) = 0$, $i \neq j$.

A goodness-of-fit measure of how well the first t principal components represent the r original variables in the lower-dimensional space is given by the ratio

$$\frac{\lambda_{t+1} + \cdots + \lambda_r}{\lambda_1 + \cdots + \lambda_r} \tag{7.16}$$

which is the proportion of the total variation in the input variables that is explained by the last $r - t$ principal components. If the first t principal components explain a large proportion of the total variation in $\mathbf{X}$, then the ratio (7.16) should be small.

Actually, more is true. Not only do $\boldsymbol{\mu}^{(t)}$, $\mathbf{A}^{(t)}$, and $\mathbf{B}^{(t)}$ minimize the scalar criterion (7.8), but also they simultaneously minimize all the eigenvalues of the $(r \times r)$-matrix

$$\boldsymbol{\Psi}^{(t)} = \mathrm{E}\{(\mathbf{X} - \boldsymbol{\mu} - \mathbf{ABX})(\mathbf{X} - \boldsymbol{\mu} - \mathbf{ABX})^\tau\}, \tag{7.17}$$

thereby also minimizing any function of those eigenvalues, such as their sum (trace of (7.17) and, hence, (7.8)) and their product (determinant of (7.17)). We can see this as follows. From (6.80), setting $\mathbf{Y} \equiv \mathbf{X}$, $s = r$, and $\boldsymbol{\Gamma} = \mathbf{I}_r$, we have that

$$\begin{aligned} \boldsymbol{\Psi}^{(t)} &\geq \boldsymbol{\Sigma}_{XX} - \boldsymbol{\Sigma}_{X,ABX} \boldsymbol{\Sigma}_{ABX,ABX}^{-1} \boldsymbol{\Sigma}_{ABX,X} \\ &= \boldsymbol{\Sigma}_{XX} - \mathbf{D}, \end{aligned} \tag{7.18}$$

where

$$\mathbf{D} = \boldsymbol{\Sigma}_{XX} \mathbf{B}^\tau \mathbf{A}^\tau (\mathbf{AB}\boldsymbol{\Sigma}_{XX}\mathbf{B}^\tau\mathbf{A}^\tau)^{-1} \mathbf{AB}\boldsymbol{\Sigma}_{XX}. \tag{7.19}$$

Note that the $(r \times r)$-matrix $\mathbf{D}$ has rank at most t $(\leq r)$. We wish to find $\boldsymbol{\mu}$, $\mathbf{A}$, and $\mathbf{B}$ to minimize the jth largest eigenvalue of $\mathbf{D}$. From the

Courant–Fischer Min-Max theorem (see Section 3.2.10),

$$
\begin{aligned}
\lambda_j(\boldsymbol{\Sigma}_{XX} - \mathbf{D}) &= \min_{\mathbf{L}:\operatorname{rank}(\mathbf{L}) \leq j-1} \max_{\boldsymbol{\alpha}:\mathbf{L}\boldsymbol{\alpha}=0} \frac{\boldsymbol{\alpha}^\tau(\boldsymbol{\Sigma}_{XX} - \mathbf{D})\boldsymbol{\alpha}}{\boldsymbol{\alpha}^\tau\boldsymbol{\alpha}} \\
&\geq \min_{\mathbf{L}} \max_{\boldsymbol{\alpha}:\mathbf{L}\boldsymbol{\alpha}=0, \mathbf{D}\boldsymbol{\alpha}=0} \frac{\boldsymbol{\alpha}^\tau\boldsymbol{\Sigma}_{XX}\boldsymbol{\alpha}}{\boldsymbol{\alpha}^\tau\boldsymbol{\alpha}} \\
&= \min_{\mathbf{L}} \max_{\boldsymbol{\alpha}:(\mathbf{L}|\mathbf{D})\boldsymbol{\alpha}=0} \frac{\boldsymbol{\alpha}^\tau\boldsymbol{\Sigma}_{XX}\boldsymbol{\alpha}}{\boldsymbol{\alpha}^\tau\boldsymbol{\alpha}} \\
&\geq \min_{\mathbf{L},\mathbf{D}} \max_{\boldsymbol{\alpha}:(\mathbf{L}|\mathbf{D})\boldsymbol{\alpha}=0} \frac{\boldsymbol{\alpha}^\tau\boldsymbol{\Sigma}_{XX}\boldsymbol{\alpha}}{\boldsymbol{\alpha}^\tau\boldsymbol{\alpha}} \\
&= \lambda_{t+j}(\boldsymbol{\Sigma}_{XX}), \tag{7.20}
\end{aligned}
$$

because $\operatorname{rank}((\mathbf{L}|\mathbf{D})) \leq j - 1 + t$. Thus,

$$
\lambda_j(\boldsymbol{\Phi}^{(t)}) \geq \lambda_{j+t}(\boldsymbol{\Sigma}_{XX}). \tag{7.21}
$$

By plugging in the above $\boldsymbol{\mu}^{(t)}$, $\mathbf{A}^{(t)}$, and $\mathbf{B}^{(t)}$ into the expression for $\boldsymbol{\Psi}^{(t)}$, it follows immediately that the minimum value of $\lambda_j(\boldsymbol{\Psi}^{(t)})$ is actually given by $\lambda_{t+j}(\boldsymbol{\Sigma}_{XX})$.

7.2.4 PCA as a Variance-Maximization Technique

In the original derivation of principal components (Hotelling, 1933). the coefficient vectors,

$$
\mathbf{b}_j = (b_{j1}, b_{j2}, \ldots, b_{jr})^\tau, \quad j = 1, 2, \ldots, t, \tag{7.22}
$$

in (7.5) were chosen in a sequential manner so that the variances of the derived variables ($\operatorname{var}\{\xi_j\} = \mathbf{b}_j^\tau \boldsymbol{\Sigma}_{XX}\mathbf{b}_j$) are arranged in descending order subject to the normalizations $\mathbf{b}_j^\tau \mathbf{b}_j = 1$, $j = 1, 2, \ldots, t$, and that they are uncorrelated with previously chosen derived variables ($\operatorname{cov}(\xi_i, \xi_j) = \mathbf{b}_i^\tau \boldsymbol{\Sigma}_{XX}\mathbf{b}_j = 0$, $i < j$).

The first principal component, ξ_1, is obtained by choosing the r coefficients, $\mathbf{b}_1$, for the linear projection ξ_1, so that the variance of ξ_1 is a maximum. A unique choice of $\{\xi_j\}$ is obtained through the normalization constraint $\mathbf{b}_j^\tau \mathbf{b}_j = 1$, for all $j = 1, 2, \ldots, t$. Form the function

$$
f(\mathbf{b}_1) = \mathbf{b}_1^\tau \boldsymbol{\Sigma}_{XX}\mathbf{b}_1 - \lambda_1(1 - \mathbf{b}_1^\tau \mathbf{b}_1), \tag{7.23}
$$

where λ_1 is a Lagrangian multiplier. Differentiating $f(\mathbf{b}_1)$ with respect to $\mathbf{b}_1$ and setting the result equal to zero for a maximum yields

$$
\frac{\partial f(\mathbf{b}_1)}{\partial \mathbf{b}_1} = 2(\boldsymbol{\Sigma}_{XX} - \lambda_1 \mathbf{I}_r)\mathbf{b}_1 = \mathbf{0}. \tag{7.24}
$$

This is a set of r simultaneous equations. If $\mathbf{b}_1 \neq \mathbf{0}$, then λ_1 must be chosen to satisfy the determinantal equation

$$|\boldsymbol{\Sigma}_{XX} - \lambda_1 \mathbf{I}_r| = 0. \tag{7.25}$$

Thus, λ_1 has to be the largest eigenvalue of $\boldsymbol{\Sigma}_{XX}$, and $\mathbf{b}_1$ the eigenvector, $\mathbf{v}_1$, associated with λ_1.

The second principal component, ξ_2, is then obtained by choosing a second set of coefficients, $\mathbf{b}_2$, for the next linear projection, ξ_2, so that the variance of ξ_2 is largest among all linear projections of $\mathbf{X}$ that are also uncorrelated with ξ_1 above. The variance of ξ_2 is $\text{var}(\xi_2) = \mathbf{b}_2^\top \boldsymbol{\Sigma}_{XX} \mathbf{b}_2$, and this has to be maximized subject to the normalization constraint $\mathbf{b}_2^\top \mathbf{b}_2 = 1$ and orthogonality constraint $\mathbf{b}_1^\top \mathbf{b}_2 = 0$. Form the function

$$f(\mathbf{b}_2) = \mathbf{b}_2^\top \boldsymbol{\Sigma}_{XX} \mathbf{b}_2 + \lambda_2(1 - \mathbf{b}_2^\top \mathbf{b}_2) + \mu \mathbf{b}_1^\top \mathbf{b}_2, \tag{7.26}$$

where λ_2 and μ are the Lagrangian multipliers. Differentiating $f(\mathbf{b}_2)$ with respect to $\mathbf{b}_2$ and setting the result equal to zero for a maximum yields

$$\frac{\partial f(\mathbf{b}_1)}{\partial \mathbf{b}_1} = 2(\boldsymbol{\Sigma}_{XX} - \lambda_2 \mathbf{I}_r)\mathbf{b}_2 + \mu \mathbf{b}_1 = \mathbf{0}. \tag{7.27}$$

Premultiplying this derivative by $\mathbf{b}_1^\top$ and using the orthogonality and normalization constraints, we have that $2\mathbf{b}_1^\top \boldsymbol{\Sigma}_{XX} \mathbf{b}_2 + \mu = \mathbf{0}$. Premultiplying the equation $(\boldsymbol{\Sigma}_{XX} - \lambda_1 \mathbf{I}_r)\mathbf{b}_1 = \mathbf{0}$ by $\mathbf{b}_2^\top$ yields $\mathbf{b}_2^\top \boldsymbol{\Sigma}_{XX} \mathbf{b}_1 = 0$, whence $\mu = 0$. Thus, λ_2 has to satisfy $(\boldsymbol{\Sigma}_{XX} - \lambda_2 \mathbf{I}_r)\mathbf{b}_2 = \mathbf{0}$. This means that λ_2 is the second largest eigenvalue of $\boldsymbol{\Sigma}_{XX}$, and the coefficient vector $\mathbf{b}_2$ for the second principal component is the eigenvector, $\mathbf{v}_2$, associated with λ_2.

In this sequential manner, we obtain the remaining sets of coefficients for the principal components $\xi_3, \xi_4, \ldots, \xi_r$, where the ith principal component ξ_i is obtained by choosing the set of coefficients, $\mathbf{b}_i$, for the linear projection ξ_i so that ξ_i has the largest variance among all linear projections of $\mathbf{X}$ that are also uncorrelated with $\xi_1, \xi_2, \ldots, \xi_{i-1}$. The coefficients of these linear projections are given by the ordered sequence of eigenvectors $\{\mathbf{v}_j\}$, where $\mathbf{v}_j$ associated with the jth largest eigenvalue, λ_j, of $\boldsymbol{\Sigma}_{XX}$.

7.2.5 Sample Principal Components

In practice, we estimate the principal components using n independent observations, $\{\mathbf{X}_i, \ i = 1, 2, \ldots, n\}$, on $\mathbf{X}$. We estimate $\boldsymbol{\mu}_X$ by

$$\widehat{\boldsymbol{\mu}}_X = \bar{\mathbf{X}} = n^{-1} \sum_{i=1}^{n} \mathbf{X}_i. \tag{7.28}$$

As before, let $\mathbf{X}_{ci} = \mathbf{X}_i - \bar{\mathbf{X}}$, $i = 1, 2, \ldots, n$, and set $\mathcal{X}_c = (\mathbf{X}_{c1}, \cdots, \mathbf{X}_{cn})$ to be an $(r \times n)$-matrix. We estimate $\boldsymbol{\Sigma}_{XX}$ by the sample covariance matrix,

$$\widehat{\boldsymbol{\Sigma}}_{XX} = n^{-1}\mathbf{S} = n^{-1}\mathcal{X}_c \mathcal{X}_c^\top. \tag{7.29}$$

The ordered eigenvalues of $\widehat{\boldsymbol{\Sigma}}_{XX}$ are denoted by $\widehat{\lambda}_1 \geq \widehat{\lambda}_2 \geq \ldots \geq \widehat{\lambda}_r \geq 0$, and the eigenvector associated with the jth largest sample eigenvalue $\widehat{\lambda}_j$ is the jth sample eigenvector $\widehat{\mathbf{v}}_j$, $j = 1, 2, \ldots, r$.

We estimate $\mathbf{A}^{(t)}$ and $\mathbf{B}^{(t)}$ by

$$\widehat{\mathbf{A}}^{(t)} = (\widehat{\mathbf{v}}_1, \cdots, \widehat{\mathbf{v}}_t) = \widehat{\mathbf{B}}^{(t)\tau}, \tag{7.30}$$

where $\widehat{\mathbf{v}}_j$ is the jth sample eigenvector of $\widehat{\boldsymbol{\Sigma}}_{XX}$, $j = 1, 2, \ldots, t$ ($t \leq r$). The best rank-t reconstruction of $\mathbf{X}$ is given by

$$\widehat{\mathbf{X}}^{(t)} = \bar{\mathbf{X}} + \widehat{\mathbf{C}}^{(t)}(\mathbf{X} - \bar{\mathbf{X}}), \tag{7.31}$$

where

$$\widehat{\mathbf{C}}^{(t)} = \widehat{\mathbf{A}}^{(t)}\widehat{\mathbf{B}}^{(t)} = \sum_{j=1}^{t} \widehat{\mathbf{v}}_j \widehat{\mathbf{v}}_j^{\tau} \tag{7.32}$$

is the reduced-rank regression coefficient matrix corresponding to the principal components case.

The jth *sample PC score* of $\mathbf{X}$ is given by

$$\widehat{\xi}_j = \widehat{\mathbf{v}}_j^{\tau}\mathbf{X}_c, \tag{7.33}$$

where $\mathbf{X}_c = \mathbf{X} - \bar{\mathbf{X}}$. The variance, λ_j, of the jth principal component is estimated by the sample variance $\widehat{\lambda}_j$, $j = 1, 2, \ldots, t$. A sample estimate of the measure (7.16) of how well the first t principal components represent the r original variables is given by the statistic

$$\frac{\widehat{\lambda}_{t+1} + \cdots + \widehat{\lambda}_r}{\widehat{\lambda}_1 + \cdots + \widehat{\lambda}_r}, \tag{7.34}$$

which is the proportion of the total sample variation that is explained by the last $r - t$ sample principal components.

It is hoped that the sample variances of the first few sample PCs will be large, whereas the rest will be small enough for the corresponding set of sample PCs to be omitted. A variable that does not change much (relative to other variables) in independent measurements may be treated approximately as a constant, and so omitting such low-variance sample PCs and putting all attention on high-variance sample PCs is, therefore, a convenient way of reducing the dimensionality of the data set.

The exact distribution of the eigenvalues of the random matrix $\mathcal{X}\mathcal{X}^{\tau} \sim \mathcal{W}_r(n, \mathbf{I}_r)$ was discovered independently and simultaneously in 1939 by Fisher, Girshick, Hsu, and Roy and in 1951 by Mood and has the form,

$$p(x_1, \ldots, x_r) = c_{r,n} \prod_{j=1}^{r} [w(x_j)]^{1/2} \prod_{j<k} (x_j - x_k), \tag{7.35}$$

where $x_1 \geq x_2 \geq \cdots \geq x_r$ are the ordered eigenvalues of $\mathcal{XX}^\tau$, $w(x) = x^{n-r-1}e^{-x}$ is the weight function for the Laguerre family of orthogonal polynomials, and $c_{r,n}$ is a normalizing constant dependent upon r and n. For a proof, see, for example, Anderson (1984, Section 13.3). The second product in (7.35) involving the pairwise differences of eigenvalues is the Jacobian term, also known as the *Vandermonde determinant* (Johnstone, 2006). In the case when the population eigenvalues are not all equal, the exact distribution of the sample eigenvalues is known (James, 1960) but is extremely complicated.

When the dimensionality, r, is very large, maybe even larger than the sample size n, then the exact distribution result (7.35) does not hold. In such situations, *random-matrix theory* has proved to be very useful in providing asymptotic results; see, e.g., Johnstone (2001, 2006). As before, suppose $\mathcal{XX}^\tau \sim \mathcal{W}_r(n, \mathbf{I}_r)$. The *empirical distribution function* computes the proportion of sample eigenvalues that are less than a given value of k,

$$G_r(k) = \frac{1}{r}\#\{x_j \leq k\}. \tag{7.36}$$

It can be shown that if $r/n \to \gamma \in (0, \infty)$, then, $G_r(k) \to G(k)$ a.s., where the limiting distribution $G(k)$ has density $g(k) = G'(k)$, and

$$g(k) = \frac{\sqrt{(b_+ - k)(k - b_-)}}{2\pi\gamma k}, \qquad b_\pm = (1 \pm \sqrt{\gamma})^2. \tag{7.37}$$

This so-called *Quarter-Circle Law* is due to Marčenko and Pastur (1967); it also holds in more general situations.

In Figure 7.2, we display the density $g(k)$ for $\gamma = 1/4$ and $\gamma = 1$. The larger is r/n, the more spread out is the limiting density. When $r = n/4$, the density is concentrated on the interval $[\frac{1}{4}, \frac{9}{4}]$, and when $r = n$, the density is spread out over the interval $[0, 4]$.

7.2.6 How Many Principal Components to Retain?

Probably the main question asked while carrying out a PCA is how many principal components to retain. Because the criterion for a good projection in PCA is a high variance for that projection, we should only retain those principal components with large variances. The question, therefore, boils down to one involving the magnitudes of the eigenvalues of $\widehat{\boldsymbol{\Sigma}}_{XX}$: how small can an eigenvalue be while still regarding the corresponding principal component as significant?

Scree Plot: The sample eigenvalues from a PCA are ordered from largest to smallest. It is usual to plot the ordered sample eigenvalues against their order number; such a display is called a "scree plot" (Cattell, 1966), after the break between a mountainside and a collection of boulders usually found

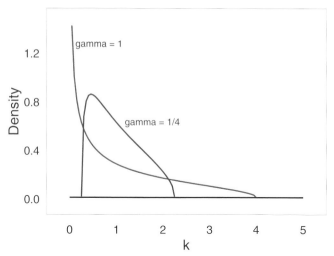

FIGURE 7.2. *Density $g(k)$ of eigenvalues of a Wishart matrix in the limiting case when $r/n \to \gamma \in (0, \infty)$. The two curves correspond to the values $\gamma = 1/4$ and $\gamma = 1$. The larger r/n, the more spread out are the eigenvalues.*

at its base. If the largest few sample eigenvalues dominate in magnitude, with the remaining sample eigenvalues very small, then the scree plot will exhibit an "elbow" in the plot corresponding to the division into "large" and "small" values of the sample eigenvalues. The order number at which the elbow occurs can be used to determine how many principal components to retain. It is usually recommended to retain those PCs up to the elbow and also the first PC following the elbow. A related popular criterion for use when an elbow may not be present in the scree plot is to use a cutoff point of 90% of total variance.

What would a scree plot look like for the eigenvalues of the covariance matrix of Gaussian data? We display two scenarios, where the only difference is the sample size. Generate an $(r \times n)$-matrix $\mathcal{Z}$ all of whose entries are iid $\mathcal{N}(0, 1)$, let $\mathbf{D}$ be an $(r \times r)$ diagonal matrix, and set $\mathcal{X} = \mathbf{D}\mathcal{Z}$. Let $\widehat{\mathbf{\Sigma}}_{XX} = n^{-1}\mathcal{X}\mathcal{X}^{\tau}$ be an $(r \times r)$ covariance matrix. Let $r = 30$ and set $\mathbf{D}^2 = \text{Diag}(12, 11, 10, 9, 8, 7, 3, 3, 3, \cdots, 3)$. Then, $\mathcal{X}\mathcal{X}^{\tau} \sim \mathcal{W}_r(n, \mathbf{D}^2)$. The scree plot of the eigenvalues of $\widehat{\mathbf{\Sigma}}_{XX}$ in the case that $n = 300$ is given in the left panel of Figure 7.3, where there is an elbow at 7. Now, suppose $n = 30$. Then, the scree plot of the eigenvalues is given in the right panel of Figure 7.3 and shows no discernible elbow. This example suggests that the relationship between n and r can determine whether or not the scree plot is useful in determining how many PCs to retain.

In the food nutrition example, the eigenvalues of the covariance matrix of the transformed data are given in Table 7.1. The scree plot of these

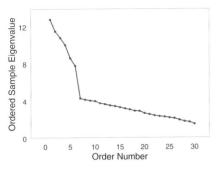
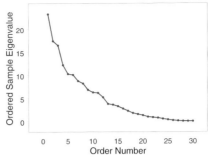

FIGURE 7.3. *Scree plots of the ordered eigenvalues of* $\widehat{\boldsymbol{\Sigma}}_{XX} = n^{-1}\mathcal{X}\mathcal{X}^{\tau}$, *where* $\mathcal{X} = \mathbf{D}\mathcal{Z}$, $\mathbf{D}$ *is a diagonal* $(r \times r)$-*matrix, and the elements of the* $(r \times n)$-*matrix* $\mathcal{Z}$ *are each independent Gaussian deviates. In this simulation,* $r = 30$ *and* $\mathbf{D}^2 = \mathrm{Diag}(12, 11, 10, 9, 8, 7, 3, 3, 3, \cdots, 3)$. *The left panel corresponds to* $n = 300$ *and has an elbow at* 7, *and the right panel corresponds to* $n = 30$ *and shows no elbow.*

eigenvalues, which is given in the left panel of Figure 7.4, shows no elbow. This may be explained by the fact that the leading PC explains only a 44% share of the total variance, there is no really dominant group of eigenvalues, and it takes four PCs to pass 90% of total variance.

PC Rank Trace: The problem of deciding how many principal components to retain is equivalent to obtaining a useful estimate of the rank of the regression coefficient matrix $\mathbf{C}$ in the principal components case. So, if we can obtain a good estimate of the rank, we should have a solution to this problem.

We saw in Chapter 6 that the rank trace plots the loss of information when approximating the full-rank regression by a sequence of reduced-rank regressions having increasing ranks. When the true rank of the regression, t_0, say, is reached, the points in the rank trace plot following that rank (i.e., ranks $t_0 + 1, \ldots, r$) should cease to change significantly from both the point for t_0 and the full-rank point (rank r).

In the principal components case, the expressions for the points in the rank trace simplify greatly and are very simple to compute. It is not difficult to show (see Exercise 7.6) that

$$\Delta\widehat{\mathbf{C}}^{(t)} = \left(1 - \frac{t}{r}\right)^{1/2}, \tag{7.38}$$

$$\Delta\widehat{\boldsymbol{\Sigma}}_{\mathcal{E}\mathcal{E}}^{(t)} = \left(\frac{\widehat{\lambda}_{t+1}^2 + \cdots + \widehat{\lambda}_r^2}{\widehat{\lambda}_1^2 + \cdots + \widehat{\lambda}_r^2}\right)^{1/2}, \tag{7.39}$$

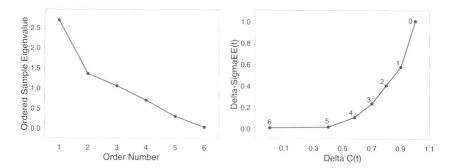

FIGURE 7.4. *Food nutrition example. Left panel: Scree plot. Right panel: PC rank-trace plot with values of t placed next to the plotted point. The scree plot for the sample covariance matrix of the transformed data does not offer any advice on the number of principal components to retain, whereas the rank trace plot suggests retaining 4 or 5 principal components. The modified version of Kaiser's rule recommends retaining three PCs.*

$t = 0, 1, 2, \ldots, r$. Comparing (7.39) with (7.34), we see that we are again looking at the smallest $r - t$ sample eigenvalues (although this time they are each squared). A plot of (7.39) against (7.38) is called a *PC rank trace plot* (Izenman, 1980). All the information regarding the dimensionality of the regression is, therefore, contained in the residual covariance matrices and not in the regression coefficients. Furthermore, the $r + 1$ plotted points decrease monotonically from $(1, 1)$ to $(0, 0)$. We assess the rank t of $\mathbf{C}$ by $\widehat{t}$, the smallest integer value between 1 and r at which an "elbow" can be detected in the PC rank trace plot.

In Figure 7.4, the right panel shows the PC rank trace plot for the sample covariance matrix of the food nutrition data. We assess the rank from the rank-trace plot as $\widehat{t} = 4$ or 5.

Kaiser's Rule: When dealing with the PCA of a sample correlation matrix, Kaiser (1960) suggested (in the context of exploratory factor analysis) that only those principal components be retained whose eigenvalues exceed unity. This decision guideline is based upon the argument that because the total variation of all r standardized variables is equal to r, it follows that a principal component should account for at least the average variation of a single standardized variable. This rule is popular but controversial; there is evidence that the cutoff value of 1 is too high. A modified rule retains all PCs whose eigenvalues of the sample correlation matrix exceed 0.7.

For the food nutrition data, the eigenvalues of the sample correlation matrix are 2.6486, 1.3301, 1.0201, 0.6801, 0.2665, and 0.00546. Three of these eigenvalues are greater than 0.7, and so the modified version of Kaiser's rule says that we should retain the first three principal components.

7.2.7 Graphical Displays

For diagnostic and data analytic purposes, it is usual to plot the first sample PC scores against the second sample PC scores,

$$(\widehat{\xi}_{i1}, \widehat{\xi}_{i2}), \quad i = 1, 2, \ldots, n, \tag{7.40}$$

where $\widehat{\xi}_{ij} = \widehat{\mathbf{v}}_j^\tau \mathbf{X}_i$, $i = 1, 2, \ldots, n$, $j = 1, 2$. A more general graphical tool for displaying the sample PC scores associated with the largest few sample eigenvalues (variances) is the scatterplot matrix, in which all possible pairs of variables are plotted in two dimensions.

See Figure 7.1 for a graphical display of the first two PCs of the food nutrition data and Figure 7.6 for a graphical display of the first three PCs of the pendigits data.

A three-dimensional scatterplot of the first three sample PC scores is also strongly recommended, especially if a "brush and spin" feature is available.

7.2.8 Example: Face Recognition Using Eigenfaces

In this example, we apply PCA to a single face photographed under $n = 11$ illumination and expression conditions; see Figure 2.4. Recall from Section 2.3.3 that each face, as a picture image, starts out as a (320×243)-matrix of intensity values, which are quantized to 8-bit grayscale (0–255, with 0 as black and 255 as white), and then translated into a stacked vector of length $r = 77,760$.

From a PCA of the n r-vectors, $\mathbf{X}_1, \ldots, \mathbf{X}_n$, of stacked images, we compute the first t PC scores, $\widehat{\boldsymbol{\xi}}_1^{(t)}, \ldots, \widehat{\boldsymbol{\xi}}_n^{(t)}$, where $\widehat{\boldsymbol{\xi}}_i^{(t)} = \widehat{\mathbf{B}}^{(t)} \mathbf{X}_i = (\widehat{\xi}_{i1}, \cdots, \widehat{\xi}_{it})^\tau$ is a t-vector, $1 \leq t \leq r$. It is usual to plot the points $(\widehat{\xi}_{i1}, \widehat{\xi}_{i2})$, $i = 1, 2, \ldots, n$, and annotate the scatterplot with face identifiers. Faces corresponding to the same individual should project to points very close to each other in the scatterplot, whereas faces corresponding to different individuals should project to more distant points. Also, faces of the same individual with very similar poses should be plotted close to each other, whereas different poses should be plotted far away from each other.

The best rank-t reconstruction of the ith original face is obtained by computing $\widehat{\mathbf{X}}_i^{(t)} = \bar{\mathbf{X}} + \widehat{\mathbf{C}}^{(t)}(\mathbf{X}_i - \bar{\mathbf{X}})$, $i = 1, 2, \ldots, n$, where $\bar{\mathbf{X}}$ is the "average" face given by (7.28) and $\widehat{\mathbf{C}}^{(t)}$ is given by (7.33). The average of all the faces can be seen in the left panel of Figure 7.5. If the r-vectors $\widehat{\mathbf{X}}_1^{(t)}, \ldots, \widehat{\mathbf{X}}_n^{(t)}$ are unstacked and displayed as images, they each have the appearance of a "ghostly" face. The reconstructed face image improves as we increase

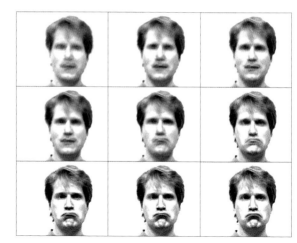

FIGURE 7.5. *The cumulative effect of the nine principal components, adding one PC at a time, for eigenface 6 ("sad"). The sad face starts to appear by the fifth PC. The average eigenface is given in the left panel.*

t. Each face image in the data set can be represented exactly as a linear combination of all r such ghostly faces or eigenfaces, or approximately as a linear combination of the first t eigenfaces, which are ordered by decreasing eigenvalues.

In the right panel of Figure 7.5, we see the effect of increasing the number of principal components on the reconstruction of face 6 ("sad"). The first eigenface is fuzzy but recognizable as a face. Adding PCs increases the sharpness of the image, and the "sad" face starts to emerge at eigenface 5.

7.2.9 Invariance and Scaling

A shortcoming of PCA is that the principal components are not invariant under rescalings of the initial variables. In other words, a PCA is sensitive to the units of measurement of the different input variables. Standardizing (centering and then scaling) the **X**-variables,

$$\mathbf{Z} \leftarrow (\operatorname{diag}\{\widehat{\boldsymbol{\Sigma}}_{XX}\})^{-1/2}(\mathbf{X} - \widehat{\boldsymbol{\mu}}_X), \qquad (7.41)$$

is equivalent to carrying out PCA using the correlation (rather than the covariance) matrix. When using the correlation matrix, the total variation of the standardized variables is r, the trace of the correlation matrix. The lack of scale invariance implies that a PCA using the correlation matrix may be very different from a similar analysis using the corresponding covariance matrix, and no simple relationship exists between the two sets of results. In the initial formulation and application of PCA, we note that Hotelling

(1933), who was dealing with a battery of test scores, extracted principal components from the correlation matrix of the data.

Standardization in the PCA context has its advantages. In some fields, standardization is customary. In heterogeneous situations, where the units of measurement of the input variables are not commensurate or the ranges of values of the variables differ considerably, standardization is especially relevant. If the variables have heterogeneous variances, it is a good idea to standardize the variables before carrying out PCA because the variables with the greatest variances will tend to overwhelm the leading principal components with the remaining variables contributing very little.

On statistical inference grounds, standardization is usually regarded as a nuisance because it complicates the distributional theory. Indeed, the asymptotic distribution theory for the eigenvalues and eigenvectors of a sample correlation matrix turns out to be extremely difficult to derive. Furthermore, certain simplifications, such as pretending that the sample correlation matrix has the same distributional properties as the sample covariance matrix, tend not to work and, hence, lead to incorrect inference results for principal components.

7.2.10   Example: Pen-Based Handwritten Digit Recognition

These data[2] were obtained from 44 writers, each of whom handwrote 250 examples of the digits $0, 1, 2, \ldots, 9$ in a random order (Alimoglu, 1995). The digits were written inside boxes of 500×500 pixel resolution on a pressure-sensitive tablet with an integrated LCD screen. The subjects were monitored only during the first entry screens. Each screen contained five boxes with the digits to be written displayed above. Subjects were told to write only inside these boxes. If they made a mistake or were unhappy with their writing, they were instructed to clear the contents of a box by using an on-screen button. Unknown to the writers, the first 10 digits were ignored as writers became familiar with the input device.

The raw data on each of $n = 10,992$ handwritten digits consisted of a sequence, (x_t, y_t), $t = 1, 2, \ldots, T$, of tablet coordinates of the pen at fixed time intervals of 100 milliseconds, where x_t and y_t were integers in the range 0–500. These data were then normalized to make the representations invariant to translation and scale distortions. The new coordinates were such that the coordinate that had the maximum range varied between 0 and 100. Usually x_t stays in this range, because most integers are taller than they are wide. Finally, from the normalized trajectory of each handwritten

---

[2]These data are available in the file **pendigits** on the book's website. The description was obtained from **www.ics.uci.edu/~learn/databases/pendigits**.

digit, 8 regularly spaced measurements, (x_t, y_t), were chosen by spatial resampling, which gave a total of $r = 16$ input variables.

A PCA of the correlation matrix (i.e., the covariance matrix of normalized variables) reveals that the variances of the first five principal component (PC) scores are larger than unity: 4.717, 3.229, 2.577, 1.230, 1.063; thus, the first five PCs together explain about 80% of the total variation, 16, in the data. A reduction in dimensionality from 16 to 5, therefore, retains a substantial amount of the total variation. Scatterplots of the first three PC scores, which explain about 66% of the total variation, are displayed in Figure 7.6, where the points are colored by type of digit.

From these three 2D scatterplots, we can make the following observations: the majority of handwritten examples of each digit cluster together, although there is a great deal of overlapping of clusters; each scatterplot has a distinctive shape, with strong suggestions of circular or torus-like appearance; and there appears to be lots of outlying points. A 3D-rotating scatterplot of the first three principal components reveals a hollow, hemispherical point configuration with crab-like arms.

7.2.11   Functional PCA

In some situations, we may need to analyze data consisting of functions or curves. Although such *functional data* are often time-dependent, we do not assume that time itself plays a special role. In fact, functional data from different and independent individuals may be recorded at different sets of time points, and in each of those instances, the data may not be equally spaced. In such cases, it is advantageous to view an individual's functional observations as a continuously defined record observed at a set of discrete points, so that a single data point is the entire function (rather than each observed data value). In other cases, we may be able to view independent replications of the entire curve.

Given a set of sample curves from a number of individuals, where each curve represents repeated measurements on the same individual, we may wish to characterize the main features of those curves. One method of doing this is through a functional version of PCA (see, e.g., Ramsay and Silverman, 1997, Chapters 6 and 7). Because we are observing curves rather than individual values, the vector-valued observations $\mathbf{X}_1, \ldots, \mathbf{X}_n$ are replaced by the univariate functions $X_1(t), \ldots, X_n(t)$, where t may indicate time, but in general is to be thought of as a continuous index varying within a closed interval $[0, T]$.

In functional PCA, each sample curve is considered to be an independent realization of a univariate stochastic process $X(t)$ (having possibly cyclical or periodic form) with smooth mean function $E\{X(t)\} = \mu(t)$ and

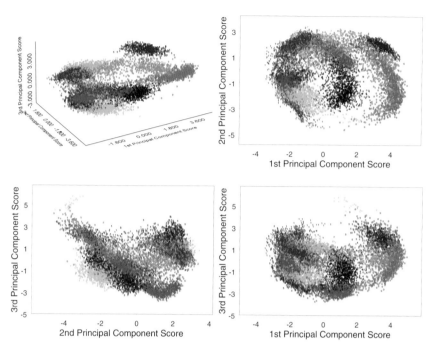

FIGURE 7.6. *Scatterplots of the first three principal components (PCs) of the correlation matrix from the* `pendigits` *data, where $r = 16$ and $n = 10,992$. The top-left panel displays the scatterplot of the first three principal component scores. The top-right panel shows the first and second PCs, the bottom-right panel shows the first and third PCs, and the bottom-left panel shows the second and third PCs. The 10 digits are shown by the following colors: green (0), brown (1), light blue (2), light magenta (3), purple (4), blue (5), light red (6), light green (7), orange (8), and light cyan (9).*

covariance function

$$\mathrm{cov}\{X(s), X(t)\} = \sigma(s, t). \qquad (7.42)$$

By a spectral decomposition of the covariance function, we can express σ as an orthogonal expansion (in the L_2 sense) in terms of its eigenvalues $\{\lambda_j\}$ and associated *eigenfunctions* $\{V_j(t)\}$, so that

$$\sigma(s, t) = \sum_{j=1}^{\infty} \lambda_j V_j(s) V_j(t), \qquad (7.43)$$

where the eigenvalues quickly tend to zero and the first few eigenfunctions are slowly varying. The covariance function σ is positive-definite and, hence, we can take the eigenvalues to be nonnegative and ordered: $\lambda_1 \geq \lambda_2 \geq \cdots \geq 0$. The goal is to determine the primary components of functional variation in $\sigma(s,t)$, where the eigenvalues indicate the amount of total variance attributed to each component.

A random curve can then be expressed as

$$X(t) = \mu(t) + \sum_{j=1}^{\infty} \xi_j V_j(t), \tag{7.44}$$

where the coefficient

$$\xi_j = \int [X(t) - \mu(t)]V_j(t)dt \tag{7.45}$$

is a scalar random variable (called the *jth functional PC score*) with $\mathrm{E}\{\xi_j\} = 0$, $\mathrm{var}\{\xi_j\} = \lambda_j$, $\sum_j \lambda_j < \infty$, and $\mathrm{cov}\{\xi_j, \xi_k\} = 0$, $j \neq k$. The eigenfunctions $\{V_j(t)\}$ (called *PC functions*) satisfy

$$\int [V_j(t)]^2 dt = 1, \quad \int V_j(t)V_k(t)dt = 0, j \neq k, \tag{7.46}$$

where the integrals are taken over $[0, T]$, which may be periodic. The expansion (7.42) is the well-known *Karhunen–Loève expansion* of $X(t)$. Thus, $X(t) - \mu(t)$ may be thought of as a finite sum of orthogonal curves each having uncorrelated random amplitudes.

Although a scientific phenomenon may be viewed as functional, in reality we typically only have a finite amount of knowledge about that phenomenon through sampling. Consequently, estimates of the mean curve $\mu(t)$ and the covariance function σ are based upon a collection of n sample curves, $X_1(t), \ldots, X_n(t)$, where $X_i(t) = \mu(t) + \sum_j \xi_{ij} V_j(t)$ is the *i*th individual curve. The *k*th point on the *i*th curve is denoted by $X_{ik} = X_i(t_k)$.

We briefly mention possible estimation procedures and refer the interested reader to the excellent books by Ramsay and Silverman on this topic. One approach to analyzing such data is, first, to smooth each individual sample curve (e.g., using spline methods or local-linear smoothers), and then apply functional PCA assuming that the smooth curves are the completely observed curves. This gives a set of eigenvalues $\{\widehat{\lambda}_j\}$ and (smooth) eigenfunctions $\{\widehat{V}_j(t)\}$ extracted from the sample covariance matrix of the smoothed data. The first and second estimated eigenfunctions are then graphed with a view to examining the extent and location of individual curve variation.

Other approaches to functional PCA have been developed, including the use of roughness penalties and regularization, which optimize the selection of smoothing parameter and choice of the number of PCs simultaneously rather than separately in two stages.

7.2.12   What Can Be Gained from Using PCA?

The short answer is that it depends on what we are trying to accomplish and the nature of the application in question. PCA is a linear technique built for several purposes: it enables us, first, to decorrelate the original variables in the study, regardless of whether $r < n$ or $n < r$; second, to carry out *data compression*, where we pay decreasing attention to the numerical accuracy by which we encode the sequence of principal components; third, to reconstruct the original input data using a reduced number of variables according to a least-squares criterion; and fourth, to identify potential clusters in the data.

In certain applications, PCA can be misleading. PCA is heavily influenced when there are outliers in the data (e.g., in computer vision, images can be corrupted by noisy pixels), and such considerations have led to the construction of robust PCA. In other situations, the linearity of PCA may be an obstacle to successful data reduction and compression, and so in Chapter 16, we consider nonlinear versions of PCA.

7.3   Canonical Variate and Correlation Analysis

Canonical variate and correlation analysis (CVA or CCA) (Hotelling, 1936) is a method for studying linear relationships between two vector variates, which we denote by $\mathbf{X} = (X_1, \cdots, X_r)^\tau$ and $\mathbf{Y} = (Y_1, \cdots, Y_s)^\tau$. As such, it has been used to solve theoretical and applied problems in econometrics, business (primarily, finance and marketing), psychometrics, geography, education, ecology, and atmospheric sciences (e.g., weather prediction).

Hotelling applied CVA to the relationship between a set of two reading test scores (X_1 = reading speed, X_2 = reading power) and a set of two arithmetic test scores (Y_1 = arithmetic speed, Y_2 = arithmetic power) obtained from 140 fourth-grade children, so that $r = s = 2$.

7.3.1   Canonical Variates and Canonical Correlations

We assume that

$$\left(\begin{array}{c} \mathbf{X} \\ \mathbf{Y} \end{array} \right) \tag{7.47}$$

is a collection of $r + s$ variables partitioned into two disjoint subcollections, where $\mathbf{X}$ and $\mathbf{Y}$ are jointly distributed with mean vector and covariance matrix given by

$$\mathrm{E}\left\{ \left(\begin{array}{c} \mathbf{X} \\ \mathbf{Y} \end{array} \right) \right\} = \left(\begin{array}{c} \boldsymbol{\mu}_X \\ \boldsymbol{\mu}_Y \end{array} \right), \tag{7.48}$$

$$E\left\{\left(\begin{array}{c} \mathbf{X} - \boldsymbol{\mu}_X \\ \mathbf{Y} - \boldsymbol{\mu}_Y \end{array}\right)\left(\begin{array}{c} \mathbf{X} - \boldsymbol{\mu}_X \\ \mathbf{Y} - \boldsymbol{\mu}_Y \end{array}\right)^{\tau}\right\} = \left(\begin{array}{cc} \boldsymbol{\Sigma}_{XX} & \boldsymbol{\Sigma}_{XY} \\ \boldsymbol{\Sigma}_{YX} & \boldsymbol{\Sigma}_{YY} \end{array}\right), \qquad (7.49)$$

respectively, where $\boldsymbol{\Sigma}_{XX}$ and $\boldsymbol{\Sigma}_{YY}$ are both assumed to be nonsingular.

CVA seeks to replace the two sets of correlated variables, $\mathbf{X}$ and $\mathbf{Y}$, by t pairs of new variables,

$$(\xi_i, \omega_i), \quad i = 1, 2, \ldots, t, \quad t \leq \min(r, s), \qquad (7.50)$$

where

$$\left.\begin{array}{l} \xi_j = \mathbf{g}_j^{\tau}\mathbf{X} = g_{1j}X_1 + g_{2j}X_2 + \cdots + g_{rj}X_r \\[2mm] \omega_j = \mathbf{h}_j^{\tau}\mathbf{Y} = h_{1j}Y_1 + h_{2j}Y_2 + \cdots + h_{sj}Y_s \end{array}\right\} \qquad (7.51)$$

$j = 1, 2, \ldots, t$, are linear projections of $\mathbf{X}$ and $\mathbf{Y}$, respectively. The jth pair of coefficient vectors, $\mathbf{g}_j = (g_{1j}, \cdots, g_{rj})^{\tau}$ and $\mathbf{h}_j = (h_{1j}, \cdots, h_{rj})^{\tau}$, are chosen so that

- the pairs $\{(\xi_j, \omega_j)\}$ are ranked in importance through their correlations,

$$\rho_j = \mathrm{corr}\{\xi_j, \omega_j\} = \frac{\mathbf{g}_j^{\tau}\boldsymbol{\Sigma}_{XY}\mathbf{h}_j}{(\mathbf{g}_j^{\tau}\boldsymbol{\Sigma}_{XX}\mathbf{g}_j)^{1/2}(\mathbf{h}_j^{\tau}\boldsymbol{\Sigma}_{YY}\mathbf{h}_j)^{1/2}}, \quad j = 1, 2, \ldots, t, \qquad (7.52)$$

which are listed in descending order of magnitude: $\rho_1 \geq \rho_2 \geq \cdots \geq \rho_t$.

- ξ_j is uncorrelated with all previously derived ξ_k:

$$\mathrm{cov}\{\xi_j, \xi_k\} = \mathbf{g}_j^{\tau}\boldsymbol{\Sigma}_{XX}\mathbf{g}_k = 0, \quad k < j. \qquad (7.53)$$

- ω_j is uncorrelated with all previously derived ω_k:

$$\mathrm{cov}\{\omega_j, \omega_k\} = \mathbf{h}_j^{\tau}\boldsymbol{\Sigma}_{YY}\mathbf{h}_k = 0, \quad k < j. \qquad (7.54)$$

The pairs (7.44) are known as *the first t pairs of canonical variates of* $\mathbf{X}$ *and* $\mathbf{Y}$ and their correlations (7.45) as *the t largest canonical correlations*.

The CVA technique ensures that every bit of correlation is wrung out of the original $\mathbf{X}$ and $\mathbf{Y}$ variables and deposited in an orderly fashion into pairs of new variables, (ξ_j, ω_j), $j = 1, 2, \ldots, t$, which have a special correlation structure. If the notion of correlation is regarded as the primary determinant of information in the system of variables, then CVA is a major tool for reducing the dimensionality of the original two sets of variables.

7.3.2   Example: COMBO-17 Galaxy Photometric Catalogue

The data for this example consist of a subset of a public catalogue of a large number of astronomical objects (e.g., stars, galaxies, quasars) with

TABLE 7.2. *Variables used to analyze 3,438 galaxies from the Chandra Deep Field South area of the sky. The variables are divided into $r = 23$ X-variables and $s = 6$ Y-variables.*

X-variables
UjMag, BjMag, VjMag, usMag, gsMag, rsMag,
UbMag, BbMag, VbMag, S280Mag,
W420F_E, W462F_E, W485F_D, W518F_E, W571F_S,
W604F_E, W646F_D, W696F_E, W753F_E, W815F_S,
W856F_D, W914F_D, W914F_E
Y-variables
Rmag, ApD_Rmag, mu_max, MC_z, MC_z_ml, chi2red

brightness measurements in 17 passbands covering the range 350–930 nm (Wolf, Meisenheimer, Kleinheinrich, Borch, Dye, Gray, Wisotski, Bell, Rix, Cimatti, Hasinger, and Szokoly, 2004).[3] All objects in the catalogue are found in the Chandra Deep Field South, one of the most popularly studied areas of the sky. Figure 7.7 shows a high-resolution composite image of the Chandran Deep Field South, based upon images obtained in 2003 with the Wide Field Imager on the ground-based 2.2-m MPG/ESO telescope located at the European Southern Observatory (ESO) on La Silla, Chile. The image displays more than 100,000 galaxies, several thousand stars, and hundreds of quasars. COMBO-17 ("Classifying Objects by Medium-Band Observations in 17 filters") is an international collaboration project whose mission is to study the evolution of galaxies.

This particular subset of the data set consists of the $n = 3,438$ objects in the Chandra Deep Field South that are classified as "Galaxies" and for which there are no missing values for any of the 65 variables (24 observations were omitted because of missing data). We also omitted five redundant variables and all error variables in the data set; the 29 remaining variables were then divided into a group of $r = 23$ X-variables and a group of $s = 6$ Y-variables, which are listed in Table 7.2.

Of the Y-variables, Rmag is the total R-band magnitude (magnitudes are inverted logarithmic measures of brightness), ApD_Rmag is the aperture difference of Rmag, mu_max is the central surface brightness in Rmag, MC_z is the mean redshift in the distribution $p(z)$, MC_z_ml is the peak

---

[3]The complete catalogue of 63,501 astronomical objects can be obtained from the website `vizier.u-strasbg.fr/viz-bin/VizieR-4` or from the COMBO-17 website `www.mpia.de/COMBO/combo_index.html`. The data set used in this example is a subset and can be downloaded from `astrostatistics.psu.edu/datasets/COMBO17.html`. The author thanks Donald Richards for very helpful discussions on this data set.

FIGURE 7.7. *High-resolution three-color composite image of the Chandra Deep Field South, obtained in January 2003 with the Wide Field Imager camera on the 2.2m MPG/ESO telescope at the European Southern Observatory (ESO), La Silla, Chile. This image is based upon a total exposure time of nearly 50 hours and displays more than 100,000 galaxies, several thousand stars, and hundreds of quasars. Source:* www.eso.org/public/outreach/press-rel/pr-2003/phot-02-03.html.

of the redshift distribution $p(z)$, and chi2red is the reduced χ^2-value of the best-fitting template.

Of the X-variables, UjMag, BjMag, VjMag, usMag, gsMag, rsMag, UbMag, BbMag, VbMag, and S280Mag are all absolute magnitudes of the galaxy in 10 bands. The first nine of these magnitudes are very highly correlated with each other, with all pairwise correlations greater than 0.93. They are based upon the measured magnitudes and the redshifts and represent the intrinsic luminosities of the galaxies. The other variables are the observed brightnesses in 13 bands in sequence from 420 nm in the ultraviolet to 915 nm in the far red; these variables are also highly correlated with each other, with correlations decreasing as distance between bands increases.

The pairwise plots of all six pairs of canonical variates of the COMBO-17 data are displayed in Figure 7.8. The canonical correlations are, in decreasing order of magnitude, 0.942, 0.538, 0.077, 0.037, 0.030, and 0.020; two of

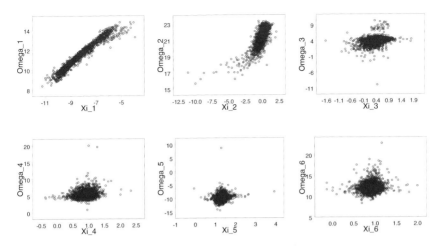

FIGURE 7.8. *Pairwise canonical variate plots of COMBO-17 galaxy data. There are n = 3,438 galaxies with r = 23 and s = 6 variables. Top-left panel: First pair of canonical variates (CVs), canonical correlation (CC) = 0.942. Top-center panel: Second pair of CVs, CC = 0.538. Top-right panel: Third pair of CVs, CC = 0.077. Bottom-left panel: Fourth pair of CVs, CC = 0.037. Bottom-center panel: Fifth pair of CVs, CC = 0.030. Bottom-right panel: Sixth pair of CVs, CC = 0.020. For the jth CV plot, ξ_j is plotted on the horizontal axis, and ω_j is plotted on the vertical axis.*

these correlations are large, whereas the rest are very small. We also see many outliers in these plots. For example, galaxy Nr = 3605 is prominent in all six plots, and galaxies Nr = 3033, 3277, and 6423 are prominent in at least three plots.

7.3.3   Least-Squares Optimality of CVA

Let the $(t \times r)$-matrix $\mathbf{G}$ and the $(t \times s)$-matrix $\mathbf{H}$, with $1 \leq t \leq \min(r, s)$, be such that $\mathbf{X}$ and $\mathbf{Y}$ are linearly projected into new vector variates,

$$\boldsymbol{\xi} = \mathbf{GX}, \quad \boldsymbol{\omega} = \mathbf{HY}, \tag{7.55}$$

respectively. Consider the problem of finding $\boldsymbol{\nu}$, $\mathbf{G}$, and $\mathbf{H}$ so that

$$\mathbf{HY} \approx \boldsymbol{\nu} + \mathbf{GX} \tag{7.56}$$

in some least-squares sense. More precisely, we wish to find $\boldsymbol{\nu}$, $\mathbf{G}$, and $\mathbf{H}$ to minimize the $(t \times t)$-matrix,

$$\mathrm{E}\{(\mathbf{HY} - \boldsymbol{\nu} - \mathbf{GX})(\mathbf{HY} - \boldsymbol{\nu} - \mathbf{GX})^{\tau}\}, \tag{7.57}$$

where we assume that the covariance matrix of $\boldsymbol{\omega}$ is $\boldsymbol{\Sigma}_{\omega\omega} = \mathbf{H}\boldsymbol{\Sigma}_{YY}\mathbf{H}^{\tau} = \mathbf{I}_t$.

Fix the matrix $\mathbf{H}$ and minimize the error criterion (7.57) first with respect to $\boldsymbol{\nu}$ and $\mathbf{G}$. We set $\boldsymbol{\omega}_c = \boldsymbol{\omega} - \boldsymbol{\mu}_\omega = \boldsymbol{\omega} - \mathbf{G}\boldsymbol{\mu}_X$, and write $\boldsymbol{\omega} - \boldsymbol{\nu} - \mathbf{G}\mathbf{X}$ as $\boldsymbol{\omega}_c + (\mathbf{H}\boldsymbol{\mu}_Y - \boldsymbol{\nu} - \mathbf{G}\boldsymbol{\mu}_X) - \mathbf{G}\mathbf{X}_c$, where $\mathbf{X}_c = \mathbf{X} - \boldsymbol{\mu}_X$. Then,

$$
\begin{aligned}
\min_{\boldsymbol{\nu},\mathbf{G}} &\mathrm{E}\{(\boldsymbol{\omega} - \boldsymbol{\nu} - \mathbf{G}\mathbf{X})(\boldsymbol{\omega} - \boldsymbol{\nu} - \mathbf{G}\mathbf{X})^\tau\} \\
&\geq \min_{\mathbf{G}} \mathrm{E}\{(\boldsymbol{\omega}_c - \mathbf{G}\mathbf{X}_c)(\boldsymbol{\omega}_c - \mathbf{G}\mathbf{X}_c)^\tau\} \\
&= \mathrm{tr}\{\boldsymbol{\Sigma}_{\omega\omega} - \boldsymbol{\Sigma}_{\omega X}\boldsymbol{\Sigma}_{XX}^{-1}\boldsymbol{\Sigma}_{X\omega}\} \\
&\quad + \min_{\mathbf{G}} \mathrm{tr}\{(\mathbf{G}\boldsymbol{\Sigma}_{XX}^{1/2} - \boldsymbol{\Sigma}_{\omega X}\boldsymbol{\Sigma}_{XX}^{-1/2})(\mathbf{G}\boldsymbol{\Sigma}_{XX}^{1/2} - \boldsymbol{\Sigma}_{\omega X}\boldsymbol{\Sigma}_{XX}^{-1/2})^\tau\} \\
&\geq \mathrm{tr}\{\boldsymbol{\Sigma}_{\omega\omega} - \boldsymbol{\Sigma}_{\omega X}\boldsymbol{\Sigma}_{XX}^{-1}\boldsymbol{\Sigma}_{X\omega}\} \\
&= \mathrm{tr}\{\mathbf{H}\boldsymbol{\Sigma}_{YY}\mathbf{H}^\tau - \mathbf{H}\boldsymbol{\Sigma}_{YX}\boldsymbol{\Sigma}_{XX}^{-1}\boldsymbol{\Sigma}_{XY}\mathbf{H}^\tau\} \\
&= t - \sum_{j=1}^{t} \lambda_j(\mathbf{H}\boldsymbol{\Sigma}_{YX}\boldsymbol{\Sigma}_{XX}^{-1}\boldsymbol{\Sigma}_{XY}\mathbf{H}^\tau),
\end{aligned} \tag{7.58}
$$

where the first inequality becomes an equality iff $\boldsymbol{\nu} = \mathbf{H}\boldsymbol{\mu}_Y - \mathbf{G}\boldsymbol{\mu}_X$, and the second inequality becomes an equality iff $\mathbf{G} = \boldsymbol{\Sigma}_{\omega X}\boldsymbol{\Sigma}_{XX}^{-1} = \mathbf{H}\boldsymbol{\Sigma}_{YX}\boldsymbol{\Sigma}_{XX}^{-1}$.

Now set $\mathbf{U}^\tau = \mathbf{H}\boldsymbol{\Sigma}_{YY}^{1/2}$, so that $\mathbf{U}^\tau\mathbf{U} = \mathbf{I}_t$. Then, by the Poincaré Separation Theorem (see Section 3.2.10), (7.58) becomes

$$
t - \sum_{j=1}^{t} \lambda_j(\mathbf{U}^\tau\mathbf{R}\mathbf{U}) \geq t - \sum_{j=1}^{t} \lambda_j(\mathbf{R}),
$$

where

$$
\mathbf{R} = \boldsymbol{\Sigma}_{YY}^{-1/2}\boldsymbol{\Sigma}_{YX}\boldsymbol{\Sigma}_{XX}^{-1}\boldsymbol{\Sigma}_{XY}\boldsymbol{\Sigma}_{YY}^{-1/2}, \tag{7.59}
$$

with equality only when the columns of $\mathbf{U}$ are the eigenvectors associated with the first t eigenvalues of $\mathbf{R}$.

To summarize: The $\boldsymbol{\nu}$, $\mathbf{G}$, and $\mathbf{H}$ that minimize (7.57) are given by

$$
\boldsymbol{\nu}^{(t)} = \mathbf{H}^{(t)}\boldsymbol{\mu}_Y - \mathbf{G}^{(t)}\boldsymbol{\mu}_X, \tag{7.60}
$$

$$
\mathbf{G}^{(t)} = \begin{pmatrix} \mathbf{v}_1^\tau \\ \vdots \\ \mathbf{v}_t^\tau \end{pmatrix} \boldsymbol{\Sigma}_{YY}^{-1/2}\boldsymbol{\Sigma}_{YX}\boldsymbol{\Sigma}_{XX}^{-1} = \begin{pmatrix} \lambda_1\mathbf{u}_1^\tau \\ \vdots \\ \lambda_t\mathbf{u}_t^\tau \end{pmatrix} \boldsymbol{\Sigma}_{XX}^{-1/2}, \tag{7.61}
$$

$$
\mathbf{H}^{(t)} = \begin{pmatrix} \mathbf{v}_1^\tau \\ \vdots \\ \mathbf{v}_t^\tau \end{pmatrix} \boldsymbol{\Sigma}_{YY}^{-1/2}, \tag{7.62}
$$

respectively, where $\mathbf{u}_j$ is the eigenvector associated with the jth largest eigenvalue λ_j^2 of

$$
\mathbf{R}^* = \boldsymbol{\Sigma}_{XX}^{-1/2}\boldsymbol{\Sigma}_{XY}\boldsymbol{\Sigma}_{YY}^{-1}\boldsymbol{\Sigma}_{YX}\boldsymbol{\Sigma}_{XX}^{-1/2}, \tag{7.63}
$$

and $\mathbf{v}_j$ is the eigenvector associated with the jth largest eigenvalue λ_j^2 of $\mathbf{R}$ in (7.59), $j = 1, 2, \ldots, t$.

Let $\mathbf{g}_j^\tau = (g_{1j}, \ldots, g_{rj})$ and $\mathbf{h}_j^\tau = (h_{1j}, \ldots, h_{sj})$ be the jth rows of $\mathbf{G}^{(t)}$ and $\mathbf{H}^{(t)}$, respectively, $j = 1, 2, \ldots, t$. The r-vector $\mathbf{g}_j$ and the s-vector $\mathbf{h}_j$ are generally assumed to have unit length; that is, $\mathbf{g}_j^\tau \mathbf{g}_j = \mathbf{h}_j^\tau \mathbf{h}_j = 1$, $j = 1, 2, \ldots, t$. The jth *pair of canonical variates scores*, (ξ_j, ω_j), is given by

$$\xi_j = \mathbf{g}_j^\tau \mathbf{X}, \quad \omega_j = \mathbf{h}_j^\tau \mathbf{Y}, \tag{7.64}$$

where

$$\mathbf{g}_j = \boldsymbol{\Sigma}_{XX}^{-1} \boldsymbol{\Sigma}_{XY} \boldsymbol{\Sigma}_{YY}^{-1/2} \mathbf{v}_j = \lambda_j \boldsymbol{\Sigma}_{XX}^{-1/2} \mathbf{u}_j, \tag{7.65}$$

$$\mathbf{h}_j = \boldsymbol{\Sigma}_{YY}^{-1/2} \mathbf{v}_j, \tag{7.66}$$

$j = 1, 2, \ldots, t$. The covariance matrix of the canonical variates scores,

$$\boldsymbol{\xi}^{(t)} = \mathbf{G}^{(t)} \mathbf{X}, \quad \boldsymbol{\omega}^{(t)} = \mathbf{H}^{(t)} \mathbf{Y}, \tag{7.67}$$

is given by

$$\mathrm{cov}\{\boldsymbol{\xi}^{(t)}, \boldsymbol{\omega}^{(t)}\} = \begin{pmatrix} \boldsymbol{\Lambda} & \boldsymbol{\Lambda} \\ \boldsymbol{\Lambda} & \mathbf{I}_t \end{pmatrix}, \tag{7.68}$$

where

$$\boldsymbol{\Lambda} = \begin{pmatrix} \lambda_1^2 & 0 & \cdots & 0 \\ 0 & \lambda_2^2 & \cdots & 0 \\ \vdots & \vdots & \ddots & \vdots \\ 0 & 0 & \cdots & \lambda_t^2 \end{pmatrix}, \tag{7.69}$$

and the correlation matrix is

$$\mathrm{corr}\{\boldsymbol{\xi}^{(t)}, \boldsymbol{\omega}^{(t)}\} = \begin{pmatrix} \mathbf{I}_t & \boldsymbol{\Lambda}^{1/2} \\ \boldsymbol{\Lambda}^{1/2} & \mathbf{I}_t \end{pmatrix}. \tag{7.70}$$

If we set $\rho_j = \lambda_j$, $j = 1, 2, \ldots, t$, then, (7.68) shows us that

- $\mathrm{corr}\{\xi_j, \xi_k\} = \delta_{jk}$, $j, k = 1, 2, \ldots, t$,

- $\mathrm{corr}\{\xi_j, \omega_k\} = \rho_j \delta_{jk}$, $j, k = 1, 2, \ldots, t$,

- $\mathrm{corr}\{\omega_j, \omega_k\} = \delta_{jk}$, $j, k = 1, 2, \ldots, t$,

where δ_{jk} is the *Kronecker delta* (i.e., $\delta_{jj} = 1, \delta_{jk} = 0, j \neq k$).

We can, therefore, view the coefficients, $\{g_{ij}\}$ and $\{h_{ij}\}$, of the linear combinations (7.51) as being chosen in the following sequential manner. The first pair (ξ_1, ω_1) has the largest possible correlation ρ_1 among all such linear combinations of $\mathbf{X}$ and $\mathbf{Y}$. The second pair, (ξ_2, ω_2), has the largest possible correlation ρ_2 among all linear combinations of $\mathbf{X}$ and $\mathbf{Y}$ in which ξ_2 is uncorrelated with ξ_1 and ω_2 is uncorrelated with ω_1. The jth pair, (ξ_j, ω_j), has the largest possible correlation ρ_j among all linear combinations of $\mathbf{X}$ and $\mathbf{Y}$ in which ξ_j is uncorrelated with $\xi_l, \xi_2, \ldots, \xi_{j-1}$ and ω_j is uncorrelated with $\omega_l, \omega_2, \ldots, \omega_{j-1}$. See Exercise 7.1. It follows that

$$1 > \rho_1 > \rho_2 > \rho_3 > \cdots > \rho_t > 0. \tag{7.71}$$

The pairs of canonical variates, (ξ_j, ω_j), $j = 1, 2, \ldots, t$, are usually arranged in computer output in the form of two groups, $\xi_l, \xi_2, \ldots, \xi_t$ and $\omega_l, \omega_2, \ldots, \omega_t$. The correlation, ρ_j, between ξ_j and ω_j is called *the canonical correlation coefficient associated with the jth pair of canonical variates*, $j = 1, 2, \ldots, t$.

7.3.4 Relationship of CVA to RRR

Compare the expressions (7.60), (7.61), and (7.62) with those of the reduced-rank regression solutions, (6.86), (6.87), and (6.88).

When $\mathbf{\Gamma} = \mathbf{\Sigma}_{YY}^{-1}$, the matrices $\mathbf{B}^{(t)}$ in (6.88) and $\mathbf{G}^{(t)}$ in (7.61) are identical. Furthermore, the matrices $\mathbf{A}^{(t)}$ in (6.87) and $\mathbf{H}^{(t)}$ in (7.62) are related by

$$\mathbf{H}^{(t)}\mathbf{A}^{(t)}\mathbf{H}^{(t)} = \mathbf{H}^{(t)}, \quad \mathbf{A}^{(t)}\mathbf{H}^{(t)}\mathbf{A}^{(t)} = \mathbf{A}^{(t)}. \tag{7.72}$$

Thus, $\mathbf{A}^{(t)}$ is a g-inverse of $\mathbf{H}^{(t)}$, and vice versa. That is,

$$\mathbf{H}^{(t)} = \mathbf{A}^{(t)-}. \tag{7.73}$$

Thus, in a least-squares sense,

$$\mathbf{A}^{(t)-}\mathbf{Y} \approx \mathbf{A}^{(t)-}\boldsymbol{\mu}^{(t)} + \mathbf{B}^{(t)}\mathbf{X}. \tag{7.74}$$

When $t = s$, two further relations hold,

$$(\mathbf{A}^{(s)}\mathbf{H}^{(s)})^\tau = \mathbf{A}^{(s)}\mathbf{H}^{(s)}, \quad (\mathbf{H}^{(s)}\mathbf{A}^{(s)})^\tau = \mathbf{H}^{(s)}\mathbf{A}^{(s)}. \tag{7.75}$$

Hence, in the full-rank case only, $\mathbf{H}^{(s)} = \mathbf{A}^{(s)+}$, the unique Moore–Penrose generalized inverse of $\mathbf{A}^{(s)}$ (see Section 3.2.7). Also, $\boldsymbol{\nu}^{(s)} = \mathbf{A}^{(s)+}\boldsymbol{\mu}^{(s)}$. *Computationally, the CVA solution, $\boldsymbol{\nu}^{(t)}$, $\mathbf{G}^{(t)}$, and $\mathbf{H}^{(t)}$, can be obtained directly from the RRR solution, $\boldsymbol{\mu}^{(t)}$, $\mathbf{A}^{(t)}$, and $\mathbf{B}^{(t)}$ (and, of course, vice versa).*

This relationship allows us to carry out a CVA using reduced-rank regression (RRR) routines. Moreover, the number t of pairs of canonical variates with nonzero canonical correlations is equal to the rank t of the regression coefficient matrix $\mathbf{C}$. This is a very important point. We have shown that the pairs of canonical variates can be computed using a multivariate RRR routine. Instead of having an isolated methodology for dealing with two sets of correlated variables (as Hotelling developed), we can incorporate canonical variate analysis as an integral part of multivariate regression methodology.

The reduced-rank regression coefficient matrix corresponding to CVA is given by

$$\mathbf{C}_{CVA}^{(t)} = \boldsymbol{\Sigma}_{YY}^{1/2} \left(\sum_{j=1}^{t} \mathbf{v}_j \mathbf{v}_j^{\tau} \right) \boldsymbol{\Sigma}_{YY}^{-1/2} \boldsymbol{\Sigma}_{YX} \boldsymbol{\Sigma}_{XX}^{-1}, \qquad (7.76)$$

where $\mathbf{v}_j$ is the eigenvector associated with the jth largest eigenvalue λ_j of $\mathbf{R}$.

Because the $(s \times s)$-matrix $\mathbf{R}$ plays such a major role in CVA, the following special cases may aid in its interpretation.

- When $s = 1$, $\mathbf{R}$ reduces to the *squared multiple correlation coefficient* (also called the *population coefficient of determination*) of Y with the best linear predictor of Y using $X_1, X_2, \ldots, X_r$,

$$\mathbf{R} = \rho_{Y.X_1,\cdots,X_r}^2 = \frac{\boldsymbol{\sigma}_{YX}^{\tau} \boldsymbol{\Sigma}_{XX}^{-1} \boldsymbol{\sigma}_{XY}}{\sigma_Y^2}, \qquad (7.77)$$

where σ_Y^2 is the variance of Y and $\boldsymbol{\sigma}_{XY}$ is the r-vector of covariances of Y with $\mathbf{X}$.

- When $r = s = 1$, $\mathbf{R}$ is the *squared correlation coefficient* between Y and X,

$$\mathbf{R} = \rho^2 = \frac{\sigma_{XY}^2}{\sigma_X^2 \sigma_Y^2}, \qquad (7.78)$$

where σ_X^2 and σ_Y^2 are the variances of X and Y, respectively, and σ_{XY} is the covariance between X and Y.

The jth canonical correlation coefficient, ρ_j, can, therefore, be interpreted as the multiple correlation coefficient of either ξ_j with $\mathbf{Y}$ or ω_j with $\mathbf{X}$. Using a multiple regression analogy, we can interpret ρ_j either as that proportion of the variance of ξ_j that is attributable to its linear regression on $\mathbf{Y}$ or as that proportion of the variance of ω_j that is attributable to its linear regression on $\mathbf{X}$.

7.3.5 CVA as a Correlation-Maximization Technique

Hotelling's approach to CVA maximized correlations between linear combinations of $\mathbf{X}$ and of $\mathbf{Y}$. Consider, again, the arbitrary linear projections $\xi = \mathbf{g}^{\tau} \mathbf{X}$ and $\omega = \mathbf{h}^{\tau} \mathbf{Y}$, where, for the sake of convenience and with no loss of generality, we assume that $E(\mathbf{X}) = \boldsymbol{\mu}_X = \mathbf{0}$ and $E(\mathbf{Y}) = \boldsymbol{\mu}_Y = \mathbf{0}$. Then, both ξ and ω have zero means. We further assume that they both have unit variances; that is, $\mathbf{g}^{\tau} \boldsymbol{\Sigma}_{XX} \mathbf{g} = 1$ and $\mathbf{h}^{\tau} \boldsymbol{\Sigma}_{YY} \mathbf{h} = 1$.

The first step is to find the vectors $\mathbf{g}$ and $\mathbf{h}$ such that the random variables ξ and ω have maximal correlation,

$$\mathrm{corr}(\xi, \omega) = \mathbf{g}^{\tau} \boldsymbol{\Sigma}_{XY} \mathbf{h}, \qquad (7.79)$$

among all such linear functions of $\mathbf{X}$ and $\mathbf{Y}$. To find $\mathbf{g}$ and $\mathbf{h}$ to maximize (7.79), we set

$$f(\mathbf{g}, \mathbf{h}) = \mathbf{g}^\tau \boldsymbol{\Sigma}_{XY} \mathbf{h} - \frac{1}{2} \lambda (\mathbf{g}^\tau \boldsymbol{\Sigma}_{XX} \mathbf{g} - 1) - \frac{1}{2} \mu (\mathbf{h}^\tau \boldsymbol{\Sigma}_{YY} \mathbf{h} - 1), \quad (7.80)$$

where λ and μ are Lagrangian multipliers. Differentiate $f(\mathbf{g}, \mathbf{h})$ with respect to $\mathbf{g}$ and $\mathbf{h}$, and then set both partial derivatives equal to zero:

$$\frac{\partial f}{\partial \mathbf{g}} = \boldsymbol{\Sigma}_{XY} \mathbf{h} - \lambda \boldsymbol{\Sigma}_{XX} \mathbf{g} = \mathbf{0}, \quad (7.81)$$

$$\frac{\partial f}{\partial \mathbf{h}} = \boldsymbol{\Sigma}_{YX} \mathbf{g} - \mu \boldsymbol{\Sigma}_{YY} \mathbf{h} = \mathbf{0}. \quad (7.82)$$

Multiplying (7.81) on the left by $\mathbf{g}^\tau$ and (7.82) on the left by $\mathbf{h}^\tau$, we obtain

$$\mathbf{g}^\tau \boldsymbol{\Sigma}_{XY} \mathbf{h} - \lambda \mathbf{g}^\tau \boldsymbol{\Sigma}_{XX} \mathbf{g} = 0, \quad (7.83)$$

$$\mathbf{h}^\tau \boldsymbol{\Sigma}_{YX} \mathbf{g} - \mu \mathbf{h}^\tau \boldsymbol{\Sigma}_{YY} \mathbf{h} = 0, \quad (7.84)$$

respectively, whence, the correlation between ξ and ω satisfies

$$\mathbf{g}^\tau \boldsymbol{\Sigma}_{XY} \mathbf{h} = \lambda = \mu. \quad (7.85)$$

Rearranging terms in (7.83), and then substituting λ for μ into (7.84), we get that

$$-\lambda \boldsymbol{\Sigma}_{XX} \mathbf{g} + \boldsymbol{\Sigma}_{XY} \mathbf{h} = \mathbf{0}, \quad (7.86)$$

$$\boldsymbol{\Sigma}_{YX} \mathbf{g} - \lambda \boldsymbol{\Sigma}_{YY} \mathbf{h} = \mathbf{0}. \quad (7.87)$$

Premultiplying (7.86) by $\boldsymbol{\Sigma}_{YX} \boldsymbol{\Sigma}_{XX}^{-1}$, then substituting (7.87) into the result, and rearranging terms gives

$$(\boldsymbol{\Sigma}_{YX} \boldsymbol{\Sigma}_{XX}^{-1} \boldsymbol{\Sigma}_{XY} - \lambda^2 \boldsymbol{\Sigma}_{YY}) \mathbf{h} = \mathbf{0}. \quad (7.88)$$

which is equivalent to

$$(\boldsymbol{\Sigma}_{YY}^{-1/2} \boldsymbol{\Sigma}_{YX} \boldsymbol{\Sigma}_{XX}^{-1} \boldsymbol{\Sigma}_{XY} \boldsymbol{\Sigma}_{YY}^{-1/2} - \lambda^2 \mathbf{I}_s) \mathbf{h} = \mathbf{0}. \quad (7.89)$$

For there to be a nontrivial solution to this equation, the following determinant has to be zero:

$$|\boldsymbol{\Sigma}_{YY}^{-1/2} \boldsymbol{\Sigma}_{YX} \boldsymbol{\Sigma}_{XX}^{-1} \boldsymbol{\Sigma}_{XY} \boldsymbol{\Sigma}_{YY}^{-1/2} - \lambda^2 \mathbf{I}_s| = 0. \quad (7.90)$$

It can be shown that the determinant in (7.90) is a polynomial in λ^2 of degree s, having s real roots, $\lambda_1^2 \geq \lambda_2^2 \geq \cdots \geq \lambda_s^2 \geq 0$, say, which are the eigenvalues of

$$\mathbf{R} = \boldsymbol{\Sigma}_{YY}^{-1/2} \boldsymbol{\Sigma}_{YX} \boldsymbol{\Sigma}_{XX}^{-1} \boldsymbol{\Sigma}_{XY} \boldsymbol{\Sigma}_{YY}^{-1/2} \quad (7.91)$$

with associated eigenvectors $\mathbf{v}_1, \mathbf{v}_2, \ldots, \mathbf{v}_s$. The maximal correlation between ξ and ω would, therefore, be achieved if we took $\lambda = \lambda_1$, the largest eigenvalue of $\mathbf{R}$. The resultant choice of coefficients $\mathbf{g}$ and $\mathbf{h}$ of ξ and ω, respectively, are given by the vectors

$$\mathbf{g}_1 = \Sigma_{XX}^{-1}\Sigma_{XY}\Sigma_{YY}^{-1/2}\mathbf{v}_1, \quad \mathbf{h}_1 = \Sigma_{YY}^{-1/2}\mathbf{v}_1; \tag{7.92}$$

compare with (7.65) and (7.66). In other words, the first pair of canonical variates is given by (ξ_1, ω_1), where $\xi_1 = \mathbf{g}_1^\tau\mathbf{X}$ and $\omega_1 = \mathbf{h}_1^\tau\mathbf{Y}$, and their correlation is $\text{corr}(\xi_1, \omega_1) = \mathbf{g}_1^\tau\Sigma_{XY}\mathbf{h}_1 = \lambda_1$.

Given (ξ_1, ω_1), let $\xi = \mathbf{g}^\tau\mathbf{X}$ and $\omega = \mathbf{h}^\tau\mathbf{Y}$ denote a second pair of arbitrary linear projections with unit variances. We require (ξ, ω) to have maximal correlation among all such linear combinations of $\mathbf{X}$ and $\mathbf{Y}$, respectively, which are also uncorrelated with (ξ_1, ω_1). This last condition translates into $\mathbf{g}^\tau\Sigma_{XX}\mathbf{g}_1 = \mathbf{h}^\tau\Sigma_{YY}\mathbf{h}_1 = 0$. Furthermore, by (7.86) and (7.87), we require

$$\text{corr}(\xi, \omega_1) = \mathbf{g}^\tau\Sigma_{XY}\mathbf{h}_1 = \lambda_1\mathbf{g}^\tau\Sigma_{XX}\mathbf{g}_1 = 0, \tag{7.93}$$

$$\text{corr}(\omega, \xi_1) = \mathbf{h}^\tau\Sigma_{YX}\mathbf{g}_1 = \lambda_1\mathbf{h}^\tau\Sigma_{YY}\mathbf{h}_1 = 0. \tag{7.94}$$

We choose $\mathbf{g}$ and $\mathbf{h}$ to maximize (7.79) subject to the above conditions. Set

$$\begin{aligned} f(\mathbf{g}, \mathbf{h}) &= \mathbf{g}^\tau\Sigma_{XY}\mathbf{h} - \frac{1}{2}\lambda(\mathbf{g}^\tau\Sigma_{XX}\mathbf{g} - 1) - \frac{1}{2}\mu(\mathbf{h}^\tau\Sigma_{YY}\mathbf{h} - 1) \\ &\quad + \eta\mathbf{g}^\tau\Sigma_{XX}\mathbf{g}_1 + \nu\mathbf{h}^\tau\Sigma_{YY}\mathbf{h}_1, \end{aligned} \tag{7.95}$$

where λ, μ, η, and ν are Lagrangian multipliers. Differentiate $f(\mathbf{g}, \mathbf{h})$ with respect to $\mathbf{g}$ and $\mathbf{h}$, and then set both partial derivatives equal to zero:

$$\frac{\partial f}{\partial \mathbf{g}} = \Sigma_{XY}\mathbf{h} - \lambda\Sigma_{XX}\mathbf{g} + \eta\Sigma_{XX}\mathbf{g}_1 = \mathbf{0}, \tag{7.96}$$

$$\frac{\partial f}{\partial \mathbf{h}} = \Sigma_{YX}\mathbf{g} - \mu\Sigma_{YY}\mathbf{h} + \nu\Sigma_{YY}\mathbf{h}_1 = \mathbf{0}. \tag{7.97}$$

Multiplying (7.96) on the left by $\mathbf{g}^\tau$ and (7.97) on the left by $\mathbf{h}^\tau$, and taking note of (7.93) and (7.94), these equations reduce to (7.86) and (7.87), respectively. We, therefore, take the second pair of canonical variates to be (ξ_2, ω_2), where

$$\mathbf{g}_2 = \Sigma_{XX}^{-1}\Sigma_{XY}\Sigma_{YY}^{-1/2}\mathbf{v}_2, \quad \mathbf{h}_2 = \Sigma_{YY}^{-1/2}\mathbf{v}_2, \tag{7.98}$$

and their correlation is $\text{corr}(\xi_2, \omega_2) = \mathbf{g}_2^\tau\Sigma_{XY}\mathbf{h}_2 = \lambda_2$.

We continue this sequential procedure, deriving eigenvalues and eigenvectors, until no further solutions can be found. This gives us sets of coefficients for the pairs of canonical variates, $(\xi_1, \omega_1), (\xi_2, \omega_2), \ldots, (\xi_k, \omega_k)$, $k = \min(r, s)$, where the ith pair of canonical variates (ξ_i, ω_i) is obtained by choosing the coefficients $\mathbf{g}_i$ and $\mathbf{h}_i$ such that (ξ_i, ω_i) has the largest correlation among all pairs of linear combinations of $\mathbf{X}$ and $\mathbf{Y}$ that are also uncorrelated with all previously derived pairs, $(\xi_j, \omega_j), j = 1, 2, \ldots, i - 1$.

7.3.6 Sample Estimates

Thus, $\mathbf{G}$ and $\mathbf{H}$ are estimated by

$$\widehat{\mathbf{G}}^{(t)} = \begin{pmatrix} \widehat{\mathbf{v}}_1^{\tau} \\ \vdots \\ \widehat{\mathbf{v}}_t^{\tau} \end{pmatrix} \widehat{\boldsymbol{\Sigma}}_{YY}^{-1/2} \widehat{\boldsymbol{\Sigma}}_{YX} \widehat{\boldsymbol{\Sigma}}_{XX}^{-1} = \begin{pmatrix} \widehat{\lambda}_1 \widehat{\mathbf{u}}_1^{\tau} \\ \vdots \\ \widehat{\lambda}_t \widehat{\mathbf{u}}_t^{\tau} \end{pmatrix} \widehat{\boldsymbol{\Sigma}}_{XX}^{-1/2}, \tag{7.99}$$

$$\widehat{\mathbf{H}}^{(t)} = \begin{pmatrix} \widehat{\mathbf{v}}_1^{\tau} \\ \vdots \\ \widehat{\mathbf{v}}_t^{\tau} \end{pmatrix} \widehat{\boldsymbol{\Sigma}}_{YY}^{-1/2}, \tag{7.100}$$

respectively, where $\widehat{\mathbf{u}}_j$ is the eigenvector associated with the jth largest eigenvalue $\widehat{\lambda}_j^2$ of the $(r \times r)$ symmetric matrix

$$\widehat{\mathbf{R}}^* = \widehat{\boldsymbol{\Sigma}}_{XX}^{-1/2} \widehat{\boldsymbol{\Sigma}}_{XY} \widehat{\boldsymbol{\Sigma}}_{YY}^{-1} \widehat{\boldsymbol{\Sigma}}_{YX} \widehat{\boldsymbol{\Sigma}}_{XX}^{-1/2}, \tag{7.101}$$

$j = 1, 2, \ldots, t$, and $\widehat{\mathbf{v}}_j$ is the eigenvector associated with the jth largest eigenvalue $\widehat{\lambda}_j^2$ of the $(s \times s)$ symmetric matrix

$$\widehat{\mathbf{R}} = \widehat{\boldsymbol{\Sigma}}_{YY}^{-1/2} \widehat{\boldsymbol{\Sigma}}_{YX} \widehat{\boldsymbol{\Sigma}}_{XX}^{-1} \widehat{\boldsymbol{\Sigma}}_{XY} \widehat{\boldsymbol{\Sigma}}_{YY}^{-1/2}, \tag{7.102}$$

$j = 1, 2, \ldots, t$. The jth row of $\widehat{\boldsymbol{\xi}} = \widehat{\mathbf{G}}^{(t)} \mathbf{X}$ and the jth row of $\widehat{\boldsymbol{\omega}} = \widehat{\mathbf{H}}^{(t)} \mathbf{Y}$ together form the jth *pair of sample canonical variates* $(\widehat{\xi}_j, \widehat{\omega}_j)$ given by

$$\widehat{\xi}_j = \widehat{\mathbf{g}}_j^{\tau} \mathbf{X}, \quad \widehat{\omega}_j = \widehat{\mathbf{h}}_j^{\tau} \mathbf{Y}, \tag{7.103}$$

with values (or *canonical variate scores*) of

$$\widehat{\xi}_{ij} = \widehat{\mathbf{g}}_j^{\tau} \mathbf{X}_i, \quad \widehat{\omega}_{ij} = \widehat{\mathbf{h}}_j^{\tau} \mathbf{Y}_i, \quad i = 1, 2, \ldots, n, \tag{7.104}$$

where

$$\widehat{\mathbf{g}}_j^{\tau} = \widehat{\mathbf{v}}_j^{\tau} \widehat{\boldsymbol{\Sigma}}_{YY}^{-1/2} \widehat{\boldsymbol{\Sigma}}_{YX} \widehat{\boldsymbol{\Sigma}}_{XX}^{-1} = \widehat{\lambda}_j \widehat{\mathbf{u}}_j^{\tau} \widehat{\boldsymbol{\Sigma}}_{XX}^{-1/2} \tag{7.105}$$

is the jth row of $\widehat{\mathbf{G}} = \widehat{\mathbf{G}}^{(t)}$ and

$$\widehat{\mathbf{h}}_j^{\tau} = \mathbf{v}_j^{\tau} \widehat{\boldsymbol{\Sigma}}_{YY}^{-1/2} \tag{7.106}$$

is the jth row of $\widehat{\mathbf{H}} = \widehat{\mathbf{H}}^{(t)}$. The *sample canonical correlation coefficient* for the jth pair of sample canonical variates, $(\widehat{\xi}_j, \widehat{\omega}_j)$, is given by

$$\widehat{\rho}_j = \widehat{\lambda}_j = \frac{\widehat{\mathbf{g}}_j^{\tau} \widehat{\boldsymbol{\Sigma}}_{XY} \widehat{\mathbf{h}}_j}{(\widehat{\mathbf{g}}_j^{\tau} \widehat{\boldsymbol{\Sigma}}_{XX} \widehat{\mathbf{g}}_j)^{1/2} (\widehat{\mathbf{h}}_j^{\tau} \widehat{\boldsymbol{\Sigma}}_{YY} \widehat{\mathbf{h}}_j)^{1/2}}, \quad j = 1, 2, \ldots, t, \tag{7.107}$$

It is usually hoped that the first t pairs of sample canonical variates will be the most important, exhibiting a major proportion of the correlation present in the data, whereas the remainder can be neglected without losing too much information concerning the correlational structure of the data. Thus, only those pairs of canonical variates with high canonical correlations should be retained for further analysis.

An estimator of the rank-t regression coefficient matrix corresponding to the canonical variates case is given by

$$\widehat{\mathbf{C}}^{(t)} = \widehat{\boldsymbol{\Sigma}}_{YY}^{1/2} \left(\sum_{j=1}^{t} \widehat{\mathbf{v}}_j \widehat{\mathbf{v}}_j^{\tau} \right) \widehat{\boldsymbol{\Sigma}}_{YY}^{-1/2} \widehat{\boldsymbol{\Sigma}}_{YX} \widehat{\boldsymbol{\Sigma}}_{XX}^{-1}, \qquad (7.108)$$

where $\widehat{\mathbf{v}}_j$ is the eigenvector associated with the jth largest eigenvalue of $\widehat{\mathbf{R}}$, $j = 1, 2, \ldots, s$. When $\mathbf{X}$ and $\mathbf{Y}$ are jointly Gaussian, the asymptotic distribution of $\widehat{\mathbf{C}}^{(t)}$ in (7.108) is available (Izenman, 1975).

The exact distribution of the sample canonical correlations when $\mathbf{X}$ and $\mathbf{Y}$ are jointly Gaussian and some of the population canonical correlations are nonzero is extremely complicated, having the form of a hypergeometric function of two matrix arguments (Constantine, 1963; James, 1964). In the null case, when $\mathbf{X}$ and $\mathbf{Y}$ are independent and all the population canonical correlations are zero, the exact density of the squares of the nonzero sample canonical correlations is given by

$$p(x_1, \ldots, x_t) = c_{r,s,n} \prod_{j=1}^{s} [w(x_j)]^{1/2} \prod_{j<k} (x_j - x_k), \qquad (7.109)$$

where the $x_1, \geq x_2 \geq \cdots \geq x_s$ are the eigenvalues of $\mathbf{R}$, $w(x) = x^{r-s-1}(1 - x)^{n-r-s-2}$ is the weight function corresponding to the Jacobi family of orthogonal polynomials, and $c_{r,s,n}$ is a normalization constant depending upon r, s, and n. For details, see, for example, Anderson (1984, Section 13.4). The second product in (7.106) is the Jacobian, also known as the *Vandermonde determinant* (Johnstone, 2006). Asymptotic distribution results are also available when the first t canonical correlations are positive, smaller than unity, and distinct.

7.3.7 *Invariance*

Unlike principal component analysis, canonical correlations are invariant under simultaneous nonsingular linear transformation of the random vectors $\mathbf{X}$ and $\mathbf{Y}$. Suppose we consider linear transformations of $\mathbf{X}$ and $\mathbf{Y}$:

$$\mathbf{X} \rightarrow \mathbf{DX}, \quad \mathbf{Y} \rightarrow \mathbf{FY}, \qquad (7.110)$$

where the $(r \times r)$-matrix $\mathbf{D}$ and the $(s \times s)$-matrix $\mathbf{F}$ are nonsingular. Then, the canonical correlations of $\mathbf{DX}$ and $\mathbf{FY}$ are identical to those of $\mathbf{X}$ and $\mathbf{Y}$. See Exercise 7.11. A consequence of this result is that a CVA using the covariance matrix will yield the *same* canonical correlations as a CVA using the corresponding correlation matrix.

7.3.8   How Many Pairs of Canonical Variates to Retain?

Because the question of how many pairs of canonical variates to retain is equivalent to determining the rank t of the regression coefficient matrix $\mathbf{C}^{(t)}$ in a reduced-rank regression for CVA, we approach this problem as a rank-determination problem. Although $\mathbf{X}$ and $\mathbf{Y}$ are treated symmetrically in CVA, the RRR formulation turns CVA into a supervised learning technique. Prediction error can be used as a measure of how good $\mathbf{X}$ is in predicting $\mathbf{Y}$ using cross-validation. In the case of the rank trace, no reductions of the expressions for the coordinates of the plotted points can be obtained for the CV case as we were able to do for the PC case. The CV rank trace can have points plotted on the exterior to the unit square, and the sequence of points may not be monotonically decreasing; we can, however, introduce a regularization parameter into the rank-trace computations to keep the plotted points within the unit square.

7.4   Projection Pursuit

Projection pursuit (PP) was motivated by the desire to discover "interesting" low-dimensional (typically, one- or two-dimensional) linear projections of high-dimensional data (Friedman and Tukey, 1974). The Gaussian distribution, which has always occupied a central place in statistical theory and application, turns out to be "least interesting" when dealing with low-dimensional projections of multivariate data. This is due to the fact that each of the marginals of a multivariate Gaussian distribution is Gaussian and that most low-dimensional projections of high-dimensional data look approximately Gaussian-distributed (Diaconis and Freedman, 1984). We should, therefore, not expect to see unusual patterns or structure in low-dimensional projections of highly multivariate data.

PP was originally driven by the desire to expose specific non-Gaussian features (variously referred to as "local concentration," "clusters of distinct groups," "clumpiness," or "clottedness") of the data. An exhaustive search for such features is clearly impossible, and so the search was automated. Indexes of interestingness were created and optimized numerically in an attempt to imitate how users instinctively (by eye) choose interesting projections. This formulation was later replaced by a search for projections that are as far from Gaussianity as possible.

The general strategy behind PP consists of the following two-step process:

1. Set up a *projection index* $\mathcal{I}$ to judge the merit of a particular one- or two-dimensional (or sometimes three-dimensional) projection of a given set of multivariate data.

2. Use an *optimization algorithm* to find the global and local extrema of that projection index over all m-dimensional projections of the data ($m = 1, 2$ or 3).

For a given m, the optimization step determines the most informative m-dimensional projection of the data. A graphical display of the projections is the output of choice in practice.

7.4.1 Projection Indexes

Huber (1985) argues that projection indexes should be chosen to possess certain computational and analytical properties, especially that of affine invariance (location and scale invariance). Examples of affine invariant indexes include absolute cumulants (e.g., skewness and kurtosis), and Shannon negative entropy (*negentropy*), both of which are nonnegative in general, but are equal to zero if the underlying distribution is Gaussian. If, however, the data are centered and sphered (having mean zero and covariance matrix the identity), then there is no reason to require affine invariance of the projection index because every projection of the sphered data inherits its properties (i.e., also has mean zero and covariance matrix the identity).

A special case of PP occurs when the projection index is the *variance*, $\mathrm{var}(Y) = \mathbf{w}^\tau \mathbf{\Sigma}_{XX} \mathbf{w}$, of the unit-length projection $Y = \mathbf{w}^\tau \mathbf{x}$. In this case, maximizing the variance with respect to $\mathbf{w}$ reduces PP to PCA, and the resulting projections are the leading principal components of $\mathbf{X}$. Bolton and Krzanowski (1999) show that maximizing the variance is equivalent to minimizing the corresponding Gaussian log-likelihood; in other words, the projection is most interesting (in a variance sense) when $\mathbf{X}$ is least likely to be Gaussian.

Cumulant-Based Index

The absolute value of kurtosis, $|\kappa_4(Y)|$, of the one-dimensional projection $Y = \mathbf{w}^\tau \mathbf{X}$ has been widely used as a measure of non-Gaussianity of Y. It has value zero for a Gaussian variable and is positive for a non-Gaussian variable. An unbiased estimate of $\kappa_4(Y)$ is given by the so-called *k-statistic* k_4 (see, e.g., Kendall and Stuart, 1969, p. 280). Although $\kappa_4(Y)$ is affine invariant and fast to compute, it is not robust, and outliers can have a pretty drastic effect on estimates of $|\kappa_4(Y)|$.

In fact, maximizing or minimizing the kurtosis, $\kappa_4(Y)$, of projected data Y with respect to direction $\mathbf{w}$ has been advocated as a way of detecting multivariate outliers (Gnanadesikan and Kettenring, 1972; Peña and Prieto, 2001). A maximal value of kurtosis would result from a small, concentrated amount of outlier contamination, whereas a minimal value of kurtosis would be due to a large amount of contamination.

Polynomial-Based Indexes

Let $Y = \mathbf{w}^\tau \mathbf{X}$ denote a continuous random variable having probability density function $p_Y(y)$. Polynomial-based projection indexes take the general form of weighted versions of integrated squared error,

$$\mathcal{I}(Y) = \int [\phi(y) - p_Y(y)]^2 w(y) dy, \qquad (7.111)$$

where $w(y)$ is a given weight function on $\Re$. Examples of $w(y)$ include $w(y) = 1/\phi(y), 1$, and $\phi(y)$, where $\phi(y)$ is the standard Gaussian density with zero mean and unit variance.

Now, Y is standard Gaussian with density $\phi(y)$ iff $U = 2\Phi(Y) - 1$ is uniformly distributed on the interval $[-1, 1]$, where $\Phi(Y) = \int_{-\infty}^{Y} \phi(y) dy$ (see Exercise 7.12). Hence, the integrated squared error between the density of U, $p_U(u)$, say, and the uniform density,

$$\mathcal{I}_F(Y) = \int_{-1}^{1} [p_U(u) - \tfrac{1}{2}]^2 du = \int_{-1}^{1} [p_U(u)]^2 du - \tfrac{1}{2}, \qquad (7.112)$$

can be used as a projection index (Friedman, 1987). The idea is that the further $p_U(u)$ is from the uniform density, the further Y would be from Gaussianity. It turns out that this index, if transformed back to the original scale, can be reexpressed as (7.111) with $w(y) = 1/\phi(y)$, assuming $p_Y(y)/[\phi(y)]^{1/2}$ is square-integrable. For heavy-tailed $p_Y(y)$, $\mathcal{I}_F(Y)$ can be infinite, and so will not be very useful as a projection index. It can be shown that $\mathcal{I}_F(Y)$ can be approximated by

$$\mathcal{I}_F(Y) \approx \frac{[\kappa_3(Y)]^2}{12} + \frac{[\kappa_4(Y)]^2}{48}, \qquad (7.113)$$

which is the moment-based projection index of Jones and Sibson (1987).

Interestingly enough, it turns out that outliers in projected data are not unusual. In simulation experiments using a moment-based index similar to (7.113) (see Friedman and Johnstone's discussions of Jones and Sibson, 1987), outliers were observed to appear repeatedly in projections of even well-behaved multivariate Gaussian data. Furthermore, there is no obvious way to robustify (7.113).

Another possibility is take $w(y) = 1$ in $\mathcal{I}(Y)$ (Hall, 1989). It is not difficult to show that Hall's index, $\mathcal{I}_H(Y)$, can be approximated by

$$\mathcal{I}_H^0(Y) \propto (\mathrm{E}\{\phi(Y)\} - \mathrm{E}\{\phi(Z)\})^2, \tag{7.114}$$

where Z is standard Gaussian and $\mathrm{E}\{\phi(Z)\} = (2\pi^{1/2})^{-1}$. Hall's index (and its two-dimensional analogue) appears to identify projections of the data that have a "hole" in their center (Cook, Buja, Cabrera, and Hurley, 1995).

Taking $w(y) = \phi(y)$ in $\mathcal{I}(Y)$ puts more weight around the center of the distribution, rather than at the tails (Cook, Buja, and Cabrera, 1993). It can be shown that $\mathcal{I}_{CBC}(Y)$ can also be approximated by (7.114). We shall see a generalized form of (7.114) again when we discuss independent component analysis.

Two-dimensional projection indexes are generally built by simple extensions of their one-dimensional versions. Suppose $\mathbf{X}$ has been centered and sphered as before. Let $\mathbf{Y} = (Y_1, Y_2)^\tau$ be a bivariate projection of $\mathbf{X}$, where $Y_1 = \mathbf{w}_1^\tau \mathbf{X}$ and $Y_2 = \mathbf{w}_2^\tau \mathbf{X}$. We want to find $\mathbf{w}_1$ and $\mathbf{w}_2$ so that Y_1 and Y_2 are uncorrelated (i.e., $\mathbf{w}_1^\tau \mathbf{w}_2 = 0$) and that the joint distribution, $p_Y(y_1, y_2)$, of (Y_1, Y_2) has some interesting structure. Furthermore, we require the projections to have equal variances (i.e., $\mathbf{w}_1^\tau \mathbf{w}_1 = \mathbf{w}_2^\tau \mathbf{w}_2 = 1$). In this case, the bivariate Gaussian density, $\phi(y_1, y_2)$, is deemed the least-interesting two-dimensional structure.

Shannon Negentropy

The *entropy* of a random variable, which was introduced by Claude E. Shannon in 1948, has become a valuable concept in information theory. The entropy of the random variable Y gives us a notion of how much information is contained in Y. Essentially, entropy is largest when Y has greatest variance (i.e., when Y is most unpredictable). If Y is a continuous random variable with probability density function $p_Y(y)$, then the (differential) entropy $\mathcal{H}(Y)$ of Y is defined by

$$\mathcal{H}(Y) = - \int p_Y(y) \log p_Y(y) dy. \tag{7.115}$$

Among all random variables having equal variance, the largest value of $\mathcal{H}(Y)$ occurs when Y has a Gaussian distribution (Rao, 1965, p. 132). Small values of $\mathcal{H}(Y)$ occur when the distribution of Y is concentrated on specific values. Huber (1985) had the idea of using differential entropy as a measure of non-Gaussianity and, hence, as a projection index.

If we normalize $\mathcal{H}(Y)$ so that it has the value zero for a Gaussian variable and otherwise is always nonnegative, we arrive at *negentropy* defined by

$$\mathcal{J}(Y) = \mathcal{H}(Z) - \mathcal{H}(Y), \tag{7.116}$$

where Z is a Gaussian random variable having the same variance as Y. If $Z \sim \mathcal{N}(0,1)$, it is easy to show that $\mathcal{H}(Z) = \frac{1}{2}[1 + \log 2\pi] \approx 1.419$. Jones and Sibson (1987) derive an efficient projection index based upon $\mathcal{J}(Y)$.

7.4.2 Optimizing the Projection Index

Given a projection index, the next step is to optimize that index, if possible using an algorithm with high speed and low memory requirements. Researchers have preferred different types of optimizing algorithms, including steepest ascent and genetic algorithms. In fact, projection indexes are notorious for getting trapped in numerous local maxima. Getting trapped repeatedly in suboptimal local maxima can delay convergence to the global maximum. It is important, therefore, to use a numerical optimizer that has the ability to avoid such local maxima.

7.5   Visualizing Projections Using Dynamic Graphics

Graphical methods are vital tools for exploring multivariate data. Most statistical graphics methods in common use today can be classified as *static graphics*, such as scatterplots, scatterplot matrices, and displays of projection pursuit results. Additional details from statistic displays can be visualized by using a range of colors or different shapes, characters, or symbols for various levels or characteristics of the data.

Innovative and more informative *dynamic graphics* were devised by John W. Tukey during the early 1970s for visually searching for low-dimensional structure within multivariate data. Such searches were enhanced by the development of custom-designed computer hardware and software (PRIM-9) to carry out the operations of *picturing* ("an ability to look at data from several different directions in multidimensional space"), *rotation* ("at a minimum, the ability to turn the data so that it can be viewed from any direction that is chosen"), *masking* ("the ability to select suitable subregions of the multidimensional space for consideration"), and *isolation* ("the ability to select any subsample of the data points for consideration") in up to 9 dimensions (Fisherkeller, Friedman, and Tukey, 1974).

The graphical data analysis concepts embedded in PRIM-9 have been upgraded and enhanced by the XGOBI/GGOBI data visualization system (Swayne, Cook, and Buja, 1998; Cook, Buja, Cabrera, and Hurley, 1995). Examples of the types of dynamic graphics included in the XGOBI/GGOBI system are

- The *grand tour* (Asimov, 1985) of data recorded on an r-dimensional set of variables, **X**, seeks to generate a continuous sequence of low-dimensional projections of the **X**-data, where projections are visualized in one, two, or three dimensions and are designed to be representative of all possible projections of the data.

- The *correlation tour* of data recorded on two nonoverlapping sets of variables, an r-dimensional set **X** and an s-dimensional set **Y**, seeks to generate a continuous sequence of one-dimensional projections of the **X**-data and of the **Y**-data in order to display the amount of correlation in those projections.

The grand tour can be regarded as a dynamic version of PCA and the correlation tour as a dynamic version of CVA. The main problem is the huge number of potentially interesting projections. Some guidance is, therefore, needed. For both tours, "interesting" projections can be automatically selected by optimizing one of the objective functions associated with projection pursuit methods. The objective functions discussed above are included in a menu of indexes in the XGOBI/GGOBI system.

7.6   Software Packages

PCA is included in R, S-PLUS, SAS, SPSS, MATLAB, and MINITAB. CVA (or CCA) is usually confused with linear discriminant analysis (see, e.g., Venables and Ripley, 2002, p. 332), which is a special case of CVA (see Chapter 8). CVA — in the sense of this chapter — is not included in most major software packages.

PCA and CVA are included as special cases of multivariate reduced-rank regression in the RRR+MULTANL package, which can be downloaded from the book's website. Versions of this package are available for use with R, S-PLUS, and MATLAB.

Bibliographical Notes

Classical descriptions of PCA and CVA can be found in any text on multivariate analysis; in particular, we recommend Anderson (1984, Chapters 11 and 12) and Seber (1984, Chapter 5) for theoretical treatments and Gnanadesikan (1977, Chapters 2 and 3) and Lattin, Carroll, and Green (2003, Chapters 4 and 9) for applied viewpoints. Detailed treatments of PCA can be found in Jackson (2003) and Jolliffe (1986). The relationships between multivariate reduced-rank regression and PCA and CVA can be found in Izenman (1975).

The original concept of projection pursuit was formulated by Kruskal (1969, 1972), but it was Friedman and Tukey (1974) who gave it the catchy name. The development of PP was based upon the experience (and frustrations) of working with an interactive computer graphics program called PRIM-9 (Fisherkeller, Friedman, and Tukey, 1974), which was the first program to use operations such as picturing, rotation, isolation, and masking for visually exploring multivariate data in up to 9 dimensions. The highpoint of PRIM-9 was a 25-minute movie taken in 1974 of Tukey analyzing high-dimensional particle physics data. Friedman and Stuetzle (2002) give an historical account of the origins and development of PRIM-9 and PP. The XGOBI/GGOBI computer graphics programs are the improved and enhanced descendants of PRIM-9. PP has recently been rediscovered by researchers in independent component analysis (see Chapter 15).

Exercises

7.1 Generate a random sample of size $n = 100$ from a three-dimensional ($r = 3$) Gaussian distribution, where one of the variables has very high variance (relative to the other two). Carry out PCA on these data using the covariance matrix and the correlation matrix. In each case, find the eigenvalues and eigenvectors, draw the scree plot, compute the PC scores, and plot all pairwise PC scores in a matrix plot. Compare results.

7.2 Carry out a RRR on the data from Exercise 7.1 using the PCA formulation (i.e., $\mathbf{Y} = \mathbf{X}$, $\mathbf{\Gamma} = \mathbf{I}_r$). Compute the rank trace and determine the number of principal components to retain. Compare results with those of Exercise 7.1.

7.3 In the file `turtles.txt`, there are three variables, length, width, and height, of the carapaces of 48 painted turtles, 24 female and 24 male. Take logarithms of all three variables. Estimate the mean vector and covariance matrix of the male turtles and of the female turtles separately. Find the eigenvalues and eigenvectors of each estimated covariance matrix and carry out a PCA of each data set. Find an expression for the volume of a turtle carapace for males and for females. (Hint: use the fact that the variables are logarithms of the original measurements.) Compare volumes of male and female carapaces.

7.4 In the pen-based handwritten digit recognition (`pendigits`) example of Section 7.2.1, compute the variance of each of the 16 variables and show that they are very similar. Then, carry out PCA using the covariance matrix. How many PCs explain 80% and 90% of the total variation in the data? Display the first three PCs using pairwise scatterplots as in Figure 7.1. Do you see any differences between a covariance-based and a correlation-based PCA for this example?

7.5 For the `pendigits` data, draw the scree plot and the rank trace plot. How many PCs would you retain based upon each plot? Do you get the same answer from both plots?

7.6 For the principal components case, show that the points in the PC rank trace are given by (7.38) and (7.39).

7.7 The file `SwissBankNotes.txt` consists of six variables measured on 200 old Swiss 1,000-franc bank notes. The first 100 are genuine and the second 100 are counterfeit. The six variables are length of the bank note, height of the bank note, measured on the left, height of the bank note, measured on the right, distance of inner frame to the lower border, distance of inner frame to the upper border, and length of the diagonal. Carry out a PCA of the 100 genuine bank notes, of the 100 counterfeit bank notes, and of all 200 bank notes combined. Do you notice any differences in the results?

7.8 In Section 5.5, condition number and condition indices were discussed as a means of detecting and identifying ill-conditioned data and collinearity in regression problems. How would such measures help in PCA or CVA? Compute these various statistics for the `pendigits` data.

7.9 Carry out a PCA of Fisher's `iris` data. These data consist of 50 observations on each of three species of iris: *Iris setosa*, *Iris versicolor*, and *Iris virginica*. The four measured variables are sepal length, sepal width, petal length, and petal width. Ignore the species labels. Compute the PC scores and plot all pairwise sets of PC scores in a matrix plot. Explain your results, taking into consideration the species labels.

7.10 Consider an $(r \times r)$ correlation matrix with the same correlation, ρ, say, in the off-diagonal entries. Find the eigenvalues and eigenvectors of this matrix when $r = 2, 3, 4$. Generalize your results to any r variables. As examples, set $\rho = 0.1, 0.3, 0.5, 0.7, 0.9$.

7.11 Show that the set of canonical variates is invariant under simultaneous nonsingular linear transformations of the random vectors $\mathbf{X}$ and $\mathbf{Y}$.

7.12 Let $r = s = 2$ and suppose the equicorrelation model holds for $\mathbf{X}$ and $\mathbf{Y}$. Then, $\mathbf{\Sigma}_{XX} = \mathbf{\Sigma}_{YY} = \begin{pmatrix} 1 & \rho \\ \rho & 1 \end{pmatrix}$ and $\mathbf{\Sigma}_{XY} = \begin{pmatrix} \rho & \rho \\ \rho & \rho \end{pmatrix}$. Find the canonical correlations and the canonical variates. Generalize your results to general r and s. Find the matrix $\mathbf{R}$ and the RRR solutions for $t = 1, 2$.

7.13 For the `COMBO-17` galaxy data, compute a rank-2 multivariate RRR of $\mathbf{Y}$ on $\mathbf{X}$ in which $\mathbf{\Gamma} = \mathbf{\Sigma}_{YY}^{-1}$ for the CV situation. Compute the multivariate residuals from the regression, plot them in any way you regard as interesting, and try to find the outliers mentioned in the example.

7.14 Show that Y is standard Gaussian with density $\phi(y)$ iff $U = 2\Phi(Y) - 1$ is uniformly distributed on the interval $[-1, 1]$, where $\Phi(Y) = \int_{-\infty}^{Y} \phi(y) dy$.

7.15 Draw the density of the eigenvalues of a Wishart matrix, $\mathcal{X}\mathcal{X}^T \sim \mathcal{W}_r(n, \mathbf{I}_r)$, where $r/n \to \gamma \in (0, \infty)$, for γ equal to 0.2, 0.5, 1, 4, 9, 16.

8
Linear Discriminant Analysis

8.1 Introduction

Suppose we are given a learning set $\mathcal{L}$ of multivariate observations (i.e., input values in $\Re^r$), and suppose each observation is known to have come from one of K *predefined* classes having similar characteristics. These classes may be identified, for example, as species of plants, levels of credit worthiness of customers, presence or absence of a specific medical condition, different types of tumors, views on Internet censorship, or whether an e-mail message is spam or non-spam. To distinguish the known classes from each other, we associate a unique *class label* (or output value) with each class; the observations are then described as *labeled observations*.

In each of these situations, there are two main goals:

Discrimination: Use the information in a learning set of labeled observations to construct a *classifier* (or *classification rule*) that will separate the predefined classes as much as possible.

Classification: Given a set of measurements on a new *unlabeled* observation, use the classifier to predict the class of that observation.

A classifier is a combination of the input variables. In the machine learning literature, discrimination and classification are described as *supervised*

A.J. Izenman, *Modern Multivariate Statistical Techniques*,
doi: 10.1007/978-0-387-78189-1_8,
© Springer Science+Business Media, LLC 2008

learning techniques; together, they are also referred to as tasks of *class prediction*.

Whether these goals are at all achievable depends upon the information provided by the input variables. When there are two classes (i.e., $K = 2$), we need only one classifier, and when there are more than two classes, we need at least two (and at most $K - 1$) classifiers to differentiate between the classes and to predict the class of a future observation.

Consider the following medical diagnosis example. If a patient enters the emergency room with severe stomach pains and symptoms consistent with both food poisoning and appendicitis, a decision has to be made as to which illness is more likely for that patient; only then can the patient be treated. For this example, the problem is that the appropriate treatment for one cause of illness is the opposite treatment for the other: appendicitis requires surgery, whereas food poisoning does not, and an incorrect diagnosis could lead to a fatal result. In light of the results from the clinical tests, the physician has to decide upon a course of treatment to maximize the likelihood of success. If the combination of test results points in a particular direction, surgery is recommended; otherwise, the physician recommends a non-surgical treatment. A classifier is constructed from past experience based upon the test results of previously treated patients (the learning set). The more reliable the classifier, the greater the chance for a successful diagnostic outcome for a future patient.

Similarly, a credit card company or a bank uses loan histories of past customers to decide whether a new customer would be a good or bad credit risk; a post office uses handwriting samples of a large number of individuals to design an automated method for distinguishing between different handwritten digits and letters; molecular biologists use gene expression data to distinguish between known classes of tumors; political scientists use frequencies of word usage to identify the authorship of different political tracts; and a person who uses e-mail would certainly like to have a filter that recognizes whether a message is spam or not.

In this chapter, we focus upon the most basic type of classifier: a *linear* combination of the input variables. This problem has been of interest to statisticians since R.A. Fisher introduced the *linear discriminant function* (Fisher, 1936).

8.1.1   Example: Wisconsin Diagnostic Breast Cancer Data

Breast cancer is the second largest cause of cancer deaths among women. Three methods of diagnosing breast cancer are currently available: mammography; fine needle aspirate (FNA) with visual interpretation; and surgical biopsy. Although biopsies are the most accurate in distinguishing

TABLE 8.1. *Ten variables for the Wisconsin breast cancer study.*

radius	Radius of an individual nucleus
texture	Variance of gray levels inside the boundary of the nucleus
peri	Distance around the perimeter of the nucleus
area	Area of the nucleus
smooth	Smoothness of the contour of a nucleus as measured by the local variation of radial segments
comp	A measure of the compactness of a cell nucleus using the formula $(\text{peri})^2/\text{area}$
scav	Severity of concavities or indentations in a cell nucleus using a size measurement that emphasizes small indentations
ncav	Number of concave points or indentations in a cell nucleus
symt	Symmetry of a cell nucleus
fracd	Fractal dimension (of the boundary) of a cell

malignant lumps from benign ones, they are invasive, time consuming, and costly.

A computer imaging system has recently been developed at the University of Wisconsin-Madison (Street, Wolberg, and Mangasarian, 1993; Mangasarian, Street, and Wolberg, 1995) with the goal of developing a procedure that diagnoses FNAs with very high accuracy. A small-gauge needle is used to extract a fluid sample (i.e., FNA) from a patient's breast lump or mass (detected by self-examination and/or mammography); the FNA is placed on a glass slide and stained to highlight the nuclei of the constituent cells; an image from the FNA is transferred to a workstation by a video camera mounted on a microscope; and the exact boundaries of the nuclei are determined.

Ten variables of the nucleus of each cell are computed from fluid samples. They are listed in Table 8.1. The variables are constructed so that larger values would typically indicate a higher likelihood of malignancy. For each image consisting of 10–40 nuclei, the mean value (mv), extreme value (i.e., largest or worst value, biggest size, most irregular shape) (ev), and standard deviation (sd) of each of these cellular features are computed, resulting in a total of 30 real-valued variables. The 30 variables are

(1) radius.mv, (2) texture.mv, (3) peri.mv, (4) area.mv, (5) smooth.mv, (6) comp.mv, (7) scav.mv, (8) ncav.mv, (9) symt.mv, (10) fracd.mv, (11) radius.sd, (12) texture.sd, (13) peri.sd, (14) area.sd, (15) smooth.sd, (16) comp.sd, (17) scav.sd, (18) ncav.sd, (19) symt.sd, (20) fracd.sd, (21) radius.ev, (22) texture.ev, (23) peri.ev, (24) area.ev, (25) smooth.ev, (26) comp.ev, (27) scav.ev, (28) ncav.ev, (29) symt.ev, (30) fracd.ev.

Because all 30 variables consist of nonnegative measurements with skewed histograms, we took natural logarithms of each variable before analyzing the data. Data values of zero were replaced by the value 0.001 prior to transforming. When we refer to variables in this example, we mean the transformed variables.

The data set[1] consists of 569 cases (images), of which 212 were diagnosed as malignant (confirmed by biopsy) and 357 as benign (confirmed by biopsy or by subsequent periodic medical examinations). Many pairs of the 30 variables are highly correlated; for example, 19 correlations are between 0.8 and 0.9, and 25 correlations are greater than 0.9 (six of which are greater than 0.99). The problem is how best to separate the malignant from the benign lumps (without performing surgery); a secondary problem is how to do this using as few variables as possible.

To discriminate between the benign and malignant lumps, a linear discriminant function (LDF) can be derived by estimating the coefficients for an optimal linear combination of the 30 input variables. From the resulting LDF, we compute a score for each of the 569 tumors, and we then separate the scores by group.

Histograms of the scores on the LDF for the benign (group 0) and malignant (group 1) tumors are displayed in the left panel of Figure 8.1, and kernel density estimates of the scores of the two groups (group 0 is the left curve and group 1 is the right curve) are displayed in the right panel. We can see a certain amount of overlap in the distribution of the LDF of the two groups, showing that perfect discrimination between benign and malignant tumors cannot be attained using the LDF with these data.

8.2 Classes and Features

We assume that the population $\mathcal{P}$ is partitioned into K unordered classes, groups, or subpopulations, which we denote by $\Pi_1, \Pi_2, \ldots, \Pi_K$. Each item in $\mathcal{P}$ is classified into one (and only one) of those classes. Measurements on a sample of items are to be used to help assign future unclassified items to one of the designated classes. The random r-vector $\mathbf{X}$, given by

$$\mathbf{X} = (X_1, \cdots, X_r)^\tau, \tag{8.1}$$

represents the r measurements on an item (i.e., $\mathbf{X} \in \Re^r$). The variables $X_1, X_2, \ldots, X_r$ are likely to be chosen because of their suspected ability

---

[1]The original data can be found in the file wdbc at the book's website and in the file breast-cancer-wisconsin/wdbc.data at the website http://www.ics.uci.edu/pub/machine-learning-databases/.

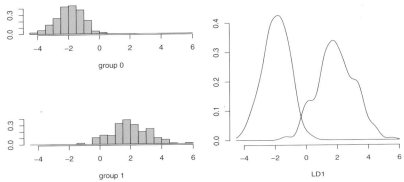

FIGURE 8.1. *Left panel: Histograms of the scores on the (first) linear discriminant function of the* wdbc *data set. Upper panel shows the histogram for the benign images (group 0) and the lower panel shows the histogram for the malignant images (group 1). Right panel: Kernel density estimates of the two sets of scores on the (first) linear discriminant function (LD1).*

to distinguish between the K classes. The variables in (8.1) are called *discriminating* or *feature variables*, and the vector $\mathbf{X}$ is the *feature vector*.

It may sometimes be appropriate to include in an analysis the additional classes of Π_D and Π_O to signify that decisions could not be made due to either an element of *doubt* in the assignment or indications that certain items constitute *outliers* and could not possibly belong to any of the designated classes.

8.3   Binary Classification

Consider, first, the *binary* classification problem $(K = 2)$ where we wish to discriminate between two classes Π_1 and Π_2, such as the "malignant" and "benign" tumors in the breast cancer example.

8.3.1   Bayes's Rule Classifier

Let

$$P(\mathbf{X} \in \Pi_i) = \pi_i, \quad i = 1, 2, \tag{8.2}$$

be the *prior probabilities* that a randomly selected observation $\mathbf{X} = \mathbf{x}$ belongs to either Π_1 or Π_2. Suppose also that the conditional multivariate probability density of $\mathbf{X}$ for the ith class is

$$P(\mathbf{X} = \mathbf{x} | \mathbf{X} \in \Pi_i) = f_i(\mathbf{x}), \quad i = 1, 2. \tag{8.3}$$

We note that there is no requirement that the $\{f_i(\cdot)\}$ be continuous; they could be discrete or be finite mixture distributions or even have singular covariance matrices. From (8.2) and (8.3), Bayes's theorem yields the *posterior probability*,

$$p(\Pi_i|\mathbf{x}) = P(\mathbf{X} \in \Pi_i|\mathbf{X} = \mathbf{x}) = \frac{f_i(\mathbf{x})\pi_i}{f_1(\mathbf{x})\pi_1 + f_2(\mathbf{x})\pi_2}, \tag{8.4}$$

that the observed $\mathbf{x}$ belongs to Π_i, $i = 1, 2$.

For a given $\mathbf{x}$, a reasonable classification strategy is to assign $\mathbf{x}$ to that class with the higher posterior probability. This strategy is called the *Bayes's rule classifier*. In other words, we assign $\mathbf{x}$ to Π_1 if

$$\frac{p(\Pi_1|\mathbf{x})}{p(\Pi_2|\mathbf{x})} > 1, \tag{8.5}$$

and we assign $\mathbf{x}$ to Π_2 otherwise. The ratio $p(\Pi_1|\mathbf{x})/p(\Pi_2|\mathbf{x})$ is referred to as the "odds-ratio" that Π_1 rather than Π_2 is the correct class given the information in $\mathbf{x}$. Substituting (8.4) into (8.5), the Bayes's rule classifier assigns $\mathbf{x}$ to Π_1 if

$$\frac{f_1(\mathbf{x})}{f_2(\mathbf{x})} > \frac{\pi_2}{\pi_1}, \tag{8.6}$$

and to Π_2 otherwise. On the boundary $\{\mathbf{x} \in R^r | f_1(\mathbf{x})/f_2(\mathbf{x}) = \pi_2/\pi_1\}$, we randomize (e.g., by tossing a fair coin) between assigning $\mathbf{x}$ to either Π_1 or Π_2.

8.3.2 Gaussian Linear Discriminant Analysis

We now make the Bayes's rule classifier more specific by following Fisher's (1936) assumption that both multivariate probability densities in (8.3) are multivariate Gaussian (see Section 3.3.2) having arbitrary mean vectors and a common covariance matrix. That is, we take $f_1(\cdot)$ to be a $\mathcal{N}_r(\boldsymbol{\mu}_1, \boldsymbol{\Sigma}_1)$ density and $f_2(\cdot)$ to be a $\mathcal{N}_r(\boldsymbol{\mu}_2, \boldsymbol{\Sigma}_2)$ density, and we make the homogeneity assumption that $\boldsymbol{\Sigma}_1 = \boldsymbol{\Sigma}_2 = \boldsymbol{\Sigma}_{XX}$.

The ratio of the two densities is given by

$$\frac{f_1(\mathbf{x})}{f_2(\mathbf{x})} = \frac{\exp\{-\frac{1}{2}(\mathbf{x} - \boldsymbol{\mu}_1)^\tau \boldsymbol{\Sigma}_{XX}^{-1}(\mathbf{x} - \boldsymbol{\mu}_1)\}}{\exp\{-\frac{1}{2}(\mathbf{x} - \boldsymbol{\mu}_2)^\tau \boldsymbol{\Sigma}_{XX}^{-1}(\mathbf{x} - \boldsymbol{\mu}_2)\}}, \tag{8.7}$$

where the normalization factors $(2\pi)^{-r/2}|\boldsymbol{\Sigma}_{XX}|^{-1/2}$ in both numerator and denominator cancel due to the equal covariance matrices of both classes. Taking logarithms (a monotonically increasing function) of (8.7), we have that

$$\log_e \frac{f_1(\mathbf{x})}{f_2(\mathbf{x})} = (\boldsymbol{\mu}_1 - \boldsymbol{\mu}_2)^\tau \boldsymbol{\Sigma}_{XX}^{-1} \mathbf{x} - \frac{1}{2}(\boldsymbol{\mu}_1 - \boldsymbol{\mu}_2)^\tau \boldsymbol{\Sigma}_{XX}^{-1}(\boldsymbol{\mu}_1 + \boldsymbol{\mu}_2) \tag{8.8}$$

$$= (\boldsymbol{\mu}_1 - \boldsymbol{\mu}_2)^\tau \boldsymbol{\Sigma}_{XX}^{-1}(\mathbf{x} - \bar{\boldsymbol{\mu}}), \tag{8.9}$$

where $\bar{\boldsymbol{\mu}} = (\boldsymbol{\mu}_1 + \boldsymbol{\mu}_2)/2$. The second term in the right-hand side of (8.8) can be written as

$$(\boldsymbol{\mu}_1 - \boldsymbol{\mu}_2)^\tau \boldsymbol{\Sigma}_{XX}^{-1}(\boldsymbol{\mu}_1 + \boldsymbol{\mu}_2) = \boldsymbol{\mu}_1^\tau \boldsymbol{\Sigma}_{XX}^{-1}\boldsymbol{\mu}_1 - \boldsymbol{\mu}_2^\tau \boldsymbol{\Sigma}_{XX}^{-1}\boldsymbol{\mu}_2. \qquad (8.10)$$

It follows that

$$L(\mathbf{x}) = \log_e \left\{ \frac{f_1(\mathbf{x})\pi_1}{f_2(\mathbf{x})\pi_2} \right\} = b_0 + \mathbf{b}^\tau \mathbf{x} \qquad (8.11)$$

is a linear function of $\mathbf{x}$, where

$$\mathbf{b} = \boldsymbol{\Sigma}_{XX}^{-1}(\boldsymbol{\mu}_1 - \boldsymbol{\mu}_2) \qquad (8.12)$$

$$b_0 = -\frac{1}{2}\{\boldsymbol{\mu}_1^\tau \boldsymbol{\Sigma}_{XX}^{-1}\boldsymbol{\mu}_1 - \boldsymbol{\mu}_2^\tau \boldsymbol{\Sigma}_{XX}^{-1}\boldsymbol{\mu}_2\} + \log_e(\pi_2/\pi_1). \qquad (8.13)$$

Thus, we assign $\mathbf{x}$ to Π_1 if the logarithm of the ratio of the two posterior probabilities is greater than zero; that is,

$$\text{if } L(\mathbf{x}) > 0, \text{ assign } \mathbf{x} \text{ to } \Pi_1. \qquad (8.14)$$

Otherwise, we assign $\mathbf{x}$ to Π_2. Note that on the boundary $\{\mathbf{x} \in R^r | L(\mathbf{x}) = 0\}$, the resulting equation is linear in $\mathbf{x}$ and, therefore, defines a hyperplane that divides the two classes. The rule (8.14) is generally referred to as *Gaussian linear discriminant analysis* (LDA).

The part of the function $L(\mathbf{x})$ in (8.11) that depends upon $\mathbf{x}$,

$$U = \mathbf{b}^\tau \mathbf{x} = (\boldsymbol{\mu}_1 - \boldsymbol{\mu}_2)^\tau \boldsymbol{\Sigma}_{XX}^{-1}\mathbf{x}, \qquad (8.15)$$

is known as *Fisher's linear discriminant function (LDF)*. Fisher actually derived the LDF using a nonparametric argument that involved no distributional assumptions. He looked for that linear combination, $\mathbf{a}^\tau \mathbf{X}$, of the feature vector $\mathbf{X}$ that separated the two classes as much as possible. In particular, he showed that $\mathbf{a} \propto \boldsymbol{\Sigma}_{XX}^{-1}(\boldsymbol{\mu}_1 - \boldsymbol{\mu}_2)$ maximized the squared difference of the two class means of $\mathbf{a}^\tau \mathbf{X}$ relative to the within-class variation of that difference (see Exercise 8.3).

Total Misclassification Probability

The LDF partitions the *feature space* $\Re^r$ into disjoint *classification regions* R_1 and R_2. If $\mathbf{x}$ falls into region R_1, it is classified as belonging to Π_1, whereas if $\mathbf{x}$ falls into region R_2, it is classified into Π_2. We now calculate the probability of misclassifying $\mathbf{x}$.

Misclassification occurs either if $\mathbf{x}$ is assigned to Π_2, but actually belongs to Π_1, or if $\mathbf{x}$ is assigned to Π_1, but actually belongs to Π_2. Define

$$\Delta^2 = (\boldsymbol{\mu}_1 - \boldsymbol{\mu}_2)^\tau \boldsymbol{\Sigma}_{XX}^{-1}(\boldsymbol{\mu}_1 - \boldsymbol{\mu}_2) \qquad (8.16)$$

to be the *squared Mahalanobis distance* between Π_1 and Π_2. Then,

$$E(U|\mathbf{X} \in \Pi_i) = \mathbf{b}^\tau \boldsymbol{\mu}_i = (\boldsymbol{\mu}_1 - \boldsymbol{\mu}_2)^\tau \boldsymbol{\Sigma}_{XX}^{-1} \boldsymbol{\mu}_i \qquad (8.17)$$

and

$$\text{var}(U|\mathbf{X} \in \Pi_i) = \mathbf{b}^\tau \boldsymbol{\Sigma}_{XX} \mathbf{b} = \Delta^2, \qquad (8.18)$$

for $i = 1, 2$. The *total misclassification probability* is, therefore,

$$P(\Delta) = P(\mathbf{X} \in R_2|\mathbf{X} \in \Pi_1)\pi_1 + P(\mathbf{X} \in R_1|\mathbf{X} \in \Pi_2)\pi_2, \qquad (8.19)$$

where

$$
\begin{aligned}
P(\mathbf{X} \in R_2|\mathbf{X} \in \Pi_1) &= P(L(\mathbf{X}) < 0|\mathbf{X} \in \Pi_1) \\
&= P\left(Z < -\frac{\Delta}{2} - \frac{1}{\Delta}\log_e \frac{\pi_2}{\pi_1}\right) \\
&= \Phi\left(-\frac{\Delta}{2} - \frac{1}{\Delta}\log_e \frac{\pi_2}{\pi_1}\right) \qquad (8.20)
\end{aligned}
$$

and

$$
\begin{aligned}
P(\mathbf{X} \in R_1|\mathbf{X} \in \Pi_2) &= P(L(\mathbf{X}) > 0|\mathbf{X} \in \Pi_2) \\
&= P\left(Z > \frac{\Delta}{2} - \frac{1}{\Delta}\log_e \frac{\pi_2}{\pi_1}\right) \\
&= \Phi\left(-\frac{\Delta}{2} + \frac{1}{\Delta}\log_e \frac{\pi_2}{\pi_1}\right). \qquad (8.21)
\end{aligned}
$$

In calculating these probabilities, we use the fact that $L(\mathbf{X}) = b_0 + U$, and then standardize U by setting

$$Z = \frac{U - E(U|\mathbf{X} \in \Pi_i)}{\sqrt{\text{var}(U|\mathbf{X} \in \Pi_i)}} \sim \mathcal{N}(0, 1).$$

In (8.20) and (8.21), $\Phi(\cdot)$ is the cumulative standard Gaussian distribution function. If $\pi_1 = \pi_2 = 1/2$, then

$$P(\mathbf{X} \in R_2|\mathbf{X} \in \Pi_1) = P(\mathbf{X} \in R_1|\mathbf{X} \in \Pi_2) = \Phi(-\Delta/2),$$

and, hence, $P(\Delta) = 2\Phi(-\Delta/2)$.

A graph of $P(\Delta)$ against Δ shows a downward-sloping curve, as one would expect; it has the value 1 when $\Delta = 0$ (i.e., the two populations are identical) and tends to zero as Δ increases. In other words, the greater the distance between the two population means, the less likely one is to misclassify $\mathbf{x}$.

Sampling Scenarios

Usually, the $2r + r(r+1)/2$ distinct parameters in $\boldsymbol{\mu}_1$, $\boldsymbol{\mu}_2$, and $\boldsymbol{\Sigma}_{XX}$ will be unknown, but can be estimated from learning data on $\mathbf{X}$. Assume, then, that we have available independent learning samples from the two classes Π_1 and Π_2. Let $\{\mathbf{X}_{1j}\}$ be a learning sample of size n_1 taken from Π_1 and let $\{\mathbf{X}_{2j}\}$ be a learning sample of size n_2 taken from Π_2.

The following different scenarios are possible when sampling from population $\mathcal{P}$:

1. *Conditional sampling*, where a sample of fixed size $n = n_1 + n_2$ is randomly selected from $\mathcal{P}$, and at a fixed $\mathbf{x}$ there are $n_i(\mathbf{x})$ observations from Π_i, $i = 1, 2$. This sampling scenario often appears in bioassays.

2. *Mixture sampling*, where a sample of fixed size $n = n_1 + n_2$ is randomly selected from $\mathcal{P}$ so that n_1 and n_2 are randomly selected. This is quite common in discrimination studies.

3. *Separate sampling*, where a sample of fixed size n_i is randomly selected from Π_i, $i = 1, 2$, and $n = n_1 + n_2$. Overall, this is the most popular scenario.

In all three cases, ML estimates of b_0 and $\mathbf{b}$ can be obtained (Anderson, 1982).

Sample Estimates

The ML estimates of $\boldsymbol{\mu}_i$, $i = 1, 2$, and $\boldsymbol{\Sigma}_{XX}$ are given by

$$\widehat{\boldsymbol{\mu}}_i = \bar{\mathbf{X}}_i = n_i^{-1} \sum_{j=1}^{n_i} \mathbf{X}_{ij}, \quad i = 1, 2, \tag{8.22}$$

$$\widehat{\boldsymbol{\Sigma}}_{XX} = n^{-1} \mathbf{S}_{XX}, \tag{8.23}$$

respectively, where

$$\mathbf{S}_{XX} = \mathbf{S}_{XX}^{(1)} + \mathbf{S}_{XX}^{(2)}, \tag{8.24}$$

and

$$\mathbf{S}_{XX}^{(i)} = \sum_{j=1}^{n_i} (\mathbf{X}_{ij} - \bar{\mathbf{X}}_i)(\mathbf{X}_{ij} - \bar{\mathbf{X}}_i)^{\tau}, \quad i = 1, 2, \tag{8.25}$$

where $n = n_1 + n_2$. If we wish to compute an unbiased estimator of $\boldsymbol{\Sigma}_{XX}$, we can divide $\mathbf{S}_{XX}$ in (8.24) by its degrees of freedom $n - 2 = n_1 + n_2 - 2$ (rather than by n) to make $\widehat{\boldsymbol{\Sigma}}_{XX}$.

The prior probabilities, π_1 and π_2, may be known or can be closely approximated in certain situations from past experience. If π_1 and π_2 are unknown, they can be estimated by

$$\widehat{\pi}_i = \frac{n_i}{n}, \quad i = 1, 2, \tag{8.26}$$

respectively. Substituting these estimates into $L(\mathbf{x})$ in (8.11) yields

$$\widehat{L}(\mathbf{x}) = \widehat{b}_0 + \widehat{\mathbf{b}}^{\tau} \mathbf{x}, \tag{8.27}$$

where

$$\widehat{\mathbf{b}} = \widehat{\boldsymbol{\Sigma}}_{XX}^{-1}(\bar{\mathbf{X}}_1 - \bar{\mathbf{X}}_2) \qquad (8.28)$$

$$\widehat{b}_0 = -\frac{1}{2}\{\bar{\mathbf{X}}_1^\tau \widehat{\boldsymbol{\Sigma}}_{XX}^{-1}\bar{\mathbf{X}}_1 - \bar{\mathbf{X}}_2^\tau \widehat{\boldsymbol{\Sigma}}_{XX}^{-1}\bar{\mathbf{X}}_2\} + \log_e \frac{n_1}{n} - \log_e \frac{n_2}{n} \qquad (8.29)$$

are the ML estimates of $\mathbf{b}$ and b_0, respectively. The classification rule assigns $\mathbf{x}$ to Π_1 if $\widehat{L}(\mathbf{x}) > 0$, and assigns $\mathbf{x}$ to Π_2 otherwise.

The second term of $\widehat{L}(\mathbf{x})$,

$$\widehat{\mathbf{b}}^\tau \mathbf{x} = (\bar{\mathbf{X}}_1 - \bar{\mathbf{X}}_2)^\tau \widehat{\boldsymbol{\Sigma}}_{XX}^{-1} \mathbf{x}, \qquad (8.30)$$

estimates Fisher's LDF. For large samples ($n_i \to \infty$, $i = 1, 2$), the distribution of $\widehat{\mathbf{b}}$ in (8.28) is Gaussian (Wald, 1944). This result allows us to study the separation of two given training samples, as well as the assumptions of normality and covariance matrix homogeneity, by drawing a histogram or normal probability plot of the LDF evaluated for every observation in the training samples. Nonparametric density estimates of the LDF scores for each class are especially useful in this regard; see, for example, Figure 8.1.

Example: Wisconsin Breast Cancer Data (Continued)

For the Wisconsin Diagnostic Breast Cancer Data, we estimate the priors π_1 and π_2 by $\widehat{\pi}_1 = n_1/n = 357/569 = 0.6274$ and $\widehat{\pi}_2 = n_2/n = 212/569 = 0.3726$, respectively. The coefficients of the LDF are estimated by first computing $\bar{\mathbf{X}}_1$, $\bar{\mathbf{X}}_2$, and the pooled covariance matrix $\widehat{\boldsymbol{\Sigma}}_{XX}$, and then using (8.28). The results are given in Table 8.2.

The leave-one-out cross-validation (CV/n) procedure drops one observation from the data set, reestimates the LDF from the remaining $n - 1$ observations, and then classifies the omitted observation; the procedure is repeated 569 times for each observation in the data set. The *confusion table* for classifying the 569 observations is given in Table 8.3. In this table, the row totals are the true classifications, and the column totals are the predicted classifications using Fisher's LDF and leave-one-out cross-validation.

From Table 8.3, we see that LDA leads to too many malignant tumors being misdiagnosed as "benign": of the 212 malignant tumors, 192 are correctly classified and 20 are not; and of the 357 benign tumors, 353 are correctly classified and 4 are not. The misclassification rate for Fisher's LDF in this example is, therefore, estimated by CV/n as $24/569 = 0.042$, or 4.2%.

For comparison, the apparent error rate (i.e., the error rate obtained by classifying each observation using the LDF, then dividing the number of misclassified observations by n) is given by $19/569 = 0.033$, or 3.3%, which is clearly an overly optimistic estimate of the LDF misclassification rate.

TABLE 8.2. *Estimated coefficients of Fisher's linear discriminant function for the Wisconsin diagnostic breast cancer data. All variables are logarithms of the original variables.*

Variable	Coeff.	Variable	Coeff.	Variable	Coeff.
radius.mv	−30.586	radius.sd	−2.630	radius.ev	6.283
texture.mv	−0.317	texture.sd	−0.602	texture.ev	2.313
peri.mv	35.215	peri.sd	0.262	peri.ev	−3.176
area.mv	−2.250	area.sd	−3.176	area.ev	−1.913
smooth.mv	0.327	smooth.sd	0.139	smooth.ev	1.540
comp.mv	−2.165	comp.sd	−0.398	comp.ev	0.528
scav.mv	1.371	scav.sd	0.047	scav.ev	−1.161
ncav.mv	0.509	ncav.sd	0.953	ncav.ev	−0.947
symt.mv	−1.223	symt.sd	−0.530	symt.ev	2.911
fracd.mv	−3.585	fracd.sd	−0.521	fracd.ev	4.168

8.3.3 LDA via Multiple Regression

The above results on LDA can also be obtained using multiple regression. We create an indicator variable Y showing which observations fall into which class, and then regress that Y on the feature vector $\mathbf{X}$. Let

$$Y = \begin{cases} y_1 & \text{if } \mathbf{X} \in \Pi_1 \\ y_2 & \text{if } \mathbf{X} \in \Pi_2 \end{cases} \tag{8.31}$$

be the *class labels* and let

$$\mathcal{Y} = (y_1 \mathbf{1}_{n_1}^\tau \;\vdots\; y_2 \mathbf{1}_{n_2}^\tau) \tag{8.32}$$

be the $(1 \times n)$ row vector whose components are the values of Y for all n observations. Let

$$\mathcal{X} = (\mathcal{X}_1 \;\vdots\; \mathcal{X}_2) \tag{8.33}$$

be an $(r \times n)$-matrix, where $\mathcal{X}_1$ is the $(r \times n_1)$-matrix of observations from Π_1 and $\mathcal{X}_2$ is the $(r \times n_2)$-matrix of observations from Π_2.

TABLE 8.3. *Confusion table for the Wisconsin Diagnostic Breast Cancer Data. Row totals are the true classifications and column totals are predicted classifications using leave-one-out cross-validation.*

	Predicted benign	Predicted malignant	Row total
True benign	353	4	357
True malignant	20	192	212
Column total	373	196	569

Let

$$\mathcal{X}_c = \mathcal{X} - \bar{\mathcal{X}} = \mathcal{X}\mathbf{H}_n \tag{8.34}$$

$$\mathcal{Y}_c = \mathcal{Y} - \bar{\mathcal{Y}} = \mathcal{Y}\mathbf{H}_n, \tag{8.35}$$

where $\mathbf{H}_n = \mathbf{I}_n - n^{-1}\mathbf{J}_n$ is the "centering matrix" and $\mathbf{J}_n = \mathbf{1}_n\mathbf{1}_n^\tau$ is an $(n \times n)$-matrix of ones.

If we regress the row vector $\mathcal{Y}_c$ on the matrix $\mathcal{X}_c$, the OLS estimator of the multiple regression coefficient vector $\boldsymbol{\beta}$ is given by

$$\widehat{\boldsymbol{\beta}}^\tau = \mathcal{Y}_c\mathcal{X}_c^\tau(\mathcal{X}_c\mathcal{X}_c^\tau)^{-1}. \tag{8.36}$$

We have the following cross-product matrices:

$$\mathcal{X}_c\mathcal{X}_c^\tau = \mathbf{S}_{XX} + k\mathbf{d}\mathbf{d}^\tau, \tag{8.37}$$

$$\mathcal{Y}_c\mathcal{X}_c^\tau = k(y_1 - y_2)\mathbf{d}^\tau, \tag{8.38}$$

$$\mathcal{Y}_c\mathcal{Y}_c^\tau = k(y_1 - y_2)^2, \tag{8.39}$$

where

$$\mathbf{d} = n_1^{-1}\mathcal{X}_1\mathbf{1}_{n_1} - n_2^{-1}\mathcal{X}_2\mathbf{1}_{n_2} = \bar{\mathbf{X}}_1 - \bar{\mathbf{X}}_2, \tag{8.40}$$

$$\mathbf{S}_{XX} = \mathcal{X}_1\mathbf{H}_{n_1}\mathcal{X}_1^\tau + \mathcal{X}_2\mathbf{H}_{n_2}\mathcal{X}_2^\tau, \tag{8.41}$$

and $k = n_1n_2/n$. See (8.23). Thus,

$$\begin{aligned}
\widehat{\boldsymbol{\beta}}^\tau &= k(y_1 - y_2)\mathbf{d}^\tau(\mathbf{S}_{XX} + k\mathbf{d}\mathbf{d}^\tau)^{-1} \\
&= k(y_1 - y_2)\mathbf{d}^\tau\mathbf{S}_{XX}^{-1}(\mathbf{I}_r + k\mathbf{d}\mathbf{d}^\tau\mathbf{S}_{XX}^{-1})^{-1}.
\end{aligned} \tag{8.42}$$

From the matrix result (3.4), setting $\mathbf{A} = \mathbf{I}_r$, $\mathbf{u} = k\mathbf{d}$, and $\mathbf{v}^\tau = \mathbf{d}^\tau\mathbf{S}_{XX}^{-1}$, we have that

$$\begin{aligned}
(\mathbf{I}_r + k\mathbf{d}\mathbf{d}^\tau\mathbf{S}_{XX}^{-1})^{-1} &= \mathbf{I}_r - \frac{k\mathbf{d}\mathbf{d}^\tau\mathbf{S}_{XX}^{-1}}{1 + k\mathbf{d}^\tau\mathbf{S}_{XX}^{-1}\mathbf{d}}. \\
&= \frac{\mathbf{I}_r}{1 + k\mathbf{d}^\tau\mathbf{S}_{XX}^{-1}\mathbf{d}},
\end{aligned}$$

whence,

$$\widehat{\boldsymbol{\beta}} = \left(\frac{k(y_1 - y_2)}{n - 2 + T^2}\right)\widehat{\boldsymbol{\Sigma}}_{XX}^{-1}\mathbf{d}, \tag{8.43}$$

where $\widehat{\boldsymbol{\Sigma}}_{XX} = \mathbf{S}_{XX}/(n-2)$ and

$$T^2 = k\mathbf{d}^\tau\widehat{\boldsymbol{\Sigma}}_{XX}^{-1}\mathbf{d} = \frac{n_1n_2}{n}(\bar{\mathbf{X}}_1 - \bar{\mathbf{X}}_2)^\tau\widehat{\boldsymbol{\Sigma}}_{XX}^{-1}(\bar{\mathbf{X}}_1 - \bar{\mathbf{X}}_2) \tag{8.44}$$

is *Hotelling's T^2 statistic*, which is used for testing the hypothesis that $\boldsymbol{\mu}_1 = \boldsymbol{\mu}_2$. Assuming multivariate normality,

$$\left(\frac{n - r - 1}{r(n - 2)}\right)T^2 \sim F_{r,n-r-1} \tag{8.45}$$

when this hypothesis is correct (see, e.g., Anderson, 1984, Section 5.3.4).

Note that $D^2 = \mathbf{d}^\tau \widehat{\boldsymbol{\Sigma}}_{XX}^{-1} \mathbf{d}$ is proportional to an estimate of Δ^2 (see (8.16)). From (8.28) and (8.43), it follows that

$$\widehat{\boldsymbol{\beta}} \propto \widehat{\boldsymbol{\Sigma}}_{XX}^{-1}(\bar{\mathbf{X}}_1 - \bar{\mathbf{X}}_2) = \widehat{\mathbf{b}}. \tag{8.46}$$

where the proportionality constant is $n_1 n_2 (y_1 - y_2)/n(n_1 + n_2 - 2 + T^2)$. This fact was first noted by Fisher (1936). Thus, we can obtain Fisher's estimated LDF (8.28) (up to a constant of proportionality) through multiple regression using an indicator response variable.

How should we choose the values y_1 and y_2? Four different choices are given in Table 8.4. In choosing the values of y_1 and y_2, researchers were initially concerned about ease of computation. The only part of $\widehat{\boldsymbol{\beta}}$ in (8.43) that depends upon y_1 and y_2 is $y_1 - y_2$. Thus, Fisher wanted $y_1 - y_2 = 1$ and $\bar{Y} = 0$; Bishop wanted $k(y_1 - y_2) = n$; Ripley wanted $\bar{Y} = 0$ and the total sum of squares $n_1 y_1^2 + n_2 y_2^2 = n$; and Lattin, Carroll, and Green wanted $\mathcal{Y}_c \mathcal{X}_c^\tau = \mathbf{d}^\tau$. With the public availability of high-speed computers, more simplistic choices are used, including $(y_1, y_2) = (1, 0)$ or $(1, -1)$. Fortunately, it does not matter which values of (y_1, y_2) we pick: these different choices of (y_1, y_2) yield $\widehat{\boldsymbol{\beta}}$s that are proportional to each other.

Example: Wisconsin Diagnostic Breast Cancer Data (Continued)

When we regress Y (1 if the patient's tumor is malignant and 0 otherwise) on each of the 30 (log-transformed) variables one at a time, all but four of the coefficients are declared to be significant. (A coefficient is "significant" at the 5% level if its absolute t-ratio is greater than the value 2.0 and is nonsignificant otherwise.)

At the other extreme, regressing Y on all 30 variables results in only eight significant coefficients. Table 8.5 gives the multiple regression of Y on the 30 (log-transformed) variables. The estimated coefficients in this table are proportional to those given in Table 8.2 for the LDF. The ordered magnitudes of the ratio of estimated coefficient to its estimated standard error for all 30 variables is displayed in Figure 8.2.

Such conflicting behavior is probably due to high pairwise correlations among the variables: 19 correlations are between 0.8 and 0.9, and 25 correlations are greater than 0.9 (six of which are greater than 0.99).

8.3.4 Variable Selection

High-dimensional data often contain pairs of highly correlated variables, which introduce collinearity into discrimination and classification problems. So, variable selection becomes a priority. The connection between Fisher's

TABLE 8.4. *Proposed values of (y_1, y_2) for LDA via multiple regression.*

Author(s)	(y_1, y_2)
Fisher (1936)	$(n_2/n, -n_1/n)$
Bishop (1995, p. 109)	$(n/n_1, -n/n_2)$
Ripley (1996, p. 102)	$\pm(-(n_2/n_1)^{1/2}, (n_1/n_2)^{1/2})$
Lattin et al (2003, p. 437)	$(1/n_1, -1/n_2)$

LDF and multiple regression provides us with a vehicle for selecting important discriminating variables. Thus, the variable selection techniques of FS and BE stepwise procedures, C_p, LARS, and Lasso can all be used in the discrimination context as well as in regression; see Exercise 8.10.

8.3.5 Logistic Discrimination

We see from (8.11) and the fact that $p(\Pi_2|\mathbf{x}) = 1 - p(\Pi_1|\mathbf{x})$ at $\mathbf{X} = \mathbf{x}$, that the posterior probability density satisfies

$$\text{logit } p(\Pi_1|\mathbf{x}) = \log_e\left(\frac{p(\Pi_1|\mathbf{x})}{1 - p(\Pi_1|\mathbf{x})}\right) = \beta_0 + \boldsymbol{\beta}^\tau\mathbf{x}, \qquad (8.47)$$

which has the form of a *logistic regression model*. The logistic approach to discrimination assumes that the log-likelihood ratio (8.11) can be modeled as a linear function of $\mathbf{x}$. Inverting the relationship (8.47), we have that

$$p(\Pi_1|\mathbf{x}) = \frac{e^{L(\mathbf{x})}}{1 + e^{L(\mathbf{x})}}, \qquad (8.48)$$

$$p(\Pi_2|\mathbf{x}) = \frac{1}{1 + e^{L(\mathbf{x})}}, \qquad (8.49)$$

where

$$L(\mathbf{x}) = \beta_0 + \boldsymbol{\beta}^\tau\mathbf{x}. \qquad (8.50)$$

We can write (8.48) as

$$p(\Pi_1|\mathbf{x}) = \frac{1}{1 + e^{-L(\mathbf{x})}} = \sigma(L(\mathbf{x})), \qquad (8.51)$$

say, where $\sigma(u) = 1/(1 + e^{-u})$ in (8.51) is a sigmoid function ("S-shaped") (see Figure 8.3), taking values of $u \in \mathcal{R}$ onto $(0, 1)$.

Maximum-Likelihood Estimation

In light of (8.50), we now write $p(\Pi_1|\mathbf{x})$ as $p_1(\mathbf{x}, \beta_0, \boldsymbol{\beta})$, and similarly for $p_2(\mathbf{x}, \beta_0, \boldsymbol{\beta})$. Thus, instead of first estimating $\boldsymbol{\mu}_1$, $\boldsymbol{\mu}_2$, and $\boldsymbol{\Sigma}_{XX}$ as we did

TABLE 8.5. *Multiple regression results for linear discriminant analysis on the Wisconsin diagnostic breast cancer data. All variables are logarithms of the original variables. Y is taken to be 1 if the patient's tumor is malignant and 0 if benign. Listed are the estimated regression coefficients, their respective estimated standard errors, and the Z-ratio of those two values. The multiple R^2 is 0.777 and the F-statistic is 62.43 on 30 and 538 degrees of freedom.*

	Coeff.	S.E.	Ratio
(Intercept)	-14.348	3.628	-3.955
radius.mv	-6.168	2.940	-2.098
texture.mv	-0.064	0.217	-0.294
peri.mv	7.102	2.385	2.978
area.mv	-0.454	1.654	-0.274
smooth.mv	0.066	0.233	0.284
comp.mv	-0.437	0.162	-2.690
scav.mv	0.277	0.104	2.669
ncav.mv	0.103	0.094	1.096
symt.mv	-0.247	0.167	-1.473
fracd.mv	-0.723	0.353	-2.047
radius.sd	-0.530	0.277	-1.915
texture.sd	-0.122	0.080	-1.527
peri.sd	0.053	0.131	0.405
area.sd	0.691	0.271	2.555
smooth.sd	0.028	0.074	0.377
comp.sd	-0.080	0.100	-0.800
scav.sd	0.010	0.096	0.100
ncav.sd	0.192	0.098	1.970
symt.sd	-0.107	0.085	-1.255
fracd.sd	-0.105	0.069	-1.516
radius.ev	1.267	1.922	0.659
texture.ev	0.467	0.283	1.647
peri.ev	-0.641	0.800	-0.801
area.ev	-0.386	1.012	-0.381
smooth.ev	0.311	0.259	1.200
comp.ev	0.106	0.173	0.617
scav.ev	-0.234	0.135	-1.730
ncav.ev	-0.191	0.126	-1.517
symt.ev	0.587	0.209	2.816
fracd.ev	0.841	0.255	3.292

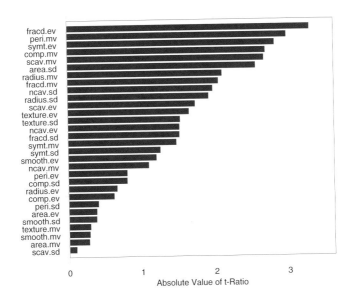

FIGURE 8.2. *Multiple regression results for linear discriminant analysis on the Wisconsin diagnostic breast cancer data. All input variables are logarithms of the original variables. Listed are the variable names on the vertical axis and the absolute value of the t-ratio for each variable on the horizontal axis. The variables are listed in descending order of their absolute t-ratios.*

in (8.24) and (8.25) in order to estimate β_0 and the coefficient vector $\boldsymbol{\beta}$, we can estimate β_0 and $\boldsymbol{\beta}$ directly through (8.47).

We define a response variable Y that identifies the population to which $\mathbf{X}$ belongs,

$$Y = \left\{ \begin{array}{ll} 1 & \text{if } \mathbf{X} \in \Pi_1 \\ 0 & \text{otherwise.} \end{array} \right. \tag{8.52}$$

The values of Y are the class labels. Conditional on $\mathbf{X}$, the Bernoulli random variable Y has $P(Y = 1) = \pi_1$ and $P(Y = 0) = 1 - \pi_1 = \pi_2$. Thus, we are interested in modeling binary data, and the usual way we do this is through *logistic regression*.

Given n observations, $(\mathbf{X}_i, Y_i), i = 1, 2, \ldots, n$, on $(\mathbf{X}, Y)$, the conditional likelihood for $(\beta_0, \boldsymbol{\beta})$ can be written as

$$\mathcal{L}(\beta_0, \boldsymbol{\beta}) = \prod_{i=1}^{n} (p_1(\mathbf{x}_i, \beta_0, \boldsymbol{\beta}))^{y_i} (1 - p_1(\mathbf{x}_i, \beta_0, \boldsymbol{\beta}))^{1-y_i}, \tag{8.53}$$

whence, the conditional log-likelihood is

$$\ell(\beta_0, \boldsymbol{\beta}) \quad = \quad \sum_{i=1}^{n} \{y_i \log_e p_1(\mathbf{x}_i, \beta_0, \boldsymbol{\beta}) + (1 - y_i) \log_e (1 - p_1(\mathbf{x}_i, \beta_0, \boldsymbol{\beta}))\}$$

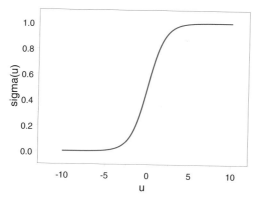

FIGURE 8.3. *Graph of* $\sigma(u) = 1/(1+e^{-u})$, *the logistic sigmoid activation function. For* $|u|$ *small,* $\sigma(u)$ *is very close to linear.*

$$= \sum_{i=1}^{n} \left\{ y_i(\beta_0 + \boldsymbol{\beta}^\tau \mathbf{x}_i) - \log_e(1 + e^{\beta_0 + \boldsymbol{\beta}^\tau \mathbf{x}_i}) \right\}. \qquad (8.54)$$

The ML estimates, $(\widetilde{\beta}_0, \widetilde{\boldsymbol{\beta}})$, of $(\beta_0, \boldsymbol{\beta})$ are obtained by maximizing $\ell(\beta_0, \boldsymbol{\beta})$ with respect to β_0 and $\boldsymbol{\beta}$. The maximization algorithm boils down to an iterative version of a weighted least-squares procedure in which the weights and the responses are updated at each iteration step. The details of the iteratively reweighted least-squares algorithm are given below.

The maximum-likelihood estimates $(\widetilde{\beta}_0, \widetilde{\boldsymbol{\beta}})$ can be plugged into (8.50) to give another estimate of the LDF,

$$\widetilde{L}(\mathbf{x}) = \widetilde{\beta}_0 + \widetilde{\boldsymbol{\beta}}^\tau \mathbf{x}. \qquad (8.55)$$

The classification rule,

$$\text{if } \widetilde{L}(\mathbf{x}) > 0, \text{ assign } \mathbf{x} \text{ to } \Pi_1, \qquad (8.56)$$

otherwise, assign $\mathbf{x}$ to Π_2, is referred to as *logistic discriminant analysis*. We note that maximizing (8.54) will not, in general, yield the same estimates for β_0 and $\boldsymbol{\beta}$ as we found in (8.28) and (8.29) for Fisher's LDF.

An equivalent classification procedure is to use $\widetilde{L}(\mathbf{x})$ in (8.55) to estimate the probability $p(\Pi_1|\mathbf{x})$ in (8.48). Substituting $\widetilde{L}(\mathbf{x})$ into (8.48) yields the estimate

$$\widetilde{p}(\Pi_1|\mathbf{x}) = \frac{e^{\widetilde{L}(\mathbf{x})}}{1 + e^{\widetilde{L}(\mathbf{x})}}, \qquad (8.57)$$

so that $\mathbf{x}$ is assigned to Π_1 if $\widetilde{p}(\Pi_1|\mathbf{x})$ is greater than some cutoff value, say 0.5, and $\mathbf{x}$ is assigned to Π_2 otherwise.

Iteratively Reweighted Least-Squares Algorithm

It will be convenient (temporarily) to redefine the r-vectors $\mathbf{x}_i$ and $\boldsymbol{\beta}$ as the following $(r+1)$-vectors: $\mathbf{x}_i \leftarrow (1, \mathbf{x}_i^\tau)^\tau$, and $\boldsymbol{\beta} \leftarrow (\beta_0, \boldsymbol{\beta}^\tau)^\tau$. Thus, $\beta_0 + \boldsymbol{\beta}^\tau \mathbf{x}_i$ can be written more compactly as $\boldsymbol{\beta}^\tau \mathbf{x}_i$. We also write $p_1(\mathbf{x}_i, \beta_0, \boldsymbol{\beta})$ as $p_1(\mathbf{x}_i, \boldsymbol{\beta})$ and $\ell(\beta_0, \boldsymbol{\beta})$ as $\ell(\boldsymbol{\beta})$.

Differentiating (8.54) and setting the derivatives equal to zero yields the *score equations*:

$$\dot{\ell}(\boldsymbol{\beta}) = \frac{\partial \ell(\boldsymbol{\beta})}{\partial \boldsymbol{\beta}} = \sum_{i=1}^{n} \mathbf{x}_i \{y_i - p_1(\mathbf{x}_i, \boldsymbol{\beta})\} = \mathbf{0}. \tag{8.58}$$

These are $r+1$ nonlinear equations in the $r+1$ logistic parameters $\boldsymbol{\beta}$. From (8.58), we see that $n_1 = \sum_{i=1}^{n} p_1(\mathbf{x}_i, \boldsymbol{\beta})$ and, hence, also that $n_2 = \sum_{i=1}^{n} p_2(\mathbf{x}_i, \boldsymbol{\beta})$.

The nonlinear equations (8.58) are solved using an algorithm known as *iteratively reweighted least-squares* (IRLS). The second derivatives of $\ell(\boldsymbol{\beta})$ are given by the $((r+1) \times (r+1))$ Hessian matrix:

$$\ddot{\ell}(\boldsymbol{\beta}) = \frac{\partial^2 \ell(\boldsymbol{\beta})}{\partial \boldsymbol{\beta} \partial \boldsymbol{\beta}^\tau} = -\sum_{i=1}^{n} \mathbf{x}_i \mathbf{x}_i^\tau p_1(\mathbf{x}_i, \boldsymbol{\beta})(1 - p_1(\mathbf{x}_i, \boldsymbol{\beta})). \tag{8.59}$$

The IRLS algorithm is based upon using the Newton–Raphson iterative approach to finding ML estimates. Starting values of $\widetilde{\boldsymbol{\beta}}^{(0)} = \mathbf{0}$ are recommended. Then, the $(k+1)$st step in the algorithm replaces the kth iterate $\widetilde{\boldsymbol{\beta}}^{(k)}$ by

$$\widetilde{\boldsymbol{\beta}}^{(k+1)} = \widetilde{\boldsymbol{\beta}}^{(k)} - (\ddot{\ell}(\boldsymbol{\beta}))^{-1} \dot{\ell}(\boldsymbol{\beta}), \tag{8.60}$$

where the derivatives are evaluated at $\widetilde{\boldsymbol{\beta}}^{(k)}$.

Using matrix notation, we set

$$\mathcal{X} = (\mathbf{X}_1, \cdots, \mathbf{X}_n), \quad \mathcal{Y} = (Y_1, \cdots, Y_n)^\tau,$$

to be an $((r+1) \times n)$ data matrix and n-vector, respectively, and let $\mathbf{W} = \mathrm{diag}\{w_i\}$ be an $(n \times n)$ diagonal weight-matrix with ith diagonal element

$$w_i = p_1(\mathbf{x}_i, \widetilde{\boldsymbol{\beta}})(1 - p_1(\mathbf{x}_i, \widetilde{\boldsymbol{\beta}})), \quad i = 1, 2, \ldots, n.$$

The score vector of first derivatives (8.58) and the Hessian matrix (8.59) can be written as

$$\dot{\ell}(\boldsymbol{\beta}) = \mathcal{X}(\mathcal{Y} - \mathbf{p}_1), \quad \ddot{\ell}(\boldsymbol{\beta}) = -\mathcal{X} \mathbf{W} \mathcal{X}^\tau, \tag{8.61}$$

respectively, where $\mathbf{p}_1$ is the n-vector

$$\mathbf{p}_1 = (p_1(\mathbf{x}_1, \widetilde{\boldsymbol{\beta}}), \cdots, p_1(\mathbf{x}_n, \widetilde{\boldsymbol{\beta}}))^{\tau}. \tag{8.62}$$

Then, (8.60) can be written as:

$$
\begin{aligned}
\widetilde{\boldsymbol{\beta}}^{(k+1)} &= \widetilde{\boldsymbol{\beta}}^{(k)} + (\mathcal{X}\mathbf{W}\mathcal{X}^{\tau})^{-1}\mathcal{X}(\mathbf{y} - \mathbf{p}_1) \\
&= (\mathcal{X}\mathbf{W}\mathcal{X}^{\tau})^{-1}\mathcal{X}\mathbf{W}\{\mathcal{X}^{\tau}\widetilde{\boldsymbol{\beta}}^{(k)} + \mathbf{W}^{-1}(\mathbf{y} - \mathbf{p}_1)\} \\
&= (\mathcal{X}\mathbf{W}\mathcal{X}^{\tau})^{-1}\mathcal{X}\mathbf{W}\mathbf{z},
\end{aligned}
\tag{8.63}
$$

where

$$\mathbf{z} = \mathcal{X}^{\tau}\widetilde{\boldsymbol{\beta}}^{(k)} + \mathbf{W}^{-1}(\mathbf{y} - \mathbf{p}_1) \tag{8.64}$$

is an n-vector. The ith element of $\mathbf{z}$ is given by

$$z_i = \mathbf{x}_i^{\tau}\widetilde{\boldsymbol{\beta}}^{(k)} + \frac{y_i - p_1(\mathbf{x}_i, \widetilde{\boldsymbol{\beta}}^{(k)})}{p_1(\mathbf{x}_i, \widetilde{\boldsymbol{\beta}}^{(k)})(1 - p_1(\mathbf{x}_i, \widetilde{\boldsymbol{\beta}}^{(k)}))}. \tag{8.65}$$

The update (8.63) has the form of a generalized least-squares estimator (see Exercise 5.17) with $\mathbf{W}$ as the diagonal matrix of weights, $\mathbf{z}$ as the response vector, and $\mathcal{X}$ as the data matrix. Note that $\mathbf{p}_1 = \mathbf{p}_1^{(k)}$, $\mathbf{z} = \mathbf{z}^{(k)}$, and $\mathbf{W} = \mathbf{W}^{(k)}$ have to be updated at every step in the algorithm because they each depend upon $\widetilde{\boldsymbol{\beta}}^{(k)}$. Furthermore, the update formula (8.63) assumes that the $((r + 1) \times (r + 1))$-matrix $\mathcal{X}\mathbf{W}\mathcal{X}^{\tau}$ can be inverted, a condition that will be violated in applications where $n < r + 1$.

Despite the fact that convergence of the IRLS algorithm to the maximum of $\ell(\boldsymbol{\beta})$ cannot be guaranteed, the algorithm does converge for most practical situations. We refer the reader to Thisted (1988, Section 4.5.6) for a detailed discussion of IRLS and its properties. The algorithm is used extensively in fitting generalized linear models (see, e.g., McCullagh and Nelder, 1989, Section 2.5).

Example: Wisconsin Diagnostic Breast Cancer Data (Continued)

Carrying out a logistic regression on all 30 transformed variables in the Wisconsin diagnostic breast cancer study results in huge values for both the estimated regression coefficients and their estimated standard errors. This, in turn, yields tiny values for all 30 t-ratios. This situation is caused by the high collinearity present in the data.

To reduce the number of variables, we apply BE stepwise regression to these data. Table 8.6 lists the parameter estimates and their estimated standard errors for a final model consisting of nine variables. Most of the pairwise correlations between these nine variables are quite moderate, with the only correlations greater than 0.8 being those of 26 (ncav.mv) with 29 (scav.ev) and 6 (comp.mv).

TABLE 8.6. *BE stepwise logistic regression results for the Wisconsin diagnostic breast cancer data.*

	Coeff.	S.E.	Ratio
(Intercept)	−66.251	19.504	−3.397
smooth.mv	15.179	7.469	2.032
comp.mv	−14.774	4.890	−3.022
ncav.mv	10.476	3.377	3.102
texture.sd	−6.963	2.304	−3.022
area.sd	12.943	3.070	4.216
fracd.sd	−5.476	1.754	−3.122
texture.ev	23.224	5.753	4.036
scav.ev	4.986	1.568	3.180
fracd.ev	17.166	5.912	2.904

8.3.6 Gaussian LDA or Logistic Discrimination?

Theoretical and empirical comparisons have been carried out between Gaussian LDA and logistic discriminant analysis. Some of the differences are the following:

1. The conditional log-likelihood (8.54) is valid under general exponential family assumptions on $f(\cdot)$ (which includes the multivariate Gaussian model with common covariance matrix). This suggests that logistic discrimination is more *robust* to nonnormality than Gaussian LDA.

2. Simulation studies have shown that when the Gaussian distributional assumptions or the common covariance matrix assumption are not satisfied, logistic discrimination performs much better.

3. Sensitivity to gross outliers can be a problem for Gaussian LDA, whereas outliers are reduced in importance in logistic discrimination, which essentially fits a sigmoidal function (rather than a linear function).

4. Logistic discriminant analysis is asymptotically less efficient than is Gaussian LDA because the latter is based upon full ML rather than conditional ML.

5. At the point when we would expect good discrimination to take place, logistic discrimination requires a much larger sample size than does Gaussian LDA to attain the same (asymptotic) error rate distribution (Efron, 1975), and this result extends to LDA using an exponential family with plug-in estimates.

8.3.7 Quadratic Discriminant Analysis

How is the classification rule (8.14) affected if the covariance matrices of the two Gaussian populations are not equal to each other? That is, if $\Sigma_1 \neq \Sigma_2$. In this case, (8.8) becomes

$$\log_e \frac{f_1(\mathbf{x})}{f_2(\mathbf{x})} =$$

$$c_0 - \frac{1}{2}\{(\mathbf{x} - \boldsymbol{\mu}_1)^\tau \Sigma_1^{-1}(\mathbf{x} - \boldsymbol{\mu}_1) - (\mathbf{x} - \boldsymbol{\mu}_2)^\tau \Sigma_2^{-1}(\mathbf{x} - \boldsymbol{\mu}_2)\} \quad (8.66)$$

$$= c_1 - \frac{1}{2}\mathbf{x}^\tau(\Sigma_1^{-1} - \Sigma_2^{-1})\mathbf{x} + (\boldsymbol{\mu}_1^\tau\Sigma_1^{-1} - \boldsymbol{\mu}_2^\tau\Sigma_2^{-1})\mathbf{x}, \quad (8.67)$$

where c_0 and c_1 are constants that depend only upon the parameters $\boldsymbol{\mu}_1$, $\boldsymbol{\mu}_2$, Σ_1, and Σ_2. The log-likelihood ratio (8.67) has the form of a quadratic function of $\mathbf{x}$. In this case, set

$$Q(\mathbf{x}) = \beta_0 + \boldsymbol{\beta}^\tau\mathbf{x} + \mathbf{x}^\tau\boldsymbol{\Omega}\mathbf{x}, \quad (8.68)$$

where

$$\boldsymbol{\Omega} = -\frac{1}{2}(\Sigma_1^{-1} - \Sigma_2^{-1}) \quad (8.69)$$

$$\boldsymbol{\beta} = \Sigma_1^{-1}\boldsymbol{\mu}_1 - \Sigma_2^{-1}\boldsymbol{\mu}_2 \quad (8.70)$$

$$\beta_0 = -\frac{1}{2}\left\{\log_e\frac{|\Sigma_1|}{|\Sigma_2|} + \boldsymbol{\mu}_1^\tau\Sigma_1^{-1}\boldsymbol{\mu}_1 - \boldsymbol{\mu}_2^\tau\Sigma_2^{-1}\boldsymbol{\mu}_2\right\} - \log_e(\pi_2/\pi_2). \quad (8.71)$$

Note that $\boldsymbol{\Omega}$ is an $(r \times r)$ symmetric matrix. The classification rule is to assign $\mathbf{x}$ to Π_1 if (8.67) is greater than $\log_e(\pi_2/\pi_1)$; that is,

$$\text{if } Q(\mathbf{x}) > 0, \text{ assign } \mathbf{x} \text{ to } \Pi_1, \quad (8.72)$$

and assign $\mathbf{x}$ to Π_2 otherwise.

The function $Q(\mathbf{x})$ of $\mathbf{x}$ is called a *quadratic discriminant function (QDF)* and the classification rule (8.72) is referred to as *quadratic discriminant analysis* (QDA). The boundary $\{\mathbf{x} \in R^r | Q(\mathbf{x}) = 0\}$ that divides the two classes is a quadratic function of $\mathbf{x}$.

An approximation to the boundaries obtained by QDA can be obtained using an LDA approach that enlists the aid of the linear terms, squared terms, and all pairwise products of the feature variables. For example, if we have two feature variables X_1 and X_2, then "quadratic LDA" would use X_1, X_2, X_1^2, X_2^2, and X_1X_2 in the linear discriminant function with $r = 5$.

Maximum-Likelihood Estimation

If the $r(r + 3)$ distinct parameters in $\boldsymbol{\mu}_1$, $\boldsymbol{\mu}_2$, Σ_1, and Σ_2 are all unknown, and π_1 and π_2 are also unknown (1 additional parameter), they can be estimated using learning samples as above, with the exception of the covariance matrices, where the ML estimator of $\boldsymbol{\Sigma}_i$ is

$$\widehat{\boldsymbol{\Sigma}}_i = n_i^{-1} \sum_{j=1}^{n_i} (\mathbf{X}_{ij} - \bar{\mathbf{X}}_i)(\mathbf{X}_{ij} - \bar{\mathbf{X}}_i)^\tau, \quad i = 1, 2. \qquad (8.73)$$

Substituting the obvious estimators into $Q(\mathbf{x})$ in (8.68) gives us

$$\widehat{Q}(\mathbf{x}) = \widehat{\beta}_0 + \widehat{\boldsymbol{\beta}}^\tau \mathbf{x} + \mathbf{x}^\tau \widehat{\boldsymbol{\Omega}} \mathbf{x}, \qquad (8.74)$$

where

$$\widehat{\boldsymbol{\Omega}} = -\frac{1}{2}(\widehat{\boldsymbol{\Sigma}}_1^{-1} - \widehat{\boldsymbol{\Sigma}}_2^{-1}), \qquad (8.75)$$

$$\widehat{\boldsymbol{\beta}} = \widehat{\boldsymbol{\Sigma}}_1^{-1} \bar{\mathbf{X}}_1 - \widehat{\boldsymbol{\Sigma}}_2^{-1} \bar{\mathbf{X}}_2 \qquad (8.76)$$

$$\widehat{\beta}_0 = -\widehat{c}_1 - \log_e \frac{n_2}{n} + \log_e \frac{n_1}{n}, \qquad (8.77)$$

and where $\widehat{c}_1$ is the estimated version of the first term in (8.67).

Because the classifier $\widehat{Q}(\mathbf{x})$ depends upon the inverses of both $\widehat{\boldsymbol{\Sigma}}_1$ and $\widehat{\boldsymbol{\Sigma}}_2$, it follows that if either n_1 or n_2 is smaller than r, then $\widehat{\boldsymbol{\Sigma}}_i$ ($i = 1$ or 2, as appropriate) will be singular and QDA will fail.

8.4   Examples of Binary Misclassification Rates

In this section, we compare the two-class discriminant analysis methods LDA and QDA on a number of well-known data sets.[2] These data sets, which are listed in Table 8.7, are

BUPA liver disorders These data are the results of blood tests considered to be sensitive to liver disorders arising from excessive alchohol consumption. The first five variables are all blood tests: mcv (mean corpuscular volume), alkphos (alkaline phosphotase), sgpt (alamine aminotransferase), sgot (aspartate aminotransferase), and gammagt (gamma-glutamyl transpeptidase); the sixth variable is drinks (number of half-pint equivalents of alchoholic beverages drunk per day). All patients are males: 145 subjects in class 1 and 200 in class 2.

Ionosphere These are radar data collected by a system of 16 high-frequency phased-array antennas in Goose Bay, Labrador, with a total transmitted power of the order 6.4 kilowatts. The targets were free electrons

---

[2]These data sets can be found in the files ionosphere, bupa, sonar, and spambase on the book's website. More details can be found in the UCI Machine Learning Repository at archive.ics.uci.edu/ml/datasets.html.

in the ionosphere. The two classes are "Good" for radar returns that show evidence of some type of structure in the ionosphere and "Bad" for those that do not and whose signals pass through the ionosphere. The received electromagnetic signals are complex-valued and were processed using an autocorrelation function whose arguments are the time of a pulse and the pulse number. There were 17 pulse numbers, which are described by two measurements per pulse number. One variable (#2) was removed from the data set because its value for all observations was zero.

Sonar Sonar signals are bounced off a metal cylinder (representing a mine) or a roughly cylindrical rock at various aspect angles and under various conditions. There are 111 observations obtained by bouncing sonar off a metal cylinder and 97 obtained from the rock. The transmitted sonar signal is a frequency-modulated chirp, rising in frequency. The data set contains signals ontained from a variety of aspect angles, spanning 90 degrees for the cylinder and 180 degrees for the rock. Each observation is a set of 60 numbers in the range 0–1, where each number represents the energy within a particular frequency band, integrated over a certain period of time.

Spambase This data set derives from a collection of spam e-mails (unsolicited commercial e-mail, which came from a postmaster and individuals who had filed spam) and non-spam e-mails (which came from filed work and personal e-mails). Most of the variables indicate whether a particular word or character was frequently occurring in the e-mail: 48 variables have the form "word_freq_WORD," that gives the percentage of the words in the e-mail which match WORD; 6 variables have the form "word_freq_CHAR," that gives the percentage of characters in the e-mail which match CHAR; and 3 "run-length" variables, measuring the average length, length of longest, and sum of length of uninterupted sequences of consecutive capital letters. There are 1813 spam (39.4%) and 2788 non-spam observations in the data set.

Table 8.7 lists the CV misclassification rates for LDA and QDA for each data set. These two-class data sets have quite varied CV misclassification rates and, in three out of the five data sets (the exceptions are the ionosphere and sonar data sets), LDA is a better classifier than QDA.

Figure 8.4 displays the kernel density estimates of the class-conditional scores of the linear discriminant function (LD1) for the binary classification data sets spambase, ionosphere, sonar, and bupa. These data sets are arranged in order of LDA misclassification rates, from smallest to largest. The less overlap between the two density estimates, the smaller the misclassification rate; the greater the overlap between the two density estimates, the larger the misclassification rate.

TABLE 8.7. *Summary of data sets with two classes. Listed are the sample size (n), number of variables (r), and number of classes (K). Also listed for each data set are leave-one-out cross-validation (CV/n) misclassification rates for linear discriminant analysis (LDA) and quadratic discriminant analysis (QDA). The data sets are listed in increasing order of LDA misclassification rates.*

Data Set	n	r	K	LDA	QDA
Breast cancer (wdbc)	569	30	2	0.042	0.062
Spambase	4601	57	2	0.113	0.170
Ionosphere	351	33	2	0.137	0.128
Sonar	208	60	2	0.245	0.240
BUPA liver disorders	345	6	2	0.301	0.406

8.5 Multiclass LDA

Assume now that the population of interest is divided into $K > 2$ nonoverlapping (disjoint) classes. For example, in a database made publicly available by the U.S. Postal Service, each item is a (16×16) pixel image of a digit extracted from a real-life zip code that is handwritten onto an envelope. The database consists of thousands of these handwritten digits, each of which is viewed as a point in an input space of 256 dimensions. The classification problem is to assign each digit to one of the 10 classes $0, 1, 2, \ldots, 9$.

We could carry out $\binom{K}{2}$ different two-class linear discriminant analyses, where we set up a sequence of "one class *versus* the rest" classification scenarios. Such a solution does not work because it would produce regions that do not belong to any of the K classes considered (see Exercise 8.13).

Instead, the two-class methodology carries over in a straightforward way to the multiclass situation. Specifically, we wish to partition the sample space into K nonoverlapping regions $R_1, R_2, \ldots, R_K$, such that an observation $\mathbf{x}$ is assigned to class Π_i if $\mathbf{x} \in R_i$. The partition is to be determined so that the total misclassification rate is a minimum.

Text Categorization

A note of caution is in order here: not all multiclass classification problems fit this description. Text categorization is an important example. At the simplest level of information processing, we save and categorize files, e-mail messages, and URLs; in more complicated activities, we assign news items, computer FAQs, security information, author identification, junk mail identification, and so on, to predefined categories. For example, about 810,000 documents of newswire stories in the Reuters Business Briefing database RCV1 (Lewis, Yang, Rose, and Li, 2004) are assigned by topic

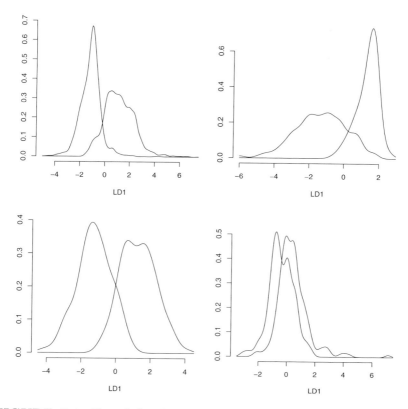

FIGURE 8.4. *Kernel density estimates of the class-conditional scores for the linear discriminant function (LD1) for the following two-class data sets:* spambase *(upper-left panel).* ionosphere *(upper-right panel).* sonar *(lower-left panel).* bupa *(lower-right panel). The amount of overlap in the density estimates is directly related to the estimated misclassification rate between the data in the two groups.*

into 103 categories. The classification problem is to assign each document to a topic based solely upon the textual content of that document (represented as a vector of words). Because documents can be assigned to more than one topic, text categorization does not fit the standard description of a classification problem.

8.5.1 Bayes's Rule Classifier

Let

$$\mathrm{Prob}(\mathbf{X} \in \Pi_i) = \pi_i, \quad i = 1, 2, \ldots, K, \tag{8.78}$$

be the prior probabilities of a randomly selected observation $\mathbf{X}$ belonging to each of the different classes in the population, and let

$$\text{Prob}(\mathbf{X} = \mathbf{x} | \mathbf{X} \in \Pi_i) = f_i(\mathbf{x}), \quad i = 1, 2, \ldots, K, \tag{8.79}$$

be the multivariate probability density for each class. The resulting posterior probability that an observed $\mathbf{x}$ belongs to the ith class is given by

$$p(\Pi_i | \mathbf{x}) = \text{Prob}(\mathbf{X} \in \Pi_i | \mathbf{X} = \mathbf{x}) = \frac{f_i(\mathbf{x})\pi_i}{\sum_{k=1}^{K} f_k(\mathbf{x})\pi_k}, \quad i = 1, 2, \ldots, K. \tag{8.80}$$

The *Bayes's rule classifier* for K classes assigns $\mathbf{x}$ to that class with the highest posterior probability. Because the denominator of (8.80) is the same for all $\Pi_i, i = 1, 2, \ldots, K$, we assign $\mathbf{x}$ to Π_i if

$$f_i(\mathbf{x})\pi_i = \max_{1 \leq j \leq K} f_j(\mathbf{x})\pi_j. \tag{8.81}$$

If the maximum in (8.81) does not uniquely define a class assignment for a given $\mathbf{x}$, then use a random assignment to break the tie between the appropriate classes.

Thus, $\mathbf{x}$ gets assigned to Π_i if $f_i(\mathbf{x})\pi_i > f_j(\mathbf{x})\pi_j$, for all $j \neq i$, or, equivalently, if $\log_e(f_i(\mathbf{x})\pi_i) > \log_e(f_j(\mathbf{x})\pi_j)$, for all $j \neq i$. The Bayes's rule classifier can be defined in an equivalent form by pairwise comparisons of posterior probabilities. We define the "log-odds" that $\mathbf{x}$ is assigned to Π_i rather than to Π_j as follows:

$$L_{ij}(\mathbf{x}) = \log_e \left\{ \frac{p(\Pi_i | \mathbf{x})}{p(\Pi_j | \mathbf{x})} \right\} = \log_e \left\{ \frac{f_i(\mathbf{x})\pi_i}{f_j(\mathbf{x})\pi_j} \right\}. \tag{8.82}$$

Then, we assign $\mathbf{x}$ to Π_i if $L_{ij}(\mathbf{x}) > 0$ for all $j \neq i$. We define *classification regions*, $R_1, R_2, \ldots, R_K$, as those areas of $\Re^r$ such that

$$\begin{aligned} R_i &= \{ \mathbf{x} \in \Re^r | L_{ij}(\mathbf{x}) > 0, j = 1, 2, \ldots, K, j \neq i \}, \\ &\qquad i = 1, 2, \ldots, K. \end{aligned} \tag{8.83}$$

This argument can be made more specific by assuming for the ith class Π_i that $f_i(\cdot)$ is the $\mathcal{N}_r(\boldsymbol{\mu}_i, \boldsymbol{\Sigma}_i)$ density, where $\boldsymbol{\mu}_i$ is an r-vector and $\boldsymbol{\Sigma}_i$ is an $(r \times r)$ covariance matrix, $i = 1, 2, \ldots, K$. We further assume that the covariance matrices for the K classes are identical, $\boldsymbol{\Sigma}_1 = \cdots = \boldsymbol{\Sigma}_K$, and equal to a common covariance matrix $\boldsymbol{\Sigma}_{XX}$.

Under these multivariate Gaussian assumptions, the log-odds of assigning $\mathbf{x}$ to Π_i (as opposed to Π_j) is a linear function of $\mathbf{x}$,

$$L_{ij}(\mathbf{x}) = b_{0ij} + \mathbf{b}_{ij}^\tau \mathbf{x}, \tag{8.84}$$

where

$$\mathbf{b}_{ij} = (\boldsymbol{\mu}_i - \boldsymbol{\mu}_j)^\tau \boldsymbol{\Sigma}_{XX}^{-1} \tag{8.85}$$

$$b_{0ij} = -\frac{1}{2}\{\boldsymbol{\mu}_i^\tau \boldsymbol{\Sigma}_{XX}^{-1} \boldsymbol{\mu}_i - \boldsymbol{\mu}_j^\tau \boldsymbol{\Sigma}_{XX}^{-1} \boldsymbol{\mu}_j\} + \log_e(\pi_i/\pi_j). \tag{8.86}$$

Because $L_{ij}(\mathbf{x})$ is linear in $\mathbf{x}$, the regions $\{R_i\}$ in (8.83) partition r-dimensional space by means of hyperplanes.

Maximum-Likelihood Estimates

Typically, the mean vectors and common covariance matrix will all be unknown. In that case, we estimate the $Kr + r(r+1)/2$ distinct parameters by taking learning samples from each of the K classes. Thus, from the ith class, we take n_i observations, $\mathbf{X}_{ij}$, $j = 1, 2, \ldots, n_i$, on the r-vector (8.1), that are then collected into the data matrix,

$$\overset{r \times n_i}{\mathcal{X}_i} = (\mathbf{X}_{i1}, \cdots, \mathbf{X}_{i,n_i}), \quad i = 1, 2, \ldots, K. \tag{8.87}$$

Let $n = \sum_{i=1}^{K} n_i$ be the total number of observations. The K data matrices (8.87) are then arranged into a single data matrix $\mathcal{X}$ which has the form

$$\begin{aligned} \overset{r \times n}{\mathcal{X}} &= (\overset{r \times n_1}{\mathcal{X}_1} \vdots \cdots \vdots \overset{r \times n_K}{\mathcal{X}_K}) \\ &= (\mathbf{X}_{11}, \cdots, \mathbf{X}_{1,n_1}, \cdots, \mathbf{X}_{K1}, \cdots, \mathbf{X}_{K,n_K}). \end{aligned} \tag{8.88}$$

The mean of each variable for the ith class is given by the r-vector,

$$\bar{\mathbf{X}}_i = n_i^{-1} \mathcal{X}_i \mathbf{1}_{n_i} = n_i^{-1} \sum_{j=1}^{n_i} \mathbf{X}_{ij} \quad i = 1, 2, \ldots, K, \tag{8.89}$$

and these K vectors are arranged into the matrix,

$$\overset{r \times n}{\bar{\mathcal{X}}} = (\underbrace{\bar{\mathbf{X}}_1, \cdots, \bar{\mathbf{X}}_1}_{n_1}, \cdots, \underbrace{\bar{\mathbf{X}}_K, \cdots, \bar{\mathbf{X}}_K}_{n_K}). \tag{8.90}$$

Let

$$\overset{r \times n}{\mathcal{X}_c} = \mathcal{X} - \bar{\mathcal{X}} = (\mathcal{X}_1 \mathbf{H}_{n_1} \vdots \cdots \vdots \mathcal{X} \mathbf{H}_{n_K}), \tag{8.91}$$

where $\mathbf{H}_{n_j}$ is the $(n_j \times n_j)$ "centering matrix." Then, we compute

$$\overset{r \times r}{\mathbf{S}_{XX}} = \mathcal{X}_c \mathcal{X}_c^\tau = \sum_{i=1}^{K} \sum_{j=1}^{n_i} (\mathbf{X}_{ij} - \bar{\mathbf{X}}_i)(\mathbf{X}_{ij} - \bar{\mathbf{X}}_i)^\tau. \tag{8.92}$$

Now, consider the following standard decomposition,

$$\mathbf{X}_{ij} - \bar{\mathbf{X}} = (\mathbf{X}_{ij} - \bar{\mathbf{X}}_i) + (\bar{\mathbf{X}}_i - \bar{\mathbf{X}}), \tag{8.93}$$

TABLE 8.8. *Multivariate decomposition of the total covariance matrix for* K *classes* $\Pi_1, \Pi_2, \ldots, \Pi_K$, *when a random learning sample of* n_i *observations is drawn from* Π_i, $i = 1, 2, \ldots, K$.

Source of Variation	df	Sum of Squares Matrix
Between classes	$K - 1$	$\mathbf{S}_{\mathrm{B}} = \sum_{i=1}^{K} n_i (\bar{\mathbf{X}}_i - \bar{\mathbf{X}})(\bar{\mathbf{X}}_i - \bar{\mathbf{X}})^\tau$
Within classes	$n - K$	$\mathbf{S}_{\mathrm{W}} = \sum_{i=1}^{K} \sum_{j=1}^{n_i} (\mathbf{X}_{ij} - \bar{\mathbf{X}}_i)(\mathbf{X}_{ij} - \bar{\mathbf{X}}_i)^\tau$
Total	$n - 1$	$\mathbf{S}_{\mathrm{tot}} = \sum_{i=1}^{K} \sum_{j=1}^{n_i} (\mathbf{X}_{ij} - \bar{\mathbf{X}})(\mathbf{X}_{ij} - \bar{\mathbf{X}})^\tau$

for the jth observation within the ith class, where

$$
\bar{\mathbf{X}} = n^{-1} \mathcal{X} \mathbf{1}_n = n^{-1} \sum_{i=1}^{K} \sum_{j=1}^{n_i} \mathbf{X}_{ij} = (\bar{X}_1, \cdots, \bar{X}_r)^\tau \qquad (8.94)
$$

is the overall mean vector ignoring class identifiers. Postmultiplying each side of (8.93) by their respective transposes, multiplying out the right-hand side, then summing over all n observations, and noting that the cross-product term vanishes (see Exercise 8.3), we arrive at the well-known multivariate analysis of variance (MANOVA) identity,

$$
\mathbf{S}_{\mathrm{tot}} = \mathbf{S}_{\mathrm{B}} + \mathbf{S}_{\mathrm{W}}, \qquad (8.95)
$$

where $\mathbf{S}_{\mathrm{tot}}$, $\mathbf{S}_{\mathrm{B}}$, and $\mathbf{S}_{\mathrm{tot}}$ are given in Table 8.8.

Thus, the total covariance matrix of the observations, $\mathbf{S}_{\mathrm{tot}}$, having $n - 1$ degrees of freedom and calculated by ignoring class identity, is partitioned into a part representing the *between-class covariance matrix*, $\mathbf{S}_B$, having $K - 1$ degrees of freedom, and another part representing the *pooled within-class covariance matrix*, $\mathbf{S}_W (= \mathbf{S}_{XX})$, having $n - K$ degrees of freedom. An unbiased estimator of the common covariance matrix, $\mathbf{\Sigma}_{XX}$, of the K classes is, therefore, given by

$$
\widehat{\mathbf{\Sigma}}_{XX} = (n - K)^{-1} \mathbf{S}_W = (n - K)^{-1} \mathbf{S}_{XX}. \qquad (8.96)
$$

If we let $f_i(\mathbf{x}) = f_i(\mathbf{x}, \boldsymbol{\eta}_i)$, where $\boldsymbol{\eta}_i$ is an r-vector of unknown parameters, and assume that the $\{\pi_i\}$ are known, the posterior probabilities (8.80) are estimated by

$$\widehat{p}(\Pi_i|\mathbf{x}) = \frac{f_i(\mathbf{x},\widehat{\boldsymbol{\eta}}_i)\pi_i}{\sum_{j=1}^{K} f_j(\mathbf{x},\widehat{\boldsymbol{\eta}}_j)\pi_j}, \quad i = 1, 2, \ldots, K, \qquad (8.97)$$

where $\widehat{\boldsymbol{\eta}}_i$ is an estimate of $\boldsymbol{\eta}_i$. The classification rule, therefore, assigns $\mathbf{x}$ to Π_i if

$$f_i(\mathbf{x},\widehat{\boldsymbol{\eta}}_i)\pi_i = \max_{1 \leq j \leq K} f_j(\mathbf{x},\widehat{\boldsymbol{\eta}}_j)\pi_j , \qquad (8.98)$$

which is often referred to as the *plug-in classifier*.

If the $\{f_i(\cdot)\}$ are multivariate Gaussian densities and $\boldsymbol{\eta}_i = (\boldsymbol{\mu}_i, \boldsymbol{\Sigma}_{XX})$, then, the sample version of $L_{ij}(\mathbf{x})$ is given by

$$\widehat{L}_{ij}(\mathbf{x}) = \widehat{b}_{0ij} + \widehat{\mathbf{b}}_{ij}^{\tau}\mathbf{x}, \qquad (8.99)$$

where

$$\widehat{\mathbf{b}}_{ij} = (\bar{\mathbf{X}}_i - \bar{\mathbf{X}}_j)^{\tau}\widehat{\boldsymbol{\Sigma}}_{XX}^{-1} \qquad (8.100)$$

$$\widehat{b}_{0ij} = -\frac{1}{2}\{\bar{\mathbf{X}}_i^{\tau}\widehat{\boldsymbol{\Sigma}}_{XX}^{-1}\bar{\mathbf{X}}_i - \bar{\mathbf{X}}_j^{\tau}\widehat{\boldsymbol{\Sigma}}_{XX}^{-1}\bar{\mathbf{X}}_j\} + \log_e\left\{\frac{n_i}{n}\right\} - \log_e\left\{\frac{n_j}{n}\right\}, \quad (8.101)$$

where we have estimated the prior π_i by the proportionality estimate, $\widehat{\pi}_i = n_i/n$, $i = 1, 2, \ldots, K$. The classification rule reduces to:

$$\text{Assign } \mathbf{x} \text{ to } \Pi_i \text{ if } \widehat{L}_{ij}(\mathbf{x}) > 0, \ j = 1, 2, \ldots, K, \ j \neq i. \qquad (8.102)$$

In other words, we assign $\mathbf{x}$ to that class Π_i with the largest value of $\widehat{L}_{ij}(\mathbf{x})$.

In the event that the covariance matrices cannot be assumed to be equal, estimates of the mean vectors are obtained using (8.89), and the ith class covariance matrix, $\boldsymbol{\Sigma}_i$, is estimated by its maximum-likelihood estimate,

$$\widehat{\boldsymbol{\Sigma}}_i = n_i^{-1} \sum_{j=1}^{n_i} (\mathbf{X}_{ij} - \bar{\mathbf{X}}_i)(\mathbf{X}_{ij} - \bar{\mathbf{X}}_i)^{\tau}, \quad i = 1, 2, \ldots, K. \qquad (8.103)$$

There are $Kr + Kr(r+1)/2$ distinct parameters that have to be estimated, and, if r is large, this is a huge increase over carrying out LDA. The resulting quadratic discriminant analysis (QDA) is similar to that of the two-class case if we make our decisions based upon comparisons of $\log_e f_i(\mathbf{x})$, $i = 1, 2, \ldots, K - 1$, with $\log_e f_K(\mathbf{x})$, say.

8.5.2 *Multiclass Logistic Discrimination*

The logistic discrimination method extends to the case of more than two classes. Setting $u_i = \log_e\{f_i(\mathbf{x})\pi_i\}$, we can express (8.80) in the form

$$p(\Pi_i|\mathbf{x}) = \frac{e^{u_i}}{\sum_{k=1}^{K} e^{u_k}} = \sigma_i, \qquad (8.104)$$

say. In the statistical literature, (8.104) is known as a *multiple logistic model*, whereas in the neural network literature, it is known as a *normalized exponential* (or *softmax*) *activation function*. Because we can write

$$\sigma_i = \frac{1}{1 + e^{-w_i}}, \tag{8.105}$$

where $w_i = u_i - \log\{\sum_{k \neq i} e^{u_k}\}$, σ_i is a generalization of the logistic sigmoid activation function (Figure 8.2).

Suppose we arbitrarily designate the last class (Π_K) to be a reference class and assume Gaussian distributions with common covariance matrices. Then, we define

$$L_i(\mathbf{x}) = u_i - u_K = b_{0i} + \mathbf{b}_i^\tau \mathbf{x}, \tag{8.106}$$

where

$$\mathbf{b}_i = (\boldsymbol{\mu}_i - \boldsymbol{\mu}_K)^\tau \boldsymbol{\Sigma}_{XX}^{-1} \tag{8.107}$$

$$b_{0i} = -\frac{1}{2}\{\boldsymbol{\mu}_i^\tau \boldsymbol{\Sigma}_{XX}^{-1} \boldsymbol{\mu}_i - \boldsymbol{\mu}_K^\tau \boldsymbol{\Sigma}_{XX}^{-1} \boldsymbol{\mu}_K\} + \log_e\{\pi_i/\pi_K\}. \tag{8.108}$$

If we divide the numerator and denominator of (8.104) by e^{u_K} and use (8.106), the posterior probabilities can be written as

$$p(\Pi_i|\mathbf{x}) = \frac{e^{L_i(\mathbf{x})}}{1 + \sum_{k=1}^{K-1} e^{L_k(\mathbf{x})}}, \quad i = 1, 2, \ldots, K-1, \tag{8.109}$$

$$p(\Pi_K|\mathbf{x}) = \frac{1}{1 + \sum_{k=1}^{K-1} e^{L_k(\mathbf{x})}} \tag{8.110}$$

If we write $f_i(\mathbf{x}) = f_i(\mathbf{x}, \boldsymbol{\eta}_i)$, where $\boldsymbol{\eta}_i$ is an r-vector of unknown parameters, then we estimate $\boldsymbol{\eta}_i$ by $\widehat{\boldsymbol{\eta}}_i$ and $f_i(\mathbf{x})$ by $\widehat{f}_i(\mathbf{x}) = f_i(\mathbf{x}, \widehat{\boldsymbol{\eta}}_i)$. As before, we assign $\mathbf{x}$ to that class that maximizes $f_i(\mathbf{x}, \widehat{\boldsymbol{\eta}}_i)$, $i = 1, 2, \ldots, K$. This classification rule is known as *multiple logistic discrimination*.

8.5.3 *LDA via Reduced-Rank Regression*

We now generalize to the multiclass case the idea for the two-class case ($K = 2$), in which we showed that the LDF can be obtained (up to a proportionality constant) by using multiple regression with a single indicator variable as the response variable.

In the multiclass case, we take the response variables to be a set of distinct indicator variables whose number is one fewer than the number of classes. If we know which observations fall into the first $K - 1$ classes, then the remaining observations automatically fall into the Kth class, and so we do not need an additional indicator variable to document that fact. The observations in the Kth class are instead each specified by a zero variable.

Some have used the Kth class (which could actually be any class, not just the last one) as a reference class to which all other classes may be compared.

As in the two-class case, the indicator variables are taken to be response variables. We now show that multiclass LDA is a special case of canonical variate analysis, which, as we saw in Chapter 7, is itself a special case of multivariate reduced-rank regression. It is for this reason that many authors refer to LDA as canonical variate analysis.

Identifying Classes Using Indicator Variables

In the following development, we set $K = s + 1$, where s is to be the number of output variables. Each observation in (8.88) is associated with its corresponding class by defining an *indicator response s-vector* $\mathbf{Y}_{ij}$, which has a 1 in the ith position if the jth observation r-vector, $\mathbf{X}_{ij}$, comes from Π_i, and zeroes in all other positions, $j = 1, 2, \ldots, n_i$, $i = 1, 2, \ldots, s + 1$. In other words, if $\mathbf{Y}_{ij} = (Y_{ijk})$, then, $Y_{ijk} = 1$ if $k = i$ and $Y_{ijk} = 0$ otherwise.

For the ith class Π_i, we have the matrix,

$$
\overset{s \times n_i}{\mathcal{Y}_i} = (\mathbf{Y}_{i1}, \ldots, \mathbf{Y}_{i,n_i}) = \begin{pmatrix} 0 & \cdots & 0 \\ \vdots & & \vdots \\ 1 & \cdots & 1 \\ \vdots & & \vdots \\ 0 & \cdots & 0 \end{pmatrix}, \tag{8.111}
$$

in which all n_i columns are identical, $i = 1, 2, \ldots, s+1$. Thus, the *indicator response matrix* $\mathcal{Y}$ is given by

$$
\begin{aligned}
\overset{s \times n}{\mathcal{Y}} &= (\overset{s \times n_1}{\mathcal{Y}_1} \vdots \cdots \vdots \overset{s \times n_{s+1}}{\mathcal{Y}_{s+1}}) \\
&= (\mathbf{Y}_{11}, \ldots, \mathbf{Y}_{1,n_1}, \ldots, \mathbf{Y}_{s+1,1}, \ldots, \mathbf{Y}_{s+1,n_{s+1}}) \\
&= \begin{pmatrix} 1 & \cdots & 1 & \cdots & 0 & \cdots & 0 & 0 & \cdots & 0 \\ \vdots & & \vdots & & \vdots & & \vdots & \vdots & & \vdots \\ 0 & \cdots & 0 & \cdots & 1 & \cdots & 1 & 0 & \cdots & 0 \end{pmatrix}. \tag{8.112}
\end{aligned}
$$

Each column of $\mathcal{Y}$ has a single 1 with the exception of the last set of n_{s+1} columns, whose every entry is equal to zero.

The s-vector of row means of $\mathcal{Y}$ is given by

$$
\bar{\mathbf{Y}} = n^{-1} \mathcal{Y} \mathbf{1}_n = (n_1/n, \cdots, n_s/n)^\tau. \tag{8.113}
$$

The ith component of $\bar{\mathbf{Y}}$ estimates the *prior probability*, π_i, that a randomly selected observation belongs to Π_i; that is, $\hat{\pi}_i = n_i/n$, $i = 1, 2, \ldots, s$, and $\hat{\pi}_{s+1} = n_{s+1}/n$. Let

$$
\overset{s \times n}{\bar{\mathcal{Y}}} = (\bar{\mathbf{Y}}, \ldots, \bar{\mathbf{Y}}) \tag{8.114}
$$

denote the matrix whose columns are n copies of the s-vector (8.113), and let

$$\overset{s \times n}{\mathcal{Y}_c} = \mathcal{Y} - \bar{\mathcal{Y}} = \mathcal{Y}\mathbf{H}_n, \tag{8.115}$$

where $\mathbf{H}_n$ is the $(n \times n)$ centering matrix. Then, the entries of $\mathcal{Y}_c$ are either $1 - (n_i/n)$ or $-n_i/n$. The cross-product matrix

$$\overset{s \times s}{\mathbf{S}_{YY}} = \mathcal{Y}_c\mathcal{Y}_c^\tau = \operatorname{diag}\{n_1, \ldots, n_s\} - n\bar{\mathbf{Y}}\bar{\mathbf{Y}}^\tau \tag{8.116}$$

has ith diagonal entry $n_i(1 - n_i/n)$ and off-diagonal entry $-n_i n_{i'}/n$ for the ith row and i'th column, $i \neq i'$, $i, i' = 1, 2, \ldots, s$. We invert $\mathbf{S}_{YY}$ to get

$$\mathbf{S}_{YY}^{-1} = \operatorname{diag}\{n_1^{-1}, \ldots, n_s^{-1}\} + n_s^{-1}\mathbf{J}_s, \tag{8.117}$$

where $\mathbf{J}_s = \mathbf{1}_s\mathbf{1}_s^\tau$ is an $(s \times s)$-matrix of 1s.

Generating Canonical Variates

We now have all the ingredients to carry out a canonical variate analysis of $\mathcal{X}$ and $\mathcal{Y}$. The central computation involves the eigenvalues and associated eigenvectors $(\widehat{\lambda}_j, \widehat{\mathbf{v}}_j)$, $j = 1, 2, \ldots, s$, of the matrix,

$$\overset{s \times s}{\widehat{\mathbf{R}}} = \mathbf{S}_{YY}^{-1/2}\mathbf{S}_{YX}\mathbf{S}_{XX}^{-1}\mathbf{S}_{XY}\mathbf{S}_{YY}^{-1/2}, \tag{8.118}$$

where

$$\overset{s \times r}{\mathbf{S}_{XY}} = \mathcal{X}_c\mathcal{Y}_c^\tau = (n_1(\bar{\mathbf{X}}_1 - \bar{\mathbf{X}}), \cdots, n_s(\bar{\mathbf{X}}_s - \bar{\mathbf{X}})) = \mathbf{S}_{YX}^\tau. \tag{8.119}$$

We recall the following fact from Section 7.3. The jth largest eigenvalue, $\widehat{\lambda}_j^*$, and associated eigenvector, $\widehat{\mathbf{v}}_j^*$, of the matrix

$$\overset{r \times r}{\widehat{\mathbf{R}}^*} = \mathbf{S}_{XX}^{-1/2}\mathbf{S}_{XY}\mathbf{S}_{YY}^{-1}\mathbf{S}_{YX}\mathbf{S}_{XX}^{-1/2} \tag{8.120}$$

are related to those of $\widehat{\mathbf{R}}$ by

$$\widehat{\lambda}_j = \widehat{\lambda}_j^*, \tag{8.121}$$

$$\widehat{\mathbf{v}}_j = \mathbf{S}_{YY}^{-1/2}\mathbf{S}_{YX}\mathbf{S}_{XX}^{-1/2}\widehat{\mathbf{v}}_j^*, \tag{8.122}$$

$j = 1, 2, \ldots, \min(r, s)$. Notice that $\widehat{\mathbf{R}}^*$ depends upon $\mathcal{Y}_c$ through the projection matrix

$$\overset{n \times n}{\mathbf{P}_\mathbf{y}} = \mathcal{Y}_c^\tau\mathbf{S}_{YY}^{-1}\mathcal{Y}_c \tag{8.123}$$

onto the columns of $\mathcal{Y}_c$. So, for any set of vectors that spans $\mathcal{Y}_c$, $\widehat{\mathbf{R}}^*$ will be unchanged.

We rescale $\widehat{\mathbf{v}}_j^*$ by setting

$$\boldsymbol{\gamma}_j = \mathbf{S}_{XX}^{-1/2}\widehat{\mathbf{v}}_j^* \tag{8.124}$$

$$= \widehat{\lambda}_j^{-1}\mathbf{S}_{XX}^{-1}\mathbf{S}_{XY}\mathbf{S}_{YY}^{-1/2}\widehat{\mathbf{v}}_j, \tag{8.125}$$

$j = 1, 2, \ldots, \min(r, s)$. From (8.122) and (8.125), we have that the $(r \times r)$-matrix $\mathbf{S}_B$ in Table 8.5 can be more easily expressed as

$$\overset{r \times r}{\mathbf{S}_B} = \mathbf{S}_{XY}\mathbf{S}_{YY}^{-1}\mathbf{S}_{YX} \tag{8.126}$$

(see Exercise 8.4). Writing out the jth eigenequation $\widehat{\mathbf{R}}\widehat{\mathbf{v}}_j = \widehat{\lambda}_j\widehat{\mathbf{v}}_j$, premultiplying both sides by $\mathbf{S}_{XX}^{-1/2}\mathbf{S}_{XY}\mathbf{S}_{YY}^{-1/2}$, and then using (8.126), we obtain

$$\mathbf{S}_B\boldsymbol{\gamma}_j = \widehat{\lambda}_j(\mathbf{S}_B + \mathbf{S}_W)\boldsymbol{\gamma}_j, \tag{8.127}$$

which shows that $\boldsymbol{\gamma}_j$ is the eigenvector associated with the jth largest eigenvalue $\widehat{\lambda}_j$ of the $(r \times r)$-matrix $(\mathbf{S}_B + \mathbf{S}_W)^{-1}\mathbf{S}_B$. Rearranging (8.127), we have that

$$\mathbf{S}_B\boldsymbol{\gamma}_j = \mu_j\mathbf{S}_W\boldsymbol{\gamma}_j, \tag{8.128}$$

where

$$\mu_j = \frac{\widehat{\lambda}_j}{1 - \widehat{\lambda}_j}, \quad j = 1, 2, \ldots, \min(r, s) . \tag{8.129}$$

In other words, the eigenvalues and eigenvectors of $\widehat{\mathbf{R}}$ are equivalent to the eigenvalues and eigenvectors of $\mathbf{S}_W^{-1}\mathbf{S}_B$ (or of its symmetric version $\mathbf{S}_W^{-1/2}\mathbf{S}_B\mathbf{S}_W^{-1/2}$). In general, the $(s \times r)$-matrix $\mathbf{S}_W^{-1}\mathbf{S}_B$ has $\min(r, s) = \min(r, K - 1)$ nonzero eigenvalues. If $K \leq r$, then $\mathbf{S}_B$ will not have full rank, resulting in $r - s = r - K + 1$ zero eigenvalues.

From (7.72) and (7.73), we set

$$\widehat{\mathbf{g}}_j^\tau = \widehat{\mathbf{v}}_j^\tau\mathbf{S}_{YY}^{-1/2}\mathbf{S}_{YX}\mathbf{S}_{XX}^{-1}, \tag{8.130}$$

$$\widehat{\mathbf{h}}_j^\tau = \widehat{\mathbf{v}}_j^\tau\mathbf{S}_{YY}^{-1/2}, \tag{8.131}$$

$j = 1, 2, \ldots, t$. Then, from (7.69), we calculate the jth pair of canonical variates $(\widehat{\xi}_j, \widehat{\omega}_j)$, where

$$\widehat{\xi}_j = \widehat{\mathbf{g}}_j^\tau\mathbf{X}_c = \boldsymbol{\gamma}_j^\tau\mathbf{X}_c, \tag{8.132}$$

$$\widehat{\omega}_j = \widehat{\mathbf{h}}_j^\tau\mathbf{Y}_c = \boldsymbol{\gamma}_j^\tau\mathbf{S}_{XY}\mathbf{S}_{YY}^{-1}\mathbf{Y}_c, \tag{8.133}$$

$j = 1, 2, \ldots, t$. In (8.132), $\mathbf{X}$ is an observed r-vector, while in (8.133), $\mathbf{Y}$ is an indicator response s-vector, and $\mathbf{X}_c = \mathbf{X} - \bar{\mathbf{X}}$ and $\mathbf{Y}_c = \mathbf{Y} - \bar{\mathbf{Y}}$. The coefficient vector

$$\boldsymbol{\gamma}_j = (\gamma_{j1}, \cdots, \gamma_{jr})^\tau \tag{8.134}$$

is the *jth discriminant vector*, $j = 1, 2, \ldots, \min(r, s)$.

The *first LDF* evaluated at $\mathbf{X}_c$ is given by

$$\widehat{\xi}_1 = \boldsymbol{\gamma}_1^\tau \mathbf{X}_c \tag{8.135}$$

and has the property that, among all such linear combinations of the xs, it alone can discriminate best between the K classes. The *second LDF* is given by

$$\widehat{\xi}_2 = \boldsymbol{\gamma}_2^\tau \mathbf{X}_c \tag{8.136}$$

and is the best discriminator between the K classes among all such linear combinations of the xs that are uncorrelated with $\widehat{\xi}_1$. The *jth LDF*,

$$\widehat{\xi}_j = \boldsymbol{\gamma}_j^\tau \mathbf{X}_c, \tag{8.137}$$

is the best discriminator between the K classes among all those linear combinations of $\mathbf{X}_c$ that are also uncorrelated with $\widehat{\xi}_1, \widehat{\xi}_2, \ldots, \widehat{\xi}_{j-1}$.

There are at most $\min(r, K - 1)$ such linear discriminant functions. One problem is to determine the smallest number $t < \min(r, s)$ of linear discriminant functions that discriminates most efficiently between the K classes. In practice, it is usual to take $t = 2$, so that only $\widehat{\xi}_1$ and $\widehat{\xi}_2$ are used in deciding whether sufficient discrimination has been obtained.

Graphical Display

Consider the kth observation $\mathbf{X}_{ik}$ (in Π_i) and its associated indicator response vector $\mathbf{Y}_{ik}$. We evaluate $\widehat{\xi}_j$ and $\widehat{\omega}_j$ at $\mathbf{X} = \mathbf{X}_{ik}$ and $\mathbf{Y} = \mathbf{Y}_{ik}$, respectively. Set

$$\widehat{\xi}_{jk}^{(i)} = \boldsymbol{\gamma}_j^\tau (\mathbf{X}_{ik} - \bar{\mathbf{X}}), \tag{8.138}$$

$$\widehat{\omega}_{jk}^{(i)} = \boldsymbol{\gamma}_j^\tau \mathbf{S}_{XY} \mathbf{S}_{YY}^{-1} (\mathbf{Y}_{ik} - \bar{\mathbf{Y}}), \tag{8.139}$$

$k = 1, 2, \ldots, n_i$, $i = 1, 2, \ldots, s + 1$. Then, we form the row vectors

$$\boldsymbol{\xi}_j^\tau = (\widehat{\xi}_{j1}^{(1)}, \cdots, \widehat{\xi}_{jn_1}^{(1)}, \cdots, \widehat{\xi}_{j1}^{(r+1)}, \cdots, \widehat{\xi}_{jn_{r+1}}^{(r+1)}), \tag{8.140}$$

$$\boldsymbol{\omega}_j^\tau = (\widehat{\omega}_{j1}^{(1)}, \cdots, \widehat{\omega}_{jn_1}^{(1)}, \cdots, \widehat{\omega}_{j1}^{(r+1)}, \cdots, \widehat{\omega}_{jn_{r+1}}^{(r+1)}), \tag{8.141}$$

of *jth discriminant scores*, $j = 1, 2, \ldots, \min(r, s)$. From (8.117) and (8.119), we have that

$$\mathbf{S}_{XY} \mathbf{S}_{YY}^{-1} = (\bar{\mathbf{X}}_1 - \bar{\mathbf{X}}_{s+1}, \cdots, \bar{\mathbf{X}}_s - \bar{\mathbf{X}}_{s+1}), \tag{8.142}$$

whence, from (8.138) and (8.139),

$$\widehat{\xi}_{jk}^{(i)} = \boldsymbol{\gamma}_j^\tau (\mathbf{X}_{ik} - \bar{\mathbf{X}}), \quad \widehat{\omega}_{jk}^{(i)} = \boldsymbol{\gamma}_j^\tau (\bar{\mathbf{X}}_i - \bar{\mathbf{X}}), \tag{8.143}$$

are the kth components of the jth pair of canonical variates evaluated for Π_i. But,

$$\bar{\mathbf{X}}_i - \bar{\mathbf{X}} = n_i^{-1} \sum_{k=1}^{n_i} (\mathbf{X}_{ik} - \bar{\mathbf{X}}), \qquad (8.144)$$

so that

$$\widehat{\omega}_{jk}^{(i)} = n_i^{-1} \sum_{a=1}^{n_i} \widehat{\xi}_{ja}^{(i)} = \bar{\xi}_j^{(i)}, \quad k = 1, 2, \ldots, n_i. \qquad (8.145)$$

In other words, *the canonical variates evaluated at the indicator response variables are the class averages of the canonical variates for the discriminating variables.* The $\{\widehat{\xi}_{jk}^{(i)}\}$ are called *discriminant coordinates* and the space generated by these coordinates is called the *discriminant space*. To visualize graphically whether the discriminant coordinates emphasize differences in class means, it is customary to plot the n points

$$(\widehat{\xi}_{1k}^{(i)}, \widehat{\xi}_{2k}^{(i)}), \quad k = 1, 2, \ldots, n_i, \ i = 1, 2, \ldots, s+1, \qquad (8.146)$$

on a scatterplot and, taking note of (8.145), we also plot a point representing the respective mean of each class,

$$(\widehat{\omega}_{1k}^{(i)}, \widehat{\omega}_{2k}^{(i)}), \quad k = 1, 2, \ldots, n_i, \ i = 1, 2, \ldots, s+1, \qquad (8.147)$$

superimposed on the same scatterplot.

8.6 Example: Gilgaied Soil

These data[3] were collected in a study of gilgaied soil at Meandarra, Queensland, Australia (Horton, Russell, and Moore, 1968). Three microtopographic classes based upon relative contours were classified as follows: top (>60 cm); slope (30–60 cm); and depression (<30 cm). The area was divided into four blocks, and soil samples were taken randomly within each microtopographic class at depths of 0–10, 10–30, 30–60, and 60–90 cm. See Table 8.9.

Chemical analyses on nine variables were carried out for each soil sample in the four blocks of the (3 positions × 4 depths) 12 groups, yielding a total of 48 soil samples. The variables are pH; total nitrogen (N); bulk-density (BD); total phosphorus (P); exchangeable + soluble calcium (Ca); exchangeable + soluble magnesium (Mg); exchangeable + soluble potassium (K); exchangeable + soluble sodium (Na); and conductivity of the saturation extract (cond).

---

[3]These data can be found in the file `gilgaied.soil` on the book's website.

TABLE 8.9. *Group numbers by depth and microtopographic position (T.P.) of gilgaied soil.*

T.P.	Soil Depth			
	0–10 cm	10–30 cm	30–60 cm	60–90 cm
Top	A	B	C	D
Slope	E	F	G	H
Depression	I	J	K	L

The first two LDF scores are computed using (8.143) and plotted in Figure 8.5. Also plotted on the same graph are the projected class averages (i.e., the letters A–L) of the 12 classes. We see that the projected class averages are plotted in roughly the same two-way position as given in Table 8.9 (with curvature). There is quite a bit of overlap of class points in this 2D discriminant space. In fact, the apparent error rate is $7/48 = 0.146$, and the leave-one-out CV misclassification rate is $31/48 = 0.646$. The curvature in the plot suggests that QDA may be more appropriate than LDA, but with only four observations in each class, QDA would fail. Another possible explanation is that the soil depths are not uniformly spaced; see Exercise 8.1.

8.7   Examples of Multiclass Misclassification Rates

In this section, we summarize how well LDA and QDA perform when applied to a wide variety of well-known multiclass data sets.[4] These data sets, which are listed in Table 8.10, are

Diabetes These data resulted from a study conducted at the Stanford Clinical Research Center of the relationship between chemical subclinical and overt nonketotic diabetes in non-obese adult subjects. The three primary variables are `glucose area` (a measure of glucose intolerance), `insulin area` (a measurement of insulin response to oral glucose), and `SSPG` (steady-state plasma glucose, a measure of insulin resistance). In addition, the `relative weight` and `fasting plasma glucose` were measured for each individual in the study. The three clinical classifications are overt diabetic (Class 1, 33 individuals), chemical diabetic (Class 2, 36), and normal (Class 3, 76).

---

[4]These data sets can be found at the book's website. The data and descriptions are taken from the UCI website, with the exception of `diabetes`, which originated from Andrews and Herzberg (1985, Table 36.1, pp. 215–219) and can be found in the Andrews subdirectory at the StatLib website, and `primate scapulae`, details of which can be found in Section 12.3.6.

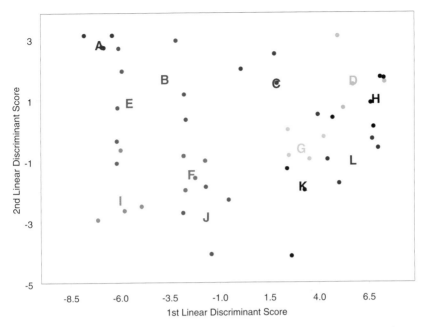

FIGURE 8.5. *LDA plot of the gilgaied soil data. There are 12 classes, A–L, and each class has four points. The projected class means are over-plotted as letters and appear in roughly the same two-way position as given in Table 8.7, albeit with some curvature.*

E-coli These data were obtained in a study of protein localization sites for 336 examples of *E. coli*. The variables are `mvg` (McGeoch's method for signal sequence recognition), `gvh` (von Heijne's method for signal sequence recognition), `lip` (von Heijne's Signal Peptidase II consensus sequence score), `chg` (presence of charge on *N*-terminus of predicted lipoproteins), `aac` (score of discriminant analysis of the amino-acid content of outer membrane and periplasmic proteins), `alm1` (score of the ALOM membrane spanning region prediction program), and `alm2` (score of the ALOM program after excluding putative cleavable signal regions from the sequence). There are 8 localization sites (classes): cp (cytoplasm, 143 examples), im (inner membrane without signal sequence, 77), pp (perisplasm, 52), imU (inner membrane, un-cleavable signal sequence, 35), om (outer membrane, 20), omL (outer membrane lipoprotein, 5), imL (inner membrane lipoprotein, 2), and imS (inner membrane, cleavable signal sequence, 2).

Forensic glass These data were collected for forensic purposes to determine whether a sample of glass is a type of "float" glass or not. There are 6 types of glass used in this data set: building windows float processed (70 examples), building windows non–float processed

(76), vehicle windows float processed (17), containers (13), tableware (9), and headlamps (29). The variables are RI (refractive index), Na (sodium), Mg (magnesium), Al (aluminum), Si (silicon), K (potassium), Ca (calcium), Ba (barium), and Fe (iron).

Iris These are Edgar Anderson's iris data made famous by R.A. Fisher. There are 150 observations made on three classes of the iris flower. The classes are *Iris setosa*, *Iris versicolour*, and *Iris virginica*, with 50 observations on each class. Four measurements (in cm) are made on each iris: sepal length, sepal width, petal length, and petal width.

Letter recognition The 26 capital letters of the English alphabet were converted into black-and-white rectangular pixel displays by using 20 different fonts, and each letter with these 20 fonts was randomly distorted to produce a file of 20,000 unique observations. Each observation was converted into 16 primitive numerical variables, which were then scaled to fit into a range of integer values of 0–15. The number of observations for each letter ranged from 734 to 813.

Pendigits These data were obtained from 44 writers, each of whom handwrote 250 examples of the digits $0, 1, 2, \ldots, 9$ in a random order. See Section 7.2.1 for a detailed description.

Primate scapulae These data consist of measurements of indices and angles on the scapulae (shoulder bones) of five genera of adult primates representing Hominoidae: gibbons (*Hylobates*), orangutangs (*Pongo*), chimpanzees (*Pan*), gorillas (*Gorilla*), and man (*Homo*). The variables are 5 indices (AD.BD, AD.CD, EA.CD, Dx.CD, and SH.ACR) and 2 angles (EAD, β). Of the 105 measurements on each variable, 16 were from *Hylobates*, 15 from *Pongo*, 20 from *Pan*, 14 from *Gorilla*, and 40 from *Homo*.

Shuttle These space-shuttle data contain 43,500 observations on 8 unidentified variables, and the observations are divided into 7 classes: Rad Flow (1), Fpv Close (2), Fpv Open (3), High (4), Bypass (5), Bpv Close (6), and Bpv Open (7). Class 1 contains about 78% of the data.

Vehicle This data set was collected by the Turing Institute, Glasgow, Scotland, in a study of how to distinguish 3D objects from a 2D image. The classes in this data set are the silhouettes of four types of *Corgi* model vehicles, an Opel Manta car (240 images), a Saab 9000 car (240), a double-decker bus (240), and a Chevrolet van (226), as viewed by a camera from many different angles and elevations. The variables are scaled variance, skewness, and kurtosis about the major/minor axes, and heuristic measures such as hollows ratio, circularity, elongatedness, rectangularity, and compactness of the silhouettes.

Wine These data are the results of a chemical analysis of 178 wines grown over the decade 1970–1979 in the same region of Italy, but derived from three different cultivars (Barolo, Grignolino, Barbera). The Barbera wines were predominately from a period that was much later than that of the Barolo and Grignolino wines. The analysis determined the quantities of 13 constituents found in each of the three types of wines: `Alcohol`, `MalicAcid`, `Ash`, `AlcAsh` (Alcalinity of Ash), `Mg` (Magnesium), `Phenols` (Total Phenols), `Flav` (Flavanoids), `NonFlavPhenols` (Non-Flavanoid Phenols) `Proa` (Proanthocyanins), `Color` (Color Intensity), `Hue`, `OD` (OD280/OD315 of Diluted Wines), and `Proline`. There are 59 Barolo wines, 71 Grignolino wines, and 48 Barbera wines.

Yeast These data were obtained in a study of protein localization sites for 1,484 examples of yeast. The variables are `mcg`, `gvh`, `alm` (see E-coli), `mit` (score of discriminant analysis of the amino-acid content of the N-terminal region, 20 residues long, of mitochondrial and non-mitochondrial proteins), `erl` (presence of HDEL substring, thought to act as a signal for retention in the endoplasmic reticulum lumen), `pox` (peroxisomal targeting signal in the C-terminus), `vac` (score of discriminant analysis of the amino-acid content of vacuolar and extracellular proteins), and `nuc` (score of discriminant analysis of nuclear localization signals of nuclear and non-nuclear proteins). There are 10 localization sites (classes): cyt (cytosolic or cytoskeletal, 463 examples), nuc (nuclear, 429), mit (mitochondrial, 244), me3 (membrane protein, no N-terminal signal, 163), me2 (membrane protein, uncleaved signal, 51), me1 (membrane protein, cleaved signal, 44), exc (extracellular, 37), vac (vacuolar, 30), pox (peroxisomal, 20), and erl (endoplasmic reticulum lumen, 5).

Table 8.10 lists the leave-one-out CV misclassification rates for LDA and QDA for each data set. The prior π_i was estimated using the proportionality estimate, $\hat{\pi}_i = n_i/n, \ i = 1, 2, \ldots, K$. These multiclass data sets have quite varied CV misclassification rates. For the `diabetes`, `glass`, `letter recognition`, `pendigits`, `vehicle`, and `wine` data sets, the QDA misclassification rate is smaller than the LDA rate, whereas the reverse happens for the `iris` and `primate scapulae` data sets. Note that if any data set has a class with fewer observations than r, then that class's estimated covariance matrix is singular, and QDA fails.

In Figure 8.6, we display the LDA plots corresponding to the six data sets `iris`, `primate.scapulae`, `shuttle`, `pendigits`, `vehicle`, and `glass`. They are arranged according to their estimated misclassification rates, as listed in Table 8.10.

We will be comparing these methods with other classification methods using the same data sets in later chapters.

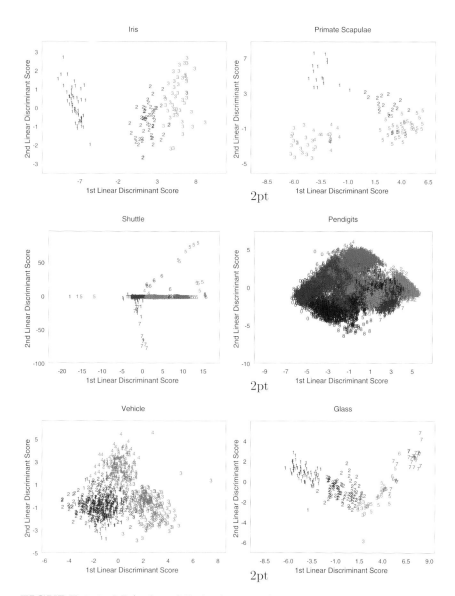

FIGURE 8.6. *LDA plot of Fisher's* iris *data*, primate.scapulae *data*, shuttle *data*, pendigits *data*, vehicle *data*, *and* glass *data*.

TABLE 8.10. *Summary of multiclass data sets. Listed are the sample size (n), number of variables (r), and number of classes (K). Also listed for each data set are leave-one-out cross-validation (CV/n) misclassification rates for linear discriminant analysis (LDA) and quadratic discriminant analysis (QDA). The data sets are ordered by size of the LDA misclassification rate. For each data set, the proportionality estimate was used for the priors. If a class has fewer than r members, QDA will fail.*

Data Set	n	r	K	LDA	QDA
Wine	178	13	3	0.011	0.006
Iris	150	4	3	0.020	0.027
Primate scapulae	105	7	5	0.029	0.057
Shuttle	43,500	8	7	0.056	
Diabetes	145	5	3	0.110	0.097
Pendigits	10,992	16	10	0.124	0.017
E-coli	336	7	8	0.128	
Vehicle	846	18	4	0.221	0.144
Letter recognition	20,000	16	26	0.298	0.114
Glass	214	9	6	0.350	0.140
Yeast	1,484	8	10	0.411	

8.8 Software Packages

All the major statistical software packages contain routines for carrying out LDA and QDA. Misclassification rates are computed in these packages by a number of methods, including the apparent error rate and cross-validation. Logistic regression is usually included within the regression methods in the packages. LDA is included as a special case of multivariate reduced-rank regression in the RRR+MULTANL package, which can be downloaded from the book's website.

Bibliographical Notes

Since Fisher (1936), LDA has seen applications in many different areas. Theoretical accounts of linear discriminant analysis may be found in Anderson (1984, Chapter 6) and Seber (1984, Chapter 6). More recent accounts are given in Ripley (1996, Chapter 3), Johnson and Wichern (1998, Chapter 11), Hastie, Tibshirani, and Friedman (2001, Chapter 4), Rencher (2002, Chapters 8, 9), and Bishop (2006, Chapter 4). A Bayesian approach is outlined in Press (1989, Chapter 7) and a nonparametric (kernel) approach in Hand (1982). The idea of using a regression model to carry out LDA can be found in Fisher (1936).

Exercises

8.1 How would you use the information in Table 8.9 to carry out a *two-way* LDA on the gilgaied soil data? Would your results change if you took into account the fact that the soil depths are not equally spaced?

8.2 Consider the `wine` data. Compute a LDA, draw a 2D-scatterplot of the first two LDF coordinates, and color-code the points by wine type. What do you notice?

8.3 Suppose $\mathbf{X}_1 \sim \mathcal{N}_r(\boldsymbol{\mu}_1, \boldsymbol{\Sigma}_{XX})$ and $\mathbf{X}_2 \sim \mathcal{N}_r(\boldsymbol{\mu}_2, \boldsymbol{\Sigma}_{XX})$ are independently distributed. Consider the statistic

$$\frac{\{\mathrm{E}(\mathbf{a}^\tau \mathbf{X}_1) - \mathrm{E}(\mathbf{a}^\tau \mathbf{X}_2)\}^2}{\mathrm{var}(\mathbf{a}^\tau \mathbf{X}_1 - \mathbf{a}^\tau \mathbf{X}_2)}$$

as a function of $\mathbf{a}$. Show that $\mathbf{a} \propto \boldsymbol{\Sigma}_{XX}^{-1}(\boldsymbol{\mu}_1 - \boldsymbol{\mu}_2)$ maximizes the statistic by using a Lagrange multiplier approach.

8.4 Consider the following alternative to QDA. Suppose you start with two variables, X_1 and X_2. Now, expand the data set by adding squares, $X_3 = X_1^2$ and $X_4 = X_2^2$, and cross-product, $X_5 = X_1 X_2$. These five variables are to be used as input to an LDA procedure. Derive the LDA boundaries from this procedure and compare them to the QDA procedure. Generalize to $r > 2$. Try this alternative procedure out on a data set of your choice.

8.5 Consider the `diabetes` data. Draw a scatterplot matrix of all five variables with different colors or symbols representing the three classes of diabetes. Do these pairwise plots suggest multivariate Gaussian distributions for each class with equal covariance matrices? Carry out an LDA and draw the 2D-scatterplot of the first two discriminating functions. Using the leave-one-out CV procedure, find the confusion table and identify those observations that are incorrectly classified based upon the LDA classification rule. Do the same for the QDA procedure.

8.6 Try the following transformation on the `iris` data. Set $X_5 = X_1/X_2$ and $X_6 = X_3/X_4$. Then, X_5 is a measure of sepal shape and X_6 is a measure of petal shape. Take logarithms of X_5 and of X_6. Plot the transformed data, and carry out an LDA on X_5 and X_6 alone. Estimate the misclassification rate for the transformed data. Do the same for the QDA procedure.

8.7 Carry out a stepwise logistic regression of the `spambase` data. Which variables are chosen to be in the final subset?

8.8 Consider The Insurance Company Benchmark data, which can be downloaded from `kdd.ics.uci.edu/databases/tic`. There are 86 variables on product-usage data and socio-demographic data derived from zip

area codes of customers of an insurance company. There is a learning set
`ticdata2000.txt` of 5,822 customers and a test set `ticeval2000.txt`
of 4,000 customers. Customers in the learning set are classified into two
classes, depending upon whether they bought a caravan insurance policy.
The problem is to predict who in the test set would be interested in buying a caravan insurance policy. Use any of the classification methods on
the learning data and then apply them to the test data. Compare your
predictions for the test set with those given in the file `tictgts2000.txt`
and estimate the test set error rate. Which variables are most useful in
predicting the purchase of a caravan insurance policy?

8.9 These data (`covertype`) were obtained from the U.S. Forest Service
and are concerned with seven different types of forest cover. The data can
be downloaded from `kdd.ics.uci.edu/databases/covertype`. There are
581,012 observations (each a 30×30 meter cell) on 54 input variables (10
quantitative variables, 4 binary wilderness areas, and 40 binary soil type
variables). Divide these data randomly into a learning set and a test set.
Use any of the methods of this chapter on the learning set to predict the
forest cover type for the test set. Estimate the test set error rate.

8.10 Consider the Wisconsin diagnostic breast cancer data. Regress Y on
each of the 30 variables, one at a time. How many coefficients are significant? Which are they? (A coefficient is declared to be "significantly different
from zero" at the 5% level if its absolute t-ratio is greater than the value 2
and is nonsignificant otherwise.) Now, regress Y on all 30 variables. How
many coefficients are significant? Which are they? Next, run the BE and
FS stepwise procedures, and the LAR and LARS-Lasso algorithms on these
data, and compare the variable subsets you obtain from these methods.

8.11 Consider the `E-coli` data. Draw a scatterplot matrix of the variables. What do you notice? Do they look Gaussian? Carry out an LDA of
the `e-coli` data by using the reduced-rank regression approach. Find the
estimated coefficients of the first two linear discriminant functions. Compute the LD scores and plot them in a scatterplot.

8.12 Consider the `yeast` data. Draw a scatterplot matrix of the data and,
if possible, draw 3D plots of various subsets of the variables and rotate the
plot ("brush and spin" in S-PLUS). What do you notice about the data?
Do they look Gaussian? Carry out an LDA of the `yeast` data by using the
reduced-rank regression approach. Find the estimated coefficients of the
first two linear discriminant functions. Compute the LD scores and plot
them in a scatterplot.

8.13 Consider the `primate.scapulae` data. Carry out five linear discriminant analyses (one for each primate species), where each analysis is of the
"one class *versus* the rest" type. Find the spatial zone (known as an *ambiguous region*) that does not correspond to any LDA assignment of a class

of primate (out of the five considered). Are the results consistent with the multiclass classification results?

8.14 Suppose LDA boundaries are found for the `primate.scapulae` data by carrying out a sequence of $\binom{5}{2} = 10$ LDA problems, each involving a distinct pair of primate species (*Hylobates* versus *Pongo*, *Gorilla* versus *Homo*, etc.). Find the *ambiguous region* that does not correspond to any LDA assignment of a class of primate (out of the five considered). Suppose we classify each primate in the data set by taking a vote based upon those boundaries. Estimate the resulting misclassification rate and compare it with the rate from the multiclass classification procedure.

9

Recursive Partitioning and Tree-Based Methods

9.1 Introduction

An algorithm known as *recursive partitioning* is the key to the nonparametric statistical method of *classification and regression trees (CART)* (Breiman, Friedman, Olshen, and Stone, 1984). Recursive partitioning is the step-by-step process by which a *decision tree* is constructed by either splitting or not splitting each node on the tree into two daughter nodes. An attractive feature of the CART methodology (or the related C4.5 methodology; Quinlan, 1993) is that because the algorithm asks a sequence of hierarchical Boolean questions (e.g., is $X_i \leq \theta_j$?, where θ_j is a threshold value), it is relatively simple to understand and interpret the results.

As we described in previous chapters, classification and regression are both supervised learning techniques, but they differ in the way their output variables are defined. For binary classification problems, the output variable, Y, is binary-valued, whereas for regression problems, Y is a continuous variable. Such a formulation is particularly useful when assessing how well a classification or regression methodology does in predicting Y from a given set of input variables $X_1, X_2, \ldots, X_r$.

In the CART methodology, the input space, $\Re^r$, is partitioned into a number of nonoverlapping rectangular ($r = 2$) or cuboid ($r > 2$) regions,

A.J. Izenman, *Modern Multivariate Statistical Techniques*,
doi: 10.1007/978-0-387-78189-1_9,
© Springer Science+Business Media, LLC 2008

each of which is viewed as homogeneous for the purpose of predicting Y. Each region, which has sides parallel to the axes of input space, is assigned a class (in a classification problem) or a constant value (in a regression problem). Such a partition corresponds to a classification or regression tree (as appropriate).

Tree-based methods, such as CART and C4.5, have been used extensively in a wide variety of fields. They have been found especially useful in biomedical and genetic research, marketing, political science, speech recognition, and other applied sciences.

9.2 Classification Trees

A classification tree is the result of asking an ordered sequence of questions, and the type of question asked at each step in the sequence depends upon the answers to the previous questions of the sequence. The sequence terminates in a prediction of the class.

The unique starting point of a classification tree is called the *root node* and consists of the entire learning set $\mathcal{L}$ at the top of the tree. A *node* is a subset of the set of variables, and it can be a terminal or nonterminal node. A *nonterminal* (or *parent*) *node* is a node that splits into two daughter nodes (a *binary split*). Such a binary split is determined by a Boolean condition on the value of a single variable, where the condition is either satisfied ("yes") or not satisfied ("no") by the observed value of that variable. All observations in $\mathcal{L}$ that have reached a particular (parent) node and satisfy the condition for that variable drop down to one of the two daughter nodes; the remaining observations at that (parent) node that do not satisfy the condition drop down to the other daughter node.

A node that does not split is called a *terminal node* and is assigned a class label. Each observation in $\mathcal{L}$ falls into one of the terminal nodes. When an observation of unknown class is "dropped down" the tree and ends up at a terminal node, it is assigned the class corresponding to the class label attached to that node. There may be more than one terminal node with the same class label. A single-split tree with only two terminal nodes is called a *stump*. The set of all terminal nodes is called a *partition* of the data.

Consider a simple example of recursive partitioning involving two input variables, X_1 and X_2. Suppose the tree diagram is given in the top panel of Figure 9.1. The possible stages of this tree are as follows: (1) Is $X_2 \leq \theta_1$? If the answer is yes, follow the left branch; if no, follow the right branch. (2) If the answer to (1) is yes, then we ask the next question: Is $X_1 \leq \theta_2$? An answer of yes yields terminal node τ_1 with corresponding region $R_1 = \{X_1 \leq \theta_2, X_2 \leq \theta_1\}$; an answer of no yields terminal node τ_2 with corresponding region $R_2 = \{X_1 > \theta_2, X_2 \leq \theta_1\}$. (3) If the answer to (1) is

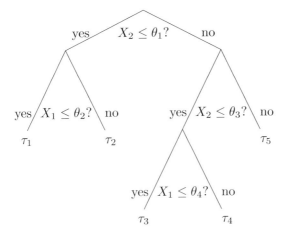

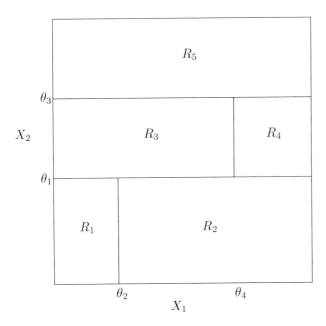

FIGURE 9.1. *Example of recursive partitioning with two input variables* X_1 *and* X_2. *Top panel shows a decision tree with five terminal nodes,* $\tau_1 - \tau_5$, *and four splits. Bottom panel shows the partitioning of* $\Re^2$ *into five regions,* $R_1 - R_5$, *corresponding to the five terminal nodes.*

no, we ask the next question: Is $X_2 \leq \theta_3$? If the answer to (3) is yes, then we ask the next question: Is $X_1 \leq \theta_4$? An answer of yes yields terminal node τ_3 with corresponding region $R_3 = \{X_1 \leq \theta_4, \theta_1 < X_2 \leq \theta_3\}$; if no, follow the right branch to terminal node τ_4 with corresponding region $R_4 = \{X_1 > \theta_4, \theta_1 < X_2 \leq \theta_3\}$. (4) If the answer to (3) is no, we arrive at terminal node τ_5 with corresponding region $R_5 = \{X_2 > \theta_3\}$. We have assumed that $\theta_2 < \theta_4$ and $\theta_1 < \theta_3$. The resulting 5-region partition of $\Re^2$ is given in the bottom panel of Figure 9.1. For a classification tree, each terminal node and corresponding region is assigned a class label.

9.2.1 Example: Cleveland Heart-Disease Data

These data[1] were obtained from a heart-disease study conducted by the Cleveland Clinic Foundation (Robert Detrano, principal investigator). For the study, the response variable is `diag` (diagnosis of heart disease: `buff` = healthy, `sick` = heart disease). There were 303 patients in the study, 164 of them healthy and 139 with heart disease.

The 13 input variables are `age` (age in years), `gender` (`male`, `fem`), `cp` (chest-pain type: `angina`=typical angina, `abnang`=atypical angina, `notang` =non-anginal pain, `asympt`=asymptomatic), `trestbps` (resting blood pressure), `chol` (serum cholesterol in mg/dl), `fbs` (fasting blood sugar < 120 mg/dl: `true`, `false`), `restecg` (resting electrocardiographic results: `norm` =normal, `abn`=having ST-T wave abnormality, `hyp`=showing probable or definite left ventricular hypertrophy by Estes's criteria), `thatach` (maximum heart rate achieved), `exang` (exercise-induced angina: `true`, `false`), `oldpeak` (ST depression induced by exercise relative to rest), `slope` (the slope of the peak exercise ST segment: `up`, `flat`, `down`), `ca` (number of major vessels (0–3) colored by flouroscopy), and `thal` (no description given: `norm`=normal, `fix`=fixed defect, `rev`=reversable effect). Of the 303 patients in the original data set, seven had missing data, and so we reduced the number of patients to 296 (160 healthy, 136 with heart disease).

The classification tree is displayed in Figure 9.2 (where we used the entropy measure as the impurity function for splitting). The root node with 296 patients is split according to whether `thal` = `norm` (163 patients) or `thal` = `fix` or `rev` (133 patients). The node with the 163 patients, which consists of 127 healthy patients and 36 patients with heart disease, is then split by whether `ca` < 0.5 (114 patients), or `ca` > 0.5 (49 patients). The node with 114 patients is declared a terminal node for `buff` because of the 102–12 majority in favor of `buff`. The node with 49 patients, which consists

---

[1]The data can be downloaded from file `cleveland.data` in the UCI repository `www.ics.uci.edu/ĩílearn/databases/heart-disease`.

of 25 healthy patients and 24 with heart disease, is split by whether `cp = abnang`, `angina`, `notang` (29 patients) or `cp = asympt` (20 patients). The node with 29 patients, which consists of 22 healthy patients and 7 with heart disease, is split by whether `age` $\leq$ 65.5 (7 patients) or `age` < 65.5 (22 patients). The node with 7 patients is declared a terminal node for `buff` because of the 7–0 majority in favor of `buff`, and the node with 22 patients, which consists of 15 healthy patients and 7 with heart disease, is split by whether `age` < 55.5 (13 patients) or `age` $\leq$ 55.5 (9 patients). The node with 13 patients is declared a terminal node for `buff` because of the 12–1 majority in favor of `buff`, and the node with 9 patients is declared a terminal node for `sick` because of the 6–3 majority in favor of `sick`. And so on.

Thus, we see that there are four paths (sequence of splits) through this tree for a patient to be declared healthy (`buff`) and five other paths for a patient to be diagnosed with heart disease (`sick`). In fact, there are 10 splits (and 11 terminal nodes) in this tree. The variables used in the tree construction are `thal`, `ca`, `cp`, `age`, `oldpeak`, `thatach`, and `exang`. The resubstitution (or apparent) error rate (i.e., the error rate obtained directly from the classification tree) is $37/296 = 0.125$ (12 `sick` patients who are classified as `buff` and 25 `buff` patients who are classified as `sick`).

9.2.2   Tree-Growing Procedure

In order to grow a classification tree, we need to answer four basic questions: (1) How do we choose the Boolean conditions for splitting at each node? (2) Which criterion should we use to split a parent node into its two daughter nodes? (3) How do we decide when a node become a terminal node (i.e., stop splitting)? (4) How do we assign a class to a terminal node?

9.2.3   Splitting Strategies

At each node, the tree-growing algorithm has to decide on which variable it is "best" to split. We need to consider every possible split over all variables present at that node, then enumerate all possible splits, evaluate each one, and decide which is best in some sense.

For a description of splitting rules, we need to make a distinction between ordinal (or continuous) and nominal (or categorical) variables.

Ordinal or Continuous Variable

For a continuous or ordinal variable, the number of possible splits at a given node is one fewer than the number of its distinctly observed values.

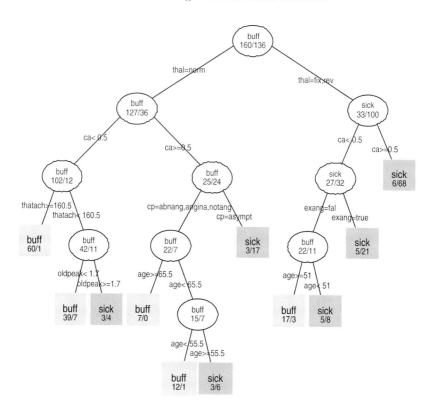

FIGURE 9.2. *Classification tree for the Cleveland heart-disease data, where the entropy measure has been used as the impurity function. The nodes (internal and terminal) are classified as* buff *(terminal nodes are colored green) or* sick *(terminal nodes are colored pink) according to the majority diagnosis of patients falling into that node. The splitting variables are displayed along the branches.*

In the Cleveland heart-disease data, we have six continuous or ordinal variables: age (40 possible splits), treatbps (48 possible splits), chol (151 possible splits), thatach (91 possible splits), ca (3 possible splits), and oldpeak (39 possible splits). The total number of possible splits from these continuous variables is, therefore, 372.

Nominal or Categorical Variable

Suppose that a particular categorical variable is defined by M distinct categories, $\ell_1, \ldots, \ell_M$. The set $\mathcal{S}$ of possible splits at that node for that variable is the set of all subsets of $\{\ell_1, \ldots, \ell_M\}$. Denote by τ_L and τ_R the left daughter-node and right daughter-node, respectively, emanating from a (parent) node τ. If we let $M = 4$, then there are $2^M - 2 = 14$ possible splits (ignoring splits where one of the daughter-nodes is empty). However, half of those splits are redundant; for example, the split $\tau_L = \{\ell_1\}$ and $\tau_R = \{\ell_2, \ell_3, \ell_4\}$ is the reverse of the split $\tau_L = \{\ell_2, \ell_3, \ell_4\}$ and $\tau_R = \{\ell_1\}$. So, the set $\mathcal{S}$ of seven distinct splits is given by the following table:

τ_L	τ_R
ℓ_1	ℓ_2, ℓ_3, ℓ_4
ℓ_2	ℓ_1, ℓ_3, ℓ_4
ℓ_3	ℓ_1, ℓ_2, ℓ_4
ℓ_4	ℓ_1, ℓ_2, ℓ_3
ℓ_1, ℓ_2	ℓ_3, ℓ_4
ℓ_1, ℓ_3	ℓ_2, ℓ_4
ℓ_1, ℓ_4	ℓ_2, ℓ_3

In general, there are $2^{M-1} - 1$ distinct splits in $\mathcal{S}$ for an M-categorical variable.

In the Cleveland heart-disease data, there are seven categorical variables: gender (1 possible split), cp (7 possible splits), fbs (1 possible split), restecg (3 possible splits), exang (1 possible split), slope (3 possible splits), and thal (3 possible splits). The total number of possible splits from these categorical variables is, therefore, 19.

Total Number of Possible Splits

We now add the number of possible splits from categorical variables (19) to the total number of possible splits from continuous variables (372) to get 391 possible splits over all 13 variables at the root node. In other words, there are 391 possible splits of the root node into two daughter nodes. So, which split is "best"?

Node Impurity Functions

To choose the best split over all variables, we first need to choose the best split for a given variable. Accordingly, we define a measure of goodness of a split.

Let $\Pi_1, \ldots, \Pi_K$ be the $K \geq 2$ classes. For node τ, we define the *node impurity function* $i(\tau)$ as

$$i(\tau) = \phi(p(1|\tau), \cdots, p(K|\tau)), \tag{9.1}$$

where $p(k|\tau)$ is an estimate of $P(\mathbf{X} \in \Pi_k|\tau)$, the conditional probability that an observation $\mathbf{X}$ is in Π_k given that it falls into node τ. In (9.1), we require ϕ to be a symmetric function, defined on the set of all K-tuples of probabilities $(p_1, \cdots, p_K)$ with unit sum, minimized at the points $(1, 0, \cdots, 0)$, $(0, 1, 0, \cdots, 0)$, $\ldots$, $(0, 0, \cdots, 0, 1)$ and maximized at the point $(\frac{1}{K}, \cdots, \frac{1}{K})$. In the two-class case ($K = 2$), these conditions reduce to a symmetric $\phi(p)$ maximized at the point $p = 1/2$ with $\phi(0) = \phi(1) = 0$.

One such function ϕ is the *entropy function*,

$$i(\tau) = -\sum_{k=1}^{K} p(k|\tau) \log p(k|\tau), \tag{9.2}$$

which is a discrete version of (7.113). When there are two classes, the entropy function reduces to

$$i(\tau) = -p \log p - (1 - p) \log(1 - p), \tag{9.3}$$

where we set $p = p(1|\tau)$. Several other ϕ-functions have also been suggested, including the *Gini diversity index*,

$$i(\tau) = \sum_{k \neq k'} p(k|\tau) p(k'|\tau) = 1 - \sum_{k} \{p(k|\tau)\}^2. \tag{9.4}$$

In the two-class case, the Gini index reduces to

$$i(\tau) = 2p(1 - p). \tag{9.5}$$

This function can be motivated by considering which quadratic polynomial satisfies the above conditions for the two-class case.

In Figure 9.3, the entropy function and the Gini index are graphed for the two-class case. For practical purposes, there is not much difference between these two types of node impurity functions. The usual default in tree-growing software is the Gini index.

Choosing the Best Split for a Variable

Suppose, at node τ, we apply split s so that a proportion p_L of the observations drops down to the left daughter-node τ_L and the remaining proportion p_R drops down to the right daughter-node τ_R. For example, suppose we have a data set in which the response variable Y has two possible values, 0 and 1. Suppose that one of the possible splits of the input variable X_j is $X_j \leq c$ vs. $X_j > c$, where c is some value of X_j. We can write down the 2×2 table in Table 9.1.

Consider, first, the parent node τ. We use the entropy function (9.3) as our impurity measure. Estimate p_L by n_{+1}/n_{++} and p_R by n_{+2}/n_{++}, and then the estimated impurity function is

$$i(\tau) = -\left(\frac{n_{+1}}{n_{++}}\right) \log_e \left(\frac{n_{+1}}{n_{++}}\right) - \left(\frac{n_{+2}}{n_{++}}\right) \log_e \left(\frac{n_{+2}}{n_{++}}\right). \tag{9.6}$$

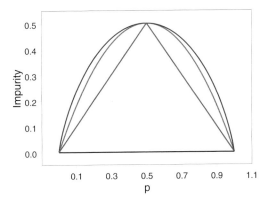

FIGURE 9.3. *Node impurity functions for the two-class case. The entropy function (rescaled) is the red curve, the Gini index is the green curve, and the resubstitution estimate of the misclassification rate is the blue curve.*

Note that $i(\tau)$ is completely independent of the type of proposed split. Now, for the daughter nodes, τ_L and τ_R. For $X_j \leq c$, we estimate p_L by n_{11}/n_{1+} and p_R by n_{12}/n_{1+}, and for $X_j > c$, we estimate p_L by n_{21}/n_{2+} and p_R by n_{22}/n_{2+}. We then compute

$$i(\tau_L) \;=\; -\left(\frac{n_{11}}{n_{1+}}\right)\log_e\left(\frac{n_{11}}{n_{1+}}\right) - \left(\frac{n_{12}}{n_{1+}}\right)\log_e\left(\frac{n_{12}}{n_{1+}}\right) \qquad (9.7)$$

$$i(\tau_R) \;=\; -\left(\frac{n_{21}}{n_{2+}}\right)\log_e\left(\frac{n_{11}}{n_{1+}}\right) - \left(\frac{n_{22}}{n_{2+}}\right)\log_e\left(\frac{n_{22}}{n_{2+}}\right). \qquad (9.8)$$

The *goodness of split s at node τ* is given by the reduction in impurity gained by splitting the parent node τ into its daughter nodes, τ_R and τ_L,

$$\Delta i(s,\tau) = i(\tau) - p_L i(\tau_L) - p_R i(\tau_R). \qquad (9.9)$$

The best split for the single variable X_j is the one that has the largest value of $\Delta i(s,\tau)$ over all $s \in S_j$, the set of possible distinct splits for X_j.

Example: Cleveland Heart-Disease Data (Continued)

Consider the first variable **age** as a possible splitting variable at the root node. There are 41 different values for **age**, and so there are 40 possible

TABLE 9.1. *Two-by-two table for a split on the variable X_j, where the response variable has value 1 or 0.*

	1	0	Row Total
$X_j \leq c$	n_{11}	n_{12}	n_{1+}
$X_j > c$	n_{21}	n_{22}	n_{2+}
Column Total	n_{+1}	n_{+2}	n_{++}

TABLE 9.2. *Two-by-two table for the split on the variable* age *in the Cleveland heart disease data: the left branch would be* age ≤ 65 *and the right branch would be* age > 65.

	Buff	Sick	Row Total
age ≤ 65	143	120	263
age > 65	17	16	33
Column Total	160	136	296

splits. We set up the 2×2 table, Table 9.2, in which age is split, for example, at 65.

Using the two-class entropy function as the impurity measure, we compute (9.7) and (9.8), respectively, for the two possible daughter nodes:

$$i(\tau_L) = -(143/263)\log_e(143/263) - (120/263)\log_e(120/263), \quad (9.10)$$
$$i(\tau_R) = -(17/33)\log_e(17/33) - (16/33)\log_e(16/33), \quad (9.11)$$

whence, $i(\tau_L) = 0.6893$ and $i(\tau_R) = 0.6927$. Furthermore, from (9.6),

$$i(\tau) = -(160/296)\log_e(160/296) - (136/296)\log_e(136/296) = 0.6899. \quad (9.12)$$

Using (9.9), the goodness of this split is given by:

$$\Delta i(s, \tau) = 0.6899 - (263/296)(0.6893) - (33/296)(0.6927) = 0.000162. \quad (9.13)$$

If we repeat these computations for all 40 possible splits for the variable age, we arrive at Figure 9.4. In the left panel, we plot $i(\tau_L)$ (blue curve) and $i(\tau_R)$ (red curve) against each of the 40 splits; for comparison, we have the constant value of $i(\tau) = 0.6899$. Note the large drop in the plot of $i(\tau_R)$ at the split age > 70. In the right panel, we plot $\Delta i(s, \tau)$ against each of the 40 splits s. The largest value of $\Delta i(s, \tau)$ is 0.04305, which corresponds to the split age ≤ 54.

Recursive Partitioning

In order to grow a tree, we start with the root node, which consists of the learning set $\mathcal{L}$. Using the "goodness-of-split" criterion for a single variable, the tree algorithm finds the best split at the root node for each of the variables, X_1 to X_r. The best split s at the root node is then defined as the one that has the largest value of (9.9) over all r single-variable best splits at that node.

In the case of the Cleveland heart-disease data, the best split at the root node (and corresponding value of $\Delta i(s, \tau)$) for each of the 13 variables is listed in Table 9.3. The largest value is 0.147 corresponding to the variable thal. So, for these data, the best split at the root node is to split the

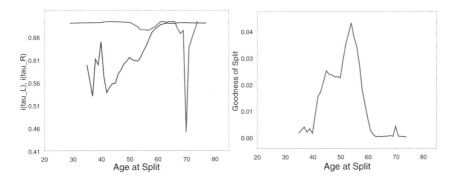

FIGURE 9.4. *Choosing the best split for the* age *variable in the Cleve-land heart-disease study. The impurity measure is the entropy function. Left panel: Plots of* $i(\tau_L)$ *(blue curve), and* $i(\tau_R)$ *(red curve) against age at split. Note the sharp dip in the* $i(\tau_R)$ *plot at the split* age > 70. *Right panel: Plot of the goodness of split* s, $\Delta i(s, \tau)$, *against age at split. The peak of this curve corresponds to the split* age ≤ 54.

variable thal according to norm vs. (fix, rev); that is, first separate the 163 normal patients from the 133 patients who have (either fixed or reversible) defects for the variable thal.

We next split each of the daughter nodes of the root node in the same way. We repeat the above computations for the left daughter node, except that we consider only those 163 patients having thal = norm, and then consider the right daughter node, except we consider only those 133 patients having thal = fix or rev. When those splits are completed, we continue to split each of the subsequent nodes. This sequential splitting process of building a tree layer-by-layer is called *recursive partitioning*. If every parent node splits into two daughter nodes, the result is called a *binary tree*. If the binary tree is grown until none of the nodes can be split any further, we say the tree is *saturated*. It is very easy in a high-dimensional classification problem to let the tree get overwhelmingly large, especially if the tree is allowed to grow until saturation.

TABLE 9.3. *Determination of the best split at the root node for the Cleve-land heart-disease data. The impurity measure is the entropy function. Each input variable is listed together with its maximum value of* $\Delta i(s, \tau)$ *over all possible splits of that variable.*

age	gender	cp	trestbps	chol	fbs	restecg
0.043	0.042	0.133	0.011	0.011	0.00001	0.015

thatach	exang	oldpeak	slope	ca	thal
0.093	0.093	0.087	0.077	0.124	0.147

One way to counter this type of situation is to restrict the growth of the tree. This was the philosophy of early tree-growers. For example, we can declare a node to be *terminal* if it fails to be larger than a certain critical size; that is, if $n(\tau) \leq n_{\min}$, where $n(\tau)$ is the number of observations in node τ and $n_{\min}$ is some previously declared minimum size of a node. Because a terminal node cannot be split into daughter nodes, it acts as a brake on tree growth; the larger the value of $n_{\min}$, the more severe the brake. Another early action was to stop a node from splitting if the largest goodness-of-split value at that node is smaller than a certain predetermined limit. These stopping rules, however, do not turn out to be such good ideas. A better approach (Breiman et al., 1984) is to let the tree grow to saturation and then "prune" it back; see Section 9.2.6.

How do we associate a class with a terminal node? Suppose at terminal node τ there are $n(\tau)$ observations, of which $n_k(\tau)$ are from class Π_k, $k = 1, 2, \ldots, K$. Then, the class which corresponds to the largest of the $\{n_k(\tau)\}$ is assigned to τ. This is called the *plurality rule*. This rule can be derived from the Bayes's rule classifier of Section 8.5.1, where we assign the node τ to class Π_i if $p(i|\tau) = \max_k p(k|\tau)$; if we estimate the prior probability π_k by $n_k(\tau)/n(\tau)$, $k = 1, 2, \ldots, K$, then this boils down to the plurality rule.

9.2.4   Example: Pima Indians Diabetes Study

This Indian population lives near Phoenix, Arizona. All patients listed in this data set[2] are females at least 21 years old of Pima Indian heritage. There are two classes: `diabetic`, if the patient shows signs of diabetes according to World Health Organization criteria (i.e., if the 2-hour post-load plasma glucose was at least 200 mg/dl at any survey examination, or if found during routine medical care), and `normal`. In the original data, there were 500 `normal` subjects and 268 `diabetic` subjects.

There are eight input variables: `npregnant` (number of times pregnant), `bmi` (body mass index, (weight in kg)/(height in m)^2), `glucose` (plasma glucose concentration at 2 hours in an oral glucose tolerance test), `pedigree` (diabetes pedigree function), `diastolic.bp` (diastolic blood pressure, mm Hg), `skinfold.thickness` (triceps skin fold thickness, mm), `insulin` (2-hour serum insulin, μU/ml), and `age` (age in years). We removed any subject with a nonsense value of zero for the variables `glucose`, `bmi`, `diastolic.bp`, `skinfold.thickness`; this reduced the data set to 532 patients (from 768), with 355 `normal` subjects and 177 `diabetic` subjects.

---

[2]These data are available on the book's website (file **pima**) and are also available from the UCI website.

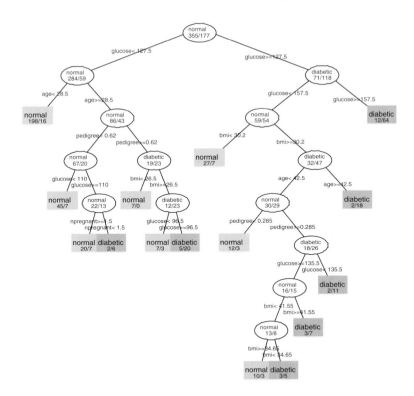

FIGURE 9.5. *A classification tree for the Pima Indians diabetes data, where the impurity measure is the Gini index. The terminal nodes are colored green for* `normal` *and pink for* `diabetic`. *The splitting variables are given on the branches of each split, and the number in each node is given as number of* `normal`/*number of* `diabetic`, *with the node classification given by the majority rule. Nodes were not split further unless they contained at least 10 subjects.*

We also did not use the variable `insulin` because it had so many zeros (374 in the original data).

A classification tree was grown for the Pima Indians diabetes data using Gini's impurity measure (9.5). The classification tree appears in Figure 9.5, where nodes are declared to be terminal if they contain fewer than 10 patients. We see 14 splits and 15 terminal nodes; a patient is declared to be normal at 8 terminal nodes and diabetic at 7 terminal nodes. The assignment of each terminal node into "normal" or "diabetic" depends upon the majority rule at that node; the numbers of normal and diabetic patients in the learning set that fall into each terminal node are displayed at that node.

9.2.5 Estimating the Misclassification Rate

Next, we compute an estimate of the within-node misclassification rate. The resubstitution estimate of the misclassification rate $R(\tau)$ of an observation in node τ is

$$r(\tau) = 1 - \max_k p(k|\tau), \tag{9.14}$$

which, for the two-class case, reduces to

$$r(\tau) = 1 - \max(p, 1 - p) = \min(p, 1 - p). \tag{9.15}$$

The resubstitution estimate (9.15) in the two-class case is graphed in Figure 9.3 (the blue curve). If $p < 1/2$, the resubstitution estimate increases linearly in p, and if $p > 1/2$, it decreases linearly in p. Because of its poor properties (e.g., nondifferentiability), (9.15) is not used much in practice.

Let T be the tree classifier and let $\widetilde{T} = \{\tau_1, \tau_2, \ldots, \tau_L\}$ denote the set of all terminal nodes of T. We can now estimate the true misclassification rate,

$$R(T) = \sum_{\tau \in \widetilde{T}} R(\tau)P(\tau) = \sum_{\ell=1}^{L} R(\tau_\ell)P(\tau_\ell) \tag{9.16}$$

for T, where $P(\tau)$ is the probability that an observation falls into node τ. If we estimate $P(\tau_\ell)$ by the proportion $p(\tau_\ell)$ of all observations that fall into node τ_ℓ, then, the resubstitution estimate of $R(T)$ is

$$R^{re}(T) = \sum_{\ell=1}^{L} r(\tau_\ell)p(\tau_\ell) = \sum_{\ell=1}^{L} R^{re}(\tau_\ell), \tag{9.17}$$

where $R^{re}(\tau_\ell) = r(\tau_\ell)p(\tau_\ell)$.

Of the 532 subjects in the Pima Indians diabetes study, the classification tree in Figure 9.5 misclassifies 29 of the 355 `normal` subjects as `diabetic`, whereas of the 177 `diabetic` patients, 46 are misclassified as `normal`. So, the resubstitution estimate is $R^{re}(T) = 75/532 = 0.141$.

The resubstitution estimate $R^{re}(T)$, however, leaves much to be desired as an estimate of $R(T)$. First, bigger trees (i.e., more splitting) have smaller values of $R^{re}(T)$; that is, $R^{re}(T') \leq R^{re}(T)$, where T' is formed by splitting a terminal node of T. For example, if a tree is allowed to grow until every terminal node contains only a single observation, then that node is classified by the class of that observation and $R^{re}(T) = 0$. Second, using only the resubstitution estimate tends to generate trees that are too big for the given data. Third, the resubstitution estimate $R^{re}(T)$ is a much-too-optimistic estimate of $R(T)$. More realistic estimates of $R(T)$ are given below.

9.2.6   Pruning the Tree

The Breiman et al. (1984) philosophy of growing trees is to grow the tree "large" and then prune off branches (from the bottom up) until the tree is the "right size." A pruned tree is a *subtree* of the original large tree. How to prune a tree, then, is the crucial part of the process. Because there are many different ways to prune a large tree, we decide which is the "best" of those subtrees by using an estimate of $R(T)$.

The pruning algorithm is as follows:

1. Grow a large tree, say, $T_{\max}$, where we keep splitting until the nodes each contain fewer than $n_{\min}$ observations;

2. Compute an estimate of $R(\tau)$ at each node $\tau \in T_{\max}$;

3. Prune $T_{\max}$ upwards toward its root node so that at each stage of pruning, the estimate of $R(T)$ is minimized.

Instead of using the resubstitution measure $R^{re}(T)$ as our estimate of $R(T)$, we modify it for tree pruning by adopting a regularization approach. Let $\alpha \geq 0$ be a *complexity parameter*. For any node $\tau \in T$, set

$$R_\alpha(\tau) = R^{re}(\tau) + \alpha. \tag{9.18}$$

From (9.18), we define a *cost-complexity pruning measure* for a tree as follows:

$$R_\alpha(T) = \sum_{\ell=1}^{L} R_\alpha(\tau_\ell) = R^{re}(T) + \alpha|\widetilde{T}|, \tag{9.19}$$

where $|\widetilde{T}| = L$ is the number of terminal nodes in the subtree T of $T_{\max}$. Think of $\alpha|\widetilde{T}|$ as a penalty term for tree size, so that $R_\alpha(T)$ penalizes $R^{re}(T)$ for generating too large a tree. For each α, we then choose that subtree $T(\alpha)$ of $T_{\max}$ that minimizes $R_\alpha(T)$:

$$R_\alpha(T(\alpha)) = \min_{T} R_\alpha(T). \tag{9.20}$$

If $T(\alpha)$ satisfies (9.20), then it is called a *minimizing subtree* (or an *optimally-pruned subtree*) of $T_{\max}$. For any α, there may be more than one minimizing subtree of $T_{\max}$.

The value of α determines the tree size. When α is very small, the penalty term will be small, and so the size of the minimizing subtree $T(\alpha)$, which will essentially be determined by $R^{re}(T(\alpha))$, will be large. For example, suppose we set $\alpha = 0$ and grow the tree $T_{\max}$ so large that each terminal node contains only a single observation; then, each terminal node takes on the class of its solitary observation, every observation is classified correctly, and $R^{re}(T_{\max}) = 0$. So, $T_{\max}$ minimizes $R_0(T)$. As we increase α, the minimizing subtrees $\dot{T}(\alpha)$ will have fewer and fewer terminal nodes. When α is very large, we will have pruned the entire tree $T_{\max}$, leaving only the root node.

Note that although α is defined on the interval $[0, \infty)$, the number of subtrees of T is finite. Suppose that, for $\alpha = \alpha_1$, the minimizing subtree is $T_1 = T(\alpha_1)$. As we increase the value of α, T_1 continues to be the minimizing subtree until a certain point, say, $\alpha = \alpha_2$, is reached, and a new subtree, $T_2 = T(\alpha_2)$, becomes the minimizing subtree. As we increase α further, the subtree T_2 continues to be the minimizing subtree until a value of α is reached, $\alpha = \alpha_3$, say, when a new subtree $T_3 = T(\alpha_3)$ becomes the minimizing subtree. This argument is repeated a finite number of times to produce a sequence of minimizing subtrees $T_1, T_2, T_3, \ldots$.

How do we get from $T_{\max}$ to T_1? Suppose the node τ in the tree $T_{\max}$ has daughter nodes τ_L and τ_R, both of which are terminal nodes. Then,

$$R^{re}(\tau) \geq R^{re}(\tau_L) + R^{re}(\tau_R) \tag{9.21}$$

(Breiman et al., 1984, Proposition 4.2). For example, in the classification tree for the Pima Indians diabetes study (Figure 9.5), the lowest subtree has a root node with 13 normals and 8 diabetics, whereas its left daughter node has 10 normals and 3 diabetics and its right daughter node has 3 normals and 5 diabetics. Thus, $R^{re}(\tau) = 8/532 > R^{re}(\tau_L) + R^{re}(\tau_R) = (3 + 3)/532 = 6/532$. If equality occurs in (9.21) at node τ, then prune the terminal nodes τ_L and τ_R from the tree. Continue this pruning strategy until no further pruning of this type is possible. The resulting tree is T_1.

Next, we find T_2. Let τ be any nonterminal node of T_1, let T_τ be the subtree whose root node is τ, and let $\tilde{T}_\tau = \{\tau'_1, \tau'_2, \ldots, \tau'_{L_\tau}\}$ be the set of terminal nodes of T_τ. Let

$$R^{re}(T_\tau) = \sum_{\tau' \in \tilde{T}_\tau} R^{re}(\tau') = \sum_{\ell'=1}^{L_\tau} R^{re}(\tau'_{\ell'}). \tag{9.22}$$

Then, $R^{re}(\tau) > R^{re}(T_\tau)$ (Breiman et al., 1984, Proposition 3.8). For example, from Figure 9.5, let τ be the nonterminal node on the right-hand side of the tree near the center of the tree having 18 normals and 26 diabetics, and let T_τ be the subtree with τ as its root node. Then, $R^{re}(\tau) = 18/532 > R^{re}(T_\tau) = (3 + 3 + 3 + 2)/532 = 11/532$. Now, set

$$R_\alpha(T_\tau) = R^{re}(T_\tau) + \alpha|\tilde{T}_\tau|. \tag{9.23}$$

As long as $R_\alpha(\tau) > R_\alpha(T_\tau)$, the subtree T_τ has a smaller cost-complexity than its root node τ, and, therefore, it pays to retain T_τ. For the previous example, we retain T_τ as long as $R_\alpha^{re}(\tau) = 18/532 + \alpha > 11/532 + 4\alpha = R_\alpha^{re}(T_\tau)$, or $\alpha < 7/(3 \cdot 532) = 0.0044$.

Substituting (9.18) and (9.23) into this condition and solving for α yields

$$\alpha < \frac{R^{re}(\tau) - R^{re}(T_\tau)}{|\widetilde{T}_\tau| - 1}. \tag{9.24}$$

So, the right-hand side of (9.24), which is positive, computes the reduction in R^{re} (due to going from a single node to the subtree with that node as root) relative to the increase in the number of terminal nodes. For $\tau \in T_1$, define

$$g_1(\tau) = \frac{R^{re}(\tau) - R^{re}(T_{1,\tau})}{|\widetilde{T}_{1,\tau}| - 1}, \quad \tau \notin \widetilde{T}(\alpha_1), \tag{9.25}$$

where $T_{1,\tau}$ is the same as T_τ. Then, $g_1(\tau)$ can be regarded as a critical value for α: as long as $g_1(\tau) > \alpha_1$, we do not prune the nonterminal nodes $\tau \in T_1$.

We define the *weakest-link node* $\widetilde{\tau}_1$ as the node in T_1 that satisfies

$$g_1(\widetilde{\tau}_1) = \min_{\tau \in T_1} g_1(\tau). \tag{9.26}$$

As α increases, $\widetilde{\tau}_1$ is the first node for which $R_\alpha(\tau) = R_\alpha(T_\tau)$, so that $\widetilde{\tau}_1$ is preferred to $T_{\widetilde{\tau}_1}$. Set $\alpha_2 = g_1(\widetilde{\tau}_1)$ and define the subtree $T_2 = T(\alpha_2)$ of T_1 by pruning away the subtree $T_{\widetilde{\tau}_1}$ (so that $\widetilde{\tau}_1$ becomes a terminal node) from T_1.

To find T_3, we find the weakest-link node $\widetilde{\tau}_2 \in T_2$ through the critical value

$$g_2(\tau) = \frac{R^{re}(\tau) - R^{re}(T_{2,\tau})}{|\widetilde{T}_{2,\tau}| - 1}, \quad \tau \in T(\alpha_2), \ \tau \notin \widetilde{T}(\alpha_2), \tag{9.27}$$

where $T_{2,\tau}$ is that part of T_τ which is contained in T_2. We set

$$\alpha_3 = g_2(\widetilde{\tau}_2) = \min_{\tau \in T_2} g_2(\tau), \tag{9.28}$$

and define the subtree T_3 of T_2 by pruning away the subtree $T_{\widetilde{\tau}_2}$ (so that $\widetilde{\tau}_2$ becomes a terminal node) from T_2. And so on for a finite number of steps.

As we noted above, there may be several minimizing subtrees for each α. How do we choose between them? For a given value of α, we call $T(\alpha)$ the *smallest minimizing subtree* if it is a minimizing subtree (i.e., satifies (9.20)) and satisfies the following condition:

$$\text{if } R_\alpha(T) = R_\alpha(T(\alpha)), \text{then } T \succ T(\alpha). \tag{9.29}$$

In (9.29), $T \succ T(\alpha)$ means that $T(\alpha)$ is a subtree of T and, hence, has fewer terminal nodes than T. This condition says that, in the event of any ties, $T(\alpha)$ is taken to be the smallest tree out of all those trees that minimize

R_α. Breiman et al. (1984, Proposition 3.7) showed that for every α, there exists a unique smallest minimizing subtree.

The above construction gives us a finite increasing sequence of complexity parameters,

$$0 = \alpha_0 < \alpha_1 < \alpha_2 < \alpha_3 < \cdots < \alpha_M, \tag{9.30}$$

which corresponds to a finite sequence of nested subtrees of $T_{\max}$,

$$T_{\max} = T_0 \succ T_1 \succ T_2 \succ T_3 \succ \cdots \succ T_M, \tag{9.31}$$

where $T_k = T(\alpha_k)$ is the unique smallest minimizing subtree for $\alpha \in [\alpha_k, \alpha_{k+1})$, and T_M is the root-node subtree. We start with T_1 and increase α until $\alpha = \alpha_2$ determines the weakest-link node $\tilde{\tau}_1$; we then prune the subtree $T_{\tilde{\tau}_1}$ with that node as root. This gives us T_2. We repeat this procedure by finding $\alpha = \alpha_3$ and the weakest-link node $\tilde{\tau}_2$ in T_2 and prune the subtree $T_{\tilde{\tau}_2}$ with that node as root. This gives us T_3. This pruning process is repeated until we arrive at T_M.

Example: Pima Indians Diabetes Study (Continued)

The sequence of seven pruned classification trees, T_k, corresponding to their critical values, α_k, are listed in Table 9.4. The tree displayed in Figure 9.5 has 14 splits (and, hence, 15 terminal nodes).

Any value of $\alpha < 0.0038$ will produce a tree with 15 terminal nodes. When $\alpha = 0.0038$, the classification tree is pruned to have 11 splits (and 12 terminal nodes), which will remain the same for all $0.0038 \leq \alpha < 0.0047$. Increasing α to 0.0047 prunes the tree to 9 splits (and 10 terminal nodes). And so on, until α is increased above 0.0883 when the tree consists only of the root node.

9.2.7   Choosing the Best Pruned Subtree

Thus far, we have constructed a finite sequence of decreasing-size subtrees $T_1, T_2, T_3, \ldots, T_M$ by pruning more and more nodes from $T_{\max}$. When do we stop pruning? Which subtree of the sequence do we choose as the "best" pruned subtree?

Choice of the best subtree depends upon having a good estimate of the misclassification rate $R(T_k)$ corresponding to the subtree T_k. Breiman et al. (1984) offered two estimation methods: use an independent test sample or use cross-validation. When the data set is very large, use of an independent test set is straightforward and computationally efficient, and is, generally, the preferred estimation method. For smaller data sets, cross-validation is preferred.

TABLE 9.4. *Pruned classification trees for the Pima Indians diabetes study. The impurity function is the Gini index. By increasing the complexity parameter α, seven classification trees, T_k, $k = 1, 2, \ldots, 6$, are derived, where the tree details are listed so that $T_k \succ T_{k+1}$; i.e., largest tree to smallest tree. Also listed for each tree are the number of terminal nodes ($|\widetilde{T}_k|$), resubstitution error (R^{re}), and 10-fold cross-validation (CV) error ($R^{CV/10}$). The $\pm$ values on the CV error are the CV standard errors ($\widehat{SE}$). The CV error estimate and its estimated standard error produce random values according to the random CV-partition of the data.*

k	α_k	$|\widetilde{T}_k|$	$R^{re}(T_k)$	$R^{CV/10}(T_k)$
1		15	0.141	0.258 ± 0.019
2	0.0038	12	0.152	0.233 ± 0.018
3	0.0047	10	0.162	0.233 ± 0.018
4	0.0069	6	0.190	0.235 ± 0.018
5	0.0085	4	0.207	0.256 ± 0.019
6	0.0188	2	0.244	0.256 ± 0.019
7	0.0883	1	0.333	0.333 ± 0.020

Independent Test Set

Randomly assign the observations in the data set $\mathcal{D}$ into a learning set $\mathcal{L}$ and a test set $\mathcal{T}$, where $\mathcal{D} = \mathcal{L} \cup \mathcal{T}$ and $\mathcal{L} \cap \mathcal{T} = \emptyset$. Suppose there are $n_{\mathcal{T}}$ observations in the test set and that they are drawn independently from the same underlying distribution as the observations in $\mathcal{L}$. Grow the tree $T_{\max}$ from the learning set only, prune it from the bottom up to give the sequence of subtrees $T_1 \succ T_2 \succ T_3 \succ \cdots \succ T_M$, and assign a class to each terminal node.

Take each of the $n_{\mathcal{T}}$ test-set observations and drop it down the subtree T_k. Each observation in $\mathcal{T}$ is then classified into one of the different classes. Because the true class of each observation in $\mathcal{T}$ is known, we estimate $R(T_k)$ by $R^{ts}(T_k)$, which is (9.19) with $\alpha = 0$; that is, $R^{ts}(T_k) = R^{re}(T_k)$, the resubstitution estimate computed using the independent test set. When the costs of misclassification are identical for each class, $R^{ts}(T_k)$ is the proportion of all test set observations that are misclassified by T_k. These estimates are then used to select the best-pruned subtree T_* by the rule

$$R^{ts}(T_*) = \min_k R^{ts}(T_k), \tag{9.32}$$

and $R^{ts}(T_*)$ is its estimated misclassification rate.

We estimate the standard error of $R^{ts}(T)$ as follows. When we drop the test set $\mathcal{T}$ down a tree T, the chance that we misclassify any one of those observations is $p^* = R(T)$. Thus, we have a binomial sampling situation with $n_{\mathcal{T}}$ Bernoulli trials and probability of success p^*. If $p = R^{ts}(T)$ is the proportion of misclassified observations in $\mathcal{T}$, then, p is unbiased for p^* and the variance of p is $p^*(1-p^*)/n_{\mathcal{T}}$. The standard error of $R^{ts}(T)$ is, therefore, estimated by

$$\widehat{\mathrm{SE}}(R^{ts}(T)) = \left\{ \frac{R^{ts}(T)(1-R^{ts}(T))}{n_{\mathcal{T}}} \right\}^{1/2}. \tag{9.33}$$

Cross-Validation

In V-fold cross-validation (CV/V), we randomly divide the data $\mathcal{D}$ into V roughly equal-size, disjoint subsets, $\mathcal{D} = \bigcup_{v=1}^{V} \mathcal{D}_v$, where $\mathcal{D}_v \cap \mathcal{D}_{v'} = \emptyset$, $v \neq v'$, and V is usually taken to be 5 or 10. We next create V different data sets from the $\{\mathcal{D}_v\}$ by taking $\mathcal{L}_v = \mathcal{D} - \mathcal{D}_v$ as the vth learning set and $\mathcal{T}_v = \mathcal{D}_v$ as the vth test set, $v = 1, 2, \ldots, V$. If the $\{\mathcal{D}_v\}$ each have the same number of observations, then each learning set will have $\left(\frac{V-1}{V}\right) \times 100$ percent of the original data set.

Grow the vth "auxilliary" tree $T_{\max}^{(v)}$ using the vth learning set $\mathcal{L}_v$, $v = 1, 2, \ldots, V$. Fix the value of the complexity parameter α. Let $T^{(v)}(\alpha)$ be the best pruned subtree of $T_{\max}^{(v)}$, $v = 1, 2, \ldots, V$. Now, drop each observation in the vth test set $\mathcal{T}_v$ down the tree $T^{(v)}(\alpha)$, $v = 1, 2, \ldots, V$. Let $n_{ij}^{(v)}(\alpha)$ denote the number of jth class observations in $\mathcal{T}_v$ that are classified as being from the ith class, $i, j = 1, 2, \ldots, K$, $v = 1, 2, \ldots, V$. Because $\mathcal{D} = \bigcup_{v=1}^{V} \mathcal{T}_v$ is a disjoint sum, the total number of jth class observations that are classified as being from the ith class is $n_{ij}(\alpha) = \sum_{v=1}^{V} n_{ij}^{(v)}(\alpha)$, $i, j = 1, 2, \ldots, K$. If we set n_j to be the number of observations in $\mathcal{D}$ that belong to the jth class, $j = 1, 2, \ldots, K$, and assume that misclassification costs are equal for all classes, then, for a given α,

$$R^{CV/V}(T(\alpha)) = n^{-1} \sum_{i=1}^{K} \sum_{j=1}^{K} n_{ij}(\alpha) \tag{9.34}$$

is the estimated misclassification rate over $\mathcal{D}$, where $T(\alpha)$ is a minimizing subtree of $T_{\max}$.

The final step in this process is to find the right-sized subtree. Breiman et al. (1984, p. 77) recommend evaluating (9.24) at the sequence of values $\alpha'_k = \sqrt{\alpha_k \alpha_{k+1}}$, where α'_k is the geometric midpoint of the interval $[\alpha_k, \alpha_{k+1})$ in which $T(\alpha) = T_k$. Set

$$R^{CV/V}(T_k) = R^{CV/V}(T(\alpha'_k)). \tag{9.35}$$

Then, select the best-pruned subtree T_* by the rule:

$$R^{CV/V}(T_*) = \min_k R^{CV/V}(T_k), \tag{9.36}$$

and use $R^{CV/V}(T_*)$ as its estimated misclassification rate.

Deriving an estimated standard error of the cross-validated estimate of the misclassification rate is more complicated than using a test set. The usual way of sidestepping issues of non-independence of the summands in (9.29) is to ignore them and pretend instead that independence holds. Actually, this approximation appears to work well in practice. See Breiman et al. (1984, Section 11.5) for details.

It is usual to take $V = 10$ for 10-fold CV. The leave-one-out CV method (i.e., $V = n$) is not recommended because the resulting auxilliary trees will be almost identical to the tree constructed from the full data set, and so nothing would be gained from this procedure.

The One-SE Rule

To overcome possible instability in selecting the best-pruned subtree, Breiman et al. (1984, Section 3.4.3) propose an alternative rule.

Let $\widehat{R}(T_*) = \min_k R(T_k)$ denote the estimated misclassification rate, calculated from either a test set (i.e., $R^{ts}(T_*)$) or cross-validation (i.e., $R^{CV/V}(T_*)$). Then, we choose the smallest tree T_{**} that satisfies the "1-SE rule," namely,

$$\widehat{R}(T_{**}) \leq \widehat{R}(T_*) + \widehat{\mathrm{SE}}(\widehat{R}(T_*)). \tag{9.37}$$

This rule appears to produce a better subtree than using T_* because it responds to the variability (through the standard error) of the cross-validation estimates.

Example: Pima Indians Diabetes Study (Continued)

For example, we apply the 1-SE rule to the Pima Indians diabetes study. From Table 9.4, the 1-SE rule yields a minimum of CV error + SE = 0.233 + 0.018 = 0.251, which leads to the choice of a classification tree with 9 splits (10 terminal nodes) based upon cross-validation. The corresponding pruned classification tree is displayed in Figure 9.6.

A diagnosis of diabetes is given to those subjects who have one of the following symptoms:

1. plasma glucose level at least 157.5;

2. plasma glucose level between 127.5 and 157.5, bmi at least 30.2, and age at least 42.5 years;

3. plasma glucose level between 127.5 and 157.6, bmi at least 30.2, age less than 42.5 years, and a pedigree at least 0.285;

4. plasma glucose level between 96.5 and 127.5, age at least 28.5 years, a pedigree at least 0.62, and bmi at least 26.5.

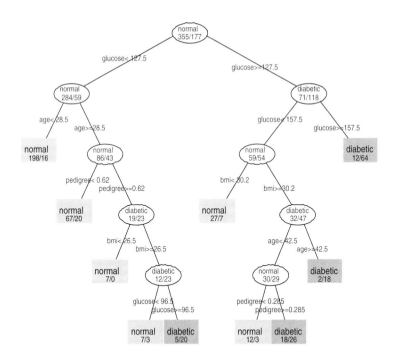

FIGURE 9.6. *A pruned classification tree for the Pima Indians diabetes data, with 9 splits and 10 terminal nodes, where the impurity measure is the Gini index. The terminal nodes are colored green for* normal *and pink for* diabetic.

This tree has a resubstitution error rate of $86/532 = 0.162$ and 10-fold CV misclassification rate of 0.233 ± 0.018.

9.2.8   Example: Vehicle Silhouettes

Consider the vehicle data[3] of Section 8.7, which were collected to study how well 3D objects could be distinguished by their 2D silhouette images. There are four classes of objects, each of which was a CORGI model vehicle: an Opel Manta car (opel, 212 images), a Saab 9000 car (saab, 217 images), a double-decker bus (bus, 218 images), and a Chevrolet van (van, 199 images), giving a total of 846 images. Each object was viewed by a camera from many different angles and elevations. The variables are scaled variance, skewness, and kurtosis about the major/minor axes, and

---

[3]These data can be found in the UCI Machine Learning Repository.

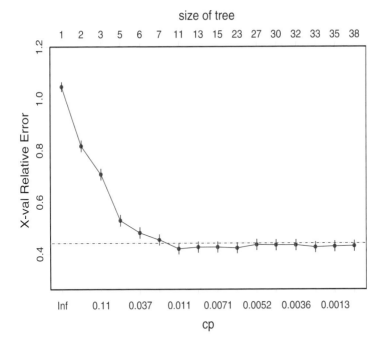

FIGURE 9.7. *Plot of 10-fold CV results of different size classification trees for the* vehicle *data. The cp-value is* α *divided by the resubstitution error rate estimate,* $R^{re}(T_0) = 628/846 = 0.742$*, for the root tree, and the vertical axis is the corresponding CV error rate also divided by* $R^{re}(T_0)$*. The vertical lines indicate* $\pm$ *two SE for each CV error estimate. The recommended tree size has cp equal to the smallest tree with the minimum CV error; in this case, 11 terminal nodes.*

heuristic measures such as `hollows ratio`, `circularity`, `elongatedness`, `rectangularity`, and `compactness` of the silhouettes.

Based upon the One–SE rule, and the resulting complexity-parameter plot in Figure 9.7, the most appropriate classification tree has 10 splits with 11 terminal nodes, with a resubstitution error rate of $0.3535 \times 0.74232 = 0.262$, and CV error rate of 0.299 ± 0.0157. In Figure 9.8, we have displayed the pruned classification tree with 10 splits and 11 terminal nodes.

9.3   Regression Trees

Suppose the data are given by $\mathcal{D} = \{(\mathbf{X}_i, Y_i), i = 1, 2, \ldots, n\}$, where the Y_i are measurements made on a continuous response variable Y, and

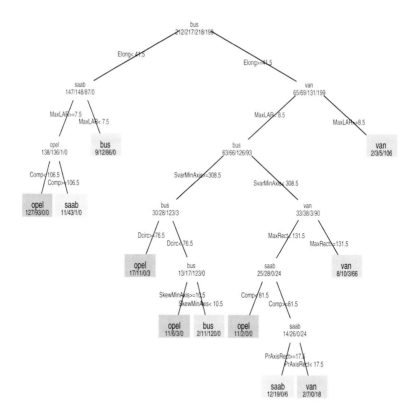

FIGURE 9.8. *A pruned classification tree for the* `vehicle` *data. There are 12 input variables, 846 observations, and four classes of vehicle models: opel (pink), saab (yellow), bus (green), and van (blue), whose numbers at each node are given by a/b/c/d, respectively, There are 10 splits and 11 terminal nodes in this tree. The resubstitution error rate is 0.262.*

the $\mathbf{X}_i$ are measurements on an input r-vector $\mathbf{X}$. We assume that Y is related to $\mathbf{X}$ as in multiple regression (see Chapter 5), and we wish to use a tree-based method to predict Y from $\mathbf{X}$.

Regression trees are constructed similarly to classification trees, and the method is generally referred to as *recursive-partitioning regression*. In a classification tree, the class of a terminal node is defined as that class that commands a plurality (a majority in the two-class case) of all the observations in that node, where ties are decided at random. In a regression tree, the output variable is set to have the constant value $Y(\tau)$ at terminal node τ. Hence, the tree can be represented as an r-dimensional histogram estimate of the regression surface, where r is the number of input variables, $X_1, X_2, \ldots, X_r$.

9.3.1   The Terminal-Node Value

How do we find $Y(\tau)$? Recall (from Chapter 5) that the resubstitution estimate of prediction error is

$$R^{re}(\widehat{\mu}) = \frac{1}{n}\sum_{i=1}^{n}(Y_i - \widehat{Y}_i)^2, \qquad (9.38)$$

where $\widehat{Y}_i = \widehat{\mu}(\mathbf{X}_i)$ is the estimated value of the predictor at $\mathbf{X}_i$. For $\widehat{Y}_i$ to be constant at each node, the predictor has to have the form

$$\widehat{\mu}(\mathbf{X}) = \sum_{\tau \in \widetilde{T}} Y(\tau)I_{[\mathbf{X}\in\tau]} = \sum_{\ell=1}^{L} Y(\tau_\ell)I_{[\mathbf{X}\in\tau_\ell]}, \qquad (9.39)$$

where $I_{[\mathbf{X}\in\tau_\ell]}$ is equal to one if $\mathbf{X} \in \tau_\ell$ and zero otherwise. For $\mathbf{X}_i \in \tau_\ell$, $R^{re}(\widehat{\mu})$ is minimized by taking $\widehat{Y}_i = \bar{Y}(\tau_\ell)$ as the constant value $Y(\tau_\ell)$, where $\bar{Y}(\tau_\ell)$ is the average of the $\{Y_i\}$ for all observations assigned to node τ_ℓ; that is,

$$\bar{Y}(\tau_\ell) = \frac{1}{n(\tau_\ell)}\sum_{\mathbf{X}_i\in\tau_\ell} Y_i, \qquad (9.40)$$

where $n(\tau_\ell)$ is the total number of observations in node τ_ℓ. Changing notation slightly to reflect the tree structure, the resubstitution estimate is

$$R^{re}(T) = \frac{1}{n}\sum_{\ell=1}^{L}\sum_{\mathbf{X}_i\in\tau_\ell}(Y_i - \bar{Y}(\tau_\ell))^2 = \sum_{\ell=1}^{L}R^{re}(\tau_\ell), \qquad (9.41)$$

where

$$R^{re}(\tau_\ell) = \frac{1}{n}\sum_{\mathbf{X}_i\in\tau_\ell}(Y_i - \bar{Y}(\tau_\ell))^2 = p(\tau_\ell)s^2(\tau_\ell), \qquad (9.42)$$

$s^2(\tau_\ell)$ is the (biased) sample variance of all the Y_i values in node τ_ℓ, and $p(\tau_\ell) = n(\tau_\ell)/n$ is the proportion of observations in node τ_ℓ. Hence, $R^{re}(T) = \sum_{\ell=1}^{L} p(\tau_\ell)s^2(\tau_\ell)$.

9.3.2   Splitting Strategy

How do we determine the type of split at any given node of the tree? We take as our splitting strategy at node $\tau \in \widetilde{T}$ the split that provides the biggest reduction in the value of $R^{re}(T)$. The reduction in $R^{re}(\tau)$ due to a split into τ_L and τ_R is given by

$$\Delta R^{re}(\tau) = R^{re}(\tau) - R^{re}(\tau_L) - R^{re}(\tau_R); \qquad (9.43)$$

the best split at τ is then the one that maximizes $\Delta R^{re}(\tau)$. The result of employing such a splitting strategy is that the best split will divide up observations according to whether Y has a small or large value; in general, where splits occur, we see either $\bar{y}(\tau_L) < \bar{y}(\tau) < \bar{y}(\tau_R)$ or its reverse with $\bar{y}(\tau_L)$ and $\bar{y}(\tau_R)$ interchanged.

We note that finding τ_L and τ_R to maximize $\Delta R^{re}(\tau)$ is equivalent to minimizing $R^{re}(\tau_L) + R^{re}(\tau_R)$. From (9.42), this boils down to finding τ_L and τ_R to solve

$$\min_{\tau_L, \tau_R} \{p(\tau_L)s^2(\tau_L) + p(\tau_R)s^2(\tau_R)\}, \tag{9.44}$$

where $p(\tau_L)$ and $p(\tau_R)$ are the proportions of observations in τ that split to τ_L and τ_R, respectively.

9.3.3 Pruning the Tree

The method for pruning a regression tree incorporates the same ideas as is used to prune a classification tree.

As before, we first grow a large tree, $T_{\max}$, by splitting nodes repeatedly until each node contains fewer than a given number of observations; that is, until $n(\tau) \leq n_{\min}$ for each $\tau \in \widetilde{T}$, where we typically set $n_{\min} = 5$.

Next, we set up an *error-complexity measure*,

$$R_\alpha(T) = R^{re}(T) + \alpha|\widetilde{T}|, \tag{9.45}$$

where $\alpha \geq 0$ is a complexity parameter. Use $R_\alpha(T)$ as the criterion for deciding when and how to split, just as we did in pruning classification trees. The result is a sequence of subtrees,

$$T_{\max} = T_0 \succ T_1 \succ T_2 \succ T_3 \succ \cdots \succ T_M, \tag{9.46}$$

and an associated sequence of complexity parameters,

$$0 = \alpha_0 < \alpha_1 < \alpha_2 < \alpha_3 < \cdots < \alpha_M, \tag{9.47}$$

such that for $\alpha \in [\alpha_k, \alpha_{k+1})$, T_k is the smallest minimizing subtree of $T_{\max}$.

9.3.4 Selecting the Best Pruned Subtree

We estimate $R(T_k)$ by using an independent test set or by cross-validation. The details follow those in Section 9.2.6.

For an independent test set, $\mathcal{T}$, an estimate of $R(T_k)$ is given by

$$R^{ts}(T_k) = \frac{1}{n_\mathcal{T}} \sum_{(\mathbf{X}_i, Y_i) \in \mathcal{T}} (Y_i - \widehat{\mu}_k(\mathbf{X}_i))^2, \tag{9.48}$$

where $n_{\mathcal{T}}$ is the number of observations in the test set and $\widehat{\mu}_k(\mathbf{X})$ is the estimated prediction function associated with subtree T_k.

For a V-fold cross-validated estimate of $R(T_k)$, we first construct the minimal error-complexity subtrees $T^{(v)}(\alpha)$, $v = 1, 2, \ldots, V$, parameterized by α. Set $\alpha'_k = \sqrt{\alpha_k \alpha_{k+1}}$ and let $\widehat{\mu}_k^{(v)}(\mathbf{x})$ denote the estimated prediction function associated with the subtree $T^{(v)}(\alpha'_k)$. The V-fold CV estimate of $R(T_k)$ is given by

$$R^{CV/V}(T_k) = n^{-1} \sum_{v=1}^{V} \sum_{(\mathbf{X}_i, Y_i) \in \mathcal{T}_v} (Y_i - \widehat{\mu}_k^{(v)}(\mathbf{X}_i))^2. \qquad (9.49)$$

We usually select $V = 10$ for a 10-fold CV estimate in which we split the learning set into 10 subsets, use 9 of those 10 subsets to grow and prune the tree, and then use the omitted subset to test the results of the tree.

Given the sequence of subtrees $\{T_k\}$, we select the smallest subtree T_{**} for which

$$\widehat{R}(T_{**}) \leq \widehat{R}(T_*) + \widehat{\mathrm{SE}}(\widehat{R}(T_*)), \qquad (9.50)$$

where $\widehat{R}(T_*) = \min_k \widehat{R}(T_k)$ is the estimated prediction error calculated using using either an independent test set (i.e., $R^{ts}(T_*)$) or cross-validation (i.e., $R^{CV/V}(T_*)$).

9.3.5 Example: 1992 Major League Baseball Salaries

As an example of a regression tree, we use data on the salaries of Major League Baseball (MLB) players for 1992 (Watnik, 1998).[4] The data consist of $n = 337$ MLB players who played at least one game in both the 1991 and 1992 seasons, excluding pitchers. The interesting aspect of these data is that a player's "value" is judged by his performance measures, which in turn could be used to determine his salary the next year or possibly to enable him to change his employer.

The output variable is the 1992 salaries (in thousands of dollars) of these players, and the input variables are the following performance measures from 1991: BA (batting average), OBP (on-base percentage), Runs (number of runs scored), Hits (number of hits), 2B (number of doubles), 3B (number of triples), HR (number of home runs), RBI (number of runs batted in), BB (number of bases on balls or walks), SO (number of strikeouts), SB (number of stolen bases), and E (number of errors made). Also included as input

---

[4]These data can be found at the website of the *Journal of Statistics Education*, www.amstat.org/publications/jse/jse_data_archive.html. Sources for these data are *CNN/Sports Illustrated*, *Sacramento Bee* (15th October 1991), *The New York Times* (19th November 1992), and the Society for American Baseball Research.

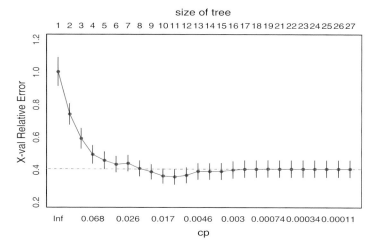

FIGURE 9.9. *Plot of 10-fold CV results of different size regression trees for 1992 baseball salary data. The cp-value is α divided by the resubstitution estimate, $R^{re}(T_0)$, for the root tree, and the vertical axis is the CV error also divided by $R^{re}(T_0)$. The vertical lines indicate $\pm$ two SE for each CV error estimate. The recommended amount of pruning is to set cp equal to the smallest tree with the minimum CV error; in this case, 11 terminal nodes.*

variables are the following four indicator variables: FAE (indicator of free-agent eligibility), FA (indicator of free agent in 1991/92), AE (indicator of arbitration eligibility), A (indicator of arbitration in 1991/92). These four variables indicated how free each player was to move to other teams. A player's BA is the ratio of number of hits to the total number of "at-bats" for that player (whether resulting in a hit or an out). The OBP is the ratio of number of hits plus the number of walks to the number of hits plus the number of walks plus the number of outs. For reference, a BA above 0.3 is very good, and an OBP above 0.4 is excellent. An RBI occurs when a runner scores as a direct result of a player's at-bat.

The plot of the CV results for this example is given in Figure 9.9, where the minimum value of the CV error occurs for a tree size of 10 terminal nodes. The pruned regression tree with 10 splits and 11 terminal nodes corresponding to the minimum 1–SE rule is given in Figure 9.10. We see from the terminal node on the right-hand side of the tree that the 14 players who score at least 46.5 runs have at least 94.5 RBIs, and are eligible for free-agency to earn the highest average salary ($3,897,214). The lowest average salary ($232,898), which is made by 108 players, is located at the terminal node on the left-hand side of the tree. We also see that performing well on at least one measure produces substantial differences in average salary. The resubstitution estimate (9.41) of prediction error for

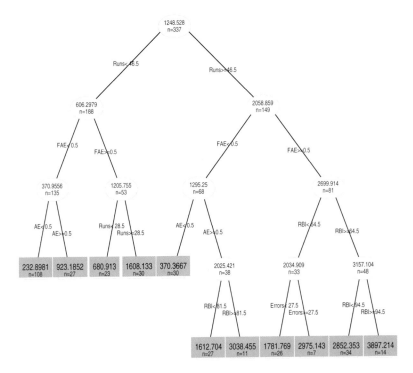

FIGURE 9.10. *Pruned regression tree for 1992 baseball salary data. The label of each node indicates the mean* `salary`, *in thousands of dollars, for the number n of players who fall into that node.*

this regression tree is $R^{re}(T) = \$341,841$, the cross-validation estimate of prediction error is \$549,217, and the cross-validation standard deviation is \$74,928. By comparison, regressing `Salary` on the 15 input variables in a multiple regression yields a residual sum of squares of \$155,032,181 and a residual mean square of \$482,966 based upon 321 df.

9.4   Extensions and Adjustments

9.4.1   Multivariate Responses

Some work has been carried out on constructing classification trees for multivariate responses, especially where each response is binary (Zhang, 1998). In such cases, the measure of within-node homogeneity at node τ for a single binary variable is generalized to a scalar-valued function of a matrix argument. Examples include $-\log|\mathbf{V}_\tau|$, where $\mathbf{V}_\tau$ is the within-node sample covariance matrix of the s binary responses at node τ, and a node-based quadratic form in $\mathbf{V}$, the covariance matrix derived from the root node. The cost-complexity of tree T is then defined as $R_\alpha(T)$ in (9.19), where $R^{re}(T)$ is a within-node homogeneity measure summed over all terminal nodes. When dealing with multivariate responses, it is clear from an applied point of view that the amount of data available for tree construction has to be very large.

9.4.2 Survival Trees

Tree-based methods for analyzing censored survival data have become very useful tools in biomedical research, where they can identify prognostic factors for predicting survival (see, e.g., Intrator and Kooperberg, 1995). The resulting trees are called *survival trees* (or *conditional inference trees*). Survival data usually take the form of time-to-death but can be more general than that, such as time to a particular event to occur. Censored survival data occur when patients live past the conclusion of the study, leave the study prematurely, or die during the period of the study from a disease not connected to the one being studied, and survival analysis has to take such conditions into account in the inference process.

When using tree-based methods to analyze censored survival data, it is necessary to choose a criterion for making splitting decisions. There are several splitting criteria, which can be divided into two types depending upon whether one prefers to use a "within-node homogeneity" measure or a "between-node heterogeneity" measure. Most applications of the former method (see, e.g., Davis and Anderson, 1989) are parametrically based; they typically incorporate a version of minus the log-likelihood loss function, where the versions differ in the loss function used and, thus, how they represent the model for the observed data likelihood within the nodes.

The first application of recursive partitioning to the analysis of censored survival data (Gordon and Olshen, 1985) used a more nonparametric approach, basing their tree-construction on within-node Kaplan-Meier estimates of the survival distribution, and then comparing those curve estimates to within-node Kaplan-Meier estimates of truly homogeneous nodes. An example of the latter method (Segal, 1988) computes the within-node Kaplan-Meier curves for the censored survival data corresponding to each of the two daughter nodes of a possible split and then applies the two-sample log-rank statistic to the Kaplan-Meier curves to measure the goodness of that split; the largest value of the log-rank statistic over all possible splits determines which split is best.

Data that fall into a particular terminal node tend to have similar experiences of survival (based upon a measure of within-node homogeneity). Survival trees can be used to partition patients into groups having similar survival results and, hence, identify common characteristics within these groups. At each terminal node of a survival tree, we compute a Kaplan-Meier estimate of the survival curve using the survival information for all patients who are members of that node and then compare the survival curves from different terminal nodes.

9.4.3   MARS

Recursive partitioning used in constructing regression trees has been generalized to a flexible class of nonparametric regression models called *multivariate adaptive regression splines (MARS)* (Friedman, 1991).

In the MARS approach, Y is related to $\mathbf{X}$ via the model $Y = \mu(\mathbf{X}) + \epsilon$, where the error term ϵ has mean zero. The regression function, $\mu(\mathbf{X})$, is taken to be a weighted sum of L basis functions,

$$\mu(\mathbf{X}) = \beta_0 + \sum_{\ell=1}^{L} \beta_\ell B_\ell(\mathbf{X}). \tag{9.51}$$

The ℓth basis function,

$$B_\ell(\mathbf{X}) = \prod_{m=1}^{M_\ell} \phi_{\ell m}(X_{q(\ell,m)}), \tag{9.52}$$

is the product of M_ℓ univariate spline functions $\{\phi_{\ell m}(X)\}$, where M_ℓ is a finite number and $q(\ell, m)$ is an index depending upon the ℓth basis function and the mth spline function. Thus, for each ℓ, $B_\ell(\mathbf{X})$ can consist of a single spline function or a product of two or more spline functions, and no input variable can appear more than once in the product. These spline functions (for ℓ odd) are often taken to be linear of the form,

$$\phi_{\ell m}(X) = (X - t_{\ell m})_+, \quad \phi_{\ell+1,m}(X) = (t_{\ell m} - X)_+, \tag{9.53}$$

where $t_{\ell m}$ is a *knot* of $\phi_{\ell m}(X)$ occurring at one of the observed values of $X_{q(\ell,m)}$, $m = 1, 2, \ldots, M_\ell$, $\ell = 1, 2, \ldots, L$. In (9.53), $(x)_+ = \max(0, x)$. If $B_\ell(\mathbf{X}) = I_{[\mathbf{X} \in \tau_\ell]}$ and $\beta_\ell = Y(\tau_\ell)$, then the regression function (9.51) is equivalent to the regression-tree predictor (9.39). Thus, whereas regression trees fit a constant at each terminal node, MARS fits more complicated piecewise linear basis functions.

Basis function are first introduced into the model (9.51) in a forwards-stepwise manner. The process starts by entering the intercept β_0 (i.e., $B_0(\mathbf{X}) = 1$) into the model, and then at each step adding one pair of terms of the form (9.53) (i.e., choosing an input variable and a knot) by minimizing an error sum of squares criterion,

$$ESS(L) = \sum_{i=1}^{n} (y_i - \mu_L(\mathbf{x}_i))^2, \tag{9.54}$$

where, for a given L, $\mu_L(\mathbf{x}_i)$ is (9.51) evaluated at $\mathbf{X} = \mathbf{x}_i$. Suppose the forwards-stepwise procedure terminates at M terms. This model is then "pruned back" by using a backwards-stepwise procedure to prevent possibly overfitting the data. At each step in the backwards-stepwise procedure, we remove one term from the model. This yields M different nested models. To choose between these M models, MARS uses a version of generalized cross-validation (GCV),

$$GCV(m) = \frac{n^{-1} \sum_{i=1}^{n} (y_i - \widehat{\mu}_m(\mathbf{x}_i))^2}{\left(1 - \frac{C(m)}{n}\right)^2}, \quad m = 1, 2, \ldots, M, \qquad (9.55)$$

where $\widehat{\mu}_m(\mathbf{x})$ is the fitted value of $\mu(\mathbf{x})$ based upon m terms, the numerator is the apparent error rate (or resubstitution error rate), and $C(m)$ is a *complexity cost function* that represents the *effective number of parameters* in the model (Craven and Wahba, 1979). The best choice of model has $m^* = \arg\min_m GCV(m)$ terms.

9.4.4   Missing Data

In some classification and regression problems, there may be missing values in the test set. Fortunately, there are a number of ways of dealing with missing data when using tree-based methods.

One obvious way is to drop a future observation with a missing data value (or values) down the tree constructed using only complete-data observations and see how far it goes. If the variable with the missing value is not involved in the construction of the tree, then the observation will drop to its appropriate terminal node, and we can then classify the observation or predict its Y value. If, on the other hand, the observation cannot drop any further than a particular internal node τ (because the next split at τ involves the variable with the missing value), we can either stop the observation at τ (Clark and Pregibon, 1992, Section 9.4.1) or force all the observations with a missing value for that variable to drop down to the same daughter node (Zhang and Singer, 1999, Section 4.8).

A method of *surrogate splits* has been proposed (Breiman et al., 1984, Section 5.3) to deal with missing data. The idea of a surrogate split at a given node τ is that we use a variable that best predicts the desired split as a substitute variable on which to split at node τ. If the best-splitting variable for a future observation at τ has a missing value at that split, we use a surrogate split at τ to force that observation further down the tree, assuming, of course, that the variable defining the surrogate split has complete data.

If the missing data occur for a nominal input variable with L levels, then we could introduce an additional level of "missing" or "NA" so that the variable now has $L + 1$ levels (Kass, 1980).

9.5   Software Packages

The original CART software is commercially available from Salford Systems. S-PLUS and R commands for classification and regression trees are discussed in Venables and Ripley (2002, Chapter 9). For the rpart library manual, which we used for the examples in this chapter, see Therneau and Atkinson (1997). Alternative software packages for carrying out tree-based classification and regression are available; they have been implemented in SAS DATA MINING, SPSS CLASSIFICATION TREES, STATISTICA, and SYSTAT, version 7. These versions differ in several aspects, including the impurity measure (typical default is the entropy function), splitting criterion, and the stopping rule.

The original MARS software is also commercially available from Salford Systems. The mars command in the mda library (Venables and Ripley, 2002, Section 8.8) in S-PLUS and R is available for fitting MARS models.

Bibliographical Notes

This chapter follows the pioneering development of CART (Classification and Regression Trees) by Breiman, Friedman, Olshen, and Stone (1984). Other treatments of the same material can be found in Clark and Pregibon (1992, Chapter 9), Ripley (1996, Chapter 7), Zhang and Singer (1999), and Hastie, Tibshirani, and Friedman (2001, Section 9.2).

Regression trees were introduced by Morgan and Sonquist (1963) using a computer program they named *Automatic Interaction Detection (AID)*. Versions of AID followed: THAID in 1973 and CHAID in 1980; CHAID is used in several computer packages that carry out tree-based methods. Comments and references on the historical development of tree-based methods are given in Ripley (1996, Section 7.4). An excellent discussion of survival trees is given by Zhang and Singer (1999). For discussions of MARS, see Hastie, Tibshirani, and Friedman (2001, Section 9.4) and Zhang and Singer (1999, Chapter 9).

Exercises

9.1 The development of classification trees in this chapter assumes that misclassifying any observation has a cost independent of the classes involved.

In many circumstances, this may be unrealistic. For example, a civilized society usually considers convicting an innocent person to be more egregious than finding a guilty person to be not guilty. Define the *misclassification cost* $c(i|j)$ as the cost of misclassifying an observation from the jth class into the ith class. Assume that $c(i|j)$ is nonnegative for $i \neq j$ and zero when $i = j$. Rewrite Sections 9.2.4, 9.2.5, and 9.2.6, taking into account the costs of misclassification.

9.2 The discussion of the way to choose the best split for a classification tree in Section 9.2 used the entropy function as the impurity measure. Use the Gini index as an impurity measure on the Cleveland heart-disease data and determine the best split for the age variable (see Table 9.2); draw the graphs of $i(\tau_l)$ and $i(\tau_R)$ for the age variable and the goodness of split (see Figure 9.3). Determine the best split for all the variables in the data set (see Table 9.3).

9.3 The full Pima Indians data (768 subjects) has a large number of missing data. In the data set, missing values are designated by zero values. How could you use those subjects having missing values for one or more variables to enhance the classification results discussed in the text?

9.4 Consider the following two examples. Both examples start out with a root node with 800 subjects of which 400 have a given disease and the other 400 do not. The first example splits the root node as follows: the left node has 300 with the disease and 100 without, and the right node has 100 with the disease and 300 without. The second example splits the root node as follows: the left node has 200 with the disease and 400 without, and the right node has 200 with the disease and 0 without. Compute the resubstitution error rate for both examples and show they are equal. Which example do you view as more useful for the future growth of the tree?

9.5 Construct the appropriate-size classification tree for the BUPA liver disorders data (see Section 8.4).

9.6 Construct the appropriate-size classification tree for the spambase data (see Section 8.4).

9.7 Construct the appropriate-size classification tree for the forensic glass data (see Section 8.7).

9.8 Construct the appropriate-size classification tree for the vehicle data (see Section 8.7).

9.9 Construct the appropriate-size classification tree for the wine data (see Section 8.7).

10
Artificial Neural Networks

10.1 Introduction

The learning technique of *artificial neural networks* (ANNs, or just *neural networks* or NNs) is the focus of this chapter. The development of ANNs evolved in periodic "waves" of research activity. ANNs were influenced by the fortunes of the fields of artificial intelligence and expert systems, which sought to answer questions such as: What makes the human brain such a formidable machine in processing cognitive thought? What is the nature of this thing called "intelligence"? And, how do humans solve problems?

These questions of "mind" and "intelligence" form the essence of *cognitive science*, a discipline that focuses on the study of interpretation and learning. "Interpretation" deals with the thought process resulting from exposure to the senses of some type of input (e.g., music, poem, speech, scientific manuscript, computer program, architectural blueprint), and "learning" deals with questions of how to learn from knowledge accumulated by studying examples having certain characteristics.

There are many different theories and models for how the mind and brain work. One such theory, called *connectionism*, uses analogues of neurons and their connections — together with the concepts of neuron firing, activation functions, and the ability to modify those connections — to form

A.J. Izenman, *Modern Multivariate Statistical Techniques*,
doi: 10.1007/978-0-387-78189-1_10,
© Springer Science+Business Media, LLC 2008

algorithms for artificial neural networks. This formulation introduces a relationship between the three notions of mind, brain, and computation, where information is processed by the brain through massively parallel computations (i.e., huge numbers of instructions processed simultaneously), unlike standard serial computations, which carry out one instruction at a time in sequential fashion.

Sophisticated types of ANNs have been used to model human intelligence, especially the ability to learn a language. These efforts include prediction of past tenses of regular and irregular English verbs (Rumelhart and McClelland, 1986b; Pinsker and Prince, 1988) and synthesis of the pronounciation of English text (Sejnowski and Rosenberg, 1987). A study involving ANNs of how the brain transforms a string of letter shapes into the meaning of a word (Hinton, Plaut, and Shallice, 1993) was instrumental in understanding the capabilities of the human brain, shedding light on specific types of impairments of the neural circuitry (e.g., surface and deep dyslexia), and in training ANNs to simulate brain damage resulting from injury or disease.

As an overly simplified model of the neuron activity in the brain, "artificial" neural networks were originally designed to mimic brain activity. Now, ANNs are treated more abstractly, as a network of highly interconnected nonlinear computing elements. The largest group of users of ANNs try to resolve problems involving machine learning, especially pattern classification and prediction. For example, problems of speech recognition, handwritten character recognition, face recognition, and robotics are important applications of ANNs. The common features to all of these types of problems are high-dimensional data and large sample sizes.

10.2   The Brain as a Neural Network

To understand how an artificial neural system can be developed, we first provide a brief description of the structure of the brain.

The largest part of the brain is the *cerebral cortex*, which consists of a vast network of interconnected cells called neurons. *Neurons* are elementary nerve cells which form the building blocks of the nervous system. In the human brain, for example, there are about 10 billion neurons of more than a hundred different types, as defined by their size and shape and by the kinds of neurochemicals they produce. A schematic diagram of a biological neuron is displayed in Figure 10.1.

The *cell body* (or *soma*) of a typical neuron contains its nucleus and two types of *processes* (or projections): *dendrites* and *axons*. The neuron receives signals from other neurons via its many dendrites, which operate as input devices. Each neuron has a single axon, a long fiber that operates as an output device; the end of the axon branches into strands, and each

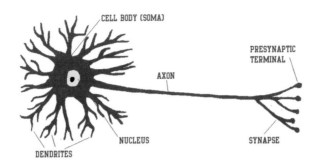

FIGURE 10.1. *Schematic view of a biological neuron.*

strand terminates in a *synapse*. Each synapse may either connect to a synapse on a dendrite or cell body of another neuron or terminate into muscle tissue. Because a neuron maintains, on average, about a thousand synaptic connections with other neurons (whereas some may have 10–20 thousand such connections), the entire collection of neurons in the brain yields an incredibly rich network of neural connections.

Neurons send signals to each other via an electrochemical process. All neurons are electrically charged due to ion concentrations inside and outside the cell. Under appropriate conditions, an activated neuron fires an electrical pulse (called an *action potential* or *spike*) of fixed amplitude and duration. The action potential travels down the axon to its endings. Each ending is swollen to form a *synaptic knob*, in which *neurotransmitters* (glutamic acid, `glu`) are stored. Neurons do not join with each other, even though they may be connected; there is a tiny gap (called the *synaptic cleft*) between the axon of the sending (or *presynaptic*) neuron and a dendrite of the receiving (or *postsynaptic*) neuron.

To send a signal to another neuron, the presynaptic neuron releases neurotransmitters across the gap to a cluster of *receptor molecules* on the dendrites of the postsynaptic neuron; these receptors act like electrical switches. When a neurotransmitter binds to one of these receptors (called an AMPA receptor), it opens up a channel into the postsynaptic neuron. Although that channel remains open for a split second, electrically charged sodium ions flood the channel, producing a local electrical disturbance (i.e., a depolarization), and start a chain reaction in which neighboring channels open up. This, in turn, sends an action potential shooting along the surface of the postsynaptic neuron toward the next neuron.

There is at least one other type of postsynaptic channel, called an NMDA glutamic acid receptor. This receptor is unusual in that it will not open unless it receives two simultaneous signals, one of which is either an electrical discarge from the postsynaptic neuron or a depolarization of its AMPA synapses, and the other is emitted by the axon from a presynaptic neuron.

When both signals arrive together, calcium ions also enter the dendrite, strengthen the synapse, and provide a mechanism for both short-term and long-term changes in the synapse. A high level of calcium released into the NMDA receptor induces *long-term potentiation (LTP)*, a form of long-term memory (lasting minutes to hours, in vitro, and hours to days and months in vivo, after which decay sets in). LTP enlarges synapses and makes them stronger, and, over time, can also change brain structure.

Note that the postsynaptic neuron may or may not fire as a result of receiving the pulse. Then, the axon shuts down for a certain amount of time (a *refractory period*) before it can fire again. To prepare the synapse for the next action potential, the synaptic cleft is cleared by *active transport* by returning the neurotransmitter to the synaptic knob of the presynaptic neuron.

Firing tends to occur randomly, but the actual rate of firing depends upon many factors. One of those factors is the status of the total input signal; this is derived from the relative strengths of the two types of synapses, namely, the *inhibitory synapses*, which prevent the neuron from firing, and the *excitatory synapses*, which push the neuron closer to firing. Depending upon whether or not the total input signal received at the synapses of a neuron exceeds some threshold limit, the neuron may fire, be in a resting state, or be in an electrically neutral state.

The brain "learns" by changing the strengths of the connections between neurons or by adding or removing such connections. Learning itself is accomplished sequentially from increasing amounts of experience.

10.3   The McCulloch–Pitts Neuron

The idea of an "artificial" neural network is usually traced back to the "computing machine" model of McCullogh and Pitts (1943), who constructed a simplified abstraction of the process of neuron activity in the human brain.

The McCulloch–Pitts neuron consists of multiple inputs (the dendrites) and a single output (the axon). The inputs are denoted by $X_1, X_2, \ldots, X_r$, and each has a value of either 0 ("off") or 1 ("on"). The signal at each input connection depends upon whether the synapse in question is excitatory or inhibitory. If any one of the synapses is inhibitory and transmits the value 1, the neuron is prevented from firing (i.e., the output is 0). If no inhibitory synapse is present, the inputs are summed to produce the *total excitation* $U = \sum_j X_j$, and then U is compared with a *threshold* value θ: if $U \geq \theta$, the output Y is 1 and the neuron fires (i.e., transmits a new signal); otherwise, Y is 0 and the neuron does not fire.

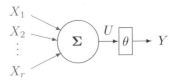

FIGURE 10.2. *McCulloch–Pitts neuron with r binary inputs, $X_1, X_2, \ldots,$*
X_r, one binary output, Y, and threshold θ.

An equivalent formulation is to say that the value of Y is determined
by the indicator function $I_{[U-\theta \geq 0]}$. Note that if $\theta > r$, the number of
inputs, the neuron will never fire. Also, if $\theta = 0$ and there are no inhibitory
synapses, the output will always have the constant value 1.

Geometrically, the input space is an r-dimensional unit hypercube, and
each of the 2^r vertices of the hypercube is associated with a specific Y-value
(either 0 or 1). For a given value of θ, the McCulloch–Pitts neuron divides
the hypercube into two half-spaces according to the hyperplane $\sum_j X_j = \theta$;
those vertices with $Y = 1$ lie on one side of the hyperplane, whereas those
with $Y = 0$ lie on the other side.

The McCulloch–Pitts neuron is usually referred to as a *threshold logic*
unit (TLU) and is displayed in Figure 10.2. It is designed to compute
simple logical functions of r arguments, where $Y = 1$ is translated as the
logical value "true" and $Y = 0$ as "false." For example, the logical functions
AND and OR for three inputs are displayed in Figure 10.3. For the logical
function AND, the neuron will fire only if all three inputs have the value 1,
whereas, for the logical function OR, the neuron will fire only if at least one
of the three inputs have the value 1. The AND and OR functions form a
basis set of logical functions. All other logical functions can be computed by
building up large networks consisting of several layers of McCulloch–Pitts

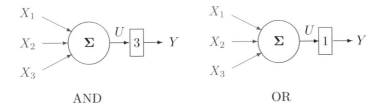

AND   OR

FIGURE 10.3. *McCulloch–Pitts neuron for the AND and OR logical*
functions with $r = 3$ binary inputs and thresholds $\theta = 3$ and $\theta = 1$, respec-
tively.

neurons. At the time, it appeared that networks of TLUs could be used to create an intelligent machine.

Although this model of a neuron was studied by many people, it is not really a good approximation of how a biological system learns. There are no adjustable parameters or weights in the network, which means that different problems can only be solved by repeatedly changing the input structure or the threshold value. Such manipulations are more complicated than adopting a flexible weighting system for the network.

10.4   Hebbian Learning Theory

At the time of the introduction of the McCulloch–Pitts neuron, little was known about how the "strength" of signals sent between neurons in the brain are changed by activity and, therefore, how learning takes place.

The next advance occurred when Donald O. Hebb, in his 1949 book *The Organization of Behavior*, summarized everything then known about how the central nervous system affects behavior and vice versa. He started out by assuming that all the neurons one needs in life are present at birth, that initial neural connections are randomly distributed, and that as we get older our neural connections multiply and become stronger. He also believed that one's perceptions, thoughts, emotions, memory, and sensations are strongly influenced by life experiences, and that such experiences leave behind a "memory trace" — via sets of interconnected neurons — which helps determine future behavior.

Using results derived from published neurophysiological experiments involving animals and humans, and from his own life observations, Hebb gave a detailed presentation of biological neurons. In particular, he formulated two new theories as to how the brain works. Building upon the ideas of Santiago Ramón y Cajal, the 1906 Nobel Laureate, Hebb's first theory focused on the nature of synaptic change and is referred to as the *Hebb learning rule* (Hebb, 1949, p. 62):

> *When an axon of cell A is near enough to excite cell B and repeatedly or persistently takes part in firing it, some growth process or metabolic change takes place in one or both cells so that A's efficiency, as one of the cells firing B, is increased.*

In other words, the strength of a synaptic connection between two neurons depends upon their associated firing history: the more often the two neurons fire together, the stronger their connection (and, by implication, the less often, the weaker their connection). The Hebb rule is time-dependent (there is an implicit ordering of events when neuron A helps to fire neuron B) and governs only what happens locally at the synapse. Any synapse that behaves according to the Hebb rule is known as a *Hebb synapse*.

The Hebb rule of *neural excitation* was later expanded (Milner, 1957) by adding the following rule of *neural inhibition*: if neuron A repeatedly or persistently sends a signal to neuron B, but B does not fire, this reduces the chance that future signals from A will entice B to fire. This inhibitory rule is necessary because otherwise the system of synaptic connections throughout the cerebral cortex would grow without limit as soon as one such connection is activated. Hebb had previously (in his 1932 M.A. thesis) incorporated the inhibitory rule into his theory but did not include it in his book.

His second theory is probably the more important idea. It was derived from a discovery by Lorente de Nó in 1944 that the brain contained closed circuits of neurons. Hebb then speculated that memory resides in the cerebral cortex in the form of overlapping clusters of thousands of highly interconnected neurons, which he called *cell assemblies*. The clusters overlap because a neuron, which has branch-like links to other neurons, can be a member of many different cell assemblies.

In Hebb's theory, a cell assembly is organized with reference to a particular sensory input and briefly acts as a closed neural circuit; sensations, thoughts, perceptions, etc., are considered different from each other if different cell assemblies are involved in the activity; and the cell assembly also retains a memory of its defining activity even after the triggering event has ceased (e.g., the memory of stubbing one's toe can remain well after the pain has subsided). Cell assemblies are thought to play an essential role in the learning process. Hebb also defined a *phase sequence* as a combination of cell assemblies that are simultaneously excited when repeatedly presented with the same sequence of stimuli.

Hebb's 1949 book was an international success; it was considered by some as ground-breaking and sensational and a starting point to build a theory of the brain. Yet it took several years before these contributions were fully recognized in the fledgling field of behavioral neuroscience. Subsequently, in the fields of psychology and neuroscience, it inspired a huge amount of research into theories of brain function and behavior. Some of Hebb's work was speculative and has since been overturned by scientific experiment and discovery. But much of it is still relevant today.

10.5   Single-Layer Perceptrons

Hebb's pioneering work on the brain led to a second wave of interest in ANNs. Frank Rosenblatt, a psychologist, had read Hebb (1949) but was not convinced that most neural connections were random and that cell assemblies could self-generate within a purely homogeneous mass of neurons. He believed that he could improve upon Hebb's work and, toward

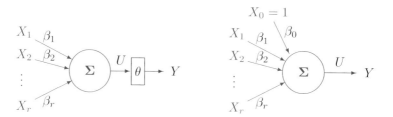

FIGURE 10.4. *Rosenblatt's single-layer perceptron with r inputs, connection weights $\{\beta_j\}$, and binary output Y. The left panel shows the perceptron with threshold θ, and the right panel shows the equivalent perceptron with bias element $\beta_0 = -\theta$ and $X_0 = 1$.*

that end, he constructed a "minimally constrained" system that he called a "perceptron" (Rosenblatt, 1958, 1962).

A *perceptron* is essentially a McCulloch–Pitts neuron, but now input X_i comes equipped with a real-valued *connection weight* β_i, $i = 1, 2, \ldots, r$. The inputs, $X_1, X_2, \ldots, X_r$ can each be binary or real-valued. Positive weights ($\beta_j > 0$) reflect excitatory synapses, and negative weights ($\beta_j < 0$) reflect inhibitory synapses. The magnitude of a weight shows the strength of the connection.

The perceptron, which is more flexible than the McCulloch–Pitts neuron for mimicking neural connections, is displayed in Figure 10.4. A weighted sum of input values, $U = \sum_j \beta_j X_j$, is computed, and the output is $Y = 1$ only if $U \geq \theta$, where θ is the threshold value; otherwise, $Y = 0$. Note that we can convert a threshold θ to 0 by introducing a *bias element* $\beta_0 = -\theta$, so that $U - \theta = \beta_0 + U$, and then comparing $U = \sum_{j=0}^{r} \beta_j X_j$ to 0, where $X_0 = 1$. If $U \geq 0$, then $Y = 1$; otherwise, $Y = 0$.

We call a function $Y \in \{0, 1\}$ *perceptron-computable* if, for a given value of θ, there exists a hyperplane that divides the input space into two half-spaces, R_1 and R_0, where R_1 corresponds to points having $Y = 1$ and R_0 to points having $Y = 0$. If the points in R_1 can be separated without error from those in R_0 by a hyperplane, we say that the two sets of points are *linearly separable*. This binary partition of input space (obtained by comparing U to the threshold value θ) enables a perceptron to predict class membership.

10.5.1 *Feedforward Single-Layer Networks*

One way of representing a network of neural interconnections is as a *directed acyclic graph (DAG)*. A *graph* is a set of *vertices* or *nodes* (representing basic computing elements) and a set of *edges* (representing the connections between the nodes), where we assume that both sets are of finite size. In a *directed graph* (or *digraph*), the edges are assigned an orientation so that numerical information flows along each edge in a particular direction. In a *feedforward network*, information flows in one direction only, from input nodes to output nodes. An *acyclic graph* is one in which no loops or feedback are allowed.

The simplest type of DAG organizes the network nodes into two separate groups: r *input nodes*, $X_1, \ldots, X_r$, and s *output nodes*, $Y_1, \ldots, Y_s$. Input nodes are also referred to as *source nodes*, *input units*, or *input variables*. No computation is carried out at these nodes. The input nodes take on values introduced by some feature external to the network. The output nodes are variously known as *sink nodes*, *neurons*, *output units*, or *output variables*. These input and output nodes can be real-valued or discrete-valued (usually, binary). Real-valued output nodes are typically scaled so that their values lie in the unit interval $[0, 1]$. Binary input and output nodes are used in the design of switching circuits; real input nodes with binary output nodes are used primarily in classification applications; and real input and output nodes are used mostly in optimization and control applications.

Despite appearances, this particular type of network is commonly called a *single-layer network* because only the output nodes involve significant amounts of computation; the input nodes, which are said to constitute a "zeroth" layer of fixed functions, involve no computation, and, hence, do not count as a layer of learnable nodes.

Every connection $X_j \rightarrow Y_\ell$ between the input nodes and the output nodes carries a *connection weight*, $\beta_{j\ell}$, which identifies the "strength" of that connection. These weights may be positive, negative, or zero; positive weights represent excitory signals, negative weights represent inhibitory signals, and zero weights represent connections that do not exist in the network.

The *architecture* (or *topology*) of the network consists of the nodes, the directed edges (with the direction of signal flow indicated by an arrow along each edge), and the connection weights.

10.5.2 Activation Functions

In the following, $\mathbf{X} = (X_1, \cdots, X_r)^\tau$ represents a random r-vector of inputs. Given $\mathbf{X}$, each output node computes an *activation value* using a linear combination of the inputs to it plus a constant; that is, for the ℓth output node or neuron, we compute the ℓth *linear activation function*,

$$U_\ell = \beta_{0\ell} + \sum_{j=1}^{r} \beta_{j\ell} X_j = \beta_{0\ell} + \mathbf{X}^\tau \boldsymbol{\beta}_\ell, \qquad (10.1)$$

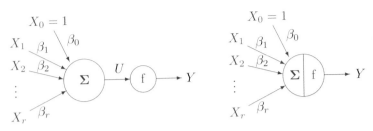

FIGURE 10.5. *Rosenblatt's single-layer perceptron with r inputs, bias element β_0, connection weights $\{\beta_j\}$, activation function f, and binary output Y. The left panel shows the perceptron with a separate computing unit for f, and the right panel shows the equivalent perceptron with a single computing unit divided into two functional parts: the addition function is written on the left and the activation function f applied to the result U of the addition is written on the right.*

where $\beta_{0\ell}$ is a constant (or *bias*) related to the threshold for the neuron to fire, and $\boldsymbol{\beta}_\ell = (\beta_{1\ell}, \cdots, \beta_{r\ell})^\tau$ is an r-vector of connection weights, $\ell = 1, 2, \ldots, s$.

In matrix notation, we can rewrite the collection of s linear activation functions (10.1) as

$$\mathbf{U} = \boldsymbol{\beta}_0 + \mathbf{BX}, \tag{10.2}$$

where $\mathbf{U} = (U_1, \cdots, U_s)^\tau$, $\boldsymbol{\beta}_0 = (\beta_{01}, \cdots, \beta_{0s})^\tau$ is an s-vector of biases, and $\mathbf{B} = (\boldsymbol{\beta}_1, \cdots, \boldsymbol{\beta}_s)^\tau$ is an $(s \times r)$-matrix of connection weights. The activation values are then each filtered through a nonlinear *threshold activation function* $f(U_\ell)$ to form the value of the ℓth output node, $\ell = 1, 2, \ldots, s$. In matrix notation,

$$\mathbf{f}(\mathbf{U}) = \mathbf{f}(\boldsymbol{\beta}_0 + \mathbf{BX}), \tag{10.3}$$

where $\mathbf{f} = (f, \cdots, f)^\tau$ is an s-vector function each of whose elements is the function f, and $\mathbf{f}(\mathbf{U}) = (f(U_1), \cdots, f(U_s))^\tau$. The simplest form of f is the identity function, $f(u) = u$. See Figure 10.5.

A partial list of activation functions is given in Table 10.1. The most interesting of these functions are the sigmoidal ("S-shaped") functions, such as the logistic and hyperbolic tangent; see Figure 8.2 for a graph of the logistic sigmoidal activation function. A *sigmoidal function* is a function $\sigma(\cdot)$ that has the following properties: $\sigma(u) \to 0$ as $u \to -\infty$ and $\sigma(u) \to 1$ as $u \to +\infty$. A sigmoidal function $\sigma(\cdot)$ is *symmetric* if $\sigma(u) + \sigma(-u) = 1$ and *asymmetric* if $\sigma(u) + \sigma(-u) = 0$. The logistic function is symmetric, whereas the tanh function is asymmetric. Note that if $f(u) = (1 + e^{-u})^{-1}$, then its derivative wrt u is $df(u)/du = e^{-u}(1 + e^{-u})^{-2} = f(u)(1 - f(u))$. The hyperbolic tangent function, $f(u) = \tanh(u)$, is a linear transformation of the logistic function (see Exercise 10.1). There is empirical evidence that

TABLE 10.1. *Examples of activation functions.*

Activation Function	$f(u)$	Range of Values
Identity, linear	u	$\Re$
Hard-limiter	$\text{sign}(u)$	$\{-1, +1\}$
Heaviside, step, threshold	$I_{[u \geq 0]}$	$\{0, 1\}$
Gaussian radial basis function	$(2\pi)^{-1/2} e^{-u^2/2}$	$\Re$
Cumulative Gaussian (sigmoid)	$\sqrt{2/\pi} \int_0^u e^{-z^2/2} dz$	$(0, 1)$
Logistic (sigmoid)	$(1 + e^{-u})^{-1}$	$(0, 1)$
Hyperbolic tangent (sigmoid)	$(e^u - e^{-u})/(e^u + e^{-u})$	$(-1, +1)$

ANN algorithms that use the tanh function converge faster than those that use the logistic function.

10.5.3 Rosenblatt's Single-Unit Perceptron

In binary classification problems, each of the n input vectors $\mathbf{X}_1, \ldots, \mathbf{X}_n$ is to be classified as a member of one of two classes, Π_1 or Π_2. For this type of application, a single-layer feedforward neural network consists of only a single output node or unit (i.e., $s = 1$).

A *single-unit perceptron* (Rosenblatt, 1958, 1962) is a single-layer feedforward network with a single output node that computes a linear combination of the input variables (e.g., $\beta_0 + \mathbf{X}^\tau \boldsymbol{\beta}$) and delivers its sign,

$$\text{sign}\{\beta_0 + \mathbf{X}^\tau \boldsymbol{\beta}\}, \tag{10.4}$$

as output, where $\text{sign}(u) = -1$ if $u < 0$, and $+1$ if $u \geq 0$. The activation function used here is the "hard-limiter" function. The output node is generally known as a *linear threshold unit*. Rosenblatt's perceptron is essentially the threshold logic unit of McCullogh and Pitts (1943) with weights.

A generalized version of the single-unit perceptron can be written as

$$f(\beta_0 + \mathbf{X}^\tau \boldsymbol{\beta}) \tag{10.5}$$

where $f(\cdot)$ is an activation function, which is usually taken to be sigmoidal.

10.5.4 The Perceptron Learning Rule

For convenience in this subsection, we make the following notational changes: $\boldsymbol{\beta} \leftarrow (\beta_0, \boldsymbol{\beta}^\tau)^\tau$ and $\mathbf{X} \leftarrow (1, \mathbf{X}^\tau)^\tau$, where both $\mathbf{X}$ and $\boldsymbol{\beta}$ are now $(r+1)$-vectors. Then, we can write $\beta_0 + \mathbf{X}^\tau \boldsymbol{\beta}$ as $\mathbf{X}^\tau \boldsymbol{\beta}$.

In the binary classification case, the single output variable takes on values $Y = \pm 1$ depending upon whether the neuron fires ($Y = +1$ if $\mathbf{X} \in \Pi_1$) or does not fire ($Y = -1$ if $\mathbf{X} \in \Pi_2$). Thus, the neuron will fire if $\mathbf{X}^\tau \boldsymbol{\beta} \geq 0$ and will not fire if $\mathbf{X}^\tau \boldsymbol{\beta} < 0$.

Suppose $\mathbf{X}_1, \ldots, \mathbf{X}_n$ are independent copies of $\mathbf{X}$, and that they are drawn from the two classes Π_1 and Π_2. Suppose, further, that these observations are *linearly separable*. That is, there exists a vector $\boldsymbol{\beta}^*$ of connection weights such that the observation vectors that belong to class Π_1 fall on one side of the hyperplane $\mathbf{X}^\tau \boldsymbol{\beta}^* = 0$, whereas the observation vectors from class Π_2 fall on the other side of the hyperplane.

As our update rule, we use a *gradient-descent algorithm*, which operates sequentially on each input vector. Such an algorithm is referred to as *on-line learning*, whereby the learning mechanism adapts quickly to correct classification errors as they occur. The input vectors are examined one at a time and classified to one of the two classes. The true class is then revealed, and the classification procedure is updated accordingly.

The algorithm proceeds by relabeling the $\{\mathbf{X}_i\}$, one at a time, so that at the hth iteration we are dealing with $\mathbf{X}_h$, $h = 1, 2, \ldots$. Set $\mathbf{X}_0 = \mathbf{0}$. The algorithm computes a sequence $\{\boldsymbol{\beta}_h\}$ of connection weights using as initial value $\boldsymbol{\beta}_0 = \mathbf{0}$. The update rule is the following:

1. If, at the hth iteration of the algorithm, the current version, $\boldsymbol{\beta}_h$, correctly classifies $\mathbf{X}_h$, we do not change $\boldsymbol{\beta}_h$ in the next iteration; that is, set $\boldsymbol{\beta}_{h+1} = \boldsymbol{\beta}_h$ if either $\mathbf{X}_h^\tau \boldsymbol{\beta}_h \geq 0$ and $\mathbf{X}_h \in \Pi_1$, or $\mathbf{X}_h^\tau \boldsymbol{\beta}_h < 0$ and $\mathbf{X}_h \in \Pi_2$.

2. If, on the other hand, the current version, $\boldsymbol{\beta}_h$, misclassifies $\mathbf{X}_h$, then we update the connection weight vector as follows: if $\mathbf{X}_h^\tau \boldsymbol{\beta}_h \geq 0$ but $\mathbf{X}_h \in \Pi_2$, then set $\boldsymbol{\beta}_{h+1} = \boldsymbol{\beta}_h - \eta \mathbf{X}_h$; if $\mathbf{X}_h^\tau \boldsymbol{\beta}_h < 0$ but $\mathbf{X}_h \in \Pi_1$, then set $\boldsymbol{\beta}_{h+1} = \boldsymbol{\beta}_h + \eta \mathbf{X}_h$, where $\eta > 0$ is the *learning-rate parameter* whose value is taken to be independent of the iteration number h.

This algorithm is popularly known as the *perceptron learning rule*. Because the value of η is irrelevant (we can always rescale $\mathbf{X}_h$ and $\boldsymbol{\beta}_h$), we set $\eta = 1$ without loss of generality.

10.5.5 Perceptron Convergence Theorem

From the update rule, it follows that $\boldsymbol{\beta}_{h+1} = \sum_{i=1}^h \mathbf{X}_i$. Assume that we have linear separability of the two classes. Suppose also that a solution vector $\boldsymbol{\beta}^*$ exists. Define

$$A = \min_{\mathbf{X}_i \in \Pi_1} \mathbf{X}_i^\tau \boldsymbol{\beta}^*, \quad B = \max_{\mathbf{X}_{*i} \in \Pi_1} \| \mathbf{X}_i \|^2 . \tag{10.6}$$

Transposing $\boldsymbol{\beta}_{h+1}$ and postmultiplying the result though by $\boldsymbol{\beta}^*$ yields

$$\boldsymbol{\beta}_{h+1}^\tau \boldsymbol{\beta}^* = \sum_{i=1}^{h} \mathbf{X}_i^\tau \boldsymbol{\beta}^* \geq hA. \tag{10.7}$$

From the Cauchy–Schwarz inequality,

$$(\boldsymbol{\beta}_{h+1}^\tau \boldsymbol{\beta}^*)^2 \leq \| \boldsymbol{\beta}_{h+1}^\tau \|^2 \| \boldsymbol{\beta}^* \|^2 . \tag{10.8}$$

Substituting (10.7) into (10.8) yields

$$\| \boldsymbol{\beta}_{h+1} \|^2 \geq \frac{h^2 A^2}{\| \boldsymbol{\beta}^* \|^2}. \tag{10.9}$$

Thus, the squared-norm of the weight vector grows at least quadratically with the number, h, of iterations.

Next, consider again the update rule, $\boldsymbol{\beta}_{k+1} = \boldsymbol{\beta}_k + \mathbf{X}_k$, at the kth iteration, where $\mathbf{X}_k \in \Pi_1$, $k = 1, 2, \ldots, h$. Then,

$$\| \boldsymbol{\beta}_{k+1} \|^2 = \| \boldsymbol{\beta}_k \|^2 + \| \mathbf{X}_k \|^2 + 2\mathbf{X}_k^\tau \boldsymbol{\beta}_k. \tag{10.10}$$

Because $\mathbf{X}_k$ has been incorrectly classified, $\mathbf{X}_k^\tau \boldsymbol{\beta}_k < 0$. It follows that,

$$\| \boldsymbol{\beta}_{k+1} \|^2 \leq \| \boldsymbol{\beta}_k \|^2 + \| \mathbf{X}_k \|^2, \tag{10.11}$$

whence,

$$\| \boldsymbol{\beta}_{k+1} \|^2 - \| \boldsymbol{\beta}_k \|^2 \leq \| \mathbf{X}_k \|^2, \tag{10.12}$$

Summing (10.12) over $k = 1, 2, \ldots, h$ yields

$$\| \boldsymbol{\beta}_{h+1} \|^2 \leq \sum_{k=1}^{h} \| \mathbf{X}_k \|^2 \leq hB. \tag{10.13}$$

Hence, the squared-norm of the weight vector grows at most linearly with the number, h, of iterations.

For large values of h, the inequalities (10.9) and (10.13) contradict each other. Thus, h cannot grow without bound. We need to find an $h_{\max}$ such that (10.9) and (10.13) both hold with equalities. In other words, $h_{\max}$ has to satisfy

$$\frac{h_{\max}^2 A^2}{\| \boldsymbol{\beta}^* \|^2} = h_{\max} B, \tag{10.14}$$

whence,

$$h_{max} = \frac{B \parallel \boldsymbol{\beta}^* \parallel^2}{A^2}. \tag{10.15}$$

We have shown the following result. Set $\eta = 1$ and $\boldsymbol{\beta}_0 = \mathbf{0}$. Then:

> For a binary classification problem with linearly separable classes, if a solution vector $\boldsymbol{\beta}^*$ exists, the algorithm will find that solution in a finite number, h_{max}, of iterations.

This is the *perceptron convergence theorem*. At the time, it was regarded as a very appealing result.

There are two difficulties implicit in this result. First, the existence of a solution vector $\boldsymbol{\beta}^*$ turns out to be crucial for the result to hold; this was made clear by Minsky and Papert (1969), who showed that there are many problems for which no perceptron solution exists.

The second difficulty derives from the fact that, even though the algorithm converges, computing h_{max} is impossible because it depends upon the solution vector $\boldsymbol{\beta}^*$, which is unknown. If the algorithm stops, we clearly have a solution. If the two classes are not linearly separable, then the algorithm will not terminate. In fact, after some large (unknown) number of iterations, the algorithm will start cycling with unknown period length. In general, if we do not know whether or not linear separability holds, we cannot reliably determine when to stop running the algorithm. If we stop the algorithm prematurely, the resulting perceptron weight vector may not generalize well for test data.

One suggested approach to this problem is to adopt a specific stopping rule whereby the algorithm is stopped after a fixed number of iterations; another approach is to make the learning-rate parameter η depend upon the iteration number (i.e., η_h) so that as the iterations proceed, the adjustments decrease in size.

10.5.6 *Limitations of the Perceptron*

Despite high initial expectations, perceptrons were found to have very limited capabilities. It was shown (Minsky and Papert, 1969) that a perceptron can learn to distinguish two classes only if the classes are linearly separable. This is not always possible as can be seen from the XOR function, which is not perceptron-computable because its input space is not linearly separable (see Exercise 10.6).

As a result, during the 1970s, research in this area was abandoned by almost everyone in that community. An additional factor to explain the absence of work on neural networks is that hardware to support neural computation did not become available until the 1980s.

10.6   Artificial Intelligence and Expert Systems

The downfall of the perceptron led to the introduction of *artificial intelligence* (AI) and *rule-based expert systems* as the main areas of research into machine intelligence. AI was viewed, first, as the study of how a human brain (or any natural intelligence) functions, and, second, as the study of how to construct an artificial intelligence (i.e., a machine that could solve problems requiring "cognition" when performed by humans). In early AI systems, problems were solved in a sequential, step-by-step fashion, by manipulating a dictionary of symbolic representations of the available knowledge on a particular subject of interest. An AI system had to store information specific to a domain of interest, use that information to solve a broad range of problems in that domain, and acquire new information from experience by solving problems in that domain.

A typical AI application was of the following type. Suppose we would like to predict the intuitive decisions made by an experienced loan officer of a bank based only on the answers given to questions on a loan application. One might first ask the loan officer to explain the value (e.g., on a 5-point scale) he or she places on the answers to each question. The points scored by an applicant on each question could be totalled and compared with some given threshold; the loan officer's decision on the loan could then be predicted based upon whether or not the applicant's total score surpassed the threshold.

This approach to predicting the decisions of a loan officer ignores possible nonlinearities in the decision-making process. For example, if the loan applicant scores high on a few specific questions, the loan officer may ignore the responses to all other questions in making a positive decision, whereas if a particular question scores low, this by itself may be sufficient to render the application unsuccessful, even though all other variables score high. Listing all the rules the loan officer can possibly use in the decision process constitutes a rule-based expert system.

Expert systems are knowledge-based systems, where "knowledge" represents a repository of data, well-known facts, specialized information, and heuristics, which experts in a field (e.g., medicine) would agree upon. Such expert systems are interactive computer programs that provide users (e.g., physicians) with computer-based consultative advice.

The earliest example of a rule-based expert system was DENDRAL, a system for identifying chemical structures from mass spectrograms. This was followed in the mid-1970s by MYCIN, which was designed to aid physicians in the diagnosis and treatment of meningitis and bacterial infections. MYCIN was made up of a "knowledge base" and an "inference engine"; the knowledge base contained information specific to the area of medical diagnosis, and the inference engine would recommend treatments to physicians who consulted the knowledge base. A generic version, known as EMYCIN ("empty" MYCIN), was then built using only the inference engine and shell, not the knowledge base. (Although never regarded by mathematicians as an AI or expert system as such, the symbolic mathematics system MAC-SYMA also emerged from the early AI world.) In the 1980s, expert systems were popularly regarded as the future of AI.

During this time, there were also ambitious attempts at AT&T Bell Laboratories to create an expert system to help users carry out statistical analyses of data. One such expert system was REX (Pregibon and Gale, 1984), which was written in the LISP language and provided rule-based guidance for simple linear regression problems. REX (short for Regression EXpert) acted as an interface between the user and a statistical software package through a flexible interactive dialogue, which only requested help when it encountered problems with the data. REX did not survive long for many reasons, including apathy due to constantly changing computational environments (Pregibon, 1991).

Despite all this activity, expert systems never lived up to their hype; they proved to be expensive, were successful only in specialized situations, and were not able to learn from their own experiences. In short, expert systems never truly possessed "cognition," which was the primary goal of AI.

The failure of AI and expert systems to come to grips with these aspects of "cognition" has been attributed to the fact that traditional computers and the human brain function very differently from each other. It was argued that AI was not providing the right environment for the emergence of a truly intelligent machine because it was not delivering a realistic model of the structure of the brain. Whereas human brains consisted of massively parallel systems of neurons, AI digital computers were serial machines; overall, the latter were incredibly slow by comparison. If one wanted to understand "cognition" (so the argument went), one should build a model based upon a detailed study of the architecture of the brain.

10.7   Multilayer Perceptrons

The most recent wave of research into ANNs arrived in the mid-1980s and has continued until the present time. Earlier suggestions of Minsky and Papert (1969) — that the limitations of the perceptron could be overcome by "layering" the perceptrons and applying nonlinear transformations prior to combining the transformed weighted inputs — were not adopted at that time due to computational limitations. Minsky and Papert's suggestions because more meaningful when high-speed computers became readily available and with the discovery of the "backpropagation" algorithm.

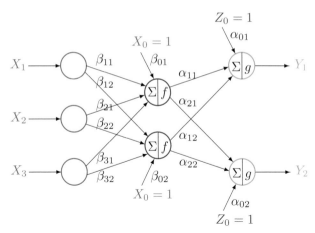

FIGURE 10.6. *Multilayer perceptron with a single hidden layer, $r = 3$ input nodes, $s = 2$ output nodes, and $t = 2$ nodes in the hidden layer. The αs and βs are weights attached to the connections between nodes, and f and g are activation functions.*

A multilayer feedforward neural network (perceptron) is a multivariate statistical technique that nonlinearly maps an input vector $\mathbf{X}=(X_1,\cdots,X_r)^{\top}$ of variables to an output vector $\mathbf{Y}=(Y_1,\cdots,Y_s)^{\top}$ of variables. Between the inputs and outputs there are also "hidden" variables arranged in layers. The hidden and output variables are traditionally called nodes, neurons, or processing units. A typical ANN is given in Figure 10.6, which has two computational layers (i.e., the hidden layer and the output layer), and $r = 3$ input nodes, $s = 2$ output nodes, and $t = 2$ nodes in the hidden layer.

ANNs can be used to model regression or classification problems. In a multiple regression situation, there is only one ($s = 1$) output variable Y and node, whereas in a multivariate regression situation, there are s output variables $Y_1, \ldots, Y_s$ and nodes. In a binary classification situation, there is only one ($s = 1$) output variable Y with value 0 or 1, whereas in a multiclass classification problem with K classes, there are $s = K - 1$ output variables $Y_1, \ldots, Y_s$ and nodes, with each Y-variable taking on the value 0 or 1.

10.7.1 Network Architecture

Multilayer perceptrons have the following architecture: r input nodes $X_1, \ldots, X_r$; one or more layers of "hidden" nodes; and s output nodes $Y_1, \ldots, Y_s$. It is usual to call each layer of hidden nodes a "hidden layer"; these nodes are not part of either the input or output of the network. If there is a single hidden layer, then the network can be described as being a "two-layer network" (the output layer being the second computational layer); in general, if there are L hidden layers, the network is described as being an $(L+1)$-layer network.

A *fully connected network* has all r input nodes connected to the nodes in the first hidden layer, all nodes in the first hidden layer connected to all nodes in the second hidden layer, ..., and all nodes in the last (Lth) hidden layer connected to all s output nodes. If some of the connections are missing, we have a *partially connected network*. We can always represent a partially connected network as a fully connected network by setting the weights of the missing connections to zero.

Given the input values, each hidden node computes an activation value by taking a weighted average of its input values and adding a constant. Similarly, each output node computes an activation value from a weighted average of the inputs to it from the hidden nodes plus a constant. The activation values are then each filtered through an *activation function* to form the output value of the neuron.

10.7.2 A Single Hidden Layer

Suppose we have a two-layer network with r input nodes (X_m, $m = 1, 2, \ldots, r$), a single layer ($L = 1$) of t hidden nodes (Z_j, $j = 1, 2, \ldots, t$), and s output nodes (Y_k, $k = 1, 2, \ldots, s$). Let β_{mj} be the weight of the connection $X_m \to Z_j$ with bias β_{0j} and let α_{jk} be the weight of the connection $Z_j \to Y_k$ with bias α_{0k}. See Figure 10.6 for a schematic diagram of a single hidden layer network with $r = 3$, $s = 2$, and $t = 2$.

Let $\mathbf{X} = (X_1, \cdots, X_r)^\tau$ and $\mathbf{Z} = (Z_1, \cdots, Z_t)^\tau$. Let $U_j = \beta_{0j} + \mathbf{X}^\tau \boldsymbol{\beta}_j$ and $V_k = \alpha_{0k} + \mathbf{Z}^\tau \boldsymbol{\alpha}_k$. Then,

$$Z_j = f_j(U_j), \quad j = 1, 2, \ldots, t, \tag{10.16}$$

$$\mu_k(\mathbf{X}) = g_k(V_k), \quad k = 1, 2, \ldots, s, \tag{10.17}$$

where $\boldsymbol{\beta}_j = (\beta_{1j}, \cdots, \beta_{rj})^\tau$ and $\boldsymbol{\alpha}_k = (\alpha_{1k}, \cdots, \alpha_{tk})^\tau$. Putting these equations together, the value of the kth output node can be expressed as

$$Y_k = \mu_k(\mathbf{X}) + \epsilon_k, \tag{10.18}$$

where

$$\mu_k(\mathbf{X}) = g_k \left(\alpha_{0k} + \sum_{j=1}^{t} \alpha_{jk} f_j \left(\beta_{0j} + \sum_{m=1}^{r} \beta_{mj} X_m \right) \right), \tag{10.19}$$

$k = 1, 2, \ldots, s$, and the $f_j(\cdot)$, $j = 1, 2, \ldots, t$, and the $g_k(\cdot)$, $k = 1, 2, \ldots, s$, are activation functions for the hidden and output layers of nodes, respectively.

The activation functions, $\{f_j(\cdot)\}$, are usually taken to be nonlinear continuous functions with sigmoidal shape (e.g., logistic or tanh functions).

The functions $\{g_k(\cdot)\}$ are often taken to be linear (in regression problems) or sigmoidal (in classification problems). The error term, ϵ_k, can be taken as Gaussian with mean zero and variance σ_k^2.

Let $s = 1$, so that we have a single output node. Suppose also that all hidden nodes in the single hidden layer have the same sigmoidal activation function $\sigma(\cdot)$. We further take the output activation function $g(\cdot)$ to be linear. Then, (10.18) reduces to $Y = \mu(\mathbf{X}) + \epsilon$, where

$$\mu(\mathbf{X}) = \alpha_0 + \sum_{j=1}^{t} \alpha_j \sigma \left(\beta_{0j} + \sum_{m=1}^{r} \beta_{mj} X_m \right), \qquad (10.20)$$

and the network is equivalent to a single-layer perceptron. If, alternatively, both $f(\cdot)$ and $g(\cdot)$ are linear, then (10.19) is just a linear combination of the inputs.

Note that sigmoidal functions play an important role in network design. They are quite flexible as activation functions and can approximate different types of other functions. For example, a sigmoidal function, $\sigma(u)$, is very close to linear when u is close to zero. Thus, we can substitute a sigmoidal function for a linear function at any hidden node while, at the same time, making the weights and bias that feed into that node very small; to compensate for the resulting scaling problem, the weights corresponding to connections emanating from that hidden node to the output node(s) are usually made much larger. Sigmoidal functions, which are smooth, monotonic functions, are especially useful for approximating discontinuous threshold functions (e.g., $I_{[u \geq 0]}$) when evaluating the gradient for a loss function of a multilayer perceptron.

We also mention the *skip-level connection*, which refers to a direct connection from input node to output node, without first passing through a hidden node. Skip-level connections can be included in the model either explicitly or through an implicit arrangement of connection weights — from input node to hidden node and then from hidden node to output node — which approximates the skip-level connection.

10.7.3   ANNs Can Approximate Continuous Functions

An important result used to motivate the use of neural networks is given by Kolmogorov's *universal approximation theorem*, which states that:

> *Any continuous function defined on a compact subset of $\Re^r$ can be uniformly approximated (in an appropriate metric) by a function of the form (10.20).*

In other words, we can approximate a continuous function by a two-layer network incorporating a single hidden layer, with a large number of hidden nodes of continuous sigmoidal nonlinearities, linear output units, and suitable connection weights. Furthermore, the closer the approximation desired, the larger the number of hidden nodes required.

Consider, for example, the Fourier series representation of the real-valued function F,

$$F(x) = \sum_{k=0}^{\infty} \{a_k \cos(kx) + b_k \sin(kx)\}, \quad x \in \Re. \tag{10.21}$$

where the $\{a_k, b_k\}$ are Fourier coefficients. The function F can be approximated by a neural network (see Exercise 10.14), which produces the approximation,

$$\widehat{F}(x) = \sum_{j=0}^{t} \alpha_j \beta_j \sin(x + \beta_{0j}). \tag{10.22}$$

The weights $\{\beta_j\}$ yield the amplitudes of the sine functions. and the constants $\{\beta_{0j}\}$ yield the phases; if, for example, we set $\beta_{0j} = \pi/2$, then $\sin(x + \beta_{0j}) = \cos(x)$, and so we do not need to include explicit cosine terms in the network. The weights $\{\alpha_j\}$ are the amplitudes of the individual Fourier terms.

The universal approximation theorem is an existence theorem: it shows, theoretically, that one can approximate an arbitrary continuous function by a single hidden-layer network. Unfortunately, it does not specify how to find that approximation; that is, how to determine the weights and the number, t, of nodes in the hidden layer (a problem known as *network complexity*). It also assumes that we know the continuous function being approximated and that the available set of hidden nodes is of unlimited size. Furthermore, the theorem is not an optimality result: it does not show that a single hidden layer is the best-possible multilayer network for carrying out the approximation.

10.7.4 More than One Hidden Layer

We can express (10.19) in matrix notation as follows:

$$\boldsymbol{\mu}(\mathbf{X}) = \mathbf{g}(\boldsymbol{\alpha}_0 + \mathbf{A}\mathbf{f}(\boldsymbol{\beta}_0 + \mathbf{B}\mathbf{X})), \tag{10.23}$$

where $\mathbf{B} = (\beta_{ij})$ is a $(t \times r)$-matrix of weights between the input nodes and the hidden layer, $\mathbf{A} = (\alpha_{jk})$ is an $(s \times t)$-matrix of weights between the hidden layer and the output layer, $\boldsymbol{\beta}_0 = (\beta_{01}, \cdots, \beta_{0t})^{\tau}$, and $\boldsymbol{\alpha}_0 = (\alpha_{01}, \cdots, \alpha_{0s})^{\tau}$; also, $\mathbf{f} = (f_1, \cdots, f_t)^{\tau}$ and $\mathbf{g} = (g_1, \cdots, g_s)^{\tau}$ are the vectors of nonlinear activation functions. In (10.23), the notation $\mathbf{h}(\mathbf{U})$ represents the vector $(h_1(U_1), \cdots, h_t(U_t))^{\tau}$, where $\mathbf{h} = (h_1, \cdots, h_t)^{\tau}$ is a vector of functions and $\mathbf{U} = (U_1, U_2, \cdots, U_t)^{\tau}$ is a random vector. Note, however, that $\boldsymbol{\mu}(\mathbf{X}) = (\mu_1(\mathbf{X}), \cdots, \mu_s(\mathbf{X}))^\tau$. Clearly, this representation permits straightforward extensions to more than one hidden layer.

An important special case of (10.23) occurs when the $\{f_j\}$ and the $\{g_k\}$ are each taken to be identity functions. In that case, (10.23) reduces to the multivariate reduced-rank regression model, $\boldsymbol{\mu}(\mathbf{X}) = \boldsymbol{\mu} + \mathbf{ABX}$, where $\boldsymbol{\mu} = \boldsymbol{\alpha}_0 + \mathbf{A}\boldsymbol{\beta}_0$. We could use the $(s \times r)$ weight-matrix $\mathbf{C} = \mathbf{AB}$ for a single-layer network (i.e., no hidden layer) and the results would be identical. The results change only when we use nonlinear activation functions at the hidden nodes.

Thus, a neural network with r input nodes, a single hidden layer with t nodes, s output nodes, and sigmoidal activation functions at the hidden nodes can be viewed as a nonlinear generalization of multivariate reduced-rank regression.

10.7.5   Optimality Criteria

Let the $(st + rt + t + s)$-vector $\boldsymbol{\omega}$ consist of the parameters of a fully connected network — the connection weights (elements of the matrices $\mathbf{A}$ and $\mathbf{B}$) and the biases (the vectors $\boldsymbol{\alpha}_0$ and $\boldsymbol{\beta}_0$). To estimate $\boldsymbol{\omega}$ in either binary classification (where outputs are either 0 or 1) or multivariate regression problems (where outputs are real-valued), it is customary to minimize the error sum of squares (ESS):

$$ESS(\boldsymbol{\omega}) = \sum_{i=1}^{n} \| \mathbf{Y}_i - \widetilde{\mathbf{Y}}_i \|^2, \tag{10.24}$$

with respect to the elements of $\boldsymbol{\omega}$, where

$$\| \mathbf{Y}_i - \widetilde{\mathbf{Y}}_i \|^2 = (\mathbf{Y}_i - \widetilde{\mathbf{Y}}_i)^\tau (\mathbf{Y}_i - \widetilde{\mathbf{Y}}_i) = \sum_{k \in \mathcal{K}} (Y_{i,k} - \widetilde{Y}_{i,k})^2, \tag{10.25}$$

and $\mathcal{K}$ is the set of output nodes. In binary classification problems, there is a single output node. In (10.25), $\mathbf{Y}_i = (Y_{i,k})$ is the value of the true (or "target") output s-vector, $\widetilde{\mathbf{Y}}_i = (\widetilde{Y}_{i,k})$ is the value of the fitted output s-vector, and $\widetilde{Y}_{i,k} = \mu_k(\mathbf{X}_i) = \mu_k(\mathbf{X}_i, \boldsymbol{\omega})$ is the fitted value at the kth output node corresponding to the ith input r-vector $\mathbf{X}_i$, $k \in \mathcal{K}$, $i = 1, 2, \ldots, n$.

For multiclass classification problems, where each observation belongs to one of $K > 2$ possible classes, there are usually K output nodes, one for each class. In this case, an error criterion is minus the logarithm of the conditional-likelihood function,

$$E(\boldsymbol{\omega}) = -\sum_{i=1}^{n} \sum_{k \in \mathcal{K}} Y_{i,k} \log \widetilde{Y}_{i,k}, \quad \widetilde{Y}_{i,k} = \frac{e^{V_{i,k}}}{\sum_{\ell \in \mathcal{K}} e^{V_{i,\ell}}}, \tag{10.26}$$

where $Y_{i,k} = 1$ if $\mathbf{X}_i \in \Pi_k$ and zero otherwise, and $V_{i,k} = \alpha_{0,k} + \mathbf{Z}_i^\tau \boldsymbol{\alpha}_k$ is the value of V_k for the ith input vector $\mathbf{X}_i$. This criterion is equivalent to the

Kullback–Leibler deviance (or *cross-entropy*), and $\widetilde{Y}_{i,k}$, which is known as the *softmax* function, is the multiclass generalization of the logistic function.

Because the fitted value, $\widetilde{Y}_{i,k}$, is a nonlinear function of $\boldsymbol{\omega}$, it follows that both the ESS and E criteria are nonlinear functions of $\boldsymbol{\omega}$. The $\boldsymbol{\omega}$ that minimizes $ESS(\boldsymbol{\omega})$ or $E(\boldsymbol{\omega})$ is not available in explicit form and, therefore, has to be found using a nonlinear optimization algorithm. The most popular numerical method for estimating the network parameters is the "backpropagation" of errors algorithm.

10.7.6 The Backpropagation of Errors Algorithm

The backpropagation algorithm (Werbos, 1974) efficiently computes the first derivatives of an error function wrt the network weights $\{\alpha_{kj}\}$ and $\{\beta_{jm}\}$. These derivatives are then used to estimate the weights by minimizing the error function through an iterative gradient-descent method.

To simplify the description of the algorithm, we treat the network as a single-hidden-layer network. All the details we present here can be generalized to a network having more than one hidden node. We denote by $\mathcal{M}$ the set of r input nodes, $\mathcal{J}$ the set of t hidden nodes, and $\mathcal{K}$ the set of s output nodes, so that $m \in \mathcal{M}$ indexes an input node, $j \in \mathcal{J}$ indexes a hidden node, and $k \in \mathcal{K}$ indexes an output node. In other words, $m \to j \to k$. As before, the input r-vectors are indexed by $i = 1, 2, \ldots, n$.

We start at the kth output node. Denote the *error signal* at that node by

$$e_{i,k} = Y_{i,k} - \widetilde{Y}_{i,k}, \quad k \in \mathcal{K}, \tag{10.27}$$

and the error sum of squares (usually referred to as the *error function*) at that node by

$$E_i = \frac{1}{2} \sum_{k \in \mathcal{K}} e_{i,k}^2 = \frac{1}{2} \sum_{k \in \mathcal{K}} (Y_{i,k} - \widetilde{Y}_{i,k})^2, \quad i = 1, 2, \ldots, n. \tag{10.28}$$

The optimizing criterion is the error sum of squares (ESS) for the entire data set; that is, the error function (10.28) averaged over all data in the learning set:

$$ESS = \frac{1}{n} \sum_{i=1}^{n} E_i = \frac{1}{2n} \sum_{i=1}^{n} \sum_{k \in \mathcal{K}} e_{i,k}^2. \tag{10.29}$$

The learning problem is to minimize ESS wrt the connection weights, $\{\alpha_{i,kj}\}$ and $\{\beta_{i,jm}\}$. Because each derivative of ESS wrt those weights is a sum over the learning set of data of the derivatives of E_i, $i = 1, 2, \ldots, n$, it suffices to minimize each E_i separately.

In the following description of the backpropagation algorithm, it may be helpful to refer to Figure 10.7.

input nodes

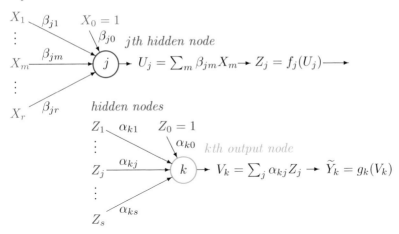

FIGURE 10.7. *Schematic diagram of the backpropagation of errors algorithm for a single-hidden-layer ANN. The top diagram relates the input nodes to the jth hidden node, and the bottom diagram relates the hidden nodes to the kth output node. To simplify notation, all reference to the ith input vector has been dropped.*

For the ith input vector, let

$$V_{i,k} = \sum_{j \in \mathcal{J}} \alpha_{kj} Z_{i,j} = \alpha_{k0} + \mathbf{Z}_i^\tau \boldsymbol{\alpha}_k, \quad k \in \mathcal{K}, \tag{10.30}$$

be a weighted sum of inputs from the set of hidden units to the kth output node, where

$$\mathbf{Z}_i = (Z_{i,1}, \ldots, Z_{i,t})^\tau, \quad \boldsymbol{\alpha}_k = (\alpha_{k1}, \ldots, \alpha_{kt})^\tau, \tag{10.31}$$

and $Z_{i,0} = 1$. Then, the corresponding output is

$$\widetilde{Y}_{i,k} = g_k(V_{i,k}), \quad k \in \mathcal{K}, \tag{10.32}$$

where $g_k(\cdot)$ is an output activation function, which we assume is differentiable.

The backpropagation algorithm is an iterative gradient-descent-based algorithm. Using randomly chosen initial values for the weights, we search for that direction that makes the error function smaller.

Consider the weights $\alpha_{i,kj}$ from the jth hidden node to the kth output node. Let $\boldsymbol{\alpha}_i = (\boldsymbol{\alpha}_{i,1}^\tau, \cdots, \boldsymbol{\alpha}_{i,s}^\tau)^\tau = (\alpha_{i,kj})$ to be the ts-vector of all the hidden-layer-to-output-layer weights at the ith iteration. Then, the update rule is

$$\boldsymbol{\alpha}_{i+1} = \boldsymbol{\alpha}_i + \Delta\boldsymbol{\alpha}_i, \tag{10.33}$$

where

$$\Delta\boldsymbol{\alpha}_i = -\eta\frac{\partial E_i}{\partial\boldsymbol{\alpha}_i} = \left(-\eta\frac{\partial E_i}{\partial\alpha_{i,jh}}\right) = (\Delta\alpha_{i,kj}). \tag{10.34}$$

Similar update equations hold also for $\alpha_{i,k0}$. In (10.34), the learning parameter η specifies how large each step should be in the iterative process. If η is too large, the iterations will move rapidly toward a local minimum, but may possibly overshoot it, whereas if η is too small, the iterations may take a long time to get anywhere near a local minimum.

Using the chain rule for differentiation, we have that

$$\begin{aligned}
\frac{\partial E_i}{\partial\alpha_{i,kj}} &= \frac{\partial E_i}{\partial e_{i,k}} \cdot \frac{\partial e_{i,k}}{\partial\widetilde{Y}_{i,k}} \cdot \frac{\partial\widetilde{Y}_{i,k}}{\partial V_{i,k}} \cdot \frac{\partial V_{i,k}}{\partial\alpha_{i,kj}} \\
&= e_{i,k} \cdot (-1) \cdot g'_k(V_{i,k}) \cdot Z_{i,j} \\
&= -e_{i,k}g'_k(\alpha_{i,k0} + \mathbf{Z}_i^\tau\boldsymbol{\alpha}_{i,k})Z_{i,j}.
\end{aligned} \tag{10.35}$$

This can also be expressed as

$$\frac{\partial E_i}{\partial\alpha_{i,jh}} = -\delta_{i,k}Z_{i,j}, \tag{10.36}$$

where

$$\delta_{i,k} = -\frac{\partial E_i}{\partial\widetilde{Y}_{i,k}} \cdot \frac{\partial\widetilde{Y}_{i,k}}{\partial V_{i,k}} = e_{i,k}g'_k(V_{i,k}) \tag{10.37}$$

is the *sensitivity* (or *local gradient*) of the ith observation at the kth output node. The expression for $\delta_{i,k}$ is the product of two terms associated with the kth node: the error signal $e_{i,k}$ and the derivative, $g'_k(V_{i,k})$, of the activation function. The gradient-descent update to $\alpha_{i,kj}$ is given by

$$\alpha_{i+1,kj} = \alpha_{i,kj} - \eta\frac{\partial E_i}{\partial\alpha_{i,kj}} = \alpha_{i,kj} + \eta\delta_{i,k}Z_{i,j}, \tag{10.38}$$

where η is the learning rate parameter of the backpropagation algorithm.

The next part of the backpropagation algorithm is to derive an update rule for the connection from the mth input node to the jth hidden node. At the ith iteration, let

$$U_{i,j} = \sum_{m\in\mathcal{M}} \beta_{i,jm}X_{i,m} = \beta_{i,j0} + \mathbf{X}_i^\tau\boldsymbol{\beta}_{i,j}, \quad j\in\mathcal{J}, \tag{10.39}$$

be the weighted sum of inputs to the jth hidden node, where

$$\mathbf{X}_i = (X_{i,1},\cdots,X_{i,r})^\tau, \quad \boldsymbol{\beta}_{i,j} = (\beta_{i,j1},\cdots,\beta_{i,jr})^\tau, \tag{10.40}$$

and $X_{i,0} = 1$. The corresponding output is

$$Z_{i,j} = f_j(U_{i,j}), \tag{10.41}$$

where $f_j(\cdot)$ is the activation function, which we assume is differentiable, at the jth hidden node. Let $\boldsymbol{\beta}_i = (\boldsymbol{\beta}_{i,1}^\tau, \cdots, \boldsymbol{\beta}_{i,t}^\tau)^\tau = (\beta_{i,jm})$ be the ith iteration of the $(r+1)t$-vector of all the input-layer-to-hidden-layer weights. Then, the update rule is

$$\boldsymbol{\beta}_{i+1} = \boldsymbol{\beta}_i + \Delta\boldsymbol{\beta}_i, \tag{10.42}$$

where

$$\Delta\boldsymbol{\beta}_i = -\eta\frac{\partial E_i}{\partial\boldsymbol{\beta}_i} = \left(-\eta\frac{\partial E_i}{\partial\beta_{i,jm}}\right) = (\Delta\beta_{i,jm}). \tag{10.43}$$

Again, similar update formulas hold for the bias terms $\beta_{i,j0}$. Using the chain rule, we have that

$$\frac{\partial E_i}{\partial\beta_{i,kj}} = \frac{\partial E_i}{\partial Z_{i,j}} \cdot \frac{\partial Z_{i,j}}{\partial U_{i,j}} \cdot \frac{\partial U_{i,j}}{\partial\beta_{i,kj}}. \tag{10.44}$$

The first term on the rhs is

$$\begin{aligned}
\frac{\partial E_i}{\partial z_{i,j}} &= \sum_{k\in\mathcal{K}} e_{i,k} \cdot \frac{\partial e_{i,k}}{\partial Z_{i,j}} \\
&= \sum_{k\in\mathcal{K}} e_{i,k} \cdot \frac{\partial e_{i,k}}{\partial V_{i,k}} \cdot \frac{\partial V_{i,k}}{\partial Z_{i,j}} \\
&= -\sum_{k\in\mathcal{K}} e_{i,k} \cdot g_k'(V_{ij}) \cdot \alpha_{i,kj} \\
&= -\sum_{k\in\mathcal{K}} \delta_{i,k}\alpha_{i,kj},
\end{aligned} \tag{10.45}$$

whence, from (10.44),

$$\frac{\partial E_i}{\partial\beta_{i,kj}} = -\sum_{k\in\mathcal{K}} e_{i,k}g_k'(\alpha_{i,k0} + \mathbf{Z}_i^\tau\boldsymbol{\alpha}_{i,k})\alpha_{i,kj}f_j'(\beta_{i,j0} + \mathbf{X}_i^\tau\boldsymbol{\beta}_{i,j})X_{i.m} \tag{10.46}$$

Putting (10.37) and (10.45) together, we have that

$$\delta_{i,j} = f_j'(U_{i,j})\sum_{k\in\mathcal{K}} \delta_{i,k}\alpha_{i,kj}. \tag{10.47}$$

This expression for $\delta_{i,j}$ is the product of two terms: the first term, $f_j'(U_{i,j})$, is the derivative of the activation function $f_j(\cdot)$ evaluated at the jth hidden node; the second term is a weighted sum of the $\delta_{i,k}$ (which requires knowledge of the error $e_{i,k}$ at the kth output node) over all output nodes, where the kth weight, $\alpha_{i,kj}$, is the connection weight of the jth hidden node to the kth output node. Thus, $\delta_{i,j}$ at the jth hidden node depends upon the $\{\delta_{i,k}\}$ from all the output nodes.

The gradient-descent update to $\beta_{i,jm}$ is given by

$$\beta_{i+1,jm} = \beta_{i,jm} - \eta\frac{\partial E_i}{\partial\beta_{i,jm}} = \beta_{i,jm} + \eta\delta_{i,j}X_{i,m}, \tag{10.48}$$

where η is the learning rate parameter of the backpropagation algorithm.

The backpropagation algorithm is defined by (10.38) and (10.48). These update formulas identify two stages of computation in this algorithm: a "feedforward pass" stage and a "backpropagation pass" stage. After an initialization step in which all connection weights are assigned values, we have the following stages in the algorithm:

Feedforward pass Inputs enter the node from the left and emerge from the right of the node; the output from the node is computed as (10.30) and (10.31), and the results are passed, from left to right, through the layers of the network.

Backpropagation pass The network is run in reverse order, layer by layer, starting at the output layer:

1. The error (10.27) is computed at the kth output node and then multiplied by the derivative of the activation function to give the sensitivity $\delta_{i,k}$ at that output node (10.37); the weights, $\{\alpha_{i,kj}\}$, feeding into the output nodes are updated by using (10.38).

2. We use (10.47) to compute the sensitivity $\delta_{i,j}$ at the jth hidden node; and, then, we use (10.48) to update the weights, $\{\beta_{i,jm}\}$, feeding into the hidden nodes.

This iterative process is repeated until some suitable stopping time.

10.7.7 Convergence and Stopping

There is no proof that the backpropagation algorithm always converges. In fact, experience has shown that the algorithm is a slow learner, the estimates may be unstable, there may exist many local minima, and convergence is not assured in practice. There have been many explanations of why this should happen.

One possible reason is that the backpropagation algorithm is a first-order approximation to the method of steepest-descent and, hence, is a version of *stochastic approximation*. As the algorithm tries to find the minimum along fairly flat regions of the surface of the error criterion, it takes many iterations to significantly reduce the error criterion; in other, highly curved regions, the algorithm may miss the minimum entirely. Another possible reason (Hwang and Ding, 1997) is that, for any ANN, instability and convergence problems may be partly caused by the "unidentifiability" of the parameter vector $\boldsymbol{\omega}$; for example, certain elements of $\boldsymbol{\omega}$ can be permuted without changing the value of $\mu(\mathbf{X})$ in (10.20).

Because of the slow progression of the backpropagation algorithm, which is both frustrating and expensive, overfitting the network has been (according to ANN folklore) accidentally avoided by stopping the algorithm prior to convergence (usually referred to as *early stopping*). Other researchers prefer to continue running the algorithm until the weights stabilize (e.g., the normed difference between successive iterates is smaller than some acceptable bound) or until the error criterion is at (or close to) a minimum. Another practical strategy is to increase the value of η to produce faster convergence, but that action could also result in oscillations.

10.8 Network Design Considerations

When fitting an ANN, the user is faced with a number of algorithmic details that need to be resolved as part of the design of the network. In this section, we discuss a collection of problems often referred to as *network complexity*.

10.8.1 Learning Modes

The most popular methods of running the backpropagation algorithm are the "on-line," "stochastic," and "batch" learning modes.

In *on-line mode*, each observation $(\mathbf{x}_i, y_i)$, $i = 1, 2, \ldots, n$, is dropped down the network in sequential fashion, one at a time, and adjustments are made to the estimates of the connection weights each time. The iteration steps (10.38) and (10.48) give an on-line update of the weights. Thus, $(\mathbf{x}_1, y_1)$ is dropped down the network first. The feedforward and backpropagation stages of the algorithm are immediately carried out, yielding updated initial values of the connection weights. Next, we drop $(\mathbf{x}_2, y_2)$ down the network, whence the feedforward and backpropagation stages are again carried out, resulting in further updated values of the connection weights. This procedure is repeated once and only once for every observation in the entire learning set, until the last observation $(\mathbf{x}_n, y_n)$ is dropped down the network and the connection weights are updated. The process then stops.

A variation on on-line learning is *stochastic learning*, where an observation is chosen at random from the learning set, dropped down the network, and the parameter values are updated using (10.38) and (10.48). As in on-line learning, each observation is dropped down the network once and only once, but in random order.

In *batch mode*, all n observations in the learning set (referred to as an *epoch*) are dropped down the network in any order. After all the observations are entered, the weights are updated by summing the derivatives over the entire learning set; that is, for the ith epoch, the updates are

$$\alpha_{i+1,jk} = \alpha_{i,jk} + \eta \sum_{h=1}^{n} \delta_{h,k} z_{h,j}, \qquad (10.49)$$

$$\beta_{i+1,jm} = \beta_{i,jm} + \eta \sum_{h=1}^{n} \delta_{h,j} x_{h,m}, \qquad (10.50)$$

$h = 1, 2, \ldots$. This entire process is repeated, epoch by epoch, until ESS becomes smaller than some preset value.

On-line learning tends to be preferred to batch learning: on-line learning is generally faster, particularly when there are many similar data values (*redundancy*) in the learning set; it can adapt better to nonstandard conditions of the data (e.g., nonstationarity); and it can more easily escape from local minima. Moreover, batch learning in very high-dimensional situations can cause computational difficulties (e.g., memory problems, cost considerations), especially when it comes to deriving the matrices $\mathbf{A}$ and $\mathbf{B}$ in (10.23).

10.8.2   Input Scaling

Inputs are often measured in widely differing scales, which may affect the relative contribution of each input to the resulting analysis. This is a common concern in data analysis. The same problem occurs when fitting an ANN. In general, it is a good idea, prior to fitting an ANN to data, to scale each input variable. A number of ways have been suggested to accomplish this objective, including (1) scale the data to the interval $[0, 1]$; (2) scale the data to $[-1, 1]$ or to $[-2, 2]$; or (3) standardize each input variable to have zero mean and unit standard deviation.

ANN theory does not require the input data to lie in $[0, 1]$; in fact, scaling to $[0, 1]$ may not be a good choice and that it is better to center the input data around zero. This implies that options (2) and (3) should be preferred to option (1). These latter two scaling options may enable an ANN to be run more efficiently and may help to avoid getting bogged down in local extrema.

If a weight-decay penalty is to be incorporated as part of the optimization process (see Section 10.8.5), then it makes sense to scale or standardize each input variable. When the data are split into learning and test sets, then the same scaling or standardization transformation applied to the learning set should also be applied to the test set. Note that the standardization transformation can only be used for stochastic or batch learning; it cannot be used for on-line learning, where the data are presented to the network one observation at a time.

10.8.3 How Many Hidden Nodes and Layers?

One of the main problems in designing a network is to determine how many hidden nodes and layers to include in the network; this, in turn, determines how many parameters are needed to model the data. The central principle here is that of Ockham's razor: keep the model as simple as possible while maintaining its ability to generalize well.

One way of choosing the number of hidden nodes is by employing cross-validation (CV). However, the presence of multiple local minima at each iteration, which result in quite different performances, can confuse the issue of deciding which solution should be used for each round of CV. Most applications of ANN determine the number of hidden nodes and layers either from the context of the problem or by trial-and-error.

10.8.4 Initializing the Weights

As with any numerical and iterative method, the backpropagation algorithm requires a choice of starting values to estimate the parameters (i.e., connection weights and biases) of the network. In general, we initialize the network by using small (close to zero), random-generated (uniformly distributed with small variance) starting values for the parameter estimates.

10.8.5 Overfitting and Network Pruning

Building a neural network can easily yield a model with a huge number of parameters. If we try to estimate all those parameters optimally by waiting for the algorithm to converge, this can lead to severe overfitting. We would like to reduce (as much as possible) the size of the network while retaining (as much as possible) its good performance characteristics.

Setting parameters to zero. One way to counter overfitting is to set some connection weights to zero, a method known as *network pruning* or, more delightfully, *optimal brain surgery*, because of the notion that ANNs try to approximate brain activity (Hassibi, Stork, Wolff, and Wanatabe, 1994). If, however, a parameter (connection weight) in the model is set to zero and the inputs are close to being collinear, then the standard errors for the remaining estimated parameters could be significantly affected; thus, it is not generally recommended to set more than one connection weight to zero (Ripley, 1996, p. 169), a strategy that defeats the objective of reducing network size.

Shrinking parameters toward zero. Another approach is to "shrink" the magnitudes of network parameters toward zero by incorporating regularization into the criterion. In such a formulation, we minimize

$$ESS_\lambda(\boldsymbol{\omega}) = ESS(\boldsymbol{\omega}) + \lambda p(\boldsymbol{\omega}), \tag{10.51}$$

where $\lambda \geq 0$ is a *regularization parameter* and $p(\cdot)$ is the *penalty function*. The term $\lambda p(\boldsymbol{\omega})$ is known as the *complexity term*. The regularization parameter λ measures the relative importance of $ESS(\boldsymbol{\omega})$ to $p(\boldsymbol{\omega})$, and is usually estimated by cross-validation.

There are two popular assignments of penalty functions in this ANN context. The simplest regularizer is *weight-decay*, whose penalty is defined by

$$p(\boldsymbol{\omega}) = \| \boldsymbol{\omega} \|^2 = \sum_\ell \omega_\ell^2, \tag{10.52}$$

where ω_ℓ is equal to α_{jm} or β_{kj}, as appropriate, and the summation is taken over all weight connections in the network (Hinton, 1987). In this case, λ is referred to as the *weight-decay parameter*. A more elaborate penalty function is the *weight-elimination penalty*, given by

$$p(\boldsymbol{\omega}) = \sum_\ell \frac{(\omega_\ell/W)^2}{1 + (\omega_\ell/W)^2}, \tag{10.53}$$

where W is a preassigned free parameter (Weigend, Rumelhart, and Huberman, 1991), such as $W = \| \boldsymbol{\omega} \|^2$. If, for some ℓ, $|\omega_\ell| \ll W$, the contribution of that connection weight to (10.53) is deemed negligible and the connection may be eliminated; if $|\omega_\ell| \gg W$, then that connection weight contributes a significant amount to (10.53) and, hence, should be retained in the network. When using penalty function (10.52) or (10.53), it is usual to start with $\lambda = 0$, which allows the network weights to be unconstrained, and then adjust that solution by increasing the value of λ in small increments.

Reducing dimensionality of input data. The user can also apply principal component analysis to the input data, thereby reducing the number of inputs, and then estimate the parameters of the resulting reduced-size ANN.

10.9   Example: Detecting Hidden Messages in Digital Images

Steganography ("covered writing," from the Greek) is "the art and science of communicating in a way which hides the existence of the communication" (Kahn, 1996). It is a method for hiding messages in different types

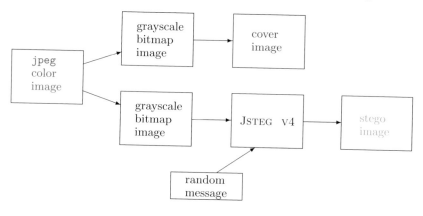

FIGURE 10.8. *Flow chart for the steganography example.*

of media, such as webpage HTML text, Microsoft Word documents, executable and dynamic link library files, digital audio files, and digital image files (`bmp, gif, jpg`). Reasons for hiding messages include the need for copyright protection of digital media (audio, image, and video), for Internet security and privacy, and to provide "stealth" military and intelligence communication.

There are many ways in which information can be hidden in digital media, including least significant bit (lsb) embedding, digital watermarking, and wavelet decomposition algorithms. A major disadvantage to lsb insertion is that it is vulnerable to slight image manipulation, such as cropping and compression. See Petitcolas, Anderson, and Kuhn (1999) for a survey.

In this example, 1,000 color `jpeg` images consisting of a mixture of various science fiction environments (including indoors, outdoors, outer space), characters, and images with special effects, were obtained from the *Star Trek* website.[1] These color images were converted into grayscale bitmap images to remove any existing digital watermarks or other hidden identifiers and cropped to a central 640×480 pixel area. These grayscale bitmap images were then duplicated to form two sets of the same 1,000 images. One set of grayscale images was decompressed to produce 1,000 "cover images." The second set was used to hide messages of random strings of characters of sufficient length (2–3 KB). Using the software package JSTEG V4,[2] 1,000

---

[1] The *Star Trek* website is www.startrek.com. The author thanks Joseph Jupin for use of the data that formed the basis for his 2004 report *Steganography* at the website astro.temple.edu/~joejupin/Steganography.pdf.

[2] Derek Upham's JSTEG V4 is available at ftp.funet.fi/pub/crypt/steganography.

"stego images" were formed. A flow chart of the steganographic process is given in Figure 10.8.

The next step is to extract from the 1,000 cover images and the 1,000 stego images a common set of variables. To identify images that contain a hidden message, we use a methodology based upon the wavelet decomposition of digital images (Farid, 2001). First, we compute a multiresolution analysis of each set of 1,000 images using *quadrature mirror filters*. For each such set, this creates orthonormal basis functions that partition the frequency space into m resolution levels and three orientations — horizontal, vertical, and diagonal. At each resolution level, separable low-pass and high-pass filters are applied along the image axes, which generate low-pass, vertical, horizontal, and diagonal subbands. Additional resolution levels are created by recursively filtering the low-pass subband.

Hiding messages in a digital image often leads to a significant change in the statistical properties of the wavelet decomposition of that image. Given an image decomposition, we compute two sets of statistical moments: (1) the mean, variance, skewness, and kurtosis of the subband coefficients at each of the three orientations and at resolution levels $1, 2, \ldots, m - 1$; (2) the same statistics, but computed from the residuals of the optimal linear predictor of coefficient magnitudes and the true coefficient magnitudes for each of the three orientation subbands at each level. This creates a total of $24(m - 1)$ variables for each image decomposition. In our example, a four-level ($m = 4$), three-orientation decomposition scheme results in a 72-dimensional vector of the moment statistics of estimated coefficients and residuals for each image.

From each set of 1,000 images, 500 images are randomly selected, but no duplicate images are taken. The resulting 1,000 images constitute our data set. The problem is to distinguish the stego images from the cover images.

We randomly divided the data from the 1,000 images into a learning set (650) and a test set (350). The learning set consists of 322 stego images and 328 cover images, and the test set consists of 178 stego images and 172 cover images. The learning set was standardized and an ANN was fit with a single hidden layer, varying the decay parameter λ between 0.0001 and 0.9, and varying the number of nodes in the hidden layer from 1 to 10. Each of these fitted models was used to predict the two classes (cover or stego) for the data in the test set, which had previously been standardized using the same scaling obtained from the learning set.

This fitting and prediction strategy is repeated 10 times using randomly generated starting values for each combination of λ and number of hidden nodes; the misclassification rates were averaged for each such combination. Figure 10.9 shows parallel boxplots of the individual results for $\lambda = 0.01$ (left panel) and 0.5 (right panel). Notice the high variability for $\lambda = 0.01$

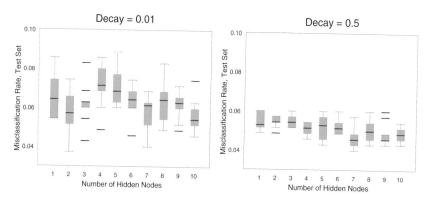

FIGURE 10.9. *Steganography example: parallel boxplots for the misclassification rate of the test set for a neural network with a single hidden layer and number of hidden nodes as displayed, and decay parameter $\lambda = 0.01$ (left panel) and 0.5 (right panel). A randomly generated start was used to fit each such model, and this was repeated 10 times for each number of hidden nodes.*

compared with $\lambda = 0.5$. The smallest average misclassification rate for the test set is 0.0463, which is obtained for $\lambda = 0.5$ and seven hidden nodes.

10.10   Examples of Fitting Neural Networks

In Table 10.2, we list the estimated misclassification rates of neural network models applied to data sets detailed in Chapter 8. The misclassification rates are estimated here by randomly dividing each data set into two subsets, a learning set (2/3) and a test set (1/3). With certain exceptions, each learning set was first standardized by subtracting the mean of each input variable and then dividing the result by the standard deviation of that variable. The same standardization was also applied to the input variables in the test set. The exceptions to this standardization are those data sets whose values fall in $[0, 1]$ (E-coli, Yeast), $[-1, 1]$ (Ionosphere), or $[0, 100]$ (Pendigits), where no transformations are made.

For each learning set, we set up a neural network model with a single hidden layer of between 0 and 10 nodes and decay parameter λ ranging from 0.00001 to 0.1. A set of initial weights is randomly generated to fit the ANN model to the learning set, the fitted ANN model is then applied to the test set, and the misclassification rate computed. This is repeated 10 times, and the resulting misclassification rates are averaged to produce the "TestSetER" in Table 10.2.

TABLE 10.2. *Summary of artificial neural network (ANN) models with a single hidden layer fitted to data sets for binary and multiclass classification. Listed are the sample size (n), number of variables (r), and number of classes (K). Also listed for each data set is the number of observations in the learning set (2/3) and in the test set (1/3) and the test-set error (misclassification) rate computed from the average of 10 random initial starts. Each learning set was standardized, and the same standardization was used for the test set (with the exception of Ionosphere, where the input values fall into [−1, 1], and E-coli, Yeast, and Pendigits, whose values fall in [0, 1]). The data sets are listed in increasing order of LDA misclassification rates (see Tables 8.5 and 8.7).*

Data Set	n	r	K	Learn	Test	TestSetER
Breast cancer (logs)	569	30	2	379	190	0.0174
Spambase	4,601	57	2	3,067	1,534	0.0669
Ionosphere	351	33	2	234	117	0.0863
Sonar	208	60	2	138	70	0.1571
BUPA liver disorders	345	6	2	230	115	0.3183
Wine	178	13	3	118	60	0.0167
Iris	150	4	3	100	50	0.0420
Primate scapulae	105	7	5	70	35	0.0114
Shuttle	58,000	8	7	43,500	14,500	0.0002
Diabetes	145	5	3	95	50	0.0020
Pendigits	10,992	16	10	7,328	3,664	0.0251
E-coli	336	7	8	224	112	0.1161
Vehicle	846	18	4	564	282	0.1897
Letter recognition	20,000	16	26	13,000	7,000	0.0987
Glass	214	9	6	143	71	0.2056
Yeast	1,484	8	10	989	495	0.4026

We see that a single hidden-layer ANN model fits some data sets better than others. Comparing Table 10.2 with Tables 8.5 and 8.7 (ANN misclassification rates are computed using an independent test set, whereas LDA and QDA used 10-fold CV), a single-hidden-layer ANN model fares better than LDA for the spambase, ionosphere, sonar, primate scapulae, shuttle, diabetes, pendigits, e-coli, vehicle, glass, and yeast data, whereas LDA comes out ahead for the breast cancer, BUPA liver, wine, and iris data. The misclassification rate for the letter-recognition data is significantly reduced if there are a large number of hidden nodes (20 or more).

10.11   Related Statistical Methods

Alternative approaches to statistical curve-fitting, such as projection-pursuit regression and generalized additive models, try to address a more general functional form than linearity. Although these methods are closely related in appearance to the ANN model, their computations are carried out in completely different ways.

10.11.1   Projection-Pursuit Regression

Consider the input r-vector $\mathbf{X}$ and a single output variable Y (i.e., $s = 1$). Suppose the model is

$$Y = \mu(\mathbf{X}) + \epsilon, \tag{10.54}$$

where $\mu(\mathbf{X}) = \mathrm{E}\{Y|\mathbf{X}\}$ is the regression function, and the errors ϵ are independent of $\mathbf{X}$ and have $\mathrm{E}(\epsilon) = 0$ and $\mathrm{var}(\epsilon) = \sigma^2$. The goal is to estimate $\mu(\mathbf{X})$. For example, suppose $r = 2$ and $\mu(\mathbf{X}) = X_1 X_2$; we can write $\mu(\mathbf{X}) = \frac{1}{4}(X_1 + X_2)^2 - \frac{1}{4}(X_1 - X_2)^2$, which is the sum of squares of the projections, $\mathbf{X}^\tau \boldsymbol{\beta}_1 = (X_1, X_2)(1, 1)^\tau$ and $\mathbf{X}^\tau \boldsymbol{\beta}_2 = (X_1, X_2)(1, -1)^\tau$. So, a regression surface can be approximated by a sum of nonlinear functions, $\{f_j\}$, of projections $\mathbf{X}^\tau \boldsymbol{\beta}_j$.

This idea is implemented in *projection-pursuit regression (PPR)* (Friedman and Stuetzle, 1981), where the regression function is taken to be

$$\mu(\mathbf{X}) = \alpha_0 + \sum_{j=1}^{t} f_j(\beta_{0j} + \mathbf{X}^\tau \boldsymbol{\beta}_j), \tag{10.55}$$

where α_0, $\{\beta_{0j}\}$, $\{\boldsymbol{\beta}_j = (\beta_{1j}, \cdots, \beta_{rj})^\tau\}$, and the $\{f_j(\cdot)\}$ are the unknown parameters of the model. This is the sum of t nonlinearly transformed linear projections of the r input variables, where t is a user-chosen parameter, and has the same form as a two-layer feedforward perceptron for a single output variable (see (10.20)). Parallel to the discussion in Section 10.5.3, it has been shown that any smooth function of $\mathbf{X}$ can be well-approximated by (10.55), where the approximation improves as t gets large enough (Diaconis and Shahshahani, 1984). It is worth noting that as we increase t, it becomes more and more difficult to interpret the fitted functions and coefficients in the PPR solution.

The linear combinations, $\beta_{0j} + \mathbf{X}^\tau \boldsymbol{\beta}_j$, $j = 1, 2, \ldots, t$, are linear projections of the inputs $\mathbf{X}$ onto t different hyperplanes, and the activation functions $f_j(\cdot)$, $j = 1, 2, \ldots, t$, are (possibly, different) smooth but unknown functions; we assume that the $\{f_j(\cdot)\}$ are each normalized to have zero mean and unit variance. These t nonlinearly transformed projections are then linearly combined to produce $\mu(\mathbf{X})$ in (10.55). The components $f_j(\beta_{0j} + \mathbf{X}^\tau \boldsymbol{\beta}_j)$, $j = 1, 2, \ldots, t$, are often referred to as *ridge functions* in r dimensions; the name derives from the fact that, in two-dimensional input space (i.e., $r = 2$), a peaked $f_j(\cdot)$ produces output with a ridge in the graph.

When there is more than one output variable, the output can be represented as a multiresponse s-vector, $\mathbf{Y} = (Y_1, \cdots, Y_s)^\tau$. Then, each component of the regression function, $\mu(\mathbf{X}) = (\mu_1(\mathbf{X}), \cdots, \mu_s(\mathbf{X}))^\tau$, where $\mu_k(\mathbf{X}) = \mathrm{E}\{Y_k|\mathbf{X}\}$, can be written in the form,

$$\mu_k(\mathbf{X}) = \alpha_{0k} + \sum_{j=1}^{t} \alpha_{jk} f_j(\beta_{0j} + \mathbf{X}^\tau \boldsymbol{\beta}_j), \quad k = 1, 2, \ldots, s, \qquad (10.56)$$

where the $f_j(\cdot)$, $j = 1, 2, \ldots, t$, are taken to be a common set of arbitrarily smooth functions having zero mean and unit variance. Models such as (10.56) are referred to as SMART (smooth multiple additive regression technique) (Friedman, 1984).

Let $\boldsymbol{\alpha} = (\alpha_0, \alpha_1, \cdots, \alpha_t)^\tau$ and $\boldsymbol{\beta}_j = (\beta_{0j}, \beta_{1j}, \cdots, \beta_{rj})^\tau$, $j = 1, 2, \ldots, t$, be each of unit length. Given data, $\{(\mathbf{X}_i, Y_i), i = 1, 2, \ldots, n\}$, the $(t(r+2)+1)$-vector $\boldsymbol{\omega} = (\boldsymbol{\alpha}^\tau, \{\boldsymbol{\beta}_j^\tau\}_{j=1}^t)^\tau$ of parameters of the PPR single-output model (10.55) can be estimated by minimizing the error sum-of-squares,

$$ESS(\boldsymbol{\omega}) = \sum_{i=1}^{n} \left\{ Y_i - \alpha_0 - \sum_{j=1}^{t} \alpha_j f_j(\beta_{0j} + \mathbf{X}_i^\tau \boldsymbol{\beta}_j) \right\}^2, \qquad (10.57)$$

for nonlinear activation functions $\{f_j(\cdot)\}$, which are also determined from the data.

The function $ESS(\boldsymbol{\omega})$ is minimized in stages, and the parameters are estimated in sequential fashion: first, the $\{\alpha_j\}$ are fitted by linear least-squares; next, the $\{f_j(\cdot)\}$ are found using one-dimensional scatterplot smoothers, and finally, the $\{\beta_{kj}\}$ are fitted by nonlinear least-squares (e.g., Gauss–Newton). Scatterplot smoothers used to estimate the PPR functions $\{f_j(\cdot)\}$ include *supersmoother* (or *variable span smoother*) (Friedman and Stuetzle, 1981), *Hermitian polynomials* (Hwang, Li, Maechler, Martin, and Schimert, 1992), and *smoothing splines* (Roosen and Hastie, 1994). These steps to minimizing (10.57) are then iterated until some stopping criterion is satisfied. Stopping too early produces an increased bias for the estimate, and waiting too long produces an enlarged variance. Typically, the process is stopped when successive iterative values of the residual sum of squares, $RSS(\widehat{\boldsymbol{\omega}})$, become small and stable. In certain examples, the amount of computation involved in finding a PPR solution could be quite large and expensive.

10.11.2 Generalized Additive Models

An *additive model* in $\mathbf{X} = (X_1, \cdots, X_r)^\tau$ is a regression model that is additive in the inputs. Specifically, we assume that $Y = \mu(\mathbf{X}) + \epsilon$, where the regression function, $\mu(\mathbf{X}) = \mathrm{E}\{Y|\mathbf{X}\}$, has the form,

$$\mu(\mathbf{X}) = \alpha_0 + \sum_{j=1}^{r} f_j(X_j), \qquad (10.58)$$

and the error ϵ is independent of $\mathbf{X}$. If $f_j(X_j) = \beta_j X_j$, then the additive model reduces to the standard multiple regression model. The key aspect of an additive model is that interactions between input variables (e.g., $X_i X_j$) are not allowed as part of the model. If simple interactions are thought to be important, we can introduce into an additive model additional terms constructed as the products $X_i X_j$, $f_{ij}(X_i X_j)$, or $\widehat{f}_i(X_i) \cdot \widehat{f}_j(X_j)$, where $\widehat{f}_i(\cdot)$ and $\widehat{f}_j(\cdot)$ are the functions obtained from fitting the additive model.

The $\{f_j(\cdot)\}$ are typically taken to be nonlinear transformations of the input variables. For example, we could transform the input variables by using logarithmic, square-root, reciprocal, or power transformations, where the choice would depend upon what we know or suspect about each input variable. In general, it is more useful if we take the $\{f_j(\cdot)\}$ to be a set of smooth, but otherwise unspecified, functions, which are centered so that $\mathrm{E}\{f_j(X_j)\} = 0$, $j = 1, 2, \ldots, r$.

To estimate $\mu(\mathbf{X})$, the strategy is to estimate each $f_j(\cdot)$ separately. Estimation is based upon a *backfitting* algorithm (Friedman and Stuetzle, 1981). The key is the identity, $\mathrm{E}\{Y - \alpha_0 - \sum_{k \neq j} f_k(X_k) | X_j\} = f_j(X_j)$. Given observations $\{(\mathbf{x}_i, y_i), i = 1, 2, \ldots, n\}$ on $(\mathbf{X}, Y)$, we estimate α_0 by $\widehat{\alpha}_0 = \bar{y}$ and use the most current function estimates $\{\widehat{f}_k, k \neq j\}$ to update $\widehat{f}_j$ by a curve obtained by smoothing the "partial residuals," $y_i - \widehat{\alpha}_0 - \sum_{k \neq j} \widehat{f}_k(x_{ki})$, against x_{ji}, $i = 1, 2, \ldots, n$. This update procedure is applied by cycling through the $\{X_j\}$ until convergence of the smoothed partial residuals. The smoothing step uses a scatterplot smoother such as a cubic regression spline, which is a set of piecewise cubic polynomials joined together at a sequence of knots and which satisfy certain continuity conditions at the knots. There are many other possible smoothing techniques, including kernel estimates and spline smoothers. In practice, the choice of smoother used depends upon the degree of "smoothness" desired.

Generalized additive models (GAMs) (Hastie and Tibshirani, 1986) extend both the class of additive models (10.58) and the class of generalized linear models (McCullagh and Nelder, 1989). The generalized additive model is usually written in the form,

$$h(\mu) = \alpha_0 + \sum_{j=1}^{r} f_j(X_j), \tag{10.59}$$

where $\mu = \mu(\mathbf{X})$ and $h(\mu)$ is a specified *link function*. Maximum-likelihood estimates of the parameter α_0 and the functions $f_1, f_2, \ldots, f_r$ are obtained in a nonparametric fashion by maximizing a penalized log-likelihood function using a *local scoring* procedure (a version of the IRLS algorithm described in Section 9.3.5, where we fit a weighted additive model rather than a weighted linear regression), which is equivalent to a version of the Newton–Raphson algorithm.

A popular example of $h(\mu)$ is the so-called *logistic link function*, $h(\mu) = \log\{\mu/(1-\mu)\}$, which is used to model binary output. If we apply the logistic link function to (10.59), then the GAM can be inverted and re-expressed as follows:

$$\mu(\mathbf{X}) = g\left(\alpha_0 + \sum_{j=1}^{r} f_j(X_j)\right), \qquad (10.60)$$

where $g(x) = (1 + e^{-x})^{-1}$. In this particular form, we see that the GAM is closely related to a neural network with logistic (sigmoid) activation function (see Exercise 10.6).

10.12 Bayesian Learning for ANN Models

Bayesian treatments of neural networks have been quite successful. As usual, $(\mathbf{X}_1, Y_1), \ldots, (\mathbf{X}_n, Y_n)$ is the learning set of data. We assume the inputs, $\mathbf{X}_1, \ldots, \mathbf{X}_n$, are given and so are omitted from any probability calculation, and the outputs, $D = \{Y_1, \ldots, Y_n\}$, constitute the data to be modeled. For this exposition, we assume a single output value Y; the results generalize to multiple outputs $\mathbf{Y}$ in a straightforward way.

An ANN model is specified by its network architecture $\mathcal{A}$ (i.e., the number of layers, number of nodes within each layer, and the activation functions) and the vector of all network parameters $\boldsymbol{\omega}$ (i.e., all connection weights and biases). Let Q be the total number of elements in the vector $\boldsymbol{\omega}$. We assume that the architecture $\mathcal{A}$ is given and, hence, does not enter the probability calculations; if different architectures are to be compared, then the influence of $\mathcal{A}$ would have to be taken into account in the calculations. In some Bayesian models, $\mathcal{A}$ is included as part of the definition of $\boldsymbol{\omega}$.

Denote the likelihood function of the parameters given the data by $p(D|\boldsymbol{\omega})$ and let $p(\boldsymbol{\omega})$ denote the prior distribution of the parameters in the model. The likelihood function gives us an idea of the extent to which the observed data D can be predicted using the parameters $\boldsymbol{\omega}$. Note that it is a function of the parameters, not the data. The likelihood function of the parameters conditional upon the data is the probability of the data given the parameters, but where the data D are fixed and the parameters $\boldsymbol{\omega}$ are variable. The prior distribution displays whatever knowledge and information we have about the parameters in the model before we observe the data.

The complexity of the model is governed by the use of a *hyperprior*, a joint distribution on the parameters of the prior distribution; the parameters of the hyperprior distribution are called *hyperparameters*. Much of Bayesian inference in ANNs uses vague (non-informative) priors for the hyperparameters; such hyperpriors represent our lack of specific knowledge about any prior parameters needed to describe the model.

From Bayes's theorem, the posterior distribution of the parameters given the data is given by

$$p(\boldsymbol{\omega}|D) = \frac{p(D|\boldsymbol{\omega})p(\boldsymbol{\omega})}{p(D)}, \tag{10.61}$$

where $p(D) = \int p(D|\boldsymbol{\omega}')p(\boldsymbol{\omega}')d\boldsymbol{\omega}'$ operates as a normalization factor to ensure that $\int p(\boldsymbol{\omega}|D)d\boldsymbol{\omega} = 1$. Note that $p(D)$ should be interpreted as $p(D|\mathcal{A})$, not as the probability of obtaining that particular set of data D. Usually, the best we can hope for is that inference based upon the posterior is *robust* (i.e., fairly insensitive) to the choice of prior.

In this section, we give brief descriptions of two popular techniques for estimating the parameters $\boldsymbol{\omega}$ in an ANN: *Laplace's method* for deriving *maximum à posteriori (MAP)* estimates (MacKay, 1991) and *Markov chain Monte Carlo (MCMC)* methods (Neal, 1996). Exact analytical Bayesian computations are infeasible for neural networks, and so approximations offer the only way of obtaining a solution in practice.

10.12.1   Laplace's Method

Predictions can be obtained by calculating the maximum (i.e., mode) of the posterior distribution (*MAP estimation*). As such, it is the Bayesian equivalent of maximum likelihood. In our discussion of this technique, we consider models for regression and classification networks separately.

Regression Networks

Suppose the output Y corresponding to input $\mathbf{X} = \mathbf{x}$ is generated by a Gaussian distribution with mean $y(\mathbf{x}, \boldsymbol{\omega})$ and known variance σ^2. Then, assuming that $\{Y_i\}$ are iid copies of Y, the likelihood function, $L_D(\boldsymbol{\omega})$, of the parameters given the data is given by

$$L_D(\boldsymbol{\omega}) = p(D|\boldsymbol{\omega}) = \frac{e^{-\kappa E_D(\boldsymbol{\omega})}}{c_D(\kappa)}, \tag{10.62}$$

where

$$E_D(\boldsymbol{\omega}) = \frac{1}{2}\sum_{i=1}^{n}(y_i - y(\mathbf{x}_i, \boldsymbol{\omega}))^2 \tag{10.63}$$

is the error sum-of-squares, $\kappa = 1/\sigma^2$ is a (known) hyperparameter,

$$c_D(\kappa) = \int e^{-\kappa E_D(\boldsymbol{\omega})}dD = (2\pi/\kappa)^{n/2} \tag{10.64}$$

is the normalization factor, and $\int dD = \int dy_1 \cdots dy_n$.

We take the prior distribution over the parameters to be the Gaussian density,

$$p(\boldsymbol{\omega}) = \frac{e^{-\lambda E_Q(\boldsymbol{\omega})}}{c_Q(\lambda)}, \qquad (10.65)$$

where

$$E_Q(\boldsymbol{\omega}) = \frac{1}{2} \parallel \boldsymbol{\omega} \parallel^2 = \frac{1}{2} \sum_{q=1}^{Q} \omega_q^2, \qquad (10.66)$$

ω_q is equal to α_{jk}, β_{ij}, α_{0k}, or β_{0j} as appropriate, λ is a hyperparameter (which we assume to be known), and $c_Q(\lambda) = (2\pi/\lambda)^{Q/2}$ is the normalization factor. We note that other types of priors for ANN modeling have been used; these include the Laplacian prior (i.e., (10.65) with $E_Q(\boldsymbol{\omega}) = \sum_q |w_q|$) and entropy-based priors (Buntine and Weigend, 1991).

Multiplying (10.62) by (10.65) and using (10.61), we get the posterior distribution of the parameters,

$$p(\boldsymbol{\omega}|D) = \frac{e^{-S(\boldsymbol{\omega})}}{c_S(\lambda, \kappa)}, \qquad (10.67)$$

where

$$\begin{aligned} S(\boldsymbol{\omega}) &= \kappa E_D(\boldsymbol{\omega}) + \lambda E_Q(\boldsymbol{\omega}) \\ &= \kappa \sum_{i=1}^{n} (y_i - y(\mathbf{x}_i, \boldsymbol{\omega}))^2 + \lambda \sum_{q=1}^{Q} \omega_q^2 \end{aligned} \qquad (10.68)$$

and the normalization factor, $c_S(\lambda, \kappa) = \int e^{-S(\boldsymbol{\omega})} d\boldsymbol{\omega}$, is an integration that cannot be evaluated explicitly. To find the maximum of the posterior distribution, we can minimize $-\log_e p(\boldsymbol{\omega}|D)$ wrt $\mathbf{w}$. Because c_S is independent of $\boldsymbol{\omega}$, it suffices to minimize $S(\boldsymbol{\omega})$. The value of $\boldsymbol{\omega}$ that maximizes the posterior probability $p(\boldsymbol{\omega}|D)$ (or, equivalently, minimizes $S(\boldsymbol{\omega})$) is regarded as the *most probable* value of $\boldsymbol{\omega}$ and is denoted by the *MAP estimate* $\boldsymbol{\omega}_{\mathrm{MP}}$. It can be found by an appropriate gradient-based optimization algorithm. The network corresponding to the parameter values $\boldsymbol{\omega}_{\mathrm{MP}}$ is referred to as the *most-probable regression network*.

From (10.68), we see that $S(\boldsymbol{\omega})$ is a constant (κ) times the error sum-of-squares of learning-set predictions plus a complexity term composed of a weight-decay penalty and regularization parameter λ. Because $S(\boldsymbol{\omega})$ has a form very similar to (10.51) and (10.52), the MAP approach can be used to determine λ in the weight-decay penalty for network pruning. Some simple arguments lead to a suggested range of 0.001 to 0.1 for exploratory values of λ (Ripley, 1996, Section 5.5). It is for this reason that MAP estimation has been characterized as "a form of maximum penalized likelihood estimation" (Neal, 1996, p. 6) rather than as a Bayesian method.

Rather than having to work with the form of the posterior density just derived, we can make the following useful approximation, known as *Laplace's method* or *approximation* (Laplace, 1774/1986). Suppose that $\boldsymbol{\omega}_{\mathrm{MP}}$ is the location of a mode of $p(\boldsymbol{\omega}|D)$. Consider the following Taylor-series expansion of $S(\boldsymbol{\omega})$ around $\boldsymbol{\omega}_{\mathrm{MP}}$:

$$S(\boldsymbol{\omega}) \approx S(\boldsymbol{\omega}_{\mathrm{MP}}) + \frac{1}{2}(\boldsymbol{\omega} - \boldsymbol{\omega}_{\mathrm{MP}})^{\tau}\mathbf{A}(\boldsymbol{\omega} - \boldsymbol{\omega}_{\mathrm{MP}}), \qquad (10.69)$$

where $\mathbf{A} = \partial^2 S(\boldsymbol{\omega})/\partial\boldsymbol{\omega}^2|_{\boldsymbol{\omega}=\boldsymbol{\omega}_{\mathrm{MP}}}$, is the $(Q \times Q)$ Hessian matrix (assumed to be positive-definite) of second-order derivatives evaluated at $\boldsymbol{\omega} = \boldsymbol{\omega}_{\mathrm{MP}}$. Substituting (10.69) into the numerator of (10.67), we can approximate $p(\boldsymbol{\omega}|D)$ by

$$\widetilde{p}(\boldsymbol{\omega}|D) = \frac{e^{-S(\boldsymbol{\omega}_{\mathrm{MP}})}}{c_S^*(\lambda)}\, e^{-\frac{1}{2}\Delta\boldsymbol{\omega}^{\tau}\mathbf{A}\Delta\boldsymbol{\omega}}, \qquad (10.70)$$

where $\Delta\boldsymbol{\omega} = \boldsymbol{\omega} - \boldsymbol{\omega}_{\mathrm{MP}}$ and the denominator (i.e., the normalizing factor) is equal to

$$c_S^*(\lambda) = (2\pi)^{Q/2}|\mathbf{A}|^{-1/2}e^{-S(\boldsymbol{\omega}_{\mathrm{MP}})}. \qquad (10.71)$$

Thus, we can approximate $p(\boldsymbol{\omega}|D)$ by

$$\widetilde{p}(\boldsymbol{\omega}|D) = (2\pi)^{-Q/2}|\mathbf{A}|^{1/2}e^{-\frac{1}{2}\Delta\boldsymbol{\omega}^{\tau}\mathbf{A}\Delta\boldsymbol{\omega}}, \qquad (10.72)$$

which is the multivariate Gaussian density, $\mathcal{N}_Q(\boldsymbol{\omega}_{\mathrm{MP}}, \mathbf{A}^{-1})$, with mean vector $\boldsymbol{\omega}_{\mathrm{MP}}$ and covariance matrix $\mathbf{A}^{-1}$. This approximation is reinforced by an asymptotic result that a posterior density converges (as $n \to \infty$) to a Gaussian density whose variance collapses to zero (Walker, 1969). Note that the Gaussian approximation $\widetilde{p}(\boldsymbol{\omega}|D)$ is different from $p(\boldsymbol{\omega}_{\mathrm{MP}}|D)$, the posterior density corresponding to the most-probable network.

For any new input vector $\mathbf{x}$, we can now write down an expression for the *predictive distribution* of a new output Y from a regression network using the learning data D:

$$p(y|\mathbf{x}, D) = \int p(y|\mathbf{x}, \boldsymbol{\omega})p(\boldsymbol{\omega}|D)d\boldsymbol{\omega}, \qquad (10.73)$$

where $p(\boldsymbol{\omega}|D)$ is the posterior density of the parameters derived above. This integral cannot be computed because of all the nonlinearities involved in the network.

To overcome this impass, we use the Gaussian approximation (10.72) to the posterior and assume that $p(y|\mathbf{x}, D)$ is a univariate Gaussian density with mean $y(\mathbf{x}, \boldsymbol{\omega})$ and variance $1/\nu$. Then, (10.73) is approximated by

$$\widetilde{p}(y|\mathbf{x}, D) \propto \int e^{-\frac{\nu}{2}(y-y(\mathbf{x},\boldsymbol{\omega}))^2 - \frac{1}{2}\Delta\boldsymbol{\omega}^{\tau}\mathbf{A}\Delta\boldsymbol{\omega}}d\boldsymbol{\omega}. \qquad (10.74)$$

We next assume that $y(\mathbf{x}, \boldsymbol{\omega})$ can be approximated by a Taylor-series expansion around $\boldsymbol{\omega}_{\mathrm{MP}}$,

$$y(\mathbf{x}, \boldsymbol{\omega}) \approx y(\mathbf{x}, \boldsymbol{\omega}_{\mathrm{MP}}) + \mathbf{g}^\tau \Delta \boldsymbol{\omega}, \tag{10.75}$$

where $\mathbf{g} = \partial y / \partial \boldsymbol{\omega}|_{\boldsymbol{\omega}_{\mathrm{MP}}}$ is the gradient. Set $y_{\mathrm{MP}} = y(\mathbf{x}, \boldsymbol{\omega}_{\mathrm{MP}})$. Substituting (10.75) into (10.74) and evaluating the resulting integral, we find that $p(y|\mathbf{x}, D)$ can be approximated by the Gaussian density,

$$\widetilde{p}(y|\mathbf{x}, D) = \frac{1}{(2\pi\sigma_y^2)^{1/2}} \, e^{-(y-y_{\mathrm{MP}})^2 / 2\sigma_y^2}, \tag{10.76}$$

with mean y_{MP} and variance $\sigma_y^2 = \frac{1}{\nu} + \mathbf{g}^\tau \mathbf{A}^{-1} \mathbf{g}$ (see Exercise 10.10). This result can be used to derive approximate confidence bounds on the most-probable output y_{MP}.

So far, we have assumed the hyperparameters κ and λ are known. But, in practice, this is a highly unlikely scenario. In a fully hierarchical-Bayesian approach to this problem, we would incorporate the hyperparameters into the model and then integrate over all parameters and hyperparameters. However, such integrations are not possible analytically, and so another approach has to be taken.

To deal with unknown κ and λ within a Bayesian framework, two different approaches to this problem have been proposed: (1) integrating out the hyperparameters analytically and then using numerical methods to estimate the most-probable parameter values (Buntine and Weigend, 1991); (2) estimating the hyperparameter values by maximizing something called "evidence" (MacKay, 1992a). These two approaches have attracted a certain amount of controversy (see, e.g., Wolpert, 1993; MacKay, 1994).

Analytically integrating out the hyperparameters. The first method involves supplying prior densities for the hyperparameters, then integrating them out (a method called *marginalization*), and finally applying numerical methods to determine $\boldsymbol{\omega}_{\mathrm{MP}}$. Thus, we can write

$$\begin{aligned} p(\boldsymbol{\omega}|D) &= \int \int p(\boldsymbol{\omega}, \kappa, \lambda|D) d\kappa d\lambda \\ &= \int \int p(\boldsymbol{\omega}|\kappa, \lambda, D) p(\kappa, \lambda|D) d\kappa d\lambda. \end{aligned} \tag{10.77}$$

Now, we use Bayes's theorem for each term in the integrand: $p(\boldsymbol{\omega}|\kappa, \lambda, D) = p(D|\boldsymbol{\omega}, \kappa, \lambda) p(\boldsymbol{\omega}|\kappa, \lambda) / p(D|\kappa, \lambda) = p(D|\boldsymbol{\omega}, \kappa) p(\boldsymbol{\omega}|\lambda) / p(D|\kappa, \lambda)$, because the likelihood does not depend upon λ and the prior does not depend upon κ; similarly, $p(\kappa, \lambda|D) = p(D|\kappa, \lambda) p(\kappa, \lambda) / p(D) = p(D|\kappa, \lambda) p(\kappa) p(\lambda) / p(D)$, where we have assumed that the two hyperparameters, κ and λ, are distributed independently of each other. We take these (improper) priors to be defined over $(0, \infty)$ as $p(\kappa) = 1/\kappa$ and $p(\lambda) = 1/\lambda$. The integral (10.77) reduces to

$$p(\boldsymbol{\omega}|D) = \frac{1}{p(D)} \int \int p(D|\boldsymbol{\omega},\kappa)p(\boldsymbol{\omega}|\lambda)p(\kappa)p(\lambda)d\kappa d\lambda. \qquad (10.78)$$

This integral can be divided up into the product of two integrals and re-expressed as (10.61). Here,

$$
\begin{aligned}
p(\boldsymbol{\omega}) &= \int p(\boldsymbol{\omega}|\lambda)p(\lambda)d\lambda \\
&= \int \frac{e^{-\lambda E_Q(\boldsymbol{\omega})}}{c_Q(\lambda)} \frac{1}{\lambda} d\lambda \\
&= \pi^{-Q/2} \int \lambda^{Q/2-1} e^{-\lambda E_Q(\boldsymbol{\omega})} d\lambda. \qquad (10.79)
\end{aligned}
$$

Using the value of a gamma integral (see, e.g., Casella and Berger, 1990, p. 100), we have that (10.79) reduces to

$$p(\boldsymbol{\omega}) = \frac{\Gamma(Q/2)}{(\pi E_Q(\boldsymbol{\omega}))^{Q/2}}. \qquad (10.80)$$

Similarly, we obtain

$$p(D|\boldsymbol{\omega}) = \int p(D|\boldsymbol{\omega},\kappa)p(\kappa)d\kappa = \frac{\Gamma(n/2)}{(\pi E_D(\boldsymbol{\omega}))^{n/2}}. \qquad (10.81)$$

Multiplying (10.80) and (10.81) to get the posterior density, taking the negative logarithm of the result, and simplifying, we get

$$-\log_e p(\boldsymbol{\omega}|D) = \frac{n}{2}\log_e E_D(\boldsymbol{\omega}) + \frac{Q}{2}\log_e E_Q(\boldsymbol{\omega}) + \text{constant}, \qquad (10.82)$$

where the constant does not depend upon $\boldsymbol{\omega}$. We differentiate (10.82) wrt $\boldsymbol{\omega}$,

$$\frac{d}{d\boldsymbol{\omega}}\{-\log_e p(\boldsymbol{\omega}|D)\} = \kappa\frac{d}{d\boldsymbol{\omega}}\{E_D(\boldsymbol{\omega})\} + \lambda\frac{d}{d\boldsymbol{\omega}}\{E_Q(\boldsymbol{\omega})\}, \qquad (10.83)$$

to find its minimum, where

$$\kappa = n/2E_D(\boldsymbol{\omega}), \quad \lambda = Q/2E_Q(\boldsymbol{\omega}). \qquad (10.84)$$

This result is next used in a nonlinear optimization algorithm in which the values of κ and λ are sequentially updated to find the most-probable parameters $\boldsymbol{\omega}_{\text{MP}}$, and then a multivariate Gaussian approximation to the posterior density is obtained centered around $\boldsymbol{\omega}_{\text{MP}}$.

Maximizing the evidence. Another method for dealing with unknown κ and λ is to maximize the "evidence" of the model, $p(D|\kappa,\lambda)$, which can be expressed as

$$
\begin{aligned}
p(D|\kappa, \lambda) &= \int p(D|\boldsymbol{\omega}, \kappa, \lambda) p(\boldsymbol{\omega}|\kappa, \lambda) d\boldsymbol{\omega} \\
&= \int p(D|\boldsymbol{\omega}, \kappa) p(\boldsymbol{\omega}|\lambda) d\boldsymbol{\omega} \\
&= (c_D(\kappa) c_Q(\lambda))^{-1} \int e^{-S(\boldsymbol{\omega})} d\boldsymbol{\omega} \\
&= \frac{c_S(\kappa, \lambda)}{c_D(\kappa) c_Q(\lambda)},
\end{aligned}
\tag{10.85}
$$

where $S(\boldsymbol{\omega})$ is given by (10.68). As usual, it is easier to maximize the logarithm of (10.85),

$$
\begin{aligned}
\log_e p(D|\kappa, \lambda) &= -\kappa E_D(\boldsymbol{\omega}_{\mathrm{MP}}) - \lambda E_Q(\boldsymbol{\omega}_{\mathrm{MP}}) - \frac{1}{2} \log_e |\mathbf{A}| \\
&\quad + \frac{n}{2} \log_e(\kappa) + \frac{Q}{2} \log_e(\lambda) - \frac{Q}{2} \log_e(2\pi).
\end{aligned}
\tag{10.86}
$$

We maximize this expression in two steps: first, fix κ and differentiate (10.86) wrt λ, set the result to zero, and solve for a maximum; next, fix λ and differentiate (10.86) wrt κ, set the result equal to zero, and solve for a maximum. These manipulations yield the following formulas (MacKay, 1992b):

$$
\lambda^* = \frac{\gamma}{2 E_Q(\boldsymbol{\omega}_{\mathrm{MP}})}
\tag{10.87}
$$

$$
\kappa^* = \frac{n - \gamma}{2 E_D(\boldsymbol{\omega}_{\mathrm{MP}})},
\tag{10.88}
$$

where

$$
\gamma = \sum_{q=1}^{Q} \frac{\eta_q}{\eta_q + \lambda^*},
\tag{10.89}
$$

and the $\{\eta_q\}$ are the eigenvalues of $\mathbf{A}^{-1}$.

Thus, we set initial values for κ^* and λ^* by sampling from their respective prior densities and determine $\boldsymbol{\omega}_{\mathrm{MP}}$ by applying a suitable nonlinear optimization algorithm to $S(\boldsymbol{\omega})$; during the progress of these iterations, the values of κ^* and λ^* are sequentially updated using (10.87)–(10.89): an initial λ_0^* gives a γ_0 using (10.89), which yields λ_1^* from (10.86) and κ_1^* from (10.88); the new λ_1^* is fed back into (10.89) to provide a new γ_1, which, in turn, gives λ_2^* and κ_2^*, and so on. These steps in the algorithm should be repeated a large number of times each time using different initial values for the parameter vector $\boldsymbol{\omega}$.

We note that this computational technique of dealing with hyperparameters is equivalent to the *empirical Bayes* (Carlin and Louis, 2000, Chapter 3) or *type II maximum-likelihood (ML-II)* approach to prior selection (Berger, 1985, Section 3.5.4).

Multiple modes. A major problem in practice, however, is that it is not generally realistic to assume that the posterior density has only a single mode. From experience of fitting Bayesian models to nonlinear networks, we find it more reasonable to assume that there will be multiple local maxima of the posterior density (see, e.g., Ripley, 1994a, p. 452, who, in a particular example, found at least 22 distinct local modes). As usual in such situations, one should try to identify as many of the distinct local maxima as possible by running the optimization algorithm using a large number of randomly chosen starting points for the parameters.

A potentially better modeling strategy for multiple modes is to use an approximation to the posterior based upon a mixture of multivariate Gaussian densities, where the component densities are assumed to have minimal overlap; each component density is centered at a different local mode of the posterior $p(\boldsymbol{\omega}|D)$, and the inverse of its covariance matrix is matched to the Hessian of the logarithm of the posterior density at the mode (MacKay, 1992a). Although some work has been carried out on Gaussian mixture models for neural networks (see, e.g., Buntine and Weigend, 1991; Ripley, 1994b), more research is needed on this topic.

Classification Networks

If the problem involves classifying data into one of two classes, Π_1 or Π_2, then the output variable Y is binary, taking on the value 1 (for Π_1) or 0 (for Π_2). The network output $y(\mathbf{x}, \boldsymbol{\omega}) = p(Y = 1|\mathbf{x}, \boldsymbol{\omega})$ is the conditional probability that the particular input vector $\mathbf{X} = \mathbf{x}$ is a member of Π_1.

The probability that $Y_i = 1$ is

$$p(Y_i = 1|\mathbf{x}_i, \boldsymbol{\omega}) = (y(\mathbf{x}_i, \boldsymbol{\omega}))^{y_i}(1 - y(\mathbf{x}_i, \boldsymbol{\omega}))^{1-y_i}. \tag{10.90}$$

The likelihood function of the parameters $\boldsymbol{\omega}$ (given the data D) is

$$p(D|\boldsymbol{\omega}) = \prod_{i=1}^{n} p(Y_i = 1|\mathbf{x}_i, \boldsymbol{\omega}) = e^{-\ell_D(\boldsymbol{\omega})}, \tag{10.91}$$

where

$$\ell_D(\boldsymbol{\omega}) = -\sum_{i=1}^{n}\{y_i \log_e y(\mathbf{x}_i, \boldsymbol{\omega}) + (1 - y_i) \log_e(1 - y(\mathbf{x}_i, \boldsymbol{\omega}))\} \tag{10.92}$$

is the negative log-likelihood function. Again, the network's architecture $\mathcal{A}$ is assumed to be given. Note that, compared to (10.62) for regression networks, (10.91) has neither a hyperparameter κ nor a denominator $c_D(\kappa)$.

For a prior on the parameters, we use the Gaussian density (10.65), which is proportional to $e^{-\lambda E_Q(\boldsymbol{\omega})}$.

Assuming the $\{Y_i\}$ are iid copies of Y, the posterior density (10.61) is

$$p(\boldsymbol{\omega}|D) = \frac{e^{-S(\boldsymbol{\omega})}}{c_S(\lambda)}, \tag{10.93}$$

where

$$S(\boldsymbol{\omega}) = \ell_D(\boldsymbol{\omega}) + \lambda E_Q(\boldsymbol{\omega}), \tag{10.94}$$

λ is, again, the regularization parameter (also known as a *weight-decay regularizer*), and $c_S(\lambda)$ is the normalization factor. Finding $\boldsymbol{\omega}$ to maximize the posterior distribution is equivalent to minimizing $S(\boldsymbol{\omega})$. The value of $\boldsymbol{\omega}$ that maximizes the posterior distribution is denoted by $\boldsymbol{\omega}_{\mathrm{MP}}$.

We can now find the probability that the input vector, $\mathbf{X} = \mathbf{x}$, is a member of class Π_1 (i.e., $Y = 1$). MacKay (1992b) suggests that if $f(\cdot)$ is one of the activation functions in Table 10.1 and $u = u(\mathbf{x}, \boldsymbol{\omega})$, then,

$$
\begin{aligned}
p(Y = 1|\mathbf{x}, D) &= \int p(Y = 1|u)p(u|\mathbf{x}, D)du \\
&= \int f(u)p(u|\mathbf{x}, D)du
\end{aligned}
\tag{10.95}
$$

provides a better estimate of the class probability than $y(\mathbf{x}, \boldsymbol{\omega}_{\mathrm{MP}})$. To evaluate this integral, MacKay first expands u in a Taylor series,

$$u(\mathbf{x}, \boldsymbol{\omega}) \approx u(\mathbf{x}, \boldsymbol{\omega}_{\mathrm{MP}}) + \mathbf{g}(\mathbf{x})^\tau \Delta\boldsymbol{\omega}, \tag{10.96}$$

where $\mathbf{g}(\mathbf{x}) = \partial u(\mathbf{x}, \boldsymbol{\omega})/\partial \boldsymbol{\omega}|_{\boldsymbol{\omega}_{\mathrm{MP}}}$ and $\Delta\boldsymbol{\omega} = \boldsymbol{\omega} - \boldsymbol{\omega}_{\mathrm{MP}}$. Thus,

$$
\begin{aligned}
p(u|\mathbf{x}, D) &= \int p(u|\mathbf{x}, \boldsymbol{\omega})p(\boldsymbol{\omega}|D)d\boldsymbol{\omega} \\
&= \int \delta(u - u_{\mathrm{MP}} - \mathbf{g}(\mathbf{x})^\tau \Delta\boldsymbol{\omega})p(\boldsymbol{\omega}|D)d\boldsymbol{\omega},
\end{aligned}
\tag{10.97}
$$

where $u_{\mathrm{MP}} = u(\mathbf{x}, \boldsymbol{\omega}_{\mathrm{MP}})$ and δ is the Dirac delta-function. This result implies that if we use Laplace's method and approximate the posterior density $p(\boldsymbol{\omega}|D)$ in (10.93) by the multivariate Gaussian density,

$$\widetilde{p}(\boldsymbol{\omega}|D) \propto e^{-\frac{1}{2}\Delta\boldsymbol{\omega}^\tau \mathbf{A}\Delta\boldsymbol{\omega}}, \tag{10.98}$$

where $\mathbf{A}$ is the (local) Hessian matrix, then, u is Gaussian,

$$p(u|\mathbf{x}, D) \propto e^{-(u - u_{\mathrm{MP}})^2/2\nu^2}, \tag{10.99}$$

with mean u_{MP} and variance

$$\nu^2 = \mathbf{g}(\mathbf{x})^\tau \mathbf{A}^{-1}\mathbf{g}(\mathbf{x}). \tag{10.100}$$

When f is sigmoidal and $p(u|\mathbf{x}, D)$ is Gaussian, the integral (10.95) does not have an analytic solution. MacKay (1992b) suggests the following simple approximation for (10.95):

$$\widetilde{p}(Y = 1|\mathbf{x}, D) = f(\alpha(\nu)u_{\mathrm{MP}}), \tag{10.101}$$

where $\alpha(\nu) = (1 + (\pi\nu^2/8))^{-1/2}$. Note that the probability (10.101) is not the same as $y(\mathbf{x}, \boldsymbol{\omega}_{\mathrm{MP}})$.

10.12.2 Markov Chain Monte Carlo Methods

As we have seen, the main computational difficulty in applying Bayesian methods involves the evaluation of complicated high-dimensional integrals. For example, the predictive distribution of the output value Y^* of a new test case $(\mathbf{X}^*, Y^*)$, given the learning data, $\mathcal{L} = \{(\mathbf{X}_1, Y_1), \ldots, (\mathbf{X}_n, Y_n)\}$, is given by

$$p(y^*|\mathbf{x}^*, \mathcal{L}) = \int p(y^*|\mathbf{x}^*, \boldsymbol{\omega})p(\boldsymbol{\omega}|\mathcal{L})d\boldsymbol{\omega}. \tag{10.102}$$

If we are to estimate Y^* in a regression model using squared-error as our loss function, then, the best predictor is the expectation of the predictive distribution (10.102),

$$\mathrm{E}\{Y^*|\mathbf{x}^*, \mathcal{L}\} = \int p(\mathbf{x}^*, \boldsymbol{\omega})p(\boldsymbol{\omega}|\mathcal{L})d\boldsymbol{\omega}. \tag{10.103}$$

Problems of approximating the posterior density or its expectation have been summarized well by Neal (1996, Section 1.2).

A recent popular and highly successful addition to the Bayesian's toolkit is a method known as *Markov chain Monte Carlo (MCMC)*, which is actually a collection of related computational techniques designed for simulating from nonstandard multivariate distributions (see, e.g., Gilks, Richardson, and Spiegelhalter, 1996; Robert and Casella, 1999). It was proposed as a method for estimating the predictive distributions of regression and classification network parameters and their expectations by Neal (1996).

The essential idea behind MCMC is to approximate the desired integration by simulating from the joint probability distribution of all the model parameters and hyperparameters. Thus, we, first, use a *Monte Carlo method* to draw a sample of B values, $\boldsymbol{\omega}^{(1)}, \ldots, \boldsymbol{\omega}^{(B)}$, from the predictive density (10.99), where $\boldsymbol{\omega}$ now includes all weights, biases, and hyperparameters; then, we approximate the expectation (10.103) by

$$\widehat{y}^* = \frac{1}{B}\sum_{b=1}^{B} p(\mathbf{x}^*, \boldsymbol{\omega}^{(b)}). \tag{10.104}$$

When the predictive density is complicated, as it is in nonlinear neural network applications, then the sequence of generated values, $\{\boldsymbol{\omega}^{(b)}\}$, has to be viewed as a dependent sequence.

One way of generating such a dependent sequence is by using an ergodic *Markov chain* with stationary distribution $P = p(\mathbf{x}, \boldsymbol{\omega})$. A Markov chain is defined on a sequence of states, $\boldsymbol{\omega}^{(b)}$, by an *initial distribution* for the startup state, $\boldsymbol{\omega}^{(0)}$, of the chain and a set of *transition probabilities*, $\{Q(\boldsymbol{\omega}^{(b)} | \boldsymbol{\omega}^{(b-1)})\}$, for a future state, $\boldsymbol{\omega}^{(b)}$, to succeed the current state, $\boldsymbol{\omega}^{(b-1)}$. The distribution P is called *stationary* (or *invariant*) if it remains the same for all states in the sequence that follow the bth state. If a stationary distribution P exists and is unique, then the Markov chain is called *ergodic* and its stationary distribution P is known as the *equilibrium distribution*. If we can find an ergodic Markov chain that has equilibrium distribution P, then it does not matter from which initial state we start the chain, convergence of the sequence will always be to P. In such a case, we can estimate (10.103) wrt P by using (10.104).

Because the members of the sequence $\{\boldsymbol{\omega}^{(b)}\}$ are dependent, we need a much larger value of B than if the sequence consisted of independent values. At the beginning, the iterates will look like the starting values, $\boldsymbol{\omega}^{(0)}$, and then, after a long time, the Markov chain will settle down. To take this into account, the first B_0 iterates are considered as the "burn-in" period; these values are discarded as not resembling the equilibrium distribution P, and only the subsequent $B - B_0$ values are regarded as essentially independent observations from P to be used for predictive purposes.

The two most popular methods for MCMC are Gibbs sampling and the Metropolis algorithm. Both (and variations of those themes) have been used extensively in mathematical physics, chemistry, biology, statistics, and image restoration.

The *Gibbs sampler* (Geman and Geman, 1984) can be applied when sampling from any distribution defined by a vector, $\boldsymbol{\omega} = (\omega_1, \cdots, \omega_Q)^\tau$, $Q \geq 2$, of parameters. Considering these parameters as random variables, we assume that all one-dimensional conditional distributions of the form $p(\omega_q | \{\omega_i, i \neq q\}), q = 1, 2, \ldots, Q$, are available to be sampled. The entire set of these conditional distributions is (under mild conditions) sufficient to determine the joint distribution and all its margins. Given a vector of starting values $\boldsymbol{\omega}^{(0)}$, we define a Markov chain by generating $\boldsymbol{\omega}^{(b)}$ from $\boldsymbol{\omega}^{(b-1)}$ according to the algorithm in Table 10.3, where we use notation from Besag, Green, Higdon, and Mengersen (1995). This process generates a sequence (or trajectory) of the chain, $\boldsymbol{\omega}^{(0)}, \boldsymbol{\omega}^{(1)}, \ldots, \boldsymbol{\omega}^{(b)}, \ldots$, and, as b gets larger and larger (after a long enough "burn-in" period), the vector $\boldsymbol{\omega}^{(b)}$ becomes approximately distributed as the desired P.

The *Metropolis algorithm* (Metropolis, Rosenbluth, Rosenbluth, Teller, and Teller, 1953) introduces a *candidate* or *proposal density*, f, whose form depends upon the current state; one generates a *candidate state*, $\boldsymbol{\omega}^*$, from f, and then decides whether or not to "accept" that candidate state. If the candidate state is accepted, it becomes the next state in the Markov chain; otherwise, it remains at the current state. See Table 10.4. The iterative

TABLE 10.3. *The Gibbs sampler.*

---

1. Let $w_1^{(0)}, \ldots, w_Q^{(0)}$ be starting values. Define

$$\boldsymbol{w}_{-q} = \{w_j, j \neq q\} = \{w_1, w_2, \ldots, w_{q-1}, w_{q+1}, \ldots, w_Q\}.$$

2. For $b = 1, 2, \ldots$:

$$\text{draw } w_q^{(b)} \sim p_q(w_q | w_{-q}^{(b-1)}), \quad q = 1, 2, \ldots, Q.$$

3. Continue the 2nd step until the joint distribution of $w_1^{(b)}, \ldots, w_Q^{(b)}$ stabilizes.

---

process moves from the current state, $\boldsymbol{w}^{(b-1)}$, to the next state, $\boldsymbol{w}^{(b)}$, corresponding to a higher-density region of $p(\boldsymbol{w}|\mathcal{L})$, whereas it rejects a percentage of those steps that move to lower-density regions of $p(\boldsymbol{w}|\mathcal{L})$. Note that the candidate densities may change from step to step; typically, the candidate density f is selected to be a member of a family of distributions, such as Gaussian densities centered at $\boldsymbol{w}^{(b-1)}$.

Unfortunately, neither the Gibbs sampler nor the Metropolis algorithm are recommended for sampling from the posterior distribution of a neural network model. Because of the huge numbers of parameters involved and the nonlinearity of the model, such MCMC procedures are either computationally infeasible or are very slow for this type of application.

TABLE 10.4. *The Metropolis algorithm.*

---

1. Let $\boldsymbol{w}^{(0)}$ be starting values. Let $p(\boldsymbol{w}|\mathcal{L})$ be the joint posterior density of $\boldsymbol{w}$.

2. For $b = 1, 2, \ldots$:

   (i) Draw a candidate state, $\boldsymbol{w}^*$, from a proposal density f, *which depends upon the current state*; i.e., $\boldsymbol{w}^* \sim f(\cdot, \boldsymbol{w}^{(b-1)})$.

   (ii) Compute the ratio $r = p(\boldsymbol{w}^*|\mathcal{L})/p(\boldsymbol{w}^{(b-1)}|\mathcal{L})$.

   (iii) (a) If $r \geq 1$, accept the candidate state and set $\boldsymbol{w}^{(b)} = \boldsymbol{w}^*$.

        (b) Otherwise, accept the candidate state with probability r or reject it with probability $1 - r$. If the candidate state is rejected, set $\boldsymbol{w}^{(b)} = \boldsymbol{w}^{(b-1)}$.

3. Continue the 2nd step until the joint distribution of $\boldsymbol{w}^{(b)}$ stabilizes.

To overcome these difficulties, Neal (1996, Chapter 3) successfully implemented a combination procedure based upon the *hybrid Monte Carlo algorithm* of Duane, Kennedy, Pendleton, and Roweth (1987). Neal's procedure separates the hyperparameters from the network parameters (i.e., weights and biases) and alternates their updates: the Gibbs sampler is used for updating the hyperparameters, and the hybrid Monte Carlo algorithm, an elaborate version of the Metropolis algorithm, is used to update the network parameters.

10.13   Software Packages

S-PLUS and R (Venables and Ripley, 2002, Sections 8.8–8.10) have commands to carry out neural networks (nnet), projection pursuit regression (ppr), and generalized additive models (gam). MATLAB has a Neural Network Toolbox with tools for designing, implementing, visualizing, and simulating neural networks. WEKA (Waikato Environment for Knowledge Analysis) is a collection of open-source machine-learning algorithms for data-mining tasks, including neural network modeling, from the University of Waikato, Hamilton, New Zealand (Witten and Frank, 2005). WEKA is downloadable from www.cs.waikato.ac.nz/ml/weka.

Gibbs sampling can be used to simulate from almost any probability model through BUGS (Bayesian inference Using Gibbs Sampling), WIN-BUGS, and OPENBUGS software, which is downloadable from

www.mrc-bsu.cam.ac.uk/bugs/.

OPENBUGS can be run from R in Windows.

Bibliographical Notes

Groundbreaking work on the neural biology of the brain appeared in the book Hebb (1949), which was reprinted in 2002 with additional material. The historical remarks in this chapter about Hebb were adapted from Milner (1993), the edited volume by Jusczyk and Klein (1980), and the excellent individual articles by Sejnowski, Milner, Kolb, Tees, and Hinton in the February 2003 issue of *Canadian Psychology*. Also highly recommended is the fascinating book by Calvin and Ojemann (1994), who use conversations between an epileptic patient and his surgeon to carry out a learning tour of the cerebral cortex.

There are many good treatments of artificial neural networks. Books include MacKay (2003, Part V), Hastie, Tibshirani, and Friedman (2001, Chapter 11), Duda, Hart, and Stork (2001, Chapters 6 and 7), Vapnik (2000), Fine (1999), Haykin (1999), Ripley (1996, Chapter 5), Rojas (1996), and Bishop (1995). Statistical perspectives of neural networks can be found in the articles by Ripley (1994a), Cheng and Titterington (1994), and Stern (1996).

The universal approximation theorem derives from the work of Kolmogorov (1957), Sprecher (1965), and others, who showed that a continuous function could have an exact representation in terms of the superposition of a few functions of one variable. Dissatisfaction with these representations for motivating neural networks led to a variety of approximation results (e.g., Cybenko, 1989; Funahashi, 1989; Hornick, Stinchcombe, and White, 1989).

The *backpropagation algorithm* (also referred to as the *generalized delta rule*) was independently discovered by several researchers at the same time. Werbos (1974) had published the basic idea of backpropagation for general networks in his doctoral dissertation, which was written during the "quiet" period of neural networks. As fate would have it, the idea lay dormant until the mid-1980s when Parker (1985) and LeCun (1985) independently rediscovered versions of the algorithm. The paper by Rumelhart, Hinton, and Williams (1986) and an expanded version, Rumelhart and McClelland (1986a), enabled the algorithm to be given wide attention. An excellent discussion of the backpropagation algorithm from the point of view of a graph-labeling problem is given by Rojas (1996, Chapter 7).

The paper by Huber (1985) and the discussion following give an excellent description of PPR and its advantages and disadvantages. Additive models and generalized additive models are described in detail in the monograph by Hastie and Tibshirani (1990). A Bayesian backfitting algorithm for fitting additive models is given by Hastie and Tibshirani (2000).

Bayesian modeling of neural networks can be found in Bishop (2006, Section 5.7), Titterington (2004), MacKay (2003, Chapter 41), Lampinen and Vehtari (2001), Fine (1999, Section 6.2), Barber and Bishop (1998), Ripley (1996, Section 5.5), Bishop (1995, Chapter 10), and Cheng and Titterington (1994).

An excellent reference to Laplace's method is Tierney and Kadane (1986), who showed how it could be used to approximate posterior expectations and, therefore, how important the method is for Bayesian computation. See also Kass, Tierney, and Kadane (1988), Bernardo and Smith (1994, Section 5.5.1), and Carlin and Louis (2000, Section 5.2.2).

Markov chain Monte Carlo (MCMC) is currently a very active field of research within the Bayesian statistical community. Books that discuss MCMC include MacKay (2003, Chapter 29), Carlin and Louis (2000, Chapter 5), Robert and Casella (1999), Neal (1996), Gilks, Richardson, and Spiegelhalter (1996), and Gelman, Carlin, Stern, and Rubin (1995, Chapter 11). Survey articles on MCMC include Cowles and Carlin (1996) and Besag, Green, Higdon, and Mengersen (1995). See also the November 2001 and February 2004 issues of *Statistical Science*. The Gibbs sampler was first used as an MCMC method by Geman and Geman (1984) in the context of image restoration. Its introduction to the statistical community is due to Gelfand and Smith (1990), who broadened its appeal considerably.

The field of neural networks is now regarded by many as part of a larger field known as *softcomputing* (due to L.A. Zadeh), which includes such topics as *fuzzy logic* (e.g., computing with words), *evolutionary computing* (e.g., genetic algorithms), *probabilistic computing* (e.g., Bayesian learning, statistical reasoning, belief networks), and *neurocomputing*. The primary goal of soft computing is to create a new AI that will reflect the workings of the human mind. According to Zadeh, this is to be accomplished using computing tools and methods that exploit a tolerance for imprecision, uncertainty, partial truth, and approximation in order to achieve robustness and a low-cost solution.

Exercises

10.1 Let $\phi(x) = a \tanh(bx)$ be the hyperbolic tangent activation function, where a and b are constants. Show that $\phi(x) = 2a\psi(bx) - a$, where $\psi(x) = (1 + e^{-x})^{-1}$ is the logistic activation function.

10.2 Show that the logistic function is symmetric, whereas the tanh function is asymmetric.

10.3 Show that the Gaussian cumulative distribution function, $\Phi(x) = (2\pi)^{-1/2} \int_{-\infty}^{x} e^{-u^2/2} du$, is a sigmoidal function.

10.4 Show that $\psi(x) = (2/\pi) \tan^{-1}(x)$ is a sigmoidal function.

10.5 For $r = 3$ inputs, draw the hyperplane in the unit cube corresponding to the McCulloch–Pitts neuron for the logical OR function.

10.6 (The XOR Problem.) Consider four points, (X_1, X_2), at the corners of the unit square: $(0,0), (0,1), (1,0), (1,1)$. Suppose that $(0,0)$ and $(1,1)$ are in class 1, whereas $(0,1)$ and $(1,0)$ are in class 2. The XOR problem is to construct a network that classifies the four points correctly. By setting $Y = 1$ to points in class 1 and $Y = 0$ to points in class 2 (or vice versa), show algebraically that a straight line cannot separate the two classes of points and, hence, that a perceptron with no hidden nodes is not an appropriate network for this problem.

10.7 (The XOR Problem, cont.) Consider a fully connected network with two input nodes (X_1, X_2), two hidden nodes (Z_1, Z_2), and a single output node (Y). Let $\beta_{11} = \beta_{12} = 1$ be the connection weights from X_1 to Z_1 and Z_2, respectively; let $\beta_{01} = 1.5$ be the bias at hidden node 1; let $\beta_{21} =$

$\beta_{22} = 1$ be the connection weights from X_2 to Z_1 and Z_2, respectively; and let $\beta_{02} = 0.5$ be the bias at hidden node 2. Next, let $\alpha_1 = -2$ and $\alpha_2 = 1$ be the connection weights from Z_1 to Y and from Z_2 to Y, respectively, with bias $\alpha_0 = 0.5$. Draw the network graph. Find the linear boundaries as defined by the two hidden nodes; in the unit square, draw the boundaries and identify which class, 0 or 1, corresponds to each region of the unit square. Show that this network solves the XOR problem. Find another solution to this problem using different weights and biases.

10.8 Write a computer program to carry out the backpropagation algorithm as detailed in Section 10.7.6 for the squared-error loss function, and then apply it to a classification data set of your choice.

10.9 Study the correspondences between a single hidden layer neural network (10.18) and a generalized additive model (10.54).

10.10 Prove that

$$\int e^{-\frac{1}{2}\mathbf{z}^\tau \mathbf{Bz} + \mathbf{h}^\tau \mathbf{z}} d\mathbf{z} = (2\pi)^{Q/2} |\mathbf{B}|^{-1/2} e^{\frac{1}{2}\mathbf{h}^\tau \mathbf{B}^{-1}\mathbf{h}}.$$

10.11 Prove (10.74). (Hint: Use Exercise 10.10 with $\mathbf{z} = \Delta\boldsymbol{\omega}$, $\mathbf{B} = \mathbf{A} + \nu\mathbf{gg}^\tau$, and $\mathbf{h} = -\nu(y - y_{\text{MP}})\mathbf{g}$. Then, multiply numerator and denominator by $\mathbf{g}^\tau (\mathbf{I} + \nu\mathbf{A}^{-1}\mathbf{gg}^\tau)\mathbf{g}$, and simplify.)

10.12 Use the logistic function as the sigmoid activation function $g(\cdot)$ and a linear function $f(\cdot)$ to derive the computational expressions for the backpropagation algorithm. Discuss the properties of this particular algorithm.

10.13 Use the cross-entropy loss function to derive the appropriate computational expressions for the backpropagation algorithm. Program the resulting algorithm, use it with a data set of your choice, and compare its output with that obtained from the squared-error loss function.

10.14 Construct a network diagram based upon the sine function that will approximate the function $F(x)$ in (10.21) by $\widehat{F}(x)$ in (10.22).

10.15 Suppose we construct a neural network with no hidden layer, just input and output nodes. Let X_j be the jth input, $j = 1, 2, \ldots, r$, and let $Y = f(\beta_0 + \mathbf{X}^\tau\boldsymbol{\beta})$ denote the output, where $f(u) = (1 + e^{-u})^{-1}$, $\mathbf{X} = (X_1, \cdots, X_r)^\tau$, and $\boldsymbol{\beta} = (\beta_1, \cdots, \beta_r)^\tau$ is an r-vector of weights. Show that the decision boundary of this network is linear. If there are two input variables (i.e., $r = 2$), draw the corresponding decision boundary.

10.16 Fit a neural network to the `gilgaied soil` data set from Section 8.6. How could the two-way format of the data be taken into account in a neural network model?

10.17 Fit a neural network to the Cleveland `heart-disease` data from Section 9.2.1. Compare results with that given by using a classification tree.

10.18 Fit a neural network to the Pima Indians diabetic data set `pima` from Section 9.2.4. Compare results with that given by using a classification tree.

10.19 Fit a regression neural network to the 1992 Major League Baseball Salaries data from Section 9.3.5. Compare results with that given by using a regression tree.

10.20 Write a computer program to implement projection pursuit regression and use it to fit the 1992 Major League Baseball Salaries data.

10.21 Consider a regression neural network in which the outputs are identical to the inputs. Generate input data from a suitable multivariate Gaussian distribution and use that same data as outputs. Fit a neural networks model to these data and comment on your results. What is the relationship between this network analysis and principal component analysis?

10.22 In the discussion of Bayesian neural networks (Section 10.12), the binary classification problem was addressed. Redo the section on Bayesian classification networks using Laplace's approximation method so that now there are more than two classes.

10.23 Take any classification data set and divide it up into a learning set and an independent test set. Change the value of one observation on one input variable in the learning set so that that value is now a univariate outlier. Fit separate single-hidden-layer neural networks to the original learning-set data and to the learning-set data with the outlier. Comment on the effect of the outlier on the fit and on its effect on classifying the test set. Shrink the value of that outlier toward its original value and evaluate when the effect of the outlier on the fit vanishes. How far away must the outlier move from its original value that significant changes to the network coefficient estimates occur?

11
Support Vector Machines

11.1 Introduction

Fisher's linear discriminant function (LDF) and related classifiers for binary and multiclass learning problems have performed well for many years and for many data sets. Recently, a brand-new learning methodology, *support vector machines (SVMs)*, has emerged (Boser, Guyon, and Vapnik, 1992), which has matched the performance of the LDF and, in many instances, has proved to be superior to it.

Development and implementation of algorithms for SVMs are currently of great interest to theoretical researchers and applied scientists in machine learning, data mining, and bioinformatics. Huge numbers of research articles, tutorials, and textbooks have been published on the topic, and annual workshops, new research journals, courses, and websites are now devoted to the subject. SVMs have been successfully applied to classification problems as diverse as handwritten digit recognition, text categorization, cancer classification using microarray expression data, protein secondary-structure prediction, and cloud classification using satellite-radiance profiles.

SVMs, which are available in both linear and nonlinear versions, involve optimization of a convex loss function under given constraints and so are unaffected by problems of local minima. This gives SVMs quite a strong

A.J. Izenman, *Modern Multivariate Statistical Techniques*,
doi: 10.1007/978-0-387-78189-1_11,
© Springer Science+Business Media, LLC 2008

competitive advantage over methods such as neural networks and decision trees. SVMs are computed using well-documented, general-purpose, mathematical programming algorithms, and their performance in many situations has been quite remarkable. Even in the face of massive data sets, extremely fast and efficient software is being designed to compute SVMs for classification.

By means of the new technology of *kernel methods*, SVMs have been very successful in building highly nonlinear classifiers. The kernel method enables us to construct linear classifiers in high-dimensional feature spaces that are nonlinearly related to input space and to carry out those computations in input space using very few parameters. SVMs have also been successful in dealing with situations in which there are many more variables than observations.

Although these advantages hold in general, we have to recognize that there will always be applications in which SVMs can get beaten in performance by a hand-crafted classification method.

In this chapter, we describe the linear and nonlinear SVM as solutions of the binary classification problem. The nonlinear SVM incorporates nonlinear transformations of the input vectors and uses the kernel trick to simplify computations. We describe a variety of kernels, including string kernels for text categorization problems. Although the SVM methodology was built specifically for binary classification, we discuss attempts to extend that methodology to multiclass classification. Finally, although the SVM methodology was originally designed to solve classification problems, we discuss how the SVM methodology has been defined for regression situations.

11.2 Linear Support Vector Machines

Assume we have available a learning set of data,

$$\mathcal{L} = \{(\mathbf{x}_i, y_i) : i = 1, 2, \ldots, n\}, \tag{11.1}$$

where $\mathbf{x}_i \in \Re^r$ and $y_i \in \{-1, +1\}$. The *binary classification problem* is to use $\mathcal{L}$ to construct a function $f : \Re^r \to \Re$ so that

$$C(\mathbf{x}) = \text{sign}(f(\mathbf{x})) \tag{11.2}$$

is a classifier. The *separating function* f then classifies each new point $\mathbf{x}$ in a test set $\mathcal{T}$ into one of two classes, Π_+ or Π_-, depending upon whether $C(\mathbf{x})$ is $+1$ (if $f(\mathbf{x}) \geq 0$) or -1 (if $f(\mathbf{x}) < 0$), respectively. The goal is to have f assign all positive points in $\mathcal{T}$ (i.e., those with $y = +1$) to Π_+ and all negative points in $\mathcal{T}$ $(y = -1)$ to Π_-. In practice, we recognize that 100% correct classification may not be possible.

11.2.1 The Linearly Separable Case

First, consider the simplest situation: suppose the positive $(y_i = +1)$ and negative $(y_i = -1)$ data points from the learning set $\mathcal{L}$ can be separated by a hyperplane,

$$\{\mathbf{x} : f(\mathbf{x}) = \beta_0 + \mathbf{x}^\tau \boldsymbol{\beta} = 0\}, \tag{11.3}$$

where $\boldsymbol{\beta}$ is the *weight vector* with Euclidean norm $\| \boldsymbol{\beta} \|$, and β_0 is the *bias*. (Note: $b = -\beta_0$ is the *threshold*.) If this hyperplane can separate the learning set into the two given classes without error, the hyperplane is termed a *separating hyperplane*. Clearly, there is an infinite number of such separating hyperplanes. How do we determine which one is the best?

Consider any separating hyperplane. Let d_- be the shortest distance from the separating hyperplane to the nearest negative data point, and let d_+ be the shortest distance from the same hyperplane to the nearest positive data point. Then, the *margin* of the separating hyperplane is defined as $d = d_- + d_+$. If, in addition, the distance between the hyperplane and its closest observation is maximized, we say that the hyperplane is an *optimal separating hyperplane* (also known as a *maximal margin classifier*).

If the learning data from the two classes are linearly separable, there exists β_0 and $\boldsymbol{\beta}$ such that

$$\beta_0 + \mathbf{x}_i^\tau \boldsymbol{\beta} \geq +1, \quad \text{if } y_i = +1, \tag{11.4}$$

$$\beta_0 + \mathbf{x}_i^\tau \boldsymbol{\beta} \leq -1, \quad \text{if } y_i = -1. \tag{11.5}$$

If there are data vectors in $\mathcal{L}$ such that equality holds in (11.4), then these data vectors lie on the hyperplane H_{+1}: $(\beta_0 - 1) + \mathbf{x}^\tau \boldsymbol{\beta} = 0$; similarly, if there are data vectors in $\mathcal{L}$ such that equality holds in (11.5), then these data vectors lie on the hyperplane H_{-1}: $(\beta_0 + 1) + \mathbf{x}^\tau \boldsymbol{\beta} = 0$. Points in $\mathcal{L}$ that lie on either one of the hyperplanes H_{-1} or H_{+1}, are said to be *support vectors*. See Figure 11.1. The support vectors typically consist of a small percentage of the total number of sample points.

If $\mathbf{x}_{-1}$ lies on the hyperplane H_{-1}, and if $\mathbf{x}_{+1}$ lies on the hyperplane H_{+1}, then,

$$\beta_0 + \mathbf{x}_{-1}^\tau \boldsymbol{\beta} = -1, \quad \beta_0 + \mathbf{x}_{+1}^\tau \boldsymbol{\beta} = +1. \tag{11.6}$$

The difference of these two equations is $\mathbf{x}_{+1}^\tau \boldsymbol{\beta} - \mathbf{x}_{-1}^\tau \boldsymbol{\beta} = 2$, and their sum is $\beta_0 = -\frac{1}{2}\{\mathbf{x}_{+1}^\tau \boldsymbol{\beta} + \mathbf{x}_{-1}^\tau \boldsymbol{\beta}\}$. The perpendicular distances of the hyperplane $\beta_0 + \mathbf{x}^\tau \boldsymbol{\beta} = 0$ from the points $\mathbf{x}_{-1}$ and $\mathbf{x}_{+1}$ are

$$d_- = \frac{|\beta_0 + \mathbf{x}_{-1}^\tau \boldsymbol{\beta}|}{\| \boldsymbol{\beta} \|} = \frac{1}{\| \boldsymbol{\beta} \|}, \quad d_+ = \frac{|\beta_0 + \mathbf{x}_{+1}^\tau \boldsymbol{\beta}|}{\| \boldsymbol{\beta} \|} = \frac{1}{\| \boldsymbol{\beta} \|}, \tag{11.7}$$

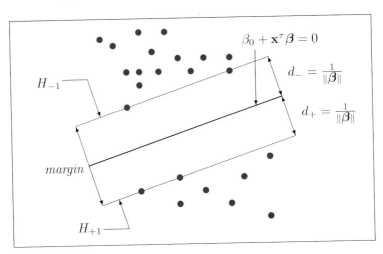

FIGURE 11.1. *Support vector machines: the linearly separable case. The red points correspond to data points with $y_i = -1$, and the blue points correspond to data points with $y_i = +1$. The separating hyperplane is the line $\beta_0 + \mathbf{x}^\tau \boldsymbol{\beta} = 0$. The support vectors are those points lying on the hyperplanes H_{-1} and H_{+1}. The margin of the separating hyperplane is $d = 2/ \parallel \boldsymbol{\beta} \parallel$.*

respectively (see Exercise 11.1). So, the margin of the separating hyperplane is $d = 2/ \parallel \boldsymbol{\beta} \parallel$.

The inequalities (11.4) and (11.5) can be combined into a single set of inequalities,

$$y_i(\beta_0 + \mathbf{x}_i^\tau \boldsymbol{\beta}) \geq +1, \quad i = 1, 2, \ldots, n. \tag{11.8}$$

The quantity $y_i(\beta_0 + \mathbf{x}_i^\tau \boldsymbol{\beta})$ is called the *margin of* $(\mathbf{x}_i, y_i)$ *with respect to the hyperplane* (11.3), $i = 1, 2, \ldots, n$. From (11.6), we see that $\mathbf{x}_i$ is a support vector with respect to the hyperplane (11.3) if its margin equals one; that is, if

$$y_i(\beta_0 + \mathbf{x}_i^\tau \boldsymbol{\beta}) = 1. \tag{11.9}$$

The support vectors in Figure 11.1 are identified (with circles around them). The empirical distribution of the margins of all the observations in $\mathcal{L}$ is called the *margin distribution of a hyperplane with respect to* $\mathcal{L}$. The minimum of the empirical margin distribution is the *margin of the hyperplane with respect to* $\mathcal{L}$.

The problem is to find the optimal separating hyperplane; namely, find the hyperplane that maximizes the margin, $2/ \parallel \boldsymbol{\beta} \parallel$, subject to the conditions (11.8). Equivalently, we wish to find β_0 and $\boldsymbol{\beta}$ to

$$minimize \quad \frac{1}{2} \parallel \boldsymbol{\beta} \parallel^2, \tag{11.10}$$

$$subject \ to \quad y_i(\beta_0 + \mathbf{x}_i^\tau \boldsymbol{\beta}) \geq 1, \quad i = 1, 2, \ldots, n. \tag{11.11}$$

This is a convex optimization problem: minimize a quadratic function subject to linear inequality constraints. Convexity ensures that we have a global minimum wthout local minima. The resulting optimal separating hyperplane is called the *maximal* (or *hard*) *margin solution*.

We solve this problem using Lagrangian multipliers. Because the constraints are $y_i(\beta_0 + \mathbf{x}_i^\tau \boldsymbol{\beta}) - 1 \geq 0$, $i = 1, 2, \ldots, n$, we multiply the constraints by positive Lagrangian multipliers and subtract each such product from the objective function (11.10) to form the *primal!functional*,

$$F_P(\beta_0, \boldsymbol{\beta}, \boldsymbol{\alpha}) = \frac{1}{2} \parallel \boldsymbol{\beta} \parallel^2 - \sum_{i=1}^{n} \alpha_i\{y_i(\beta_0 + \mathbf{x}_i^\tau \boldsymbol{\beta}) - 1\}, \qquad (11.12)$$

where

$$\boldsymbol{\alpha} = (\alpha_1, \cdots, \alpha_n)^\tau \geq \mathbf{0} \qquad (11.13)$$

is the n-vector of (nonnegative) Lagrangian coefficients. We need to minimize F with respect to the *primal variables* β_0 and $\boldsymbol{\beta}$, and then maximize the resulting minimum-F with respect to the *dual variables* $\boldsymbol{\alpha}$.

The Karush–Kuhn–Tucker conditions give necessary and sufficient conditions for a solution to a constrained optimization problem. For our primal problem, β_0, $\boldsymbol{\beta}$, and $\boldsymbol{\alpha}$ have to satisfy:

$$\frac{\partial F_P(\beta_0, \boldsymbol{\beta}, \boldsymbol{\alpha})}{\partial \beta_0} = -\sum_{i=1}^{n} \alpha_i y_i = 0, \qquad (11.14)$$

$$\frac{\partial F_P(\beta_0, \boldsymbol{\beta}, \boldsymbol{\alpha})}{\partial \boldsymbol{\beta}} = \boldsymbol{\beta} - \sum_{i=1}^{n} \alpha_i y_i \mathbf{x}_i = 0, \qquad (11.15)$$

$$y_i(\beta_0 + \mathbf{x}_i^\tau \boldsymbol{\beta}) - 1 \geq 0, \qquad (11.16)$$

$$\alpha_i \geq 0, \qquad (11.17)$$

$$\alpha_i\{y_i(\beta_0 + \mathbf{x}_i^\tau \boldsymbol{\beta}) - 1\} = 0, \qquad (11.18)$$

for $i = 1, 2, \ldots, n$. The condition (11.18) is known as the *Karush–Kuhn–Tucker complementarity condition*.

Solving equations (11.14) and (11.15) yields

$$\sum_{i=1}^{n} \alpha_i y_i = 0, \qquad (11.19)$$

$$\boldsymbol{\beta}^* = \sum_{i=1}^{n} \alpha_i y_i \mathbf{x}_i. \qquad (11.20)$$

Substituting (11.19) and (11.20) into (11.12) yields the minimum value of $F_P(\beta_0, \boldsymbol{\beta}, \boldsymbol{\alpha})$, namely,

$$F_D(\boldsymbol{\alpha}) = \frac{1}{2} \parallel \boldsymbol{\beta}^* \parallel^2 - \sum_{i=1}^{n} \alpha_i\{y_i(\beta_0^* + \mathbf{x}_i^\tau \boldsymbol{\beta}^*) - 1\}$$

$$= \frac{1}{2}\sum_{i=1}^{n}\sum_{j=1}^{n}\alpha_i\alpha_j y_i y_j (\mathbf{x}_i^\tau \mathbf{x}_j) - \sum_{i=1}^{n}\sum_{j=1}^{n}\alpha_i\alpha_j y_i y_j (\mathbf{x}_i^\tau \mathbf{x}_i) + \sum_{i=1}^{n}\alpha_i$$

$$= \sum_{i=1}^{n}\alpha_i - \frac{1}{2}\sum_{i=1}^{n}\sum_{j=1}^{n}\alpha_i\alpha_j y_i y_j (\mathbf{x}_i^\tau \mathbf{x}_j), \tag{11.21}$$

where we used (11.18) in the second line. Note that the primal variables have been removed from the problem. The expression (11.21) is usually referred to as the *dual functional* of the optimization problem.

We next find the Lagrangian multipliers $\boldsymbol{\alpha}$ by maximizing the dual functional (11.21) subject to the constraints (11.17) and (11.19). The constrained maximization problem (the "Wolfe dual") can be written in matrix notation as follows. Find $\boldsymbol{\alpha}$ to

$$maximize \quad F_D(\boldsymbol{\alpha}) = \mathbf{1}_n^\tau \boldsymbol{\alpha} - \frac{1}{2}\boldsymbol{\alpha}^\tau \mathbf{H}\boldsymbol{\alpha} \tag{11.22}$$

$$subject\ to \quad \boldsymbol{\alpha} \geq \mathbf{0},\ \boldsymbol{\alpha}^\tau \mathbf{y} = \mathbf{0}, \tag{11.23}$$

where $\mathbf{y} = (y_1, \cdots, y_n)^\tau$ and $\mathbf{H} = (H_{ij})$ is a square $(n \times n)$-matrix with $H_{ij} = y_i y_j (\mathbf{x}_i^\tau \mathbf{x}_j)$. If $\widehat{\boldsymbol{\alpha}}$ solves this optimization problem, then

$$\widehat{\boldsymbol{\beta}} = \sum_{i=1}^{n}\widehat{\alpha}_i y_i \mathbf{x}_i \tag{11.24}$$

yields the optimal weight vector. If $\widehat{\alpha}_i > 0$, then, from (11.18), $y_i(\beta_0^* + \mathbf{x}_i^\tau \boldsymbol{\beta}^*) = 1$, and so $\mathbf{x}_i$ is a support vector; for all observations that are not support vectors, $\widehat{\alpha}_i = 0$. Let $sv \subset \{1, 2, \ldots, n\}$ be the subset of indices that identify the support vectors (and also the nonzero Lagrangian multipliers). Then, the optimal $\boldsymbol{\beta}$ is given by (11.24), where the sum is taken only over the support vectors; that is,

$$\widehat{\boldsymbol{\beta}} = \sum_{i \in sv}\widehat{\alpha}_i y_i \mathbf{x}_i. \tag{11.25}$$

In other words, $\widehat{\boldsymbol{\beta}}$ is a linear function only of the support vectors $\{\mathbf{x}_i, i \in sv\}$. In most applications, the number of support vectors will be small relative to the size of $\mathcal{L}$, yielding a *sparse* solution. In this case, the support vectors carry all the information necessary to determine the optimal hyperplane.

The primal and dual optimization problems yield the same solution, although the dual problem is simpler to compute and, as we shall see, is simpler to generalize to nonlinear classifiers. Finding the solution involves standard convex quadratic-programming methods, and so any local minimum also turns out to be a global minimum.

Although the optimal bias $\widehat{\beta}_0$ is not determined explicitly by the optimization solution, we can estimate it by solving (11.18) for each support vector and then averaging the results. In other words, the estimated bias of the optimal hyperplane is given by

$$\widehat{\beta}_0 = \frac{1}{|sv|} \sum_{i \in sv} \left(\frac{1 - y_i \mathbf{x}_i^T \widehat{\boldsymbol{\beta}}}{y_i} \right),$$
(11.26)

where $|sv|$ is the number of support vectors in $\mathcal{L}$.

It follows that the optimal hyperplane can be written as

$$\begin{aligned} \widehat{f}(\mathbf{x}) &= \widehat{\beta}_0 + \mathbf{x}^T \widehat{\boldsymbol{\beta}} \\ &= \widehat{\beta}_0 + \sum_{i \in sv} \widehat{\alpha}_i y_i (\mathbf{x}^T \mathbf{x}_i). \end{aligned}$$
(11.27)

Clearly, only support vectors are relevant in computing the optimal separating hyperplane; observations that are not support vectors play no role in determining the hyperplane and are, thus, irrelevant to solving the optimization problem. The *classification rule* is given by

$$C(\mathbf{x}) = \operatorname{sign}\{\widehat{f}(\mathbf{x})\}.$$
(11.28)

If $j \in sv$, then, from (11.27),

$$y_j \widehat{f}(\mathbf{x}_j) = y_j \widehat{\beta}_0 + \sum_{i \in sv} \widehat{\alpha}_i y_i y_j (\mathbf{x}_j^T \mathbf{x}_i) = 1.$$
(11.29)

Hence, the squared-norm of the weight vector $\widehat{\boldsymbol{\beta}}$ of the optimal hyperplane is

$$\begin{aligned} \| \widehat{\boldsymbol{\beta}} \|^2 &= \sum_{i \in sv} \sum_{j \in sv} \widehat{\alpha}_i \widehat{\alpha}_j y_i y_j (\mathbf{x}_i^T \mathbf{x}_j) \\ &= \sum_{j \in sv} \widehat{\alpha}_j y_j \sum_{i \in sv} \widehat{\alpha}_i y_i (\mathbf{x}_i^T \mathbf{x}_j) \\ &= \sum_{j \in sv} \widehat{\alpha}_j (1 - y_j \widehat{\beta}_0) \\ &= \sum_{j \in sv} \widehat{\alpha}_j. \end{aligned}$$
(11.30)

The third line used (11.29) and the fourth line used (11.19). It follows from (11.30) that the optimal hyperplane has maximum margin $2/ \| \widehat{\boldsymbol{\beta}} \|$, where

$$\frac{1}{\| \widehat{\boldsymbol{\beta}} \|} = \left(\sum_{j \in sv} \widehat{\alpha}_j \right)^{-1/2}.$$
(11.31)

11.2.2   The Linearly Nonseparable Case

In real applications, it is unlikely that there will be such a clear linear separation between data drawn from two classes. More likely, there will be some overlap. We can generally expect some data from one class to infiltrate the region of space perceived to belong to the other class, and vice versa. The overlap will cause problems for any classification rule, and, depending upon the extent of the overlap, we should expect that some of the overlapping points will be misclassified.

The *nonseparable case* occurs if either the two classes are separable, but not linearly so, or that no clear separability exists between the two classes, linearly or nonlinearly. One reason for overlapping classes is the high noise level (i.e., large variances) of one or both classes. As a result, one or more of the constraints will be violated.

The way we cope with overlapping data is to create a more flexible formulation of the problem, which leads to a *soft-margin solution*. To do this, we introduce the concept of a nonnegative *slack variable*, ξ_i, for each observation, $(\mathbf{x}_i, y_i)$, in $\mathcal{L}$, $i = 1, 2, \ldots, n$. See Figure 11.2 for a two-dimensional example. Let

$$\boldsymbol{\xi} = (\xi_1, \cdots, \xi_n)^\tau \geq \mathbf{0}. \tag{11.32}$$

The constraints (11.11) now become $y_i(\beta_0 + \mathbf{x}_i^\tau\boldsymbol{\beta}) + \xi_i \geq 1$ for $i = 1, 2, \ldots, n$. Data points that obey these constraints have $\xi_i = 0$. The classifier now has to find the optimal hyperplane that controls both the margin, $2/\|\boldsymbol{\beta}\|$, and some computationally simple function of the slack variables, such as

$$g_\sigma(\boldsymbol{\xi}) = \sum_{i=1}^{n} \xi_i^\sigma, \tag{11.33}$$

subject to certain constraints. The usual values of σ are 1 ("1-norm") or 2 ("2-norm"). Here, we discuss the case of $\sigma = 1$; for $\sigma = 2$, see Exercise 11.2.

The *1-norm soft-margin optimization problem* is to find β_0, $\boldsymbol{\beta}$, and $\boldsymbol{\xi}$ to

$$minimize \quad \frac{1}{2}\|\boldsymbol{\beta}\|^2 + C\sum_{i=1}^{n}\xi, \tag{11.34}$$

$$subject\ to \quad \xi_i \geq 0, \quad y_i(\beta_0 + \mathbf{x}_i^\tau\boldsymbol{\beta}) \geq 1 - \xi_i, \quad i = 1, 2, \ldots, n, \tag{11.35}$$

where $C > 0$ is a *regularization parameter*. C takes the form of a tuning constant that controls the size of the slack variables and balances the two terms in the minimizing function.

Form the primal functional, $F_P = F_P(\beta_0, \boldsymbol{\beta}, \boldsymbol{\xi}, \boldsymbol{\alpha}, \boldsymbol{\eta})$, where

$$F_P = \frac{1}{2}\|\boldsymbol{\beta}\|^2 + C\sum_{i=1}^{n}\xi_i - \sum_{i=1}^{n}\alpha_i\{y_i(\beta_0 + \mathbf{x}_i^\tau\boldsymbol{\beta}) - (1 - \xi_i)\} - \sum_{i=1}^{n}\eta_i\xi_i, \tag{11.36}$$

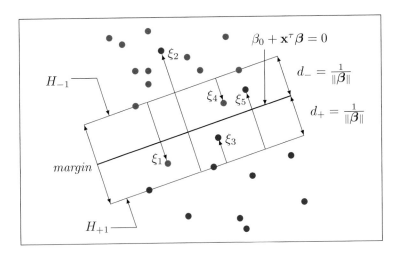

FIGURE 11.2. *Support vector machines: the nonlinearly separable case. The red points correspond to data points with $y_i = -1$, and the blue points correspond to data points with $y_i = +1$. The separating hyperplane is the line $\beta_0 + \mathbf{x}^T\boldsymbol{\beta} = 0$. The support vectors are those circled points lying on the hyperplanes H_{-1} and H_{+1}. The slack variables ξ_1 and ξ_4 are associated with the red points that violate the constraint of hyperplane H_{-1}, and points marked by $\xi_2, \xi_3,$ and ξ_5 are associated with the blue points that violate the constraint of hyperplane H_{+1}. Points that satisfy the constraints of the appropriate hyperplane have $\xi_i = 0$.*

with $\boldsymbol{\alpha} = (\alpha_1, \cdots, \alpha_n)^\tau \geq \mathbf{0}$ and $\boldsymbol{\eta} = (\eta_1, \cdots, \eta_n)^\tau \geq \mathbf{0}$. Fix $\boldsymbol{\alpha}$ and $\boldsymbol{\eta}$, and differentiate F_P with respect to β_0, $\boldsymbol{\beta}$, and $\boldsymbol{\xi}$:

$$\frac{\partial F_P}{\partial \beta_0} = -\sum_{i=1}^{n} \alpha_i y_i, \tag{11.37}$$

$$\frac{\partial F_P}{\partial \boldsymbol{\beta}} = \boldsymbol{\beta} - \sum_{i=1}^{n} \alpha_i y_i \mathbf{x}_i, \tag{11.38}$$

$$\frac{\partial F_P}{\partial \xi_i} = C - \alpha_i - \eta_i, \quad i = 1, 2, \ldots, n. \tag{11.39}$$

Setting these derivatives equal to zero and solving yields

$$\sum_{i=1}^{n} \alpha_i y_i = 0, \quad \boldsymbol{\beta}^* = \sum_{i=1}^{n} \alpha_i y_i \mathbf{x}_i, \quad \alpha_i = C - \eta_i. \tag{11.40}$$

Substituting (11.37) into (11.33) gives the dual functional,

$$F_D(\boldsymbol{\alpha}) = \sum_{i=1}^{n} \alpha_i - \frac{1}{2} \sum_{i=1}^{n} \sum_{j=1}^{n} \alpha_i \alpha_j y_i y_j (\mathbf{x}_i^\tau \mathbf{x}_j), \tag{11.41}$$

which, remarkably, is the same as (11.18) for the linearly separable case. From the constraints $C - \alpha_i - \eta_i = 0$ and $\eta_i \geq 0$, we have that $0 \leq \alpha_i \leq C$. In addition, we have the Karush–Kuhn–Tucker conditions:

$$y_i(\beta_0 + \mathbf{x}_i^\tau \boldsymbol{\beta}) - (1 - \xi_i) \geq 0 \tag{11.42}$$

$$\xi_i \geq 0, \tag{11.43}$$

$$\alpha_i \geq 0, \tag{11.44}$$

$$\eta_i \geq 0, \tag{11.45}$$

$$\alpha_i \{ y_i(\beta_0 + \mathbf{x}_i^\tau \boldsymbol{\beta}) - (1 - \xi_i) \} = 0, \tag{11.46}$$

$$\xi_i(\alpha_i - C) = 0, \tag{11.47}$$

for $i = 1, 2, \ldots, n$. From (11.47), a slack variable, ξ_i, can be nonzero only if $\alpha_i = C$. The Karush–Kuhn–Tucker complementarity conditions, (11.46) and (11.47), can be used to find the optimal bias β_0.

We can write the dual maximization problem in matrix notation as follows. Find $\boldsymbol{\alpha}$ to

$$maximize \quad F_D(\boldsymbol{\alpha}) = \mathbf{1}_n^\tau \boldsymbol{\alpha} - \frac{1}{2} \boldsymbol{\alpha}^\tau \mathbf{H} \boldsymbol{\alpha} \tag{11.48}$$

$$subject\ to \quad \boldsymbol{\alpha}^\tau \mathbf{y} = 0, \ \mathbf{0} \leq \boldsymbol{\alpha} \leq C \mathbf{1}_n. \tag{11.49}$$

The only difference between this optimization problem and that for the linearly separable case, (11.22) and (11.23), is that, here, the Lagrangian coefficients α_i, $i = 1, 2, \ldots, n$, are each bounded above by C; this upper bound restricts the influence of each observation in determining the solution. This type of constraint is referred to as a *box constraint* because $\boldsymbol{\alpha}$ is constrained by the box of side C in the positive orthant. From (11.49), we see that the *feasible region* for the solution to this convex optimization problem is the intersection of the hyperplane $\boldsymbol{\alpha}^\tau \mathbf{y} = 0$ with the box constraint $\mathbf{0} \leq \boldsymbol{\alpha} \leq C \mathbf{1}_n$. If $C = \infty$, then the problem reduces to the hard-margin separable case.

If $\widehat{\boldsymbol{\alpha}}$ solves this optimization problem, then,

$$\widehat{\boldsymbol{\beta}} = \sum_{i \in sv} \widehat{\alpha}_i y_i \mathbf{x}_i \tag{11.50}$$

yields the optimal weight vector, where the set sv of support vectors contains those observations in $\mathcal{L}$ which satisfy the constraint (11.42).

11.3 Nonlinear Support Vector Machines

So far, we have discussed methods for constructing a linear SVM classifier. But what if a linear classifier is not appropriate for the data set in question? Can we extend the idea of linear SVM to the nonlinear case? The key to constructing a nonlinear SVM is to observe that the observations in $\mathcal{L}$ only enter the dual optimization problem through the inner products $\langle \mathbf{x}_i, \mathbf{x}_j \rangle = \mathbf{x}_i^\tau \mathbf{x}_j$, $i, j = 1, 2, \ldots, n$.

11.3.1 Nonlinear Transformations

Suppose we transform each observation, $\mathbf{x}_i \in \Re^r$, in $\mathcal{L}$ using some nonlinear mapping $\mathbf{\Phi} : \Re^r \rightarrow \mathcal{H}$, where $\mathcal{H}$ is an $N_\mathcal{H}$-dimensional feature space. The nonlinear map $\mathbf{\Phi}$ is generally called the *feature map* and the space $\mathcal{H}$ is called the *feature space*. The space $\mathcal{H}$ may be very high-dimensional, possibly even infinite dimensional. We will generally assume that $\mathcal{H}$ is a Hilbert space of real-valued functions on $\Re$ with inner product $\langle \cdot, \cdot \rangle$ and norm $\| \cdot \|$.

Let

$$\mathbf{\Phi}(\mathbf{x}_i) = (\phi_1(\mathbf{x}_i), \cdots, \phi_{N_\mathcal{H}}(\mathbf{x}_i))^\tau \in \mathcal{H}, \quad i = 1, 2, \ldots, n. \qquad (11.51)$$

The transformed sample is then $\{\mathbf{\Phi}(\mathbf{x}_i), y_i\}$, where $y_i \in \{-1, +1\}$ identifies the two classes. If we substitute $\mathbf{\Phi}(\mathbf{x}_i)$ for $\mathbf{x}_i$ in the development of the linear SVM, then data would only enter the optimization problem by way of the inner products $\langle \mathbf{\Phi}(\mathbf{x}_i), \mathbf{\Phi}(\mathbf{x}_j) \rangle = \mathbf{\Phi}(\mathbf{x}_i)^\tau \mathbf{\Phi}(\mathbf{x}_j)$. The difficulty in using nonlinear transformations in this way is computing such inner products in high-dimensional space $\mathcal{H}$.

11.3.2 The "Kernel Trick"

The idea behind nonlinear SVM is to find an optimal separating hyperplane (with or without slack variables, as appropriate) in high-dimensional feature space $\mathcal{H}$ just as we did for the linear SVM in input space. Of course, we would expect the dimensionality of $\mathcal{H}$ to be a huge impediment to constructing an optimal separating hyperplane (and classification rule) because of the curse of dimensionality. The fact that this does not become a problem in practice is due to the "kernel trick," which was first applied to SVMs by Cortes and Vapnik (1995).

The so-called kernel trick is a wonderful idea that is widely used in algorithms for computing inner products of the form $\langle \Phi(\mathbf{x}_i), \Phi(\mathbf{x}_j) \rangle$ in feature space $\mathcal{H}$. The trick is that instead of computing these inner products in $\mathcal{H}$, which would be computationally expensive because of its high dimensionality, we compute them using a nonlinear kernel function, $K(\mathbf{x}_i, \mathbf{x}_j) = \langle \Phi(\mathbf{x}_i), \Phi(\mathbf{x}_j) \rangle$, in input space, which helps speed up the computations. Then, we just compute a *linear* SVM, but where the computations are carried out in some other space.

11.3.3 Kernels and Their Properties

A *kernel* K is a function $K : \Re^r \times \Re^r \to \Re$ such that, for all $\mathbf{x}, \mathbf{y} \in \Re^r$,

$$K(\mathbf{x}, \mathbf{y}) = \langle \Phi(\mathbf{x}), \Phi(\mathbf{y}) \rangle. \tag{11.52}$$

The kernel function is designed to compute inner-products in $\mathcal{H}$ by using only the original input data. Thus, wherever we see the inner product $\langle \Phi(\mathbf{x}), \Phi(\mathbf{y}) \rangle$, we substitute the kernel function $K(\mathbf{x}, \mathbf{y})$. The choice of K implicitly determines both Φ and $\mathcal{H}$. The big advantage to using kernels as inner products is that if we are given a kernel function K, then we do not need to know the explicit form of Φ.

We require that the kernel function be symmetric, $K(\mathbf{x}, \mathbf{y}) = K(\mathbf{y}, \mathbf{x})$, and satisfy an inequality, $[K(\mathbf{x}, \mathbf{y})]^2 \leq K(\mathbf{x}, \mathbf{x}) K(\mathbf{y}, \mathbf{y})$. derived from the Cauchy–Schwarz inequality. If $K(\mathbf{x}, \mathbf{x}) = 1$ for all $\mathbf{x} \in \Re^r$, this implies that $\|\Phi(\mathbf{x})\|_{\mathcal{H}} = 1$. A kernel K is said to have the *reproducing property* if, for any $f \in \mathcal{H}$,

$$\langle f(\cdot), K(\mathbf{x}, \cdot) \rangle = f(\mathbf{x}). \tag{11.53}$$

If K has this property, we say it is a *reproducing kernel*. K is also called the *representer of evaluation*. In particular, if $f(\cdot) = K(\cdot, \mathbf{x})$, then,

$$\langle K(\mathbf{x}, \cdot), K(\mathbf{y}, \cdot) \rangle = K(\mathbf{x}, \mathbf{y}). \tag{11.54}$$

Let $\mathbf{x}_1, \ldots, \mathbf{x}_n$ be any set of n points in $\mathcal{R}^r$. Then, the $(n \times n)$-matrix $\mathbf{K} = (K_{ij})$, where $K_{ij} = K(\mathbf{x}_i, \mathbf{x}_j)$, $i, j = 1, 2, \ldots, n$, is called the *Gram* (or *kernel*) matrix of K with respect to $\mathbf{x}_1, \ldots, \mathbf{x}_n$. If the Gram matrix $\mathbf{K}$ satisfies $\mathbf{u}^\tau \mathbf{K} \mathbf{u} \geq 0$, for any n-vector $\mathbf{u}$, then it is said to be *nonnegative-definite* with nonnegative eigenvalues, in which case we say that K is a nonnegative-definite kernel[1] (or *Mercer kernel*).

If K is a specific Mercer kernel on $\mathcal{R}^r \times \mathcal{R}^r$, we can always construct a unique Hilbert space $\mathcal{H}_K$, say, of real-valued functions for which K is its reproducing kernel. We call $\mathcal{H}_K$ a (real) *reproducing kernel Hilbert space* (rkhs). We write the inner-product and norm of $\mathcal{H}_K$ by $\langle \cdot, \cdot \rangle_{\mathcal{H}_K}$ (or just $\langle \cdot, \cdot \rangle$ when K is understood) and $\| \cdot \|_{\mathcal{H}_K}$, respectively.

11.3.4 Examples of Kernels

An example of a kernel is the *inhomogeneous polynomial kernel of degree d*,

$$K(\mathbf{x}, \mathbf{y}) = (\langle \mathbf{x}, \mathbf{y} \rangle + c)^d, \quad \mathbf{x}, \mathbf{y} \in \Re^r, \tag{11.55}$$

---

[1] In the machine-learning literature, nonnegative-definite matrices and kernels are usually referred to as positive-definite matrices and kernels, respectively.

TABLE 11.1. *Kernel functions, $K(\mathbf{x}, \mathbf{y})$, where $\sigma > 0$ is a scale parameter, $a, b, c \geq 0$, and d is an integer. The Euclidean norm is $\|\mathbf{x}\|^2 = \mathbf{x}^\tau \mathbf{x}$.*

Kernel	$K(\mathbf{x}, \mathbf{y})$
Polynomial of degree d	$(\langle \mathbf{x}, \mathbf{y} \rangle + c)^d$
Gaussian radial basis function	$\exp\left\{ -\frac{\|\mathbf{x}-\mathbf{y}\|^2}{2\sigma^2} \right\}$
Laplacian	$\exp\left\{ -\frac{\|\mathbf{x}-\mathbf{y}\|}{\sigma} \right\}$
Thin-plate spline	$\left(\frac{\|\mathbf{x}-\mathbf{y}\|}{\sigma} \right)^2 \log_e \left\{ \frac{\|\mathbf{x}-\mathbf{y}\|}{\sigma} \right\}$
Sigmoid	$\tanh(a\langle \mathbf{x}, \mathbf{y} \rangle + b)$

where c and d are parameters. The homogeneous form of the kernel occurs when $c = 0$ in (12.55). If $d = 1$ and $c = 0$, the feature map reduces to the identity. Usually, we take $c > 0$. A simple nonlinear map is given by the case $r = 2$ and $d = 2$. If $\mathbf{x} = (x_1, x_2)^\tau$ and $\mathbf{y} = (y_1, y_2)^\tau$, then,

$$K(\mathbf{x}, \mathbf{y}) = (\langle \mathbf{x}, \mathbf{y} \rangle + c)^2 = (x_1 y_1 + x_2 y_2 + c)^2 = \langle \Phi(\mathbf{x}), \Phi(\mathbf{y}) \rangle,$$

where $\Phi(\mathbf{x}) = (x_1^2, x_2^2, \sqrt{2}x_1 x_2, \sqrt{2c}x_1, \sqrt{2}x_2, c)^\tau$ and similarly for $\Phi(\mathbf{y})$. In this example, the function $\Phi(\mathbf{x})$ consists of six features ($\mathcal{H} = \Re^6$), all monomials having degree at most 2. For this kernel, we see that c controls the magnitudes of the constant term and the first-degree term.

In general, there will be $\dim(\mathcal{H}) = \binom{r+d}{d}$ different features, consisting of all monomials having degree at most d. The dimensionality of $\mathcal{H}$ can rapidly become very large: for example, in visual recognition problems, data may consist of 16×16 pixel images (so that each image is turned into a vector of dimension $r = 256$); if $d = 2$, then $\dim(\mathcal{H}) = 33,670$, whereas if $d = 4$, we have $\dim(\mathcal{H}) = 186,043,585$.

Other popular kernels, such as the *Gaussian radial basis function (RBF)*, the *Laplacian kernel*, the *thin-plate spline kernel*, and the *sigmoid kernel*, are given in Table 11.1. Strictly speaking, the sigmoid kernel is not a kernel (it satisfies Mercer's conditions only for certain values of a and b), but it has become very popular in that role in certain situations (e.g., two-layer neural networks).

The Gaussian RBF, Laplacian, and thin-plate spline kernels are examples of *translation-invariant* (or *stationary*) *kernels* having the general form

$K(\mathbf{x}, \mathbf{y}) = k(\mathbf{x} - \mathbf{y})$, where $k : \Re^r \to \Re$. The polynomial kernel is an example of a nonstationary kernel. A stationary kernel $K(\mathbf{x}, \mathbf{y})$ is *isotropic* if it depends only upon the distance $\delta = \|\mathbf{x} - \mathbf{y}\|$, i.e., if $K(\mathbf{x}, \mathbf{y}) = k(\delta)$, scaled to have $k(0) = 1$.

It is not always obvious which kernel to choose in any given application. Prior knowledge or a search through the literature can be helpful. If no such information is available, the best approach is to try either a Gaussian RBF, which has only a single parameter (σ) to be determined, or a polynomial kernel of low degree ($d = 1$ or 2). If necessary, more complicated kernels can then be applied to compare results.

String Kernels for Text Categorization

Text categorization is the assignment of natural-language text (or hypertext) documents into a given number of predefined categories based upon the content of those documents (see Section 2.2.1). Although manual categorization of text documents is currently the norm (e.g., using folders to save files, e-mail messages, URLs, etc.), some text categorization is automated (e.g., filters for spam or junk mail to help users cope with the sheer volume of daily e-mail messages). To reduce costs of text categorization tasks, we should expect a greater degree of automation to be present in the future.

In text-categorization problems, *string kernels* have been proposed based upon ideas derived from bioinformatics (see, e.g., Lodhi, Saunders, Shawe-Taylor,Cristianini, and Watkins, 2002).

Let $\mathcal{A}$ be a finite alphabet. A "string"

$$s = s_1 s_2 \cdots s_{|s|} \tag{11.56}$$

is a finite sequence of elements of $\mathcal{A}$, including the empty sequence, where $|s|$ denotes the length of s. We call u a *subsequence* of s (written $u = s(\mathbf{i})$) if there are indices $\mathbf{i} = (i_1, i_2, \cdots, i_{|u|})$, with $1 \leq i_1 < \cdots < i_{|u|} \leq |s|$, such that $u_j = s_{i_j}, j = 1, 2, \ldots, |u|$. If the indices $\mathbf{i}$ are contiguous, we say that u is a *substring* of s. The length of u in s is

$$\ell(\mathbf{i}) = i_{|u|} - i_1 + 1, \tag{11.57}$$

which is the number of elements of s overlaid by the subsequence u. For example, let s be the string "cat" ($s_1 = c, s_2 = a, s_3 = t, |s| = 3$), and consider all possible 2-symbol sequences, "ca," "ct," and "at," derived from s. For the string $u = $ ca, we have that $u_1 = c = s_1, u_2 = a = s_2$, whence, $u = s(\mathbf{i})$, where $\mathbf{i} = (i_1, i_2) = (1, 2)$. Thus, $\ell(\mathbf{i}) = 2$. Similarly, for the subsequence $u = $ ct, $u_1 = c = s_1, u_2 = t = s_3$, whence, $\mathbf{i} = (i_1, i_2) = (1, 3)$, and $\ell(\mathbf{i}) = 3$. Also, the subsequence $u = $ at has $u_1 = a = s_2, u_2 = t = s_3$, whence, $\mathbf{i} = (2, 3)$, and $\ell(\mathbf{i}) = 2$.

If $D = \mathcal{A}^m$ is the set of all finite strings of length at most m from $\mathcal{A}$, then, the feature space for a string kernel is $\Re^D$. The feature map Φ_u, operating on a string $s \in \mathcal{A}^m$, is characterized in terms of a given string $u \in \mathcal{A}^m$. To deal with noncontiguous subsequences, define $\lambda \in (0, 1)$ as the *drop-off rate* (or *decay factor*); we use λ to weight the interior gaps in the subsequences. The degree of importance we put into a contiguous subsequence is reflected in how small we take the value of λ. The value $\Phi_u(s)$ is computed as follows: identify all subsequences (indexed by $\mathbf{i}$) of s that are identical to u; for each such subsequence, raise λ to the power $\ell(\mathbf{i})$; and then sum the results over all subsequences. Because $\lambda < 1$, larger values of $\ell(\mathbf{i})$ carry less weight than smaller values of $\ell(\mathbf{i})$. We write

$$\Phi_u(s) = \sum_{\mathbf{i}:u=s(\mathbf{i})} \lambda^{\ell(\mathbf{i})}, \quad u \in \mathcal{A}^m. \tag{11.58}$$

In our example above, $\Phi_{ca}(\text{cat}) = \lambda^2$, $\Phi_{ct}(\text{cat}) = \lambda^3$, and $\Phi_{at}(\text{cat}) = \lambda^2$.

Two documents are considered to be "similar" if they have many subsequences in common: the more subsequences they have in common, the more similar they are deemed to be. Note that the degree of contiguity present in a subsequence determines the weight of that substring in the comparison; the closer the subsequence is to a contiguous substring, the more it should contribute to the comparison.

Let s and t be two strings. The kernel associated with the feature maps corresponding to s and t is given by the sum of inner products for all *common* substrings of length m,

$$\begin{aligned} K_m(s, t) &= \sum_{u \in \mathcal{D}} \langle \Phi_u(s), \Phi_u(t) \rangle \\ &= \sum_{u \in \mathcal{D}} \sum_{\mathbf{i}:u=s(\mathbf{i})} \sum_{\mathbf{j}:u=s(\mathbf{j})} \lambda^{\ell(\mathbf{i})+\ell(\mathbf{j})}. \end{aligned} \tag{11.59}$$

The kernel (11.59) is called a *string kernel* (or a *gap-weighted subsequences kernel*). For the example, let t be the string "car" ($t_1 = \text{c}, t_2 = \text{a}, t_3 = \text{r}$, $|t| = 3$). Note that the strings "cat" and "car" are both substrings of the string "cart." The three 2-symbol substrings of t are "ca," "cr," and "ar." For these substrings, we have that $\Phi_{ca}(\text{car}) = \lambda^2, \Phi_{cr}(\text{car}) = \lambda^3$, and $\Phi_{ar}(\text{car}) = \lambda^2$. The inner product (11.62) is given by $K_2(\text{cat}, \text{car}) = \langle \Phi_{ca}(\text{cat}), \Phi_{ca}(\text{car}) \rangle = \lambda^4$.

The feature maps in feature space are usually normalized to remove any bias introduced by document length. This is equivalent to normalizing the kernel (11.59),

$$K_m^*(s, t) = \frac{K_m(s, t)}{\sqrt{K_m(s, s)K_m(t, t)}}. \tag{11.60}$$

For our example, $K_2(\text{cat}, \text{cat}) = \langle \Phi_{\text{ca}}(\text{cat}), \Phi_{\text{ca}}(\text{cat}) \rangle + \langle \Phi_{\text{ct}}(\text{cat}), \Phi_{\text{ct}}(\text{cat}) \rangle + \langle \Phi_{\text{at}}(\text{cat}), \Phi_{\text{at}}(\text{cat}) \rangle = \lambda^6 + 2\lambda^4$, and, similarly, $K_2(\text{car}, \text{car}) = \lambda^6 + 2\lambda^4$, whence, $K_2^*(\text{cat}, \text{car}) = \lambda^4/(\lambda^6 + 2\lambda^4) = 1/(\lambda^2 + 2)$.

The parameters of the string kernel (11.59) are m and λ. The choices of $m = 5$ and $\lambda = 0.5$ have been found to perform well on segments of certain data sets (e.g., on subsets of the Reuters-21578 data) but do not fare as well when applied to the full data set.

11.3.5 Optimizing in Feature Space

Let K be a kernel. Suppose, first, that the observations in $\mathcal{L}$ are linearly separable in the feature space corresponding to the kernel K. Then, the dual optimization problem is to find $\boldsymbol{\alpha}$ and β_0 to

$$\text{maximize}\quad F_D(\boldsymbol{\alpha}) = \mathbf{1}_n^\tau \boldsymbol{\alpha} - \frac{1}{2}\boldsymbol{\alpha}^\tau \mathbf{H}\boldsymbol{\alpha} \tag{11.61}$$

$$\text{subject to}\quad \boldsymbol{\alpha} \geq \mathbf{0},\ \boldsymbol{\alpha}^\tau \mathbf{y} = 0, \tag{11.62}$$

where $\mathbf{y} = (y_1, \cdots, y_n)^\tau$, $\mathbf{H} = (H_{ij})$, and

$$H_{ij} = y_i y_j K(\mathbf{x}_i, \mathbf{x}_j) = y_i y_j K_{ij}, \quad i, j = 1, 2, \ldots, n. \tag{11.63}$$

Because K is a kernel, the Gram matrix $\mathbf{K} = (K_{ij})$ is nonnegative-definite, and so is the matrix $\mathbf{H}$ with elements (11.63). Hence, the functional $F_D(\boldsymbol{\alpha})$ is convex (see Exercise 11.8). So, there is a unique solution to this constrained optimization problem. If $\widehat{\boldsymbol{\alpha}}$ and $\widehat{\beta}_0$ solve this problem, then, the SVM decision rule is $\text{sign}\{\widehat{f}(\mathbf{x})\}$, where

$$\widehat{f}(\mathbf{x}) = \widehat{\beta}_0 + \sum_{i \in sv} \widehat{\alpha}_i y_i K(\mathbf{x}, \mathbf{x}_i) \tag{11.64}$$

is the optimal separating hyperplane in the feature space corresponding to the kernel K.

In the nonseparable case, using the kernel K, the dual problem of the 1-norm soft-margin optimization problem is to find $\boldsymbol{\alpha}$ to

$$\text{maximize}\quad F_D^*(\boldsymbol{\alpha}) = \mathbf{1}_n^\tau \boldsymbol{\alpha} - \frac{1}{2}\boldsymbol{\alpha}^\tau \mathbf{H}\boldsymbol{\alpha} \tag{11.65}$$

$$\text{subject to}\quad \mathbf{0} \leq \boldsymbol{\alpha} \leq C\mathbf{1}_n,\ \boldsymbol{\alpha}^\tau \mathbf{y} = 0, \tag{11.66}$$

where $\mathbf{y}$ and $\mathbf{H}$ are as above. For an optimal solution, the Karush–Kuhn–Tucker conditions, (11.42)–(11.47), must hold for the primal problem. So, a solution, $\boldsymbol{\alpha}$, to this problem has to satisfy all those conditions. Fortunately, it suffices to check a simpler set of conditions: we have to check that $\boldsymbol{\alpha}$

satisfies (11.66) and that (11.42) holds for all points where $0 \le \alpha_i < C$ and $\xi_i = 0$, and also for all points where $\alpha_i = C$ and $\xi_i \ge 0$.

11.3.6 Grid Search for Parameters

We need to determine two parameters when using a Gaussian RBF kernel, namely, the cost, C, of violating the constraints and the kernel parameter $\gamma = 1/\sigma^2$. The parameter C in the box constraint can be chosen by searching a wide range of values of C using either CV (usually, 10-fold) on $\mathcal{L}$ or an independent validation set of observations. In practice, it is usual to start the search by trying several different values of C, such as 10, 100, 1,000, 10,000, and so on. A initial grid of values of γ can be selected by trying out a crude set of possible values, say, 0.00001, 0.0001, 0.001, 0.01, 0.1, and 1.0.

When there appears to be a minimum CV misclassification rate within an interval of the two-way grid, we make the grid search finer within that interval. Armed with a two-way grid of values of (C, γ), we apply CV to estimate the generalization error for each cell in that grid. The (C, γ) that has the smallest CV misclassification rate is selected as the solution to the SVM classification problem.

11.3.7 Example: E-mail or Spam?

This example (spambase) was described in Section 8.4, where we applied LDA and QDA to a collection of 4,601 messages, comprising 1,813 spam e-mails and 2,788 non-spam e-mails. There are 57 variables (attributes) and each message is labeled as one of the two classes email or spam.

Here we apply nonlinear SVM (R package libsvm) using a Gaussian RBF kernel to the 4,601 messages. The SVM solution depends upon the cost C of violating the constraints and the variance, σ^2, of the Gaussian RBF kernel. After applying a trial-and-error method, we used the following grid of values for C amd $\gamma = 1/\sigma^2$:

$C = 10, 80, 100, 200, 500, 1,000$,

$\gamma = 0.00001(0.00001)0.0001(0.0001)0.002(0.001)0.01(0.01)0.04$.

In Figure 11.3, we plot the values of the 10-fold CV misclassification rate against the values of γ listed above, where each curve (connected set of points) represents a different value of C. For each C, we see that the CV/10 misclassification curves have similar shapes: a minimum value for γ very close to zero, and for values of γ away from zero, the curve trends upwards. In this initial search, we find a minimum CV/10 misclassification rate of 8.06% at $(C, \gamma) = (500, 0.0002)$ and $(1,000, 0.0002)$. We see that the general

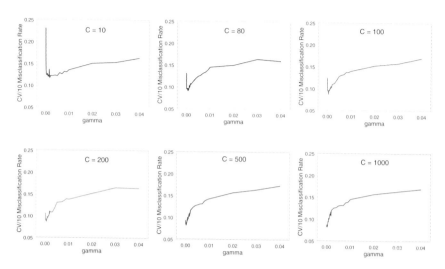

FIGURE 11.3. *SVM cross-validation misclassification rate curves for the spambase data. Initial grid search for the minimum 10-fold CV misclassification rate using $0.00001 \leq \gamma \leq 0.04$. The curves correspond to $C = 10$ (dark blue), 80 (brown), 100 (green), 200 (orange), 500 (light blue), and 1,000 (red). Within this intial grid search, the minimum CV/10 misclassification rate is 8.06%, which occurs at $(C, \gamma) = (500, 0.0002)$ and $(1,000, 0.0002)$.*

level of the misclassification rate tends to decrease as C increases and γ decreases together.

A detailed investigation of $C > 1000$ and γ close to zero reveals a minimum CV/10 misclassification rate of 6.91% at $C = 11,000$ and $\gamma = 0.00001$, corresponding to the following 10 CV estimates of the true classification rate:

0.9043, 0.9478, 0.9304, 0.9261, 0.9109,

0.9413, 0.9326, 0.9500. 0.9326, 0.9328.

This solution has 931 support vectors (482 e-mails, 449 spam), which means that a large percentage (79.8%) of the messages (82.7% of the e-mails and 75.2% of the spam) are not support points. Of the 4,601 messages, 2,697 e-mails and 1,676 spam are correctly classified (228 misclassified), yielding an apparent error rate of 4.96%.

This example turns out to be more computationally intensive than are the other binary-classification examples discussed in this chapter. Although the value of γ has very little effect on the speed of computating the 10-fold CV error rate, the speed of computation does depend upon C: as we increase the value of C, the speed of computation slows down considerably.

TABLE 11.2. *Summary of support vector machine (SVM) application to data sets for binary classification. Listed are the sample size (n), number of variables (r), and number of classes (K). Also listed for each data set is the 10-fold cross-validation (CV/10) misclassification rates corresponding to the best choice of (C, γ) for the SVM. The data sets are listed in increasing order of LDA misclassification rates (see Table 8.5).*

Data Set	n	r	K	SVM–CV/10
Breast cancer (logs)	569	30	2	0.0158
Spambase	4601	57	2	0.0691
Ionosphere	351	33	2	0.0427
Sonar	208	60	2	0.1010
BUPA liver disorders	345	6	2	0.2522

Also worth noting is that for fixed γ, increasing C reduces the number of support vectors and the apparent error rate. We cannot make similar general statements about fixed C and increasing γ; however, for fixed C, we generally see that the number of support vectors tends to increase (but not always) with increasing γ.

The nonlinear SVM is clearly a better classifier for this example than is LDA or QDA, whose leave-one-out CV misclassification rate is around 11% for LDA and 17% for QDA, but the amount of computational work involved in the grid search for the SVM solution is much greater and, hence, a lot more expensive.

11.3.8 Binary Classification Examples

We apply the SVM algorithm to the binary classification examples of Section 8.4: the log-transformed breast cancer data, the ionosphere data, the BUPA liver disorders data, the sonar data, and the spambase data. Except for spambase, computations for these examples were very fast.

In Table 11.2, we list the minimum 10-fold CV misclassification rate for each data set. Comparing these results to those of LDA (see Table 8.5, where we used leave-one-out CV), we see that SVM produces remarkable decreases in misclassification rates: the breast cancer rate decreased from 11.3% to 1.58%, the spambase rate decreased from 11.3% to 6.91%, the ionosphere rate decreased from 13.7% to 4.27%, the sonar rate decreased from 24.5% to 10.1%, and the BUPA liver disorders rate decreased from 30.1% to 25.22%.

11.3.9 SVM as a Regularization Method

The SVM classifier can also be regarded as the solution to a particular regularization problem. Let $f \in \mathcal{H}_K$, the reproducing kernel Hilbert space

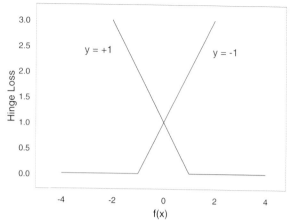

FIGURE 11.4. *Hinge loss function* $(1 - yf(\mathbf{x}))_+$ *for* $y = -1$ *and* $y = +1$.

(rkhs) associated with the kernel K, with $\| f \|_{\mathcal{H}_K}^2$ the squared-norm of f in $\mathcal{H}_K$.

Consider the classification error, $y_i - f(\mathbf{x}_i)$, where $y_i \in \{-1, +1\}$. Then,

$$|y_i - f(\mathbf{x}_i)| = |y_i(1 - y_i f(\mathbf{x}_i))| = |1 - y_i f(\mathbf{x}_i)| = (1 - y_i f(\mathbf{x}_i))_+, \quad (11.67)$$

$i = 1, 2, \ldots, n$, where $(x)_+ = \max\{x, 0\}$. The quantity $(1 - y_i f(\mathbf{x}_i))_+$, which could be zero if all $\mathbf{x}_i$ are correctly classified, is called the *hinge loss function* and is displayed in Figure 11.4. The hinge loss plays a vital role in SVM methodology; indeed, it has been shown to be Bayes consistent for classification in the sense that minimizing the loss function yields the Bayes rule (Lin, 2002). The hinge loss is also related to the misclassification loss function $I_{[y_i C(\mathbf{x}_i) \leq 0]} = I_{[y_i f(\mathbf{x}_i) \leq 0]}$. When $f(\mathbf{x}_i) = \pm 1$, the hinge loss is twice the misclassification loss; otherwise, the ratio of the two losses depends upon the sign of $y_i f(\mathbf{x}_i)$.

We wish to find a function $f \in \mathcal{H}_K$ to minimize a penalized version of the hinge loss. Specifically, we wish to find $f \in \mathcal{H}_K$ to

$$minimize \quad \frac{1}{n} \sum_{i=1}^{n} (1 - y_i f(\mathbf{x}_i))_+ + \lambda \| f \|_{\mathcal{H}_K}^2, \quad (11.68)$$

where $\lambda > 0$. In (11.69), the first term, $n^{-1} \sum_{i=1}^{n} (1 - y_i f(\mathbf{x}_i))_+$, measures the distance of the data from separability, and the second term, $\lambda \| f \|_{\mathcal{H}_K}^2$, penalizes overfitting. The tuning parameter λ balances the trade-off between estimating f (the first term) and how well f can be approximated (the second term). After the minimizing f has been found, the SVM classifier is $C(\mathbf{x}) = \text{sign}\{f(\mathbf{x})\}$, $\mathbf{x} \in \mathcal{R}^r$.

The optimizing criterion (11.68) is nondifferentiable due to the shape of the hinge-loss function. Fortunately, we can rewrite the problem in a slightly different form and thereby solve it.

We start from the fact that every $f \in \mathcal{H}$ can be written uniquely as the sum of two terms:

$$f(\cdot) = f^{\|}(\cdot) + f^{\perp}(\cdot) = \sum_{i=1}^{n} \alpha_i K(\mathbf{x}_i, \cdot) + f^{\perp}(\cdot), \qquad (11.69)$$

where $f^{\|} \in \mathcal{H}_K$ is the projection of f onto the subspace $\mathcal{H}_K$ of $\mathcal{H}$ and $f^{\perp}$ is in the subspace perpendicular to $\mathcal{H}_K$; that is, $\langle f^{\perp}(\cdot), K(\mathbf{x}_i, \cdot) \rangle_{\mathcal{H}} = 0$, $i = 1, 2, \ldots, n$. We can write $f(\mathbf{x}_i)$ via the reproducing property as follows:

$$f(\mathbf{x}_i) = \langle f(\cdot), K(\mathbf{x}_i, \cdot) \rangle = \langle f^{\|}(\cdot), K(\mathbf{x}_i, \cdot) \rangle + \langle f^{\perp}(\cdot), K(\mathbf{x}_i, \cdot) \rangle. \quad (11.70)$$

Because the second term on the rhs is zero, then,

$$f(\mathbf{x}) = \sum_{i=1}^{n} \alpha_i K(\mathbf{x}_i, \mathbf{x}), \qquad (11.71)$$

independent of $f^{\perp}$, where we used (11.69) and $\langle K(\mathbf{x}_i, \cdot), K(\mathbf{x}_j, \cdot) \rangle_{\mathcal{H}_K} = K(\mathbf{x}_i, \mathbf{x}_j)$. Now, from (11.69),

$$
\begin{aligned}
\| f \|_{\mathcal{H}_K}^2 &= \left\| \sum_i \alpha_i K(\mathbf{x}_i, \cdot) + f^{\perp} \right\|_{\mathcal{H}_K}^2 \\
&= \left\| \sum_i \alpha_i K(\mathbf{x}_i, \cdot) \right\|_{\mathcal{H}_K}^2 + \| f^{\perp} \|_{\mathcal{H}_K}^2 \\
&\geq \left\| \sum_i \alpha_i K(\mathbf{x}_i, \cdot) \right\|_{\mathcal{H}_K}^2, \qquad (11.72)
\end{aligned}
$$

with equality iff $f^{\perp} = 0$, in which case any $f \in \mathcal{H}_K$ that minimizes (11.68) admits a representation of the form (11.71). This important result is known as the *representer theorem* (Kimeldorf and Wahba, 1971); it says that the minimizing f (which would live in an infinite-dimensional rkhs if, for example, the kernel is a Gaussian RBF) can be written as a linear combination of a reproducing kernel evaluated at each of the n data points.

From (11.72), we have that $\| f \|_{\mathcal{H}_K}^2 = \sum_i \sum_j \alpha_i \alpha_j K(\mathbf{x}_i, \mathbf{x}_j) = \| \boldsymbol{\beta} \|^2$, where $\boldsymbol{\beta} = \sum_{i=1}^{n} \alpha_i \boldsymbol{\Phi}(\mathbf{x}_i)$. If the space $\mathcal{H}_K$ consists of linear functions of the form $f(\mathbf{x}) = \beta_0 + \boldsymbol{\Phi}(\mathbf{x})^{\tau} \boldsymbol{\beta}$ with $\| f \|_{\mathcal{H}_K}^2 = \| \boldsymbol{\beta} \|^2$, then the problem of finding f in (11.68) is equivalent to one of finding β_0 and $\boldsymbol{\beta}$ to

$$minimize \quad \frac{1}{n} \sum_{i=1}^{n} (1 - y_i(\beta_0 + \boldsymbol{\Phi}(\mathbf{x}_i)^{\tau}\boldsymbol{\beta}))_+ + \lambda \| \boldsymbol{\beta} \|^2 . \qquad (11.73)$$

Then, (11.68), which is nondifferentiable due to the hinge loss function, can be reformulated in terms of solving the 1-norm soft-margin optimization problem (11.34)–(11.35).

11.4   Multiclass Support Vector Machines

Often, data are derived from more than two classes. In the multiclass situation, $\mathbf{X} \in \Re^r$ is a random r-vector chosen for classification purposes and $Y \in \{1, 2, \ldots, K\}$ is a class label, where K is the number of classes. Because SVM classifiers are formulated for only two classes, we need to know if (and how) the SVM methodology can be extended to distinguish between $K > 2$ classes. There have been several attempts to define such a multiclass SVM strategy.

11.4.1   Multiclass SVM as a Series of Binary Problems

The standard SVM strategy for a multiclass classification problem (over K classes) has been to reduce it to a series of binary problems. There are different approachs to this strategy:

One-versus-rest: Divide the K-class problem into K binary classifica-
tion subproblems of the type "kth class" vs. "not kth class," $k = 1, 2, \ldots, K$. Corresponding to the kth subproblem, a classifier $\widehat{f}_k$ is constructed in which the kth class is coded as positive and the union of the other classes is coded as negative. A new $\mathbf{x}$ is then assigned to the class with the largest value of $\widehat{f}_k(\mathbf{x})$, $k = 1, 2, \ldots, K$, where $\widehat{f}_k(\mathbf{x})$ is the optimal SVM solution for the binary problem of the kth class versus the rest.

One-versus-one: Divide the K-class problem into $\binom{K}{2}$ comparisons of all pairs of classes. A classifier $\widehat{f}_{jk}$ is constructed by coding the jth class as positive and the kth class as negative, $j, k = 1, 2, \ldots, K$, $j \neq k$. Then, for a new $\mathbf{x}$, aggregate the votes for each class and assign $\mathbf{x}$ to the class having the most votes.

Even though these strategies are widely used in practice to resolve multiclass SVM classification problems, one has to be cautious about their use.

In Table 11.3, we report the CV/10 misclassification rates for one-versus-one multiclass SVM applied to the same data sets from Section 8.7. Also listed in Table 11.3 are the values of (C, γ) that yield the minimum misclassification rate for each data set. It is instructive to compare these rates with those in Table 8.7, where we used LDA and QDA. We see that for

TABLE 11.3. *Summary of support vector machine (SVM) "one-versus-one" classification results for data sets with more than two classes. Listed are the sample size (n), number of variables (r), and number of classes (K). Also listed for each data set is the 10-fold cross-validation (CV/10) misclassification rates corresponding to the best choice of (C, γ). The data sets are listed in increasing order of LDA misclassification rates (Table 8.7).*

Data Set	n	r	K	SVM–CV/10	C	γ
Wine	178	13	3	0.0169	10^6	8×10^{-8}
Iris	150	4	3	0.0200	100	0.002
Primate scapulae	105	7	5	0.0286	100	0.0002
Shuttle	43,500	8	7	0.0019	10	0.0001
Diabetes	145	5	3	0.0414	100	0.000009
Pendigits	10,992	16	10	0.0031	10	0.0001
E-coli	336	7	8	0.1280	10	1.0
Vehicle	846	18	4	0.1501	600	0.00005
Letter recognition	20,000	16	26	0.0183	50	0.04
Glass	214	9	6	0.0093	10	0.001
Yeast	1,484	8	10	0.3935	10	7.0

the shuttle, diabetes, pendigits, vehicle, letter recognition, glass, and yeast data sets, the SVM method performs better than does the LDA method; for the iris, primate scapulae, and e-coli data sets, the SVM and LDA methods perform about the same; and LDA performs better than does SVM for the wine data set. Thus, neither one-versus-one SVM nor LDA performs uniformly best for all of these data sets.

The one-versus-rest approach is popular for carrying out text categorization tasks, where each document may belong to more than one class. Although it enjoys the optimality property of the SVM method for each binary subproblem, it can yield a different classifier than the Bayes optimal classifier for the multiclass case. Furthermore, the classification success of the one-versus-rest approach depends upon the extent of the class-size imbalance of each subproblem and whether one class dominates all other classes when determining the most-probable class for each new **x**.

The one-versus-one approach, which uses only those observations belonging to the classes involved in each pairwise comparison, suffers from the problem of having to use smaller samples to train each classifier, which may, in turn, increase the variance of the solution.

11.4.2 A True Multiclass SVM

To construct a true multiclass SVM classifier, we need to consider all K classes, $\Pi_1, \Pi_2, \ldots, \Pi_K$, simultaneously, and the classifier has to reduce to the binary SVM classifier if $K = 2$. Here we describe the construction due to Lee, Lin, and Wahba (2004).

Let $\mathbf{v}_1, \ldots, \mathbf{v}_K$ be a sequence of K-vectors, where $\mathbf{v}_k$ has a 1 in the kth position and whose elements sum to zero, $k = 1, 2, \ldots, K$; that is, let

$$\mathbf{v}_1 = \left(1, -\frac{1}{K-1}, \cdots, -\frac{1}{K-1}\right)^{\tau}$$

$$\mathbf{v}_2 = \left(-\frac{1}{K-1}, 1, \cdots, -\frac{1}{K-1}\right)^{\tau}$$

$$\vdots$$

$$\mathbf{v}_K = \left(-\frac{1}{K-1}, -\frac{1}{K-1}, \cdots, 1\right)^{\tau}.$$

Note that if $K = 2$, then $\mathbf{v}_1 = (1, -1)^{\tau}$ and $\mathbf{v}_2 = (-1, 1)^{\tau}$. Every $\mathbf{x}_i$ can be labeled as one of these K vectors; that is, $\mathbf{x}_i$ has label $\mathbf{y}_i = \mathbf{v}_k$ if $\mathbf{x}_i \in \Pi_k$, $i = 1, 2, \ldots, n$, $k = 1, 2, \ldots, K$.

Next, we generalize the separating function $f(\mathbf{x})$ to a K-vector of separating functions,

$$\mathbf{f}(\mathbf{x}) = (f_1(\mathbf{x}), \cdots, f_K(\mathbf{x}))^{\tau}, \tag{11.74}$$

where

$$f_k(\mathbf{x}) = \beta_{0k} + h_k(\mathbf{x}), \quad h_k \in \mathcal{H}_K, \quad k = 1, 2, \ldots, K. \tag{11.75}$$

In (11.75), $\mathcal{H}_K$ is a reproducing-kernel Hilbert space (rkhs) spanned by the $\{K(\mathbf{x}_i, \cdot), i = 1, 2, \ldots, n\}$. For example, in the linear case, $h_k(\mathbf{x}) = \mathbf{x}^{\tau}\boldsymbol{\beta}_k$, for some vector of coefficients $\boldsymbol{\beta}_k$. We also assume, for uniqueness, that

$$\sum_{k=1}^{K} f_k(\mathbf{x}) = 0. \tag{11.76}$$

Let $\mathbf{L}(\mathbf{y}_i)$ be a K-vector with 0 in the kth position if $\mathbf{x}_i \in \Pi_k$, and 1 in all other positions; this vector represents the equal costs of misclassifying $\mathbf{x}_i$ (and allows for an unequal misclassification cost structure if appropriate). If $K = 2$ and $\mathbf{x}_i \in \Pi_1$, then $\mathbf{L}(\mathbf{y}_i) = (0, 1)^{\tau}$, while if $\mathbf{x}_i \in \Pi_2$, then $\mathbf{L}(\mathbf{y}_i) = (1, 0)^{\tau}$.

The multiclass generalization of the optimization problem (11.68) is, therefore, to find functions $\mathbf{f}(\mathbf{x}) = (f_1(\mathbf{x}), \cdots, f_K(\mathbf{x}))^{\tau}$ satisfying (11.76) which

$$minimize \quad I_{\lambda}(\mathbf{f}, \mathcal{Y}) = \frac{1}{n}\sum_{i=1}^{n}[\mathbf{L}(\mathbf{y}_i)]^{\tau}(\mathbf{f}(\mathbf{x}_i) - \mathbf{y}_i)_+ + \frac{\lambda}{2}\sum_{k=1}^{K}\| h_k \|^2, \tag{11.77}$$

where $(\mathbf{f}(\mathbf{x}_i) - \mathbf{y}_i)_+ = ((f_1(\mathbf{x}_i) - y_{i1})_+, \cdots, (f_K(\mathbf{x}_i) - y_{iK})_+)^{\tau}$ and $\mathcal{Y} = (\mathbf{y}_1, \cdots, \mathbf{y}_n)$ is a $(K \times n)$-matrix.

By setting $K = 2$, we can see that (11.77) is a generalization of (11.68). If $\mathbf{x}_i \in \Pi_1$, then $\mathbf{y}_i = \mathbf{v}_1 = (1, -1)^\tau$, and

$$
\begin{aligned}
[\mathbf{L}(\mathbf{y}_i)]^\tau (\mathbf{f}(\mathbf{x}_i) - \mathbf{y}_i)_+ &= (0, 1)((f_1(\mathbf{x}_i) - 1)_+, (f_2(\mathbf{x}_i) + 1)_+)^\tau \\
&= (f_2(\mathbf{x}_i) + 1)_+ \\
&= (1 - f_1(\mathbf{x}_i))_+,
\end{aligned}
\tag{11.78}
$$

while if $\mathbf{x}_i \in \Pi_2$, then $\mathbf{y}_i = \mathbf{v}_2 = (-1, 1)$, and

$$
[\mathbf{L}(\mathbf{y}_i)]^\tau (\mathbf{f}(\mathbf{x}_i) - \mathbf{y}_i)_+ = (f_1(\mathbf{x}_i) + 1)_+.
\tag{11.79}
$$

So, the first term (with f) in (11.68) is identical to the first term (with f_1) in (11.77) when $K = 2$. If we set $K = 2$ in the second term of (11.77), we have that

$$
\sum_{k=1}^{2} \| h_k \|^2 = \| h_1 \|^2 + \| -h_1 \|^2 = 2 \| h_1 \|^2,
\tag{11.80}
$$

so that the second terms of (11.68) and (11.77) are identical.

The function $h_k \in \mathcal{H}_K$ can be decomposed into two parts:

$$
h_k(\cdot) = \sum_{\ell=1}^{n} \beta_{\ell k} K(\mathbf{x}_\ell, \cdot) + h_k^\perp(\cdot),
\tag{11.81}
$$

where the $\{\beta_{\ell k}\}$ are constants and $h_k^\perp(\cdot)$ is an element in the rkhs orthogonal to $\mathcal{H}_K$. Substituting (11.76) into (11.77), then using (11.81), and rearranging terms, we have that

$$
f_K(\cdot) = -\sum_{k=1}^{K-1} \beta_{0k} - \sum_{k=1}^{K-1} \sum_{i=1}^{n} \beta_{ik} K(\mathbf{x}_i, \cdot) - \sum_{k=1}^{K-1} h_k^\perp(\cdot).
\tag{11.82}
$$

Because $K(\cdot, \cdot)$ is a reproducing kernel,

$$
\langle h_k, K(\mathbf{x}_i, \cdot) \rangle = h_k(\mathbf{x}_i), \quad i = 1, 2, \ldots, n,
\tag{11.83}
$$

and so,

$$
\begin{aligned}
f_k(\mathbf{x}_i) &= \beta_{0k} + h_k(\mathbf{x}_i) \\
&= \beta_{0k} + \langle h_k, K(\mathbf{x}_i, \cdot) \rangle \\
&= \beta_{0k} + \langle \sum_{\ell=1}^{n} \beta_{\ell k} K(\mathbf{x}_\ell, \cdot) + h_k^\perp(\cdot), K(\mathbf{x}_i, \cdot) \rangle \\
&= \beta_{0k} + \sum_{\ell=1}^{n} \beta_{\ell k} K(\mathbf{x}_\ell, \mathbf{x}_i).
\end{aligned}
\tag{11.84}
$$

Note that, for $k = 1, 2, \ldots, K - 1$,

$$
\begin{aligned}
\| h_k(\cdot) \|^2 &= \| \sum_{\ell=1}^{n} \beta_{\ell k} K(\mathbf{x}_\ell, \cdot) + h_k^\perp(\cdot) \|^2 \\
&= \sum_{\ell=1}^{n} \sum_{i=1}^{n} \beta_{\ell k} \beta_{ik} K(\mathbf{x}_\ell, \mathbf{x}_i) + \| h_k^\perp(\cdot) \|^2, \qquad (11.85)
\end{aligned}
$$

and, for $k = K$,

$$
\| h_K(\cdot) \|^2 = \| \sum_{k=1}^{K-1} \sum_{i=1}^{n} \beta_{ik} K(\mathbf{x}_i, \cdot) \|^2 + \| \sum_{k=1}^{K-1} h_k^\perp(\cdot) \|^2. \qquad (11.86)
$$

Thus, to minimize (11.86), we set $h_k^\perp(\cdot) = 0$ for all k.

From (11.84), the zero-sum constraint (11.76) becomes

$$
\bar{\beta}_0 + \sum_{\ell=1}^{n} \bar{\beta}_\ell K(\mathbf{x}_\ell, \cdot) = 0, \qquad (11.87)
$$

where $\bar{\beta}_0 = K^{-1} \sum_{k=1}^{K} \beta_{0k}$ and $\bar{\beta}_i = K^{-1} \sum_{k=1}^{K} \beta_{ik}$. At the n data points, $\{\mathbf{x}_i, i = 1, 2, \ldots, n\}$, (11.87) in matrix notation is given by

$$
\left(\sum_{k=1}^{K} \beta_{0k} \right) \mathbf{1}_n + \mathbf{K} \left(\sum_{k=1}^{K} \boldsymbol{\beta}_{\cdot k} \right) = \mathbf{0}, \qquad (11.88)
$$

where $\mathbf{K} = (K(\mathbf{x}_i, \mathbf{x}_j))$ is an $(n \times n)$ Gram matrix and $\boldsymbol{\beta}_{\cdot k} = (\beta_{1k}, \cdots, \beta_{nk})^\tau$. Let $\beta_{0k}^* = \beta_{0k} - \bar{\beta}_0$ and $\beta_{ik}^* = \beta_{ik} - \bar{\beta}_i$. Using (11.87), we see that the centered version of (11.84) is $f_k^*(\mathbf{x}_i) = \beta_{0k}^* + \sum_{\ell=1}^{n} \beta_{\ell k}^* K(\mathbf{x}_\ell, \mathbf{x}_i) = f_k(\mathbf{x}_i)$. Then,

$$
\sum_{k=1}^{K} \| h_k^*(\cdot) \|^2 = \sum_{k=1}^{K} \boldsymbol{\beta}_{\cdot k}^\tau \mathbf{K} \boldsymbol{\beta}_{\cdot k} - K \bar{\boldsymbol{\beta}}^\tau \mathbf{K} \bar{\boldsymbol{\beta}} \leq \sum_{k=1}^{K} \boldsymbol{\beta}_{\cdot k}^\tau \mathbf{K} \boldsymbol{\beta}_{\cdot k} = \sum_{k=1}^{K} \| h_k(\cdot) \|^2,
$$
$$(11.89)$$

where $\bar{\boldsymbol{\beta}} = (\bar{\beta}_1, \cdots, \bar{\beta}_n)^\tau$; if $\mathbf{K}\bar{\boldsymbol{\beta}} = \mathbf{0}$, the inequality becomes an equality and so $\sum_{k=1}^{K} \beta_{0k} = 0$. Thus,

$$
0 = K^2 \bar{\boldsymbol{\beta}}^\tau \mathbf{K} \bar{\boldsymbol{\beta}} = \| \sum_{i=1}^{n} (\sum_{k=1}^{K} \beta_{ik}) K(\mathbf{x}_i, \cdot) \|^2 = \| \sum_{k=1}^{K} \sum_{i=1}^{n} \beta_{ik} K(\mathbf{x}_i, \cdot) \|^2,
$$
$$(11.90)$$

whence, $\sum_{k=1}^{K} \sum_{i=1}^{n} \beta_{ik} K(\mathbf{x}_i, \mathbf{x}) = 0$, for all $\mathbf{x}$. Thus,

$$
\sum_{k=1}^{K} \left\{ \beta_{0k} + \sum_{i=1}^{n} \beta_{ik} K(\mathbf{x}_i, \mathbf{x}) \right\} = 0, \qquad (11.91)
$$

for every $\mathbf{x}$. So, minimizing (11.77) under the zero-sum constraint (11.76) only at the n data points is equivalent to minimizing (11.77) under the same constraint for every $\mathbf{x}$.

We next construct a Lagrangian formulation of the optimization problem (11.77) using the following notation. Let $\boldsymbol{\xi}_i = (\xi_{i1}, \cdots, \xi_{iK})^\tau$ be a K-vector of slack variables corresponding to $(f(\mathbf{x}_i) - y_i)_+$, $i = 1, 2, \ldots, n$, and let $(\boldsymbol{\xi}_{\cdot 1}, \cdots, \boldsymbol{\xi}_{\cdot K}) = (\boldsymbol{\xi}_1, \cdots, \boldsymbol{\xi}_n)^\tau$ be the $(n \times K)$-matrix whose kth column is $\boldsymbol{\xi}_{\cdot k}$ and whose ith row is $\boldsymbol{\xi}_i$. Let $(\mathbf{L}_1, \cdots, \mathbf{L}_K) = (\mathbf{L}(\mathbf{y}_1), \cdots, \mathbf{L}(\mathbf{y}_n))^\tau$ be the $(n \times K)$-matrix whose kth column is $\mathbf{L}_k$ and whose ith row is $\mathbf{L}(\mathbf{y}_i) = (L_{i1}, \cdots, L_{iK})$. Let $(\mathbf{y}_{\cdot 1}, \cdots, \mathbf{y}_{\cdot K}) = (\mathbf{y}_1, \cdots, \mathbf{y}_n)^\tau$ denote the $(n \times K)$-matrix whose kth column is $\mathbf{y}_{\cdot k}$ and whose ith row is $\mathbf{y}_i$.

The primal problem is to find $\{\beta_{0k}\}$, $\{\boldsymbol{\beta}_{\cdot k}\}$, and $\{\boldsymbol{\xi}_{\cdot k}\}$ to

$$minimize \quad \sum_{k=1}^K \mathbf{L}_k^\tau \boldsymbol{\xi}_{\cdot k} + \frac{n\lambda}{2} \sum_{k=1}^K \boldsymbol{\beta}_{\cdot k}^\tau \mathbf{K} \boldsymbol{\beta}_{\cdot k} \tag{11.92}$$

subject to

$$\beta_{0k} \mathbf{1}_n + \mathbf{K} \boldsymbol{\beta}_{\cdot k} - \mathbf{y}_{\cdot k} \ \leq\ \boldsymbol{\xi}_{\cdot k}, \quad k = 1, 2, \ldots, K, \tag{11.93}$$

$$\boldsymbol{\xi}_{\cdot k} \ \geq\ \mathbf{0}, \quad k = 1, 2, \ldots, K, \tag{11.94}$$

$$\left(\sum_{k=1}^K \beta_{0k}\right) \mathbf{1}_n + \mathbf{K}\left(\sum_{k=1}^K \boldsymbol{\beta}_{\cdot k}\right) \ =\ \mathbf{0}. \tag{11.95}$$

Form the primal functional $F_P = F_P(\{\beta_{0k}\}, \{\boldsymbol{\beta}_{\cdot k}\}, \{\boldsymbol{\xi}_{\cdot k}\})$, where

$$\begin{aligned}
F_P \ =\ & \sum_{k=1}^K \mathbf{L}_k^\tau \boldsymbol{\xi}_{\cdot k} + \frac{n\lambda}{2} \sum_{k=1}^K \boldsymbol{\beta}_{\cdot k}^\tau \mathbf{K} \boldsymbol{\beta}_{\cdot k} \\
& + \sum_{k=1}^K \boldsymbol{\alpha}_{\cdot k}^\tau (\beta_{0k} \mathbf{1}_n + \mathbf{K} \boldsymbol{\beta}_{\cdot k} - \mathbf{y}_{\cdot k} - \boldsymbol{\xi}_{\cdot k}) \\
& - \sum_{k=1}^K \boldsymbol{\gamma}_k^\tau \boldsymbol{\xi}_{\cdot k} + \boldsymbol{\delta}^\tau \left(\left(\sum_{k=1}^K \beta_{0k}\right)\mathbf{1}_n + \mathbf{K}\left(\sum_{k=1}^K \boldsymbol{\beta}_{\cdot k}\right) \right). \tag{11.96}
\end{aligned}$$

In (11.96), $\boldsymbol{\alpha}_{\cdot k} = (\alpha_{1k}, \cdots, \alpha_{nk})^\tau$ and $\boldsymbol{\gamma}_k$ are n-vectors of nonnegative Lagrange multipliers for the inequality constraints (11.93) and (11.94), respectively, and $\boldsymbol{\delta}$ is an n-vector of unconstrained Lagrange multipliers for the equality constraint (11.95).

Differentiating (11.96) with respect to β_{0k}, $\boldsymbol{\beta}_{\cdot k}$, and $\boldsymbol{\xi}_{\cdot k}$ yields

$$\frac{\partial F_P}{\partial \beta_{0k}} \ =\ (\boldsymbol{\alpha}_{\cdot k} + \boldsymbol{\delta})^\tau \mathbf{1}_n, \tag{11.97}$$

$$\frac{\partial F_P}{\partial \boldsymbol{\beta}_{\cdot k}} \ =\ n\lambda \mathbf{K} \boldsymbol{\beta}_{\cdot k} + \mathbf{K} \boldsymbol{\alpha}_{\cdot k} + \mathbf{K} \boldsymbol{\delta}, \tag{11.98}$$

$$\frac{\partial F_P}{\partial \boldsymbol{\xi}_{\cdot k}} = \mathbf{L}_k - \boldsymbol{\alpha}_{\cdot k} - \boldsymbol{\gamma}_k, \tag{11.99}$$

$$\boldsymbol{\alpha}_{\cdot k} \geq \mathbf{0}, \tag{11.100}$$

$$\boldsymbol{\gamma}_k \geq \mathbf{0}. \tag{11.101}$$

The Karush–Kuhn–Tucker complementarity conditions are

$$\boldsymbol{\alpha}_{\cdot k}(\beta_{0k}\mathbf{1}_n + \mathbf{K}\boldsymbol{\beta}_{\cdot k} - \mathbf{y}_{\cdot k} - \boldsymbol{\xi}_{\cdot k})^\tau = 0, \quad k = 1, 2, \ldots, K, \tag{11.102}$$

$$\boldsymbol{\gamma}_k \boldsymbol{\xi}_{\cdot k}^\tau = 0, \quad k = 1, 2, \ldots, K, \tag{11.103}$$

where, from (11.99), $\boldsymbol{\gamma}_k = \mathbf{L}_k - \boldsymbol{\alpha}_{\cdot k}$. Note that (11.102) and (11.103) are outer products of two column vectors, meaning that each of the n^2 elementwise products of those vectors are zero.

From (11.99) and (11.101), we have that $\mathbf{0} \leq \boldsymbol{\alpha}_{\cdot k} \leq \mathbf{L}_k$, $k = 1, 2, \ldots, K$. Suppose, for some i, $0 < \alpha_{ik} < L_{ik}$; then, $\gamma_{ik} > 0$, and, from (11.103), $\xi_{ik} = 0$, whence, from (11.102), $y_{ik} = \beta_{0k} + \sum_{\ell=1}^n \beta_{\ell k} K(\mathbf{x}_\ell, \mathbf{x}_i)$.

Setting the derivatives equal to zero for $k = 1, 2, \ldots, K$ yields $\boldsymbol{\delta} = -\bar{\boldsymbol{\alpha}} = -K^{-1}\sum_{k=1}^K \boldsymbol{\alpha}_{\cdot k}$ from (11.97), whence, $(\boldsymbol{\alpha}_{\cdot k} - \bar{\boldsymbol{\alpha}})^\tau \mathbf{1}_n = 0$, and, from (11.98), $\boldsymbol{\beta}_{\cdot k} = -(n\lambda)^{-1}(\boldsymbol{\alpha}_{\cdot k} - \bar{\boldsymbol{\alpha}})$, assuming that $\mathbf{K}$ is positive-definite. If $\mathbf{K}$ is not positive-definite, then $\boldsymbol{\beta}_{\cdot k}$ is not uniquely determined. Because (11.97), (11.98), and (11.99) are each zero, we construct the dual functional F_D by using them to remove a number of the terms of F_P.

The resulting dual problem is to find $\{\boldsymbol{\alpha}_{\cdot k}\}$ to

$$minimize \quad F_D = \frac{1}{2}\sum_{k=1}^K (\boldsymbol{\alpha}_{\cdot k} - \bar{\boldsymbol{\alpha}})^\tau \mathbf{K}(\boldsymbol{\alpha}_{\cdot k} - \bar{\boldsymbol{\alpha}}) + n\lambda\sum_{k=1}^K \boldsymbol{\alpha}_{\cdot k}^\tau \mathbf{y}_{\cdot k} \tag{11.104}$$

$$subject\ to \quad \mathbf{0} \leq \boldsymbol{\alpha}_{\cdot k} \leq \mathbf{L}_k, \quad k = 1, 2, \ldots, K, \tag{11.105}$$

$$(\boldsymbol{\alpha}_{\cdot k} - \bar{\boldsymbol{\alpha}})^\tau \mathbf{1}_n = 0, \quad k = 1, 2, \ldots, K. \tag{11.106}$$

From the solution, $\{\widehat{\boldsymbol{\alpha}}_{\cdot k}\}$, to this quadratic programming problem, we set

$$\widehat{\boldsymbol{\beta}}_{\cdot k} = -(n\lambda)^{-1}(\widehat{\boldsymbol{\alpha}}_{\cdot k} - \widehat{\boldsymbol{\alpha}}), \tag{11.107}$$

where $\widehat{\boldsymbol{\alpha}} = K^{-1}\sum_{k=1}^K \widehat{\boldsymbol{\alpha}}_{\cdot k}$.

The multiclass classification solution for a new $\mathbf{x}$ is given by

$$C_k(\mathbf{x}) = \arg\max_k \{\widehat{f}_k(\mathbf{x})\}, \tag{11.108}$$

where

$$\widehat{f}_k(\mathbf{x}) = \widehat{\beta}_{0k} + \sum_{\ell=1}^n \widehat{\beta}_{\ell k} K(\mathbf{x}_\ell, \mathbf{x}), \quad k = 1, 2, \ldots, K. \tag{11.109}$$

Suppose the row vector $\widehat{\boldsymbol{\alpha}}_i = (\widehat{\alpha}_{i1}, \cdots, \widehat{\alpha}_{iK}) = \mathbf{0}$ for $(\mathbf{x}_i, \mathbf{y}_i)$; then, from (11.107), $\widehat{\boldsymbol{\beta}}_i = (\widehat{\beta}_{i1}, \cdots, \widehat{\beta}_{iK}) = \mathbf{0}$. It follows that the term $\widehat{\beta}_{ik}K(\mathbf{x}_i, \mathbf{x}) = 0$, $k = 1, 2, \ldots, K$. Thus, any term involving $(\mathbf{x}_i, \mathbf{y}_i)$ does not appear in (11.109); in other words, it does not matter whether $(\mathbf{x}_i, \mathbf{y}_i)$ is or is not included in the learning set $\mathcal{L}$ because it has no effect on the solution. This result leads us to a definition of support vectors: an observation $(\mathbf{x}_i, \mathbf{y}_i)$ is called a *support vector* if $\widehat{\boldsymbol{\beta}}_i = (\widehat{\beta}_{i1}, \cdots, \widehat{\beta}_{iK}) \neq \mathbf{0}$. As in the binary SVM solution, it is in our computational best interests for there to be relatively few support vectors for any given application.

The one issue remaining is the choice of tuning parameter λ (and any other parameters involved in the computation of the kernel). A *generalized approximate cross-validation* (GACV) method is derived in Lee, Lin, and Wahba (2004) based upon an approximation to the leave-one-out cross-validation technique used for penalized-likelihood methods. The basic idea behind GACV is the following. Write (11.77) as

$$I_\lambda(\mathbf{f}, \mathcal{Y}) = n^{-1} \sum_{i=1}^n g(\mathbf{y}_i, \mathbf{f}(\mathbf{x}_i)) + J_\lambda(\mathbf{f}), \qquad (11.110)$$

where $g(\mathbf{y}_i, \mathbf{f}(\mathbf{x}_i)) = [\mathbf{L}(\mathbf{y}_i)]^\tau (\mathbf{f}(\mathbf{x}_i) - \mathbf{y}_i)_+$ and $J_\lambda(\mathbf{f}) = (\lambda/2) \sum_{i=1}^n \| h_j \|^2$. Let $\mathbf{f}_\lambda = \arg\min_{\mathbf{f}} I_\lambda(\mathbf{f}, \mathcal{Y})$ and let $\mathbf{f}_\lambda^{(-i)}$ denote that $\mathbf{f}_\lambda$ that yields the minimum of $I_\lambda(\mathbf{f}, \mathcal{Y})$ by omitting the ith observation $(\mathbf{x}_i, \mathbf{y}_i)$ from the first term in (11.110). If we write

$$g(\mathbf{y}_i, \mathbf{f}_\lambda^{(-i)}(\mathbf{x}_i)) = g(\mathbf{y}_i, \mathbf{f}_\lambda(\mathbf{x}_i)) + [g(\mathbf{y}_i, \mathbf{f}_\lambda^{(-i)}(\mathbf{x}_i)) - g(\mathbf{y}_i, \mathbf{f}_\lambda(\mathbf{x}_i))], \quad (11.111)$$

then the λ that minimizes $n^{-1} \sum_{i=1}^n g(\mathbf{y}_i, \mathbf{f}_\lambda^{(-i)}(\mathbf{x}_i))$ is found by using a suitable approximation of $D(\lambda) = n^{-1} \sum_{i=1}^n [g(\mathbf{y}_i, \mathbf{f}_\lambda^{(-i)}(\mathbf{x}_i)) - g(\mathbf{y}_i, \mathbf{f}_\lambda(\mathbf{x}_i))]$, computed over a grid of values of λ.

This solution of the multiclass SVM problem has been found to be successful in simulations and in analyzing real data. Comparisons of various multiclass classification methods, such as multiclass SVM, "all-versus-rest," LDA, and QDA, over a number of data sets show that no one classification method appears to be superior for all situations studied; performance appears to depend upon the idiosyncrasies of the data to be analyzed.

11.5 Support Vector Regression

The SVM was designed for classification. Can we extend (or generalize) the idea to regression? How would the main concepts used in SVM — convex optimization, optimal separating hyperplane, support vectors, margin, sparseness of the solution, slack variables, and the use of kernels — translate to the regression situation? It turns out that all of these concepts find

their analogues in regression analysis and they add a different view to the
topic than the views we saw in Chapter 5.

11.5.1 ϵ-Insensitive Loss Functions

In SVM classification, the margin is used to determine the amount of
separation between two nonoverlapping classes of points: the bigger the
margin, the more confident we are that the optimal separating hyperplane is
a superior classifier. In regression, we are not interested in separating points
but in providing a function of the input vectors that would track the points
closely. Thus, a regression analogue for the margin would entail forming a
"band" or "tube" around the true regression function that contains most
of the points. Points not contained within the tube would be described
through slack variables. In formulating these ideas, we first need to define
an appropriate loss function.

We define a loss function that ignores errors associated with points falling
within a certain distance (e.g., $\epsilon > 0$) of the true linear regression function,

$$\mu(\mathbf{x}) = \beta_0 + \mathbf{x}^T \boldsymbol{\beta}. \tag{11.112}$$

In other words, if the point $(\mathbf{x}, y)$ is such that $|y - \mu(\mathbf{x})| \leq \epsilon$, then the loss
is taken to be zero; if, on the other hand, $|y - \mu(\mathbf{x})| > \epsilon$, then we take the
loss to be $|y - \mu(\mathbf{x})| - \epsilon$.

With this strategy in mind, we can define the following two types of loss
function:

- $L_1^\epsilon(y, \mu(\mathbf{x})) = \max\{0, |y - \mu(\mathbf{x})| - \epsilon\}$,

- $L_2^\epsilon(y, \mu(\mathbf{x})) = \max\{0, (y - \mu(\mathbf{x}))^2 - \epsilon\}$.

The first loss function, L_1^ϵ, is called the *linear ϵ-insensitive loss function*,
and the second, L_2^ϵ, is the *quadratic ϵ-insensitive loss function*. The two
loss functions, linear (red curve) and quadratic (blue curve), are graphed
in Figure 11.5. We see that the linear loss function ignores all errors falling
within $\pm\epsilon$ of the true regression function $\mu(\mathbf{x})$ while dampening in a linear
fashion errors that fall outside those limits.

11.5.2   Optimization for Linear ϵ-Insensitive Loss

We define slack variables ξ_i and ξ_j' in the following way. If the point
$(\mathbf{x}_i, y_i)$ lies above the ϵ-tube, then $\xi_i' = y_i - \mu(\mathbf{x}_i) - \epsilon \geq 0$, whereas if the
point $(\mathbf{x}_j, y_j)$ lies below the ϵ-tube, then $\xi_j = \mu(\mathbf{x}_j) - \epsilon - y_j \geq 0$. For
points that fall outside the ϵ-tube, the values of the slack variables depend

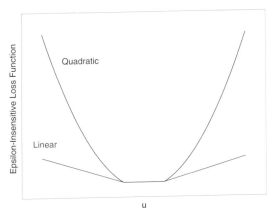

FIGURE 11.5. *The linear ϵ-insensitive loss function (red curve) and the quadratic ϵ-insensitive loss function (blue curve) for support vector regression. Plotted are $L_i(u) = \max\{0, |u|^i - \epsilon\}$ vs. u, $i = 1, 2$, where $u = y - \mu(\mathbf{x})$. For the linear loss function, the "flat" part of the curve has width 2ϵ.*

upon the shape of the loss function; for points inside the ϵ-tube, the slack variables have value zero.

For linear ϵ-insensitive loss, the primal optimization problem is to find β_0, $\boldsymbol{\beta}$, $\boldsymbol{\xi}' = (\xi_1', \cdots, \xi_n')^{\tau}$, and $\boldsymbol{\xi} = (\xi_1, \cdots, \xi_n)^{\tau}$ to

$$minimize \quad \frac{1}{2} \parallel \boldsymbol{\beta} \parallel^2 + C \sum_{i=1}^{n} (\xi_i + \xi_i') \qquad (11.113)$$

$$subject\ to \quad \begin{aligned} y_i - (\beta_0 + \mathbf{x}_i^{\tau}\boldsymbol{\beta}) &\leq \epsilon + \xi_i', \\ (\beta_0 + \mathbf{x}_i^{\tau}\boldsymbol{\beta}) - y_i &\leq \epsilon + \xi_i, \\ \xi_i' \geq 0,\ \xi_i \geq 0, &\quad i = 1, 2, \ldots, n. \end{aligned} \qquad (11.114)$$

The constant $C > 0$ exists to balance the flatness of the function μ against our tolerance of deviations larger than ϵ. Notice that because ϵ is found only in the constraints, the solution to this optimization problem has to incorporate a band around the regression function.

Form the primal Lagrangian,

$$\begin{aligned} F_P \quad = \quad &\frac{1}{2} \parallel \boldsymbol{\beta} \parallel^2 + C \sum_{i=1}^{n} (\xi_i + \xi_i') - \sum_i a_i \{y_i - (\beta_0 + \mathbf{x}_i^{\tau}\boldsymbol{\beta}) - \epsilon - \xi_i'\} \\ &- \sum_i b_i \{(\beta_0 + \mathbf{x}_i^{\tau}\boldsymbol{\beta}) - y_i - \epsilon - \xi_i\} \\ &- \sum_i c_i \xi_i' - \sum_i d_i \xi_i, \qquad (11.115) \end{aligned}$$

where $a_i, b_i, c_i,$ and d_i, $i = 1, 2, \ldots, n$, are the Lagrange multipliers. This, in turn, implies that a_i, b_i, c_i, d_i, $i = 1, 2, \ldots, n$, are all nonnegative. The derivatives are

$$\frac{\partial F_P}{\partial \beta_0} = \sum_i a_i - \sum_i b_i \tag{11.116}$$

$$\frac{\partial F_P}{\partial \beta} = \beta + \sum_i a_i \mathbf{x}_i - \sum_i b_i \mathbf{x}_i \tag{11.117}$$

$$\frac{\partial F_P}{\partial \xi_i} = C + b_i - d_i \tag{11.118}$$

$$\frac{\partial F_P}{\partial \xi_i'} = C + a_i - c_i \tag{11.119}$$

Setting these derivatives equal to zero for a stationary solution yields:

$$\boldsymbol{\beta}^* = \sum_i (b_i - a_i)\mathbf{x}_i, \tag{11.120}$$

$$\sum_i (b_i - a_i) = 0, \tag{11.121}$$

$$C + b_i - d_i = 0, \quad C + a_i - c_i = 0, \quad i = 1, 2, \ldots, n. \tag{11.122}$$

The expression (11.120) is known as the *support vector expansion* because $\boldsymbol{\beta}^*$ can be written as a linear combination of the input vectors $\{\mathbf{x}_i\}$. Setting $\boldsymbol{\beta} = \boldsymbol{\beta}^*$ in the true regression equation (11.112) gives us

$$\mu^*(\mathbf{x}) = \beta_0 + \sum_{i=1}^{n} (b_i - a_i)(\mathbf{x}^\tau \mathbf{x}_i). \tag{11.123}$$

Substituting $\boldsymbol{\beta}^*$ into the primal Lagrangian and using (11.120) and (11.121) gives us the dual problem: find $\mathbf{a} = (a_1, \cdots, a_n)^\tau$, $\mathbf{b} = (b_1, \cdots, b_n)^\tau$ to

$$maximize \quad F_D = (\mathbf{b} - \mathbf{a})^\tau \mathbf{y} - \epsilon(\mathbf{b} + \mathbf{a})^\tau \mathbf{1}_n$$

$$- \frac{1}{2}(\mathbf{b} - \mathbf{a})^\tau \mathbf{K}(\mathbf{b} - \mathbf{a}) \tag{11.124}$$

$$subject\ to \quad \mathbf{0} \leq \mathbf{a}, \mathbf{b} \leq C\mathbf{1}_n, \quad (\mathbf{b} - \mathbf{a})^\tau \mathbf{1}_n = 0, \tag{11.125}$$

where $\mathbf{K} = (\langle \mathbf{x}_i, \mathbf{x}_j \rangle)$ for linear SVM. The Karush–Kuhn–Tucker complementarity conditions state that the products of the dual variables and the constraints are all zero:

$$a_i(\beta_0 + \mathbf{x}_i^\tau \boldsymbol{\beta} - y_i - \epsilon - \xi_i) = 0, \quad i = 1, 2, \ldots, n, \tag{11.126}$$

$$b_i(y_i - \beta_0 - \mathbf{x}_i^\tau \boldsymbol{\beta} - \epsilon - \xi_i') = 0, \quad i = 1, 2, \ldots, n, \tag{11.127}$$

$$\xi_i \xi_i' = 0, \quad a_i b_i = 0, \quad i = 1, 2, \ldots, n, \tag{11.128}$$

$$(a_i - C)\xi_i = 0, \quad (b_i - C)\xi_i' = 0, \quad i = 1, 2, \ldots, n. \tag{11.129}$$

In practice, the value of ϵ is usually taken to be around 0.1.

The solution to this optimization problem produces a linear function of $\mathbf{x}$ accompanied by a band or tube of $\pm\epsilon$ around the function. Points that do not fall inside the tube are the *support vectors*.

11.5.3 Extensions

The optimization problem using quadratic ϵ-insensitive loss can be solved in a similar manner; see Exercise 11.3.

If we formulate this problem using nonlinear transformations of the input vectors, $\mathbf{x} \rightarrow \boldsymbol{\Phi}(\mathbf{x})$, to a feature space defined by the kernel $K(\mathbf{x}, \mathbf{y})$, then the stationary solution (11.120) is replaced by

$$\boldsymbol{\beta}^* = \sum_{i=1}^{n} (b_i - a_i)\boldsymbol{\Phi}(\mathbf{x}_i), \tag{11.130}$$

the inner product $\langle \mathbf{x}_i, \mathbf{x}_j \rangle = \mathbf{x}_i^T \mathbf{x}_j$ in (11.120) is replaced by the more general kernel function,

$$K(\mathbf{x}_i, \mathbf{x}_j) = \langle \boldsymbol{\Phi}(\mathbf{x}_i), \boldsymbol{\Phi}(\mathbf{x}_j) \rangle = \boldsymbol{\Phi}(\mathbf{x}_i)^T \boldsymbol{\Phi}(\mathbf{x}_j), \tag{11.131}$$

the matrix $\mathbf{K} = (K(\mathbf{x}_i, \mathbf{x}_j))$ replaces the matrix $\mathbf{K}$ in (11.124), and the SVM regression function (11.122) becomes

$$\mu^*(\mathbf{x}) = \beta_0 + \sum_{i=1}^{n} (b_i - a_i)K(\mathbf{x}, \mathbf{x}_i); \tag{11.132}$$

see Exercise 11.4. Note that $\boldsymbol{\beta}^*$ in (11.130) does not have an explicit representation as it has in (11.120).

11.6 Optimization Algorithms for SVMs

When a data set is small, general-purpose linear programming (LP) or quadratic programming (QP) optimizers work quite well to solve SVM problems; QP optimizers can solve problems having about a thousand points, whereas LP optimizers can deal with hundreds of thousands of points. With large data sets, however, a more sophisticated approach is required.

The main problem when computing SVMs for very large data sets is that storing the entire kernel in main memory dramatically slows down computation. Alternative algorithms, constructed for the specific task of overcoming such computational inefficiencies, are now available in certain SVM software.

We give only brief descriptions of some of these algorithms. The simplest procedure for solving a convex optimization problem is that of gradient ascent:

Gradient Ascent: Start with an initial estimate of the $\boldsymbol{\alpha}$-coefficient vector and then successively update $\boldsymbol{\alpha}$ one α-coefficient at a time using the steepest ascent algorithm.

A problem with this approach is that the solution for $\boldsymbol{\alpha} = (\alpha_1, \cdots, \alpha_n)^{\tau}$ has to satisfy the linear constraint $\boldsymbol{\alpha}^{\tau}\mathbf{y} = \sum_{i=1}^{n} \alpha_i y_i = 0$. Carrying out a non-trivial one-at-a-time update of each α-component (while holding the remaining αs constant at their current values) will violate this constraint, and the solution at each iteration will fall outside the feasible region. The minimum number of αs that can be changed at each iteration is two.

More complicated (but also more efficient) numerical techniques for large learning data sets are now available in many SVM software packages. Examples of such advanced techniques include "chunking," decomposition, and sequential minimal optimization. Each method builds upon certain common elements: (1) choose a subset of the learning set $\mathcal{L}$, (2) monitor closely the KKT optimality conditions to discover which points not in the subset violate the conditions, and (3) apply a suitable optimizing strategy. These strategies are

Chunking: Start with an arbitrary subset (called the "working set" or "chunk") of size 100–500 of the learning set $\mathcal{L}$; use a general LP or QP optimizer to train an SVM on that subset and keep only the support vectors; apply the resulting classifier to all the remaining data in $\mathcal{L}$ and sort the misclassified points by how badly they violate the KKT conditions; add to the support vectors found previously a predetermined number of those points that most violate the KKT conditions; iterate until all points satisfy the KKT conditions. The general optimizer and the point selection process make this algorithm slow and inefficient.

Decomposition: Similar to chunking, except that at each iteration, the size of the subset is always the same; adding new points to the subset means that an equal number of old points must be removed.

Sequential Minimal Optimization (SMO): An extreme version of the decomposition algorithm, whereby the subset consists of only two points at each iteration (see above comments related to the gradient ascent algorithm). These two αs are found at each iteration by using a heuristic argument and then updated so that the constraint $\boldsymbol{\alpha}^{\tau}\mathbf{y} = \sum_{i=1}^{n} \alpha_i y_i = 0$ is satisfied and the solution is found within the feasible region.

TABLE 11.4. *Some implementations of SVM.*

Package	Implementation
SVM^{light}	http://svmlight.joachims.org/
LIBSVM	http://csie.ntu.edu.tw/~cjlin/libsvm/
SVMTorch II	http://www.idiap.ch/machine-learning.php
SVMsequel	http://www.isi.edu/~hdaume/SVMsequel/
TinySVM	http://chasen.org/~taku/TinySVM/

A big advantage of SMO (Platt, 1999) is that the algorithm has an analytical solution and so does not need to refer to a general QP optimizer; it also does not need to store the entire kernel matrix in memory. Although more iterations are needed, SMO is much faster than the other algorithms. The SMO algorithm has been improved in many ways for use with massive data sets.

11.7 Software Packages

There are several software packages for computing SVMs. Many are available for downloading over the Internet. See Table 11.4 for a partial list. Most of these SVM packages use similar data-input formats and command lines.

The most popular SVM package is SVM^{light} by Thorsten Joachims; it is very fast and can carry out classification and regression using a variety of kernels and is used for text classification. It is often used as the basis for other SVM software packages.

The C++-based package LIBSVM by C.-C. Chang and C.-J. Lin, which carries out classification and regression, is based upon SMO and SVM^{light}, and has interfaces to MATLAB, python, perl, ruby, S-Plus (function svm in library libsvm), and R (function svm in library e1071); see Venables and Ripley (2002, pp. 344–346). SVMTorch II is an extremely fast C++ program for classification and regression that can handle more than 20,000 observations and more than 100 input variables. SVMsequel is a very fast program that handles classification problems, a variety of kernels (including string kernels), and enormous data sets. TinySVM, which supports C++, perl, ruby, python, and Java interfaces, is based upon SVM^{light}, carries out classification and regression, and can deal with very large data sets.

Bibliographical Notes

There are several excellent references on support vector machines. Our primary references include the books by Vapnik (1998, 2000), Cristianini and Shawe-Taylor (2000), Shawe-Taylor and Cristianini (2004, Chapter 7), Schölkopf and Smola (2002), and Hastie, Tibshirani, and Friedman (2001, Section 4.5 and Chapter 12) and the review articles by Burges (1998), Schölkopf and Smola (2003), and Moguerza and Munoz (2006). An excellent book on convex optimization is Boyd and Vandenberghe (2004).

Most of the theoretical work on kernel functions goes back to about the beginning of the 1900s. The idea of using kernel functions as inner products was introduced into machine learning by Aizerman, Braverman, and Rozoener (1964). Kernels were then put to work in SVM methodology by Boser, Guyon, and Vapnik (1992), who borrowed the "kernel" name from the theory of integral operators.

Our description of string kernels for text categorization is based upon Lodhi, Saunders, Shawe-Taylor, Cristianini, and Watkins (2002). See also Shawe-Taylor and Cristianini (2004, Chapter 11). For applications of SVM to text categorization, see the book by Joachims (2002) and Cristianini and Shawe-Taylor (2000, Section 8.1).

Exercises

11.1 (a) Show that the perpendicular distance of the point (h, k) to the line $f(x, y) = ax + by + c = 0$ is $\pm (ah + bk + c)/\sqrt{a^2 + b^2}$, where the sign chosen is that of c.

(b) Let $\mu(\mathbf{x}) = \beta_0 + \mathbf{x}^\tau \boldsymbol{\beta} = 0$ denote a hyperplane, where $\beta_0 \in \Re$ and $\boldsymbol{\beta} \in \Re^r$, and let $\mathbf{x}_k \in \Re^r$ be a point in the space. By minimizing $\| \mathbf{x} - \mathbf{x}_k \|^2$ subject to $\mu(\mathbf{x}) = 0$, show that the perpendicular distance from the point to the hyperplane is $|\mu(\mathbf{x}_k)|/ \| \boldsymbol{\beta} \|$.

11.2 In the support vector regression problem using a quadratic ϵ-insensitive loss function, formulate and solve the resulting optimization problem.

11.3 The "2-norm soft margin" optimization problem for SVM classification: Consider the regularization problem of minimizing $\frac{1}{2} \| \boldsymbol{\beta} \|^2 + C \sum_{i=1}^n \xi_i^2$ subject to the constraints $y_i(\beta_0 + \mathbf{x}_i^\tau \boldsymbol{\beta}) \geq 1 - \xi_i$, and $\xi \geq 0$, for $i = 1, 2, \ldots, n$.

(a) Show that the same optimal solution to this problem is reached if we remove the constraints $\xi_i \geq 0$, $i = 1, 2, \ldots, n$, on the slack variables. (Hint: What is the effect on the objective functional if this constraint is violated?)

(b) Form the primal Lagrangian F_P, which will be a function of β_0, $\boldsymbol{\beta}$, $\boldsymbol{\xi}$, and the Lagrangian multipliers $\boldsymbol{\alpha}$. Differentiate F_P wrt β_0, $\boldsymbol{\beta}$, and $\boldsymbol{\xi}$, set the results equal to zero, and solve for a stationary solution.

(c) Substitute the results from (b) into the primal Lagrangian to obtain the dual objective functional F_D. Write out the dual problem (objective functional and constraints) in matrix notation. Maximize the dual wrt $\boldsymbol{\alpha}$. Use the Karush–Kuhn–Tucker complementary conditions $\alpha_i \{ y_i (\beta_0 + \mathbf{x}_i^T \boldsymbol{\beta}) - (1 - \xi_i) \} = 0$ for $i = 1, 2, \ldots, n$.

(d) If $\boldsymbol{\alpha}^*$ is the solution to the dual problem, find $\widehat{\boldsymbol{\beta}}$ and its norm, which gives the width of the margin.

11.4 For the support vector regression problem in a feature space defined by a general kernel function K representing the inner product of pairs of nonlinearly transformed input vectors, formulate and solve the resulting optimization problem using (a) a linear ϵ-insensitive loss function and (b) a quadratic ϵ-insensitive loss function.

11.5 In the support vector regression problem, let $\epsilon = 0$. Consider the quadratic (2-norm) primal optimization problem,

$minimize \quad \lambda \| \boldsymbol{\beta} \|^2 + \sum_{i=1}^{n} \xi_i^2$

$subject \ to \quad y_i - \mathbf{x}_i^T \boldsymbol{\beta} = \xi_i, \quad i = 1, 2, \ldots, n.$

Form the Lagrangian, differentiate wrt $\boldsymbol{\beta}$ and ξ_i, $i = 1, 2, \ldots, n$, and set the results equal to zero for a stationary solution. Substitute these values into the primal functional to get the dual problem. Use $\mathbf{K}$ to represent the Gram matrix with entries either $K_{ij} = \mathbf{x}_i^T \mathbf{x}_j$ or $K_{ij} = K(\mathbf{x}_i, \mathbf{x}_j)$. Differentiate the dual functional wrt the Lagrange multipliers $\boldsymbol{\alpha}$, and set the result equal to zero. Show that this solution is related to ridge regression (see Section 5.7.4).

11.6 Let $\mathbf{x}, \mathbf{y} \in \Re^2$. Consider the polynomial kernel function, $K(\mathbf{x}, \mathbf{y}) = \langle \mathbf{x}, \mathbf{y} \rangle^2$, so that $r = 2$ and $d = 2$. Find two different maps $\boldsymbol{\Phi} : \Re^2 \to \mathcal{H}$ for $\mathcal{H} = \Re^3$.

11.7 Let $z \in \Re$ and define the $(2m + 1)$-dimensional $\boldsymbol{\Phi}$-mapping,

$$\boldsymbol{\Phi}(z) = (2^{-1/2}, \cos z, \cdots, \cos mz, \sin z, \cdots, \sin mz)^\tau.$$

Using this mapping, show that the kernel $K(x, y) = \langle \boldsymbol{\Phi}(x), \boldsymbol{\Phi}(y) \rangle$, $x, y \in \Re$, reduces to the *Dirichlet kernel* given by

$$K(x, y) = \frac{\sin((m + \frac{1}{2})\delta)}{2 \sin(\delta/2)},$$

where $\delta = x - y$.

11.8 Show that the homogeneous polynomial kernel, $K(\mathbf{x}, \mathbf{y}) = \langle \mathbf{x}, \mathbf{y} \rangle^d$, satisfies Mercer's condition (11.54).

11.9 If K_1 and K_2 are kernels and $c_1, c_2 \geq 0$ are real numbers, show that the following functions are kernels:

(a) $c_1 K_1(\mathbf{x}, \mathbf{y}) + c_2 K_2(\mathbf{x}, \mathbf{y})$;

(b) $K_1(\mathbf{x}, \mathbf{y}) K_2(\mathbf{x}, \mathbf{y})$;

(c) $\exp\{K_1(\mathbf{x}, \mathbf{y})\}$.

(Hint: In each case, you have to show that the function is nonnegative-definite.)

11.10 Prove that in finite-dimensional input space, a symmetric function $K(\mathbf{x}, \mathbf{y})$ is a kernel function iff $\mathbf{K} = (K(\mathbf{x}_i, \mathbf{x}_j))$ is a nonnegative-definite matrix with nonnegative eigenvalues. (Hint: Use the symmetry and the spectral theorem for $\mathbf{K}$ to show that K is a kernel. Then, show that for a negative eigenvalue, the squared-norm of any point $\mathbf{z} \in \mathcal{H}$ is negative, which is impossible.)

11.11 Show that the functional $F_D(\boldsymbol{\alpha})$ in (11.40) is convex; i.e., show that, for $\theta \in (0, 1)$ and $\boldsymbol{\alpha}, \boldsymbol{\beta} \in \Re^n$,

$$F_D(\theta\boldsymbol{\alpha} + (1 - \theta)\boldsymbol{\beta}) \leq \theta F_D(\boldsymbol{\alpha}) + (1 - \theta)F_D(\boldsymbol{\beta}).$$

11.12 Apply nonlinear-SVM to a binary classification data set of your choice. Make up a two-way table of values of (C, γ) and for each cell in that table compute the CV/10 misclassification rate. Find the pair (C, γ) with the smallest CV/10 misclassification rate. Compare this rate with results obtained using LDA and that using a classification tree.

12
Cluster Analysis

12.1 Introduction

Cluster analysis, which is the most well-known example of *unsupervised learning*, is a very popular tool for analyzing unstructured multivariate data. Within the data-mining community, cluster analysis is also known as *data segmentation*, and within the machine-learning community, it is also known as *class discovery*. The methodology consists of various algorithms each of which seeks to organize a given data set into homogeneous subgroups, or "clusters." There is no guarantee that more than one such group can be found; however, in any practical application, the underlying hypothesis is that the data form a heterogeneous set that should separate into natural groups familiar to the domain experts.

Clustering is a statistical tool for those who need to arrange large quantities of multivariate data into natural groups. For example, marketers use demographics and consumer profiles in an attempt to segment the marketplace into small, homogeneous groups so that promotional campaigns may be carried out more efficiently; biologists divide organisms into hierarchical orders in order to describe the notion of biological diversity; financial managers categorize corporations into different types based upon relevant financial characteristics; archaeologists group artifacts (e.g., broaches) found in

A.J. Izenman, *Modern Multivariate Statistical Techniques*,
doi: 10.1007/978-0-387-78189-1_12,
© Springer Science+Business Media, LLC 2008

graves in order to understand movements of ancient peoples; physicians use medical records to cluster patients for treatment diagnosis; and audiologists use repeated utterances of specific words by different speakers to provide a basis for speaker recognition. There are many other similar examples,

Cluster analysis resembles methods for classifying items; yet the two data analytic methods are philosophically different from each other. First, in classification, it is known a priori how many classes or groups are present in the data and which items are members of which class or group; in cluster analysis, the number of classes is unknown and so is the membership of items into classes. Second, in classification, the objective is to classify new items (possibly in the form of a test set) into one of the given classes based upon experience obtained using a learning set of data; clustering falls more into the framework of exploratory data analysis, where no prior information is available regarding the class structure of the data. Third, classification deals almost exclusively with classifying observations, whereas clustering can be applied to clustering observations or variables or both observations and variables simultaneously, depending upon the context.

Methods for clustering items (either observations or variables) depend upon how similar (or dissimilar) the items are to each other. Similar items are treated as a homogeneous class or group, whereas dissimilar items form additional classes or groups. Much of the output of a cluster analysis is visual, with the results displayed as scatterplots, trees, dendrograms, silhouette plots, and heatmaps.

12.1.1 What Is a Cluster?

This is a difficult question to answer mainly because there is no universally accepted definition of exactly what constitutes a cluster. As a result, the various clustering methods usually do not produce identical or even similar solutions.

A cluster is generally thought of as a group of *items* (*objects, points*) in which each item is "close" (in some appropriate sense) to a central item of a cluster and that members of different clusters are "far away" from each other. In a sense, then, clusters can be viewed as "high-density regions" of some multidimensional space (Hartigan, 1975). Such a notion seems fine on the surface if clusters are to be thought of as convex elliptical regions.

However, it is not difficult to conceive of situations in which natural clusterings of items do not follow this pattern. When the dimension of a space is large enough, these multidimensional items, plotted as points in that space, may congregate in clusters that curve and twist around each other; even if the various swarms of points are non-overlapping (which is unlikely), the oddly shaped configurations of points may be almost impossible to detect and identify using current techniques.

12.1.2 Example: Old Faithful Geyser Eruptions

The data for this example[1] is a set of 107 bivariate observations, that were taken from a study of the eruptions of Old Faithful Geyser in Yellowstone National Park, Wyoming (Weisberg, 1985, p. 231). A geyser is a hot spring which occasionally becomes unstable and erupts hot water and steam into the air. Old Faithful Geyser is the most famous of all geysers and is an extremely popular tourist attraction. The variables measured are duration of eruption (X_1) and waiting time until the next eruption (X_2), both recorded in minutes, for all eruptions of Old Faithful Geyser between 6 a.m. and midnight, 1–8 August 1978. Prior to clustering, one could argue that there are two or three possible clusters in the data.

Because the two variables are measured on very different scales (the standard deviations of X_1 and X_2 being approximately 1 and 13, respectively), the derived clusters (using any clustering algorithm) are completely determined by X_2, the interval between eruptions; the observations are divided into clusters by straight-line boundaries parallel to the horizontal axis. Without standardizing both variables, we cannot obtain a realistic partitioning of the data. So, for this example, we standardize the variables prior to clustering.

The results of this clustering study, where we set the number of clusters to be two or three for each method, are displayed in Figure 12.1. The most interesting result is that "perfect" clustering (according to our intuition) for both two and three clusters is accomplished only by the single-linkage, hierarchical agglomerative method (see first row of Figure 12.1). If we use the single-linkage results as the gold standard, we see that average-linkage and complete-linkage methods (second row), which produced the same results for two and three clusters, had one incorrect allocation for two clusters and three incorrect allocations for three clusters. Although both of the nonhierarchical clustering methods, pam and K-means (third row), had perfect clustering for two clusters, they performed poorly for three clusters, where they both had 45 incorrectly allocations.

12.2 Clustering Tasks

There are numerous ways of clustering a data set of n independent measurements on each of r correlated variables.

Clustering Observations: When we speak about "clustering," we usually think of clustering the n observations into groups, where the

---

[1] The data can be found in the file geyser on the book's website.

$K = 2$ $K = 3$

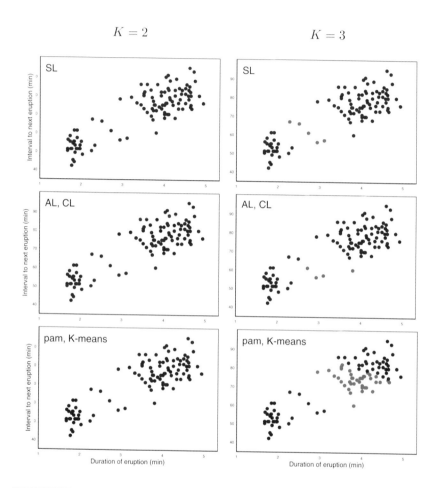

FIGURE 12.1. *Clustering results for Old Faithful Geyser data. The scatterplots in the left column panels are solutions for $K = 2$ classes, with red and blue as the two cluster colors. The scatterplots in the right column panels are solutions for $K = 3$ classes, with red, green, and blue as the three cluster colors. The first row is the single-linkage (SL) solutions, the second row is both average-linkage (AL) and complete-linkage (CL) solutions, the third row is both* pam *and K-means solutions.*

number, K, of groups is unknown and has to be determined from the data. When analyzing microarray data, the observations may be, for example, tissue samples, disease types, or experimental conditions, and so this task is often referred to as "clustering samples."

Clustering Variables: We may wish to partition the p variables into K distinct groups, where the number K is unknown and has to be determined from the data. A group may be determined by using only one variable; however, most clusters will be formed using several variables. These clusters should be far enough apart (in some sense) that groupings are easily identifiable. Each cluster of variables may later be replaced by a single variable representative of that cluster. When analyzing microarray data, the variables are genes, and so we refer to this task as "gene clustering."

Two-Way Clustering: Instead of clustering the variables or the observations separately, it might in certain circumstances be more appropriate to cluster them both simultaneously. Two-way clustering is known by different names, such as "block clustering" or "direct clustering." This goal is especially appropriate in microarray studies, where it is desired to cluster genes and tissue samples at the same time to show which subset of genes is most closely related to which subset of disease types.

NOTE: Because many of the clustering algorithms can be applied to observations or variables (or both simultaneously), it will often be convenient in this chapter to use the generic word "item" when a distinction between observation or variable is unnecessary.

12.3 Hierarchical Clustering

There are two types of hierarchical clustering methods: *agglomorative* and *divisive*. Agglomerative clustering algorithms, often called "bottom-up" methods, start with each item being its own cluster; then, clusters are successively merged, until only a single cluster remains. Divisive clustering algorithms, often called "top-down" methods, do the opposite: they start with all items as members of a single cluster; then, that cluster is split into two separate clusters, and so on for every successive cluster, until each item is its own cluster. Most attention in the clustering literature has been on agglomerative methods; however, arguments have been made that divisive methods can provide more sophisticated and robust clusterings.

12.3.1 Dendrogram

The end result of all hierarchical clustering methods is a *dendrogram* (i.e., hierarchical tree diagram), where the k-cluster solution is obtained by merging some of the clusters from the $(k + 1)$-cluster solution. The dendrogram may be drawn horizontal or vertical, depending upon user choice or software decision; both types give the same information. In this discussion, we assume a vertical dendrogram.

The dendrogram allows the user to read off the "height" of the linkage criterion at which items or clusters or both are combined together to form a new, larger cluster. Items that are similar to each other are combined at low heights, whereas items that are more dissimilar are combined higher up the dendrogram. Thus, it is the difference in heights that defines how close items are to each other. The greater the distance between heights at which clusters are combined, the more readily we can identify substantial structure in the data.

A partition of the data into a specified number of groups can be obtained by "cutting" the dendrogram at an appropriate height. If we draw a horizontal line on the dendrogram at a given height, then the number, K, of vertical lines cut by that horizontal line identifies a K-cluster solution; the intersection of the horizontal line and one of those K vertical lines then represents a cluster, and the items located at the end of all branches below that intersection constitute the members of the cluster.

Unlike the vertical distances, which are crucial in defining a solution, the horizontal distances between items are irrelevant; the software that draws a dendrogram is generally written so that the dendrogram can be easily interpreted. For large data sets, however, this goal becomes impossible.

12.3.2 Dissimilarity

The basic tool for hierarchical clustering is a measure of the *dissimilarity* or *proximity* (i.e., distance) of one item relative to another item. Which definition of distance is used in any given application is often a matter of subjective choice. Let $\mathbf{x}_i, \mathbf{x}_j \in \Re^r$. Dissimilarities usually satisfy the following three properties:

1. $d(\mathbf{x}_i, \mathbf{x}_j) \geq 0$;

2. $d(\mathbf{x}_i, \mathbf{x}_i) = 0$;

3. $d(\mathbf{x}_j, \mathbf{x}_i) = d(\mathbf{x}_i, \mathbf{x}_j)$.

Such dissimilarities are termed *metric* or *ultrametric* according to whether they satisfy a fourth property. A metric dissimilarity satisfies

4a. $d(\mathbf{x}_i, \mathbf{x}_j) \leq d(\mathbf{x}_i, \mathbf{x}_k) + d(\mathbf{x}_k, \mathbf{x}_j)$, and an ultrametric dissimilarity satisfies

4b. $d(\mathbf{x}_i, \mathbf{x}_j) \leq \max\{d(\mathbf{x}_i, \mathbf{x}_k), d(\mathbf{x}_j, \mathbf{x}_k)\}$.

Ultrametric dissimilarities can be displayed graphically by a dendrogram.

There are several ways to define a dissimilarity, the most popular being *Euclidean distance* and *Manhattan city-block distance.*

Let $\mathbf{x}_i = (x_{i1}, \cdots, x_{ir})^\tau$ and $\mathbf{x}_j = (x_{j1}, \cdots, x_{jr})^\tau$ denote two points in $\Re^r$. Then, these dissimilarity measures are defined as follows:

Euclidean: $d(\mathbf{x}_i, \mathbf{x}_j) = [(\mathbf{x}_i - \mathbf{x}_j)^\tau (\mathbf{x}_i - \mathbf{x}_j)]^{1/2} = \left[\sum_{k=1}^{r} (x_{ik} - x_{jk})^2\right]^{1/2}$.

Manhattan: $d(\mathbf{x}_i, \mathbf{x}_j) = \sum_{k=1}^{r} |x_{ik} - x_{jk}|$.

Minkowski: $d_m(\mathbf{x}_i, \mathbf{x}_j) = \left[\sum_{k=1}^{r} |x_{ik} - x_{jk}|^m\right]^{1/m}$.

In some applications, *squared-Euclidean distance* is used. Minkowski distance includes as special cases Euclidean distance ($m = 2$) and Manhattan distance ($m = 1$).

These dissimilarity measures are all computed using raw data, not standardized data. Standardization is usually recommended when the variability of the variables is quite different: a larger variability will have a more pronounced affect upon the clustering procedure than will a variable with relatively low variability.

A dissimilarity measure used for clustering variables is

1-correlation: $d(\mathbf{x}_i, \mathbf{x}_j) = 1 - \rho_{ij} = 1 - s_{ij}/s_i s_j$,

where $-1 \leq \rho_{ij} \leq 1$ is the correlation between the pair of variables X_i and X_j. Here, $s_{ij} = \sum_{k=1}^{r} (x_{ik} - \bar{x}_i)(x_{jk} - \bar{x}_j)$, $s_i = [\sum_{k=1}^{r} (x_{ik} - \bar{x}_i)^2]^{1/2}$, $s_2 = [\sum_{k=1}^{r} (x_{jk} - \bar{x}_j)^2]^{1/2}$, and $\bar{x}_\ell = r^{-1} \sum_{\ell=1}^{r} x_{\ell k}$, $\ell = i, j$. A relatively large absolute value of ρ_{ij} suggests the variables are "close" to each other, whereas a small correlation ($\rho_{ij} \approx 0$) suggests the variables are "far away" from each other. Thus, $1 - \rho_{ij}$ is taken as a measure of "dissimilarity" between the variables.

Given n observations, $\mathbf{x}_1, \ldots, \mathbf{x}_n \in \Re^r$, the starting point of any hierarchical clustering procedure is to compute the pairwise dissimilarities between observations and then arrange them into a symmetric, $(n \times n)$ *proximity matrix,* $\mathbf{D} = (d_{ij})$, where $d_{ij} = d(\mathbf{x}_i, \mathbf{x}_j)$, with zeroes along the diagonal. If we are clustering variables, the proximity matrix $\mathbf{D} = (d_{ij})$ is a symmetric, $(r \times r)$-matrix with ijth dissimilarity $d_{ij} = 1 - \rho_{ij}$.

12.3.3 Agglomerative Nesting (agnes)

Table 12.1 lists the algorithm for agglomerative hierarchical clustering. The most popular of these clustering methods are referred to as *single-linkage* (or *nearest-neighbor*), *complete-linkage* (or *farthest-neighbor*), and a compromise between these two, *average-linkage* methods. Each of these clustering methods is defined by the way in which two clusters (which may be single items) are combined or "joined" to form a new, larger cluster. Single linkage uses a minimum-distance metric between clusters, complete linkage uses a greatest-distance metric, and average linkage computes the average distance between all pairs of items within the two different clusters, one item from each cluster. There is also a weighted version of average linkage, where the weights reflect the (possibly disparate) sizes of the clusters in question.

No one of these algorithms is uniformly best for all clustering problems. Whereas the dendrograms from single-linkage and complete-linkage methods are invariant under monotone transformations of the pairwise dissimilarities, this property does not hold for the average-linkage method. Single-linkage often leads to long "chains" of clusters, joined by singleton points near each other, a result that does not have much appeal in practice, whereas complete-linkage tends to produce many small, compact clusters. Average linkage is dependent upon the size of the clusters, whereas single and complete linkage, which depend only upon the smallest or largest dissimilarity, respectively, do not.

12.3.4 A Worked Example

To understand agglomerative hierarchical clustering, we give a detailed analysis of a small example. Consider the following $n = 8$ bivariate points:

$$\mathbf{x}_1 = (1, 3)^\tau, \mathbf{x}_2 = (2, 4)^\tau, \mathbf{x}_3 = (1, 5)^\tau, \mathbf{x}_4 = (5, 5)^\tau,$$
$$\mathbf{x}_5 = (5, 7)^\tau, \mathbf{x}_6 = (4, 9)^\tau, \mathbf{x}_7 = (2, 8)^\tau, \mathbf{x}_8 = (3, 10)^\tau.$$

A scatterplot of these points is given in Figure 12.2 (top-left panel). Using Euclidean distance, the upper-triangular portion of the symmetric, (8×8)-matrix $\mathbf{D}^{(1)}$ is as follows:

	1	2	3	4	5	6	7	8
1	0	1.414	2.000	4.472	5.657	6.708	5.099	7.280
2		0	1.414	3.162	4.243	5.385	4.000	6.083
3			0	4.000	4.472	5.000	3.162	5.385
4				0	2.000	4.123	4.243	5.385
5					0	2.236	3.162	3.606
6						0	2.236	1.414
7							0	2.236
8								0

TABLE 12.1. *Algorithm for agglomerative hierarchical clustering.*

---

1. Input: $\mathcal{L} = \{\mathbf{x}_i, i = 1, 2, \ldots, n\}$, n = number of clusters, each cluster of which contains one item.

2. Compute $\mathbf{D} = (d_{ij})$, the $(n \times n)$-matrix of dissimilarities between the n clusters, where $d_{ij} = d(\mathbf{x}_i, \mathbf{x}_j)$, $i, j = 1, 2, \ldots, n$.

3. Find the smallest dissimilarity, say, d_{IJ}, in $\mathbf{D} = \mathbf{D}^{(1)}$. Merge clusters I and J to form a new cluster IJ.

4. Compute dissimilarities, $d_{IJ,K}$, between the new cluster IJ and all other clusters $K \neq IJ$. These dissimilarities depend upon which linkage method is used. For all clusters $K \neq I, J$, we have the following linkage options:

   Single linkage: $d_{IJ,K} = \min\{d_{I,K}, d_{J,K}\}$.

   Complete linkage: $d_{IJ,K} = \max\{d_{I,K}, d_{J,K}\}$.

   Average linkage: $d_{IJ,K} = \sum_{i \in IJ} \sum_{k \in K} d_{ik}/(N_{IJ} N_K)$,

   where N_{IJ} and N_K are the numbers of items in clusters IJ and K, respectively.

5. Form a new $((n-1) \times (n-1))$-matrix, $\mathbf{D}^{(2)}$, by deleting rows and columns I and J and adding a new row and column IJ with dissimilarities computed from step 4.

6. Repeat steps 3, 4, and 5 a total of $n - 1$ times. At the ith step, $\mathbf{D}^{(i)}$ is a symmetric $((n-i+1) \times (n-i+1))$-matrix, $i = 1, 2, \ldots, n$. At the last step $(i = n)$, $\mathbf{D}^{(n)} = 0$, and all items are merged together into a single cluster.

7. Output: List of which clusters are merged at each step, the value (or *height*) of the dissimilarity of each merge, and a dendrogram to summarize the clustering procedure.

Single Linkage. The smallest dissimilarity is $d_{12} = d_{23} = d_{68} = 1.414$. We choose to merge $\mathbf{x}_2$ and $\mathbf{x}_3$ to form the new cluster "23." We next compute new dissimilarities, $d_{23,K} = \min\{d_{2K}, d_{3K}\}$ for $K = 1, 4, 5, 6, 7, 8$. The (7×7)-matrix $\mathbf{D}^{(2)}$ is given by the following:

	1	23	4	5	6	7	8
1	0	1.414	4.472	5.657	6.708	5.099	7.280
23		0	3.162	4.243	5.000	3.162	5.385
4			0	2.000	4.123	4.243	5.385
5				0	2.236	3.162	3.606
6					0	2.236	1.414
7						0	2.236
8							0

The smallest dissimilarity is $d_{1,23} = d_{68} = 1.414$. We choose to merge $\mathbf{x}_1$ with the "23" cluster, producing a new cluster "123." We next compute new dissimilarities, $d_{123,K} = \min\{d_{12,K}, d_{3K}\}$ for $K = 4, 5, 6, 7, 8$. The

(6×6)-matrix $\mathbf{D}^{(3)}$ is as follows:

	123	4	5	6	7	8
123	0	3.162	4.243	5.000	3.162	5.385
4		0	2.000	4.123	4.243	5.385
5			0	2.236	3.162	3.606
6				0	2.236	1.414
7					0	2.236
8						0

The smallest dissimilarity is $d_{68} = 1.414$, and so we merge $\mathbf{x}_6$ and $\mathbf{x}_8$ to form the new cluster "68." We compute new dissimilarities, $d_{68,K} = \min\{d_{6K}, d_{8K}\}$ for $K = 123, 4, 5, 7$. This gives us the (5×5)-matrix $\mathbf{D}^{(4)}$,

	123	4	5	6	7
123	0	3.162	4.243	5.000	3.162
4		0	2.000	4.123	4.243
5			0	2.236	3.162
68				0	2.236
7					0

The smallest dissimilarity is $d_{45} = 2.0$, and so we merge $\mathbf{x}_4$ and $\mathbf{x}_5$ to form the new cluster "45." We compute new dissimilarities, $d_{45,K} = \min\{d_{4K}, d_{5K}\}$ for $K = 123, 68, 7$. This gives the (4×4)-matrix $\mathbf{D}^{(5)}$,

	123	45	6	7
123	0	3.162	5.000	3.162
45		0	2.236	4.243
68			0	2.236
7				0

The smallest dissimilarity is $d_{45,68} = d_{68,7} = 2.236$. We choose to merge the cluster "68" with $\mathbf{x}_7$ to produce the new cluster "678." The new dissimilarities, $d_{678,K} = \min\{d_{68,K}, d_{7K}\}$ for $K = 123, 45$, yield the matrix $\mathbf{D}^{(6)}$,

	123	45	678
123	0	3.162	3.162
45		0	2.236
678			0

The smallest dissimilarity is $d_{45,678} = 2.236$, so the next merge is the cluster "45" with the cluster "678." The matrix $\mathbf{D}^{(7)}$ is

	123	45678
123	0	3.162
45678		0

The last merge is cluster "123" with cluster "45678," and the merging dissimilarity is $d_{123,45678} = 3.162$. The dendrogram is displayed in the top-right panel of Figure 12.2.

Complete Linkage. Complete linkage uses the same idea as single linkage, but instead of taking the smallest dissimilarity as the distance measure between clusters, we take the largest such dissimilarity. From $\mathbf{D}^{(1)}$ given

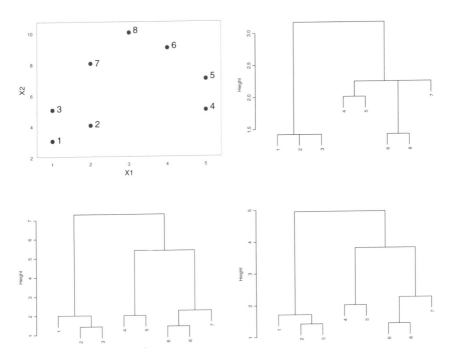

FIGURE 12.2. *Agglomerative hierarchical clustering for worked example using Euclidean distance. Top-left panel: Scatterplot of eight bivariate points. Other panels show dendrograms showing hierarchical clusters and value of Euclidean distance at merge points. Top-right panel: Single linkage. Bottom-left panel: Complete linkage. Bottom-right panel: Average linkage.*

previously, we merge $\mathbf{x}_2$ and $\mathbf{x}_3$ to form the "23" cluster at height 1.414, as before. Using Euclidean distance (but omitting square-roots in the presentation), the upper-triangular portion of the (7×7)-matrix $\mathbf{D}^{(2)}$ is as follows:

	1	23	4	5	6	7	8
1	0	2.0	4.472	5.657	6.708	5.099	7.280
23		0	4.000	4.472	5.385	4.000	6.083
4			0	2.000	4.123	4.243	5.385
5				0	2.236	3.162	3.606
6					0	2.236	1.414
7						0	2.236
8							0

The smallest dissimilarity is $d_{68} = 1.414$. We merge $\mathbf{x}_6$ and $\mathbf{x}_8$ to form a new cluster "68." We compute new dissimilarities, $d_{68,K} = \max\{d_{6K}, d_{8K}\}$

for $K = 1, 23, 4, 5, 7$. This gives us a (6×6)-matrix $\mathbf{D}^{(3)}$,

	1	23	4	5	68	7
1	0	2.000	4.472	5.657	7.280	5.099
23		0	4.000	4.472	6.083	4.000
4			0	2.000	4.123	4.243
5				0	2.236	3.162
68					0	2.236
7						0

The smallest dissimilarity is $d_{1,23} = d_{45} = 2.0$. We choose to merge the cluster "23" with $\mathbf{x}_1$ to form a new cluster "123." We compute new dissimilarities, $d_{123,K} = \max\{d_{12,K}, d_{3K}\}$ for $K = 4, 5, 68, 7$. This gives us a new (5×5)-matrix $\mathbf{D}^{(4)}$,

	123	4	5	68	7
123	0	4.472	5.657	7.280	5.099
4		0	2.000	5.385	4.243
5			0	3.606	3.162
68				0	2.236
7					0

The smallest dissimilarity is $d_{45} = 2.0$. We merge $\mathbf{x}_4$ and $\mathbf{x}_5$ to form a new cluster "45." We compute dissimilarities, $d_{45,K} = \max\{d_{4K}, d_{5K}\}$ for $K = 123, 68, 7$. This gives us a new (4×4)-matrix $\mathbf{D}^{(5)}$,

	123	45	68	7
123	0	5.657	7.280	5.099
45		0	5.385	4.243
68			0	2.236
7				0

The smallest dissimilarity is $d_{68,7} = 2.236$. We merge cluster "68" with $\mathbf{x}_7$ to form the new cluster "678." New dissimilarities $d_{678,K} = \max\{d_{68,K}, d_{7K}\}$ are computed for $K = 123, 45$ to give the new (3×3)-matrix $\mathbf{D}^{(6)}$,

	123	45	678
123	0	5.657	7.280
45		0	5.385
678			0

The last steps merge the clusters "45" and "678" with a merging value of $d_{45,678} = 5.385$, and then the clusters "123" and "45678" with a merging value of $d_{123,45678} = 7.280$. The dendrogram is displayed in the bottom-left panel of Figure 12.2.

Average Linkage. For average linkage, the distance between two clusters is found by computing the average dissimilarity of each item in the first cluster to each item in the second cluster.

We start with the matrix $\mathbf{D}^{(1)}$. The smallest dissimilarity is $d_{12} = \sqrt{2} = 1.414$, and so we merge $\mathbf{x}_1$ and $\mathbf{x}_2$ to form cluster "12." We compute dissimilarities between the cluster "12" and all other points using the average distance, $d_{12,K} = (d_{1K} + d_{2K})/2$, for $K = 3, 4, 5, 6, 7, 8$. For example,

$d_{12,3} = (d_{13} + d_{23})/2 = (\sqrt{4} + \sqrt{2})/2 = 1.707$. The matrix $\mathbf{D}^{(2)}$ is given by

	12	3	4	5	6	7	8
12	0	1.707	3.817	4.950	6.047	4.550	6.681
3		0	4.000	4.472	5.000	3.162	5.385
4			0	2.000	4.123	4,243	5.385
5				0	2.236	3.162	3.606
6					0	2.236	1.414
7						0	2.236
8							0

The smallest dissimilarity is $d_{68} = 1.414$, and so we merge $\mathbf{x}_6$ and $\mathbf{x}_8$ to form the new cluster "68." We compute dissimilarities between the cluster "68" and all other points and clusters using the average distance, $d_{68,12} = (d_{16} + d_{26} + d_{18} + d_{28})/4 = 6.364$, and $d_{68,K} = (d_{6K} + d_{8K})/2$, for $K = 3, 4, 5, 7$. The matrix $\mathbf{D}^{(3)}$ is

	12	3	4	5	68	7
12	0	1.707	3.817	4.950	6.364	4.550
3		0	4.000	4.472	5.193	3.162
4			0	2.000	4.754	4,243
5				0	2.921	3.162
68					0	2.236
7						0

The smallest dissimilarity is $d_{12,3} = 1.707$, and so we merge $\mathbf{x}_3$ and the cluster "12" to form the new cluster "123." We compute dissimilarities between the cluster "123" and all other points using the average distance, $d_{123,68} = (d_{16} + d_{18} + d_{26} + d_{28} + d_{36} + d_{38})/6 = 5.974$ and $d_{123,K} = (d_{1K} + d_{2K} + d_{3K})/3$, for $K = 4, 5, 7$. This gives the matrix $\mathbf{D}^{(4)}$:

	123	4	5	68	7
123	0	3.878	4.791	5.974	4.087
4		0	2.000	4.754	4.243
5			0	2.921	3.162
68				0	2.236
7					0

The smallest dissimilarity is $d_{45} = 2.0$, and so we merge $\mathbf{x}_4$ and $\mathbf{x}_5$ to form the new cluster "45." We compute dissimilarities between the cluster "45" and the other clusters as before. This gives the matrix $\mathbf{D}^{(5)}$:

	123	45	68	7
123	0	4.334	5.974	4.087
45		0	3.837	3.702
68			0	2.236
7				0

The smallest dissimilarity is $d_{68,7} = 2.236$, and so we merge $\mathbf{x}_7$ and the cluster "68" to form the new cluster "678." This gives the matrix $\mathbf{D}^{(6)}$:

	123	45	678
123	0	4.334	5.345
45		0	3.792
678			0

The smallest dissimilarity is $d_{45,678} = 3.782$, and so we merge the two clusters "45" and "678" to form a new cluster "45678." We merge the last two clusters and compute their dissimilarity $d_{123,45678} = 4.940$. The dendrogram is displayed in the bottom-right panel of Figure 12.2.

12.3.5 Divisive Analysis (diana)

The most-used divisive hierarchical clustering procedure is that proposed by MacNaughton-Smith, Williams, Dale, and Mockett (1964).

The idea is that at each step, the items are divided into a "splinter" group (say, cluster A) and the "remainder" (say, cluster B). The splinter group is initiated by extracting that item that has the largest average dissimilarity from all other items in the data set; that item is set up as cluster A. Given this separation of the data into A and B, we next compute, for each item in cluster B, the following two quantities: (1) the average dissimilarity between that item and all other items in cluster B, and (2) the average dissimilarity between that item and all items in cluster A. Then, we compute the difference (1)–(2) for each item in B. If all differences are negative, we stop the algorithm. If any of these differences are positive (indicating that the item in B is closer on average to cluster A than to the other items in cluster B), we take the item in B with the largest positive difference, move it to A, and repeat the procedure. This algorithm provides a binary split of the data into two clusters A and B. This same procedure can then be used to obtain binary splits of each of the clusters A and B separately.

The dendrogram corresponding to divisive hierarchical clustering of the worked example is displayed in Figure 12.3. Compare the result with that of the various agglomerative hierarchical clustering options in Figure 12.2. The major difference we see is that $\mathbf{x}_4$ is now included in the cluster with items $\mathbf{x}_1, \mathbf{x}_2$, and $\mathbf{x}_3$, rather than in the other cluster.

12.3.6 Example: Primate Scapular Shapes

This example is a small part of a much larger study (Ashton, Oxnard, and Spence, 1965) on measurements of the scapulae (shoulder bones) from 30 genera covering most of the primate order. The data[2] used in this example consist of measurements on the scapulae of five genera of adult primates

---

[2]The author thanks Charles Oxnard and Rebecca German for providing him with these data. The data can be found in the file primate.scapulae on the book's website.

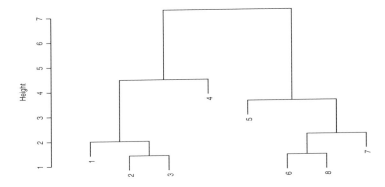

FIGURE 12.3. *Divisive hierarchical clustering for the worked example using Euclidean distance.*

representing Hominoidea; that is, gibbons (*Hylobates*), orangutans (*Pongo*), chimpanzees (*Pan*), gorillas (*Gorilla*), and man (*Homo*).

The measurements consist of indices and angles that are related to scapular shape, but not to functional meaning. Other studies showed that gender differences for such measurements were not statistically significant, and so no attempt was made by the authors of the study to divide the specimens by gender. Interest centered upon determining the extent to which these scapular shape measurements could be useful in classifying living primates.

There are eight variables in this data set, of which the first five (AD.BD, AD.CD, EA.CD, Dx.CD, and SH.ACR) are indices and the last three (EAD, β, and γ) are angles. Of the 105 measurements on each variable, 16 were taken on *Hylobates* scapulae, 15 on *Pongo* scapulae, 20 on *Pan* scapulae, 14 on *Gorilla* scapulae, and 40 on *Homo* scapulae. The angle γ was not available for *Homo* and, thus, was not used in this example.

Agglomerative and divisive hierarchical methods were employed for clustering the scapulae data using all five indices and two of the angles (EAD and β). Figure 12.4 shows dendrograms from the single-linkage, average-linkage, and complete-linkage agglomerative hierarchical methods and the dendrogram from the divisive hierarchical method. Although five clusters can be identified for each dendrogram, the single-linkage dendrogram, which shows long, stringy clusters, has a very different shape than do the other three dendrograms.

We can see that certain primates are separated from the others. In particular, primates 6, 18, 20, 55, and 102 stand out in the agglomerative dendrograms, and primate 3 also stands out in the single-linkage dendrogram.

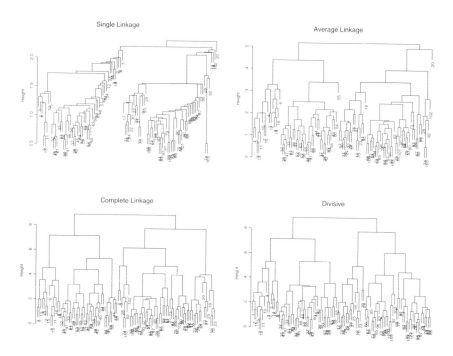

FIGURE 12.4. *Dendrograms from hierarchical clustering of the primate scapulae data. Upper-left panel: single linkage. Upper-right panel: average linkage. Lower-left panel: complete linkage. Lower-right panel: divisive.*

When an isolated observation appears high enough up in a dendrogram, it becomes a cluster of size one and, hence, plays the role of an outlier in the data. In fact, single linkage for five clusters produces three clusters each of size one (primates 3, 20, and 102), and average linkage produces one cluster of size one (primate 20). We see from Figure 12.4 that single-linkage and average-linkage clustering algorithms tend to have more isolated observations than do either the complete-linkage or divisive clustering algorithms.

12.4   Nonhierarchical or Partitioning Methods

Nonhierarchical clustering methods (also known as *partitioning methods*) simply split the data items into a predetermined number K of groups or clusters, where there is no hierarchical relationship between the K-cluster solution and the $(K + 1)$-cluster solution; that is, the K-cluster solution is not the initial step for the $(K + 1)$-cluster solution. Given K, we seek to partition the data into K clusters so that the items within each cluster are similar to each other, whereas items from different clusters are quite dissimilar.

One sledgehammer method of nonhierarchical clustering would conceivably involve as a first step the total enumeration of all possible groupings of the items. Then, using some optimizing criterion, the grouping that is chosen as "best" would be that partition that optimized the criterion. Clearly, for large data sets (e.g., microarray data used for gene clustering), such a method would rapidly become infeasible, requiring incredible amounts of computer time and storage. As a result, all available clustering techniques are iterative and work on only a very limited amount of enumeration. Thus, nonhierarchical clustering methods, which do not need to store large proximity matrices, are computationally more efficient than are hierarchical methods.

This category of clustering methods includes all of the partitioning methods, (e.g., *K-means, partitioning around medoids*) and *mode-searching* (or *bump-hunting*) methods using parametric mixtures or nonparametric density estimates.

12.4.1   K-Means Clustering (kmeans)

The popular K-means algorithm (MacQueen, 1967) is listed in Table 12.2. Because it is extremely efficient, it is often used for large-scale clustering projects. Note that the K-means algorithm needs access to the original data.

The K-means algorithm starts either by assigning items to one of K predetermined clusters and then computing the K cluster centroids, or by pre-specifying the K cluster centroids. The pre-specified centroids may be randomly selected items or may be obtained by cutting a dendrogram at an appropriate height. Then, in an iterative fashion, the algorithm seeks to minimize ESS by reassigning items to clusters. The procedure stops when no further reassignment reduces the value of ESS.

The solution (a configuration of items into K clusters) will typically not be unique; the algorithm will only find a local minimum of ESS. It is recommended that the algorithm be run using different initial random assignments of the items to K clusters (or by randomly selecting K initial centroids) in order to find the lowest minimum of ESS and, hence, the best clustering solution based upon K clusters.

For the worked example, the K-means clustering solutions for $K = 2, 3, 4$ are listed in Table 12.3. For $K = 2$, ESS=23.5; for $K = 3$, ESS=8.67; and for $K = 4$, ESS=5.67. Note that, in general, we expect ESS to be a monotonically decreasing function of K, unless the solution for a given value of K turns out to be a local minimum.

TABLE 12.2. *Algorithm for K-means clustering.*

1. Input: $\mathcal{L} = \{\mathbf{x}_i, i = 1, 2, \ldots, n\}$, K = number of clusters.

2. Do one of the following:

   - Form an initial random assignment of the items into K clusters and, for cluster k, compute its current centroid, $\bar{\mathbf{x}}_k$, $k = 1, 2, \ldots, K$.

   - Pre-specify K cluster centroids, $\bar{\mathbf{x}}_k$, $k = 1, 2, \ldots, K$.

3. Compute the squared-Euclidean distance of each item to its current cluster centroid:

$$\text{ESS} = \sum_{k=1}^{K} \sum_{c(i)=k} (\mathbf{x}_i - \bar{\mathbf{x}}_k)^\tau (\mathbf{x}_i - \bar{\mathbf{x}}_k),$$

where $\bar{\mathbf{x}}_k$ is the kth cluster centroid and $c(i)$ is the cluster containing $\mathbf{x}_i$.

4. Reassign each item to its nearest cluster centroid so that ESS is reduced in magnitude. Update the cluster centroids after each reassignment.

5. Repeat steps 3 and 4 until no further reassignment of items takes place.

12.4.2 Partitioning Around Medoids (pam)

This clustering method (Vinod, 1969) is a modification of the *K-medoids clustering* algorithm. Although similar to K-means clustering, this algorithm searches for K "representative objects" (or *medoids*) — rather than the centroids — among the items in the data set, and a dissimilarity-based distance is used instead of squared-Euclidean distance. Because it minimizes a sum of dissimilarities instead of a sum of (squared) Euclidean distances, the method is more robust to data anomolies such as outliers and missing values.

This algorithm starts with the proximity matrix $\mathbf{D} = (d_{ij})$, where $d_{ij} = d(\mathbf{x}_i, \mathbf{x}_j)$, either given or computed from the data set, and an initial configuration of the items into K clusters. Using $\mathbf{D}$, we find that item (called a *representative object* or *medoid*) within each cluster that minimizes the total dissimilarity to all other items within its cluster. In the K-medoids algorithm, the centroids of steps 2, 3, and 4 in the K-means algorithm (Table 12.2) are replaced by medoids, and the objective function ESS is replaced by ESS_{Kmed}. See Table 12.4 (steps 1, 2, 3, and 4a) for the K-medoids algorithm.

The *partitioning around medoids* (pam) modification of the K-medoids algorithm (Kaufman and Rousseeuw, 1990, Section 2.4) introduces a swapping strategy by which the medoid of each cluster is replaced by another item in that cluster, but only if such a swap reduces the value of the

TABLE 12.3. *K-means clustering solutions (K = 2, 3, 4) for the worked example.*

K	k	Indexes	Centroid	Within-Cluster SS
2	1	1,2,3,4	(3.5, 8.5)	13.5
	2	5,6,7,8	(2.25, 4.25)	10.0
3	1	1,2,3	(1.33, 4.0)	2.67
	2	4,5	(5.0, 6.0)	2.0
	3	6,7,8	(3.0, 9.0)	4.0
4	1	1,2,3	(1.33, 4.0)	2.67
	2	4,5	(5.0, 6.0)	2.0
	3	6,8	(3.5, 9.5)	1.0
	4	7	(2.0, 8.0)	0.0

objective function. The `pam` algorithm is listed in Table 12.4 (steps 1, 2, 3, and 4b).

A disadvantage of both the K-medoids and the `pam` algorithms is that, although they run well on small data sets, they are not efficient enough to use for clustering large data sets.

12.4.3 Fuzzy Analysis (fanny)

The idea behind *fuzzy clustering* is that items to be clustered can be assigned probabilities of belonging to each of the K clusters (Kaufman and Rousseeuw, 1990, Section 4.4). Let u_{ik} denote the *strength of membership* of the ith item for the kth cluster. For the ith item, we require that the $\{u_{ik}\}$ behave like probabilities; that is, $u_{ik} \geq 0$, for all i and $k = 1, 2, \ldots, K$, and $\sum_{k=1}^{K} u_{iv} = 1$ for each i. This contrasts with the partitioning methods of `kmeans` or `pam`, where each item is assigned to one and only one cluster.

Given a proximity matrix $\mathbf{D} = (d_{ij})$ and number of clusters K, the unknown membership strengths, $\{u_{ik}\}$, are found by minimizing the objective function,

$$\sum_{k=1}^{K} \frac{\sum_i \sum_j u_{ik}^2 u_{jk}^2 d_{ij}}{2 \sum_\ell u_{\ell k}^2}. \tag{12.1}$$

The objective function is minimized subject to the nonnegativity and unit sum restrictions by using an iterative algorithm.

For the worked example, the solution (after 90 iterations) is given in Table 12.5, where the most likely cluster memberships are as follows: cluster 1: items 1, 2, 3; cluster 2: items 4, 5; cluster 3: items 6, 7, 8. The minimum of the objective function is 3.428.

TABLE 12.4. *Algorithms for K-medoid and partitioning-around-medoids clustering.*

1. Input: proximity matrix $\mathbf{D} = (d_{ij})$; K = number of clusters.

2. Form an initial assignment of the items into K clusters.

3. Locate the *medoid* for each cluster. The medoid of the kth cluster is defined as that item in the kth cluster that minimizes the total dissimilarity to all other items within that cluster, $k = 1, 2, \ldots, K$.

4a. For K-medoids clustering:

- For the kth cluster, reassign the i_kth item to its nearest cluster medoid so that the objective function,

$$\text{ESS}_{\text{med}} = \sum_{k=1}^{K} \sum_{c(i)=k} d_{ii_k},$$

is reduced in magnitude, where $c(i)$ is the cluster containing the ith item.

- Repeat step 3 and the reassignment step until no further reassignment of items takes place.

4b. For partitioning-around-medoids clustering:

- For each cluster, swap the medoid with the non-medoid item that gives the largest reduction in ESS_{med}.

- Repeat the swapping process over all clusters until no further reduction in ESS_{med} takes place.

12.4.4 Silhouette Plot

A useful feature of partitioning methods based upon the proximity matrix $\mathbf{D}$ (e.g., kmeans, pam, and fanny) is that the resulting partition of the data can be graphically displayed in the form of a *silhouette plot* (Rousseeuw, 1987).

Suppose we are given a particular clustering, $\mathcal{C}_K$, of the data into K clusters. Let $c(i)$ denote the cluster containing the ith item. Let a_i be the average dissimilarity of that ith item to all other members of the same cluster $c(i)$. Also, let c be some cluster other than $c(i)$, and let $d(i, c)$ be the average dissimilarity of the ith item to all members of c. Compute $d(i, c)$ for all clusters c other than $c(i)$. Let $b_i = \min_{c \neq c(i)} d(i, c)$. If $b_i = d(i, C)$, then, cluster C is called the *neighbor* of data point i and is regarded as the second-best cluster for the ith item.

TABLE 12.5. *Fuzzy clustering for the worked example with $K = 3$. The boldfaced entries show the most probable cluster memberships for each item.*

	Cluster k		
i	1	2	3
1	**0.799**	0.117	0.083
2	**0.828**	0.107	0.065
3	**0.735**	0.146	0.119
4	0.116	**0.790**	0.094
5	0.102	**0.715**	0.183
6	0.072	0.146	**0.782**
7	0.196	0.239	**0.565**
8	0.064	0.097	**0.839**

The ith *silhouette value* (or *width*) is given by

$$s_i(\mathcal{C}_K) = s_{iK} = \frac{b_i - a_i}{\max\{a_i, b_i\}}, \tag{12.2}$$

so that $-1 \leq s_{iK} \leq 1$. Large positive values of s_{iK} (i.e., $a_i \approx 0$) indicate that the ith item is well-clustered, large negative values of s_{iK} (i.e., $b_i \approx 0$) indicate poor clustering, and $s_{iK} \approx 0$ (i.e., $a_i \approx b_i$) indicates that the ith item lies between two clusters. If $\max_i\{s_{iK}\} < 0.25$, this indicates either that there are no definable clusters in the data or that, even if there are, the clustering procedure has not found it. Negative silhouette widths tend to attract attention: the items corresponding to these negative values are considered to be borderline allocations; they are neither well-clustered nor are they assigned by the clustering process to an alternative cluster.

A *silhouette plot* is a bar plot of all the $\{s_{iK}\}$ after they are ranked in decreasing order, where the length of the ith bar is s_{iK}. For the worked example, where we used the **pam** clustering method with $K = 3$ clusters, the silhouette plot is displayed in Figure 12.5.

The *average silhouette width*, $\bar{s}_K$, is the average of all the $\{s_{iK}\}$. For the worked example with $K = 3$, the overall average silhouette width is $\bar{s}_3 = 0.51$. (For $K = 2$, $\bar{s}_2 = 0.44$, and for $K = 4$, $\bar{s}_4 = 0.41$.) The statistic $\bar{s}_K$ has been found to be a very useful indicator of the merit of the clustering $\mathcal{C}_K$. The average silhouette width has also been used to choose the value of K by finding K to maximize $\bar{s}_K$.

As a clustering diagnostic, Kaufman and Rousseeuw defined the *silhouette coefficient*, $SC = \max_K\{\bar{s}_K\}$, and gave subjective interpretations of its value:

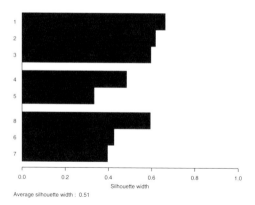

Silhouette width

Average silhouette width : 0.51

FIGURE 12.5. *Silhouette plot for the worked example using the partitioning around medoids (*pam*) clustering method with $K = 3$ clusters.*

SC	Interpretation
0.71–1.00	A strong structure has been found
0.51–0.70	A reasonable structure has been found
0.26–0.50	The structure is weak and could be artificial
≤ 0.25	No substantial structure has been found

12.4.5 Example: Landsat Satellite Image Data

Since 1972, Landsat satellites orbiting the Earth have used a combination of scanning geometry, satellite orbit, and Earth rotation to collect high-resolution multispectral digital information for detecting and monitoring different types of land surface cover characteristics. The Landsat data in this example were generated from a Landsat Multispectral Scanner (MSS) image database used in the European STATLOG Project for assessing machine-learning methods.[3] The following description of the data is taken from the STATLOG website:

> One frame of Landsat MSS imagery consists of four digital images of the same scene in different spectral bands. Two of these are in the visible region (corresponding approximately to green and red regions of the visible spectrum) and two are in the (near) infrared. Each pixel is an 8-bit word, with 0

---

[3]These data, which are available in the file satimage at the book's website, can also be downloaded from http://www.niaad.liacc.up.pt/old/statlog/. For information on the Landsat satellites, see http://edc.usgs.gov/guides/landsat_mss.html.

TABLE 12.6. *Comparison of results of different clustering algorithms applied to the Landsat image data. The data consist of six groups of 4,435 observations measured on 36 variables. Prior to clustering, all variables were standardized. The six derived clusters are designated A–F. The agglomerative hierarchical clustering methods are single-linkage (SL), average-linkage (AL), and complete-linkage (CL), and the nonhierarchical methods are K-means and partitioning around mediods (pam). Each column in this table gives the cluster sizes distributed among the six clusters, ordered from largest cluster (A) to smallest cluster (F).*

Cluster	SL	AL	CL	K-Means	pam
A	4,428	2,203	1,717	1,420	999
B	2	1,764	1,348	1,134	937
C	1	370	885	763	790
D	1	57	266	694	708
E	1	23	162	242	613
F	1	18	57	182	388

corresponding to black and 255 to white. The spatial resolution of a pixel is about 80m×80m. Each image contains 2,340×3,380 such pixels. The data set is a (tiny) sub-area of a scene, consisting of 82×100 pixels. Each line of the data corresponds to a 3×3 square neighborhood of pixels completely contained within the 82×100 sub-area. Each line contains the pixel values in the four spectral bands of each of the 9 pixels in the 3×3 neighborhood.

The 36 variables are arranged in groups of four spectral bands (1, 2, 3, 4) covering each pixel of the 3×3 neighborhood (top-left (TL), top-center (TC), top-right (TR); center-left (CL), center-center (CC), center-right (CR); bottom-left (BL), bottom-center (BC), bottom-right (BR)). The center pixel (CC) of each of 4,435 neighborhoods is classified into one of six classes: 1. red soil (1,072), 2. cotton crop (479), 3. gray soil (961), 4. damp gray soil (415), 5. soil with vegetation stubble (470), and 7. very damp gray soil (1038). There is no class 6. Although we do not use these classifications in the clustering algorithms, we can compare our results with the true classifications.

The results of five clustering methods (we specified six clusters for each method) are given in Table 12.6. We see that of the agglomerative hierarchical clustering methods, single-linkage (SL) puts almost all the observations into a single cluster, whereas average-linkage (AL) and complete-linkage (CL) are somewhat better at distributing the observations among the six clusters. K-means is better still, but pam is closest to the true configuration of the data. The pam silhouette plot for six clusters is given in Figure 12.6 and the average silhouette width is 0.32.

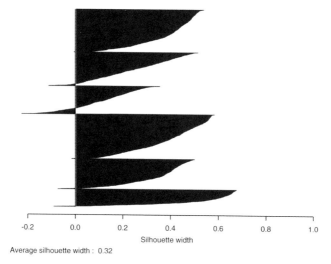

Silhouette width

Average silhouette width : 0.32

FIGURE 12.6. *Silhouette plot for the Landsat image example using the partitioning around medoids (*pam*) clustering method with $K = 6$ clusters.*

The largest four eigenvalues of the (36×36) correlation matrix of the Landsat data are 18.68, 14.08, 1.61, and 0.91, respectively. Kaiser's rule says that we should retain only those PCs whose eigenvalues are greater than unity; in this case, we retain the first three PCs. In Figure 12.7, we display a scatterplot of the first two PC scores of the Landsat data. The six clusters of points (corresponding to Table 12.6) found using the pam algorithm are each identified by their color. The scatterplot of the PC scores appears to be wedge-shaped, with three primary "rods." The "bottom" rod is divided into three distinct bands, consisting of clusters A (dark blue), C (red), and B (green); the "middle" rod is similarly divided up into three distinct bands of clusters D (orange), E (light blue), and some B (green); and the "top" rod only consists of cluster F (brown). There are also many points in the scatterplot that fall between the rods.

The picture becomes more interpretable if we look at a 3D scatterplot of the first three PC scores (not shown here), especially if we use a rotation/spin operation as is available in S–PLUS or R. Rotating the 3D plot shows a tripod-like structure, with the top of the tripod being cluster B and the three rods being the three legs of the tripod. We can compute a *confusion table*, Table 12.7, which details how many neighborhoods from each class are allocated to the various clusters. From Table 12.7, we see that one leg consists of clusters of primarily different types of gray soil (A, C, and B); the second leg consists of clusters of primarily red soil (D and E); and the third leg consists of a cluster of cotton crop (F). Image neighborhoods classified by Landsat as soil with vegetation stubble appear mostly within clusters B and E.

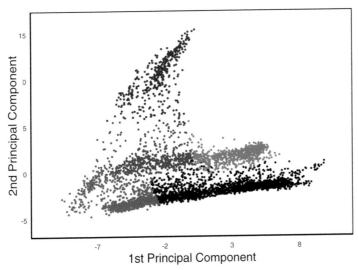

FIGURE 12.7. *Scatterplot of first two principal components of the Landsat image data, with points colored to identify the clusters found in the data. The six derived clusters are A. dark blue; B. green; C. red; D. orange; E. light blue; F. brown.*

12.5 Self-Organizing Maps (SOMs)

The *self-organizing map (SOM)* algorithm (Kohonen, 1982) has its roots in artificial neural networks and has also been likened to methods such as multidimensional scaling (MDS; see Chapter 14) and K-means clustering. It is also referred to as a *Kohonen self-organizing feature map*. The original motivation for SOMs was expressed in terms of an artificial neural network

TABLE 12.7. *The confusion table showing results of the* pam *clustering algorithm applied to the Landsat image data. The six derived clusters are designated A–F. The entry in the ith row and jth column shows the number of neighborhoods classified by Landsat into the ith image-type and allocated to the jth cluster.*

Class	A	B	C	D	E	F	Total
1	22	0	11	651	388	0	1,072
2	0	1	10	8	72	388	479
3	883	1	63	14	0	0	961
4	78	18	307	4	7	0	415
5	0	249	48	31	142	0	470
7	15	668	351	0	4	0	1,038
Total	999	937	790	708	613	388	4,435

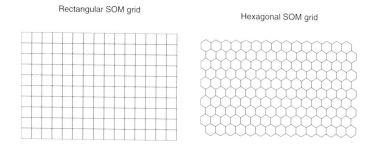

FIGURE 12.8. *Displays of* 10×15 *rectangular and hexagonal SOM grids.*

for modeling the human brain, and much of the literature still uses the image of neurons in describing the building blocks of a SOM.

SOMs have been applied to clustering problems in fields as diverse as geographical information systems, bioinformatics, medical research, physical anthropology, natural language processing, document retrieval systems, and ecology. Its primary use is in reducing high-dimensional data to a lower-dimensional nonlinear manifold, usually two or three dimensions, and in displaying graphically the results of such data reduction. In a SOM, the aim is to map the projected data to discrete interconnected *nodes*, where each node represents a grouping or cluster of relatively homogeneous points.

12.5.1   The SOM Algorithm

Two versions of the SOM algorithm are available: an "on-line" version, in which items are presented to the algorithm in sequential fashion (one at a time, possibly in random order), and a "batch" version, in which all the data are presented together at one time. Both algorithms are due to Kohonen.

The end product of the SOM algorithm (after a large number of iteration steps) is a graphical image called a *SOM plot*. The SOM plot is displayed in output space and consists of a *grid* (or network) of a large number of interconnected *nodes* (or *artificial neurons*). In two dimensions, the nodes are typically arranged as a square, rectangular, or hexagonal grid. See Figure 12.8. For visualization reasons, an hexagonal grid is preferred.

In a two-dimensional rectangular grid, for example, the set of rows is $\mathcal{K}_1 = \{1, 2, \ldots, K_1\}$ and the set of columns is $\mathcal{K}_2 = \{1, 2, \ldots, K_2\}$, where K_1 (the *height*) and K_2 (the *width*) are chosen by the user. Then, a node is defined by its coordinates, $(\ell_1, \ell_2) \in \mathcal{K}_1 \times \mathcal{K}_2$. The total number of nodes, $K = K_1 K_2$, is usually chosen by trial and error, initially much larger than the suspected number of clusters in the data. After an initial SOM analysis, one can reconfigure the SOM by reducing the number of row and column nodes. It will be convenient to map the collection of nodes into an ordered sequence, so that the node $(\ell_1, \ell_2) \in \mathcal{K}_1 \times \mathcal{K}_2$ is relabeled as the index $k = (\ell_1 - 1)K_2 + \ell_2 \in \mathcal{K}$, where $\mathcal{K} = \{1, 2, \ldots, K\}$.

The SOM algorithm has much in common with K-means clustering. In K-means clustering, items assigned to a particular cluster are averaged to obtain a "cluster centroid" (or "representative" of that cluster), which is subsequently updated. With this in mind, we associate with the kth node in a SOM plot a *representative* in input space, $\mathbf{m}_k \in \Re^r$, $k \in \mathcal{K}$. Representatives have also been called *synaptic weight vectors, prototypes, codebook vectors, reference vectors,* and *model vectors*. It is usual to initialize the process by setting the components of $\mathbf{m}_k$, $k \in \mathcal{K}$, to be random numbers.

12.5.2 *On-line Versions*

At the first step of the on-line SOM algorithm, we set up the *map size* (i.e., select K_1 and K_2) and initialize all representatives $\{\mathbf{m}_k\}$ so that they each consist of random values.

At each subsequent step of the algorithm, an input vector $\mathbf{X}$ is randomly selected from the data set and standardized so that each component variable of $\mathbf{X}$ has zero mean and variance one. In this way, no component variable has undue influence on the results just because it has a large variance or absolute value. We then present $\mathbf{X}$ to the SOM algorithm.

We compute the Euclidean distance between $\mathbf{X}$ and each representative and find that node whose representative yields the smallest distance to $\mathbf{X}$. If

$$k^* = \arg\min_k \{\| \mathbf{X} - \mathbf{m}_k \|\}, \tag{12.3}$$

where $\| \cdot \|$ denotes Euclidean norm, then the representative $\mathbf{m}_{k^*}$ is declared the "winner," and k^* is referred to as the *best-matching unit (BMU)* or *winning node* for the input vector $\mathbf{X}$.

Next, we look at those nodes that are "neighbors" of the winning node. A node $k' \in \mathcal{K}$ is defined to be a *grid neighbor* of the node $k \in \mathcal{K}$ if the Euclidean distance between $\mathbf{m}_k$ and $\mathbf{m}_{k'}$ is smaller than a given threshold c. The set of nodes, $\mathcal{N}_c(k^*)$, which are grid neighbors of the winning node k^*, is called the *neighborhood set* for that node. We then update the representatives corresponding to each grid neighbor of the winning node k^* (including k^* itself) so that each $\mathbf{m}_k$, $k \in \mathcal{N}_c(k^*)$, is closer to $\mathbf{X}$; the simplest way of doing this is to use the uniformly weighted update formula,

$$\mathbf{m}_k \leftarrow \mathbf{m}_k + \alpha(\mathbf{X} - \mathbf{m}_k), \quad k \in \mathcal{N}_c(k^*), \tag{12.4}$$

where $0 < \alpha < 1$ is a *learning-rate factor*. For $k \notin \mathcal{N}_c(k^*)$, we set $\alpha = 0$, so that $\mathbf{m}_k$, $k \notin \mathcal{N}_c(k^*)$, remains unchanged. This process, which is repeated a large number of times, runs through the collection of input vectors one at a time. A useful "rule of thumb" is to run the algorithm steps for at least 500 times the number of nodes (Kohonen, 2001, p. 112).

A "distance-weighted" version of (12.4) is probably the more popular strategy,

$$\mathbf{m}_k \leftarrow \mathbf{m}_k + \alpha h_k(\mathbf{X} - \mathbf{m}_k), \quad k \in \mathcal{N}_c(k^*), \tag{12.5}$$

where the *neighborhood function* h depends upon how close the neighboring representatives are to $\mathbf{m}_{k^*}$. Those representatives that are neighbors of $\mathbf{m}_{k^*}$ are adjusted, but not by as much as is $\mathbf{m}_{k^*}$; the further a neighbor is from $\mathbf{m}_{k^*}$, the less of an adjustment is made. The h-function takes the value one when the distance is zero and becomes progressively smaller as the distances become larger. For $k \notin \mathcal{N}_c(k^*)$, we set $h_k = 0$. The most-popular h-function is the multivariate Gaussian kernel function,

$$h_k = \exp\left\{-\frac{\|\mathbf{m}_k - \mathbf{m}_{k^*}\|^2}{2\sigma^2}\right\} I_{[k \in \mathcal{N}_c(k^*)]}, \tag{12.6}$$

where $\sigma > 0$ is the *neighborhood radius*.

Values of c, α, and σ are provided by the user but may change during the sequential process. In the on-line process, c is shrunk during the first 1,000 or so observations from, say, an initial value of C (chosen by the user) to 1. If we take the threshold value c to be so small that each neighborhood contains only a single point, then we lose the dependencies between representatives, which would be independently updated, and the SOM algorithm reduces to an on-line version of K-means clustering, where K is the total number of nodes. The value of α decreases from a large initial value of just less than 1 to a value slightly greater than zero over the same observation span. Three forms of the learning rate, $\alpha(t)$, as a function of the iteration number t are used:

linear: $\alpha(t) = \alpha_0(1 - t/T)$;

power: $\alpha(t) = \alpha_0(0.005/\alpha_0)^{t/T}$;

inverse: $\alpha(t) = \alpha_0/(1 + 100t/T)$,

where α_0 is the initial learning rate and T is the total number of iterations. In Figure 12.9, the functions $\alpha(t)$ are drawn for the `linear`, `power`, and `inverse` forms, where we have taken $\alpha_0 = 0.5$ and $T = 100$. Like α, σ in (12.6) is also taken to decrease monotonically.

12.5.3 Batch Version

The batch SOM algorithm is significantly faster than the on-line version. As before, we first make an initial choice of representatives $\{\mathbf{m}_k\}$. For the kth node, we list all those items $\mathbf{X}_i$ whose $\mathbf{m}_{k^*} \in \mathcal{N}_c(k)$. Then, we

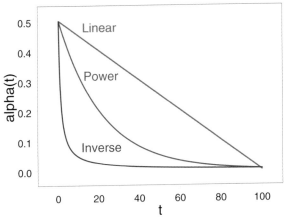

FIGURE 12.9. *Graphs of the on-line SOM learning-rate $\alpha(t)$ as a function of the iteration number t for the* `linear`, `power`, *and* `inverse` *forms, where the initial learning rate $\alpha_0 = 0.5$ and the total number of iterations is $T = 100$.*

update $\mathbf{m}_k$ by averaging the items obtained from the previous step of the algorithm, where we might use a weighted average, with weights $\{h_{ik^*}\}$ given by (12.6). Finally, repeat the process a few times.

In a batch SOM display, the nodes are drawn as circles, and the data points that are mapped to a node are then randomly plotted within the circle corresponding to that particular node; see Figure 12.10, which presents a SOM display of the Landsat data. This can be a very useful graphical display for showing the interrelated structure of the (often high-dimensional) representatives in a 2D plot, together with the input points that are mapped to each representative.

If each data point has a unique identifier, such as a gene description, then it is not difficult to determine the identities of the data points that are captured by each node. In many clustering problems, however, individual points do not have unique identifiers; so, instead, class membership can be used as a plotting symbol in the SOM plot, as in Figure 12.10. From a SOM plot, cluster patterns should be visible.

12.5.4 Unified-Distance Matrix

A different type of visualization of the cluster structure of a SOM is a *U-matrix*, where U stands for "unified distance" (Ultsch and Siemon, 1990). Each entry in a *U*-matrix is the Euclidean distance (in input space) between neighboring representatives. For example, if we have a map with one row of five nodes with representatives $\{\mathbf{m}_1, \mathbf{m}_2, \mathbf{m}_3, \mathbf{m}_4, \mathbf{m}_5\}$, then the

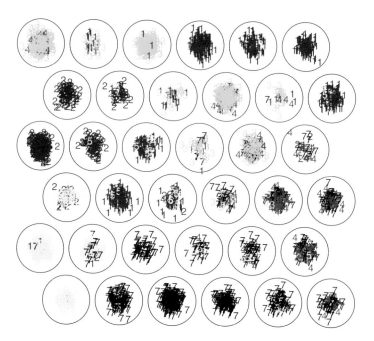

FIGURE 12.10. *A 6×6 hexagonal batch-SOM plot of the Landsat satellite image data. The circles correspond to nodes, and the projected points are plotted randomly within the appropriate circle to which they were deemed closest. The six classes of vegetation are used as plotting symbols (1=red, 2=blue, 3=turquoise, 4=purple, 5=yellow, 7=black).*

U-matrix is a (1×9)-vector,

$$U = (u_1, u_{12}, u_2, u_{23}, u_3, u_{34}, u_4, u_{45}, u_5), \qquad (12.7)$$

where $u_{ij} = \| \mathbf{m}_i - \mathbf{m}_j \|$ is the Euclidean distance between neighboring representatives, and u_i is a representative-specific value; for example, $u_3 = (u_{23} + u_{34})/2$ is the average distance from that representative to all neighboring representatives. A small value in a U-matrix indicates that the SOM nodes are close together in input space, whereas a large value indicates that the SOM nodes, even though they are neighbors in output space, are quite far apart in input space. Thus, the U-matrix provides a useful guide to the underlying probability density function of $\mathbf{X}$ projected onto two dimensions.

Rather than displaying these U-matrix values as a 3D landscape (with low valleys showing clusters and high ridges showing separations between clusters), it is usual instead to discretize the distance values and then color-code them in a 2D colormap, where the colors show the gradations in values. In the SOM TOOLBOX for MATLAB, for example, large distances in the U-matrix are colored as yellow and red and indicate a cluster border, whereas

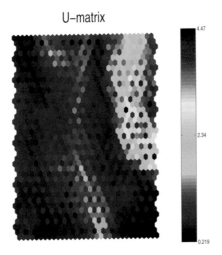

FIGURE 12.11. *The U-matrix from the batch SOM with hexagonal grids for the Landsat satellite image data.*

small distances are colored as blue and indicate items in the same cluster. Figure 12.11 displays the U-matrix with an hexagonal grid for the Landsat image data, where a number of clusters are visible.

A *hierarchical SOM (HSOM)* is a tree of maps (U-matrices), where the "lower" maps on the tree act as a preprocessing stage to the "higher" maps. As we climb up the hierarchy, the information becomes more abstract. HSOMs have been successfully used in the development of bibliographic information retrieval tools. For example, a "document map" has been created for organizing astronomical text documents (Lesteven, Poinçot, and Murtagh, 2001). Using more than 10,300 articles published in several leading astronomy journals, the authors selected 269 keywords, each of which appeared in at least five different articles. By clicking on an individual node in the map, information about the articles located at that node can be retrieved. From this information, the user can then access article content (title, authors, abstract, and the on-line full paper).

12.5.5 Component Planes

An additional useful visualization tool is a colormap of the various component planes. In general, the "components" are the individual input variables that make up **X**.

Figure 12.12 shows the 36 component planes for the Landsat data. Because these data have an easily visualized physical structure, the component planes are arranged into four groups of nine images (corresponding to the four spectral bands and the nine positions). The component planes

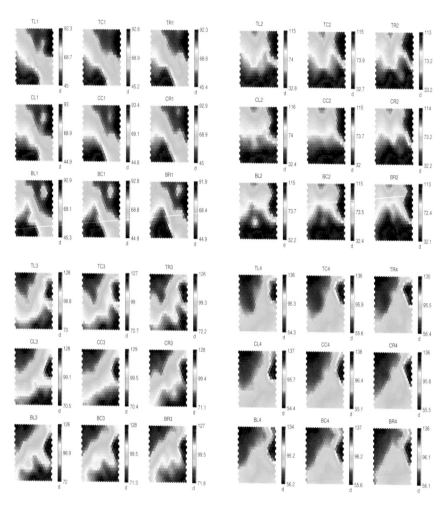

FIGURE 12.12. *Colormaps of the 36 component planes from the batch-SOM algorithm with hexagonal grids for the Landsat image data. The component planes are arranged into four groups (corresponding to the four spectral bands, 1, 2, 3, and 4), each group having nine component planes (corresponding to the nine positions (TL, TC, TR; CL, CC, CR; BL, BC, BR, where T is top position, C is center, B is bottom, L is left, C is center, R is right) in the 3×3 pixel neighborhoods.*

show that the variable values differ substantially between the four spectral bands. Within each set of 3×3 pixel neighborhoods, the component planes show some differences, but those differences are not as significant as between spectral bands. In this example, the component planes have given us a good view of the differences in measurement of each of the four spectral bands.

The U-matrix and component planes derived from SOMs have been applied to the visualization of gene clusters derived from microarray data (see, e.g., Tomayo, Slonim, Mesirov, Zhu, Kitareewan, Dmitrovsky, Lander, and Golub, 1999). In particular, if the genes are expressed at different points in time or at different temperatures, then the component planes, which can be thought of as "slices" of the U-matrix, show the cluster structure obtained at each timepoint or temperature.

12.6 Clustering Variables

We can use the same clustering methods for variables as we used for clustering observations, the main difference being the measure of distance between variables. For clustering variables, we generally use a distance metric based upon the correlation matrix for the r variables. The correlations provide a reasonable measure of "closeness" between pairs of variables. Those pairs of variables with relatively large correlations can be thought of as being "close" to each other; those pairs for which the corresponding correlations are small are considered to be "far away" from each other.

If we standardize each of the r variables to have zero mean and unit variance, then it is not difficult to show that

$$\frac{1}{2(n-1)} \sum_{i=1}^{n} (X_{ji} - X_{ki})^2 = 1 - \rho_{jk}, \tag{12.8}$$

where ρ_{jk} is the correlation between variables X_j and X_k. This shows us that using squared Euclidean distance, $\sum_i (X_{ji} - X_{ki})^2$, is equivalent to using $1 - \rho_{jk}$ as a dissimilarity measure. Either distance metric enables us to utilize any of the hierarchical or nonhierarchical/partitioning clustering methods discussed above, and the graphical output can be a dendrogram or a silhouette plot as appropriate.

12.6.1 Gene Clustering

The most popular use of variable clustering has been in clustering the thousands or tens of thousands of genes measured using a microarray experiment. Concern over the enormous volume of biological information in an organism's genome has led to the idea of grouping together those genes with similar expression patterns. This type of clustering is referred to as *gene clustering*, where, in addition to the usual hierarchical and partitioning methods, some specialized methods have been developed.

In gene clustering, the $(r \times n)$ data matrix $\mathcal{X} = (X_{ij})$ contains the gene-expression data derived from a microarray experiment, where i indexes the row (gene), j indexes the column (tissue sample), and X_{ij} is, for example, the intensity log-ratio of the abundance of the ith gene in the experimental sample relative to some reference sample; in other words, X_{ij} is a measurement of how strongly the ith gene is expressed in the jth sample. Because X_{ij} is the log of a ratio, it follows that those ratios with values between 0 and 1 will yield negative X_{ij}, whereas those ratios greater than 1 will yield positive X_{ij}. For typical microarray experiments, $r \gg n$, so that matrix $\mathcal{X}$ will be "vertically long and skinny."

12.6.2 Principal-Component Gene Shaving

Suppose our goal is to discover a gene cluster that has high variability across samples. Let $\mathcal{S}_k$ denote the set of (row) indices of a cluster of k genes. Consider the jth tissue sample (i.e., jth column of $\mathcal{X}$) and compute the average gene-expression over the k genes for that sample,

$$\bar{X}_{j,\mathcal{S}_k} = \frac{1}{k} \sum_{i \in \mathcal{S}_k} X_{ij}, \quad j = 1, 2, \ldots, n. \tag{12.9}$$

The variance of the $\bar{X}_{j,\mathcal{S}_k}$, $j = 1, 2, \ldots, n$, is given by

$$\text{var}\{\bar{X}_{\mathcal{S}_k}\} = \frac{1}{n} \sum_{j=1}^{n} (\bar{X}_{j,\mathcal{S}_k} - \bar{X}_{\mathcal{S}_k})^2, \tag{12.10}$$

where

$$\bar{X}_{\mathcal{S}_k} = \frac{1}{n} \sum_{j=1}^{n} \bar{X}_{j,\mathcal{S}_k} = \frac{1}{kn} \sum_{j=1}^{n} \sum_{i \in \mathcal{S}_k} X_{ij}. \tag{12.11}$$

Given all possible clusters of size k, we can search for that cluster $\mathcal{S}_k$ with the highest $\text{var}\{\bar{X}_{\mathcal{S}_k}\}$. Unfortunately, such a search procedure is computationally infeasible because it entails evaluating $\binom{r}{k}$ different subsets, which gets big very quickly for r large, as would be common in gene clustering.

Gene shaving (Hastie, Tibshirani, Eisen, Alzadeh, Levy, Staudt, Chan, Botstein, and Brown, 2000) has been proposed as a method for clustering genes, where the primary goal is to identify small subsets (i.e., clusters) of highly correlated ("coherent") genes that vary as much as possible between samples. This method differs from those described previously in that genes are allowed to be included as members of more than one cluster.

Consider the linear combination,

$$Z_j = \mathbf{a}^\tau \mathbf{X}_j = \sum_{i=1}^{r} a_i X_{ij}, \qquad (12.12)$$

of the jth column gene expressions, where $\mathbf{X}_j = (X_{1j}, \cdots, X_{rj})^\tau$, $\mathbf{a} = (a_1, \cdots, a_r)^\tau$, the $\{a_i\}$ are positive, negative, or zero weights, and $\sum_{i=1}^{r} a_i^2 = 1$. For example, for given k, we could set $a_i = \pm 1/\sqrt{k}$ for $i \in \mathcal{S}_k$, and zero otherwise. We wish to find the coefficients $\{a_i\}$ such that the variance of Z_j is maximized.

The solution is given by the first principal component (PC1) of the r rows of $\mathcal{X}$. The $\min(r-1, n)$ principal components of $\mathcal{X}$ are referred to as *eigen-genes*. The individual genes may be ordered according to the magnitude (from largest to smallest in absolute value) of their respective coefficients in the first eigen-gene PC1; we expect that many of the coefficients in PC1 will be close to zero. We could threshold those "near-zero" coefficients (i.e., set the coefficient value equal to zero if it is smaller than a prespecified limit), thereby removing those particular genes from the cluster, but, from experience with simulations, we can do better.

As a selection process for weeding out unimportant genes, we instead compute the inner product (or correlation) of each gene with PC1 and "shave off" (i.e., remove) those genes (rows of $\mathcal{X}$) with the $100\alpha\%$ smallest absolute inner products (e.g., $\alpha = 0.1$). This shaving process decreases the size of the set of available genes, say to k_1 genes. From the reduced subset of k_1 rows, we recompute the first principal component, which, in turn, is shaved to a subset of, say, k_2 rows. This iteration is repeated until a finite sequence of nested gene clusters, $\mathcal{S}_r \supset \mathcal{S}_{k_1} \supset \mathcal{S}_{k_2} \supset \cdots \supset \mathcal{S}_1$, is obtained, where $\mathcal{S}_k$ denotes the set of indices of a cluster of k genes.

The next step is to decide on k and $\mathcal{S}_k$. For a given value of k, define the following ANOVA-type decomposition of the total variance,

$$V_T = \frac{1}{kn} \sum_{i \in \mathcal{S}_k} \sum_{j=1}^{n} (X_{ij} - \bar{X}_{\mathcal{S}_k})^2 = V_B + V_W, \qquad (12.13)$$

where

$$V_B = \frac{1}{n} \sum_{j=1}^{n} (\bar{X}_{j,\mathcal{S}_k} - \bar{X}_{\mathcal{S}_k})^2, \qquad (12.14)$$

$$V_W = \frac{1}{n} \sum_{j=1}^{n} \left[\frac{1}{k} \sum_{i \in \mathcal{S}_k} (X_{ij} - \bar{X}_{j,\mathcal{S}_k})^2 \right] \qquad (12.15)$$

are the between-variance and within-variance, respectively. A natural statistic is

$$R^2(\mathcal{S}_k) = \frac{V_B}{V_T} \times 100\% = \frac{V_B/V_W}{1 + V_B/V_W} \times 100\%, \qquad (12.16)$$

which is the percentage of the total variance explained by the gene cluster $\mathcal{S}_k$. The larger the value of R^2, the more coherent the gene cluster.

Hastie et al. now determine the cluster size k by a permutation argument applied to the R^2-value in (12.16). The "significance" of the R^2-value is judged by comparing it with its expectation computed under a suitable reference null distribution; in this case, the reference distribution assumes the rows and columns of $\mathcal{X}$ are independent. Randomly permute the elements of each row of $\mathcal{X}$ to get $\mathcal{X}^*$. Do this B times to get $\mathcal{X}^{*b}$, $b = 1, 2, \ldots, B$. Apply the shaving algorithm to $\mathcal{X}^{*b}$, that gives $\mathcal{S}_k^{*b}$, and then compute $R^2(\mathcal{S}_k^{*b})$, $b = 1, 2, \ldots, B$.

The *gap statistic* (Tibshirani, Walther, and Hastie, 2001) is defined as

$$\mathrm{Gap}(k) = R^2(\mathcal{S}_k) - \overline{R^2(\mathcal{S}_k^*)}, \qquad (12.17)$$

where $\overline{R^2(\mathcal{S}_k^*)}$ is the average of all the $\{R^2(\mathcal{S}_k^{*b}), b = 1, 2, \ldots, B\}$. We choose that value, $\hat{k}$, of k (and, hence, $\mathcal{S}_{\hat{k}}$) which results in the maximum gap; that is, $\hat{k} = \arg\max_k \mathrm{Gap}(k)$. A useful graphical technique is to plot the *gap curve*, which is a plot of $\mathrm{Gap}(k)$ against cluster size k. Set $\hat{k} = \hat{k}^{(1)}$.

After determining the number, $\hat{k}^{(1)}$, of genes and their identities, we look for a second gene cluster. Before we do that, we need to remove the effects of the first cluster of genes. Hastie et al. apply an orthogonalization trick: first, compute the first *supergene*, $\bar{\mathbf{X}}^{(1)} = (\bar{X}_1^{(1)}, \cdots, \bar{X}_r^{(1)})^\tau$, an r-vector of average genes corresponding to the first cluster $\mathcal{S}_{\hat{k}^{(1)}}$, where $\bar{X}_j^{(1)} = \sum_{i \in \mathcal{S}_{\hat{k}^{(1)}}} X_{ij} / \hat{k}^{(1)}$, $j = 1, 2, \ldots, r$; second, orthogonalize $\mathcal{X}$ by regressing each row of $\mathcal{X}$ on the supergene $\bar{\mathbf{X}}^{(1)}$ and replacing the rows of $\mathcal{X}$ by the residuals from each such regression. This gives us the matrix $\mathcal{X}_1$. Rerun the shaving algorithm on $\mathcal{X}_1$ and then use the gap statistic to obtain $\hat{k}^{(2)}$, the second gene cluster $\mathcal{S}_{\hat{k}^{(2)}}$, and the second supergene $\bar{\mathbf{X}}^{(2)}$. This process is applied repeatedly a total of t times, where t is prespecified, by modifying $\mathcal{X}$ and $\bar{\mathbf{X}}$ at each step; at the kth step, $\mathcal{X}$ is orthogonal to all the previously obtained supergenes $\bar{\mathbf{X}}^{(\ell)}$, $\ell = 1, 2, \ldots, k - 1$.

One of the main steps in the gene-shaving process is the use of the gap statistic to determine the cluster size k. Hastie et al. report good results for the gap statistic when the clusters are well-separated. However, there is evidence that the gap statistic tends to overestimate the number of clusters (Dudoit and Fridlyand, 2002; Simon et al., 2003, p. 151).

After identifying each gene cluster, the rows of $\mathcal{X}$ can be reordered to display those gene clusters more explicitly. The tissue samples (columns of

$\mathcal{X}$) can also be reordered according to either the average gene expression of each column of $\mathcal{X}$ or some external covariate reflecting additional information, such as tissue type or cancer class. A supervised version of gene shaving (Hastie et al., 2000) has been developed, which, for example, is able to identify gene clusters that are closely associated with patient survival times.

12.6.3 Example: Colon Cancer Data

We apply PC gene-shaving to the colon cancer microarray data described in Section 2.2.1. The microarray data consist of expression levels of 92 genes obtained from a microarray study on 62 colon tissue samples. The gene-expression heatmap for the colon cancer data is displayed in Figure 2.1. Figure 12.13 shows the gap curves for the first four clusters derived using the gene-shaving algorithm. For each cluster, the value of k at which the gap curve attains its maximum is chosen to be the estimated size of the cluster. The estimated cluster sizes for the first four clusters are 41, 15, 6, and 19, respectively. The four heatmaps for those gene clusters are displayed in Figure 12.14, where the samples are ordered by the values of the column averages; each panel gives the values of the total variance V_T, the between-variance V_B, the ratio V_B/V_W, and $R^2 = V_B/V_T \times 100\%$, the percentage of the total variance explained by that cluster. The largest R^2 value was that of the third cluster at 64.8%.

The four clusters in Figure 12.14 display different patterns of gene expression. The first cluster has an interesting feature in that the genes split into two equal-sized subgroups: for a given tissue sample, when the "upper" subgroup of genes are strongly upregulated (red color), the "lower" subgroup are strongly downregulated (green color), and vice versa. Furthermore, the red/green split depends upon whether the sample is a tumor sample or a normal sample. The second and third clusters of genes have the same overall appearance: in both, the tumor samples (mostly located on the right of the heatmap) tend to be upregulated, whereas normal samples (mostly located on the left of the heatmap) tend to be downregulated. The reds and greens of the fourth cluster are somewhat more randomly sprinkled around the heatmap, although there are pockets of adjacent cells (e.g., the top few rows and a portion of the right-hand side) that seem to share similar expression patterns.

12.7 Block Clustering

So far, our focus has been on clustering observations (cases, samples) or variables separately. Now, we consider the problem of clustering observations and variables simultaneously.

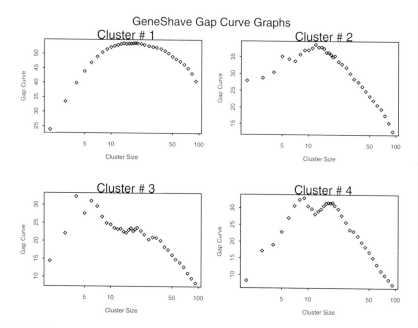

FIGURE 12.13. *Gap curves for the first four clusters of colon cancer data. The gap estimate of cluster size is that value of k for which the gap curve is a maximum. The estimated cluster sizes are first cluster (top-left panel), 41; second cluster (top-right panel), 15; third cluster (bottom-left panel), 6; and fourth cluster (bottom-right panel), 19.*

The simplest way to do this is to apply a hierarchical clustering method to rows and columns separately. Figure 12.15 displays the heatmap of the colon cancer data, where rows and columns have been rearranged through separate hierarchical clustering algorithms. We see a partition of the heatmap into blocks of mainly reds or greens. The rearrangement of rows (colon tissue samples) does not correspond to the known division into tumor samples and normal samples.

Block clustering, also known as *direct clustering* (Hartigan, 1972), produces a simultaneous reordering of the rows and columns of the $(r \times n)$ data matrix $\mathcal{X} = (X_{ij})$ so that the data matrix is partitioned into K submatrices or "data clusters." As an example, Hartigan (1974) clustered the voting records of 126 nations on 50 selected issues at the United Nations, where each vote was coded as 1 (= yes), 2 (= abstain), 3 (= no), 5 (= absent), or 0 (= unknown), and the "absents" are treated as missing data. To motivate the two-way clustering, a natural problem was whether "blocs" of countries exist that vote alike on "blocs" of questions that arise from the same issue.

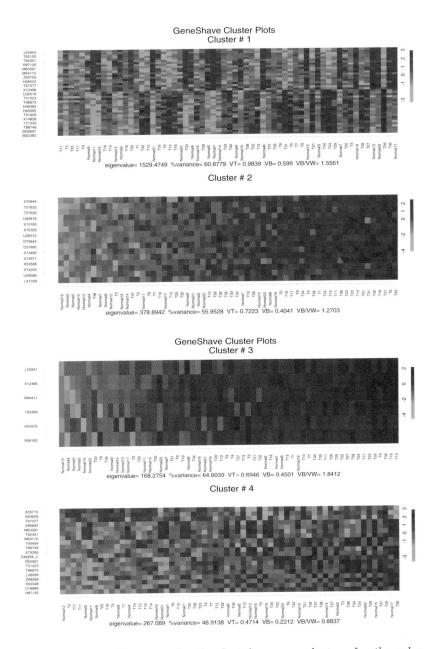

FIGURE 12.14. *Heatmaps for the first four gene clusters for the colon cancer data, where each cluster size is determined by the maximum of that gap curve. The genes are the rows and the samples are the columns. The samples are ordered by the values of the column averages.*

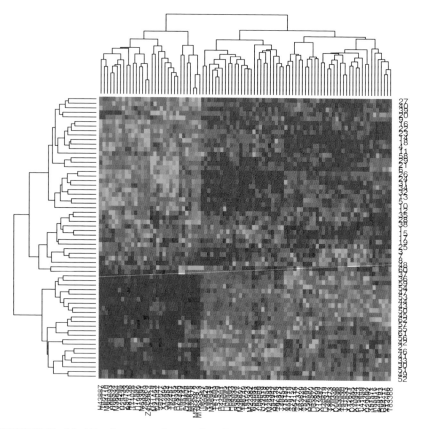

FIGURE 12.15. *Separate hierarchical clustering of rows (colon tissue samples) and columns (genes) of the colon cancer data.*

In block clustering, each entry in the data matrix appears in one and only one data cluster, and each data cluster corresponds to a particular "row cluster" and a particular "column cluster." The block-clustering algorithm given in Table 12.8 partitions the rows and columns of $\mathcal{X}$ into homogeneous, disjoint blocks (i.e., where the elements of each block can be closely approximated by the same value) so that the row clusters and column clusters are hierarchically arranged to form row and column dendrograms, respectively.

12.8 Two-Way Clustering of Microarray Data

For clustering gene expression data, it can be argued that creating *disjoint* blocks of genes and samples may be an over-simplification of the situation. Biological systems are notoriously complicated, and interrelations between these systems may result from some genes possessing multiple

TABLE 12.8. *Hartigan's block-clustering algorithm.*

---

1. Start with all data in a single block (i.e., $K = 1$).

2. Let $B_1, B_2, \ldots, B_K$ denote a partition of the rows and columns of $\mathcal{X}$ into K blocks (or data clusters), where $B_k = (\mathcal{R}_k, \mathcal{C}_k)$ consists of a set, $\mathcal{R}_k$, of r_k rows and a set, $\mathcal{C}_k$, of c_k columns of $\mathcal{X}$, $k = 1, 2, \ldots, K$.

3. Within the kth block B_k, compute $\bar{X}_k$, the average of all the X_{ij} within that block. Approximate $\mathcal{X}$ by the matrix $\widehat{\mathcal{X}} = (\widehat{X}_{ij})$, where the $\widehat{X}_{ij} = \bar{X}_k$ are constant within block B_k. Compute $\mathrm{ESS} = \sum_{k=1}^{K} \sum_{(i,j) \in B_k} (X_{ij} - \bar{X}_k)^2$, the total within-block variance.

4. At the hth step, there will be h blocks, $B_1, B_2, \ldots, B_k, \ldots, B_h$. Suppose we destroy B_k by splitting it into two subblocks, B'_k and B''_k, either by splitting the rows or the columns. Consider a row-split of the block $B_k = (\mathcal{R}_k, \mathcal{C}_k)$. Suppose $\mathcal{R}_k$ *contains* a previous row-split of a different block $B_\ell = (\mathcal{R}_\ell, \mathcal{C}_\ell)$ into $B'_\ell = (\mathcal{R}'_\ell, \mathcal{C}'_\ell)$ and $B''_\ell = (\mathcal{R}''_\ell, \mathcal{C}''_\ell)$. Then, the only row-split allowable for B_k is a *fixed split* given by $\mathcal{R}'_k = \mathcal{R}'_\ell$ and $\mathcal{R}''_k = \mathcal{R}''_\ell$. Similarly for column splits. A *free split* is a split in which no such restrictions are specified.

5. The reduction in ESS due to row-splitting B_k into B'_k and B''_k is given by

$$\Delta\mathrm{ESS} = c_k r'_k [\bar{X}(B'_k) - \bar{X}(B_k)]^2 + c_k r''_k [\bar{X}(B''_k) - \bar{X}(B_k)]^2,$$

where $\bar{X}(B)$ denotes the average of $\mathcal{X}$ over the block B.

6. At each step, compute $\Delta\mathrm{ESS}$ for each (row or column) split of all existing blocks. Choose that split that maximizes $\Delta\mathrm{ESS}$.

7. Stop when any further splitting leads to $\Delta\mathrm{ESS}$ becoming too small or when the number of blocks K becomes too large.

---

functions. Hence, it may be more realistic to accept the idea that certain clusters should naturally overlap each other. Furthermore, similarities between related genes and between related samples may be more complex due to gene-sample interaction effects.

12.8.1 *Biclustering*

With this in mind, the *biclustering* approach (Cheng and Church, 2000) seeks to divide the $(r \times n)$-matrix $\mathcal{X} = (X_{ij})$ of gene-expression data into a pre-specified number of "biclusters," which do not have to be disjoint. Each bicluster corresponds to a subset of the genes and a subset of the samples that possess a high degree of similarity. So, certain rows and columns of $\mathcal{X}$ will appear in several biclusters. The basic idea is to determine in a sequential fashion one bicluster at a time.

A bicluster is defined as a submatrix, $\mathcal{X}(\mathcal{I}, \mathcal{J})$, of $\mathcal{X}$, where $\mathcal{I}$ is a subset of $n_{\mathcal{I}}$ rows and $\mathcal{J}$ is a subset of $n_{\mathcal{J}}$ columns in $\mathcal{X}$. Consider the expression level X_{ij}, $i \in \mathcal{I}, j \in \mathcal{J}$. If we model the bicluster by an additive two-way analysis of variance (ANOVA) model, then we can write

$$X_{ij} \approx \mu + \alpha_i + \beta_j, \quad i \in \mathcal{I}, j \in \mathcal{J}, \tag{12.18}$$

where μ is the overall mean effect, α_i represents the effect of the ith row, β_j the effect of the jth column, and, for uniqueness, we assume that $\sum_{i \in \mathcal{I}} \alpha_i = \sum_{j \in \mathcal{J}} \beta_j = 0$. Least-squares estimates of μ, α_i, and β_j are given by

$$\widehat{\mu} = \bar{X}_{..}, \quad \widehat{\alpha}_i = \bar{X}_{i.} - \bar{X}_{..}, \quad \widehat{\beta}_j = \bar{X}_{.j} - \bar{X}_{..}, \tag{12.19}$$

where

$$\bar{X}_{i.} = n_{\mathcal{J}}^{-1} \sum_{j \in \mathcal{J}} X_{ij}, \quad \bar{X}_{.j} = n_{\mathcal{I}}^{-1} \sum_{i \in \mathcal{I}} X_{ij} \tag{12.20}$$

$$\bar{X}_{..} = (n_{\mathcal{I}} n_{\mathcal{J}})^{-1} \sum_{i \in \mathcal{I}} \sum_{j \in \mathcal{J}} X_{ij}. \tag{12.21}$$

The least-squares *residual* at X_{ij} is defined as

$$\widehat{e}_{ij} = X_{ij} - \widehat{\mu} - \widehat{\alpha}_i - \widehat{\beta}_j = X_{ij} - \bar{X}_{i.} - \bar{X}_{.j} + \bar{X}_{..}, \quad i \in \mathcal{I}, j \in \mathcal{J}. \tag{12.22}$$

Let

$$RSS(\mathcal{I}, \mathcal{J}) = \sum_{i \in \mathcal{I}} \sum_{j \in \mathcal{J}} \widehat{e}_{ij}^2 \tag{12.23}$$

be the residual sum of squares for the bicluster. The objective function is

$$H(\mathcal{I}, \mathcal{J}) = \frac{RSS(\mathcal{I}, \mathcal{J})}{n_{\mathcal{I}} n_{\mathcal{J}}}, \tag{12.24}$$

which is proportional to the residual mean square $RMS(\mathcal{I}, \mathcal{J})$ for the bicluster; that is, $RMS = [(n_{\mathcal{I}} - 1)(n_{\mathcal{J}} - 1)/n_{\mathcal{I}} n_{\mathcal{J}}]H$. The aim is to find a row set $\mathcal{I}$ and a column set $\mathcal{J}$ such that $H(\mathcal{I}, \mathcal{J})$ has a small value.

A bicluster is constructed by sequentially deleting one or multiple rows or columns at a time from $\mathcal{X}$, where the choice is determined at each step so as to achieve the largest decrease in the value of H. Deleting rows or columns will reduce the value of H. A similar result allows one to add some rows or columns without increasing H. Like all greedy algorithms, this algorithm needs a threshold value; it is usual to fix a maximum-acceptable threshold $\delta \geq 0$ for the value of H while running the algorithm.

As each bicluster is found, the elements of $\mathcal{X}$ corresponding to that bicluster are replaced by random numbers (so that no recognizable pattern from that bicluster is retained that could be correlated with future biclusters), and the next bicluster is sought. The random numbers are sampled from a uniform density over a range appropriate for the given application.

12.8.2 Plaid Models

Plaid models (Lazzeroni and Owen, 2002) form a family of models for carrying out block-clustering, in which sums of "layers" of two-way ANOVA models are fitted to gene-expression data. As such, it generalizes the biclustering approach. Each "layer" is formed by a subset of the rows and columns and can be viewed as a two-way clustering of the elements of the data matrix, except that genes can be members of different layers or of none of them. Hence, overlapping clusters (i.e., layers) are allowed.

There are several different types of plaid models, some more detailed than others. Consider the following simple model,

$$X_{ij} \approx \mu_0 + \sum_{k=1}^{K} \mu_k \rho_{ik} \kappa_{jk}. \qquad (12.25)$$

In this model, μ_0 represents the expression level for the *background layer*, μ_k represents the expression level in the kth layer, and ρ_{ik} and κ_{jk} are two indicators whose value is 1 if the subscripts are equal and 0 otherwise. Thus, $\rho_{ik} = 1$ (or 0) indicates the presence (or absence) of the ith gene in the kth gene-layer, whereas $\kappa_{jk} = 1$ (or 0) indicates the presence (or absence) of the jth sample in the kth sample-layer. The expression level μ_k is said to be *upregulated* if $\mu_k > 0$ and *downregulated* if $\mu_k < 0$.

Requiring each gene and each sample to be in exactly one cluster would mean that $\sum_k \rho_{ik} = 1$ for every i, and $\sum_k \kappa_{jk} = 1$ for every j, respectively. To allow overlapping levels, these constraints would have to be relaxed: for example, we could set $\sum_k \rho_{ik} \geq 2$ for some i, or $\sum_k \kappa_{jk} \geq 2$ for some j. We would also need to recognize that there may be genes or samples that do not belong naturally to any layer; for such genes, $\sum_k \rho_{ik} = 0$, and for such samples, $\sum_k \kappa_{jk} = 0$. In general, we do not need to impose any restrictions on the $\{\rho_{ik}\}$ and $\{\kappa_{jk}\}$.

A more general ANOVA-type model is given by

$$X_{ij} \approx \mu_0 + \sum_{k=1}^{K} (\mu_k + \alpha_{ik} + \beta_{jk}) \rho_{ik} \kappa_{jk}, \qquad (12.26)$$

where α_{ik} and β_{jk} measure the effects of the ith row (genes) and jth column (samples), respectively, in the kth layer. To avoid overparameterization, we require $\sum_i \rho_{ik} \alpha_{ik} = \sum_j \kappa_{jk} \beta_{jk} = 0$, $k = 1, 2, \ldots, K$. The description of model (12.26) as a "plaid" model derives from the visual appearance of the fitted heatmap of $\mu_k + \alpha_{ik} + \beta_{jk}$, where we see the row-stripes of the $\{\rho_{ik}\}$ and the column-stripes of the $\{\kappa_{jk}\}$.

Let $\theta_{ijk} = \mu_k + \alpha_{ik} + \beta_{jk}$, $k = 1, 2, \ldots, K$. Then, we can write the plaid model (12.26) as

$$X_{ij} \approx \theta_{ij0} + \sum_{k=1}^{K} \theta_{ijk} \rho_{ik} \kappa_{jk}. \qquad (12.27)$$

To estimate the parameters $\{\theta_{ijk}\}$ in (12.27), we minimize the criterion,

$$Q = \frac{1}{2} \sum_{i=1}^{r} \sum_{j=1}^{n} \left(X_{ij} - \theta_{ij0} - \sum_{k=1}^{K} \theta_{ijk} \rho_{ik} \kappa_{jk} \right)^2, \tag{12.28}$$

with respect to $\{\theta_{ijk}\}, \{\rho_{ik}\}, \{\kappa_{jk}\}$, where $\rho_{ik}, \kappa_{jk} \in \{0, 1\}$. Given the number of layers K, this optimization problem quickly becomes computationally infeasible (each gene and each sample can be in or out of each layer, and so there are $(2^r - 1)(2^n - 1)$ possible combinations of genes and samples).

To overcome this problem, the minimization of Q is turned into an iterative process, where we add one layer at a time. Suppose we have already fitted $K - 1$ layers, and we need to identify the Kth layer by minimizing Q. If we let

$$Z_{ij} = X_{ij} - \theta_{ij0} - \sum_{k=1}^{K-1} \theta_{ijk} \rho_{ik} \kappa_{jk} \tag{12.29}$$

denote the "residual" remaining after the first $K - 1$ layers, then we can write Q as

$$Q = \frac{1}{2} \sum_{i=1}^{r} \sum_{j=1}^{n} (Z_{ij} - \theta_{ijK} \rho_{iK} \kappa_{jK})^2 \tag{12.30}$$

$$= \frac{1}{2} \sum_{i=1}^{r} \sum_{j=1}^{n} (Z_{ij} - (\mu_K + \alpha_{iK} + \beta_{jK}) \rho_{iK} \kappa_{jK})^2. \tag{12.31}$$

We wish to minimize Q subject to the identifying conditions

$$\sum_{i=1}^{r} \alpha_{iK} \rho_{iK}^2 = \sum_{j=1}^{n} \beta_{jK} \kappa_{jK}^2 = 0. \tag{12.32}$$

From (12.31) and (12.32), we set up the usual Lagrangian multipliers, differentiate wrt μ_K, α_{iK}, and β_{jK}, set the derivatives equal to zero, and solve. The results give:

$$\mu_K^* = \frac{\sum_i \sum_j Z_{ij} \rho_{iK} \kappa_{jK}}{(\sum_i \rho_{iK}^2)(\sum_j \kappa_{jK}^2)} \tag{12.33}$$

$$\alpha_{iK}^* = \frac{\sum_j (Z_{ij} - \mu_K \rho_{iK} \kappa_{jK}) \kappa_{jK}}{\rho_{iK}(\sum_j \kappa_{jK}^2)} \tag{12.34}$$

$$\beta_{jK}^* = \frac{\sum_i (Z_{ij} - \mu_K \rho_{iK} \kappa_{jK}) \rho_{iK}}{\kappa_{jK}(\sum_i \rho_{iK}^2)}. \tag{12.35}$$

Given the values of $\rho_{iK}^{(s-1)}$ and $\kappa_{jK}^{(s-1)}$ from the $(s-1)$st iteration, we use (12.33)–(12.35) to update $\theta_{ijK}^{(s)}$ at the sth iteration. Note that updating

α_{iK}^* only requires data for the ith gene, and updating β_{jK}^* only requires data for the jth sample; hence, the resulting iterations are very fast.

Given values for θ_{ijK}, the update formulas for ρ_{iK} and κ_{jK} are found by differentiating (12.14) wrt ρ_{iK} and κ_{jK}, setting the results equal to zero, and solving. This gives:

$$\rho_{iK}^* = \frac{\sum_j Z_{ij}\theta_{ijK}\kappa_{jK}}{\sum_j \theta_{ijK}^2\kappa_{jK}^2} \tag{12.36}$$

$$\kappa_{jK}^* = \frac{\sum_i Z_{ij}\theta_{ijK}\rho_{iK}}{\sum_i \theta_{ijK}^2\rho_{iK}^2}. \tag{12.37}$$

So, set the initial values of all the ρs and the κs to be in $(0,1)$ (say, make them all equal to 0.5). Then, given values of $\theta_{ijK}^{(s)}$ and $\kappa_{jK}^{(s-1)}$, we use (12.20) to update $\rho_{iK}^{(s)}$. Similarly, given values of $\theta_{ijK}^{(s)}$ and $\rho_{iK}^{(s-1)}$, we use (12.21) to update $\kappa_{jK}^{(s)}$. The trick is to keep ρ and κ away from 0 and 1 early in the iteration process, but to force ρ and κ toward 0 and 1 late in the process. At convergence, the estimated parameters for the kth layer are denoted by $\widehat{\mu}_k, \widehat{\alpha}_{ik}$, and $\widehat{\beta}_{jk}$, $k = 1, 2, \ldots, K$.

The absolute values of the row effects, $|\widehat{\mu}_k + \widehat{\alpha}_{ik}|$, and the column effects, $|\widehat{\mu}_k + \widehat{\beta}_{jk}|$, for the kth layer $(k = 1, 2, \ldots, K)$ can each be ordered to show which genes and samples are most affected by the biological conditions of that layer. Within the kth layer, genes are *upregulated* if $\widehat{\mu}_k + \widehat{\alpha}_{ik} > 0$, whereas genes with $\widehat{\mu}_k + \widehat{\alpha}_{ik} < 0$ are said to be *downregulated*. The "size" or "importance" of the kth layer is indicated by the value of

$$\sigma_k^2 = \sum_{i=1}^n \sum_{j=1}^r \rho_{ij}^*\kappa_{jk}^*\theta_{ijk}^2, \tag{12.38}$$

and this quantity is used in a permulation argument by Lazzeroni and Owen to choose the number of layers K.

12.8.3 Example: Leukemia (ALL/AML) Data

The data for this example[4] are obtained from a study of two types of acute leukemias — acute lymphoblastic leukemia (ALL) and acute myeloid leukemia (AML) (Golub et al, 1999). The leukemia data, which consist of gene expression levels for 7,219 probes from 6,817 human genes, were

---

[4]The leukemia data can be found in the file ALL_AML_Merge.txt on the book's website. The data are available in the BIOCONDUCTOR R package golubEsets, and the preprocessing code is in the BIOCONDUCTOR R package multtest, both of which can be downloaded from the website http://www.bioconductor.org.

derived using Affymetrix high-density oligonucleotide arrays. There are 72 mRNA samples made up of 47 ALL samples (38 B-cell and 9 T-cell) and 25 AML samples extracted from bone marrow (BM) or from peripheral blood (PB).

The leukemia data were preprocessed following the methods of Golub et al. (see Dudoit, Fridlyand, and Speed, 2002): (1) a floor and ceiling of 100 and 16,000, respectively, were set for the expression levels; (2) any gene that has low variability (i.e., any gene with either $\max / \min \leq 5$ or $\max - \min \leq$ 500) over all tissue samples was excluded; (3) the remaining expression levels were transformed using a logarithmic (base-10) transformation; (4) the preprocessed leukemia data were standardized by centering (mean 0) and scaling (variance 1) each of the mRNA samples across rows (genes). This left a data array, $\mathcal{X} = (X_{gi})$, consisting of 3,571 rows (genes) by 72 columns (mRNA samples), where X_{gi} denotes the expression level for the gth gene in the ith mRNA sample.

We applied the plaid model to the leukemia data. Our strategy consisted of (1) four shuffles in the stopping rule; (2) a common sign for $\mu + \alpha_i$ and for $\mu + \beta_j$ within each layer; and (3) any row (or column) released from a layer if being part of a layer failed to reduce its sum of squares by at least 0.51. The algorithm stopped after finding 11 layers, each containing α_i and β_j components. After the 11th layer, the algorithm failed to find a layer that retained any rows under the release criterion.

Table 12.9 shows the composition of each of the 11 layers. We see that layer 4 is completely composed of AML samples, layer 5 consists of only ALL B-cell samples, and layers 3 and 11 contain only ALL samples. All other layers are mixed ALL and AML samples. Only 55 of the 72 samples are contained in the 11 layers, so that 17 samples were not included in any layer. The biggest percentage omission is for the ALL T-cell samples with 5 out of 9 samples not included; 9 of the 38 ALL B-cell samples and 3 of the 25 AML samples are omitted.

Table 12.10 gives the estimated column effects, $\widehat{\mu}_k + \widehat{\beta}_{jk}$, in the first 8 layers; notice that the signs of each column effect are the same within each layer. We see a pattern of similar mRNA samples appearing in the odd layers 1, 3, 5, 7, and 11, and in the even layers 2, 4, 6, and 8. These odd-even patterns, however, are switched in layers 9 and 10.

While we see from Table 12.9 that the number of samples in the different layers is about the same, the number of genes decreases from more than 200 in the first few layers to a much smaller number in each of the last few layers. About half of the genes in each of the first two layers are the same, whereas a third of the genes in layer 3 are present in layer 4 and vice versa. The amount of gene overlap in the other layers is negligible.

TABLE 12.9. *Plaid analysis of the leukemia data. Composition of each layer by the number of genes (rows) and number of samples (columns), and the number of ALL B-cells, ALL T-cells, and AML samples in each layer.*

Layer	Genes	Samples	ALL-B	ALL-T	AML
1	230	14	12	0	2
2	222	16	9	1	6
3	265	13	12	1	0
4	238	19	0	0	19
5	61	14	14	0	0
6	13	16	3	2	11
7	15	13	11	0	2
8	3	17	6	2	9
9	11	17	5	1	11
10	5	14	13	0	1
11	10	10	9	1	0

12.9 Clustering Based Upon Mixture Models

So far, our treatment of clustering has been algorithmic; rather than creating clustering methods based upon a statistical model with stochastic elements (so that the the full force of the traditional statistical inference framework could be applied), we have used nonstochastic methods whose computational solution in each case is an iterative algorithm, which is a general optimization routine for the treatment of incomplete data. The EM algorithm has been found to be especially valuable for clustering data in problems from machine learning, computer vision, vector quantization, image restoration, and market segmentation.

Suppose $\mathbf{X} \sim p(\cdot|\boldsymbol{\psi})$, where $\boldsymbol{\psi}$ is an unknown parameter vector. The *complete-data likelihood* is given by

$$\mathcal{L}(\boldsymbol{\psi}|\mathbf{X}) = p(\mathbf{X}|\boldsymbol{\psi}). \tag{12.39}$$

Now, suppose some components of $\mathbf{X}$ are missing. We can write

$$\mathbf{X} = (\mathbf{X}_{obs}^{\tau}, \mathbf{X}_{mis}^{\tau})^{\tau}, \tag{12.40}$$

where $\mathbf{X}_{obs}$ is the observed part of $\mathbf{X}$, and $\mathbf{X}_{mis}$ is the missing part of $\mathbf{X}$. If the probability that a particular variable is unobserved depends only upon $\mathbf{X}_{obs}$ and not on $\mathbf{X}_{mis}$, then the *observed-data likelihood* is obtained by integrating $\mathbf{X}_{mis}$ out of the complete-data likelihood,

$$\mathcal{L}_{obs}(\boldsymbol{\psi}|\mathbf{X}_{obs}) = \int p(\mathbf{X}_{obs}, \mathbf{X}_{mis}|\boldsymbol{\psi}) \, d\mathbf{X}_{mis}. \tag{12.41}$$

TABLE 12.10. *Plaid analysis of the leukemia data. Estimated column effects $(\widehat{\mu} + \widehat{\beta}_j)$ for the first 8 layers. Samples whose estimated effects do not appear in a column are not included in that layer.*

Sample	1	2	3	4	5	6	7	8
ALLT 3		0.72				0.53		
ALLB 4								0.63
ALLB 5	−1.04		1.15		−0.63			
ALLT 6						0.66		0.84
ALLB 7		0.81						1.37
ALLB 8		1.09				0.74		1.58
ALLB 13	−0.86		1.10		−0.68			
ALLT 14			0.61					
ALLB 15	−1.19		1.07		−0.82		−0.96	
ALLB 16			0.63					
ALLB 19					−0.51			
ALLB 20	−1.24		1.39		−0.99			
ALLB 21	−0.81		1.47					
ALLB 22		0.65						
ALLT 23								0.49
ALLB 24			0.96					
ALLB 27		1.54				0.70		1.54
AML 28				−0.65		0.47		0.67
AML 29	−0.77						−0.85	
AML 30				−0.79				
AML 31				−0.54				0.70
AML 32				−0.70		0.71		
AML 33		0.86		−1.13		0.78		0.60
AML 34		0.69		−0.70		0.84		
AML 35		1.06		−0.62				
AML 36				−0.96		0.69		
AML 37				−0.92		0.88		0.39
AML 38		0.67		−0.84				0.93
ALLB 39		0.72						0.96
ALLB 40		0.86						
ALLB 41	−1.09		1.08		−0.63		−0.78	
ALLB 43							−1.25	
ALLB 44	−0.72				−0.43		−0.63	
ALLB 45	−0.74				−0.41		−0.75	
ALLB 46	−0.80				−0.47		−0.89	
ALLB 47			0.63		−0.60			
ALLB 48	−0.74		1.25		−0.78		−0.69	
ALLB 49		1.29						1.07
AML 50				−0.85		0.93		1.31
AML 51				−0.85		0.97		
AML 53				−0.94		0.77		0.85
ALLB 56		0.85				0.63		
AML 58		1.04		−0.78		0.77		0.68
ALLB 59					−0.36		−0.67	
AML 61		0.71		−0.59				0.58
AML 62				−0.60				
AML 63				−0.68				
AML 64		1.06		−0.82		0.76		
AML 65				−0.49				
AML 66	−1.04						−0.71	
ALLB 68	−1.26		1.19		−0.74		−1.01	
ALLB 69	−1.04		0.90		−0.76		−0.83	
ALLB 70							−0.53	

TABLE 12.11. *The EM algorithm.*

1. Input: $\widehat{\psi}^{(0)}$ = initial guess for the parameter vector ψ.
2. Let $\mathbf{X} = (\mathbf{X}_{obs}^{\tau}, \mathbf{X}_{mis}^{\tau})^{\tau}$ represent the "complete" data, where $\mathbf{X}_{obs}$ and $\mathbf{X}_{mis}$ are the portions of $\mathbf{X}$ which are observed and missing, respectively.
3. For m=0,1,2,..., iterate between the following two steps:
   - *E-step*: Compute

$$Q(\psi \mid \widehat{\psi}^{(m)}) = \mathrm{E}\left\{\ell(\psi|\mathbf{X}) \mid \mathbf{X}_{obs}, \widehat{\psi}^{(m)}\right\}$$

   as a function of ψ.
   - *M-step*: Find $\widehat{\psi}^{(m+1)} = \arg\max_{\psi} Q(\psi \mid \widehat{\psi}^{(m)})$.
4. Stop when convergence of the log-likelihood is attained.

The MLE for ψ based upon the observed data $\mathbf{X}_{obs}$ is the ψ that maximizes $\mathcal{L}_{obs}(\psi|\mathbf{X}_{obs})$. Unfortunately, a direct attack on this problem usually fails.

The EM algorithm is tailor-made for this type of problem. It is a two-step iterative process, incorporating an *expectation step* (E-step) with a *maximization step* (M-step); see Table 12.11 for the algorithmic details. The E-step computes the conditional expectation of the complete-data log-likelihood given the observed data and the current parameter estimate, and the M-step updates the parameter estimate by maximizing the conditional expectation from the E-step.

Because $p((\mathbf{X}_{mis}|\mathbf{X}_{obs}, \psi) = p(\mathbf{X}_{obs}, \mathbf{X}_{mis}|\psi)/p(\mathbf{X}_{obs}|\psi)$, the observed-data log-likelihood is

$$\begin{aligned}
\ell(\psi|\mathbf{X}_{obs}) &= \log p(\mathbf{X}_{obs}|\psi) \\
&= \ell(\psi|\mathbf{X}) - \log p(\mathbf{X}_{mis}|\mathbf{X}_{obs}, \psi),
\end{aligned} \tag{12.42}$$

where $\ell(\psi|\mathbf{X})$ is the complete-data log-likelihood, which may be easy to compute, and $\log p(\mathbf{X}_{mis}|\mathbf{X}_{obs}, \psi)$ is the part of the complete-data log-likelihood due to the missing data. Taking expectations of (12.39) wrt the conditional density $p(\mathbf{X}_{mis}|\mathbf{X}_{obs}, \psi')$, where ψ' is a current value of ψ, yields

$$\ell(\psi|\mathbf{X}_{obs}) = Q(\psi|\psi') - H(\psi|\psi'), \tag{12.43}$$

where

$$\begin{aligned}
Q(\psi|\psi') &= \int \ell(\psi|\mathbf{X})p(\mathbf{X}_{mis}|\mathbf{X}_{obs}, \psi')d\mathbf{X}_{mis} \\
&= \mathrm{E}\{\ell(\psi|\mathbf{X})|\mathbf{X}_{obs}, \psi'\},
\end{aligned} \tag{12.44}$$

and

$$H(\boldsymbol{\psi}|\boldsymbol{\psi}') = \int \log p(\mathbf{X}_{mis}|\mathbf{X}_{obs}, \boldsymbol{\psi}) p(\mathbf{X}_{mis}|\mathbf{X}_{obs}, \boldsymbol{\psi}') d\mathbf{X}_{mis}$$
$$= \mathrm{E}\{\log p(\mathbf{X}_{mis}|\mathbf{X}_{obs}, \boldsymbol{\psi})|\mathbf{X}_{obs}, \boldsymbol{\psi}'\}. \qquad (12.45)$$

If we now set

$$h(\mathbf{X}_{mis}) = \frac{p(\mathbf{X}_{mis}|\mathbf{X}_{obs}, \boldsymbol{\psi})}{p(\mathbf{X}_{mis}|\mathbf{X}_{obs}, \boldsymbol{\psi}')}, \qquad (12.46)$$

then,

$$H(\boldsymbol{\psi}|\boldsymbol{\psi}') - H(\boldsymbol{\psi}'|\boldsymbol{\psi}') = \mathrm{E}\{\log h(\mathbf{X}_{mis})|\mathbf{X}_{obs}, \boldsymbol{\psi}'\}$$
$$\leq \mathrm{E}\{h(\mathbf{X}_{mis}|\mathbf{X}_{obs}, \boldsymbol{\psi}')\} - 1$$
$$= 0, \qquad (12.47)$$

where we have used the inequality $\log x \leq x - 1$. Thus, $H(\boldsymbol{\psi}|\boldsymbol{\psi}') \leq H(\boldsymbol{\psi}'|\boldsymbol{\psi}')$.

From (12.43), the difference in $\ell(\boldsymbol{\psi}|\mathbf{X}_{obs})$ at the mth and $(m+1)$st iterations is

$$\ell(\boldsymbol{\psi}^{(m+1)}|\mathbf{X}_{obs}) - \ell(\boldsymbol{\psi}^{(m)}|\mathbf{X}_{obs})$$
$$\geq Q(\boldsymbol{\psi}^{(m+1)}|\boldsymbol{\psi}^{(m)}) - Q(\boldsymbol{\psi}^{(m)}|\boldsymbol{\psi}^{(m)}) \geq 0, \qquad (12.48)$$

where we have used (12.44) and the fact that the EM algorithm finds $\boldsymbol{\psi}^{(m+1)}$ to make $Q(\boldsymbol{\psi}^{(m+1)}|\boldsymbol{\psi}^{(m)}) > Q(\boldsymbol{\psi}^{(m)}|\boldsymbol{\psi}^{(m)})$. Thus, *the log-likelihood function increases at each iteration* (more accurately, it does not decrease). From this result, it can be shown that (under reasonably mild regularity conditions) convergence of the log-likelihood, at least to a local maximum, is ensured by this iterative process (Wu, 1983). Note, however, that local convergence of the log-likelihood does not automatically imply local convergence of the parameter estimates, although the latter convergence holds under additional regularity conditions.

The EM algorithm possesses reliable convergence properties and low cost per iteration, does not require much storage space, and is easy to program. Yet, it can be extremely slow to converge if there are many missing data and if the size of the data set is large. (We note that some effort has been made to speed up the EM algorithm.) Furthermore, because convergence is guaranteed only to a local maximum, and because likelihood surfaces often possess many local maxima, it is usually necessary to run the EM algorithm using different random starts to try to find a global maximum of the likelihood function.

12.9.1   The EM Algorithm for Finite Mixtures

In mixture problems, if we knew which observations belonged to which group or class, then we could divide up the data by class and then estimate the parameters of each component density separately. Not knowing the class labels means that the labels and the parameters have to be estimated simultaneously.

One of the first applications of the EM algorithm was to the finite mixtures problem. The "trick" here is to introduce a K-vector of dummy variables,

$$\mathbf{X}_{i,mis} = (X_{i1,mis}, \cdots, X_{iK,mis})^{\tau}, \tag{12.49}$$

where

$$X_{ik,mis} = \begin{cases} 1 & \text{if } \mathbf{X}_{i,obs} \in \Pi_k \\ 0 & \text{otherwise} \end{cases} \tag{12.50}$$

$k = 1, 2, \ldots, K$, and use it to augment the ith observation, $\mathbf{X}_{i,obs}$, to produce a "complete" data vector,

$$\mathbf{X}_i = (\mathbf{X}_{i,obs}^{\tau}, \mathbf{X}_{i,mis}^{\tau})^{\tau}, \quad i = 1, 2, \ldots, n. \tag{12.51}$$

This idea of creating "missing data" for this problem as indicators of the unknown class labels was a key innovation of Dempster, Laird, and Rubin (1977).

Assume now that $\mathbf{X}_{i,mis}$ is iid according to a single draw from a K-class multinomial distribution with probabilities $\pi_k = \text{Prob}\{\mathbf{X}_{i,obs} \in \Pi_k\}$, $k = 1, 2, \ldots, K$. That is,

$$\mathbf{X}_{i,mis} \overset{iid}{\sim} \text{Mult}_K(1, \boldsymbol{\pi}), \quad i = 1, 2, \ldots, n, \tag{12.52}$$

where $\boldsymbol{\pi} = (\pi_1, \ldots, \pi_K)^{\tau}$. Hence,

$$\mathbf{X}_{i,obs}|\mathbf{X}_{i,mis} \sim \prod_{k=1}^{K} [f_k(\mathbf{X}_{i,obs}|\boldsymbol{\theta}_k)]^{X_{ik,mis}}. \tag{12.53}$$

From (13.49) and (13.50), the complete-data log-likelihood is

$$\begin{aligned} \ell(\boldsymbol{\psi}|\mathbf{X}) &= \ell(\{\boldsymbol{\theta}_k\}, \{\pi_k\}, \{X_{ik,mis}\}|\mathbf{X}) \\ &= \sum_{i=1}^{n} \sum_{k=1}^{K} X_{ik,mis} \log\{\pi_k f_k(\mathbf{X}_{i,obs}|\boldsymbol{\theta}_k)\}. \end{aligned} \tag{12.54}$$

The E-step computes $Q(\boldsymbol{\psi}|\widehat{\boldsymbol{\psi}}^{(m)})$ by replacing each dummy variable $X_{ik,mis}$ in (12.54) by its conditional expectation,

$$\widehat{X}_{ik,mis}^{(m)} = \mathrm{E}\{X_{ik,mis}|\mathbf{X}_{i,obs}, \widehat{\boldsymbol{\psi}}^{(m)}\}, \tag{12.55}$$

where $\widehat{\boldsymbol{\psi}}^{(m)}$ is the current estimate of $\boldsymbol{\psi}$. In other words, at the mth iteration, $X_{ik,mis}$ is estimated by the posterior probability that $\mathbf{X}_{i,obs} \in \Pi_k$; from Section 9.5.1, this is

$$\widehat{X}_{ik,mis}^{(m)} = \frac{\widehat{\pi}_k^{(m)} f_k(\mathbf{X}_{i,obs}|\widehat{\boldsymbol{\theta}}_k^{(m)})}{\sum_{j=1}^{K} \widehat{\pi}_j^{(m)} f_j(\mathbf{X}_{i,obs}|\widehat{\boldsymbol{\theta}}_j^{(m)})}. \tag{12.56}$$

The M-step then takes the probabilities of class membership provided by the E-step, inserts them into (12.54) in place of $X_{ik,mis}$, and updates the parameter values from the E-step by maximizing (12.54) wrt $\{\pi_k\}, \{\boldsymbol{\theta}_k\}$. The M-step for the mixture proportions $\{\pi_k\}$ is given by

$$\widehat{\pi}_k^{(m+1)} = n^{-1} \sum_{i=1}^{n} \widehat{X}_{ik,mis}^{(m)}, \quad k = 1, 2, \ldots, K. \tag{12.57}$$

The M-step for the parameter vector $\boldsymbol{\psi}$ depends upon the context. The E-step and M-step are iterated as many times as it is necessary to achieve convergence of the log-likelihood. *The ML determination of the class of the ith observation is then the class corresponding to the largest value of* $\widehat{X}_{ik,mis}, k = 1, 2, \ldots, K.$

Consider, for example, a mixture of the two univariate Gaussian densities $\phi(x|\boldsymbol{\theta}_1)$ and $\phi(x|\boldsymbol{\theta}_2)$, where the parameter vectors are $\boldsymbol{\theta}_1 = (\mu_1, \sigma_1^2)^\tau$ and $\boldsymbol{\theta}_2 = (\mu_2, \sigma_2^2)^\tau$, and the mixture proportions are $\pi_1 = 1 - \pi$ and $\pi_2 = \pi$. We also drop the subscript k. The E-step (13.56) reduces to

$$\widehat{X}_{i,mis}^{(m)} = \frac{\widehat{\pi}^{(m)} \phi(X_{i,obs}|\widehat{\boldsymbol{\theta}}_2^{(m)})}{(1 - \widehat{\pi}^{(m)})\phi(X_{i,obs}|\widehat{\boldsymbol{\theta}}_1^{(m)}) + \widehat{\pi}^{(m)} \phi(X_{i,obs}|\widehat{\boldsymbol{\theta}}_2^{(m)})}, \tag{12.58}$$

where $\widehat{\pi}^{(m)} = n^{-1} \sum_{i=1}^{n} \widehat{X}_{i,mis}^{(m)}$. By maximizing (13.54) while fixing $X_{ik,mis} = \widehat{X}_{ik,mis}^{(m)}$, the M-step yields the estimates

$$\widehat{\mu}_1^{(m+1)} = \frac{\sum_{i=1}^{n}(1 - \widehat{X}_{i,mis}^{(m)})X_{i,obs}}{\sum_{i=1}^{n}(1 - \widehat{X}_{i,mis}^{(m)})}, \tag{12.59}$$

$$(\widehat{\sigma}_1^2)^{(m+1)} = \frac{\sum_{i=1}^{n}(1 - \widehat{X}_{i,mis}^{(m)})(X_{i,obs} - \widehat{\mu}_1^{(m+1)})^2}{\sum_{i=1}^{n}(1 - \widehat{X}_{i,mis}^{(m)})}, \tag{12.60}$$

$$\widehat{\mu}_2^{(m+1)} = \frac{\sum_{i=1}^{n} \widehat{X}_{i,mis}^{(m)} X_{i,obs}}{\sum_{i=1}^{n} \widehat{X}_{i,mis}^{(m)}}, \tag{12.61}$$

$$(\widehat{\sigma}_2^2)^{(m+1)} = \frac{\sum_{i=1}^{n} \widehat{X}_{i,mis}^{(m)}(X_{i,obs} - \widehat{\mu}_2^{(m+1)})^2}{\sum_{i=1}^{n} \widehat{X}_{i,mis}^{(m)}}. \tag{12.62}$$

Experimentation with this mixture model has shown that whereas convergence of the log-likelihood may be incredibly slow, most of the progress toward convergence tends to occur during the first few iterations (Redner and Walker, 1984).

In the multivariate Gaussian mixture problem (see Exercise 12.9), the "curse of dimensionality" raises its ugly head, where the number of parameters grows quickly with the increase in dimensionality. Although PCA is often used as a first step to reduce the dimensionality, this does not help in mixtures problems because any class structure as exists may not be preserved by the principal components (Chang, 1983). Furthermore, whenever estimates of the covariance matrix become singular or nearly singular, the EM algorithm breaks down; this can happen, for example, if the mixture has too many components and at least one of those components has too few observations, or when the dimensionality is greater than the number of observations, such as occurs with microarray experiments. This is currently an area of much research (Fraley and Raftery, 2002).

12.9.2 How Many Components?

The number of components, K, is one of the most important ingredients in mixture modeling, which becomes more complicated when the value of K is unknown. As a result, much attention has been paid to this issue. By and large, attempts at formulating test criteria to decide on the number of components have not been successful.

For example, an early decision procedure was the likelihood-ratio test statistic $-2\log\lambda_k$, where λ_k is the likelihood ratio (LR) (Wolfe, 1970). The LR compares a mixture having k components with a mixture having $k+1$ components and then repeats the test for a succession of increasing values of k, each time comparing the result to a reference χ^2-distribution. The testing stops the first time that a k-mixture density is not rejected in favor of a $(k+1)$-mixture density. Recent empirical evidence indicates that this test tends to overestimate the value of K. More seriously, the regularity conditions for the χ^2 approximation do not hold in finite-mixture problems.

Several alternatives to the likelihood ratio test have since been proposed. The two most prominent approaches are a nonparametric bootstrap assessment of the number of modes in the data using a kernel density estimator with a sequence of decreasing window-widths (Silverman, 1981, 1983) and a Bayesian solution that uses the EM algorithm to fit the mixture model and then computes approximate Bayes factors to decide on K (Fraley and Raftery, 2002). Silverman's approach is promising, but there are a number of anomolies in its behavior (Izenman and Sommer, 1988). *Bayes factors* (Kass and Raftery, 1995) are ratios of high-dimensional integrals and are often impossible to compute; arguments have been made to justify BIC as approximate Bayes factors to estimate K, even though the regularity conditions for the BIC approximation do not hold for finite-mixture models.

12.10   Software Packages

Almost all the major statistical software packages contain hierarchical and non-hierarchical clustering routines for clustering observations or variables as appropriate. Software for two-way clustering methods, model-based clustering methods, and other recently developed methods have to be downloaded from the Internet.

There are two SOM methods, `batchSOM` and `SOM`, in the R package (Venables and Ripley, 2002, pp. 310–311) and a CRAN package `som` (formerly `GeneSOM`) for gene expression data. A SOM TOOLBOX for MATLAB can be downloaded free from `www.cis.hut.fi/projects/somtoolbox/`.
Another package for computing SOMs is GENECLUSTER, which can be downloaded from the website
`www-genome.wi.mit.edu/cancer/software/software.html`.
The U-matrix and component planes in Figures 13.11 and 13.12 were computed using MATLAB `somtoolbox`.

A fast algorithm for gene-shaving forms the basis for the software package GENECLUST, which can be downloaded free from
`odin.mdacc.tmc.edu/~kim/geneclust`; see Do, Broom, and Wen (2003). Software and documentation (Owen, 2000) for applying plaid models to a data array can be downloaded from
`www-stat.stanford.edu/ owen/clickwrap/plaid.html`.

Most research into model-based clustering from a Bayesian viewpoint has been carried out by Adrian Raftery and colleagues. Their S-PLUS functions `mclust` and `mclust-em` and documentation (Fraley and Raftery, 1998) can be downloaded from
`www.stat.washington.edu/raftery/Research/Mclust`.

The EMMIX software package can fit a mixture model with Gaussian or t-components (McLachlan, Peel, Basford, and Abrams, 1999) and can be downloaded from `www.jstatsoft.org`.

Bibliographical Notes

Books that focus on cluster analysis include Kaufman and Rousseeuw (1990) and Hartigan (1975). Cluster analysis can be found as a chapter of most books on multivariate analysis: Rencher (2002, Chapter 14), Lattin, Carroll, and Green (2003, Chapter 8), Johnson and Wichern (1998, Chapter 12), Seber (1984, Chapter 7). See also Ripley (1996, Section 9.3).

Books on self-organizing maps include Oja and Kaski (2003), and Kohonen (2001). There is also a Special Issue of *Neural Networks* in 2002 on New Developments in Self-Organizing Maps.

Review articles on the use of clustering in analyzing microarray data include Sebastiani, Gussoni, Kohane, and Ramoni (2003), Bryan (2004), and Chipman, Hastie, and Tibshirani (2003).

There is a huge literature on mixtures of distributions. Book references include Everitt and Hand (1981), Titterington, Smith, and Makov (1985), McLachlan and Basford (1988), and McLachlan and Peel (2000). The idea of representing a density function as a mixture of two Gaussian components was popularized by Tukey (1960) as a way of modeling outliers in data, where he assumed equal means but different variances, one variance much larger than the other.

The EM algorithm has a long and interesting history, with the earliest version published in 1926. It was named in Dempster, Laird, and Rubin (1977), who showed the monotonic behavior of the log-likelihood function and gave examples of the general applicability of the algorithm. Books that give good accounts of the EM algorithm include Hastie, Tibshirani, and Friedman (2001, Section 8.5), Schafer (1997, Chapter 3), Ripley (1996, Appendix A.2), and Little and Rubin (1987, Chapter 7). See also the edited volume by Wanatabe and Yamaguchi (2004). An excellent review of model-based clustering is given by Fraley and Raftery (2002).

Exercises

12.1 Run the clustering algorithms for the `satimage` data, but only using the center pixels (i.e., variables `CC1`, `CC2`, `CC3`, `CC4`) of each 3×3 neighborhood. Compare your results with those in Table 12.10.

12.2 Write a computer program to implement single-linkage, average-linkage, and complete-linkage agglomerative hierarchical clustering. Try it out on a data set of your choice.

12.3 Cluster the `primate.scapulae` data using single-linkage, average-linkage, and complete-linkage agglomerative hierarchical clustering methods. Find the five-cluster solutions for all three methods, which allows comparison with the true primate classifications. Find the misclassification rate for all three methods. Show that the lowest rate occurs for the complete-linkage method and the highest for the single-linkage method.

12.4 Using the leukemia (ALL/AML) data, run a SOM algorithm (either on-line or batch) to cluster the genes. Draw a SOM plot and identify the genes captured by each representative. Consult with a biologist to see whether the clusters of genes are biologically meaningful. Compute the U-matrix and the component planes. Solely on the basis of the patterns provided by the component planes, can you separate them into the three groups of ALL-B, ALL-T, and AML tissue samples?

12.5 Microarray data from the National Cancer Institute can be found in the file `ncifinal.txt` on the book's website. There are 5,244 genes and 61 samples in this data set; the samples are derived from tumors with different sites of origin: 7 breast, 5 central nervous system (CNS), 7 colon, 6 leukemia, 8 melanoma, 9 non–small-cell lung carcinoma (NSCLC), 6 ovarian, and 9 renal. There are also data from independent microarray experiments yielding 2 leukemia samples (K562) and 2 breast cancer samples (MCF7). Use the gene shaving method to cluster the genes in this data set into 8 clusters. Describe the appearance of the heatmap for each cluster, and use the gap statistic to determine the number of genes in each cluster.

12.6 Nutritional data from 961 different food items is given in the file `food.txt`, which can be downloaded from the book's website or from `http://www.ntwrks.com/~mikev/chart1.html`. For each food item, there are 7 variables: fat (grams), food energy (calories), carbohydrates (grams), protein (grams), cholesterol (milligrams), weight (grams), and saturated fat (grams). To equalize out the different types of servings of each food, first divide each variable by weight of the food item. Next, because of the wide variations in the different variables, standardize each variable. The resulting data are $\mathcal{X} = (X_{ij})$. Apply plaid models to these data. Describe your findings for each of the first 10 layers.

12.7 Establish the ML estimates (12.57), (12.59)–(12.62) for the parameters of the two-component univariate Gaussian mixture.

12.8 Using the EM algorithm, find the ML estimates of the parameters of a finite mixture of multivariate Gaussian densities with equal covariance matrice $\mathbf{\Sigma}$. Show that the ML estimate $\widehat{\mathbf{\Sigma}}^{(m)}$ has to be inverted at each iteration m, which is one of the factors slowing down the computational speed of the algorithm.

12.9 Run a batch-SOM analysis on the Wisconsin Breast-Cancer data `wbcd`. Find the "circles" representation for the data and describe how well the SOM method clusters the tumor cases into `benign` and `malignant`. Compute the U-matrix and discuss its representation for these data.

13
Multidimensional Scaling and Distance Geometry

13.1 Introduction

Imagine you have a map of a particular geographical region, which includes a number of cities and towns. Usually, such a map will be accompanied by a two-way table displaying how close a selected number of those towns and cities are to each other. Each cell of that table will show the degree of "closeness" (or *proximity*) of the row city to the column city that identifies that cell. The notion of proximity between two geographical locations is easy to understand, even though it could have different meanings: for example, proximity could be defined as straight-line distance or as shortest traveling distance.

In more general situations, proximity could be a more complicated concept. We can talk about the proximity of any two entities to each other, where by "entity" we might mean an object, a brand-name product, a nation, a stimulus, etc. The proximity of a pair of such entities could be a measure of association (e.g., the absolute value of a correlation coefficient), a confusion frequency (i.e., to what extent one entity is confused with another in an identification exercise), or some other measure of how alike (or how different) one perceives the entities. If we are studying a set of linked Internet webpages, we may be interested in visualizing a hypermedia network

A.J. Izenman, *Modern Multivariate Statistical Techniques*,
doi: 10.1007/978-0-387-78189-1_13,
© Springer Science+Business Media, LLC 2008

in which proximity would be based upon a notion of *network distance* (i.e., the number of hyperlinks needed to jump from one node to another).

The general problem of *multidimensional scaling (MDS)* essentially reverses that relationship: given only a two-way table of proximities, we wish to reconstruct the original map as closely as possible. A further wrinkle in the problem is that we also do not know the number of dimensions in which the given entities are located. So, determining the number of dimensions is another major problem to be solved.

MDS is not a single procedure but a family of different algorithms, each designed to arrive at an optimal low-dimensional configuration for a particular type of proximity data. MDS is primarily a data visualization method for identifying "clusters" of points, where points in a particular cluster are viewed as being "closer" to the other points in that cluster than to points in other clusters.

In this chapter, we describe a number of MDS methods. Specifically, we describe and illustrate classical scaling (also called "distance geometry" by those in bioinformatics) and distance scaling (divided according to whether the distances are of metric or nonmetric type). Distance scaling is also referred to as metric and nonmetric MDS. The standard treatment of classical scaling yields an eigendecomposition problem and as such is the same as PCA if the goal is dimensionality reduction. The distance scaling methods, on the other hand, use iterative procedures to arrive at a solution.

In Table 13.1, we list some of the application areas of MDS. We shall see that the essential ideas behind MDS also play prominent roles in evaluating random forests (Chapter 14) and revealing nonlinear manifolds (Chapter 16).

13.1.1 Example: Airline Distances

As a simple example of the MDS problem, consider Table 13.2, which is taken from p. 131 of the Revised 6th Edition (1995) of the *National Geographic Atlas of the World*. The table lists the airline distances (in kms) between $n = 18$ cities: Beijing, Cape Town, Hong Kong, Honolulu, London, Melbourne, Mexico, Montreal, Moscow, New Delhi, New York, Paris, Rio de Janeiro, Rome, San Francisco, Singapore, Stockholm, and Tokyo. For this application of MDS, the problem is to re-create the map that yielded the table of airline distances. Because the cities are scattered around the surface of a sphere, we should expect to recover a solution in three dimensions. Furthermore, because airplanes do not fly through the earth but over its surface, airline distances between cities do not always obey the triangle inequality and so may not be Euclidean.

We used the classical scaling method to obtain 2D and 3D maps of the MDS reconstruction, where each map has 18 points, one for each city. We

TABLE 13.1. *Some application areas and research topics in MDS.*

Psychology: Study the underlying structure of perceptions of different classes of psychological stimuli (e.g., personality traits, gender roles) or physical stimuli (e.g., human faces, everyday sounds, fragrances, colors) and create a "perceptual map" of those stimuli. Understand the psychological dimensions hidden in the data so that we can describe how proximity judgments are generated.

Marketing: Derive "product maps" of consumer choice and product preference (e.g., automobiles, beer) so that relationships between products can be discerned. Use these maps to position new products appropriately, to modify an existing product image to emphasize brand differentiation, or to design future experiments to determine what type of consumer can best discriminate between similar products and on which dimensions.

Ecology: Provide "environmental impact maps" of pollution (e.g., oil spills, sewage pollution, drilling-mud dispersal) on local communities of animals, marine species, and insects. Use such maps to develop a biological taxonomy to classify populations using morphometric or genetic data or from evolutionary theory.

Molecular Biology: Reconstruct the spatial structures of molecules (e.g., amino acids) using biomolecular conformation (3D structure). Interpret their interrelations, similarities, and differences. Construct a 3D "protein map" as a global view of the protein structure universe.

Computational Chemistry: Use a measure of molecular similarity (e.g., interatomic distance) to characterize the behavior and function of molecules derived from large collections of compounds.

Social Networks: Develop "telephone-call graphs," where the vertices are telephone numbers and the edges correspond to calls between them. Recognize instances of credit card fraud and network intrusion detection. Identify clusters in large scientific collaboration networks.

Graph Layout: Design a diagram to describe a network and the system it represents using a graph-theoretic distance (e.g., minimum-path length) between pairs of nodes or vertices. Examples include communications networks, electrical circuit diagrams, wiring diagrams, and protein-protein interaction graphs. Create graphic visualizations of digital image libraries, with images as vertices and proximities (e.g., perceptual differences) between pairs of images as edge weights.

Music: Use a measure of musical sound quality (e.g., a set of spectral components with high resolution at low frequencies to mimic the human auditory system) as input to a nonlinear distance measure to assess the similarities and differences between a variety of songs.

TABLE 13.2. *Airline distances (km) between 18 cities. Source: Atlas of the World, Revised 6th Edition, National Geographic Society, 1995, p. 131.*

	Beijing	Cape Town	Hong Kong	Honolulu	London	Melbourne
Cape Town	12947					
Hong Kong	1972	11867				
Honolulu	8171	18562	8945			
London	8160	9635	9646	11653		
Melbourne	9093	10338	7392	8862	16902	
Mexico	12478	13703	14155	6098	8947	13557
Montreal	10490	12744	12462	7915	5240	16730
Moscow	5809	10101	7158	11342	2506	14418
New Delhi	3788	9284	3770	11930	6724	10192
New York	11012	12551	12984	7996	5586	16671
Paris	8236	9307	9650	11988	341	16793
Rio de Janeiro	17325	6075	17710	13343	9254	13227
Rome	8144	8417	9300	12936	1434	15987
San Francisco	9524	16487	11121	3857	8640	12644
Singapore	4465	9671	2575	10824	10860	6050
Stockholm	6725	10334	8243	11059	1436	15593
Tokyo	2104	14737	2893	6208	9585	8159

	Mexico	Montreal	Moscow	New Delhi	New York	Paris
Montreal	3728					
Moscow	10740	7077				
New Delhi	14679	11286	4349			
New York	3362	533	7530	11779		
Paris	9213	5522	2492	6601	5851	
Rio	7669	8175	11529	14080	7729	9146
Rome	10260	6601	2378	5929	6907	1108
S.F.	3038	4092	9469	12380	4140	8975
Singapore	16623	14816	8426	4142	15349	10743
Stockholm	9603	5900	1231	5579	6336	1546
Tokyo	11319	10409	7502	5857	10870	9738

	Rio	Rome	S.F.	Singapore	Stockholm
Rome	9181				
S.F.	10647	10071			
Singapore	15740	10030	13598		
Stockholm	10682	1977	8644	9646	
Tokyo	18557	9881	8284	5317	8193

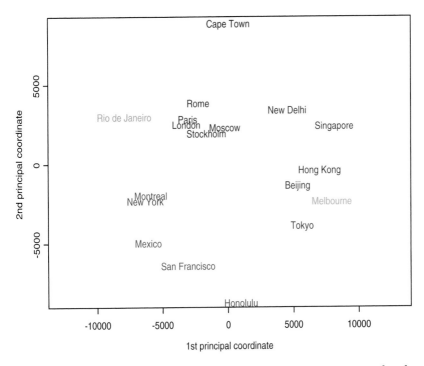

FIGURE 13.1. *Two-dimensional map of 18 world cities using the classical scaling algorithm on airline distances between those cities. The colors reflect the different continents: Asia (purple), North America (red), South America (orange), Europe (blue), Africa (brown), and Australasia (green).*

expect cities with low airline mileage between them to correspond to points in the display that are close together and cities with high airline mileage to correspond to points far apart from each other. In Figure 13.1, we display a scatterplot of the 2D solution.

The 3D solution is given in Figure 13.2. Different colors are used to label the different continents. A dynamic "brush and spin" of the 3D solution shows that the points appear to be scattered around the surface of a sphere; we also see three outliers: Melbourne, Rio de Janeiro, and Cape Town. We expect to see (and we do see) geographically related clusters of points.

Note that the points are not in their customary locations on a globe, and it may be necessary to carry out a rotation and reflection to get them into their usual positions. The computational details needed to produce Figures 13.1 and 13.2 can be found in Section 13.6.3.

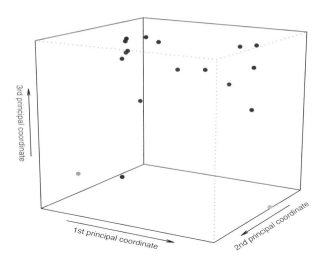

FIGURE 13.2. *Three-dimensional map of 18 world cities using the classical scaling algorithm on airline distances between those cities. The colors reflect the different continents: Asia (purple), North America (red), South America (yellow), Europe (blue), Africa (brown), and Australasia (green).*

13.2   Two Golden Oldies

The primary goal of MDS is to rearrange the entities in some optimal manner so that distances between different entities in the resulting spatial configuration correspond closely to the given proximities. The rearrangement of entities takes place in a space of specified low dimension (usually, 1, 2, or 3 dimensions), where MDS ensures that the given proximities between the entities are well-reproduced by the new configuration.

Before we get into details about the different MDS methods, we first look at a couple of classic examples that were instrumental in paving the way to a greater understanding of the power of MDS for researchers in various fields. These classic examples are the pairwise comparison of color stimuli and of Morse-code signals, where the similarity or dissimilarity of the members of each pair is evaluated by a number of subjects.

13.2.1   Example: Perceptions of Color in Human Vision

In an experiment designed to study the perceptions of color in human vision (Ekman, 1954), 14 colors differing only in their hue (i.e., wavelengths from 434 μm to 674 μm) were projected two at a time onto a screen in an all-pairs design (see Section 13.3 for definition) to 31 subjects, who rated

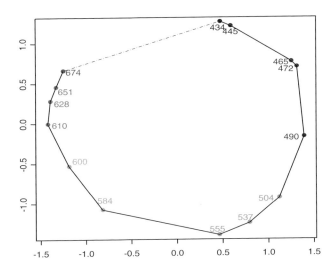

FIGURE 13.3. *Two-dimensional nonmetric MDS representation of color dissimilarities showing the "color circle." The colors correspond to the following wavelengths: 434=indigo, 445=blue, 472=blue-green, 504=green, 555=yellow-green, 600=yellow, 628=orange-yellow, 651=orange, 674=red.*

each of the possible $m = 91$ pairs on a five-point scale from 0 ("no similarity at all") to 4 ("identical"). The rating for each pair of colors was averaged over all subjects and the result divided by 4 to bring the similarity ratings into the interval $[0, 1]$. These mean similarity ratings were then collected into a (14×14) table (see Exercise 13.1), which was treated as a correlation matrix. A visual inspection of the similarities shows that the higher values cluster on the diagonal closest to the main diagonal.

A nonmetric MDS solution for the the color experiment (Shepard, 1962) essentially reproduces the well-known two-dimensional "color circle." Figure 13.3 shows a two-dimensional circular configuration of points representing the 14 colors arranged in order of their wavelengths. A one-dimensional solution would not work because a projection onto the x-axis would make points 434 and 555 lie very close to each other, whereas the dissimilarity between those two colors was one of the largest.

13.2.2   Example: Confusion of Morse-Code Signals

Morse code consists of 36 short signals of dots and dashes (26 letters of the alphabet and the digits 0–9). In a study of the extent of confusion over these different codes (Rothkopf, 1957), the 36 Morse-code signals were acoustically presented by machine in pairs to 598 subjects who had no knowledge of Morse code; each pair of signals was presented twice (e.g.,

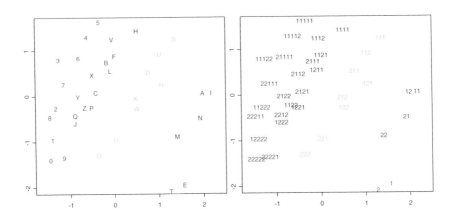

FIGURE 13.4. *Two-dimensional nonmetric MDS representation of Morse-code dissimilarities. The left panel shows the configuration of letters and numbers, and the right panel shows the corresponding Morse code. A "beep" is a dot or a dash. A dot (short beep) is coded as a "1" and a dash (long beep) is coded as a "2." Colors are used to distinguish between code lengths: one beep (purple), two beeps (brown), three beeps (green), four beeps (red), and five beeps (blue).*

A then B, and B then A), and the subjects had to determine whether the members of each pair were the same or different. The results of this experiment yielded 1,260 proximities (instead of the usual $m = 630$) due to asymmetric results from the repeated and inverted presentation of each paired signal. The proximities are given in Exercise 13.2.

A two-dimensional nonmetric MDS solution (Shepard, 1963) is displayed in Figure 13.4. For ease in visualization, dots and dashes are coded by using a "1" for a dot and a "2" for a dash. The graph shows the complexity of the signals. We see that the horizontal axis accounts for code length (i.e., the total number of dots and dashes in the Morse-code symbol) and the vertical axis accounts for the fraction of dots (i.e., ratio of number of dots to code length).

A reanalysis of the MDS solution to the Morse-code data (Buja and Swayne, 2002; Buja, Swayne, Littman, and Hofmann, 2002) using XGvis, an interactive data visualization system for MDS calculations based upon the XGobi package, found evidence that code length and fraction of dots are slightly confounded: long codes that have many dots are more often confused with shorter codes that have many dashes, and vice versa, thereby suggesting a confusion effect due to the physical duration of the code. Furthermore, two additional dimensions were suggested by the graphical analysis: a dummy dimension for the codes of length one and a dummy dimension for initial exposure position (i.e., a dot or dash in the starting position) for the long codes.

13.3 Proximity Matrices

The focus on *pairwise* comparisons of entities is fundamental to MDS. The "closeness" of two entities is measured by a *proximity* measure, which can be defined in a number of different ways. On the one hand, a proximity can be a continuous measure of how physically close one entity is to another (i.e., a *bona fide* distance measure, as in the airline distances example) or it could be a subjective judgment recorded on an ordinal scale, but where the scale is sufficiently well-calibrated as to be considered continuous.

In other cases, especially in studies of perception, a proximity will not be quantitative but will be a subjective rating of *similarity* (or *dissimilarity*) recorded on a pair of entities. A similarity rating is designed to indicate how "close" a pair of entities are to each other, whereas a dissimilarity rating shows the opposite, how unalike are the pair.

In many types of experiments, proximity data are obtained from a group of subjects, each of whom make similarity (or dissimilarity) judgments on all possible $m = \binom{n}{2} = \frac{1}{2}n(n-1)$ unordered pairs of n entities. This type of experiment is said to have an *all-pairs design* (Ramsay, 1982). For example, the color stimuli and Morse-code experiments both followed all-pairs designs. It is unusual for such an experiment to be repeated with the same group of subjects (due to boredom, fatigue, or memory of previous responses), although designs have been constructed to present fewer than all possible pairs to each subject.

It is irrelevant whether we use similarities or dissimilarities as our measure of proximity between two entities. In other words, "closeness" of one entity to another could be measured by a small or large value. The only thing that matters when carrying out MDS is that there should be a monotonic relationship (either increasing or decreasing) between the "closeness" of two entities and the corresponding similarity or dissimilarity value. Anyway, we usually convert similarities into dissimilarities through a monotonically decreasing transformation.

Consider a particular collection of n entities. Let δ_{ij} represent the dissimilarity of the ith entity to the jth entity. We arrange the m dissimilarities, $\{\delta_{ij}\}$, into an $(m \times m)$ square matrix,

$$\boldsymbol{\Delta} = (\delta_{ij}), \tag{13.1}$$

called a *proximity matrix*. The proximity matrix is usually displayed as a lower-triangular array of nonnegative entries, with the understanding that the diagonal entries are all zeroes and that the upper-triangular array is a mirror image of the given lower-triangle (i.e., the matrix is symmetric). In other words, for all $i, j = 1, 2, \ldots, n$,

$$\delta_{ij} \geq 0, \quad \delta_{ii} = 0, \quad \delta_{ji} = \delta_{ij}. \tag{13.2}$$

In order for a dissimilarity measure to be regarded as a *metric* distance, we also require that δ_{ij} satisfy the *triangle inequality*,

$$\delta_{ij} \leq \delta_{ik} + \delta_{kj}, \quad \text{for all } k. \tag{13.3}$$

In some applications (such as the Morse-code example described above), we should not expect symmetry; in such cases, adjustments (e.g., setting $\delta_{ij} \leftarrow \frac{1}{2}(\delta_{ij} + \delta_{ji})$ to form a symmetrized version of $\mathbf{\Delta}$) can be made.

13.4   Comparing Protein Sequences

There are about 100,000 different proteins in the human body, and they provide the internal structure of cells and tissues. Proteins are macromolecules and carry out important bodily functions, including supporting cell structure (skin, tendons, hair, nails, bone), protecting against infection from bacteria and viruses (antibodies, immune system), aiding movement (muscles), transporting materials (hemoglobin for oxygen), and regulating control (enzymes, hormones, metabolism, insulin) of the body. Nearly all of these proteins have a similar chemical structure and, in some instances, even share a common evolutionary origin.

Of major interest in the study of molecular biology is the notion of a spatial "protein map," which would show how existing protein families relate to one another, structurally and functionally. One would hope that such a map would yield important insight into the evolutionary origins of existing protein structures. In this way, researchers might be able to predict the functions of newly discovered proteins from their spatial locations and proximities to other proteins in the map, where we would expect neighboring proteins to have very similar biochemical properties. This also raises the issue of whether a protein map can help justify classifications of proteins into empirically determined classes, such as the four primary classes (α, β, α/β, and $\alpha + \beta$) of proteins as defined by the Structural Classification System of Proteins (SCOP).

13.4.1   Optimal Sequence Alignment

The argument used to compute the proximity of two proteins centers on the idea that amino acids can be altered by random mutations over a long period of evolution. Mutations of a protein sequence can take various

TABLE 13.3. *The 20 amino acids (and their 3-letter and 1-letter abbreviations).*

Alanine (`ala`, A), Arginine (`arg`, R), Asparagine (`asn`, N), Aspartic acid (`asp`, D), Cysteine (`cys`, C), Glutamine (`gln`, Q), Glutamic acid (`glu`, E), Glycine (`gly`, G), Histidine (`his`, H), Isoleucine (`ile`, I), Leucine (`leu`, L), Lysine (`lys`, K), Methionine (`met`, M), Phenylalanine (`phe`, F), Proline (`pro`, P), Serine (`ser`, S), Threonine (`thr`, T), Tryptophan (`trp`, W), Tyrosine (`tyr`, Y), Valine (`val`, V)

forms, such as the deletion or insertion of amino acids, or swapping similar amino acids for ones already in the sequence. For an evolving organism to survive, the structure and functionality of the most important segments of its protein sequences would have to be preserved (or even be improved). Thus, researchers try to understand the evolutionary process of proteins by studying relationships between their respective amino acid sequences.

The comparison problem is complicated by the fact that each sequence is actually a "word" composed of a string of letters selected from a 20-letter alphabet; see Table 13.3. It is a nontrivial task to compute a similarity value between two sequences that have different lengths and different amino acid distributions. The trick here is to align the two sequences (or segments of each of them) so that as many letters in one sequence can be "matched" with the corresponding letters in the other sequence. The extent to which matching occurs will have some bearing on how related (or unrelated) we consider the sequences to be.

There are several methods for carrying out *sequence alignment*. These are generally divided into global and local methods. *Global alignment* tries to align all the letters in the two entire sequences assuming that the two sequences are very similar from beginning to end, whereas *local alignment* assumes that the two sequences are highly similar only over short segments of letters. Alignment methods use *dynamic programming* algorithms as the primary tool (Needleman and Wunsch, 1970; Smith and Waterman, 1981). For searching the huge databases available today, local methods, such as BLAST (Altschul, Gish, Miller, Myers, and Lipman, 1990) and FASTA (Pearson and Lipman, 1988), which use more heuristic-type techniques, have become popular because of their extremely fast computation times, even though their solutions may be slightly suboptimal.

A sequence alignment is declared to be "optimal" if it maximizes an alignment score. For a particular alignment of two sequences, an *alignment score* is the sum of a number of terms, each term comparing an element from the first sequence and a corresponding element in the same position from the second sequence, where an element is either an amino acid or a "gap." When the amino acids in a given position are identical in both

TABLE 13.4. *The BLOSUM62 amino acid substitution matrix. The rows correspond to the amino acids in one protein sequence and the columns correspond to the amino acids in another sequence. At a given position in an alignment of the two sequences, the substitution score of the aligned amino acids is given in the appropriate cell of the matrix. The diagonal entries (in blue) show the scores applied to identities, whereas off-diagonal positive scores are given in red.*

	A	C	D	E	F	G	H	I	K	L	M	N	P	Q	R	S	T	V	W	Y
A	4	0	-2	-1	-2	0	-2	-1	-1	-1	-1	-2	-1	-1	-1	1	0	0	-3	-2
C	0	9	-3	-4	-2	-3	-3	-1	-3	-1	-1	-3	-3	-3	-3	-1	-1	-1	-2	-2
D	-2	-3	6	2	-3	-1	-1	-3	-1	-4	-3	1	-1	0	-2	0	-1	-3	-4	-3
E	-1	-4	2	5	-3	-2	0	-3	1	-3	-2	0	-1	2	0	0	-1	-2	-3	-2
F	-2	-2	-3	-3	6	-3	-1	0	-3	0	0	-3	-4	-3	-3	-2	-2	-1	1	3
G	0	-3	-1	-2	-3	6	-2	-4	-2	-4	-3	0	-2	-2	-2	0	-2	-3	-2	-3
H	-2	-3	-1	0	-1	-2	8	-3	-1	-3	-2	1	-2	0	0	-1	-2	-3	-2	2
I	-1	-1	-3	-3	0	-4	-3	4	-3	2	1	-3	-3	-3	-3	-2	-1	3	-3	-1
K	-1	-3	-1	1	-3	-2	-1	-3	5	-2	-1	0	-1	1	2	0	-1	-2	-3	-2
L	-1	-1	-4	-3	0	-4	-3	2	-2	4	2	-3	-3	-2	-2	-2	-1	1	-2	-1
M	-1	-1	-3	-2	0	-3	-2	1	-1	2	5	-2	-2	0	-1	-1	-1	1	-1	-1
N	-2	-3	1	0	-3	0	1	-3	0	-3	-2	6	-2	0	0	1	0	-3	-4	-2
P	-1	-3	-1	-1	-4	-2	-2	-3	-1	-3	-2	-2	7	-1	-2	-1	-1	-2	-4	-3
Q	-1	-3	0	2	-3	-2	0	-3	1	-2	0	0	-1	5	1	0	-1	-2	-2	-1
R	-1	-3	-2	0	-3	-2	0	-3	2	-2	-1	0	-2	1	5	-1	-1	-3	-3	-2
S	1	-1	0	0	-2	0	-1	-2	0	-2	-1	1	-1	0	-1	4	1	-2	-3	-2
T	0	-1	-1	-1	-2	-2	-2	-1	-1	-1	-1	0	-1	-1	-1	1	5	0	-2	-2
V	0	-1	-3	-2	-1	-3	-3	3	-2	1	1	-3	-2	-2	-3	-2	0	4	-3	-1
W	-3	-2	-4	-3	1	-2	-2	-3	-3	-2	-1	-4	-4	-2	-3	-3	-2	-3	11	2
Y	-2	-2	-3	-2	3	-3	2	-1	-2	-1	-1	-2	-3	-1	-2	-2	-2	-1	2	7

sequences, we say that an *identity* has occurred and give it a high positive score. When two different amino acids are present at the same position in an alignment, we call it a *substitution* and give it a score that could be negative, zero, or positive.

To each possible pairing of amino acids (one from each sequence, at the same position in the alignment), we assign a *substitution score*, which gives a quantitative measure of the "cost" of replacing one amino acid by another. The substitution scores for all 210 possible pairs of amino acids are collected together to form a symmetric, (20×20) *substitution matrix*, which is used to measure the closeness of the two sequences. One of the most popular substitution matrices is BLOSUM62 (BLOcks SUbstitution Matrix; see Table 13.4), which assumes that no more than 62% of the letters in the two sequences are identical (Henikoff and Henikoff, 1996).

A *gap* (or *indel*) is an empty space (denoted by a "-") introduced into an alignment to compensate for an *insertion* or a *deletion* of an amino acid in one sequence relative to the other. A gap is penalized by assigning to it a large value (the *gap score*, usually set by the user), which is then subtracted from the alignment score. There are two types of gap penalties, one for starting (or opening) a gap and another for extending the gap; typically, the latter is considered to be more serious than is the former, so that opening a gap merits a smaller penalty than does extending that gap. Gap-scoring methods usually define the gap penalty as $q + rk$, where q and r are chosen by the user; the *gap open penalty* uses $k = 1$ and the *gap extension penalty* uses $k = 2, 3, \ldots$.

The alignment score s is the sum of the identity and substitution scores, minus the gap score. Implicitly, we are assuming that the score for a particular position in the alignment is independent of scores derived from neighboring positions (Karlin and Altschul, 1990); such an assumption appears to be reasonable for protein sequences. The optimal alignment between two sequences (including gaps) corresponds to that alignment with the highest alignment score.

In general, given n proteins from some database, let s_{ij} be the alignment score between the ith and jth protein, $i, j = 1, 2, \ldots, n$. Because closely related proteins will have a high alignment score, the alignment score is a similarity and so has to be transformed into a dissimilarity using $\delta_{ij} = s_{\max} - s_{ij}$, where $s_{\max}$ is the largest alignment score among all $m = n(n-1)/2$ protein pairs. The proximity matrix is then given by $\boldsymbol{\Delta} = (\delta_{ij})$.

13.4.2 Example: Two Hemoglobin Chains

Suppose we wish to compare the *hemoglobin alpha chain* protein (Swiss-Prot database code HBA_HUMAN, AC# P69905/P019122) having length 141 with the related *hemoglobin beta chain* protein (Swiss-Prot database code HBB_HUMAN, AC# P68871/P02023) having length 146. Both of these human proteins transport oxygen from the lungs to the various peripheral tissues. HBA gives blood its red color, and defects in HBB are the cause of sickle cell anemia.

To compare these proteins, we use the BLOSUM62 matrix and the gap scoring method with $q = 12, r = 4$. The SIM algorithm (Huang and Miller, 1991), which is a local similarity program using dynamic programming techniques, finds that the optimal alignment over 145 amino acids is:

```
LSPADKTNVKAAWGKVGAHAGEYGAEALERMFLSFPTTKTYFPHF------DLSH
L+P +K+ V A WGKV  +  E G EAL R+ + +P T+ +F  F        D
LTPEEKSAVTALWGKV--NVDEVGGEALGRLLVVYPWTQRFFESFGDLSTPDAVM

GSAQVKGHGKKVADALTNAVAHVDDMPNALSALSDLHAHKLRVDPVNFKLLSHCL
G+ +VK HGKKV  A ++ +AH+D++     + LS+LH  KL VDP NL+LL + L
GNPKVKAHGKKVLGAFSDGLAHLDNLKGTFATLSELHCDKLHVDPENFRLLGNVL

LVTLAAHLPAEFTPAVHASLDKFLASVSTVLTSKY
+  LA H   EFTP V A+  K +A V+ L  KY
VCVLAHHFGKEFTPPVQAAYQKVVAGVANALAHKY}
```

The first line is a portion of the HBA_HUMAN protein sequence, and the third line is a portion of HBB_HUMAN. The sequences have been "locally" aligned (with gaps). Looking at the middle line, we see 86 positive substitution scores (the 25 "+"s and the 61 identities). The alignment score is $s = 259$. For different values of q and r, we would obtain different optimal alignments and alignment scores.

13.5   String Matching

The problem of comparing different protein sequences is closely related to a more general class of problems involving the matching of different strings of letters, characters, or symbols drawn from a common alphabet $\mathcal{A}$. The alphabet could be binary $\{0, 1\}$, decimal $\{0, 1, 2, \ldots, 9\}$, English language $\{A, B, C, \ldots, Z\}$, the four DNA bases $\{A, C, G, T\}$, or the 20 amino acids. In *pattern matching*, we study the problem of finding a given pattern (typically, a collection of strings described in terms of some alphabet $\mathcal{A}$) within a body of text. If a pattern is a single string, the problem is called *string matching*. We can imagine, for example, a string-matching problem in which we need to know whether a particular word or phrase can be found within a given sentence, paragraph, article, or book.

String matching is used extensively in text-processing applications; in particular, it is used in searching a document for a word, phrase, or an arbitrary string of letters; designing spell-checkers; predicting unknown words when writing in a second language; and name-retrieval systems in genealogical research. The UNIX programming environment (Kernighan and Pike, 1984), for example, employs various string- and pattern-matching algorithms (e.g., awk, diff, and grep), and the PERL language was designed specifically to possess powerful string-matching capabilities. The related problems of string- and pattern-matching have obvious implications for the design of an Internet search engine (e.g., Google™, www.google.com), where the text is the union of all linked webpages (Brin and Page, 1998).

String matching techniques are needed in many different applications, including matching melodies in large databases of digital music (Uitdenbogerd and Zobel, 1999), dating trees by the sequence of rings they contain (Wenk, 1999), and comparing different speech pronunciations in computational linguistics (Nerbonne, Heeringa, and Kleiweg, 1999).

13.5.1   Edit Distance

A popular numerical measure of the similarity between two strings is *edit distance* (also called *Levenshtein distance*), which was adapted from methods used to compare two different protein sequences. The usual definition of edit distance is the fewest number of editing operations (insertions, deletions, substitutions) which would be needed to transform one string into the other. An *insertion* inserts a letter into the sequence, a *deletion* deletes a letter from the sequence, and a *substitution* replaces one letter in the sequence by another letter. Identities (or matches) are not counted in the distance measure. In some definitions of edit distance, each editing operation is assigned a nonnegative cost, that reduces to the above definition if each editing operation has unit cost. The sequence of editing operations that achieves the minimum edit distance will probably not be unique.

An early application of edit distance was to comparative biochemistry (de Duve, 1984, p. 354), where it was used to construct a *phylogenetic tree* — a diagram laying out a possible evolutionary history — of a single protein. The resulting proximity matrix shows the number of amino-acid substitutions in the protein cytochrome *c* from 25 different species, including mammals and other vertebrates, invertebrates, plants, and fungi. The entries in the matrix show the fewest number of nucleotide substitutions in DNA (according to the genetic code) needed to account for the observed amino-acid replacements.

13.5.2 Example: Employee Careers at Lloyds Bank

An unusual example of string matching using edit distance is that of analyzing changes in employee careers over a given period of time. The careers of two individuals can be compared by determining the fewest number of changes necessary to transform one career into the other (Abbott and Hrycak, 1990). Each type of change incurs a cost, and the total cost of transforming one career into another is the sum of all such costs.

One fascinating study looked at a large database of employee information from Lloyds Bank, one of England's oldest and largest banks, during the period 1890–1970 (Stovel, Savage, and Bearman, 1996). The authors were interested in tracing how "static, status-based employment arrangements" of the early 1900s had been replaced, less than two generations later, by "highly-dynamic, achievement-oriented careers" within large bureaucratic organizations, such as the British banking system and, in particular, Lloyds Bank.

The available data give every job held by each employee of the bank. The data are described by a rectangular array, where each row corresponds to a different employee and the columns (variables) record the various jobs held by that employee over the number of years of the study. In this particular study, job termination (resignation, death, firing) is coded by type of termination, and each year the employee is absent from bank employment is coded as 999. These termination codes and 999s are not used in the matching algorithm.

An employee's job at Lloyds Bank is characterized by three factors: branch size and type (1=small rural, 2=large rural, 3=small urban, 4=large urban [London], 5=specialist head office, and 6=head office) and job category (1=clerk, 2=senior clerk, 3=regular manager, and 4=specialist manager). Thus, there are $6 \times 4 = 24$ branch-position categories of jobs, where a job would be characterized as "years@branch·position." For example, a 45-year career at Lloyds might be summarized as $\{15@11, 6@22, 24@23\}$, which translates into 15 years as a clerk in a small rural branch, then a move to a large rural branch where he spent 6 years as a senior clerk and 24 years as a regular manager.

In this example, we reanalyze two data sets on the careers of Lloyds' employees. The data sets consist of sequential employment records for random samples of $n = 80$ employees drawn from two different cohorts, those who started work at Lloyds during the period 1905–1909 and those who started during 1925–1929.[1] (See also Oh and Raftery, 2001, who used only the 1905-1909 cohort data.) Each data set contains an ID variable, a variable containing the first year of the employee's employment, and $r = 71$ variables containing the sequential data of the employment history of each employee.

A (24×24) substitution matrix with branch-position categories forming its rows and columns was constructed from the entire collection of employee records (Stovel, Savage, and Bearman, 1996, Table A3). The entries in the substitution matrix represent costs; they range from 0.5 to 6.5 and reflect the notion that unlikely changes are costly and frequent changes are inexpensive. The cost of an insertion or deletion was fixed at the maximum substitution cost, 6.5. The career records, the substitution matrix, and the edit-distance method were then used by an alignment algorithm to construct an (80×80) non-Euclidean proximity matrix for each cohort of employees.

13.6   Classical Scaling and Distance Geometry

The airline-distances example (see Section 13.1.1) illustrates the classical scaling method of MDS. Suppose we are given n points $\mathbf{X}_1, \ldots, \mathbf{X}_n \in \Re^r$. From these points, we compute an $(n \times n)$ proximity matrix $\mathbf{\Delta} = (\delta_{ij})$ of dissimilarities, where

$$\delta_{ij} = \|\mathbf{X}_i - \mathbf{X}_j\| = \left\{ \sum_{k=1}^{r} (X_{ik} - X_{jk})^2 \right\}^{1/2} \qquad (13.4)$$

---

[1] The author thanks Katherine Stovel for kindly providing him with the employment data on the two cohorts of Lloyds' employees and for the corresponding two (80×80) proximity matrices. The data can be found in the files samp05 and samp25, and the proximity matrices in the files samp05d and samp25d, all on the book's website.

is the dissimilarity between the points $\mathbf{X}_i = (X_{ik})$ and $\mathbf{X}_j = (X_{jk})$; these dissimilarities are the Euclidean distances between all $m = \frac{1}{2}n(n-1)$ pairs of points in that space.

Actually, there is no requirement that the $\{\delta_{ij}\}$ be Euclidean distances; they can be any kind of distances. For example, the *Minkowski* or L_p *distance* is given by

$$\delta_{ij} = \left\{ \sum_{k=1}^{r} |X_{ik} - X_{jk}|^p \right\}^{1/p}, \tag{13.5}$$

where $p \geq 1$ is set by the user. When $p = 1$, we have the *city-block* or *Manhattan distance*, and when $p = 2$, we have Euclidean distance.

13.6.1 From Dissimilarities to Principal Coordinates

From (13.4), we note that

$$\delta_{ij}^2 = \|\mathbf{X}_i\|^2 + \|\mathbf{X}_j\|^2 - 2\mathbf{X}_i^\tau \mathbf{X}_j. \tag{13.6}$$

Let $b_{ij} = \mathbf{X}_i^\tau \mathbf{X}_j = -\frac{1}{2}(\delta_{ij}^2 - \delta_{i0}^2 - \delta_{j0}^2)$, where $\delta_{i0}^2 = \|\mathbf{X}_i\|^2$ is the squared distance from the point $\mathbf{x}_i$ to the origin. Summing (13.6) over i and over j yields the following identities:

$$n^{-1} \sum_i \delta_{ij}^2 = n^{-1} \sum_i \delta_{i0}^2 + \delta_{j0}^2 \tag{13.7}$$

$$n^{-1} \sum_j \delta_{ij}^2 = \delta_{i0}^2 + n^{-1} \sum_j \delta_{j0}^2 \tag{13.8}$$

$$n^{-2} \sum_i \sum_j \delta_{ij}^2 = 2n^{-1} \sum_i \delta_{i0}^2. \tag{13.9}$$

Substituting (13.7)–(13.9) into (13.6) and simplifying, we get

$$b_{ij} = a_{ij} - a_{i\cdot} - a_{\cdot j} + a_{\cdot\cdot}, \tag{13.10}$$

where $a_{ij} = -\frac{1}{2}\delta_{ij}^2$, and the usual "dot" notation is used, $a_{i\cdot} = n^{-1} \sum_j a_{ij}^2$, $a_{\cdot j} = n^{-1} \sum_i a_{ij}^2$, and $a_{\cdot\cdot} = n^{-2} \sum_i \sum_j a_{ij}^2$. If we set $\mathbf{A} = (a_{ij})$ to be the matrix of squared dissimilarities and $\mathbf{B} = (b_{ij})$, then $\mathbf{A}$ and $\mathbf{B}$ are related through $\mathbf{B} = \mathbf{HAH}$, where $\mathbf{H} = \mathbf{I}_n - n^{-1}\mathbf{J}_n$ is a centering matrix and $\mathbf{J}_n$ is an $(n \times n)$-matrix of ones. The matrix $\mathbf{B}$ is said to be a "doubly centered" version of $\mathbf{A}$.

In the dimensionality-reduction aspect of MDS, we wish to find a t-dimensional representation, $\mathbf{Y}_1, \ldots, \mathbf{Y}_n \in \Re^t$ (referred to as *principal coordinates*), of those r-dimensional points (with $t < r$), such that the interpoint distances in t-space "match" those in r-space. When dissimilarities are defined as Euclidean interpoint distances, this type of "classical" MDS is equivalent to PCA in that the principal coordinates are identical to the scores of the first t principal components of the $\{\mathbf{X}_i\}$.

Typically, in *classical scaling* (Torgerson, 1952, 1958) we are not given the $\{\mathbf{X}_i\} \subset \Re^r$; instead, we are given only the dissimilarities $\{\delta_{ij}\}$ through the $(n \times n)$ proximity matrix $\mathbf{\Delta}$. Using $\mathbf{\Delta}$, we form $\mathbf{A}$, and then $\mathbf{B}$. Motivation for classical scaling comes from a least-squares argument similar to the one employed for PCA; see Section 7.2.2. The idea is to find a matrix $\mathbf{B}^* = (b_{ij}^*)$ with rank at most t that minimizes $\operatorname{tr}\{(\mathbf{B} - \mathbf{B}^*)^2\} = \sum_i \sum_j (b_{ij} - b_{ij}^*)^2$. It can be shown (Mardia, 1978) that if $\{\lambda_k\}$ are the eigenvalues of $\mathbf{B}$ and $\{\lambda_k^*\}$ are the eigenvalues of $\mathbf{B}^*$, then the minimum of $\operatorname{tr}\{(\mathbf{B}-\mathbf{B}^*)^2\}$ is given by $\sum_{k=1}^n (\lambda_k - \lambda_k^*)^2$, where $\lambda_k^* = \max(\lambda_k, 0)$ for $k = 1, 2, \ldots, t$, and zero otherwise. Because of the rank constraint, at least $n-t$ of the eigenvalues of $\mathbf{B}^*$ have to be zero. If any of the eigenvalues of $\mathbf{B}$ are negative, a suitable constant can be added to the dissimilarities, or the negative eigenvalues can be ignored. The first t principal coordinates, as defined by the classical scaling algorithm in Table 13.5, are taken to be the required projections in t-dimensional space.

The classical scaling algorithm is based upon an eigendecomposition of the matrix $\mathbf{B}$. This eigendecomposition produces $\mathbf{Y}_1, \ldots, \mathbf{Y}_n \in \Re^t$, $t < r$, a configuration whose Euclidean interpoint distances,

$$d_{ij}^2 = \|\mathbf{Y}_i - \mathbf{Y}_j\|^2 = (\mathbf{Y}_i - \mathbf{Y}_j)^\tau (\mathbf{Y}_i - \mathbf{Y}_j), \tag{13.11}$$

match those given in the matrix $\mathbf{\Delta}$. The classical scaling algorithm automatically sets the mean $\bar{\mathbf{Y}}$ of all n points in the configuration to be the origin in $\Re^t$. To see this, we note that because $\mathbf{H1}_n = \mathbf{0}$, we have that $\mathbf{B1}_n = \mathbf{0}$, whence, $n^2 \bar{\mathbf{Y}} \bar{\mathbf{Y}} = (\mathbf{Y}^\tau \mathbf{1}_n)^\tau (\mathbf{Y}^\tau \mathbf{1}_n) = \mathbf{1}_n \mathbf{B1}_n = \mathbf{0}$, and so $\bar{\mathbf{Y}} = \mathbf{0}$.

The solution of the classical scaling problem is not unique. Consider an orthogonal transformation of two points that are obtained through the classical scaling algorithm: $\mathbf{Y}_i \to \mathbf{PY}_i$ and $\mathbf{Y}_j \to \mathbf{PY}_j$, where $\mathbf{P}$ is an orthogonal matrix. Then, $\mathbf{PY}_i - \mathbf{PY}_j = \mathbf{P}(\mathbf{Y}_i - \mathbf{Y}_j)$, whence,

$$\|\mathbf{P}(\mathbf{Y}_i - \mathbf{Y}_j)\|^2 = (\mathbf{Y}_i - \mathbf{Y}_j)^\tau \mathbf{P}^\tau \mathbf{P}(\mathbf{Y}_i - \mathbf{Y}_j) = \|\mathbf{Y}_i - \mathbf{Y}_j\|^2. \tag{13.12}$$

So, a common orthogonal transformation of the points in the configuration found by classical scaling yields a different solution of the classical scaling problem.

13.6.2 Assessing Dimensionality

One way of determining the dimensionality of the resulting configuration is to look at the eigenvalues of $\mathbf{B}$.

The usual strategy is to plot the ordered eigenvalues (or some function of them) against dimension and then identify a dimension at which the

TABLE 13.5. *The classical scaling algorithm.*

---

1. Given an $(n \times n)$-matrix of interpoint distances $\mathbf{\Delta} = (\delta_{ij})$, form the $(n \times n)$-matrix $\mathbf{A} = (a_{ij})$, where $a_{ij} = -\frac{1}{2}\delta_{ij}^2$.

2. Form the "doubly centered," symmetric, $(n \times n)$-matrix $\mathbf{B} = \mathbf{HAH}$, where $\mathbf{H} = \mathbf{I}_n - n^{-1}\mathbf{J}_n$ and $\mathbf{J}_n = \mathbf{1}_n\mathbf{1}_n^\tau$ is an $(n \times n)$-matrix of ones.

3. Compute the eigenvalues and eigenvectors of $\mathbf{B}$. Let $\mathbf{\Lambda} = \text{diag}\{\lambda_1, \cdots, \lambda_n\}$ be the diagonal matrix of the eigenvalues of $\mathbf{B}$ and let $\mathbf{V} = (\mathbf{v}_1, \cdots, \mathbf{v}_n)$ be the matrix whose columns are the eigenvectors of $\mathbf{B}$. Then, by the spectral theorem, $\mathbf{B} = \mathbf{V\Lambda V}^\tau$.

4. If $\mathbf{B}$ is nonnegative-definite with rank $r(\mathbf{B}) = t < n$, the largest t eigenvalues will be positive and the remaining $n - t$ eigenvalues will be zero. Denote by $\mathbf{\Lambda}_1 = \text{diag}\{\lambda_1, \cdots, \lambda_t\}$ the $(t \times t)$ diagonal matrix of the positive eigenvalues of $\mathbf{B}$ and let $\mathbf{V}_1 = (\mathbf{v}_1, \cdots, \mathbf{v}_t)$ be the corresponding matrix of eigenvectors of $\mathbf{B}$. Then,
$$\mathbf{B} = \mathbf{V}_1\mathbf{\Lambda}_1\mathbf{V}_1^\tau = (\mathbf{V}_1\mathbf{\Lambda}_1^{1/2})(\mathbf{\Lambda}_1^{1/2}\mathbf{V}_1) = \mathbf{YY}^\tau,$$
where $\mathbf{Y} = \mathbf{V}_1\mathbf{\Lambda}_1^{1/2} = (\sqrt{\lambda_1}\mathbf{v}_1, \cdots, \sqrt{\lambda_t}\mathbf{v}_t) = (\mathbf{Y}_1, \cdots, \mathbf{Y}_n)^\tau$.

5. The *principal coordinates*, which are the columns, $\mathbf{Y}_1, \ldots, \mathbf{Y}_n$, of the $(t \times n)$-matrix $\mathbf{Y}^\tau$, yield the n points in t-dimensional space whose interpoint distances $d_{ij} = \|\mathbf{Y}_i - \mathbf{Y}_j\|$ are equal to the distances δ_{ij} in the matrix $\mathbf{\Delta}$.

6. If the eigenvalues of $\mathbf{B}$ are not all nonnegative, then either ignore the negative eigenvalues (and associated eigenvectors) or add a suitable constant to the dissimilarities (i.e., $\delta_{ij} \leftarrow \delta_{ij} + c$ if $i \neq j$, and unchanged otherwise) and return to step 1. If t is too large for practical purposes, then the largest $t' < t$ positive eigenvalues and associated eigenvectors of $\mathbf{B}$ can be used to construct a reduced set of principal coordinates. In this case, the interpoint distances d_{ij} approximate the δ_{ij} from the matrix $\mathbf{\Delta}$.

---

eigenvalues become "stable" (i.e., do not change perceptively). At that dimension, we may observe an "elbow" that shows where stability occurs. If $\mathbf{X}_i \in \Re^t$, $i = 1, 2, \ldots, n$, then stability in the plot should occur at dimension $t + 1$. For easier graphical interpretation of a classical scaling solution, we hope that t is small, of the order 2 or 3.

13.6.3   Example: Airline Distances (Continued)

In Table 13.6, we give the 18 eigenvalues of the matrix $\mathbf{B}$. One can see three large positive eigenvalues, eight negative eigenvalues, six smaller positive eigenvalues, and one zero eigenvalue (due to the double-centering operation). Ignoring the negative eigenvalues (which, in this case, result

TABLE 13.6. *Eigenvalues of* **B** *and the eigenvectors corresponding to the first three largest eigenvalues (in red) for the airline distances example.*

	Eigenvalues	Eigenvectors		
1	471582511	0.245	-0.072	0.183
2	316824787	0.003	0.502	-0.347
3	253943687	0.323	-0.017	0.103
4	-98466163	0.044	-0.487	-0.080
5	-74912121	-0.145	0.144	0.205
6	-47505097	0.366	-0.128	-0.569
7	31736348	-0.281	-0.275	-0.174
8	-7508328	-0.272	-0.115	0.094
9	4338497	-0.010	0.134	0.202
10	1747583	0.209	0.195	0.110
11	-1498641	-0.292	-0.117	0.061
12	145113	-0.141	0.163	0.196
13	-102966	-0.364	0.172	-0.473
14	60477	-0.104	0.220	0.163
15	-6334	-0.140	-0.356	-0.009
16	-1362	0.375	0.139	-0.054
17	100	-0.074	0.112	0.215
18	0	0.260	-0.214	0.173

TABLE 13.7. *First three principal coordinates of the 18 cities in the airline distances example.*

City	Principal Coordinates		
	1st	2nd	3rd
Beijing	5315.24	-1272.90	2920.75
Cape Town	57.63	8935.14	-5522.26
Hong Kong	7010.90	-306.52	1645.53
Honolulu	962.86	-8677.05	-1270.47
London	-3157.53	2557.96	3268.11
Melbourne	7948.29	-2283.67	-9062.28
Mexico	-6108.97	-4896.64	-2778.04
Montreal	-5912.57	-2039.70	1495.92
Moscow	-220.84	2377.27	3221.22
New Delhi	4528.94	3474.33	1751.50
New York	-6341.02	-2078.66	972.39
Paris	-3058.30	2910.08	3118.95
Rio de Janeiro	-7905.60	3067.34	-7537.69
Rome	-2262.26	3916.47	2595.85
San Francisco	-3041.92	-6341.23	-142.88
Singapore	8139.01	2470.83	-867.84
Stockholm	-1610.37	1997.61	3429.67
Tokyo	5656.51	-3810.66	2761.56

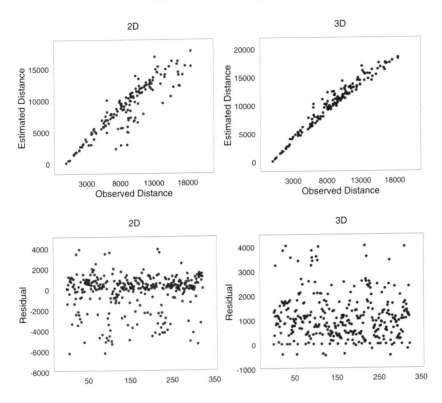

FIGURE 13.5. *Estimated and observed airline distances. The left panels show the 2D solution and the right panels show the 3D solution. The top panels show the estimated distances plotted against the observed distances, and the bottom panels show the residuals from the the fit (residual = estimated distance − observed distance) plotted against sequence number.*

from the distances not being Euclidean due to the earth's curvature), we see that the magnitudes of the three largest positive eigenvalues suggest that a 3D solution makes the most sense here for recreating the world map. As a result, we retain only the eigenvectors corresponding to the first $t = 3$ eigenvalues. In Table 13.7, we display the $n = 18$ scores of the first three principal coordinates using step 4 of the classical scaling algorithm. The 2D solution is given in Figure 13.1 and the 3D solution in Figure 13.2.

Figure 13.5 shows the estimated and observed airline distances plotted against each other for the 2D and 3D solutions. In the top-left panel, the scatterplot corresponding to the 2D solution shows that many of the observed distances are severely underestimated, with a number of them also being overestimated. In the top-right panel, the scatterplot corresponding to the 3D solution indicates a better fit to the observed distances, yet it also shows that the observed distances are consistently overestimated.

We should not really be surprised at the results in this example. The differences occur because of the fact that the estimated airline distances are taken to be Euclidean. Airline distances are measured over a curved surface rather than a flat one. We should, therefore, expect to see a certain amount of distortion when we use a Euclidean metric to estimate distances between cities distributed across the surface of a globe.

The Euclidean distance between two cities "near" each other is close to its airline distance; see, for example, the European, Asian, or North American clusters of cities in Figure 13.1, whose 2D configurations are similar to their usual geographical locations. However, when the cities are far apart from each other, maybe on opposite sides of the globe, we expect large distortions to be introduced. We see this effect in the 2D and 3D solutions, with the three cities of Cape Town, Rio de Janeiro, and Melbourne each involved in producing all the largest absolute residuals (where residual = $d_{ij} - \delta_{ij}$); see the bottom two panels of Figure 13.5, where residuals are plotted against sequence number. The largest residuals in the 3D solution are the Cape Town–Rio de Janeiro and Melbourne–Tokyo distances.

13.6.4 Example: Mapping the Protein Universe

Molecular evolution has led to the development of "families" of proteins, so that information on the shape and function of one protein can be used to predict the shape and function of another protein within the same family. Sifting through the 100,000 or so amino acid sequences to group similar proteins into families becomes more difficult when the evolutionary distances between proteins grow too large. In such cases, it is natural to turn toward comparing the three-dimensional shapes of proteins (rather than their one-dimensional amino acid sequences).

Molecular biologists would, therefore, like to obtain a global representation (i.e., a map) of the "protein structure universe," in which adjacent points represent structurally related proteins. In order to do this, biologists have been using the classical scaling algorithm under the name "distance geometry" (Havel, Kuntz, and Crippen, 1983) to construct various 2D and 3D protein maps (see, e.g., Holm and Sander, 1996; Hou, Sims, Zhang, and Kim, 2003).

In this example, we reanalyze data on 498 proteins taken from the SCOP (Structural Classification System of Proteins) database.[2] We applied the

---

[2]The author thanks Sung-Hou Kim and Jingtong Hou for kindly providing him with their list of 498 proteins and the resulting (498 × 498) proximity matrix $\boldsymbol{\Delta}$. The list of

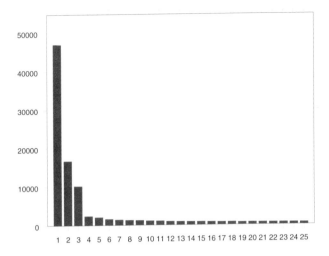

FIGURE 13.6. *The first 25 ordered eigenvalues of* **B** *obtained from the classical scaling algorithm on 498 proteins.*

classical scaling algorithm (Table 13.5) to the proximity matrix Δ. From inspection of the largest 25 eigenvalues of **B** (see Figure 13.6), we see that the first three eigenvalues are dominant, suggesting a 3D configuration is probably most appropriate.

A 2D map of the first two principal coordinate scores for the 498 proteins is given in Figure 13.7. We can clearly see three arms with four clusters of points corresponding to four of the SCOP classes. The first (red dots) arm contains 136 α-helix proteins (class 1), the second (blue dots) arm contains 92 β-sheet proteins (class 2), the third (green dots) arm consists of 94 α/β proteins (class 3, mainly parallel β-sheets), and the 176 $\alpha + \beta$ proteins (class 4, mainly antiparallel β-sheets) congregate (brown dots) at the junction of the three arms. Class 1 does not overlap with class 3 and has minimal overlap (two outlying points) with class 2; classes 2 and 3 have only two overlapping points; class 4, however, spreads and mixes with all three other classes. These results suggest that certain proteins may be misclassified by SCOP. We also notice the presence of a few outliers in the display.

A 3D map of the first three principal coordinates for the 498 proteins shows more interesting structure; see Figure 13.8. The blue points of class 1 (the α-helix proteins) and the red points of class 2 (the β-sheet proteins)

proteins may be found in the file `498.SCOP.txt` and the proximity matrix is in the file `498.matrix.txt` on the book's website.

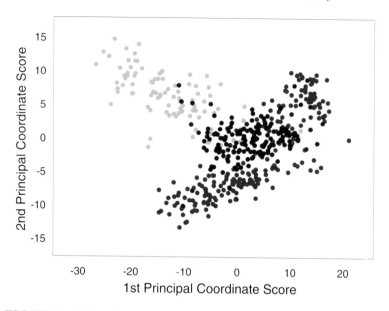

FIGURE 13.7. *Two-dimensional map of four protein classes using the classical scaling algorithm on 498 proteins. Class 1 (red dots) are α-proteins, class 2 (blue dots) are β-proteins, class 3 (green dots) are α/β-proteins, and class 4 (brown dots) are α + β-proteins.*

appear to fall along two separate axes. The black points of class 4 (the $\alpha+\beta$ proteins, a random mixture of α-helix and β-sheet proteins) jut out from the middle of those two axes and lie on the plane formed by those axes. The green points representing the proteins in class 3 (the α/β proteins) are actually scattered around a third axis, perpendicular to the other two axes. These results are very similar to those discovered by Hou, Sims, Zhang, and Kim (2003).

13.7 Distance Scaling

Given n items (or entities) and the matrix of their dissimilarities, $\boldsymbol{\Delta} = (\delta_{ij})$, we saw that the classical scaling problem is to find a configuration of points in a lower-dimensional space such that the interpoint distances $\{d_{ij}\}$ satisfy $d_{ij} \approx \delta_{ij}$. In *distance scaling*, this relationship is relaxed; we wish to find a suitable configuration for which

$$d_{ij} \approx f(\delta_{ij}), \qquad (13.13)$$

where f is some monotonic function. The function f transforms the dissimilarities into distances. The use of "metric" or "nonmetric" distance

FIGURE 13.8. *A three-dimensional map of four protein classes using the classical scaling algorithm on 498 proteins. The graph shows two separate axes, one for the blue points of the α-helix class of proteins and the other for the red points of the β-sheet class of proteins. A third axis of green points for the α/β class of proteins is also visible in both panels. Lying midway in the plane formed by the α and β axes, we see the black points of the $\alpha + \beta$ class of proteins.*

scaling depends upon the the nature of the dissimilarities. If the dissimilarities are quantitative (e.g., ratio or interval scale), we use metric distance scaling, whereas if the dissimilarities are qualitative (e.g., ordinal), we use nonmetric distance scaling. In the MDS literature, metric distance scaling is traditionally called *metric MDS* and nonmetric distance scaling is called *nonmetric MDS*.

13.8 Metric Distance Scaling

In metric distance scaling, the dissimilarities $\{\delta_{ij}\}$ are quantitative measurements, usually Euclidean, but other distance metrics are possible. The function f is usually taken to be a parametric monotonic function, such as $f(\delta_{ij}) = \alpha + \beta\delta_{ij}$, where α and β are unknown positive coefficients. In some MDS software (e.g., SAS PROC MDS), metric distance scaling is characterized in three ways: *absolute MDS* ($\alpha = 0$, $\beta = 1$), *ratio MDS* ($\alpha = 0$, $\beta > 0$), and *interval MDS* ($\alpha \geq 0$ and $\beta \geq 0$). It is worth noting that absolute MDS is not very useful in practice. If the $\{\delta_{ij}\}$ are similarities (rather than dissimilarities), then we need $\beta < 0$.

13.8.1 Metric Least-Squares Scaling

Because f is a parametric function, the distances $\{d_{ij}\}$ can be fitted to $\{f(\delta_{ij})\}$ by least-squares (LS). The result is *metric LS scaling*. If the dissimilarities are Euclidean distances and f is taken to be the identity function, then classical scaling can be viewed as an example of metric LS scaling. In fact, metric distance scaling is often regarded as synonymous with classical scaling.

A given configuration of points $\{\mathbf{Y}_{ij}\} \subset \Re^t$ can be evaluated by computing the pairwise distances $\{d_{ij}\}$ and then, for an unknown monotone function f, using the weighted loss function,

$$\mathcal{L}_f(\mathbf{Y}_1, \ldots, \mathbf{Y}_n; \mathbf{W}) = \sum_{i<j} w_{ij}(d_{ij} - f(\delta_{ij}))^2, \qquad (13.14)$$

as a goodness-of-fit criterion, where $\mathbf{W} = (w_{ij})$ is a given matrix of weights. For a specific dimensionality t, the square-root of $\mathcal{L}_f$,

$$\text{stress} = [\mathcal{L}_f(\mathbf{Y}_1, \ldots, \mathbf{Y}_n; \mathbf{W})]^{1/2}, \qquad (13.15)$$

is known as the *metric stress function*. Minimizing stress over all t-dimensional configurations $\{\mathbf{Y}_{ij}\}$ and monotone f yields an optimal metric distance scaling solution. Weighting systems include $w_{ij} = \{\sum_{k<\ell} \delta_{k\ell}^2\}^{-1}$ and $w_{ij} = \delta_{ij}^{-2}$. More general loss functions, where $g(d_{ij})$ replaces d_{ij} in (13.14), for some function g, have also been proposed.

13.8.2 Sammon Mapping

The so-called *Sammon nonlinear mapping*, which has become a popular tool for pattern recognition, is a special case of metric LS scaling, where $w_{ij} = \delta_{ij}^{-1}\{\sum_{k<\ell} \delta_{k\ell}\}^{-1}$ is used as the weighting system in (13.14) and f is the identity function (Sammon, 1969). This weighting system normalizes the squared-errors in pairwise distances by using the distance in the original space. As a result, Sammon mapping preserves the small δ_{ij}, giving them a greater degree of importance in the fitting procedure than for larger values of δ_{ij}; this can be a useful strategy if one is trying to identify clusters in the data.

Differentiating (13.14) with Sammon weights yields a set of nonlinear least-squares equations, which are then solved using an iterative numerical procedure. The usual algorithm (see, e.g., Cox and Cox, 2001, Section 2.4) starts at the classical scaling solution, then follows that up by using a pseudo-Newton iterative procedure with step size reduced by a "magic" factor, usually in the range 0.3–0.4; in some cases, this factor has to be set to a much smaller value to carry the algorithm to convergence.

13.8.3   Example: Lloyds Bank Employees

As an example of metric LS scaling, we compare the two-dimensional Sammon mapping and classical scaling solutions of the 1905–1909 and 1925–1929 cohorts of the Lloyds Bank employee data (see Section 13.5.2). Figure 13.9 displays the two types of metric MDS for each cohort. We see that whereas the plotted points for classical scaling and Sammon mapping appear to have similar patterns, with a number of well-separated clusters, the points in the Sammon map for each cohort are considerably more spread out than are those derived from classical scaling. A similar effect using different data was also noticed by Ripley (1996, p. 309).

For the 1905–1909 cohort, there are three employees (1587, 1590, 3240) who can be considered as outliers with respect to the remaining employees. The two employees 1590 and 3240 only worked at Lloyds Bank for two years (the next shortest employment tenure was 10 years) and employee 1587 worked there for 59 years (the next longest tenure was 48 years). The Sammon mapping algorithm stopped (no further iterations) at the classical scaling solution, so that the upper-left and upper-right panels of Figure 13.9 are identical.

13.8.4   Bayesian MDS

In certain situations, it may be reasonable to assume that the observed dissimilarities in the proximity matrix $\mathbf{\Delta} = (\delta_{ij})$ are tainted by measurement error. We may see this, for example, when the elements of $\mathbf{\Delta}$ are clearly measured in three dimensions, but the stress value for the three-dimensional solution is not zero as it should be; instead, it may require a much-higher dimensional solution to reduce stress down to zero. One way to incorporate measurement error into metric MDS is to adopt a more explicit modeling framework, such as a Bayesian viewpoint (Oh and Raftery, 2001).

A cautionary note: in general, it is often difficult to verify the types of distributional assumptions used in statistical modeling, and the assumptions used in Bayesian MDS are no exception.

In this model, we assume the dissimilarity, $\delta_{ij} > 0$, between entities i and j is observed with Gaussian error:

$$\delta_{ij} = \delta_{ij}^0 + \epsilon_{ij}, \tag{13.16}$$

where δ_{ij}^0 is the true dissimilarity and $\epsilon_{ij} \sim \mathcal{N}(0, \sigma^2)$, $i, j = 1, 2, \ldots, n$.

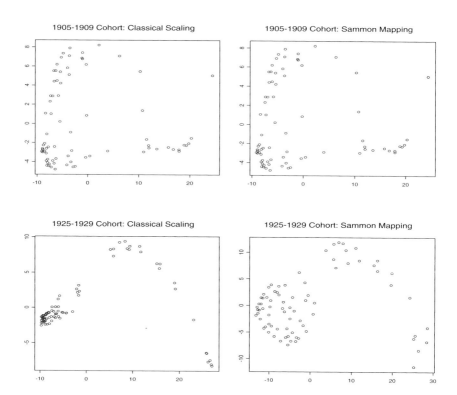

FIGURE 13.9. *Two-dimensional MDS solutions for the Lloyds Bank data. The left panels show classical scalings and the right panels show Sammon mappings. The upper panels show the 1905–1909 cohorts of Lloyds Bank employees and the lower panels show the 1925–1929 cohorts.*

Thus, given δ_{ij}^0, the observed dissimilarity δ_{ij}, which is a function of the unknown $\{\mathbf{X}_i\}$, follows the truncated Gaussian distribution,

$$\delta_{ij} \sim \mathcal{N}(\delta_{ij}^0, \sigma^2) I_{[\delta_{ij} > 0]}, \quad i \neq j, \ i, j = 1, 2, \ldots, n. \tag{13.17}$$

The likelihood function of $(\{\mathbf{X}_i\}, \sigma^2)$, given $\boldsymbol{\Delta}$, is given by

$$
\begin{aligned}
L(\{\mathbf{X}_i\}, \sigma^2 | \boldsymbol{\Delta}) &= \prod_{i < j} \frac{1}{\sqrt{2\pi\sigma^2}} \exp\left\{ -\frac{(\delta_{ij} - \delta_{ij}^0)^2}{2\sigma^2} \right\} \left\{ 1 - \Phi\left(-\frac{\delta_{ij}^0}{\sigma} \right) \right\}^{-1} \\
&\propto (\sigma^2)^{-m/2} \exp\left\{ -\frac{ESS}{2\sigma^2} - \sum_{i < j} \log \Phi\left(\frac{\delta_{ij}^0}{\sigma} \right) \right\}, \quad (13.18)
\end{aligned}
$$

where $ESS = \sum_{i < j} (\delta_{ij} - \delta_{ij}^0)^2$ is the error sum of squares, $\Phi(\cdot)$ is the standard Gaussian cdf, and $m = n(n-1)/2$ is the number of dissimilarities. The second term in the exponent of the likelihood function is the modification to the normalizing constant due to the truncation.

Next, we assume that the $\{\mathbf{X}_i\}$ are iid with a common multivariate-Gaussian prior density,

$$\mathbf{X}_i \sim \mathcal{N}_r(\mathbf{0}, \boldsymbol{\Sigma}_{XX}), \quad i = 1, 2, \ldots, n, \tag{13.19}$$

where $\boldsymbol{\Sigma}_{XX} = \text{diag}\{\lambda_1, \ldots, \lambda_r\}$. Then, the full conditional posterior density of the $\{\mathbf{X}_i\}$, that is, $\pi(\{\mathbf{X}_i\}|\sigma^2, \{\lambda_j\})$, is proportional to

$$(\sigma^2)^{-m/2} \left(\prod_{j=1}^{r} \lambda_j^{-n/2} \right) \exp\left\{ -\frac{Q_1 + Q_2}{2} - \sum_{i<j} \log \Phi\left(\frac{\delta_{ij}^0}{\sigma} \right) \right\}, \tag{13.20}$$

where $Q_1 = ESS/\sigma^2$ and $Q_2 = \sum_{i=1}^{n}(\mathbf{X}_i^\tau \boldsymbol{\Sigma}_{XX}^{-1} \mathbf{X}_i) = \sum_{j=1}^{r} \lambda_j^{-1} s_j$ are quadratic functions of the $\{\mathbf{x}_i\}$, and $s_j = \sum_{i=1}^{n} X_{ij}^2$.

We now assume that the error variance σ^2 has the (conjugate) prior

$$\sigma^2 \sim IG(a, b), \tag{13.21}$$

where $IG(a, b)$ is the inverse-gamma distribution with parameters a and b (i.e., $\pi(\sigma^2) \propto (\sigma^2)^{-(a+1)} e^{-b/\sigma^2}$, $a, b > 0$; see, e.g., Bernardo and Smith, 1994, p. 119). Similarly, the prior for λ_j is taken to be

$$\lambda_j \sim IG(\alpha, \beta_j) \tag{13.22}$$

(i.e., $\pi(\lambda_j) \propto \lambda_j^{-(\alpha+1)} e^{-\beta_j/\lambda_j}$, $\alpha, \beta_j > 0$), independently for each $j = 1, 2, \ldots, r$. Finally, the prior densities of $\{\mathbf{X}_i\}$, $\{\lambda_j\}$, and σ^2 are assumed to be independent.

The joint posterior density of $(\{\mathbf{X}_i\}, \{\lambda_j\}, \sigma^2)$, given the proximity matrix $\boldsymbol{\Delta} = (\delta_{ij})$, is

$$p(\{\mathbf{X}_i\}, \{\lambda_j\}, \sigma^2|\boldsymbol{\Delta}) = L(\{\mathbf{X}_i\}, \sigma^2|\boldsymbol{\Delta}) \cdot \pi(\{\mathbf{x}_i\}) \cdot \pi(\sigma^2) \cdot \pi(\{\lambda_j\})$$

$$\propto (\sigma^2)^{-(m/2+a+1)} \left(\prod_{j=1}^{r} \lambda_j^{-(n/2+\alpha+1)} \right) e^{-A}, \tag{13.23}$$

where

$$A = \frac{Q_1 + Q_2}{2} + \sum_{i<j} \log \Phi\left(\frac{\delta_{ij}^0}{\sigma} \right) + \frac{b}{\sigma^2} + \sum_{j=1}^{r} \frac{\beta_j}{\lambda_j}. \tag{13.24}$$

The posterior distribution (13.23) is a complicated function of the unknown quantities $(\{\mathbf{x}_i\}, \{\lambda_j\}, \sigma^2)$.

The numerical integration necessary to compute Bayes estimates of these quantities is best accomplished using Markov chain Monte Carlo (MCMC) methods. Oh and Raftery used a random-walk, Metropolis–Hastings algorithm. Initial values for the $\{\mathbf{X}_i\}$ and the other unknown parameters, σ^2 and $\{\lambda_j\}$, of the posterior distributions are taken from a classical scaling solution. For the algorithmic details, we refer the reader to the original article.

13.9 Nonmetric Distance Scaling

In many applications of MDS, dissimilarities are known only by their rank order, and the spacing between successively ranked dissimilarities is of no interest or is unavailable. This may happen because the data collected involve only ordinal information (possibly through pairwise comparisons of a set of entities). See, for example, the color stimuli and Morse code examples in Section 13.2. In these types of situations, we have no metric to deal with such comparisons.

In *nonmetric distance scaling* (also known as *ordinal MDS*), we assume that f is an arbitrary function that satisfies the monotonicity constraint $f(\delta_{ij}) \leq f(\delta_{k\ell})$ whenever $\delta_{ij} < \delta_{k\ell}$, for all $i, j, k, \ell = 1, 2, \ldots, n$. Explanations as to why f should be a monotone transformation of a dissimilarity include the following:

> *People vary remarkably in the way in which they use rating scales in general, with some showing tendencies to avoid extreme ratings, others using specific categories disproportionately often, and still others piling their judgments up against one extreme of the scale.* (Ramsay, 1988)

> *In psychophysical applications, the measuring device by which dissimilarities are observed is the human mind, which is known to perceive distances in ways that are subject to monotonic distortion. (For example, the mind has a tendency to underestimate large distances.)* (Trosset, 1998)

So, rather than using subjective judgment as a distance measure, we choose instead to construct f to preserve the *rank-order* of the dissimilarities.

13.9.1 Disparities

Suppose we have a symmetric matrix $\mathbf{\Delta} = (\delta_{ij})$ of dissimilarities (with zero diagonal entries) between a collection of n r-dimensional entities. Ignoring the diagonal entries in $\mathbf{\Delta}$ (which avoids the problem in the Morse-code example), we have $m = \frac{1}{2}n(n-1)$ dissimilarities, which we further assume can be strictly ordered from smallest to largest:

$$\delta_{i_1 j_1} < \delta_{i_2 j_2} < \cdots < \delta_{i_m j_m}, \tag{13.25}$$

where (i_1, j_1) indicates the pair of entities having the smallest dissimilarity, and (i_m, j_m) indicates the pair of entities having the largest dissimilarity.

The objective is to represent these r-dimensional entities as a configuration of n points in the lower-dimensional space $\Re^t$, where for the moment we assume that the dimensionality t is given. Denote the points in this configuration by $\mathbf{Y}_1, \ldots, \mathbf{Y}_n$ and let

$$d_{ij} = \|\mathbf{Y}_i - \mathbf{Y}_j\| = \{(\mathbf{Y}_i - \mathbf{Y}_j)^\tau (\mathbf{Y}_i - \mathbf{Y}_j)\}^{1/2} \tag{13.26}$$

be the Euclidean distance between the points $\mathbf{Y}_i$ and $\mathbf{Y}_j$, $i < j$. Nonmetric distance scaling finds a configuration such that the ordering of the distances

$$d_{i_1 j_1} < d_{i_2 j_2} < \cdots < d_{i_m j_m} \tag{13.27}$$

matches exactly the ordering of the dissimilarities in (13.25).

A plot of the configuration distances $\{d_{ij}\}$ against their rank-order will not necessarily produce a monotonically looking scatterplot, thereby violating the monotone condition (13.27). To overcome this difficulty, we approximate the $\{d_{ij}\}$ by $\{\widehat{d}_{ij}\}$, say (usually called *disparities*), which are monotonically related to the $\{d_{ij}\}$ and where

$$\widehat{d}_{i_1 j_1} \leq \widehat{d}_{i_2 j_2} \leq \cdots \leq \widehat{d}_{i_m j_m}. \tag{13.28}$$

This formulation allows for possible ties in the disparities. Think of the $\{\widehat{d}_{ij}\}$ as fitted values obtained from fitting a monotonically increasing function to the $\{d_{ij}\}$; the $\{\widehat{d}_{ij}\}$ are not themselves distances, and there may be no configuration of points $\{\mathbf{y}_i\}$ for which the $\{\widehat{d}_{ij}\}$ are its interpoint distances. These disparities, which are joined up to form a "curve," are then superimposed upon the plot of the configuration distances against rank-order. The resulting plot is usually called a *Shepard diagram*.

There are two main methods for computing nondecreasing disparities for the nonmetric distance-scaling problem. The first method, *isotonic regression* (also known as *monotonic regression*) (Kruskal, 1964b), results in a step-like function, whereas the second, *monotone splines*, yields a smoother transformation.

Isotonic Regression: A Simple Example

Consider the following artificial example with $n = 6$ entities. Suppose the rank-order of the 15 dissimilarities, $\{\delta_{ij}\}$, is given in Table 13.8 by the column marked "rank." Suppose, further, that a specific configuration yields

TABLE 13.8. *Finding the disparities by isotonic regression for an artificial example with $n = 6$ and $m = 15$. The columns I, II, III, IV, V, and VI display a sequence of trial solutions for the disparities. The cells in red indicate the active block at each trial solution. The value of S is 6.85%.*

rank	d_{ij}	I	II	III	IV	V	VI	$\widehat{d}_{ij}$
1	2.3	2.3	2.3	2.30	2.30	2.30	2.30	2.30
2	2.7	2.7	2.7	2.70	2.70	2.70	2.70	2.70
3	8.1	8.1	6.9	6.67	6.67	6.67	6.67	6.67
4	5.7	5.7	6.9	6.67	6.67	6.67	6.67	6.67
5	6.2	6.2	6.2	6.67	6.67	6.67	6.67	6.67
6	8.1	8.1	8.1	8.10	8.13	7.80	7.80	7.80
7	8.6	8.6	8.6	8.60	8.13	7.80	7.80	7.80
8	7.7	7.7	7.7	7.70	8.13	7.80	7.80	7.80
9	6.8	6.8	6.8	6.80	6.80	7.80	7.80	7.80
10	9.3	9.3	9.3	9.30	9.30	9.30	9.30	9.30
11	10.5	10.5	10.5	10.50	10.50	10.50	10.15	10.10
12	9.8	9.8	9.8	9.80	9.80	9.80	10.15	10.10
13	10.0	10.0	10.0	10.00	10.00	10.00	10.00	10.10
14	12.6	12.6	12.6	12.60	12.60	12.60	12.60	12.60
15	12.8	12.8	12.8	12.80	12.80	12.80	12.80	12.60

the estimated dissimilarities, $\{d_{ij}\}$, given in the second column. Clearly, the estimates are not rank-ordered to fit with the ranks of the dissimilarities.

We partition the estimated dissimilarities into blocks, and at each step of the algorithm one of these blocks becomes "active." A "block" is a consecutive set of dissimilarities that have to be set equal to each other to maintain monotonicity. A trial solution consists of averaging the values within the active block. Table 13.8 shows the complete sequence of trial solutions for this example to obtain the set of disparities for a single iteration.

From the second column of Table 13.8, we see that the first three d_{ij} are increasing (2.3, 2.7, 8.1). The next distance (5.7) is smaller only than the preceding 8.1, so the active block is (8.1, 5.7), whose values are averaged to get 6.9. The next distance 6.2 is smaller than the two previous 6.9s, so the active block is (6.9, 6.9, 6.2), with an average of 6.67. The two distances (8.1, 8.6) are increasing, but the next one (7.7) is smaller than the preceding two. The active block is now (8.1, 8.6, 7.7), and their average value is 8.13. The next distance (6.8) is smaller than the three 8.13s, so the active block is (8.13, 8.13, 8.13, 6.80), with an average of 7.80. The next two distances (9.3, 10.5) are increasing, but 9.8 is smaller than 10.5. So, we average the two distances (10.5, 9.8) to get 10.15. The next distance 10.0 is smaller than the two 10.15s, so we average the three values to get 10.1. The remaining distances satisfy the monotonicity requirement, and the procedure stops.

The last column of Table 13.8 shows the disparities $\{\widehat{d}_{ij}\}$. The Shepard diagram of the $\{d_{ij}\}$ and the $\{\widehat{d}_{ij}\}$ is given in the left panel of Figure 13.10. In preparation for the next step in the algorithm (i.e., updating the

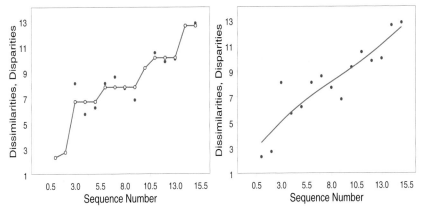

FIGURE 13.10. *Shepard diagram for the artificial example. Left panel: Isotonic regression. Right panel: Monotone spline. Horizontal axis is rank order. For the red points, the vertical axis is the dissimilarity d_{ij}, whereas for the fitted blue points, the vertical axis is the disparity $\widehat{d}_{ij}$.*

configuration), the disparities are normalized so that their sum-of-squares equals $\frac{1}{2}n(n-1)$.

Monotone Splines

As we see from the left panel of Figure 13.10, the disparities are plotted as a step-like function. We would like to make the transformation smoother than a step function while retaining the property that it be nondecreasing. We now describe a class of monotone spline transformations (Ramsay, 1988), which can be constrained to be everywhere nondecreasing and smooth. Monotone splines are constructed from polynomials defined over a grid of subintervals so that adjacent polynomials are joined up in a very smooth way at the interval boundaries.

Let $[L, U] \subset \Re$ be an interval. Define a grid in the interior of that interval by the sequence of points (or *knots*) $L = \xi_0 < \xi_1 < \cdots < \xi_q < \xi_{q+1} = U$. This grid has q interior knots, $\xi_1, \ldots, \xi_q$. Let p_i represent the rank of the ith dissimilarity, $i = 1, 2, \ldots, m$. The grid of points defines a sequence of adjacent subintervals, $[\xi_0, \xi_1], [\xi_1, \xi_2], \ldots, [\xi_q, \xi_{q+1}]$, so that each p_i falls into one of these subintervals. Within the jth subinterval $[\xi_j, \xi_{j+1}]$, the function f consists of a polynomial P_j having a given degree $k-1$ (or order k). The smoothness of f is characterized by the two polynomials P_j and P_{j+1} having equal (and, hence, continuous) derivatives up to order $k-2$ at the knot ξ_j, $i = 1, 2, \ldots, q$; that is, $(D^{i-1}P_j)(\xi_j) = (D^{i-1}P_{j+1})(\xi_j)$, $i = 1, 2, \ldots, k-1$, where $(D^{i-1}P)(\xi) = d^{i-1}P/dx^{i-1}$ evaluated at the point ξ if $i > 1$ and $(D^0P)(\xi) = P(\xi)$.

Thus, if $k = 1$, the spline is a step function discontinuous at the knots; if $k = 2$, the spline is a sequence of piecewise linear segments that join up continuously at the knots; if $k = 3$, we have piecewise quadratic segments with continuous first derivatives at the knots; and if $k = 4$, we have piecewise cubic segments with continuous first and second derivatives at the knots (usually called a *cubic spline*). Note that the number of knots and their placement play important roles in the definition of any spline function; a poor choice of knots can result in a low-quality spline fit to the data.

It can be shown that a monotone spline of degree k with q interior knots can be computed using the equation,

$$\widehat{\mathbf{d}} = b_0 \mathbf{1}_m + \mathbf{Mb}, \tag{13.29}$$

where $\mathbf{M} = (\mathbf{M}_1, \cdots, \mathbf{M}_{k+q})$ is an $(m \times (k+q))$-matrix, $\mathbf{b}$ is a $(k+q)$-vector of nonnegative weights, and b_0 is a nonnegative constant. These type of splines are also called *regression splines*. The columns $\{\mathbf{M}_j\}$ of $\mathbf{M}$ are each piecewise polynomial functions of the p_i. The first $\max\{0, j - k\}$ intervals of $\mathbf{M}_j$ are each zero and the last $\max\{0, q - j + 1\}$ intervals of $\mathbf{M}_j$ are each one. For example, suppose $k = 2$ (quadratic) and $q = 4$ interior knots. Then, $\mathbf{M}_1$ has ones in the last four intervals, $\mathbf{M}_2$ has ones in the last three intervals, $\mathbf{M}_3$ has a zero in the first interval and ones in the last two intervals, $\mathbf{M}_4$ has zeroes in the first two intervals and a one in the last interval, $\mathbf{M}_5$ has zeroes in the first three intervals, and $\mathbf{M}_6$ has zeroes in the first four intervals. The remaining intervals constitute an appropriate polynomial in the p_i with equal derivatives at the knots.

More formally, let M_{ij} denote the ith element of the jth column $\mathbf{M}_j$ of the matrix $\mathbf{M}$, $i = 0, 1, 2, \ldots, m - 1$, $j = 1, 2, \ldots, k + q$, where q is the number of interior knots. A zero-order ($k = 0$) spline has M_{ij} equal to zero if $\xi_0 \leq p_i < \xi_j$ and one if $\xi_j \leq p_i < \xi_{q+1}$. A linear ($k = 1$) spline has elements

$$M_{ij} = \begin{cases} 0 & \text{if } \xi_0 \leq p_i < \xi_{j-1} \\ a_{ij} & \text{if } \xi_{j-1} \leq p_i < \xi_j \\ 1 & \text{if } \xi_j \leq p_i < \xi_{q+1} \end{cases} \tag{13.30}$$

where $a_{ij} = (p_i - \xi_j)/(\xi_j - \xi_{j-1})$. For a quadratic spline ($k = 2$), we have that:

$$M_{ij} = \begin{cases} 0 & \text{if } \xi_0 \leq p_i < \xi_{j-2} \\ b_{ij} & \text{if } \xi_{j-2} \leq p_i < \xi_{j-1} \\ 1 - c_{ij} & \text{if } \xi_{j-1} \leq p_i < \xi_j \\ 1 & \text{if } \xi_j \leq p_i < \xi_{q+1} \end{cases} \tag{13.31}$$

where $b_{ij} = (\xi_{j-2} - p_i)^2/(\xi_{j-1} - \xi_{j-2})(\xi_j - \xi_{j-2})$ and $c_{ij} = (\xi_j - p_i)^2/(\xi_j - \xi_{j-1})(\xi_j - \xi_{j-2})$. For $j = 1$, we set $\xi_{-1} = \xi_0$, and for $j = q + 1$, we set $\xi_{q+1} = \xi_q$. In the special case that $q = m - 1$ and $k = 0$, the monotone spline (using appropriately located knots) is identical to a monotone regression transformation (see Exercise 13.4).

Thus, we can write the ith disparity (i.e., ith element of the m-vector $\widehat{\mathbf{d}}$) as the linear combination,

$$\widehat{d}_i = \sum_{j=0}^{k+q} b_j M_{ij}, \tag{13.32}$$

with nonnegative weights, $b_j \geq 0$, $j = 0, 1, 2, \ldots, k + q$, where we set $M_{i0} = 1$, $i = 1, 2, \ldots, m$. Redefine $\mathbf{M}$ to be an $(m \times (k+q+1))$-matrix with first column $\mathbf{1}_m$ to take care of the constant b_0. Then, $\mathbf{b} = (b_0, b_1, \cdots, b_{k+q})^\tau$ is the coefficient vector having nonnegative elements. The vector of disparities is defined as $\widehat{\mathbf{d}} = \mathbf{Mb}$.

We now wish to find nonnegative $\mathbf{b}$ that will solve the LS problem,

$$\mathbf{b}^* = \arg\min_{\mathbf{b} \geq \mathbf{0}} (\mathbf{d} - \mathbf{Mb})^\tau (\mathbf{d} - \mathbf{Mb}). \tag{13.33}$$

This problem can be solved using the following *alternating least-squares (ALS)* algorithm. First, fix all entries of $\mathbf{b}$ except b_j. We now choose a nonnegative b_j to minimize (13.33). Compute the "residual" $\widehat{\mathbf{e}}_j = \mathbf{d} - \sum_{k \neq j} b_k \mathbf{M}_k$. Then, $\widehat{\mathbf{e}}_j - b_j \mathbf{M}_j = \mathbf{d} - \mathbf{Mb}$, and $(\widehat{\mathbf{e}}_j - b_j \mathbf{M}_j)^\tau (\widehat{\mathbf{e}}_j - b_j \mathbf{M}_j) = \widehat{\mathbf{e}}_j^\tau \widehat{\mathbf{e}}_j + b_j^2 \mathbf{M}_j^\tau \mathbf{M}_j - 2b_j \mathbf{M}_j^\tau \widehat{\mathbf{e}}_j$, which is minimized when $b_j = \mathbf{M}_j^\tau \widehat{\mathbf{e}}_j / \mathbf{M}_j^\tau \mathbf{M}_j$. If $b_j < 0$, set $b_j = 0$. Thus, $b_j \geq 0$. Repeat this computation for every other element of $\mathbf{b}$ while keeping all other elements of $\mathbf{b}$ fixed. These steps constitute the first iteration of the ALS algorithm, which yields $\mathbf{b} \geq \mathbf{0}$. We update these values in an iterative fashion until no change is observed in $\mathbf{b}$. This algorithm has been shown to converge to a global minimum of the nonnegative LS problem.

In the right panel of Figure 13.10, we show the monotone spline fitted to the artificial data example given in Table 13.8.

13.9.2 The Stress Function

If we square the horizontal deviations, $d_{ij} - \widehat{d}_{ij}$, from a Shepard diagram and then add them up, we get a form of residual sum of squares,

$$\text{raw stress} = S^*(\mathbf{Y}_1, \ldots, \mathbf{Y}_n) = \sum_{i<j} (d_{ij} - \widehat{d}_{ij})^2, \tag{13.34}$$

which acts as a measure of goodness of fit. Although this measure is invariant under translations, reflections, and rotations (orthogonal transformations) of the $\{\mathbf{Y}_i\}$, it is not scale-invariant under stretching (or shrinking) of each of the $\{\mathbf{Y}_i\}$ by some constant k; we see that $\mathbf{Y}_i \to k\mathbf{Y}_i$ means $d_{ij} \to kd_{ij}$ and $\widehat{d}_{ij} \to k\widehat{d}_{ij}$, so that $S^* \to k^2 S^*$. Thus, raw stress can always be reduced in magnitude by scaling down (shrinking) the configuration to a single point where all the $d_{ij} = 0$. To counter this effect of scale-dependency, we normalize the measure S^* to have the general form,

$$\left\{ \sum_{i<j} w_{ij} (d_{ij} - \widehat{d}_{ij})^2 \right\}^{1/2}, \tag{13.35}$$

where the $\{w_{ij}\}$ are weights chosen by the user. The most popular normalization is where $w_{ij} = (\sum_{i<j} d_{ij}^2)^{-1}$, so that (13.34) becomes (Kruskal, 1964a)

$$\text{stress} = S(\mathbf{Y}_1, \ldots, \mathbf{Y}_n) = \left\{ \frac{\sum_{i<j} (d_{ij} - \widehat{d}_{ij})^2}{\sum_{i<j} d_{ij}^2} \right\}^{1/2}, \tag{13.36}$$

where it is understood that the summations in both the numerator and denominator of S are computed for all $i, j = 1, 2, \ldots, n$ such that $i < j$. The stress value S lies between 0 and 1.

The stress criterion S (more commonly known as *Kruskal's stress formula one* or *Stress-1*) can be interpreted as a loss function that depends upon the configuration points $\{\mathbf{Y}_i\}$ and the disparities $\{\widehat{d}_{ij}\}$ and measures how well a particular configuration fits the given dissimilarities. It is worth noting that certain authors refer to S^2 as the stress function.

A slightly different version of S (called *stress formula two* or *Stress-2*) has weights given by

$$w_{ij} = \left\{ \sum_{i<j} (d_{ij} - \bar{d})^2 \right\}^{-1}, \tag{13.37}$$

where $\bar{d}$ is the average distance. The normalization (13.37) has been used in situations where certain types of degeneracies occur. Other recommended normalizations include $w_{ij} = \widehat{d}_{ij}^{-2}$ and $w_{ij} = (\sum_{i<j} \widehat{d}_{ij})^{-1} \widehat{d}_{ij}^{-1}$ (Sammon, 1969). The *sstress* criterion, which uses squared distances and squared disparities,

$$\text{sstress} = \sum_{i<j} (d_{ij}^2 - \widehat{d}_{ij}^2)^2, \tag{13.38}$$

is the minimization criterion of choice in the MDS program ALSCAL (Takane, Young, and de Leeuw, 1977). A disadvantage of the sstress criterion (13.38) is that it emphasizes larger dissimilarities at the expense of smaller ones. More general versions of all these stress functions are available.

TABLE 13.9. *The nonmetric distance-scaling algorithm.*

1. Order the $m = \frac{1}{2}n(n-1)$ dissimilarities $\{\delta_{ij}\}$ from smallest to largest as in (13.25).

2. Fix the number t of dimensions and choose an initial configuration of points $\mathbf{Y}_i \in \Re^t$, $i = 1, 2, \ldots, n$.

3. Compute the set of distances $\{d_{ij}\}$ between all pairs of points in the initial configuration.

4. Use an isotonic regression algorithm to produce fitted values $\{\widehat{d}_{ij}\}$. Compute the initial value of stress.

5. Change the configuration of points by applying an iterative gradient search algorithm (e.g., method of steepest descent) to the stress criterion. This step will produce a new set of $\{d_{ij}\}$.

6. Use an isotonic regression algorithm to produce revised values of the $\{\widehat{d}_{ij}\}$, together with a smaller stress value.

7. Repeat steps 5 and 6 until the current configuration produces a minimum stress value, so that no further improvement in stress can take place by further reconfiguring the points.

8. Repeat the previous steps using a different value of t. Plot stress against t. Choose that value of t that gives a reasonably small value of stress and where no significant decrease in stress can result from increasing t. This is usually exhibited by an "elbow" in the plot.

13.9.3 *Fitting Nonmetric Distance-Scaling Models*

The goal here is to find a configuration of points $\{\mathbf{Y}_i\} \subset \Re^t$ that minimizes the stress value S under the monotonicity condition (13.28) for the disparities. To minimize such a nonlinear function in many variables, gradient-based optimization algorithms (e.g., method of steepest descent) have traditionally been used (Kruskal, 1964b).

Starting with an arbitrary configuration (which may be a random scatter of points having little relationship to the given dissimilarities or the configuration found from carrying out metric MDS on $\mathbf{\Delta}$), we change the locations of the points in an iterative fashion. At each iteration, we improve the configuration by finding the direction for which S decreases most quickly, and we move the points in the configuration a short step in that direction. This iterative scheme is carried out until S does not decrease significantly. The algorithm is listed in Table 13.9.

Let $\mathcal{Y}^\tau = (\mathbf{Y}_1, \cdots, \mathbf{Y}_n)$ be a $(t \times n)$-matrix whose columns are the configuration points. Let $\mathbf{y} = \text{vec}(\mathcal{Y}^\tau) = (\mathbf{Y}_1^\tau, \cdots, \mathbf{Y}_n^\tau)^\tau$ be the nt-vector obtained by placing the columns of $\mathcal{Y}^\tau$ under one another successively. Stress

$S = S(\mathbf{y})$ is now a function of $\mathbf{y}$. The method of steepest descent moves the configuration in a direction determined by the partial derivatives of S with respect to $\mathbf{y}$. Thus, given the configuration $\mathbf{y}^{(m)}$ at the mth iteration, a revised configuration at the next iteration is given by:

$$\mathbf{y}^{(m+1)} = \mathbf{y}^{(m)} - \alpha_{m+1}\mathbf{z}, \tag{13.39}$$

where α_{m+1} is the step-size at the $(m + 1)$th iteration and

$$\mathbf{z} = \frac{\partial S}{\partial \mathbf{y}} / |\frac{\partial S}{\partial \mathbf{y}}| \tag{13.40}$$

is the (normalized) gradient function. Explicit formulas for $\mathbf{z}$ were first given by Kruskal (1964b). See also Cox and Cox (2001, Section 3.2.2). Step size should be changed at each iteration to speed up the algorithm (Kruskal suggests starting in general with $\alpha_0 = 0.2$).

This gradient-based procedure has been extended and generalized in many different ways. However, there is no guarantee that any of these algorithms will find a global minimum. Indeed, it is not unusual for these algorithms to find only local minima. As a result, the best that can often be accomplished is to try different initial configurations (i.e., random starts) to check the convergence properties of the algorithm. This may be accomplished by choosing a very large step size to start the iteration process all over again whenever a local minimum is thought to have been reached. If the same solution is obtained from repeated application of the algorithm, then the common solution is probably a global minimum.

13.9.4   How Good Is an MDS Solution?

Kruskal's experience with various types of real and simulated data led him to assess the global fit of any nonmetric distance-scaling solution by various levels of stress values (Kruskal, 1964a); see Table 13.10. The distance-scaling solution for the color-stimuli example has a clear elbow at two dimensions in the scree plot and a 2D minimum-stress value of 0.023, which would classify the configuration as an excellent fit to the data.

The assessment given by Table 13.10 should be considered only as a possible guideline of how well an MDS solution fits the "true" data structure; in fact, this guideline often fails to do this, especially in situations where the data are noisy. For example, in the distance-scaling solution for the Morse-code example, the scree plot shows no elbow and the minimum-stress value for the 2D solution is 0.18, which would declare the configuration close to a poor fit to the data. In general, if the number of subjects is larger in one study than in another, then we would expect the stress value from the former study to be larger. We see this in the Morse-code example, where

TABLE 13.10 *Evaluation of "stress."*

Stress	Goodness of Fit
0.20	Poor
0.10	Fair
0.05	Good
0.025	Excellent
0.0	"Perfect"

the number of subjects is much larger than in the color-stimuli example and so would be expected to have a higher stress value.

13.9.5 How Many Dimensions?

Stress S measures the goodness of fit of a given configuration in $\Re^t$, and the configuration that best matches the dissimilarities enjoys minimum stress. Furthermore, as we increase t, we find that the minimum stress decreases. In fact, if $t \geq n - 1$, the minimum stress is exactly zero, a solution that is clearly undesirable; too large a dimensionality implies that we are including in the solution overly many noisy dimensions, which, in turn, leads to overfitting.

The goal is to choose a configuration for which t is reasonably small (typically, 2 or 3, if possible). With this consideration in mind, we compute the minimum value of stress, $S_{min}^{(t)}$, for different dimensionalities t, plot the points $(t, S_{min}^{(t)})$, and then join up the plotted points. The resulting "curve" will be monotonically decreasing from right to left, with the decrease becoming less severe as t gets larger. This "curve" is sometimes called a *scree plot* if all minimum-stress values for t from 1 to r are computed.

We choose that value of t for which the minimum stress is small and any further increase in t does not significantly decrease the minimum stress. Using an informal selection procedure also found in PCA and factor analysis, we look for a t that exhibits an "elbow" in the plot. That value of t is taken as the chosen dimensionality. The results of simulation studies with noise-perturbed distances (e.g., Spence and Graef, 1974), however, have shown that "elbows" are not all that obvious in scree plots even when noise is kept at a fairly low level.

13.10 Software Packages

Classical scaling can be carried out in S-PLUS and R by using the command `cmdscale` [in library(`mva`)] (Venables and Ripley, 2002, p. 306). Sammon mapping can be computed using the S-PLUS and R command `sammon` [in library(`MASS`)] (Venables and Ripley, 2002, p. 308) and is also available in the `SOM` toolbox in MATLAB. A `Fortran` program, `bmds`, written by Oh and Raftery to compute Bayesian MDS is available at the STATLIB website. R (version 1.9.0) contains the command `isoreg` [now in package `stats`, moved from package `modreg`] to compute isotonic regression. Kruskal's method of nonmetric distance scaling using the stress function (13.36) and isotonic regression can be carried out in S-PLUS and R by using the command `isoMDS` [in library(`MASS`)] (Venables and Ripley, 2002, p. 308).

Bibliographical Notes

Book-length descriptions of MDS include Cox and Cox (1994) and Borg and Groenen (1997).

The early development of MDS procedures was dominated by applications to psychology. Another very popular area for MDS application has been marketing, where the entities are different brand-name products, and the distance between a pair of those products gives a measure of how closely associated the two products appear to be in the eyes of consumers. Researchers in areas of molecular biology (Crippen and Havel, 1978; Havel, 1991; Glunt, Hayden, and Raydan, 1993; Basalaj and Eilbeck, 2003; Hou, Sims, Zhang, and Kim, 2003), computational chemistry (Trosset, 1998), social networks (Theus and Schonlau, 1998), and graph layout and drawing (Kruskal and Seery, 1980; Di Battista, Eades, Tamassia, and Tollis, 1994) have shown that those areas can also profit from using MDS. We note that the MDS application to network design (Kruskal and Seery, 1980) is used as part of the Isomap algorithm for nonlinear manifold learning (see Section 16.7.3), but where geodesic distance along the manifold is used instead of Euclidean distance.

Classical scaling was introduced by Torgerson (1952, 1958). Gower (1966) called it *principal coordinate analysis* because of its close resemblance to PCA. Its roots go back to the results of Eckart and Young (1936) and Young and Householder (1938). Classical scaling has variously been referred to as *Torgerson scaling*, *Torgerson–Gower scaling*, and *Torgerson–Young scaling*.

In 1995, the National Academy Commission on Physical Sciences, Mathematics, and Applications published a report entitled *Mathematical Challenges from Theoretical/Computational Chemistry*. In Chapter 3 of that report, "distance geometry" was described as "an important technique in computational chemistry" and "a key tool in the NMR spectroscopist's arsenal, providing not only the [3D] structures, but also a [quantification] of how accurately they are known."

Useful books on protein sequence alignment include Durbin, Eddy, Krogh, and Mitchison (1998) and Deonier, Tavaré, and Waterman (2005). An excellent account of BLAST can be found in the book by Korf, Yandell, and Bedell (2003), where Altschul reports that the name BLAST was originally chosen to be a pun on the name FASTA, but then morphed into its current expanded name. The SIM sequence alignment program can be found at the website `us.expasy.org/tools/sim-prot.html`.

Nonmetric MDS was formulated by Kruskal (1964a,b), who introduced the notion of stress and gave an iterative computational algorithm for carrying out MDS. Monotone splines have been used as a main ingredient in a model-based framework for statistical inference in MDS (Ramsay, 1982). We note that even though the idea of extending nonmetric MDS into a model-based methodology is very controversial (see the discussion accompanying Ramsay's article), monotone splines in MDS have been found to be a useful exploratory tool for calculating disparities.

The algorithms that are still being used for MDS are known to be very slow and inefficient and do not scale well for very large data sets. Accordingly, workshops on MDS algorithms are being held to develop new and different algorithms for MDS.

Exercises

13.1 Consider the color-stimuli experiment outlined in Section 14.2.1. The similarity ratings are given in the file `color-stimuli` on the books's website. Carry out a classical scaling of the data and show that the solution is a "color circle" ranging from violet (434 mμ) to blue (472 mμ) to green (504 mμ) to yellow (584 mμ) to red (674 mμ). Compare your solution to the nonmetric scaling solution given in Section 13.2.

13.2 Consider the Morse-code experiment outlined in Section 13.2.2. The file `Morse-code` on the book's website gives a table of the percentages of times that a signal corresponding to the row label was identified as being the same as the signal corresponding to the column label. A row of this table shows the confusion rate for that particular Morse-code signal when presented *before* each of the column signals, whereas a column of the table shows the confusion rate for that particular signal when presented *after* each of the row signals. This table of confusion rates is not symmetric and the diagonal elements are not each 100%. Now, every square matrix $\mathbf{M}$ can be decomposed uniquely into the sum of two orthogonal matrices, $\mathbf{M} = \mathbf{A} + \mathbf{B}$, where $\mathbf{A} = \frac{1}{2}(\mathbf{M} + \mathbf{M}^\tau)$ is symmetric ($\mathbf{A}^\tau = \mathbf{A}$), and $\mathbf{B} = \frac{1}{2}(\mathbf{M} - \mathbf{M}^\tau)$ is skew-symmetric ($\mathbf{B}^\tau = -\mathbf{B}$) with zero diagonal entries. Find the decomposition for the Morse-code data. Ignore that part of the Morse-code data provided by $\mathbf{B}$ and carry out a nonmetric scaling only of the symmetric part $\mathbf{A}$. Decide how many dimensions you think are appropriate for representing the data.

13.3 Let $\|\mathbf{M}\|^2 = \sum_i \sum_j M_{ij}^2$. From the decomposition in Exercise 13.2, show that $\|\mathbf{M}\|^2 = \|\mathbf{A}\|^2 + \|\mathbf{B}\|^2$. This result enables us to analyze separately the symmetric part (see Exercise 13.2) and the asymmetric part of the Morse-code data. Ignore the diagonal entries in $\mathbf{M}$, $\mathbf{A}$, and $\mathbf{B}$. Find the sum of squares of the remaining entries of all three matrices and argue why you may think that the symmetric part of the data plays a major role in the analysis, whereas the asymmetric part plays only a minor role.

13.4 Show that the dissimilarities in the matrix $\mathbf{\Delta}$ are Euclidean distances if and only if the doubly centered matrix $\mathbf{B} = \mathbf{HAH}$ is nonnegative definite, where $\mathbf{A}$ is given in the classical scaling algorithm of Table 13.5.

13.5 This exercise shows that monotone regression is a special case of monotone spline transformations. Consider a zero-order ($k = 0$) monotone spline with $q = m - 1$ interior knots (where m is the number of dissimilarities). Let $p_i = i$, $i = 0, 1, 2, \ldots, m - 1$. Let $m = 5$ and put the knots at the points $\xi_0 = 0.5, \xi_1 = 1.5, \xi_2 = 2.5, \xi_3 = 3.5, \xi_4 = 4.5, \xi_5 = 5.5$. Find the (5×4)-matrix $\mathbf{M}$ and the vector of disparities $\hat{\mathbf{d}} = b_0 \mathbf{1}_m + \mathbf{Mb}$, for any nonnegative b_i, $i = 0, 1, \ldots, m - 1$. Show that the disparities obey the same monotonicity property as they do in (13.27) for monotone regression.

13.6 In the `British-towns` file on the book's website, there is a proximity matrix of the distances between 48 towns in Great Britain. Carry out a classical scaling of these pairwise distances and construct a map of Great Britain.

13.7 In ratio MDS and interval MDS, find the LS estimates of α and β in each case, where the minimizing criterion is the weighted loss function (13.14).

14
Committee Machines

14.1 Introduction

One of the most important research topics in machine learning is the problem of how to lower the generalization error of a learning algorithm, either by reducing the bias or the variance (or both). A major complication of any attempt to reduce variance or bias (or both) is that the definitions of "bias" and "variance" of a classification rule are not as obvious as they are in regression. In fact, there have been several conflicting suggestions for the bias-variance decomposition for classification problems.

Such a desire to control bias and variance, and, hence, generalization error, is related to the idea of "instability" of a prediction or classification method. If a small perturbation of the learning set induces major changes in the resulting predictor or classifier, we say that the associated regression or classification method is *unstable*. Unstable predictors or classifiers have high variance (due to overfitting) and low bias. High bias occurs for predictors or classifiers that underfit the data. Decision trees and neural nets are, by this definition, unstable, whereas linear discriminant analysis is an example of a stable classifier with low variance and possibly high bias.

In this chapter, we show that the *instability* of a predictor or classifier (or, more generally, of any learning algorithm) is an important tool that can be

A.J. Izenman, *Modern Multivariate Statistical Techniques*,
doi: 10.1007/978-0-387-78189-1_14,
© Springer Science+Business Media, LLC 2008

used to improve the accuracy of that learning algorithm. Novel approaches to the problem of predictor instability include *bagging* and *boosting*. Both of these approaches exploit the presence of instability in order to create a more accurate learning method (i.e., predictor or classifier). By perturbing the learning set, these methods generate an *ensemble* of different *base predictors* or *base classifiers*, which are then combined into a single *combined predictor* or *combined classifier*, as appropriate. The success of such combined learning methods — called *ensemble learning* or *committee machines* — often depends upon the degree of instability of the base predictors or classifiers.

Bagging and boosting can be distinguished from each other by the manner in which their respective perturbations are generated. The bagging process (Breiman, 1996b) generates perturbations by random and independent drawings from the learning set, whereas the boosting process (Freund and Schapire, 1998) is deterministic and generates perturbations by successive reweightings of the learning set, where current weights depend upon the misclassification history of the process. Bagging was designed specifically to reduce variance, whereas boosting appears to have more of a bias-reducing flavor. Another example of a committee machine that will be described in this chapter is *random forests* (Breiman, 2001b).

14.2 Bagging

The word *bagging* is an acronym for the phrase "bootstrap aggregating" (Breiman, 1996b). Bagging was the first procedure that successfully combined an ensemble of learning algorithms to improve performance over a single such algorithm.

Bagging is most successful if the predictor is unstable. If the learning procedure is stable, the bagged predictor will not differ much from the single predictor and may even weaken its performance somewhat. However, when the learning procedure is unstable, we tend to see a significant improvement for the bagged predictor over the original unstable procedure.

As before, we denote the learning set of n observations by

$$\mathcal{L} = \{(\mathbf{X}_i, Y_i), i = 1, 2, \dots, n\}, \tag{14.1}$$

where the $\{Y_i\}$ are continuous responses (a regression problem) or unordered class labels (a classification problem). Bagging takes an ensemble of learning sets, $\{\mathcal{L}_k\}$, say, each containing n observations drawn from the same underlying distribution as those in $\mathcal{L}$, and combines the predictors from those learning sets in such a way that the resulting predictor improves upon that obtained from the single learning set $\mathcal{L}$.

The bagging procedure starts by drawing B bootstrap samples from $\mathcal{L}$. Each bootstrap sample is obtained by repeated sampling *with replacement* from $\mathcal{L}$. In other words, we place equal probabilities on the sample points (i.e., $p_i = 1/n$ on the ith observation $(\mathbf{X}_i, Y_i)$ in $\mathcal{L}$, $i = 1, 2, \ldots, n$) and then sample n times with replacement from this distribution. We denote the bootstrap samples by

$$\mathcal{L}^{*b} = \{(\mathbf{X}_i^{*b}, Y_i^{*b}), i = 1, 2, \ldots, n\}, \quad b = 1, 2, \ldots, B. \tag{14.2}$$

Some of the original learning set will appear in $\mathcal{L}^{*b}$, some will appear several times, whereas others will not appear at all. What we do next depends upon whether we are dealing with a classification or a regression problem.

14.2.1 Bagging Tree-Based Classifiers

In the classification case, $Y_i \in \{1, 2, \ldots, K\}$ is a class label attached to $\mathbf{X}_i$. We grow a classification tree $\mathcal{T}^{*b}$ from the bth bootstrap sample $\mathcal{L}^{*b}$. To reduce bias, we grow this tree very large without pruning. Suppose $(\mathbf{X}, Y)$ is independently drawn from the same joint distribution as the members in $\mathcal{L}$. We drop $\mathbf{X}$ down each of the B bootstrap trees. For each tree, when $\mathbf{X}$ falls into a terminal node associated with a particular class, we say that the tree "votes" for that class. We then predict the class of $\mathbf{X}$ by the class that receives the most number of votes over all B trees. We call this classification procedure *the majority-vote rule*.

In order to evaluate the bagging method, we need an independent test set of observations. The fact that we are sampling (with replacement) from $\mathcal{L}$ means that about 37% of the observations in $\mathcal{L}$ will not be chosen for each bootstrap sample (see Section 5.5.3). Let $\mathcal{L} - \mathcal{L}^{*b}$ denote those observations in $\mathcal{L}$ that are not selected for the bth bootstrap sample $\mathcal{L}^{*b}$. If the observation $(\mathbf{X}, Y)$ is in $\mathcal{L} - \mathcal{L}^{*b}$ (which we write as $(\mathbf{X}, Y) \notin \mathcal{L}^{*b}$), then $(\mathbf{X}, Y)$ is called an *out-of-bag (OOB) observation*. The collection of OOB observations (which we call an OOB sample) corresponding to the bootstrap sample $\mathcal{L}^{*b}$ will function as an independent test set.

The OOB approach to estimating generalization error is equivalent to using an independent test set of the same size. The OOB approach is also able to use all the data, rather than partitioning the data into a separate (and smaller) learning set and a test set, and it does not require any additional computing as is needed for cross-validation.

Suppose $(\mathbf{X}_i, Y_i) \notin \mathcal{L}^{*b}$. We drop $\mathbf{X}_i$ down the classification tree $\mathcal{T}^{*b}$ grown from $\mathcal{L}^{*b}$, and predict the class label for $\mathbf{X}_i$. This acts as a classification vote on $\mathbf{X}_i$. Suppose there are n_i ($\leq B$) trees for which $\mathbf{X}_i$ is a member of the corresponding OOB sample. Drop $\mathbf{X}_i$ down each of those n_i trees and aggregate the votes for each of the K classes. Summarize the results by the K-vector,

$$\widehat{\mathbf{p}}(\mathbf{x}_i) = (\widehat{p}_1(\mathbf{x}_i), \widehat{p}_2(\mathbf{x}_i), \cdots, \widehat{p}_K(\mathbf{x}_i))^\tau, \tag{14.3}$$

where $\widehat{p}_k(\mathbf{x}_i)$ is the proportion of the n_i trees that votes for $\mathbf{X}_i = \mathbf{x}_i$ to be a member of the kth class Π_k. The proportion $\widehat{p}_k(\mathbf{x}_i)$ is an estimate of the true probability, $p(\Pi_k|\mathbf{x}_i) = \text{Prob}(\mathbf{X} \in \Pi_k|\mathbf{X} = \mathbf{x}_i)$, that the observed $\mathbf{x}_i$ belongs to Π_k. The *OOB classifier*, $C_{\text{bag}}(\mathbf{x}_i)$, of $\mathbf{x}_i$ is then obtained by the majority-vote rule:

$$C_{\text{bag}}(\mathbf{x}_i) = \arg \max_k \{\widehat{p}_k(\mathbf{x}_i)\}. \tag{14.4}$$

That is, it assigns $\mathbf{x}_i$ to that class that enjoys the largest number of votes. We repeat this for every observation in $\mathcal{L}$. The *OOB misclassification rate*,

$$PE_{\text{bag}} = n^{-1} \sum_{i=1}^{n} I_{[C_{\text{bag}}(\mathbf{x}_i) \neq y_i]}, \tag{14.5}$$

is the proportion of times that the predicted class, $C_{\text{bag}}(\mathbf{x}_i)$, is different from the true class, $Y = y_i$, for all observations in $\mathcal{L}$, and is an unbiased estimate of generalization error.

Examples of Bagging Classification Trees

As a first example of bagging classification trees, we estimate the OOB misclassification rate for the binary classification data set spambase, which consists of 57 variables measured on 4,601 messages, each one classified as spam (1,813 messages) or e-mail (2,788 messages). If we declare every message as non-spam, we get a baseline misclassification rate of $1,813/4,601 = 0.394$.

We grew different-sized classification trees (stumps, 4-node trees, 8-node trees, and largest-possible trees) and then bagged them using $B = 10(25)200$ bootstrap replications; each combination of tree-size and B was then repeated 10 times. Figure 14.1 plots the average OOB misclassification rates for bagging different size trees against B (left panel) and parallel boxplots for bagging the largest-possible trees (right panel). We see that bagging stumps is obviously a bad idea. Otherwise, as the complexity of the tree increases, the OOB misclassification rates decrease significantly.

In Figure 14.2, we display the results of bagging classification trees for the two-class data sets BUPA liver disorders and Wisconsin diagnostic breast cancer (wdbc) and for the multiclass data sets glass (six classes) and yeast (ten classes) as parallel boxplots. For each data set, the largest-possible tree was grown, the number of bootstrap samples was varied as $B = 10, 25(25)200$, and for each B, we repeated the bagging procedure 10 times. The results, which are representative of many different data sets, show that for binary classification problems, as we increase B, the misclassification rate declines, until about $B = 50$, when it appears to stabilize. For multiclass classification problems, and especially in situations where there

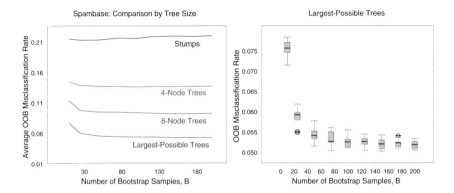

FIGURE 14.1. *Bagging classification trees for the* spambase *data. Left panel: Comparison of average profiles (over 10 repetitions) of out-of-bag (OOB) misclassification rates plotted against number of bootstrap samples (B = 10, 25(25)200) for different size trees (stumps, 4-node trees, 8-node trees, and the largest-possible trees). Notice how poorly stumps perform as base classifiers, and misclassification rates decline as tree complexity increases. Right panel: Parallel boxplots of OOB misclassification rates for the* spambase *data plotted against the number of bootstrap samples B, where largest-possible trees were grown.*

are a large number of classes, the misclassification rate tends to stabilize when B is taken to be 75–100.

14.2.2 *Bagging Regression-Tree Predictors*

In the regression case, $Y_i \in \Re$. Bagging regression-tree estimates is a very similar procedure to that applied to classification trees, but instead of using a voting mechanism to determine the predicted class of an observation, we average the predicted response values obtained from the individual regression trees.

Specifically, from the bth bootstrap sample $\mathcal{L}^{*b}$, we grow a regression tree T^{*b} and obtain the predictor $\widehat{\mu}^{*b}(\mathbf{X})$. We drop $\mathbf{X}$ down each of the B regression trees and then average the predictions,

$$\widehat{\mu}_{\text{bag}}(\mathbf{X}) = B^{-1} \sum_{b=1}^{B} \widehat{\mu}^{*b}(\mathbf{X}), \qquad (14.6)$$

to arrive at a *bagged estimate* of Y.

To evaluate the predictive abilities of a bagged regression estimate such as (14.6), we again use the OOB approach. Let $(\mathbf{X}_i, Y_i) \in \mathcal{L}$. We drop $\mathbf{X}_i$ down each of the n_i bootstrap trees whose OOB samples contain $(\mathbf{X}_i, Y_i)$. The *OOB regression estimate, $\widehat{\mu}_{\text{bag}}(\mathbf{X}_i)$, is found by averaging the n_i bootstrap*

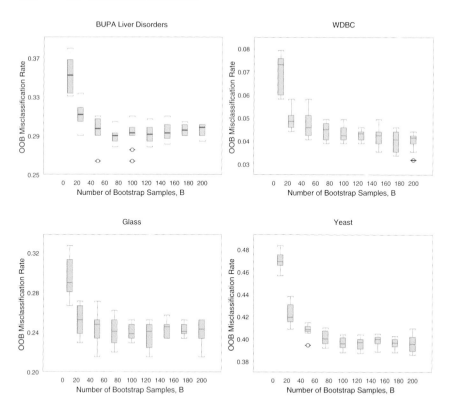

FIGURE 14.2. *Parallel boxplots of out-of-bag (OOB) misclassification rates plotted against the number of bootstrap samples B. Top-left panel: BUPA liver disorders (K = 2); top-right panel: WDBCt (K = 2); bottom-left panel: Glass (K = 6); bottom-right panel: Yeast (K = 10), where K is the number of classes. For each B = 10, 25(25)200, 10 repetitions were generated.*

predicted values; that is,

$$\widehat{\mu}_{\text{bag}}(\mathbf{X}_i) = n_i^{-1} \sum_{b \in \mathcal{N}_i} \widehat{\mu}^{*b}(\mathbf{X}_i), \qquad (14.7)$$

where $\mathcal{N}_i$ is the set of n_i bootstrap samples that do not contain $(\mathbf{X}_i, Y_i)$. We repeat this procedure for all observations in $\mathcal{L}$. We then estimate the generalization error of the bagged estimate by the *OOB error rate*,

$$PE_{\text{bag}} = n^{-1} \sum_{i=1}^{n} (Y_i - \widehat{\mu}_{\text{bag}}(\mathbf{X}_i))^2, \qquad (14.8)$$

which is computed as the mean-squared-error between the bagged estimates and their true response values.

14.3 Boosting

The underlying notion of "boosting" is to enhance the accuracy of a "weak" binary classification learning algorithm. This idea originated in a field known in machine learning as "probably approximately correct" (PAC) learning (Valiant, 1984). The first successful boosting algorithms were provided by Schapire (1990) and Freund (1995). The name derives from the idea of creating a "strong" classifier by substantially improving or "boosting" the performance of a single "weak" classifier, where improvement is obtained by combining the classification votes from an ensemble of similar classifiers.

We define a *weak* (or *base*) *classifier* to be one that correctly classifies slightly more than 50% of the time (i.e., a little better than random guessing). Boosting algorithms combine M base classifiers $C_1, C_2, \ldots, C_M$ in the following way. For an observation $\mathbf{X} = \mathbf{x}$, the *boosted classifier* is given by:

$$C_{\boldsymbol{\alpha}}(\mathbf{x}) = \text{sign}\{f_{\boldsymbol{\alpha}}(\mathbf{x})\}, \qquad (14.9)$$

where

$$f_{\boldsymbol{\alpha}}(\mathbf{x}) = \sum_{j=1}^{M} \left(\frac{\alpha_j}{\sum_{j'} \alpha_{j'}} \right) C_j(\mathbf{x}), \qquad (14.10)$$

and $\boldsymbol{\alpha} = (\alpha_1, \cdots, \alpha_M)^\tau$ is an M-vector of constant coefficients.

Suppose, for example, we wish to determine whether a particular e-mail is spam (i.e., junk e-mail) or not without actually opening it. If we decide it is spam, we delete the e-mail without looking at it; if not, we read it. Suppose we have software that automatically detects whether an e-mail contains any particular word, say, the word "money," and then classifies the e-mail as spam or not spam depending upon whether that word is or is not in the e-mail. This is an example of a weak classifier because by itself it may classify too many legitimate e-mails as spam and give the appearance of pure guessing. We could improve upon this classifier by combining it with other weak classifiers each of which detects one word thought to characterize spam, say, "free," "order," "credit," and so on. We would then expect the resulting combined classifier to be a much stronger classifier than any of them separately.

More often than not, boosting (and the other ensemble methods) is applied to classifiers derived from decision trees. The weak classifier described above is an example of a "stump" classifier, a decision tree having only a single split and two terminal nodes. In that example, the stump classifier asks only one question: Does the e-mail contain the word "money"? If it does, classify it as spam (i.e., +1); otherwise, as not spam (i.e., –1). More complicated problems may require a weak classifier to be derived from two- or three-level decision trees. A *strong classifier* has a much smaller misclassification rate using a test set of observations.

Suppose, in the spam/not spam example, we use four ($M = 4$) stump classifiers that separately use the words "money," "free," "order," and "credit" to characterize spam. Define these classifiers as follows:

$$C_1(\text{e-mail}) = \begin{cases} +1 & \text{if e-mail contains word "money"} \\ -1 & \text{otherwise} \end{cases}$$

$$C_2(\text{e-mail}) = \begin{cases} +1 & \text{if e-mail contains word "free"} \\ -1 & \text{otherwise} \end{cases}$$

$$C_3(\text{e-mail}) = \begin{cases} +1 & \text{if e-mail contains word "order"} \\ -1 & \text{otherwise} \end{cases}$$

$$C_4(\text{e-mail}) = \begin{cases} +1 & \text{if e-mail contains word "credit"} \\ -1 & \text{otherwise} \end{cases}$$

Now, linearly combine these four classifiers by using nonnegative weights summing to one. Suppose the combined classifier is

$$f(\text{e-mail}) = 0.2C_1(\text{e-mail}) + 0.1C_2(\text{e-mail}) + 0.4C_3(\text{e-mail}) + 0.3C_4(\text{e-mail}).$$

How should an e-mail having the words "money," "order," and "credit" be classified? We calculate $f(\text{e-mail}) = 0.2 - 0.1 + 0.4 + 0.3 = 0.8$. The classification is given by $\text{sign}\{f(\text{e-mail})\} = \text{sign}\{0.8\} = +1$, and so we classify the e-mail as spam.

Different versions of boosting have been applied to a wide variety of data sets with enormous success; consequently, this class of improvement algorithms has become an important research topic in both the statistics and machine learning communities. The most well-known of these boosting algorithms is ADABOOST (Freund and Schapire, 1997).

14.3.1 ADABOOST: *Boosting by Reweighting*

ADABOOST (an acronym for "adaptive boosting") is an algorithm that is designed to improve performance in binary classification problems; it is generally regarded as the first step toward a truly practical boosting procedure. Details of the algorithm are shown in Table 14.1. It is also known as "Discrete ADABOOST" (because the goal is to predict class labels). A simple generalization of ADABOOST to more than two classes is called "ADABOOST.M1." ADABOOST was originally devised with the

TABLE 14.1. ADABOOST *algorithm for binary classification.*

---

1. Input: $\mathcal{L} = \{(\mathbf{X}_i, Y_i), i = 1, 2, \ldots, n\}$, $Y_i \in \{-1, +1\}$, $i = 1, 2, \ldots, n$, $\mathcal{C} = \{C_1, C_2, \ldots, C_M\}$, T = number of iterations.

2. Initialize the weight vector: Set $\mathbf{w}_1 = (w_{11}, \cdots, w_{n1})^\tau$, where $w_{i1} = 1/n$, $i = 1, 2, \ldots, n$.

3. For $t = 1, 2, \ldots, T$:

   - Select a weak classifier $C_{j_t}(\mathbf{x}) \in \{-1, +1\}$ from $\mathcal{C}$, $j_t \in \{1, 2, \ldots, M\}$, and train it on the learning set $\mathcal{L}$, where the ith observation $(\mathbf{X}_i, Y_i)$ has (normalized) weight w_{it}, $i = 1, 2, \ldots, n$.

   - Compute the weighted prediction error:

   $$PE_t = PE(\mathbf{w}_t) = \mathrm{E}_w\{I_{[Y_i \neq C_{j_t}(\mathbf{X}_i)]}\} = \left(\frac{\mathbf{w}_t^\tau}{\mathbf{1}_n^\tau \mathbf{w}_t}\right) \mathbf{e}_t,$$

   where E_w indicates taking expectation with respect to the probability distribution of $\mathbf{w}_t = (w_{1t}, \cdots, w_{nt})^\tau$, and $\mathbf{e}_t$ is an n-vector with ith entry $[\mathbf{e}_t]_i = I_{[Y_i \neq C_{j_t}(\mathbf{X}_i)]}$.

   - Set $\beta_t = \frac{1}{2}\log\left(\frac{1-PE_t}{PE_t}\right)$.

   - Update weights:

   $$w_{i,t+1} = \frac{w_{it}}{W_t}\exp\{2\beta_t I_{[Y_i \neq C_{j_t}(\mathbf{X}_i)]}\}, \quad i = 1, 2, \ldots, n,$$

   where W_t is a normalizing constant needed to ensure that the vector $\mathbf{w}_{t+1} = (w_{1,t+1}, \cdots, w_{n,t+1})^\tau$ represents a true weight distribution over $\mathcal{L}$; that is, $\mathbf{1}_n^\tau \mathbf{w}_{t+1} = 1$.

4. Output: $\mathrm{sign}\{f(\mathbf{x})\}$, where $f(\mathbf{x}) = \sum_{t=1}^T \beta_t C_{j_t}(\mathbf{x}) = \sum_{j=1}^M \alpha_j C_j(\mathbf{x})$, and $\alpha_j = \sum_{t=1}^T \beta_t I_{[j_t=j]}$.

---

specific intention of driving the prediction error from the learning set (i.e., the *learning set error*) quickly to zero.

In the ADABOOST algorithm for binary classification, we start with a learning set $\mathcal{L} = \{(\mathbf{x}_i, y_i)\}$, where $\mathbf{x}_i$ is an r-vector of inputs and $y_i \in \{-1, +1\}$ is a class label. ADABOOST weights the observations in $\mathcal{L}$ by a weight vector, $\mathbf{w} = (w_1, w_2, \cdots, w_n)^\tau$, and these weights are recalculated at each iteration. Initially, we use equal weights for each observation in $\mathcal{L}$.

At each iteration, the algorithm selects a "weak" classifier from a very large, but finite, set $\mathcal{C}$ of all possible weak classifiers. The finiteness assumption always holds for classification problems where each classifier in $\mathcal{C}$ has a finite set of possible outputs. For example, in binary classification, at most 2^n distinct labelings can be applied to the learning set. Because $\mathcal{C}$ is finite, it is entirely possible that, in constructing the ensemble, certain of the weak classifiers in $\mathcal{C}$ will be selected more than once (i.e., the smaller the set, the more likely that repetitions will occur).

At the tth iteration of ADABOOST, we modify the weighting system so that observations misclassified in the previous iteration will be more heavily weighted in the current iteration. In this way, ADABOOST tries hard to classify correctly any previously misclassified observations.

After T iterations, we have a sequence, $C_{j_1}(\mathbf{x}), C_{j_2}(\mathbf{x}), \ldots, C_{j_T}(\mathbf{x})$, of weak classifiers, where $j_t \in \{1, 2, \ldots, M\}, t = 1, 2, \ldots, T$. If the weak classifier C_j is selected multiple times in the process of the algorithm, then the coefficient for that component in the combined classifier is the sum of those coefficients obtained at all iterations when C_j was chosen. If the classifiers are small decision trees (as they often are when boosting is applied), then the jth weak classifier can be parameterized as $C_j(\mathbf{x}; \mathbf{a}_j)$, where the parameter vector $\mathbf{a}_j$ contains information on the splitting variables, split points, and the mean at each terminal node of the jth tree.

The value of the boosted classifier $C(\mathbf{x})$ depends upon the *sign* of the linear combination, $f(\mathbf{x}) = \sum_{j=1}^{M} \alpha_j C_j(\mathbf{x})$, of the weak classifiers, where α_j is the coefficient for C_j. In other words, $C(\mathbf{x}) = +1$ if $f(\mathbf{x}) > 0$, and -1 otherwise. ADABOOST does not restrict the sum of the coefficients $\{\alpha_j\}$, which may grow to be very large; all ADABOOST assumes is that f is in the linear span of the class $\mathcal{C}$ of weak classifiers. If we restrict the coefficients to be nonnegative with a fixed sum λ, say, this produces a regularized version of ADABOOST, where λ acts as a smoothing parameter; in this case, $f \in \text{conv}(\mathcal{C})$, the convex hull of $\mathcal{C}$ (see, e.g., Lugosi and Vayatis, 2004).

14.3.2   Example: Aqueous Solubility in Drug Discovery

In order to identify high-quality candidate drugs, pharmaceutical companies need to assess the absorption, distribution, metabolism, and excretion (ADME) characteristics of compounds, including biopharaceutical properties such as aqueous solubility, permeability, metabolic stability, and in vivo pharmacokinetics. One of the most fundamental tests to perform is that of solubility of a compound in water (or a solvent mixture), which now takes place routinely prior to biological testing. In fact, "aqueous-solubility" testing now usually occurs very early within the drug discovery and development process. Moreover, the Biopharmaceutics Classification System classifies compounds based upon their solubility and other properties.

Because patients tend to prefer oral medication, the commercial viability of a candidate drug would be greatly improved if the drug were soluble in water and could be delivered orally. For compounds that are not water soluble, results from experimental in vitro screening assays (which test the ability of a compound to dissolve in water) may not be reliable or reproducible and can lead to biological problems and increased drug-development costs. In recent years, the pharmaceutical industry has seen more candidate drugs that are highly insoluble, and this has become a real problem in drug development.

This example examines a data set involving 5,631 compounds on which an in-house solubility screen was performed.[1] Based upon this screen, compounds were categorized as either "insoluble" (3,493 compounds) or "soluble" (2,138 compounds). Then, for each compound, 72 continuous, noisy structural variables were recorded. One variable (71) had a large number (14%) of missing data and so was deleted from the data set. For proprietary reasons, the identities of the variables and compounds were not made publicly available.

The 5,631 compounds were randomly separated into a learning set (2,815 compounds) and a test set (2,816 compounds). We applied the discrete ADABOOST algorithm (using an exponential loss function; see below) and the results are displayed in Figure 14.3, where we plot the misclassification rate of both the learning set and the test set.

When we use "stumps" as classification trees in this example (left panel), the misclassification rates of both the learning set and the test set continue to decline, even after 2,000 iterations of ADABOOST, where the misclassification rates are 0.2298 for the learning set and 0.2553 for the test set. When we use 16-node trees (right panel), we see a different picture: after 500 boosting iterations, the learning set has a misclassification rate of zero, reached at iteration 312, and the test set declines to about 0.205.

In Figure 14.4, we show a comparison of test-set misclassification rates using different size trees: stumps (red curve), 4-node trees (magenta), 8-node trees (blue), and 16-node trees (green). We see that using ADABOOST on stumps actually performs the worst, whereas 16-node trees perform best. However, boosting 32-node trees (not shown here) yields slightly higher test-set misclassification rates than does boosting 16-node trees.

14.3.3 Convergence Issues and Overfitting

Empirical experiments have demonstrated that ADABOOST tends to be quite resistant to overfitting: the test set error (an estimate of *generalization*

---

[1] These data are available at the book's website under the filename soldat. The data are part of the R-package ada (Culp, Johnson, and Michailidis, 2006), which implements several versions of boosting, and can be downloaded from the website www.stat.lsa.umich.edu/ǔulpm/math/ada/img.html. The description of the data set is taken from that article. The author thanks Mark V. Culp for discussions about the ada package and the soldat data set.

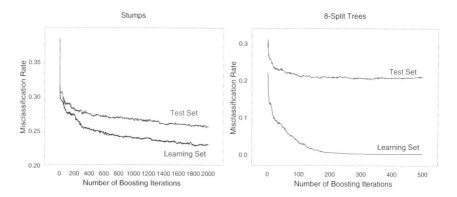

FIGURE 14.3. ADABOOST *for the solubility data (*`soldat`*). Misclassification rates for the training set (blue curve) and test set (red curve) plotted as a function of the number of boosting iterations. An exponential loss function was used with the discrete* ADABOOST *algorithm. Left panel: Stumps are used as classification trees and* ADABOOST *was run for 2,000 iterations. Right panel: 16-node trees and* ADABOOST *was run for 500 iterations.*

error) almost always continues to decline (and then levels off) as we increase the number of classifiers involved *even after the learning-set error has been driven to zero*! Recall that in a typical classification scenario, test-set error decreases for a little while and then begins to increase as the classifier becomes more and more complex. The discovery that ADABOOST is resistant to overfitting led to it being called the "most accurate, general-purpose, classification algorithm available" (Breiman, 2004).

Since ADABOOST was introduced, hundreds of articles have been published attempting to penetrate the "mysterious" secret of why it appears to be resistant to overfitting. Many explanations have been attempted, but the question still remains open. This mystery has been described as "the most important unsolved problem in machine learning" (Breiman, 2004).

This does not mean, however, that ADABOOST *never* overfits. Indeed, examples of ADABOOST have been constructed in which the test-set error increases (i.e., ADABOOST does overfit) as the number of iterations increases.

In a simulated 2D example of 150 observations drawn from each of two circular-Gaussian distributions with some overlap, Breiman (2002) reports that the test-set error decreases to a minimum after about 5,000 iterations, but then reverses direction and starts to increase. Friedman, Hastie, and Tibshirani (2000) observed much the same behavior when they applied ADABOOST to 400 observations drawn from each of two 10-dimensional,

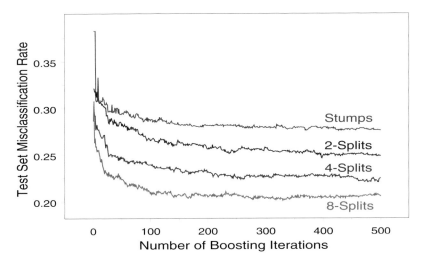

FIGURE 14.4. *Comparison of test-set misclassification rates for the solu-bility data using different size trees: stumps, 4-node trees (2-Splits), 8-node trees (4-Splits), and 16-node trees (8-Splits), where* ADABOOST *was run for 500 iterations.*

spherical-Gaussian distributions having the same mean, where the reversal occurs at about 50 iterations.

Breiman (2004) suggests that the ADABOOST process may actually consist of two stages. In the first stage (which may consist of several thousand iterations), the test-set error approaches close to the optimal Bayes error, mimicking its population (i.e., infinite sample size n) behavior. (In its population version, Breiman showed that ADABOOST is *Bayes consistent*; that is, its risk converges in probability to the Bayes risk.) If, for whatever reason, convergence fails, its test-set error then starts increasing. This second-stage behavior is not yet understood.

Further study of the convergence problem has shown that, for finite sample sizes, ADABOOST can be Bayes consistent only if it is regularized; see, for example, the articles on boosting in the February 2004 issue of *The Annals of Statistics*. One possible type of regularization for ADABOOST is that of stopping the algorithm very early (e.g., after 10 or 100 iterations), rather than letting it run forever; essentially, the argument is that overfitting will occur as soon as the classifier becomes too complicated and that continuing to run the algorithm will only produce larger misclassification rates. Jiang (2004) and Bickel and Ritov (2004) show that for any finite n, there is a stopping time t_n such that if the algorithm is stopped at t_n iterations, then ADABOOST will be Bayes consistent. The question then becomes, if the strategy is to stop ADABOOST early, how does one determine the best time to stop (i.e., an optimal t_n)? One suggested method is to use a data-based procedure, such as cross-validation. Other regularized versions of ADABOOST are discussed in Section 14.3.6.

There is empirical evidence (Mease and Wyner, 2007) that shows that early stopping may not be the panacea needed to prevent overfitting; indeed, the evidence suggests that overfitting tends to occur very early in the life of the algorithm and that running the algorithm for a much larger number of iterations actually reduces the amount of overfitting (to a level close to that of the Bayes risk) rather than increases it.

14.3.4   Classification Margins

One interesting argument put forward to explain why boosting works so well in classification problems involves the concept of a "margin" (Schapire, Freund, Bartlett, and Lee, 1998).

Let $\mathcal{C}$ be the set of all potential weak classifiers. For example, weak classifiers could be chosen from all those decision trees that have a specified number of terminal nodes. Consider a boosted classifier f of the type (14.9), where the weights, $\{\alpha_t\}$, are each nonnegative and sum to one. Then, $f \in \text{conv}(\mathcal{C})$ is a weighted average of weak classifiers from $\mathcal{C}$. If the weak classifiers are defined by a voting scheme, then the prediction is that label y that receives the highest vote from the weak classifiers.

Let $g(\mathbf{x}, y)$ denote a classifier that predicts the label y for an observation $\mathbf{x}$. Then, g predicts y iff $g(\mathbf{x}, y) > \max_{y' \neq y} g(\mathbf{x}, y')$. The *classification margin* of the labeled observation $(\mathbf{x}, y)$ is defined as

$$m(\mathbf{x}, y) = g(\mathbf{x}, y) - \max_{y' \neq y} g(\mathbf{x}, y'). \tag{14.11}$$

Thus, if y is the correct label for $\mathbf{x}$, then g misclassifies $\mathbf{x}$ iff $m(\mathbf{x}, y) < 0$. If $g(\mathbf{x}, y) = \sum_t I_{[C_{j_t}(\mathbf{x}) = y]}$ denotes the total number of votes for y obtained from all the weak classifiers, then the classification margin is the amount by which the total vote for the correct class y exceeds the highest total vote for any incorrect class. That is,

$$m(\mathbf{x}, y) = \sum_t I_{[C_{j_t}(\mathbf{x}) = y]} - \max_{y' \neq y} \left\{ \sum_t I_{[C_{j_t}(\mathbf{x}) = y']} \right\}. \tag{14.12}$$

Thus, an observation $(\mathbf{x}, y)$ is misclassified by the voting scheme iff its margin is not positive. Because the observation $(\mathbf{x}, y)$ is misclassified by the boosted classifier f only if $yf(\mathbf{x}) \leq 0$, we can think of the margin of $(\mathbf{x}, y)$ with respect to f as $m(\mathbf{x}, y) = yf(\mathbf{x})$. The margin of the boosted classifier f is the *minimum margin* over all n observations in $\mathcal{L}$.

In binary classification problems (with labels -1 and $+1$), the margin can be viewed in the following terms: the bigger the margin, the more "confidence" we have that the observation has been correctly classified. If the margin is large but negative, this tells us we are very confident that the observation has been misclassified. Small margins indicate doubtful reliance on classifications.

To assess the performance of a boosted classifier, Schapire et al. (1998) derive a probabilistic upper-bound on its generalization error. The upper bound turns out to depend upon the sum of the empirical margin distribution, $n^{-1} \sum_{i=1}^{n} I_{[y_i f(\mathbf{x}_i) \leq \delta]}$, and the *VC-dimension* of the class of boosted classifiers (Vapnik and Cheronenkis, 1971), but is independent of the number of weak classifiers being combined. From the upper-bound, they argue that the bigger the margins (over a learning set), the lower the generalization error of the classifier. They then conjecture that ADABOOST is successful because it produces large margins for the learning set.

Unfortunately, the probabilistic upper-bound only tells part of the story. Schapire et al. (1998) realized that their bound is much too loose to be useful for a majority-vote classifier. Although not asymptotic by construction (the bound is not dependent upon the size of the learning set), empirical results show that for the bound to be of any practical use, the size of the learning set would have to be huge (of the order tens of thousands). Constructing tighter upper bounds on the generalization error remains an open problem (see, e.g., Koltchinskii and Panchenko, 2002).

Breiman (1999) demonstrated also that high margins alone cannot explain the success of ADABOOST. Using a game-theoretic argument, he constructed a boosted classifier that not only had large margins (higher indeed than obtained by ADABOOST on each of a number of data sets) but also had higher generalization error in each case.

14.3.5 ADABOOST *and Maximal Margins*

So far, we have adopted a "nonoptimal" point of view (or strategy) for ADABOOST, where it is only necessary to provide a "sufficiently good" classifier at each iteration (not necessarily the best one) from the set $\mathcal{C}$ of weak classifiers; examples of nonoptimal ADABOOST include decision trees and neural networks. We can also identify an "optimal" ADABOOST strategy, where the best weak classifier is selected at each iteration from $\mathcal{C}$. This strategy has the effect of introducing an optimality step into the ADABOOST algorithm, so that, in principle, specific weak classifiers can be chosen again and again from $\mathcal{C}$.

From the above discussion, we know that ADABOOST induces "large" margins. In fact, if ρ is the maximum achievable margin, then ADABOOST produces a margin $m(\mathbf{x}, y)$ that is bounded below and above by

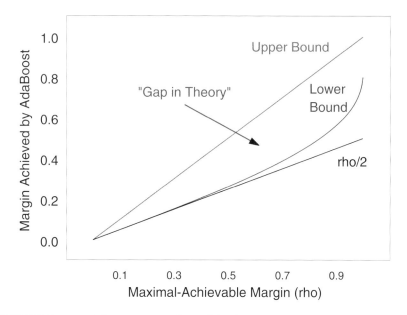

FIGURE 14.5. *The margin achieved by* ADABOOST *as a function of the maximal-achievable margin ρ. Shown are the upper bound (green line), the lower bound (red curve), and the line $\rho/2$ (blue line). The lower bound was derived by Rätsch and Warmuth (2005).*

$$\frac{\rho}{2} \leq -\frac{\log(1-\rho^2)}{\log\left(\frac{1+\rho}{1-\rho}\right)} \leq m(\mathbf{x}, y) \leq \rho \qquad (14.13)$$

(Schapire et al., 1998; Rätsch and Warmuth, 2005). In Figure 14.5, we plot the lower and upper bounds of $m(\mathbf{x}, y)$ based upon (14.13), and the line $\rho/2$. The vertical distance between the upper and lower bounds has been referred to as the "gap in theory" (Rätsch and Warmuth, 2005). The lower bound (i.e., the red curve in Figure 14.5) has been shown to be exactly tight, however, for the nonoptimal ADABOOST strategy (Rudin, Schapire, and Daubechies, 2007).

Recall that the closer a classifier gets to the maximum margin, the more confidence we have in that classifier. Even though ADABOOST was not specially designed to attain the maximum margin (margin theory came just after the introduction of ADABOOST), there is widespread belief that (as a by-product of its remarkable practical properties) ADABOOST also maximizes the margin. Rätsch and Warmuth (2005) noted, however, that empirical evidence (simulations and Figure 14.3) showed that might not always be the case.

The conjecture that ADABOOST does not always attain the maximum possible margin turns out to be true (Rudin, Daubechies, and Schapire, 2004). Because the margin does not increase monotonically as the iterations proceed, standard methods for examining convergence properties of AD-ABOOST margins are not applicable. Instead, following the remarkable work by Rudin et al., we look at the limiting performance of the sequence of weight vectors $\{\mathbf{w}_t\}$ that defines ADABOOST.

Let $\mathbf{Q} = (Q_{ij})$ be an $(n \times M)$-matrix, where $Q_{ij} = y_i C_m(\mathbf{x}_i)$, $i = 1, 2, \ldots, n$, $j = 1, 2, \ldots, M$, are the margin values. The columns of $\mathbf{Q}$ are the M weak classifiers in $\mathcal{C}$, the rows are the n observations in $\mathcal{L}$, and Q_{ij} is $+1$ if $\mathbf{x}_i$ is correctly classified by the weak classifier C_j and -1 otherwise. In applications, the values of n and M may be huge; as a result, $\mathbf{Q}$ is a matrix that is unlikely to be used in practice. However, $\mathbf{Q}$ has proved most useful in understanding certain properties of ADABOOST.

If the learning algorithm selects classifier C_{j_t}, we can write $Q_{ij_t} = I_{[y_i = C_{j_t}(\mathbf{x}_i)]} - I_{[y_i \neq C_{j_t}(\mathbf{x}_i)]} = 1 - 2I_{[y_i \neq C_{j_t}(\mathbf{x}_i)]}$. Substituting $I_{[y_i \neq C_{j_t}(\mathbf{x}_i)]} = \frac{1}{2}(1 - Q_{ij_t})$ into the weighted prediction error in Table 14.1 yields

$$PE_t = \frac{1}{2} - \frac{1}{2}r_t, \tag{14.14}$$

where

$$r_t = [\mathbf{w}_t^\tau \mathbf{Q}]_{j_t} = \sum_{i=1}^{n} w_{it} Q_{ij_t} \tag{14.15}$$

is the *edge* of C_{j_t} over $\mathcal{L}$, and where the $\mathbf{w}_t$ are normalized. Thus, r_t shows how much C_{j_t} varies from a pure-chance classifier.

The "edge" can be used to select a weak classifier from $\mathcal{C}$. Up to this point, we have not explained how to select an element of $\mathcal{C}$ at each iteration. Corresponding to the two types of ADABOOST strategies, we consider the following two selection rules:

optimal strategy: $j_t \in \arg\max_j [\mathbf{w}_t^\tau \mathbf{Q}]_j$

nonoptimal strategy: $j_t \in \{j : [\mathbf{w}_t^\tau \mathbf{Q}]_j \geq \rho\}$,

where ρ is the maximum achievable margin for $\mathbf{Q}$. In other words, the optimal strategy selects the classifier from $\mathcal{C}$ that has the largest edge, whereas the nonoptimal strategy selects any classifier in $\mathcal{C}$ whose edge is at least ρ.

The next step is to initialize the M-vector of coefficients by setting $\boldsymbol{\beta}_1 = \mathbf{0}$. Substituting PE_t from (14.14) into the expression for the coefficient β_t in Table 14.1 yields

$$\beta_t = \frac{1}{2} \log \left(\frac{1 + r_t}{1 - r_t} \right). \tag{14.16}$$

By iterating on the update formula for the weights, we have that $w_{it} \propto e^{-[\mathbf{Q}\boldsymbol{\beta}_t]_i}$, where $[\mathbf{Q}\boldsymbol{\beta}_t]_i = \sum_{j=1}^{t} y_i \beta_j C_j(\mathbf{x}_i)$ is the ith entry of the n-vector

$\mathbf{Q}\boldsymbol{\beta}_t$, and the proportionality factor does not involve the subscript i. We can update $\boldsymbol{\beta}_t$ by the formula

$$\boldsymbol{\beta}_{t+1} = \boldsymbol{\beta}_t + \boldsymbol{\delta}_{j_t} \tag{14.17}$$

where $\boldsymbol{\delta}_{j_t}$ is an M-vector with β_t in the j_tth position and zeroes in all other positions. The normalized weights $\{w_{it}\}$ can now be written as:

$$w_{it} = \frac{e^{-[\mathbf{Q}\boldsymbol{\beta}_t]_i}}{\sum_{i'=1}^{n} e^{-[\mathbf{Q}\boldsymbol{\beta}_t]_{i'}}}, \quad i = 1, 2, \ldots, n. \tag{14.18}$$

Note that the initialization step $\boldsymbol{\beta}_1 = \mathbf{0}$ yields $w_{i1} = 1/n$, $i = 1, 2, \ldots, n$.

From (14.16) and the fact that the elements of $\mathbf{Q}$ are ± 1, we have that

$$e^{-Q_{ij_t}\beta_t} = \left(\frac{1-r_t}{1+r_t}\right)^{Q_{ij_t}/2} = \left(\frac{1-Q_{ij_t}r_t}{1+Q_{ij_t}r_t}\right)^{1/2}, \tag{14.19}$$

whence, the update for the weights is given by

$$\begin{aligned} w_{i,t+1} &= \frac{w_{it}e^{-Q_{ij_t}\beta_t}}{\sum_{i'=1}^{n} w_{i't}e^{-Q_{i'j_t}\beta_t}} \\ &= \frac{w_{it}}{\sum_{i'=1}^{n} w_{i't}\left(\frac{1-Q_{i'j_t}r_t}{1+Q_{i'j_t}r_t}\right)^{1/2}\left(\frac{1+Q_{ij_t}r_t}{1-Q_{ij_t}r_t}\right)^{1/2}}. \end{aligned} \tag{14.20}$$

The first line uses (14.17); the second line divides both numerator and denominator by $\sum_{i'=1}^{n} e^{-[\mathbf{Q}\boldsymbol{\beta}_t]_{i'}}$, and then uses (14.18); and the third line uses (14.19).

To simplify the denominator of (14.20), consider the two cases, $Q_{ij_t} = -1$ and $Q_{ij_t} = +1$, separately. Then, for each case, divide the summation into two sets of indices, $\{i' : Q_{i'j_t} = -1\}$ and $\{i' : Q_{i'j_t} = +1\}$. Let $w_{-,t} = \sum_{\{i':Q_{i'j_t}=-1\}} w_{i't}$ and $w_{+,t} = 1 - w_{-,t}$. On the set $\{i' : Q_{i'j_t} = -1\}$, we have $r_t = 1 - 2w_{-,t}$, or $w_{-,t} = \frac{1}{2}(1 - r_t)$. Similarly, on the set $\{i' : Q_{i'j_t} = +1\}$, we have $w_{+,t} = \frac{1}{2}(1 + r_t)$. Simple algebra yields the update formula,

$$w_{i,t+1} = \frac{w_{it}}{1 + Q_{ij_t}r_t}, \quad i = 1, 2, \ldots, n. \tag{14.21}$$

Note that $\sum_{i=1}^{n} w_{i,t+1} = 1$. Thus, the weight vector $\mathbf{w}_t$ (at iteration t) in the ADABOOST algorithm can be expressed as a nonlinear iterated map (14.21) that connects $\mathbf{w}_{t+1}$ directly to $\mathbf{w}_t$, including renormalization.

The update formula (14.21) was discovered by Rudin, Daubechies, and Schapire (2004). The version derived here is for the nonoptimal ADABOOST strategy; the corresponding optimal strategy can be obtained by incorporating a step into the algorithm that, at each iteration, picks the weak classifier from $\mathcal{C}$ that has the largest edge and, hence, is furtherest away from a pure-chance classifier.

Rudin et al. showed that ADABOOST can be written as a dynamic system, which can be analyzed in terms of fixed points and stable limit cycles. ADABOOST is said to exhibit "cyclic behavior" if the same weak classifiers keep turning up again and again and the sequence of weight vectors keeps repeating with constant periodicity; that is, a cycle with period s (called an "s-cycle") occurs if there exists an integer s such that, at some iteration t, $\mathbf{w}_{t+s} = \mathbf{w}_t$. Large-scale simulations have shown that it is not unusual for ADABOOST to produce periodic cycles in its weight vectors. A fixed point is produced if, at some iteration t, $\mathbf{w}_{t+1} = \mathbf{w}_t$.

Using specific low-dimensional examples that are simple enough for the details to be worked out completely, Rudin et al. showed that ADABOOST does not always converge to a maximum-margin solution. Instead, AD-ABOOST may converge to a solution whose margin is *significantly below* the maximum value. It may do this for nonoptimal ADABOOST even if optimal ADABOOST converges to a maximum-margin solution. ADABOOST can also operate in chaotic mode, where the algorithm moves into and out of cyclic behavior, possibly due to a sensitivity to initial conditions.

With so much attention paid to ADABOOST and maximum margins, it was only natural for alternative boosting algorithms to be designed specifically to maximize the margin. Such algorithms include ARC-GV (Breiman, 1999) and ADABOOST* (Rätsch and Warmuth, 2002), neither of which are based upon coordinate-descent optimization. Two related algorithms, Coordinate-Ascent Boosting and Approximate Coordinate-Ascent Boosting algorithms (Rudin, Schapire, and Daubechies, 2004), do use a coordinatewise optimization method to find the classifier to maximize the margin.

14.3.6 A Statistical Interpretation of ADABOOST

Can we give a statistical interpretation of the ADABOOST algorithm? Friedman, Hastie, and Tibshirani (2000) showed that ADABOOST is equivalent to running a coordinate-descent algorithm to fit an additive, logistic-discrimination model to the learning set. That article (and the discussants) had much to say about the philosophical, statistical, and computational issues of boosting; we outline some of their development work here.

Let $\{C_j, j = 1, 2, \ldots, M\}$ be a set of M base classifiers, where each $C_j \in \{-1, +1\}$. Consider the following linear combination of those classifiers:

$$f(\mathbf{x}) = \sum_{j=1}^{M} \alpha_j C_j(\mathbf{x}), \tag{14.22}$$

where the $\{\alpha_j\}$ are constants. This has the general form of an *additive model*. Because $\alpha_j C_j(\mathbf{x}) \in \Re$, the combined binary classifier is defined as

$\text{sign}\{f(\mathbf{x})\}$. We wish to find the $\{\alpha_j\}$ and $\{C_j\}$ in (14.22) to minimize some optimality criterion.

To evaluate the classifier $f(\mathbf{x})$, we would like to minimize the number of misclassifications using a criterion based upon the usual zero-one loss function,

$$L(y, f(\mathbf{x})) = I_{[y \neq f(\mathbf{x})]} = I_{[yf(\mathbf{x}) \leq 0]}. \tag{14.23}$$

Unfortunately, this minimization problem will not work. Instead, we use a smooth, strictly convex, differentiable loss function of the random variable $Yf(\mathbf{X})$. In this case, the risk function,

$$R(f) = \mathrm{E}_{\mathbf{X},Y}\{L(Y, f(\mathbf{X}))\}, \tag{14.24}$$

is constructed using the exponential loss function,

$$L(y, f(\mathbf{x})) = e^{-yf(\mathbf{x})}, \quad y \in \{-1, +1\}. \tag{14.25}$$

In (14.24), the expectation E is a population expectation (taken over the joint distribution of $\mathbf{X}$ and Y). If $y \neq f(\mathbf{x})$, then, $yf(\mathbf{x}) \leq 0$, and so, $e^{-yf(\mathbf{x})} \geq 1$. Thus,

$$I_{[yf(\mathbf{x}) \leq 0]} \leq e^{-yf(\mathbf{x})} \tag{14.26}$$

(Schapire and Singer, 1998). It follows that the generalization error (in this case, the probability of misclassification),

$$\text{Prob}\{Y \neq f(\mathbf{X})\} = \mathrm{E}\{I_{[Yf(\mathbf{X}) \leq 0]}\} \leq R(f), \tag{14.27}$$

is bounded above by $R(f)$.

The objective now is to minimize $R(f)$. Because

$$R(f) = \mathrm{E}_{\mathbf{X}}[\, \mathrm{E}_Y\{L(Y, f(\mathbf{x}))| \mathbf{x}\} \,], \tag{14.28}$$

it suffices to carry out the minimization conditional on $\mathbf{X} = \mathbf{x}$; that is, we wish to find $f(\mathbf{x})$ to minimize

$$\ell(f(\mathbf{x})) = \mathrm{E}_Y\{L(Y, f(\mathbf{x}))| \mathbf{x}\}. \tag{14.29}$$

Plugging the exponential loss function (14.25) into (14.29), we have that

$$\mathrm{E}_Y\{e^{-Yf(\mathbf{x})}|\mathbf{x}\} = e^{f(\mathbf{x})}\text{Prob}\{Y = -1|\mathbf{x}\} + e^{-f(\mathbf{x})}\text{Prob}\{Y = 1|\mathbf{x}\}. \tag{14.30}$$

Differentiating (14.30) wrt $f(\mathbf{x})$ and setting the result equal to zero gives

$$e^{f(\mathbf{x})}\text{Prob}\{Y = -1|\mathbf{x}\} - e^{-f(\mathbf{x})}\text{Prob}\{Y = 1|\mathbf{x}\} = 0. \tag{14.31}$$

Solving for f yields

$$f(\mathbf{x}) = \frac{1}{2}\log\left(\frac{\text{Prob}\{Y = 1|\mathbf{x}\}}{\text{Prob}\{Y = -1|\mathbf{x}\}}\right), \tag{14.32}$$

which is half the log-odds of the class probabilities. Rearranging (14.32), we have

$$\text{Prob}\{Y = 1|\mathbf{x}\} = \frac{1}{1 + e^{-2f(\mathbf{x})}}, \tag{14.33}$$

$$\text{Prob}\{Y = -1|\mathbf{x}\} = \frac{1}{1 + e^{2f(\mathbf{x})}}, \tag{14.34}$$

which gives us the *logistic regression model*.

Next, consider an empirical version of the risk function (14.24) for the classifier f. In this case, we replace the expectation E by an average over the learning set; this gives us the learning-set prediction error for f:

$$PE = n^{-1} \sum_{i=1}^{n} L(y_i, f(\mathbf{x}_i)) = n^{-1} \sum_{i=1}^{n} I_{[y_i \neq f(\mathbf{x}_i)]}. \tag{14.35}$$

Because of the nonconvexity of the indicator function, this makes the problem of minimizing (14.35) wrt f a computationally difficult task. A way of avoiding this problem is to minimize instead a convex upper bound on the indicator function (see, e.g., Schapire and Singer, 1998). First, note that $I_{[y_i \neq f(\mathbf{x}_i)]} = I_{[y_i f(\mathbf{x}_i) \leq 0]}$; then, from the inequality (14.26), we have that

$$PE \leq n^{-1} \sum_{i=1}^{n} e^{-y_i f(\mathbf{x}_i)}. \tag{14.36}$$

Thus, the exponential criterion (14.36) is a differentiable upper bound on PE.

Now, let the partial sum, $f_j(\mathbf{x}_i) = \sum_{k=1}^{j} \alpha_k C_k(\mathbf{x}_i)$, of (14.22) be an additive model in the first j classifiers, $j = 1, 2, \ldots, M$. We can write f_j as an update to f_{j-1}; that is,

$$f_j(\mathbf{x}_i) = f_{j-1}(\mathbf{x}_i) + \alpha_j C_j(\mathbf{x}_i). \tag{14.37}$$

Then, the minimization problem can be formulated as

$$(\alpha_j, C_j) = \arg\min_{\alpha, C} \sum_{i=1}^{n} e^{-y_i [f_{j-1}(\mathbf{x}_i) + \alpha C(\mathbf{x}_i)]}. \tag{14.38}$$

ADABOOST solves this minimization problem using a *coordinate-descent optimization algorithm*; see Table 14.2. Friedman et al. (2000) call this procedure a *forward-stagewise* minimization (see Section 5.9.1) of an additive model; in this case, the model is an additive, logistic-regression model.

We solve (14.38) in two steps: first, by fixing α and minimizing (14.38) wrt C, and then, given C_j, minimizing the result wrt α to get α_j. Now,

$$e^{-y_i [f_{j-1}(\mathbf{x}_i) + \alpha C(\mathbf{x}_i)]} = w_{i,j-1} e^{-y_i \alpha C(\mathbf{x}_i)}, \tag{14.39}$$

TABLE 14.2. *Coordinate-descent algorithm for fitting an additive model.*

---

1. Input: $\mathcal{L} = \{(\mathbf{X}_i, Y_i), i = 1, 2, \ldots, n\}$, T = number of iterations, $h(\mathbf{x}; \boldsymbol{\theta})$ is a parametric function of $\mathbf{x}$ with unknown parameters $\boldsymbol{\theta}$.

2. Initialize: $f_0(\mathbf{x}) = 0$.

3. For $t = 1, 2, \ldots, T$:

   - Compute $(\beta_t, \boldsymbol{\theta}_t) = \arg\min_{\beta, \boldsymbol{\theta}} \sum_{i=1}^{n} L(y_i, f_{t-1}(\mathbf{x}_i) + \beta h(\mathbf{x}_i; \boldsymbol{\theta}))$.
   - Set $f_t(\mathbf{x}) = f_{t-1}(\mathbf{x}) + \beta_t h(\mathbf{x}; \boldsymbol{\theta}_t)$.

4. Output: $\widehat{f}(\mathbf{x}) = f_T(\mathbf{x}) = \sum_{t=1}^{T} \beta_t h(\mathbf{x}; \boldsymbol{\theta}_t)$.

---

where $w_{i,j-1} = e^{-y_i f_{j-1}(\mathbf{x}_i)}$ is a weight. Using the fact that $y_i C(\mathbf{x}_i) = -1$ if $y_i \neq C(\mathbf{x}_i)$ and $+1$ otherwise, the criterion is

$$e^{\alpha} \sum_{i=1}^{n} w_{i,j-1} I_{[y_i \neq C(\mathbf{x}_i)]} + e^{-\alpha} \sum_{i=1}^{n} w_{i,j-1} \{1 - I_{[y_i \neq C(\mathbf{x}_i)]}\}, \qquad (14.40)$$

which can be written as

$$e^{-\alpha} \sum_{i=1}^{n} w_{i,j-1} + (e^{\alpha} - e^{-\alpha}) \sum_{i=1}^{n} w_{i,j-1} I_{[y_i \neq C(\mathbf{x}_i)]}. \qquad (14.41)$$

The only term depending upon C is the second sum, and so, we take

$$C_j = \arg\min_{C} \left\{ \sum_{i=1}^{n} w_{i,j-1} I_{[y_i \neq C(\mathbf{x}_i)]} \right\}. \qquad (14.42)$$

Next, we substitute C_j from (14.42) into (14.41) and minimize the result wrt α. Differentiating (14.41) wrt α and setting the result equal to zero, we get

$$-e^{-\alpha_j} \sum_{i=1}^{n} w_{i,j-1} + (e^{\alpha_j} + e^{-\alpha_j}) C_j = 0, \qquad (14.43)$$

which, solving for α_j, yields

$$\alpha_j = \frac{1}{2} \log\left(\frac{1 - PE_j}{PE_j}\right), \qquad (14.44)$$

where

$$PE_j = \frac{C_j}{\sum_{i=1}^{n} w_{i,j-1}} = \sum_{i=1}^{n} \left(\frac{w_{i,j-1}}{\sum_{\ell=1}^{n} w_{\ell,j-1}}\right) I_{[y_i \neq C_j(\mathbf{x}_i)]}) \qquad (14.45)$$

is the weighted learning-set prediction error. Plugging (14.44) into (14.37) gives us the update formula for $f(\mathbf{x}_i)$:

$$f_j(\mathbf{x}_i) = f_{j-1}(\mathbf{x}_i) + \frac{1}{2} \log \left(\frac{1 - PE_j}{PE_j} \right) C_j(\mathbf{x}_i). \qquad (14.46)$$

Using (14.37), we update the weights,

$$w_{ij} = e^{-y_i f_j(\mathbf{x}_i)} = w_{i,j-1} e^{-y_i \alpha_j C_j(\mathbf{x}_i)}. \qquad (14.47)$$

From above, we have that

$$-y_i C_j(\mathbf{x}_i) = 2I_{[y_i \neq C_j(\mathbf{x}_i)]} - 1. \qquad (14.48)$$

Substituting (14.48) into (14.47) gives

$$w_{ij} = w_{i,j-1} e^{2\alpha_j I_{[y_i \neq C_j(\mathbf{x}_i)]}} e^{-\alpha_j}. \qquad (14.49)$$

We can ignore the final term $e^{-\alpha_j}$ in (14.49) as it is a multiplying constant with respect to all the weights and, thus, is removed when the weights are normalized. These results constitute the ADABOOST algorithm (see Table 14.1).

Notice that the coordinate-descent algorithm (Table 14.2) used in AD-ABOOST fits the terms in the additive model one term (or coordinate) at a time, not jointly. At each iteration, the procedure fits only a single term of the model, unlike a *stepwise* procedure, *which readjusts all currently existing terms to compensate for adding a new term*. This fitting procedure is contrary to the usual statistical way of fitting a model with many terms (see, e.g., Buja's discussion of Friedman et al., 2000). If the model has a large number of terms, as happens with ADABOOST, then coordinatewise fitting, one term at a time, makes a lot more sense — computationally — than does fitting all the terms simultaneously, even though the latter would be optimal. Coordinatewise algorithms are typically quite efficient, converge fairly rapidly, and are simple to program. Thus, although coordinatewise fitting is a suboptimal procedure, it enables ADABOOST to work successfully.

14.3.7 Some Questions About ADABOOST

Since the Friedman, Hastie, and Tibshirani (2000) paper on the statistical view of boosting appeared, much has been written on the subject, with extensions in many directions. Many studies using real and simulated data have appeared that try to examine the statistical issues discussed in the Friedman et al. paper. Simulated data have been particularly important in understanding the behavior of ADABOOST because then the joint distribution of $(\mathbf{X}, Y)$ is completely known. However, several major questions about ADABOOST have been left unanswered by Friedman et al. and other researchers. Here, we offer brief discussions of a few of these issues.

Why Does ADABOOST Work?

This is probably the most important question of interest to users of AD-ABOOST. As we have seen, ADABOOST can be viewed as algorithmically similar to an approach consisting of an amalgam of three separate components: (1) an additive logistic regression model (e.g., a linear combination of classification trees), (2) an exponential loss criterion, and (3) a coordinatewise fitting procedure. This interpretation of ADABOOST has since encouraged researchers to develop other boosting algorithms by changing either the type of smooth, convex loss function used in the basic algorithm, or the numerical fitting procedure, or both.

But this still begs the question of why ADABOOST works so well. AD-ABOOST yields very small misclassification rates (compared with other competing classifiers) over a wide variety of data sets and is (in most cases) highly resistant to overfitting. As we have already noted, not all data sets are immune to overfitting; there are a number of specially constructed examples that show that ADABOOST can indeed overfit. What Friedman et al. gave us is a useful description of a way of thinking — statistically — about ADABOOST. But they did not address the main issue of *why* AD-ABOOST is so resistant to overfitting, whether for simulated or real data.

Since the appearance of that article, many suggestions have been made as to why ADABOOST is successful in classification situations. Some researchers have pointed to the stagewise fitting machine, or to the 0–1 loss function, or to the notion of margin, but none of these explanations are really convincing. It is still an open question as to why ADABOOST works as well as it does. Specifically, we would like to know under what conditions we can expect ADABOOST not to overfit, and under what conditions we should expect ADABOOST to overfit.

How Well Can ADABOOST Estimate Conditional Class Probabilities?

On a related problem to classification, we may wish to estimate the *conditional class probability function,*

$$p(\mathbf{x}) = \mathrm{P}\{Y = 1|\mathbf{x}\}. \tag{14.50}$$

If we can estimate $p(\mathbf{x})$ well across the entire range of $\mathbf{x}$, we would then be able to obtain a solution to the classification problem by choosing an appropriate quantile q of this function to be the class boundary; that is, find q such that all cases in the region $p(\mathbf{x}) > q$ are classified as positive $(+1)$.

Building upon the connection between ADABOOST and logistic regression, Friedman, Hastie, and Tibshirani (2000, Algorithm 3) introduced the LOGITBOOST algorithm to estimate $p(\mathbf{x})$ directly using the link function (14.33); that is,

$$\widehat{p}_j(\mathbf{x}) = \frac{1}{1 + e^{-2f_j(\mathbf{x})}}, \tag{14.51}$$

where f_j is the classifier evaluated at the jth iteration and which satisfies (14.37). LOGITBOOST is a modified version of the ADABOOST algorithm that uses stagewise minimization of the binomial log-likelihood loss function (in place of exponential loss). Thus, the current estimate of the boosted classifier $f_j(\mathbf{x})$ is transformed via the link (14.51) to produce a current estimate of $p(\mathbf{x})$.

In simulations, Mease, Wyner, and Buja (2007) show that boosting classification trees (and LOGITBOOST, in particular) is not well-suited to estimating $p(\mathbf{x})$, except for estimating the median of that probability function (and other special cases). Indeed, there is empirical evidence that boosting can severely overfit the estimate of $p(\mathbf{x})$ — even when the ADABOOST classification rule performs well with no appearance of overfitting. These results throw doubt on the popularly made claim that the success of boosting is due to its close relationship to logistic regression.

Do Stumps Make the Best Base Classifiers for ADABOOST?

It has been argued (Friedman, Hastie, and Tibshirani, 2000, pp. 360–361; Hastie, Tibshirani, and Friedman, 2001, Section 10.11) that larger trees introduce higher-level interaction effects among the input variables $\mathbf{X}$. Thus, stumps represent main effects (X_j), the second level of 4 nodes represents first-order interactions $(X_j X_k)$, the third level of 8 nodes represents second-order interactions $(X_j X_k X_\ell)$, and so on. Such higher-order interactions, it is argued, then lead to overfitting. A corollary to this argument is that if we believe that the optimal Bayes risk can be closely approximated by an additive function of elements of $\mathbf{X}$, then only stumps provide an additive model. Although larger trees are not ruled out as base classifiers, stumps, in this context, are said to provide an "ideal match" and, according to this argument, are to be preferred to larger trees.

Yet, simulations have shown (Mease and Wyner, 2007) that stumps do not necessarily provide the best base classifiers for ADABOOST even if the optimal Bayes risk is additive, and that larger trees can actually be more effective. The solubility example in Section 14.3.2 shows that using stumps as base classifiers gives a relatively 'poor' performance when compared with the results from using larger trees with 4, 8, or 16 terminal nodes.

14.3.8 Gradient Boosting for Regression

We saw that one of the crucial steps in the derivation of ADABOOST was the minimization of $\ell(f(\mathbf{x}))$. Given an exponential loss criterion and an additive model, the minimization procedure led to a coordinate-descent algorithm. In an extension of that idea, Friedman (2001) showed that other boosting strategies could be obtained by using different minimizing procedures combined with different loss functions. In particular, he adapted the well-known *gradient-descent* (also known as *steepest-descent*) algorithm to derive a more general boosting procedure — which he called "gradient boosting" — primarily for regression situations.

The general minimization problem is to find $\widehat{f}$ such that

$$\widehat{f}(\mathbf{x}) = \arg\min_{f} \ell(f(\mathbf{x})). \tag{14.52}$$

For a given $\mathbf{x}$, let $f_0(\mathbf{x})$ be an initial guess and let $f_{t-1}(\mathbf{x})$ be the current approximation to $\widehat{f}(\mathbf{x})$. According to the gradient-descent algorithm, we update $f_{t-1}(\mathbf{x})$ by moving a small *step-size* $\rho_t > 0$ in the direction of the negative gradient, and evaluate the result at $f_{t-1}(\mathbf{x})$. In other words, we set

$$f_t(\mathbf{x}) = f_{t-1}(\mathbf{x}) - \rho_t g_t(\mathbf{x}), \tag{14.53}$$

where $-\rho_t g_t(\mathbf{x})$ is the best steepest-descent step direction toward $\widehat{f}(\mathbf{x})$,

$$
\begin{aligned}
g_t(\mathbf{x}) &= \left. \frac{\partial \ell(f(\mathbf{x}))}{\partial f(\mathbf{x})} \right|_{f(\mathbf{x})=f_{t-1}(\mathbf{x})} \\
&= \left. \frac{\partial E_Y\{L(Y, f(\mathbf{x})) \mid \mathbf{x}\}}{\partial f(\mathbf{x})} \right|_{f(\mathbf{x})=f_{t-1}(\mathbf{x})} \\
&= E_Y \left\{ \left. \frac{\partial L(Y, f(\mathbf{x}))}{\partial f(\mathbf{x})} \right|_{f(\mathbf{x})=f_{t-1}(\mathbf{x})} \middle| \mathbf{x} \right\}
\end{aligned}
\tag{14.54}
$$

is the *gradient* (assuming differentiation and integration can be exchanged), and the step-size (or *learning rate*) ρ_t is determined from the *line search*,

$$\rho_m = \arg\min_{\rho} E_{\mathbf{X},Y}\{L(Y, f_{t-1}(\mathbf{X}) - \rho g_t(\mathbf{X}))\}. \tag{14.55}$$

Choice of ρ_m is crucial to the steepest-descent method: too large a ρ_m may lead to overshooting the minimum and possibly divergent oscillations, whereas too small a ρ_m will slow down the search and greatly increase computation time.

The expectations in (14.54) and (14.55) are estimated using the learning set $\mathcal{L} = \{(\mathbf{x}_i, y_i), i = 1, 2, \ldots, n\}$. For example, the gradient (14.54) at $\mathbf{x}_i$

is estimated by

$$g_t(\mathbf{x}_i) = \left. \frac{\partial L(y_i, f(\mathbf{x}_i))}{\partial f(\mathbf{x}_i)} \right|_{f(\mathbf{x}_i)=f_{t-1}(\mathbf{x}_i)} \quad, \quad i = 1, 2, \ldots, n, \tag{14.56}$$

the step-size ρ_t by

$$\rho_t = \arg\min_{\rho} \sum_{i=1}^{n} L(y_i, f_{t-1}(\mathbf{x}_i) - \rho g_t(\mathbf{x}_i)), \tag{14.57}$$

and the update rule by $\mathbf{f}_t = \mathbf{f}_{t-1} - \rho_t \mathbf{g}_t$, where $\mathbf{f}_t = (f_t(\mathbf{x}_i))$, $\mathbf{f}_{t-1} = (f_{t-1}(\mathbf{x}_i))$, and $\mathbf{g}_t = (g_t(\mathbf{x}_i))$ are each n-vectors.

The important point to note here is that the gradient vector $\mathbf{g}_t$ is only defined at a very specific set $\mathcal{L}$ of n points; we cannot use this formulation to compute the gradient and step-size at any set of points not in $\mathcal{L}$.

Friedman (2001) found an ingenious way around this problem by approximating the negative gradient $-g_t(\mathbf{x})$ by a parametric function, $h(\mathbf{x}; \boldsymbol{\theta}_t)$, with parameter vector $\boldsymbol{\theta}_t$. For example, if we use a J-terminal-node regression tree as a base learner, then $h(\mathbf{x}; \boldsymbol{\theta}_t)$ takes the simple form (see Section 14.3),

$$h(\mathbf{x}; \boldsymbol{\theta}_t) = \sum_{j=1}^{J} \bar{y}_j I_{[\mathbf{x} \in R_j]}, \tag{14.58}$$

where the components of the parameter vector $\boldsymbol{\theta}_t = (\{\bar{y}_j, R_j\})^\tau$ define the entire tree: the $\{R_j\}$ are the J disjoint regions of input space and represent the terminal nodes of the tree, and the $\{\bar{y}_j\}$ are terminal-node means that define the region boundaries.

How can we choose $\boldsymbol{\theta}_t$? A simple idea is to choose $\boldsymbol{\theta}_t$ so that $h(\mathbf{x}; \boldsymbol{\theta}_t)$ is most highly correlated with the negative gradient $-g_t(\mathbf{x})$. This is a least-squares minimization problem: define n "pseudoresponses" as $\widetilde{y}_i = -g(\mathbf{x}_i)$, $i = 1, 2, \ldots, n$, and solve

$$(\boldsymbol{\theta}_t, \beta_t) = \arg\min_{\boldsymbol{\theta}, \beta} \sum_{i=1}^{n} (\widetilde{y}_i - \beta h(\mathbf{x}_i; \boldsymbol{\theta}))^2. \tag{14.59}$$

The update formula is

$$f_t(\mathbf{x}) = f_{t-1}(\mathbf{x}) + \rho_t h(\mathbf{x}; \boldsymbol{\theta}_t), \tag{14.60}$$

where ρ_t is found from the line search,

$$\rho_t = \arg\min_{\rho} \sum_{i=1}^{n} L(y_i, f_{t-1}(\mathbf{x}_i) + \rho h(\mathbf{x}_i; \boldsymbol{\theta}_t)). \tag{14.61}$$

These steps constitute the GRADIENT.BOOST algorithm (Friedman, 2001, Algorithm 1) given in Table 14.3. If each $h(\mathbf{x}; \boldsymbol{\theta}_t)$ is a J-terminal-node

TABLE 14.3. GRADIENT.BOOST *algorithm for fitting an additive model.*

---

1. Input: $\mathcal{L} = \{(\mathbf{x}_i, y_i), i = 1, 2, \ldots, n\}$, $T =$ number of iterations.

2. Initialize: $f_0(\mathbf{x}) = \arg\min_\rho \sum_{i=1}^n L(y_i, \rho)$.

3. For $t = 1, 2, \ldots, T$:

   - $\tilde{y}_i = -g_t(\mathbf{x}_i) = -\left. \frac{\partial L(y_i, f(\mathbf{x}_i))}{\partial f(\mathbf{x}_i)} \right|_{f(\mathbf{x}_i) = f_{t-1}(\mathbf{x}_i)}$, $i = 1, 2, \ldots, n$,

   - Compute $(\boldsymbol{\theta}_t, \beta_t) = \arg\min_{\boldsymbol{\theta}, \beta} \sum_{i=1}^n (\tilde{y}_i - \beta h(\mathbf{x}_i; \boldsymbol{\theta}))^2$.

   - Compute $\rho_t = \arg\min_\rho \sum_{i=1}^n L(y_i, f_{t-1}(\mathbf{x}_i) + \rho h(\mathbf{x}_i; \boldsymbol{\theta}_t))$.

   - Set $f_t(\mathbf{x}) = f_{t-1}(\mathbf{x}) + \beta_t h(\mathbf{x}; \boldsymbol{\theta}_t)$.

4. Output: $\widehat{f}(\mathbf{x}) = f_T(\mathbf{x}) = \sum_{t=1}^T \beta_t h(\mathbf{x}; \boldsymbol{\theta}_t)$.

---

regression tree, the algorithm is referred to as the GRADIENT.TREEBOOST algorithm.

14.3.9   Other Loss Functions

Several different loss functions have been proposed as alternatives to exponential loss (14.25) as part of ADABOOST or for gradient boosting. These include the following:

logistic (log)loss: $L(y, f_t(\mathbf{x})) = \log\left\{1 + e^{-2y f_t(\mathbf{x})}\right\}$, $y \in \{-1, +1\}$.

   This loss function is used in the LOGITBOOST algorithm (Friedman et al., 2000) for classification, where $f_t(\mathbf{x}) = C_t(\mathbf{x})$; see Exercise 14.3.

squared-error loss: $L(y, f_t(\mathbf{x})) = \frac{1}{2}(y - f_t(\mathbf{x}))^2$, $y \in \Re$.

   For continuous $Y \in \Re$, this loss function is used for least-squares regression boosting by the LS.BOOST (and the LS.TREEBOOST) algorithm (Friedman, 2001, Algorithm 2) and the L_2BOOST algorithm (Buhlmann and Yu, 2003); see Exercise 14.4.

absolute-error loss: $L(y, f_t(\mathbf{x})) = |y - f_t(\mathbf{x})|$, $y \in \Re$.

   This loss function is used in the LAD.TREEBOOST algorithm (Friedman, 2001, Algorithm 3) for boosting regression trees using least-absolute-deviation. The resulting procedure is robust against outliers in both input and output variables.

Huber loss: $L(y, f_t(\mathbf{x})) = \begin{cases} \frac{1}{2}(y - f_t(\mathbf{x}))^2, & \text{if } |y - f_t(\mathbf{x})| \leq \delta, \\ \delta(|y - f_t(\mathbf{x})| - \delta/2), & \text{otherwise.} \end{cases}$

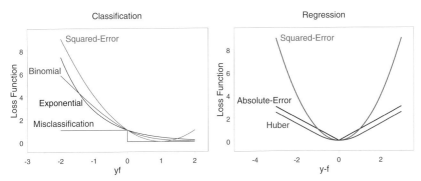

FIGURE 14.6. *A comparison of loss functions for binary classification (left panel), where the label y is -1 or $+1$, and regression (right panel), where y is real-valued. The predictor is f. Left panel: Shown are the exponential (e^{-yf}, blue curve), binomial (scaled version of $\log\{1 + e^{-2yf}\}$, red curve), squared-error ($(y - f)^2 = 2(1 - yf)$, green curve), and misclassification ($I_{[yf<0]}$, black step-function) loss functions, graphed as functions of the margin yf. Right panel: Shown are the absolute-error ($|y - f|$, blue curve), squared-error ($(y - f)^2$, green curve), and Huber (with $\delta = 1$, red curve) loss functions, graphed as functions of the error $y - f$.*

This loss function (specially constructed for Huber's theory of robust M-regression) combines features of squared-error loss and absolute-error loss (Huber, 1964), and is used by the M.TREEBOOST algorithm (Friedman, 2001, Algorithm 4). The constant δ is used to identify outlying residuals, whose loss is measured by absolute error instead of squared error.

The left panel of Figure 14.6 shows graphs of the exponential, binomial, and squared-error loss functions, each of which can be regarded as a continuous, convex approximation to misclassification loss, a step function having a loss of one for $yf < 0$ and zero for $yf > 0$. We can clearly see that squared-error loss is a poor approximation to misclassification loss. Instead of being a monotonically decreasing function of the margin yf as are the exponential, binomial, and misclassification losses, squared-error loss becomes larger the more confidently we can classify an observation (i.e., the larger the margin value)! The right panel shows graphs of absolute-error loss, squared-error loss, and the Huber loss function (with $\delta = 1$).

14.3.10 Regularization

Regularization by restricting the fitting process is popularly used as an antidote to overfitting. There are several ways to do this; one possible approach is that of model selection, whereby the number of components in the combined classifier or predictor is not allowed to get too large. Another approach shrinks the values of all coefficients in the model. As a general rule, regularization by restricting the number of components is not as effective as shrinking the coefficient values.

Regularized predictors are obtained using a penalty function $p(\boldsymbol{\alpha})$ to restrict the size of the coefficient vector $\boldsymbol{\alpha} = (\alpha_1, \cdots, \alpha_M)^\tau$ of the combined predictor $f(\mathbf{x}) = \sum_{j=1}^{M} \alpha_j C_j(\mathbf{x})$. The penalized estimates are given by the solution to the constrained minimization problem:

$$\widehat{\boldsymbol{\alpha}}_\lambda = \arg\min_{\boldsymbol{\alpha}} \left\{ \sum_{i=1}^{n} L(y_i, f(\mathbf{x}_i)) + \lambda p(\boldsymbol{\alpha}) \right\}, \tag{14.62}$$

where L is any of the loss functions listed above and $\lambda > 0$ is a small regularization parameter usually interpreted as a "learning rate." We then apply the forwards-stagewise (or a related gradient-boosting) procedure to the constrained-minimization problem. There are two types of penalty functions that have been suggested for use in the boosting context:

L_1-Penalty: The coefficients are constrained so that the sum of their absolute values,

$$p_1(\boldsymbol{\alpha}) = \sum_{j=1}^{M} |\alpha_j|, \tag{14.63}$$

is smaller than a given value. In the case of squared-error loss function for regression boosting, this L_1-penalty yields a method closely related to the Lasso algorithm and to the least-angle regression (LARS) algorithm; see Sections 5.8 and 5.9. As we noted in Section 5.8, empirical evidence suggests that the L_1-penalty works best when there are a small-to-medium number of moderate-sized true coefficients.

L_2-Penalty: This penalty function restricts the sum of squares of the coefficients,

$$p_2(\boldsymbol{\alpha}) = \sum_{j=1}^{M} \alpha_j^2. \tag{14.64}$$

When the combined learner is a convex combination of base learners and we use squared-error loss, the optimum penalized-regression predictor is the *ridge regression* estimator of Section 6.3.3.

As we vary the value of λ in the above constrained-minimization problem, we obtain a sequence of values of the components of the estimated coefficient vector $\widehat{\boldsymbol{\alpha}}_\lambda$. We then plot the trace of each coefficient $\widehat{\alpha}_{j,\lambda}$ against λ.

Friedman (2001) introduces a regularization parameter λ into the GRA-DIENT.BOOST algorithm in Table 14.3 by adding $\lambda \in (0, 1]$ to the input line, and then changing the fourth line in the for-loop to

$$f_t(\mathbf{x}) = f_{t-1}(\mathbf{x}) + \lambda \cdot \beta_t h(\mathbf{x}; \boldsymbol{\theta}_t), \tag{14.65}$$

where $h(\mathbf{x}; \boldsymbol{\theta}_t)$ is a parametric function. This particular form of regularization is also referred to as "shrinkage." The parameter λ operates (in conjunction with the number of iterations T) to find the best fit to the data. However, there is a "trade-off" between the values of λ and T in the fitting process: the best value for T is observed to be higher for smaller values of λ. From simulations, Friedman notes that the performance of the gradient-boosting method is generally enhanced (sometimes dramatically) by using as large a value of T as is computationally feasible and then setting λ to be a small (but not too small) value (e.g., $\lambda = 0.1$) so that an appropriate criterion is optimized close to the chosen value for T. Using a value of λ close to one typically produces evidence of serious overfitting.

14.3.11 Noisy Class Labels

In classification problems, *label noise* exists when the learning set contains observations with incorrect class labels. Dietterich (2000) showed that noisy labels degrade the accuracy of ADABOOST when applied to classification trees, whereas bagging appears to be quite robust against label noise.

We can create label noise by randomly selecting (without replacement) a fraction (e.g., 5%) of the observations from a data set and then changing the class label of each chosen observation using a random assignment from the set of incorrect class labels. In Section 5.6.2, we saw that, on average, about 37% of the observations from the learning set are omitted from each bootstrap sample; thus, it is likely that a large proportion of the mislabeled observations will not appear in a bootstrap sample. The omission of misclassified observations (which should behave like regression outliers) from the bootstrap sample will increase instability and, hence, improve the performance of bagging.

On the other hand, after a few iterations, ADABOOST will keep assigning large weights to the fraction of mislabeled observations because it will have difficulty classifying the "corrupted" observations, and this may, in turn, degrade performance and lead to overfitting.

When noisy class labels are present, there is empirical evidence (Krieger, Long, and Wyner, 2001) that we can improve the classifier's performance if we apply bagging following boosting (a "BB" algorithm). Specifically, we generate $B = \rho n$ bootstrap samples from the learning set ($0 < \rho < 1$), compute a boosted classifier from each bootstrap sample using M iterations, combine the B different boosted classifiers into an ensemble, and then average over the ensemble. Studies show, using real data, that the BB classifier averages out (or smoothes) the overfitting in ADABOOST and, hence, decreases test error.

14.4   Random Forests

We have seen how perturbing the learning set $\mathcal{L}$ in various ways can be used to generate an ensemble (or forest) of tree-structured classifiers. A classification tree T_k is grown for each perturbation $\mathcal{L}_k$ of the learning set, $k = 1, 2, \ldots, K$; a test set observation $\mathbf{x}$ is dropped down each tree; and the classifier predicts the class of that observation by that class that enjoys the largest number of total votes over all of the trees.

In bagging, randomization is used only in selecting the data set on which to grow each tree. An extension of this idea is *random forests* (Breiman, 2001b), where randomization adds another layer onto bagging and is a crucial part of constructing each tree. Suggestions on how to introduce randomization into tree construction include *random split selection* in which each node is split by randomly choosing one of the t best splits at that node (Dietterich, 2000) and *random input selection* in which the split at each node is decided by a random choice of subset of the r input features (Ho, 1998).

14.4.1   Randomizing Tree Construction

In random forests, we start in the same way that bagging starts, with B bootstrap samples drawn from the learning set $\mathcal{L}$, but the difference is how the trees are grown from those samples. The idea is to introduce a randomization component into tree construction so that, for the tree T^{*b}, each node is split in a random manner. Possible options for developing a randomized splitting strategy at each node include using some form of random input selection and linear combinations of inputs.

Recall that bagging applied to a tree-structured classifier reduces variance (due to aggregation) and bias (if the trees are fully grown). A random forest reduces the correlation between the tree-structured classifiers that enter into the averaging step. The algorithm is given in Table 14.4.

There are only two tuning parameters for a random forest: the number m of variables randomly chosen as a subset at each node and the number B of bootstrap samples. The procedure is relatively insensitive to a wide range of values of m and B. A good starting point is to take m as $\sqrt{r}$; if that is not sufficient, it is recommended to rerun the program with $m = 2\sqrt{r}$ and $m = 0.5\sqrt{r}$ as a way of monitoring the procedure. We have often found that values smaller than $\sqrt{r}$ yield smaller misclassification rates. The number B of bootstrap samples can be taken to be at least 1,000, and if r is very large, then B can be around 5,000.

TABLE 14.4. *Random forest classification algorithm using random input selection at each tree node.*

1. Input: $\mathcal{L} = \{(\mathbf{x}_i, y_i), i = 1, 2, \ldots, n\}$, $y_i \in \{1, 2, \ldots, K\}$, m = number of variables to be chosen at each node ($m << r$), B = number of bootstrap samples.

2. For $b = 1, 2, \ldots, B$:

   - Draw a bootstrap sample $\mathcal{L}^{*b}$ from the learning set $\mathcal{L}$.

   - From $\mathcal{L}^{*b}$, grow a tree classifier T^{*b} using random input selection: at each node, randomly select a subset m of the r input variables, and, using only the m selected variables, determine the best split at that node (using entropy or the Gini index). To reduce bias, grow the tree to a maximum depth with no pruning.

   - The tree T^{*b} generates an associated random vector $\boldsymbol{\theta}_b$, which is independent of the previous $\boldsymbol{\theta}_1, \ldots, \boldsymbol{\theta}_{b-1}$, and whose form and dimensionality are determined by context.

   - Using $\boldsymbol{\theta}_b$ and an input vector $\mathbf{x}$, define a classifier $h(\mathbf{x}, \boldsymbol{\theta}_b)$ having a single vote for the class of $\mathbf{x}$.

3. The B randomized tree-structured classifiers $\{h(\mathbf{x}, \boldsymbol{\theta}_b)\}$ are collectively called a *random forest*.

4. The observation $\mathbf{x}$ is assigned to the majority vote-getting class as determined by the random forest.

14.4.2 Generalization Error

Consider an *ensemble* (or *committee*) of B randomized tree-structured classifiers,

$$h(\mathbf{x}, \boldsymbol{\theta}_1), h(\mathbf{x}, \boldsymbol{\theta}_2), \ldots, h(\mathbf{x}, \boldsymbol{\theta}_B). \tag{14.66}$$

Define the generalization error for a random forest having B trees as

$$PE_B = P_{\mathbf{X}, Y}\{m_B(\mathbf{X}, Y) < 0\}, \tag{14.67}$$

where

$$m_B(\mathbf{X}, Y) = B^{-1} \sum_{b=1}^{B} I_{[h(\mathbf{X}, \boldsymbol{\theta}_b)=Y]} - \max_{k \neq y}\left\{ B^{-1} \sum_{b=1}^{B} I_{[h(\mathbf{X}, \boldsymbol{\theta}_b)=k]} \right\} \tag{14.68}$$

is the classification margin for the ensemble, and the probability is computed over the $(\mathbf{X}, Y)$-space. Note that if $m_B(\mathbf{X}, Y) > 0$, then the committee votes for the correct classification, whereas otherwise it does not.

Breiman (2001b) showed, using the strong law of large numbers, that, as the number of trees increases ($B \to \infty$), PE_B converges almost surely ($\{\boldsymbol{\theta}_b\}$) to the generalization error,

$$PE = P_{\mathbf{X},Y}\{m(\mathbf{X},Y) < 0\}, \tag{14.69}$$

where

$$m(\mathbf{X},Y) = P_{\boldsymbol{\Theta}}\{h(\mathbf{X},\boldsymbol{\Theta}) = Y\} - \max_{k \neq Y} P_{\boldsymbol{\Theta}}\{h(\mathbf{X},\boldsymbol{\Theta}) = k\}. \tag{14.70}$$

is defined as the *margin function* for a random forest. The margin, $m(\mathbf{X},Y)$, is the amount by which the average number of votes at $(\mathbf{X},Y)$ for the correct class exceeds the average vote for any other class. This limiting result is important: it shows that as we increase the number of trees in the forest, generalization error for a random forest converges to a limit; in other words, random forests *cannot overfit*, even if we have an infinite number of trees in the forest.

14.4.3   An Upper Bound on Generalization Error

The generalization error of a random forest can be bounded by a quantity that depends upon two parameters: a first-order parameter μ measuring the "strength" of any single tree in the forest and a second-order parameter $\bar{\rho}$ measuring the overall "correlation" between pairs of trees in the forest (Breiman, 2001b). These two parameters can be used to assess the accuracy of classifiers and the amount of dependence between them. For an accurate classification, we would like a strong classifier (large μ) with low correlation (small $\bar{\rho}$) between trees.

Consider the set of classifiers (14.66). From (14.68), define

$$\mu = E_{\mathbf{X},Y}\{m(\mathbf{X},Y)\} \tag{14.71}$$

to be the expected "strength" of the set of classifiers, which is assumed to be positive. Think of strength as a measure of accuracy of a tree in the forest. In the binary case, we see from (14.68) that $m(\mathbf{X},Y)$ can be written as

$$m(\mathbf{X},Y) = 2 \cdot P_{\boldsymbol{\Theta}}\{h(\mathbf{X},\boldsymbol{\Theta}) = Y\} - 1, \tag{14.72}$$

and the condition $\mu > 0$ translates to $E_{\mathbf{X},Y}P_{\boldsymbol{\Theta}}\{h(\mathbf{X},\boldsymbol{\Theta}) = Y\} > 0.5$; this result mimics the learning condition that a "weak" classifier is one that correctly classifies at a rate higher than 50%.

Our goal in this section is to provide an upper bound on the generalization error,

$$PE^* = P_{\mathbf{X},Y}\{|m(\mathbf{X},Y) - E_{\mathbf{X},Y}\{m(\mathbf{X},Y)\}| > \mu\}, \tag{14.73}$$

of a random forest. Applying Chebychev's inequality to (14.73), it follows that

$$PE^* \leq \frac{\mathrm{var}_{\mathbf{X},Y}\{m(\mathbf{X},Y)\}}{\mu^2}. \tag{14.74}$$

We now derive a suitable expression for the numerator of this upper bound.

Let $\widetilde{k} = \widetilde{k}(\mathbf{X},Y)$ denote the class with the most incorrect votes; that is,

$$\widetilde{k} = \arg\max_{k \neq Y} \mathrm{P}_{\Theta}\{h(\mathbf{X},\Theta) = k\}. \tag{14.75}$$

Then, from (14.70),

$$\begin{aligned} m(\mathbf{X},Y) &= \mathrm{P}_{\Theta}\{h(\mathbf{X},\Theta) = Y\} - \mathrm{P}_{\Theta}\{h(\mathbf{X},\Theta) = \widetilde{k}\} \\ &= \mathrm{E}_{\Theta}\{m^*(\mathbf{X},Y,\Theta)\}, \end{aligned} \tag{14.76}$$

where

$$m^*(\mathbf{X},Y,\boldsymbol{\theta}) = I_{[h(\mathbf{X},\boldsymbol{\theta})=Y]} - I_{[h(\mathbf{X},\boldsymbol{\theta})=\widetilde{k}]} \tag{14.77}$$

can be regarded as a "raw" margin function. Assuming that Θ and Θ' are iid,

$$[m(\mathbf{X},Y)]^2 = [\mathrm{E}_{\Theta}\{m^*(\mathbf{X},Y,\Theta)\}]^2 = \mathrm{E}_{\Theta,\Theta'}\{m^*(\mathbf{X},Y,\Theta)m^*(\mathbf{X},Y,\Theta')\}. \tag{14.78}$$

Thus, the variance function is

$$\begin{aligned} \mathrm{var}_{\mathbf{X},Y}\{m(\mathbf{X},Y)\} &= \mathrm{E}_{\Theta,\Theta'}\{\mathrm{cov}_{\mathbf{X},Y}(m^*(\mathbf{X},Y,\Theta), m^*(\mathbf{X},Y,\Theta'))\} \\ &= \mathrm{E}_{\Theta,\Theta'}\{\rho(\Theta,\Theta')\sigma(\Theta)\sigma(\Theta')\}, \end{aligned} \tag{14.79}$$

where, for fixed $\boldsymbol{\theta}$ and $\boldsymbol{\theta}'$,

$$\rho(\boldsymbol{\theta},\boldsymbol{\theta}') = \mathrm{corr}_{\mathbf{X},Y}\{m^*(\mathbf{X},Y,\boldsymbol{\theta}), m^*(\mathbf{X},Y,\boldsymbol{\theta}')\} \tag{14.80}$$

is the correlation between the raw margin functions of two different members in the forest, and, for fixed $\boldsymbol{\theta}$, $\sigma(\boldsymbol{\theta})$ is the square-root of

$$\sigma^2(\boldsymbol{\theta}) = \mathrm{var}_{\mathbf{X},Y}\{m^*(\mathbf{X},Y,\boldsymbol{\theta})\}. \tag{14.81}$$

Hence, from (14.79) and the definition of variance,

$$\mathrm{var}_{\mathbf{X},Y}\{m(\mathbf{X},Y)\} = \bar{\rho} \cdot [\mathrm{E}_{\Theta}\{\sigma(\Theta)\}]^2 \leq \bar{\rho} \cdot \mathrm{E}_{\Theta}\{\sigma^2(\Theta)\}, \tag{14.82}$$

where

$$\bar{\rho} = \frac{\mathrm{E}_{\Theta,\Theta'}\{\rho(\Theta,\Theta')\sigma(\Theta)\sigma(\Theta)\}}{\mathrm{E}_{\Theta,\Theta'}\{\sigma(\Theta)\sigma(\Theta)\}} \tag{14.83}$$

is the average correlation between all possible pairs of trees in the forest. Note that, from (14.82), we can write

$$\bar{\rho} = \frac{\mathrm{var}_{\mathbf{X},Y}\{m(\mathbf{X},Y)\}}{[\mathrm{E}_{\Theta}\{\sigma(\Theta)\}]^2}. \tag{14.84}$$

Now, from (14.81),

$$\mathrm{E}_{\boldsymbol{\Theta}}\{\sigma^2(\boldsymbol{\Theta})\} = \mathrm{E}_{\boldsymbol{\Theta}}\{\mathrm{E}_{\mathbf{X},Y}[(m^*(\mathbf{X},Y,\boldsymbol{\Theta}))^2] - [\mathrm{E}_{\mathbf{X},Y}(m^*(\mathbf{X},Y,\boldsymbol{\Theta}))]^2\}.$$
(14.85)

In the first term on the rhs, $m^*(\mathbf{X},Y,\boldsymbol{\Theta})$ is the difference of two indicator functions; see (14.77). So, $[m^*(\mathbf{X},Y,\boldsymbol{\Theta})]^2 \leq 1$. The second term on the rhs can be written as

$$\begin{aligned} \mathrm{E}_{\boldsymbol{\Theta}}\{[\mathrm{E}_{\mathbf{X},Y}(m^*(\mathbf{X},Y,\boldsymbol{\Theta}))]^2\} &\geq [\mathrm{E}_{\boldsymbol{\Theta}}(\mathrm{E}_{\mathbf{X},Y}(m^*(\mathbf{X},Y,\boldsymbol{\Theta}))]^2 \\ &= [\mathrm{E}_{\mathbf{X},Y}(\mathrm{E}_{\boldsymbol{\Theta}}(m^*(\mathbf{X},Y,\boldsymbol{\Theta}))]^2 \\ &= [\mathrm{E}_{\mathbf{X},Y}(m(\mathbf{X},Y))]^2 \\ &= \mu^2 \end{aligned}$$
(14.86)

The first line used the inequality $E(X^2) \geq [E(X)]^2$. Thus,

$$\mathrm{E}_{\boldsymbol{\Theta}}\{\sigma^2(\boldsymbol{\Theta})\} \leq 1 - \mu^2.$$
(14.87)

Substituting the inequality (14.87) into (14.82), and the result into (14.74) gives us an upper bound on generalization error for a random forest in terms of μ and $\bar{\rho}$:

$$PE^* \leq \frac{\bar{\rho}(1-\mu^2)}{\mu^2}.$$
(14.88)

This upper bound was derived by Breiman (2001b); however, the bound is generally quite loose.

Estimation of μ and $\bar{\rho}$ can be carried out as follows. Let $\mathcal{L}^{*b}$ be the bth bootstrap sample and let $h(\mathbf{x}, \boldsymbol{\theta}_b)$ be the bth classifier of $\mathbf{x}$ based upon $\mathcal{L}^{*b}$. For an OOB observation $(\mathbf{x}, y)$, let

$$\widehat{p}(\mathbf{x}, y) = \frac{\sum_b I_{[h(\mathbf{x},\boldsymbol{\theta}_b)=y;(\mathbf{x},y)\notin\mathcal{L}^{*b}]}}{\sum_b I_{[(\mathbf{x},y)\notin\mathcal{L}^{*b}]}}$$
(14.89)

denote the proportion of votes received for class y. It is an estimate of $\mathrm{P}_{\boldsymbol{\Theta}}\{h(\mathbf{x},\boldsymbol{\Theta}) = y\}$. The strength (14.71), which is the expected value of (14.70), can be estimated by:

$$\widehat{\mu} = n^{-1} \sum_{i=1}^{n} \{\widehat{p}(\mathbf{x}_i, y_i) - \widehat{p}(\mathbf{x}_i, \widetilde{y}_i)\},$$
(14.90)

where $\widetilde{y}_i = \arg\max_{y' \neq y} \widehat{p}(\mathbf{x}_i, y_i')$.

To estimate $\bar{\rho}$ in (14.84), we estimate the numerator and denominator separately. The numerator,

$$\mathrm{var}_{\mathbf{X},Y}\{m(\mathbf{X},Y)\} =$$

$$\mathrm{E}_{\mathbf{X},Y}\left\{\mathrm{P}_{\boldsymbol{\Theta}}[h(\mathbf{X},\boldsymbol{\Theta}) = Y] - \mathrm{P}_{\boldsymbol{\Theta}}[h(\mathbf{X},\boldsymbol{\Theta}) = \widetilde{Y}]\right\}^2 - \mu^2, \quad (14.91)$$

can be estimated by

$$n^{-1} \sum_{i=1}^{n} \{\widehat{p}(\mathbf{x}_i, y_i) - \widehat{p}(\mathbf{x}_i, \widetilde{y}_i)\}^2 - \widehat{\mu}^2. \tag{14.92}$$

The standard deviation is

$$\sigma(\boldsymbol{\theta}) = \{p_1 + p_2 + (p_1 - p_2)^2\}^{1/2}, \tag{14.93}$$

where $p_1 = \mathrm{E}_{\mathbf{X},Y}\{h(\mathbf{X}, \boldsymbol{\Theta}) = Y\}$ and $p_2 = \mathrm{E}_{\mathbf{X},Y}\{h(\mathbf{X}, \boldsymbol{\Theta}) = \widetilde{Y}\}$. We can estimate p_1 and p_2 for the bth OOB sample by

$$\widehat{p}_{1,b} = n_b^{-1} \sum_{(\mathbf{x}_i, y_i) \notin \mathcal{L}^{*b}} I_{[h(\mathbf{x}_i, \boldsymbol{\theta}_b) = y_i]}, \tag{14.94}$$

$$\widehat{p}_{2,b} = n_b^{-1} \sum_{(\mathbf{x}_i, y_i) \notin \mathcal{L}^{*b}} I_{[h(\mathbf{x}_i, \boldsymbol{\theta}_b) = \widetilde{y}_i]}, \tag{14.95}$$

respectively, where $n_b = \sum_{i=1}^{n} I_{[(\mathbf{x}_i, y_i) \notin \mathcal{L}^{*b}]}$ is the number of observations in the bth OOB sample. The denominator of (14.84) is found by substituting (14.94) and (14.95) into (14.93) to get an estimate of $\sigma(\boldsymbol{\theta}_b)$, and then averaging over all OOB samples:

$$\widehat{\sigma}(\boldsymbol{\theta}) = B^{-1} \sum_{b=1}^{B} \{\widehat{p}_{1,b} + \widehat{p}_{2,b} + (\widehat{p}_{1,b} - \widehat{p}_{2,b})^2\}^{1/2}. \tag{14.96}$$

14.4.4 Example: Diagnostic Classification of Four Childhood Tumors

Gene expression profiling using cDNA microarrays has become a very popular way of studying diseases. In this example, we analyze data from microarray experiments (Khan et al., 2001) on the small, round, blue-cell tumors (SRBCTs) of childhood, which include the distinct diagnostic categories of neuroblastoma (NB), rhabdomyosarcoma (RMS), non–Hodgkin lymphoma (NML), and the Ewing family of tumors (EWS). SRBCTs are so-named because of their similar appearance on routine histology; they often masquerade as each other, making correct clinical diagnosis difficult. Getting the diagnosis correct impacts directly upon the type of treatment, therapy, and prognosis the patient receives. Currently, there is no single clinical test that can discriminate between these cancers.

Gene-expression data were collected with a goal of distinguishing between the four types of SRBCT categories.[2] The data initially consisted of 83 cases

---

[2]The data are publicly available and can be downloaded from the website research.nhgri.nih.gov/microarray/Supplement.

(29 EWS, 11 BL, 18 NB, and 25 RMS) of both tumor biopsy material and cell lines measured on microarrays containing 6,567 genes. Requiring that each gene should have a certain minimal level of intensity reduced the number of genes to 2,308.

A random forest was applied to these data using 500 fully grown trees, where at each node we specified that 25 variables were to be randomly sampled (from the 2,308 variables available) as candidates for splitting. Over the 500 trees, the 83 cases were OOB the following numbers of times:

1	2	3	4	5	6	7	8	9	10
194	170	187	189	181	179	201	187	175	195
11	12	13	14	15	16	17	18	19	20
175	190	174	195	199	189	162	174	187	189
21	22	23	24	25	26	27	28	29	30
175	185	179	180	201	192	170	180	192	191
31	32	33	34	35	36	37	38	39	40
163	165	182	179	173	193	184	168	186	166
41	42	43	44	45	46	47	48	49	50
165	169	191	186	186	187	185	191	183	185
51	52	53	54	55	56	57	58	59	60
183	177	178	172	194	180	185	188	176	185
61	62	63	64	65	66	67	68	69	70
192	165	198	179	179	180	176	205	187	180
71	72	73	74	75	76	77	78	79	80
172	187	183	181	184	181	186	183	188	193
81	82	83							
181	172	193							

From these OOB instances, we obtain, for each case, the fraction of "votes" from the random forest for each disease category; each case is then classified according to the category with the highest fraction of votes received; and the OOB misclassification rate is calculated over all 83 classified cases. For this example, the results are indeed impressive: all 83 samples are correctly classified (0% OOB misclassification rate).

14.4.5 Assessing Variable Importance

If the objective is to classify new observations, it is useful to know which variables really control the classification process; in a regression situation, we need to know which subset of variables best explains the response values. We recognize, of course, that identifying which variables are important can be complicated by the existence of interactions between variables. Random forests can be used to evaluate the variables in a data set and provide a graphical display to assess the importance of each variable.

Computations are carried out one tree at a time. As before, let T^{*b} be the tree classifier constructed from the bootstrap sample $\mathcal{L}^{*b}$. First, drop the OOB observations corresponding to $\mathcal{L}^{*b}$ down the tree T^{*b}, record the resulting classifications, and compute the OOB error rate, $PE_b(OOB)$. Next, randomly permute the OOB values on the jth variable X_j while leaving the data on all other variables unchanged. If X_j is important, permuting its observed values will reduce our ability to classify successfully each of the OOB observations. Then, we drop the altered OOB observations down the tree T^{*b}, record the resulting classifications, and compute the OOB error rate, $PE_b(OOB_j)$, which should be larger than the error rate of the unaltered data. A raw T^{*b}-score for X_j can be computed by the difference between those two OOB error rates,

$$raw_b(j) = PE_b(OOB_j) - PE_b(OOB), \quad b = 1, 2, \ldots, B. \tag{14.97}$$

Finally, average the raw scores over all the B trees in the forest,

$$imp(j) = \frac{1}{B} \sum_{b=1}^{B} raw_b(j), \tag{14.98}$$

to obtain an overall measure of the importance of X_j. Call this measure the *raw permutation accuracy importance score* for the jth variable.

Assuming the B raw scores (14.97) are independent from tree to tree, we can compute a straightforward estimate of the standard error. Empirical studies using many different types of data sets show that a good case can be made for independence: indeed, scores between the trees appear to have low correlations. If this estimate of standard error is acceptable, we compute a z-score by dividing the raw score by the estimated standard error and then compute an appropriate Gaussian-based significance level for that z-score. Call this z-score the *mean decrease in accuracy* for the jth variable.

A second measure of variable importance derives from the fact that the Gini impurity index for a given parent node is larger than the value of that measure for its two daughter nodes. By averaging the (Gini) decreases in node impurities over all trees in the forest, we obtain a measure we call the *Gini importance index*.

For the example of childhood SRBCTs, the 30 most important variables for classification are displayed in Figure 14.7. The 10 variables [gene ID, gene description] that give the largest mean decrease in accuracy (left panel) are, in order of importance: 742 [*756847*, suppressin (nuclear deformed epidermal autoregulatory factor-1 (DEAF-1)-related)], 1955 [*80410*, farnesyl diphosphate synthase], 246 [*345538*, cathepsin L], 1003 [*825433*, ESTs], 1389 [*525799*, GTP cyclohydrolase I feedback regulatory protein], 509 [*37553*, protein phosphatase 2A, regulatory subunit B' (PR 53)], 2050 [*244154*, KIAA0875 protein], 2046 [*128054*, ESTs], 1799 [*196189*,

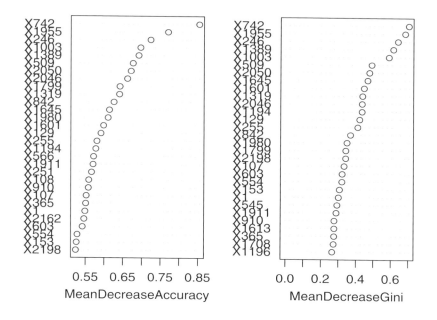

FIGURE 14.7. *Variable-importance plots for the SRBCT data.*

cytochrome b-5], and 1319 [*146868*, mitogen-activated protein kinase kinase kinase 11],

The 14 variables [gene ID] that give the largest mean decrease in the Gini index (right panel) are, in order of importance: 742 [*756847*], 1955 [*80410*], 246 [*345538*], 1389 [*525799*], 1003 [*825433*], 509 [*37553*], 2050 [*244154*], 1645 [*839374*, exostoses (multiple)-like 2], 1601 [*725188*, malate dehydrogenase 1, NAD (soluble)], 1319 [*146868*, mitogen-activated protein kinase kinase kinase 11], 2046 [*128054*], 1194 [*48285*, p53-induced protein], 129 [*298062*, troponin T2, cardiac], and 255 [*154472*, fibroblast growth factor receptor 1 (fms-related tyrosine kinase 2, Pfeiffer syndrome)].

Note that the rankings of important variables changes with the number of variables randomly chosen for splitting at each node, the initial seed for randomization, and the number of bootstrap trees in the forest.

14.4.6 Proximities for Classical Scaling

One of the most useful notions incorporated into random forests is that of computing proximities between pairs of observations. Using proximities, we can apply MDS (see Chapter 13) to the learning set to give a graphical view of data clustering in a lower-dimensional space. Proximities can also be used for imputing missing values and identifying multivariate outliers, if they are present in the data.

Suppose we construct a random forest of trees $\{T^{*b}\}$ from a learning set $\mathcal{L}$. Recall that each tree T^{*b} is unpruned and, hence, each terminal node in T^{*b} will contain only a few observations. If we drop all cases in $\mathcal{L}$ (including the OOB observations) down all the trees in the forest, how often do pairs of observations occupy the same terminal node? The answer to this question gives us a measure of "closeness" (or "proximity") of those pairs of observations to each other.

We, therefore, wish to define a similarity measure, $\text{prox}(\mathbf{x}_i, \mathbf{x}_j)$, between pairs of observations, $\mathbf{x}_i$ and $\mathbf{x}_j$, say, so that the closer $\mathbf{x}_i$ and $\mathbf{x}_j$ are to each other, the larger the value of $\text{prox}(\mathbf{x}_i, \mathbf{x}_j)$. If the two observations $\mathbf{x}_i$ and $\mathbf{x}_j$ end up at the same terminal node in T^{*b}, we increase $\text{prox}(\mathbf{x}_i, \mathbf{x}_j)$ by one. We repeat this procedure over all B trees in the forest, and then divide the frequency totals of pairwise proximities by the number, B, of trees in the forest; this gives us the proportion of all trees for which each pair of observations end up at the same terminal nodes. The results are subtracted from one to yield dissimilarities:

$$\delta_{ij} = 1 - \text{prox}(\mathbf{x}_i, \mathbf{x}_j), \quad \mathbf{x}_i, \mathbf{x}_j \in \mathcal{L}, \quad i, j = 1, 2, \ldots, n. \qquad (14.99)$$

We collect these pairwise dissimilarities into an $(n \times n)$ proximity matrix $\mathbf{\Delta} = (\delta_{ij})$, which is symmetric, positive-definite, with diagonal entries equal to zero. The proximity matrix is then used as input into the classical-scaling algorithm (see Table 13.5); this algorithm provides us with a visual comparison of the n observations in a lower-dimensional setting, where the interpoint distances between all pairs of observations are preserved (as much as possible) in the reduction to a lower-dimensional space.

A graphical display of pairs of principal coordinates (typically, the first plotted against the second) often yields a worthwhile comparison of the data in the learning set. In Figure 14.8, we show the MDS plot of proximities for the SRBCT data (left panel) and the BUPA data (right panel). We see that the SRBCT plot separates the data into four clusters correspond-ing to the four classes of SRBCTs, which probably contributes to its 0% OOB misclassification rate. The BUPA MDS plot, by contrast, shows three "arms" corresponding to the two classes in the data; the clusters have a number of overlapping points, and the OOB misclassification rate is 24.35% (500 bootstrap trees in the forest and 2 variables selected at random from the six for splitting each node in each tree), although this rate depends upon the same factors as listed at the end of the previous section.

14.4.7 Identifying Multivariate Outliers

Detecting and identifying outliers in multivariate data can be very diffi-cult, especially when the dimensionality is high. So, any procedure that is

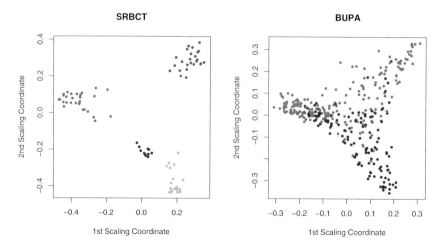

FIGURE 14.8. *MDS plots of the SRBCT data (left panel) and the BUPA data (right panel). The types of tumors in the SRBCT plot are (number of points, color): BL (11, red), EWS (29, blue), NB (18, green), and RMS (25, purple). The points in the BUPA plot correspond to Class 1 (145, red) and Class 2 (200, blue).*

successful in outlier-detection is worth its weight in gold. The proximities computed for random forests can be used to detect outliers.

The basic idea is that we identify an outlier by how far away it is from all other observations belonging to its class in the learning set. Suppose $\mathbf{x}_i \in \Pi_k$. If the proximity of, say, $\mathbf{x}_i$ to another kth-class observation, say, $\mathbf{x}_j$ is small, then it is rare for those two observations to end up at the same terminal nodes when they are simultaneously dropped down all the trees in the forest. In other words, $\mathbf{x}_i$ and $\mathbf{x}_j$ are far apart from each other iff their proximity is small. If $\mathbf{x}_i$ is far away from all the other kth-class observations in the learning set, then all the proximities, $\text{prox}(\mathbf{x}_i.\mathbf{x}_\ell)$, of $\mathbf{x}_i$ with $\mathbf{x}_\ell$, $\ell \neq i$, will be small. Breiman and Cutler (2004) suggest that a raw outlier measure for the ith observation, $\mathbf{x}_i$, in the kth class be given by

$$u_{ik} = \frac{n}{\sum_{\mathbf{x}_\ell \in \Pi_k, \ell \neq i} [\text{prox}(\mathbf{x}_i, \mathbf{x}_\ell)]^2}, \quad i = 1, 2, \ldots, n, \qquad (14.100)$$

where $k = 1, 2, \ldots, K$. Thus, if $\mathbf{x}_i$ is really an outlier for the kth class, the denominator of (14.100) will be small, so that u_{ik} will be large.

Let $m_k = \text{med}_{\mathbf{x}_\ell \in \Pi_k} \{u_{\ell k}\}$ be the median of the raw outlier measures over all kth-class observations. Then, for $k = 1, 2, \ldots, K$, a standardized version of u_{ik} is given by

$$\widetilde{u}_{ik} = \frac{u_{ik} - m_k}{\sum_{\mathbf{x}_\ell \in \Pi_k} |u_{\ell k} - m_k|}. \quad i = 1, 2, \ldots, n. \qquad (14.101)$$

SRBCT

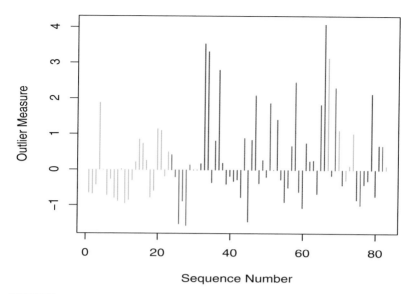

FIGURE 14.9. *Outlier plot for the SRBCT data. The types of tumors in the SRBCT plot are BL (red), EWS (blue), NB (green), and RMS (purple).*

The values of (14.101) are plotted against sequence number, with each class's values plotted using either a different symbol or color. Values of (14.101) in excess of 10 should generate concern. The SRBCT data set does not appear to have any outliers; see Figure 14.9.

14.4.8   Treating Unbalanced Classes

A major impediment to good classification in practical problems occurs when at least one of the classes (often the class of primary interest) contains only a very small proportion of the observations. Examples of such "unbalanced" situations include detection of fraudulent telephone calls, information retrieval and filtering, diagnosis of rare thyroid diseases, and detection of oil spills from satellite images (Chen, Liaw, and Breiman, 2004). In each of these examples, the result is wildly varying prediction errors for the different classes.

Classification algorithms, which focus on minimizing the overall misclassification rate, classify most observations according to the class of the majority of observations (the "majority" class); as a result, the misclassification rate will be very low, but the observations belonging to the class of primary interest (the "minority" class) will be totally misclassified. In the case of random forests, for example, the bootstrap samples will contain very few (and maybe none) of the minority class observations, and so we will see poor class prediction (i.e., high prediction error) for the minority class.

To alleviate such difficulties, various modifications to the random forest classifier were considered by Chen, Liaw, and Breiman (2004), including *balanced random forest* (BRF), where the majority class is undersampled, and *weighted random forest* (WRF), where a heavier weight is placed upon selecting the minority class in bootstrap samples in order to prevent misclassifying that class. Based upon experiments with various data sets, no real difference in prediction error has been found between BRF and WRF, although BRF turns out to be computationally more efficient.

14.5 Software Packages

Bagging for classification or regression can be carried out in R using the package `ipred` (short for Improved Predictors), which can be downloaded from an appropriate CRAN site. For bagging decision trees, `ipred` uses the package `rpart`. For R users, there are several packages that carry out boosting using ADABOOST: `ada`, `boost`, and `adabag` each use `rpart` and each can be downloaded from an appropriate CRAN site. The ADABOOST computations were carried out here using the `ada` package.

Breiman's random forest software is now a commercial product that is licenced exclusively to Salford Systems (`www.salford-systems.com`). See the URL `www.stat.berkeley.edu/users/breiman/RandomForests`. An R-interface to the random forests classifier has been written by A. Liaw and M. Wiener based upon original Fortran code written by Breiman and Cutler. R documentation and help files for version 4.4–2 are available at

`lib.stat.cmu.edu/R/CRAN/doc/packages/randomForest.pdf`.

Software to carry out bagging, boosting, and random forests is also available in other packages, such as MATLAB, WEKA, and STATISTICA.

Bibliographical Notes

Much of the material in this chapter is based upon the work of Leo Breiman, who has provided great insights into the ensemble methods of bagging, boosting, and random forests, and who has left an indelible mark on this field.

The statistical derivation of ADABOOST is adapted from the treatments in Friedman, Hastie, and Tibshirani (2000) and Hastie, Tibshirani, and Friedman (2001, Section 10.4). The algorithmic development of GRADIENT.BOOST follows the work in Friedman (2001). The division of ADABOOST

into optimal and nonoptimal strategies can be found in Rätsch and Warmuth (2002); see also Rudin, Daubechies, and Schapire (2004) and Rudin,
Schapire, and Daubechies (2007).

The section on random forests is based upon Breiman (2001b) and the
short course he and Adele Cutler gave at the *26th Symposium on the
Interface*, which was held in Baltimore, MD, in May 2004. Breiman's in-
spiration for random forests came from reading Amit and Geman (1997).
This author thanks Adele Cutler for conversations on random forests and
especially for pointing out and correcting a typographical error in Breiman
(2001, equation (8)).

Exercises

14.1 Let $Y \in \{-1, +1\}$ and let $C(\mathbf{x}) \in \{-1, +1\}$ be a classifier of $\mathbf{x}$. Show
that $Y^* = (Y + 1)/2$ is a Bernoulli variable that takes the value 0 with
probability $p(\mathbf{x}) = e^{C(\mathbf{x})}/(e^{C(\mathbf{x})} + e^{-C(\mathbf{x})})$ and the value 1 with probability
$1 - p(\mathbf{x})$. Find the binomial log-likelihood and show that it is equal to
$L(y, C(\mathbf{x})) = \log_e \{1 + e^{-2yC(\mathbf{x})}\}$.

14.2 Consider the regression situation, where Y is continuous. Assume
squared-error loss: $L(y, f_m(\mathbf{x})) = \frac{1}{2}(y - f_m(\mathbf{x}))^2$. Show that the pseudoresponses are given by $\tilde{y}_i = y_i - f_{m-1}(\mathbf{x}_i)$, $i = 1, 2, \ldots, n$, and that the
learning rate is $\rho_m = \beta_m$. Hence, show that the GRADIENT.BOOST algorithm reduces to an iterative least-squares fitting of the current residuals.

14.3 Show that (14.19) can be reduced to (14.20). Furthermore, show that
$\sum_{i=1}^n w_{i,m+1} = 1$.

14.4 Consider the following 10 two-dimensional points: the first five points,
(1, 4), (3.5, 6.5), (4.5, 7.5), (6, 6), (1.5, 1.5), belong to Class 1, and the
second five points, (8, 6.5), (3, 4.5), (4.5, 4), (8, 1.5), (2.5, 0), belong to
Class 2. Plot these points on a scatterplot using different symbols or colors to distinguish the two classes. Carry through *by hand* the ADABOOST
algorithm on these points, showing the weights at each step of the process.
Determine the final classifier and calculate its misclassification rate.

14.5 Write a program that implements ADABOOST for tree-based binary
classification. Extend your program to more than two classes.

14.6 Use ADABOOST to classify the `pima-indian-diabetes` data. Compare your results with the classification tree results. Can you do any better
with random forests?

14.7 Use a random forest to classify the `spambase` data. Repeat the analysis 100 times using different random seeds to start each replication. For each

repetition, find the OOB misclassification rate and draw the boxplot for OOB misclassification rates. Repeat this for different values of m (number of variables selected as candidates for splitting) and B (number of bootstrap trees in the forest). What can you say about the effect of m and B on the OOB misclassification rate?

14.8 Carry out the same computations as in Exercise 14.7 for the `glass` data. What do you notice about the MDS plot?

14.9 Carry out the same computations as in Exercise 14.7 for the Wisconsin Diagnostic Breast Cancer data (`wdbc`).

14.10 Run random forests 100 times on the `SRBCT` data, and each time find the 30 most-important variables. Set $B = 500$ bootstrap trees and $m = 25$, and for each run use different random seeds as starting values. You should see different sets of variables being ranked as the 30 most important for each run. Create a method for visualizing the overall ranking of the variables. Repeat these operations using different B and different m. Using the Internet, try to get some corroboration for your findings.

15

Latent Variable Models for Blind Source Separation

15.1 Introduction

Models incorporating "latent" variables have been commonplace in the social and behavioral sciences for a long time. The most popular of those models is the *factor analysis model*, in which a set of observed continuous variables is explained in terms of a much smaller set of continuous latent variables (called *factors*), and the relationship is taken to be a linear one.

Latent variables, which can be continuous or discrete, are quite different from observed variables in that they are artificial or hypothetical constructs. Latent variables are typically used to give a formal representation of ideas or concepts that cannot be well-defined or measured directly. In educational and psychometric research, for example, fuzzy concepts such as "general intelligence," "verbal ability," "ambition," "socioeconomic status," "quality of life," and "happiness" are constructed from certain observed variables that are regarded as proxies for those unobservable concepts. Moreover, it is not unusual to hear of a causal relationship between a latent variable and a set of given observable variables (e.g., "it is because of a person's high level of intelligence that he or she does so well on standardized tests").

A.J. Izenman, *Modern Multivariate Statistical Techniques*,
doi: 10.1007/978-0-387-78189-1_15,
© Springer Science+Business Media, LLC 2008

Latent variables are also known, for example, as *hidden variables* in neural network modeling and as *sources* that are statistically independent of each other in independent component analysis. Latent variables have been introduced into MCMC sampling as *auxiliary variables* and as a data-augmentation technique in missing-value problems. Latent variables are usually formed as linear combinations of observable variables for the purpose of reducing the dimensionality of a data set. Indeed, it is easier to consider a single latent variable interpreted as "quantitative ability" than to have to deal with understanding a battery of different arithmetic and mathematics test scores. As we will see, latent variables play the fundamental role of "sources" in blind source separation problems.

15.2 Blind Source Separation and the Cocktail-Party Problem

A common type of problem that arises in such diverse fields as telecommunications, sound and image processing, brain imaging, speech enhancement, predicting stock-price movements, remote sensing, biomedical engineering, and signal processing — all situations in which the data consist of multiple time series — is to find a way of solving the *blind source separation (BSS)* problem. The BSS problem involves decomposing an unknown mixture of non-Gaussian signals into its independent component signals (Cardoso, 1998). BSS is similar to the classical electrical engineering problem of source separation, but in BSS there is no prior knowledge of the signals that make up the mixture.

The best-known example of BSS is the so-called cocktail-party problem (Cherry, 1953). In this problem, m people are speaking simultaneously at a party, and each of r microphones placed in the same room at different distances from each speaker records a different mixture of the speakers' voices at n time points. The question is whether, based upon these microphone recordings, we can separate out the individual speech signals of each of the m speakers. Despite the fact that the cocktail-party problem assumes the speakers babble on independently without considering the presence of other partygoers (who usually speak in clustered groups), it does give a fairly simplistic explanation of how one can envision BSS problems.

Thus, we see that mixtures of signals occur everywhere, and it is of great interest to develop methods for separating (or "unmixing") those signals so that we can view the individual raw signals that make up that mixture. With this in mind, we describe in this chapter a general latent variable model that is proposed to solve the BSS problem. Special cases of this model include independent component analysis, (exploratory) factor analysis, and independent factor analysis.

15.3   Independent Component Analysis

Independent component analysis (ICA) is a multivariate statistical technique that seeks to uncover hidden variables in high-dimensional data. As such, it belongs to the class of latent variable models. Furthermore, because of its success in analyzing signal processing data, ICA is also regarded as a digital signal transform method.

In its most basic form, the ICA model is assumed to be a linear mixture of an unknown number of unknown hidden *source* variables, where the mixing coefficients are also unknown. A totally "blind" approach to determining both the hidden variables and the mixing coefficients solely from the observed multivariate data fails because the problem as stated is not well-defined.

To build more structure into the problem, we require the hidden variables to be mutually independent and also (with at most one exception) non-Gaussian. ICA is actually an amalgam of several related approaches to this problem, and these approaches are characterized by the types of assumptions visited upon the distributions of the independent source variables and whether or not a separate noise component should be included in the ICA model.

15.3.1   Applications of ICA

ICA has been extensively applied to the study of human brain functions. Patterns of human brain-wave activity can be viewed through noninvasive recordings made by r (usually around 20, sometimes a lot more) electrodes placed evenly around a subject's head during different periods of consciousness and sleep. The electrodes capture a mixture of brain waves from different areas of the brain. Electroencephalographic (EEG) recordings make it possible to relate certain types of behavior to changes in the electrical activity of the cerebral cortex; event-related potential (ERP) recordings are finely-tuned EEGs resulting from the stimulation of specific visual, auditory, or sensory systems; and magnetoencephalographic (MEG) recordings measure the strength of magnetic fields that are generated by cortical activity. ICA has been used successfully to separate EEG, ERP, and MEG recordings into individual (and meaningful) source signals.

ICA has also been successful in extracting three-dimensional spatial recordings (called *component maps*) from functional magnetic resonance imaging (fMRI) experiments used to study the human brain. These experiments consist of a number of trials in which subjects perform certain experimental and control psychomotor tasks. The component maps take the form of a mixture of signals from thousands of voxels (volume elements) located in each of several brain slices and measured over a given period of time. The voxel values indicate brain regions that are actively involved in the cognitive processing of the specified tasks. If the active voxels are sparsely distributed in the maps and are mostly nonoverlapping, then the maps are considered to be independent. ICA has been used to separate fMRI data into m independent component maps together with their corresponding component activation patterns.

Other applications of ICA include extracting structure from financial stock returns, mapping the cosmic microwave background anisotropy from satellite radiometric sky maps, separating out the effects of major volcanic eruptions from climate and temperature data, identifying spatial-variation patterns in manufacturing processes such as automobile assembly, Web image retrieval and classification, wireless communications and speech recognition systems, and agricultural remote sensing images. Classification of microarray gene expression profiles using ICA methods has also become a popular research issue.

15.3.2   Example: Cutaneous Potential Recordings of a Pregnant Woman

In prenatal diagnostics, it is important for a physician to be able to monitor — in a non-invasive way — the fetal heart activity of a pregnant woman so that the health and condition of the fetus can be assessed. A multichannel electrocardiogram (ECG) can be used to obtain a mixture of maternal and fetal electrical activity, including fetal heart rate and maternal heart rate; however, the maternal ECG signal is many hundreds or thousands times stronger than the fetal ECG signal, and the signals are further contaminated by respiration baseline wandering and other sources of electrical interference.

The data[1] for this example consist of 2,500 ECG points sampled at 500 Hz using 8-channel cutaneous (i.e., on the skin) potential recordings of a pregnant woman (de Lathauwer, de Moor, and Vandewalle, 2000). The 8 sets of cardiac rhythms are displayed in Figure 15.1; the 2,500 points are recorded over a period of 5 seconds, one point every 0.002 seconds. Note that the range of amplitudes increases as we go from Channel 1 to Channel 8. The first five channels (1–5) are measured near the fetus and, hence, show abdominal signals. Fetal contributions are visible in Channels 1, 2,

---

[1]These data are included in DaISy: "Database for the Identification of Systems," de Moor, B.L.R. (1997) (ed.), Department of Electrical Engineering, ESAT/SISTA, K.U. Leuven, Belgium, and can be downloaded from the website www.esat.kuleuven.ac.be/~smc/daisy/daisydata.html.

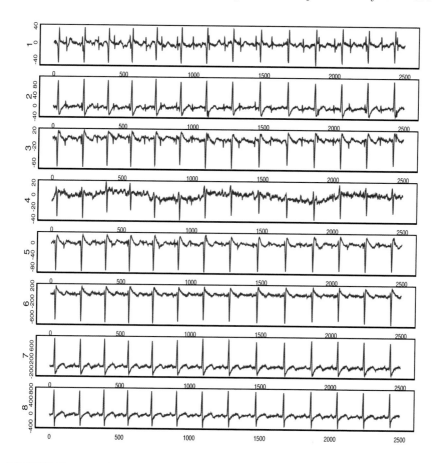

FIGURE 15.1. *Cardiac rhythms obtained from 2,500 ECG points sampled at 500 Hz using an 8-channel cutaneous potential recording of a pregnant woman.*

and 3, but their magnitudes are quite weak. The other three channels (6–8) were placed on the mother's thorax (chest), near the heart; note that the high magnitudes of the maternal ECG in the thoracic signals tend to swamp the fetal ECG signals. We illustrate the power of ICA methods for this example by reconstructing the fetal ECG from multichannel potential recordings on the mother's skin.

First, we preprocess the data by applying PCA to the sample correlation matrix; this produces 8 uncorrelated and ordered principal components whose variances decrease in magnitude. Only the first two PCs have eigenvalues greater than unity, and together they account for about 93% of the total variation of the data. For this example, we retain all 8 PCs as inputs to ICA.

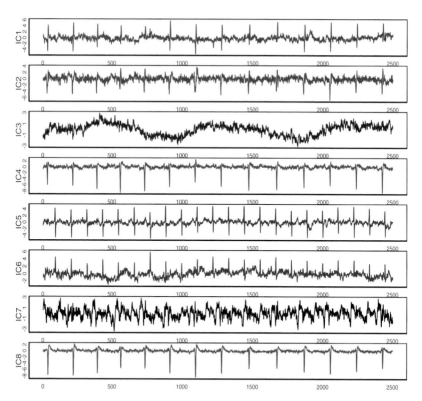

FIGURE 15.2. *Eight independent components from the ICA of the 8-channel cutaneous potential recordings of a pregnant woman. The red curves (IC1, IC2, IC4, IC8) reflect the cardiac rhythms of the mother, whereas the blue curves (IC5, IC6) reflect the cardiac rhythms of the fetus. The purple curve (IC3) shows a respiration component, and the black curve (IC7) shows the noise level of the sensors.*

We then apply the FASTICA algorithm (see Section 15.3.11) to all 8 sets of principal component scores, which, in turn, yields 8 independent components (ICs). These ICs are displayed in Figure 15.2. We see four ICs that reflect the mother's cardiac rhythm (red curves: IC1, IC2, IC4, IC8) and two ICs reflecting the fetal cardiac rhythm (blue curves: IC5, IC6). The purple curve (IC3) probably reflects a low-periodic respiration component, and IC7 displays a noise component.

15.3.3   Connection to Projection Pursuit

The technical aspects of ICA in its basic formulation are remarkably similar to those of exploratory projection pursuit (PP) (Friedman and Tukey, 1974), a methodology developed more than a decade earlier than ICA.

ICA and PP methodologies look at the same data in very different ways, yet they both use the same (or similar) computational tool (numerically optimizing an objective function) to achieve a common statistical goal of finding low-dimensional, non-Gaussian projections of the data. The differences between ICA and PP derive from the different problems they were originally built to solve.

For example, ICA was introduced to resolve a separation problem, starting with the estimation of independent components, whereas PP was designed to be an exploratory tool for data visualization, focusing on dimensionality reduction of a high-dimensional space. Furthermore, the manner in which PP and ICA extract a sequence of signals from a given collection of mixtures differs: PP extracts signals one at a time, whereas ICA can extract the entire set of signals in parallel. The PP and ICA solutions are also related: PP also makes no assumptions about the data or about independent components, as does ICA; if the ICA model holds, then the optimization process produces independent components, whereas if the model does not hold, then we obtain the PP solution.

Although much of the PP methodology has been incorporated into the ICA toolkit, there has been little cross-pollination in the other direction. Recent enhancements of the ICA model that take into account time-structure and nonlinearity of the mixing coefficients have further helped to distinguish ICA from PP.

15.3.4 Centering and Sphering

Suppose we observe a random r-vector, $\mathbf{X} = (X_1, \cdots, X_r)^\tau$, of correlated measurements with mean r-vector $E\{\mathbf{X}\} = \boldsymbol{\mu}$ and $(r \times r)$ covariance matrix $\text{cov}\{\mathbf{X}\} = \boldsymbol{\Sigma}_{XX}$. Prior to carrying out PP or ICA applications, we preprocess $\mathbf{X}$ so that its r components have commensurate scales. We do this by first *centering* $\mathbf{X}$ so that its components have zero mean, and then by *sphering* (or *whitening*) the result so that its components are uncorrelated with unit variances.

Sphering is a linear transformation that removes all traces of scale and correlation structure from $\mathbf{X}$. From the spectral decomposition of the covariance matrix, $\boldsymbol{\Sigma}_{XX} = \mathbf{U}\boldsymbol{\Lambda}\mathbf{U}^\tau$, where the columns of the orthogonal matrix $\mathbf{U}$ are the eigenvectors of $\boldsymbol{\Sigma}_{XX}$ and $\boldsymbol{\Lambda}$ is a diagonal matrix with diagonal elements the eigenvalues of $\boldsymbol{\Sigma}_{XX}$. The columns of $\mathbf{U}$ and the diagonal elements of $\boldsymbol{\Lambda}$ are ordered by the decreasing magnitudes of the eigenvalues of $\boldsymbol{\Sigma}_{XX}$.

Assume that $\boldsymbol{\mu}$ and $\boldsymbol{\Sigma}_{XX}$ are both known. Then, we can write $\boldsymbol{\Sigma}_{XX}^{-1/2} = \mathbf{U}\boldsymbol{\Lambda}^{-1/2}\mathbf{U}^\tau$, The (centered and) sphered version of $\mathbf{X}$ is given by

$$\mathbf{X} \leftarrow \boldsymbol{\Lambda}^{-1/2}\mathbf{U}^\tau(\mathbf{X} - \boldsymbol{\mu}). \tag{15.1}$$

This transformation is equivalent to computing the principal components of $\mathbf{X} - \boldsymbol{\mu}$ and then rescaling each of the principal components to have unit variance. If $\boldsymbol{\Sigma}_{XX}$ has less than full rank, only those principal components having nonzero variance would be retained (and rescaled). A benefit of sphering $\mathbf{X}$ is that it is now affine invariant, with $\boldsymbol{\mu} = \mathbf{0}$ and $\boldsymbol{\Sigma}_{XX} = \mathbf{I}_r$.

In practice, $\boldsymbol{\mu}$ and $\boldsymbol{\Sigma}_{XX}$ will be unknown. Thus, we use n independent observations, $\mathbf{X}_1, \ldots, \mathbf{X}_n$, on $\mathbf{X}$ to compute $\bar{\mathbf{X}} = n^{-1} \sum_{i=1}^{n} \mathbf{X}_i$ and $\widehat{\boldsymbol{\Sigma}}_{XX} = n^{-1} \sum_{i=1}^{n} (\mathbf{X}_i - \bar{\mathbf{X}})(\mathbf{X}_i - \bar{\mathbf{X}})^\tau = \widehat{\mathbf{U}} \widehat{\boldsymbol{\Lambda}} \widehat{\mathbf{U}}^\tau$, respectively. Centering and sphering the data using $\mathbf{X}_i \leftarrow \widehat{\boldsymbol{\Lambda}}^{-1/2} \widehat{\mathbf{U}}^\tau (\mathbf{X}_i - \bar{\mathbf{X}})$, $i = 1, 2, \ldots, n$, transform an elliptically shaped symmetric cloud of points into a spherically shaped cloud. To reduce the dimensionality of the data, only the first $J < r$ sphered variables need be retained, where J is chosen to explain a certain (high) proportion of the total variance. If outliers are present, robust versions of the sphering process can be used (see, e.g., Tukey and Tukey, 1981).

We note that the practice of sphering is somewhat controvertial. Although sphering has computational and interpretational advantages (see, e.g., Friedman, 1987), arguments have been made that the act of sphering is too closely tied to underlying unimodal (and especially Gaussian) distributions, an environment we wish to avoid (see, e.g., the comments of Gower, and Hastie and Tibshirani in the discussion of Jones and Sibson, 1987). However, we follow PP and ICA practice by assuming that the components of $\mathbf{X}$ have been preprocessed to be mutually uncorrelated, each having zero mean and unit variance.

15.3.5 The General ICA Problem

In its most general form, the ICA model assumes that $\mathbf{X}$ is generated by

$$\mathbf{X} = f(\mathbf{S}) + \mathbf{e}, \tag{15.2}$$

where $\mathbf{S} = (S_1, \cdots, S_m)^\tau$ is an (unobservable) random m-vector variate of *sources* whose components $\{S_j\}$ are independent latent variables each having zero mean, $f : \Re^m \to \Re^r$ is an unknown *mixing function*, and $\mathbf{e}$ is a zero-mean, additive, r-vector-valued component that represents measurement noise and any other type of variability that cannot be directly attributed to the sources. Independence of the sources means that each individual source signal is thought to be generated by a process unrelated to any other source signal. We assume that $E(\mathbf{S}) = \mathbf{0}$ and $cov(\mathbf{S}) = \mathbf{I}_m$, but that the distribution of $\mathbf{S}$ is otherwise unknown.

The BSS problem is to invert f and estimate $\mathbf{S}$. As it stands, this problem is ill-posed and needs some additional constraints or regularization on $\mathbf{S}$, f, and $\mathbf{e}$. If we take f to be a linear function, $f(\mathbf{S}) = \mathbf{A}\mathbf{S}$, where $\mathbf{A}$ is a "mixing" matrix, then (15.2) is described as a *linear ICA* model, whereas if f assumed to be nonlinear, then (15.2) is described as a *nonlinear ICA* model. Most applications of ICA assume no additive noise $\mathbf{e}$ and that all noise in the model is to be associated with the components of the random vector $\mathbf{S}$. Such a model is referred to as *noiseless ICA*. If $\mathbf{e}$ is included in (15.2), the model is described as *noisy ICA*.

It turns out that the noiseless ICA model with linear mixing, $\mathbf{X} = \mathbf{AS}$, can only be solved if the vector $\mathbf{S}$ with independent components is not Gaussian. We can see this by assuming the contrary. Suppose that the sources, $S_1, \ldots, S_m$, are independent and Gaussian, each with zero mean and unit variance. Their joint density is given by $q_{\mathbf{S}}(\mathbf{s}) = \prod_{j=1}^{m} q_{S_j}(s_j) = (2\pi)^{-m/2} e^{-\|\mathbf{s}\|^2/2}$, where $\|\mathbf{s}\|^2 = \sum_j s_j^2$. If the mixing matrix $\mathbf{A}$ is square ($m = r$) and, hence, orthogonal ($\mathbf{I}_r = \boldsymbol{\Sigma}_{XX} = \mathbf{AA}^\tau$, so that $\mathbf{A}^{-1} = \mathbf{A}^\tau$), then one can show that the density of $\mathbf{X} = \mathbf{AS}$ is given by $p_{\mathbf{X}}(\mathbf{x}) = (2\pi)^{-m/2} e^{-\|\mathbf{A}^\tau \mathbf{x}\|^2/2} |\det(\mathbf{A}^\tau)|$. But $\mathbf{A}$ is orthogonal, and so $\|\mathbf{A}^\tau \mathbf{x}\|^2 = \|\mathbf{x}\|^2$ and $|\det(\mathbf{A}^\tau)| = 1$. Thus, the density of $\mathbf{X}$ reduces to $p_{\mathbf{X}}(\mathbf{x}) = (2\pi)^{-m/2} e^{-\|\mathbf{x}\|^2/2}$, which is identical to the density of $\mathbf{S}$, so that the orthogonal mixing matrix $\mathbf{A}$ cannot be identified for independent Gaussian sources. Thus, it makes sense to require that, with the exception of at most one component, the remaining independent source components cannot be Gaussian distributed.

There are a number of ways of estimating this type of ICA model while ensuring that the components of $\mathbf{S}$ are as statistically independent and non-Gaussian as possible. Usually, we are in possession of n repeated r-variate observations, $\mathbf{X}_i = (X_{i1}, \cdots, X_{ir})^\tau$, $i = 1, 2, \ldots, n$, on $\mathbf{X}$, which constitute our data set. From this, our goal is to recover the m independent sources, $\mathbf{S}_i = (S_{i1}, \cdots, S_{im})^\tau$, $i = 1, 2, \ldots, n$, which generated the data through $\mathbf{X}_i = \mathbf{AS}_i$, $i = 1, 2, \ldots, n$. Several efficient computational algorithms have been created to reach this goal.

In most ICA applications, $\mathbf{X}$ is regarded as an r-vector-valued stochastic process $\mathbf{X}(t) = (X_1(t), \cdots, X_r(t))^\tau$, such as audio or music signals, EEG or MEG tracings, or seismic recordings, where t is a time or index parameter. We usually assume that $\mathbf{X}(t)$ is an unknown non-Gaussian process with zero mean. In the linear noiseless ICA model with temporally structured sources and *static mixing* (i.e., $\mathbf{A}$ is an fixed matrix of constants, non-time-varying, without trends or delays), the model is written as $\mathbf{X}(t) = \mathbf{AS}(t)$, where $\mathbf{S}(t) = (S_1(t), \cdots, S_m(t))^\tau$ is assumed to be an m-vector of *stationary sources*, $1 \leq t \leq n$. For example, in the cocktail-party problem, $S_i(t)$ is the tth sound spoken by the ith speaker ($i = 1, 2, \ldots, m$), and $X_j(t)$ is the tth acoustic recording made by the jth microphone ($j = 1, 2, \ldots, r$).

In this formulation, ICA is closely related to the *deconvolution of time series*; see, for example, Donoho (1981), who discusses at length the single-channel ($r = 1$) deconvolution problem and its application to exploratory seismology. Donoho points out that the geophysicist's technique of minimum entropy deconvolution is actually a PP method with kurtosis as the projection index. See Huber (1985, Section 18). Extensions to the multi-channel (general r) case have also been studied.

If the mixing matrix $\mathbf{A} = \mathbf{A}(t)$ is allowed to depend upon the time parameter, then we refer to the model as *dynamic mixing*. By incorporating the temporal structure of the sources into the ICA model, there is a good chance that the separation properties of the analysis can be improved. In our description of ICA models, we omit the explicit dependence of $\mathbf{X}$ on t unless specifically needed in the exposition.

15.3.6 Linear Mixing: Noiseless ICA

The simplest form of the ICA model is the linear mixing version with no additive noise, usually called the *noiseless* (or *classical*) *ICA model*. In this scenario, $\mathbf{X}$ is modeled deterministically as

$$\mathbf{X} = \mathbf{AS}, \tag{15.3}$$

where $\mathbf{S} = (S_1, \cdots, S_m)^\tau$ is a latent random m-vector of independent source components, and $\mathbf{A}$ is a full-rank $(r \times m)$ *mixing matrix* of unknown parameters. Usually, $m \leq r$. For model (15.3), where the sources have mean zero, $\mathbf{X}$ has mean zero and covariance matrix $\mathbf{AA}^\tau$. Given n iid observations on $\mathbf{X}$, the BSS (and ICA) problem is to estimate $\mathbf{A}$ and, hence, recover $\mathbf{S}$.

For a given $\mathbf{A}$ with full-rank, there exists a *separating* (or *unmixing matrix*) $\mathbf{W}$ such that the sources can be recovered exactly from the observed $\mathbf{X}$ by $\mathbf{S} = \mathbf{WX}$, where $\mathbf{W} = (\mathbf{A}^\tau \mathbf{A})^{-1} \mathbf{A}^\tau$. If the number of independent sources is equal to the number of measurements (i.e., $m = r$), then we refer to (15.3) as the *square invertible mixing model*, and, for that special case, $\mathbf{W} = \mathbf{A}^{-1}$. As we saw above, if $\mathbf{X}$ has been centered and sphered, then the resulting square mixing matrix $\mathbf{A}$ in model (15.3) is orthogonal, and so $\mathbf{W} = \mathbf{A}^\tau$.

In practice, $\mathbf{A}$ is unknown and the goal is to estimate the separating matrix and the source components based solely upon the observed $\mathbf{X}$. Given an estimate $\widehat{\mathbf{W}} = (\widehat{\mathbf{w}}_1, \cdots, \widehat{\mathbf{w}}_m)^\tau$ of the separating matrix $\mathbf{W}$, the source component vector $\mathbf{S}$ is approximated by

$$\mathbf{Y} = \widehat{\mathbf{W}}\mathbf{X}, \tag{15.4}$$

where the elements, $Y_1 = \widehat{\mathbf{w}}_1^\tau \mathbf{X}, \ldots, Y_m = \widehat{\mathbf{w}}_m^\tau \mathbf{X}$, of $\mathbf{Y}$ are taken to be statistically independent and as non-Gaussian as possible.

15.3.7 Identifiability Aspects

Given $\mathbf{X}$, the model (15.3) suffers from a certain amount of arbitrariness:

1. The original sources are ordered arbitrarily. Let $\mathbf{P}$ be an $(m \times m)$ permutation matrix (a permutation of the rows and columns of the identity matrix such that every row and column has exactly one 1). Then, the model (15.3) can be written as $\mathbf{X} = \mathbf{A}\mathbf{P}^{-1}\mathbf{P}\mathbf{S}$, where $\mathbf{A}\mathbf{P}^{-1}$ is a new mixing matrix and $\mathbf{P}\mathbf{S}$ permutes the elements of $\mathbf{S}$. In practical terms, $\mathbf{S}$ and $\mathbf{P}\mathbf{S}$ are indistinguishable.

2. The elements of $\mathbf{A}$ (and $\mathbf{S}$) have arbitrary scaling. Multiplying S_j by an arbitrary nonzero constant c_j (i.e., increasing the amplitude of that particular signal) while dividing the jth column of $\mathbf{A}$ by the same c_j, $j = 1, 2, \ldots, m$, will not change the product $\mathbf{A}\mathbf{S}$. In other words, we cannot recover the original scalings of the source signals in $\mathbf{S}$.

3. There is an arbitrary rotational factor in the matrix $\mathbf{A}$ that cannot be resolved by just observing $\mathbf{X}$. Setting $\mathbf{A}^* = \mathbf{A}\mathbf{T}$ and $\mathbf{S}^* = \mathbf{T}^\tau \mathbf{S}$, where $\mathbf{T}$ is an orthogonal matrix, we see that $\mathbf{X}^* = \mathbf{A}^*\mathbf{S}^*$ has the same mean and covariance matrix as $\mathbf{X} = \mathbf{A}\mathbf{S}$

Thus, we should expect the columns of the separating matrix $\mathbf{W}$ to be a scaled and permuted version of the true $\mathbf{W}_0$. In practice, identifiability issues are not really serious; as long as we require at most one of the components of $\mathbf{X}$ to be Gaussian, then $\mathbf{W}$ is identifiable up to scaling and permutation of its rows, and we are able to extract the independent source components.

15.3.8 Objective Functions

The general strategy behind ICA is very similar to that of PP described in Section 7.4. Note that a projection index of PP is called an *objective* (or *contrast*) *function* in ICA. In practice, objective functions should be non-negative and equal to zero iff the projections are mutually independent. In the case of PP, interest is primarily in one- and two-dimensional (and, sometimes, three-dimensional) projections, while for ICA, we would be interested in a specified number of projections (possibly $m > 3$, depending upon context).

The same projection indexes of PP (third- and fourth-order cumulants, polynomial-based indexes, and negentropy; see Section 7.4) are often used as objective functions in ICA, especially as a means of approximating the entropy $\mathcal{H}(Y)$ of $Y = \widehat{\mathbf{w}}^\tau \mathbf{X}$. The main difficulty of using such moment-based indexes arises from their well-known lack of robustness.

Researchers working with ICA now tend to use instead objective functions based upon nonpolynomial approximations of the density function to maximize the entropy $\mathcal{H}(Y)$.

15.3.9 *Nonpolynomial-Based Approximations*

Suppose $G_i(Y)$, $i = 1, 2, \ldots, N$, are different nonpolynomial functions of Y which (like Hermite polynomials) form an orthonormal system with respect to the standard Gaussian density ϕ,

$$\int \phi(y)G_i(y)G_j(y)ds = \delta_{ij}, \tag{15.5}$$

where $\delta_{ij} = 1$ or 0 according as $i = j$ or $i \neq j$, respectively, and which are orthogonal to all polynomials of up to second order,

$$\int \phi(y)G_i(y)y^k dy = 0, \quad k = 0, 1, 2. \tag{15.6}$$

The orthogonality constraints (15.5) and (15.6) can always be satisfied by using ordinary Gram–Schmidt orthonormalization. We further assume that the expectations of the first N of the $G_i(Y)$ are given by the following values:

$$E\{G_i(Y)\} = \int G_i(y)q_Y(y)dy = c_i, \quad i = 1, 2, \ldots, N. \tag{15.7}$$

Assuming also that Y has mean 0 and variance 1 yields two more constraints,

$$
\begin{aligned}
G_{N+1}(y) &= y, \quad c_{N+1} = 0, &\tag{15.8}\\
G_{N+2}(y) &= y^2, \quad c_{N+2} = 1. &\tag{15.9}
\end{aligned}
$$

If the probability density $p_Y^0(y)$ satisfies the constraints (15.5)–(15.9) and also has the largest entropy among all such densities, then it can be shown that

$$p_Y^0(y) = A \exp\left\{\sum_i a_i G_i(y)\right\}, \tag{15.10}$$

where A and the $\{a_i\}$ are constants to be determined from (15.7). If we further assume that $p_Y(y) \approx \phi(y)$, then for (15.10) to be close to $e^{-y^2/2}$, the only substantial coefficient has to be $a_{N+2} \approx -1/2$. We can rewrite (15.10) as follows:

$$
\begin{aligned}
p_Y^0(y) &= A \exp\left\{-y^2/2 + a_{N+1}y + (a_{N+2} + 1/2)y^2 + \sum_{i=1}^{N} a_i G_i(y)\right\}\\
&= \bar{A}\,\phi(y)\left(1 + a_{N+1}y + (a_{N+2} + 1/2)y^2 + \sum_{i=1}^{N} a_i G_i(y)\right), (15.11)
\end{aligned}
$$

where $\bar{A} = (2\pi)^{1/2}A$ and where we used the approximation $e^\epsilon \approx 1 + \epsilon$. Furthermore,

$$1 = \int p_Y^0(y)dy = \bar{A}[1 + (a_{N+2} + 1/2)] \tag{15.12}$$

$$0 = \mathrm{E}\{Y\} = \int p_Y^0(y)y\,dy = \bar{A}a_{N+1} \tag{15.13}$$

$$1 = \mathrm{E}\{Y^2\} = \int p_Y^0(y)y^2\,dy = \bar{A}[1 + 3(a_{N+2} + 1/2)] \tag{15.14}$$

$$c_i = \int p_Y^0(y)G_i(y)\,dy = \bar{A}a_i, \quad i = 1, 2, \dots, N. \tag{15.15}$$

These equations are easily solved to give $a_i = c_i, i = 1, 2, \dots, N, a_{N+1} = 0, a_{N+2} = -1/2$, and $\bar{A} = 1$. Substituting these values into (15.11) yields

$$p_Y^0(y) = \phi(y)\left(1 + \sum_{i=1}^{N} c_i G_i(y)\right), \tag{15.16}$$

which is referred to as the *approximate maximum entropy density*. Compare this representation with that given by (15.10).

From (15.16), the entropy of Y, $\mathcal{H}(Y) = -\int p_Y(y)\log p_Y(y)\,dy$, can be approximated by

$$
\begin{aligned}
\mathcal{H}(Y) &\approx -\int p_Y^0(y)\log p_Y^0(y)\,dy \\
&= -\int \phi(y)\left(1 + \sum_{i=1}^{N} c_i G_i(y)\right)\log\left[\phi(y)\left(1 + \sum_{i=1}^{N} c_i G_i(y)\right)\right]dy \\
&\approx -\int \phi(y)\log\phi(y)\,dy - \sum_{i=1}^{N} c_i \int \phi(y)G_i(y)\log\phi(y)\,dy \\
&\quad - \int \phi(y)\left(1 + \sum_{i=1}^{N} c_i G_i(y)\right)\log\left(1 + \sum_{i=1}^{N} c_i G_i(y)\right)dy \\
&= \mathcal{H}(Z) - \sum_{i=1}^{N} c_i \int \phi(y)G_i(y)\log\phi(y)\,dy - \sum_{i=1}^{N} c_i \int \phi(y)G_i(y)\,dy \\
&\quad - \frac{1}{2}\sum_{i=1}^{N} c_i^2 \int \phi(y)G_i^2(y)\,dy - o\left(\sum_{i=1}^{N} c_i^2 \int \phi(y)G_i^2(y)\,dy\right) \\
&= \mathcal{H}(Z) - 0 - 0 - \frac{1}{2}\sum_{i=1}^{N} c_i^2 + o\left(\sum_{i=1}^{N} c_i^2\right), \tag{15.17}
\end{aligned}
$$

where we have used the conditions (15.5) and (15.6), the expansion $(1 + \epsilon)\log(1 + \epsilon) = \epsilon + \epsilon^2/2 + o(\epsilon^2)$ for ϵ small, and $Z \sim \mathcal{N}(0, 1)$. From (15.7) and (15.17), we have that

$$\mathcal{H}(Z) - \mathcal{H}(Y) = \mathcal{J}(Y) \approx \mathcal{J}_N(Y) \equiv \frac{1}{2}\sum_{i=1}^{N}(\mathrm{E}\{G_i(Y)\})^2. \tag{15.18}$$

All that remains now is to choose the functions $\{G_i(Y)\}$.

The simplest choice of these functions has $N = 1$ or $N = 2$. First, taking $N = 2$, we can make G_1 an odd function $(G_1(-y) = -G_1(y)$, reflecting symmetry *vs.* asymmetry) and G_2 an even function $(G_2(-y) = G_2(y)$, reflecting *sub-Gaussian* (negative kurtosis) *vs.* *super-Gaussian* (positive kurtosis) distributions). One can show that in this case, the approximation (15.18) boils down to

$$\mathcal{J}_2(Y) = \beta_1 \left(\text{E}\{G_1(Y)\}\right)^2 + \beta_2 \left(\text{E}\{G_2(Y)\} - \text{E}\{G_2(Z)\}\right)^2, \qquad (15.19)$$

where β_1 and β_2 are positive constants. If we take $N = 1$, the approximation becomes

$$\mathcal{J}_1(Y) = \beta \left(\text{E}\{G(Y)\} - \text{E}\{G(Z)\}\right)^2, \quad \beta > 0, \qquad (15.20)$$

for any nonquadratic *objective function* G, where $Z \sim \mathcal{N}(0, 1)$. So, we see that (15.20) generalizes the objective functions (7.111), where $G(Y) = Y^4$, and (7.112), where G is given by the standard Gaussian density ϕ.

The approximation (15.20) to negentropy is used in the R/S-Plus and C code implementation (Marchini, Heaton, and Ripley, 2003) of the FastICA algorithm (see Section 15.3.11), where $\beta = 1$. By choosing G carefully, we can do much better than (7.111), which is sensitive to outliers. In particular, the following choices of the G function are more robust performers:

- logcosh : $G(y) = \frac{1}{\alpha} \log \cosh(\alpha y)$, $1 \le \alpha \le 2$ (usually, $\alpha = 1$),

- exp : $G(y) = -e^{-y^2/2} = -(2\pi)^{1/2}\phi(y)$.

The logcosh function has been found to be good for most types of ICA problems, and the exp function is probably best for highly super-Gaussian source components where robustness is a serious consideration. The logcosh function has also been used successfully as a flexible family of Bayesian prior distributions, especially for the image reconstruction of photon emission computed tomographic data (Green, 1990; Weir and Green, 1994; Weir, 1997).

The exp function yields a version of $\mathcal{J}_1(Y)$ that is proportional to the objective functions $\mathcal{I}_H^0(Y)$ and $\mathcal{I}_{CBC}^0(Y)$ (see Section 7.4.1). An immediate consequence of this result is that the FastICA algorithm can be used for PP as a fast computational method for finding "interesting" one-dimensional projections of multivariate data, as well as for finding a single source component by ICA.

15.3.10   Mutual Information

The *relative entropy* or *Kullback–Leibler divergence* of a multivariate probability density p with respect to another multivariate probability density

q is defined as

$$KL(p \parallel q) = \int p(\mathbf{y}) \log \frac{p(\mathbf{y})}{q(\mathbf{y})} \, d\mathbf{y}$$

$$= -\mathcal{H}(\mathbf{Y}) - \int p(\mathbf{y}) \log q(\mathbf{y}) \, d\mathbf{y}, \qquad (15.21)$$

where $\mathcal{H}(\mathbf{Y})$ is the entropy of the vector $\mathbf{Y}$, and $-\int p(\mathbf{y}) \log q(\mathbf{y}) d\mathbf{y}$ is the *cross-entropy* between p and q (Cover and Thomas, 1991, Chapter 2). Note that Kullback–Leibler divergence is nonnegative,

$$KL(p \parallel q) = \mathrm{E}_p \left\{ \log \frac{p(\mathbf{y})}{q(\mathbf{y})} \right\}$$

$$\geq -\log \mathrm{E}_p \left\{ \frac{q(\mathbf{y})}{p(\mathbf{y})} \right\}$$

$$= -\log \left\{ \int q(\mathbf{y}) d\mathbf{y} \right\} = 0, \qquad (15.22)$$

and is zero if $p = q$. In (15.22), we used Jensen's inequality $\mathrm{E}\{f(x)\} \geq f(\mathrm{E}\{x\})$ for the convex function $f(x) = -\log(x)$, and E_p indicates expectation taken with respect to the density p. However, $KL(p \parallel q)$ is not a bona fide distance measure because it is not a symmetric function of p and q; that is, $KL(p \parallel q) \neq KL(q \parallel p)$.

We define the amount of *mutual information* ($\mathcal{MI}$) between the m components, $Y_1, \ldots, Y_m$, of $\mathbf{Y}$ by setting q in (15.22) to be the product of the marginal densities of $\mathbf{Y}$, $q(\mathbf{y}) = \prod_{j=1}^{m} p_j(y_j)$, where $p_j(y_j)$ is the (marginal) density of Y_j:

$$\mathcal{MI}(\mathbf{Y}) = KL(p \parallel \prod_j p_j)$$

$$= -\mathcal{H}(\mathbf{Y}) - \int p(\mathbf{y}) \log \left(\prod_{j=1}^{m} p_j(y_j) \right) d\mathbf{y}$$

$$= \sum_{j=1}^{m} \mathcal{H}(Y_j) - \mathcal{H}(\mathbf{Y}). \qquad (15.23)$$

Thus, mutual information can be regarded as the difference between the total amount of information carried by each of the components of $\mathbf{Y}$ and the information carried by the components jointly. $\mathcal{MI}(\mathbf{Y})$ is always nonnegative and is zero if and only if the components of $\mathbf{Y}$ are statistically independent (i.e., $p(\mathbf{y}) = \prod_j p_j(y_j)$).

In the square-mixing case (i.e., $m = r$), let $\mathbf{Y} = \mathbf{WX}$ be the m-vector of recovered source components, where $\mathbf{W} = (\mathbf{w}_i, \cdots, \mathbf{w}_m)^\tau$ minimizes the mutual information of the transformed components $\{S_j\}$. Then, the entropy of $\mathbf{Y} = (Y_1, \cdots, Y_m)^\tau$ is given by

$$\mathcal{H}(\mathbf{Y}) = \log|\det(\mathbf{W})| + \mathcal{H}(\mathbf{X}). \tag{15.24}$$

Assuming that each $Y_j = \mathbf{w}_j^\tau \mathbf{X}$ has zero mean and unit variance, $j = 1, 2, \ldots, m$, and that the $\{Y_j\}$ are uncorrelated, we have that $\mathrm{E}\{\mathbf{YY}^\tau\} = \mathbf{W}\mathbf{\Sigma}_{XX}\mathbf{W}^\tau = \mathbf{I}$, whence, $\det(\mathbf{W}) = [\det(\mathbf{\Sigma}_{XX})]^{-1/2}$, which does not depend upon $\mathbf{W}$. If $\mathbf{X}$ has been centered and sphered, (15.24) reduces to $\mathcal{H}(\mathbf{Y}) = \mathcal{H}(\mathbf{X})$. Thus, we can write (15.23) as

$$\mathcal{MI}(\mathbf{Y}) = c - \sum_{j=1}^{m} \mathcal{J}(Y_j), \tag{15.25}$$

where $c = m\mathcal{H}(Z) - \mathcal{H}(\mathbf{X})$ does not depend upon $\mathbf{W}$ and, hence, is constant (Z is a standard Gaussian variate). In terms of optimizing the mutual information between the m components of $\mathbf{Y}$ with respect to the square separating matrix $\mathbf{W}$, we see that mutual information is the negative of the sum of the negentropies of each of the $\{Y_j\}$. In other words, minimizing the mutual information between the components of $\mathbf{Y}$ is equivalent to maximizing the sum of the negentropies of the independent components of $\mathbf{Y}$.

15.3.11   The FastICA Algorithm

Let Y be a projection, $Y = \mathbf{w}^\tau \mathbf{X}$, of $\mathbf{X}$. The idea is to find that direction $\mathbf{w}$ that optimizes a given objective function. For example, if the variance of the projection, $\mathrm{var}(Y) = \mathbf{w}^\tau \mathbf{\Sigma}_{XX}\mathbf{w}$, where $\|\mathbf{w}\| = 1$, is taken as the objective function, then maximizing that function with respect to $\mathbf{w}$ yields the first principal component of $\mathbf{X}$. In this case, the solution is the eigenvector corresponding to the largest eigenvalue of $\mathbf{\Sigma}_{XX}$. Subsequent principal components can be sequentially extracted by maximizing projection variance within the orthogonal complement of the space spanned by previously derived eigenvectors. PCA is, therefore, a special case of ICA (but not vice versa), but whereas PCA obtains uncorrelated components, ICA yields independent components. Hence, sphering by PCA is typically used as a preprocessing tool in ICA algorithms.

In this section, we describe the FastICA algorithm that is popularly used for optimizing a given objective function and thereby extracting a single component or multiple independent components from $\mathbf{X}$.

Extracting a Single Source Component

First, consider a single ($m = 1$) source component $Y = \mathbf{w}^\tau \mathbf{X}$, where the r-vector $\mathbf{w}$ represents a direction for a one-dimensional projection. We

TABLE 15.1. *Nonquadratic density functions and their first and second derivatives to be used as input to the* `FastICA` *algorithm. Note that for the* `logcosh` *density,* $1 \leq \alpha \leq 2$.

Density	$G(y)$	$g(y) = G'(y)$	$g'(y) = G''(y)$
`logcosh`	$\frac{1}{\alpha}\log\cosh(\alpha y)$	$\tanh(\alpha y)$	$\alpha(1 - \tanh^2(\alpha y))$
`exp`	$-e^{-y^2/2}$	$ye^{-y^2/2}$	$(1 - y^2)e^{-y^2/2}$

wish to find that $\mathbf{w}$ that maximizes the approximation (15.20) to negentropy subject to the sphering constraint $\mathrm{E}\{(\mathbf{w}^\tau\mathbf{X})^2\} = \|\mathbf{w}\|^2 = 1$ on the projection. In other words, $\mathbf{w}$ is to be that direction that makes the density of the one-dimensional projection $Y = \mathbf{w}^\tau\mathbf{X}$ as far away from the Gaussian density as possible.

Because the maxima of the negentropy $\mathcal{J}(\mathbf{w}^\tau\mathbf{X})$ are typically obtained at certain maxima of $\mathrm{E}\{G(\mathbf{w}^\tau\mathbf{X})\}$, we set

$$F(\mathbf{w}) = \mathrm{E}\{G(\mathbf{w}^\tau\mathbf{X})\} - \frac{\lambda}{2}(\|\mathbf{w}\|^2 - 1), \tag{15.26}$$

where λ is the Lagrangian multiplier. To maximize (15.26), the Newton–Raphson iterative method (see, e.g., Thisted, 1988, Section 4.2.2) yields the iteration

$$\mathbf{w} \leftarrow \mathbf{w} - \left(\frac{\partial^2 F(\mathbf{w})}{\partial\mathbf{w}^2}\right)^{-1}\left(\frac{\partial F(\mathbf{w})}{\partial\mathbf{w}}\right). \tag{15.27}$$

We, thus, need to find the first and second partial derivatives of $F(\mathbf{w})$ with respect to $\mathbf{w}$.

Differentiating (15.26) with respect to $\mathbf{w}$ yields

$$\frac{\partial F(\mathbf{w})}{\partial\mathbf{w}} = \mathrm{E}(\mathbf{X}g(\mathbf{w}^\tau\mathbf{X})) - \lambda\mathbf{w}, \tag{15.28}$$

where $g = \partial G/\partial\mathbf{w}$. The stationary values of the function F are found by equating (15.28) to zero. Premultiplying both sides of the resulting equation by $\mathbf{w}^\tau$ yields

$$\lambda = \mathrm{E}(\mathbf{w}^\tau\mathbf{X}g(\mathbf{w}^\tau\mathbf{X})). \tag{15.29}$$

Differentiating (15.28) with respect to $\mathbf{w}$ gives the approximate second derivative of F,

$$\begin{aligned}\frac{\partial^2 F(\mathbf{w})}{\partial\mathbf{w}^2} &= \mathrm{E}\{(\mathbf{X}\mathbf{X}^\tau g'(\mathbf{w}^\tau\mathbf{X})\} - \lambda\mathbf{I}_r \\ &\approx \mathrm{E}\{(\mathbf{X}\mathbf{X})^\tau\}\mathrm{E}\{g'(\mathbf{w}^\tau\mathbf{X})\} - \lambda\mathbf{I}_r \\ &= (\mathrm{E}\{g'(\mathbf{w}^\tau\mathbf{X})\} - \lambda)\mathbf{I}_r, \end{aligned} \tag{15.30}$$

TABLE 15.2. `FastICA` *algorithm for determining a single source component.*

---

1. Center and whiten the data to give $\mathbf{X}$.

2. Choose an initial version of the r-vector $\mathbf{w}$ with unit norm.

3. Choose G to be any nonquadratic density with first and second partial derivatives g and g', respectively. If the choice is either the `logcosh` or `exp` density, g and g' are given in the text.

4. Let $\mathbf{w} \leftarrow \mathrm{E}(\mathbf{X}g(\mathbf{w}^{\tau}\mathbf{X})) - \mathbf{w}\mathrm{E}(g'(\mathbf{w}^{\tau}\mathbf{X}))$. In practice, the expectations are estimated using sample averages.

5. Let $\mathbf{w} \leftarrow \mathbf{w}/\|\mathbf{w}\|$.

6. Iterate between steps 4 and 5. Stop when convergence is attained.

---

where we used the fact that $\mathbf{X}$ has been sphered. Substituting (15.28) and (15.30) into (15.27), the iteration reduces to

$$\mathbf{w} \leftarrow \mathbf{w} - \frac{\mathrm{E}\{\mathbf{X}g(\mathbf{w}^{\tau}\mathbf{X})\} - \lambda\mathbf{w}}{\mathrm{E}\{g'(\mathbf{w}^{\tau}\mathbf{X})\} - \lambda} \ . \tag{15.31}$$

If we set $E_1 = \mathrm{E}\{\mathbf{X}g(\mathbf{w}_{k-1}^{\tau}\mathbf{X})\}$ and $E_2 = \mathrm{E}\{g'(\mathbf{w}_{k-1}^{\tau}\mathbf{X})\}$, then (15.31) can be written as $\mathbf{w}_k = \mathbf{w}_{k-1} - (E_1 - \lambda\mathbf{w}_{k-1})/(E_2 - \lambda)$ for the kth iteration. Multiplying both sides by $\lambda - E_2$ yields $\mathbf{w}_k(\lambda - E_2) = E_1 - \mathbf{w}_{k-1}E_2$. Because we divide $\mathbf{w}$ by its norm $\|\mathbf{w}\|$ at each step of the iterative procedure, the factor $\lambda - E_2$ can be ignored. The iteration (15.31) is, therefore, equivalent to

$$\mathbf{w} \leftarrow \mathrm{E}\{\mathbf{X}g(\mathbf{w}^{\tau}\mathbf{X})\} - \mathbf{w}\mathrm{E}\{g'(\mathbf{w}^{\tau}\mathbf{X})\}. \tag{15.32}$$

For the `logcosh` and `exp` densities, the functions g and g' are given in Table 15.1. Substituting for g and g' in (15.32) for either the `logcosh` or `exp` density as appropriate yields the `FastICA` algorithm, which is given in Table 15.2.

The values of $\mathbf{w}$ can change substantially from iteration to iteration; this is because the ICA model cannot determine the sign of $\mathbf{w}$, so that $-\mathbf{w}$ and $\mathbf{w}$ become equivalent and define the same direction. In light of this comment, "convergence" of the `FastICA` algorithm is taken to have a different meaning than usual, and is taken here to mean that successive iterative values of $\mathbf{w}$ (i.e., $\mathbf{w}_{k-1}$ and $\mathbf{w}_k$ for some k) are oriented in the same direction (i.e., $\mathbf{w}_k^{\tau}\mathbf{w}_{k-1}$ is very close to 1).

Extracting Multiple Source Components

The `FastICA` package (Hurri, Gävert, Särelä, and Hyvärinen, 1998) includes two different ways of extracting more than one independent source component. Both methods (termed "deflation" and "parallel" methods) repeatedly call the single component extraction algorithm of Table 15.2. Essentially, at each step in the algorithic cycle:

deflation: the single component routine finds a new component, that new component is orthogonalized using the Gram–Schmidt method with respect to all previously found components, and then the resulting new component is normalized.

parallel: the single component routine is carried out in parallel for each independent component to be extracted, and then a symmetric orthogonalization is carried out on all components simultaneously.

The `deflation` method extracts independent components sequentially one-at-a-time, whereas the `parallel` method extracts all the independent components at the same time. Both algorithms are listed in Table 15.3. Note that the `parallel` algorithm is used for minimizing mutual information $\mathcal{MI}(\mathbf{Y})$ because the `deflation` algorithm is not appropriate.

15.3.12 Example: Identifying Artifacts in MEG Recordings

Brain signals are very weak electrical signals. Neurons located in the brain conduct electrical activity, which, in turn, produces magnetic fields. Because magnetic signals pass unchanged through brain tissue and the skull, they can be recorded outside the head and used to identify the locations of brain activity. A MEG device is used for real-time mapping of changes in the magnetic field caused by brain activity. However, such recordings often contain artifacts due to external disturbances such as eye movements or blinks, or sensory malfunctions. It is, therefore, advisable to detect, identify, and remove such artifacts from the records. In this example, we discuss the issue of separating artifacts from true brain activity. The primary assumption here is that artifacts are an anatomically and physiologically separate process from brain activity, so that, statistically, the two types of magnetic signals generated by such processes can be considered to be independent.

In a noninvasive experiment carried out by the ICA group at the Helsinki University of Technology (Vigário, Jousmäki, Hämäläinen, Hari, and Oja, 1997), the MEG signals of a test subject were recorded in a magnetically shielded room. Measurements were taken using a whole-scalp neuromagnetometer (a helmet-shaped device; see Figure 15.3) with 122 SQUID (superconducting quantum interference device) sensors organized in pairs at 61 grid locations uniformly distributed around the head. The weak magnetic fields produced by brain activity are detected by these sensors. The

TABLE 15.3. *Two* `FastICA` *algorithms for extracting multiple indepen-dent source components.*

Deflation algorithm

1. Center and whiten the data to give $\mathbf{X}$.

2. Decide on the number, m, of independent components to be extracted.

3. For $k = 1, 2, \ldots, m,$

   - Initialize (e.g., randomly) the r-vector $\mathbf{w}_k$ to have unit norm.

   - Let $\mathbf{w}_k \leftarrow \mathrm{E}(\mathbf{X}g(\mathbf{w}_k^\top\mathbf{X})) - \mathbf{w}_k\mathrm{E}(g'(\mathbf{w}_k^\top\mathbf{X}))$ be the `FastICA` single-component update for $\mathbf{w}_k$, where g and g' are given in Table 15.1. In practice, the expectations are estimated using sample averages.

   - Use the Gram–Schmidt process to orthogonalize $\mathbf{w}_k$ with respect to the previously chosen $\mathbf{w}_1, \ldots, \mathbf{w}_{k-1}$:

$$\mathbf{w}_k \leftarrow \mathbf{w}_k - \sum_{j=1}^{k-1}(\mathbf{w}_k^\top\mathbf{w}_j)\mathbf{w}_j.$$

   - Let $\mathbf{w}_k \leftarrow \mathbf{w}_k/\|\mathbf{w}_k\|$.

   - Iterate $\mathbf{w}_k$ until convergence.

4. Set $k \leftarrow k + 1$. If $k \leq m$, return to step 3.

Parallel algorithm

1. Center and whiten the data to give $\mathbf{X}$.

2. Decide on the number, m, of independent components to be extracted.

3. Initialize (e.g., randomly) the r-vectors $\mathbf{w}_1, \ldots, \mathbf{w}_m$, each to have unit norm. Let $\mathbf{W} = (\mathbf{w}_1, \cdots, \mathbf{w}_m)^\top$.

4. Carry out a symmetric orthogonalization of $\mathbf{W}$ by $\mathbf{W} \leftarrow (\mathbf{W}\mathbf{W}^\top)^{-1/2}\mathbf{W}$.

5. For each $k = 1, 2, \ldots, m$, let $\mathbf{w}_k \leftarrow \mathrm{E}(\mathbf{X}g(\mathbf{w}_k^\top\mathbf{X})) - \mathbf{w}_k\mathrm{E}(g'(\mathbf{w}_k^\top\mathbf{X}))$ be the `FastICA` single-component update for $\mathbf{w}_k$, where g and g' are given in Table 15.1. In practice, the expectations are estimated using sample averages.

6. Carry out another symmetric orthogonalization of $\mathbf{W}$.

7. If convergence has not occurred, return to step 5.

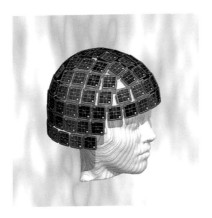

FIGURE 15.3. *Helmet-shaped device with array of sensors uniformly distributed around the head to provide MEG measurements. Source:* `ltl.tkk.fi/research/brain/head.jpg`

MEG signals were deliberately contaminated by having the test subject induce the following artifacts: (1) blink his eyes; (2) make horizontal saccades (quick, simultaneous movements of both eyes at the same time in the same direction) to simulate typical ocular (eye) artifacts; and (3) bite his teeth for as long as 20 seconds to simulate myographic (muscle) artifacts. Two more artifacts were added: (4) a piece of metal was placed next to the navel to simulate breathing artifacts; and (5) a digital watch was placed one meter away from the helmet in the shielded room to simulate a general artifact.

The data consist of $n = 17,730$ amplitudes of each of $r = 122$ MEG signals recorded over a period of 2 minutes.[2] A sample of 12 of these signals is displayed in Figure 15.4. We first used PCA to convert the MEG data into principal components with decreasing variance; see Figure 15.5 for a scree plot of the eigenvalues. Because we used the sample correlation matrix for PCA, we retained only those PCs whose eigenvalues were greater than unity, which also corresponded to an "elbow" in Figure 15.5. This reduced the 122-dimensional data to 22 PCs, which accounted for 77.8% of total variance. Next, we extracted 22 independent components from the PCs (using the parallel FASTICA algorithm). The 22 ICs are displayed in Figure 15.6.

We see certain patterns in the ICs. Counting from the top of Figure 15.6, IC1–IC10 (purple curves) show low-fequency, bump-like, overlearning artifacts (Särelä and Vigario, 2003); IC11 and IC12 (light-blue curves) show

---

[2]The data are publicly available and can be downloaded from the website `www.cis.hut.fi/projects/ica/eegmeg/MEG_data.html`.

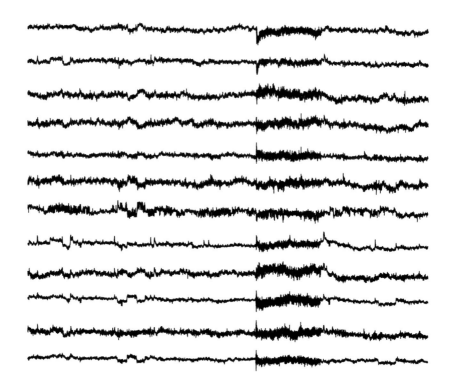

FIGURE 15.4. *Spontaneous MEG signals for a sample of 12 channels (locations) of a 122-channel whole-scalp neuromagnetometer over the frontal, temporal, and occipital areas of a test subject's scalp. Artifacts were introduced by saccades, blinking, and biting, in that order.*

horizontal eye movements, and IC13 and IC14 (green curves) show eye blinks; IC15 (red curve) represents a cardiac cycle artifact, and IC16 (dark-blue curve) shows the digital watch artifact, both signals of which are not visible in the raw data; and IC17 and IC18 (orange curves) correspond to the muscle (biting) artifact. The remaining four signals reflect noise components.

15.3.13 Maximum-Likelihood ICA

Another way of carrying out ICA is to specify a parametric distribution, $p_\mathbf{S}(\mathbf{s})$, for the latent source variables $\mathbf{S}$ and then apply the maximum-likelihood (ML) method to estimate the parameters of that distribution. In

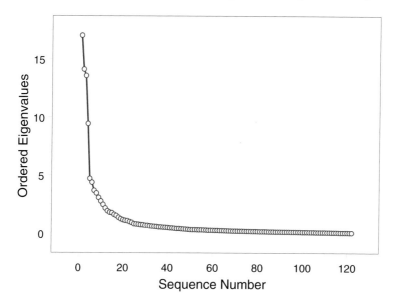

FIGURE 15.5. *Scree plot of the 122 ordered eigenvalues (variances) of the sample correlation matrix computed from the MEG data.*

this section, we describe a fixed-point algorithm (which utilizes the `FastICA` algorithm) for square mixing ($m = r$).

Suppose the density of the m-vector $\mathbf{S} = (S_1, \cdots, S_m)$ of sources is $p_{\mathbf{S}}(\mathbf{s})$, and suppose $\mathbf{X} = \mathbf{A}\mathbf{S}$, where $\mathbf{A}$ is square and nonsingular. Let $\mathbf{W} = \mathbf{A}^{-1}$. Then, the density of $\mathbf{X}$ is $p_{\mathbf{X}}(\mathbf{x}) = |\det(\mathbf{W})| p_{\mathbf{S}}(\mathbf{s})$. Because the sources are assumed to be statistically independent, then,

$$p_{\mathbf{X}}(\mathbf{x}) = |\det(\mathbf{W})| \prod_{j=1}^{m} p_{S_j}(\mathbf{w}_j^\tau \mathbf{x}), \tag{15.33}$$

where $p_{S_j}(s_j)$ is the density of S_j and $\mathbf{w}_j^\tau$ is the jth row of $\mathbf{W}$. Given n i.i.d. observations, $\mathbf{X}_1 \ldots, \mathbf{X}_n$, on $\mathbf{X}$, the log-likelihood function for $\mathbf{W}$ is

$$n^{-1} \log \mathcal{L}(\mathbf{W}|\{\mathbf{X}_i\}) = \log |\det(\mathbf{W})| + n^{-1} \sum_{i=1}^{n} \left\{ \sum_{j=1}^{m} \log p_{S_j}(\mathbf{w}_j^\tau \mathbf{X}_i) \right\}. \tag{15.34}$$

In this case, the parameters are the elements of $\mathbf{W}$. To find the ML estimator of $\mathbf{W}$, we derive a fixed-point algorithm that will maximize (15.34) numerically.

For convenience, we write "E" in the second term on the right-hand side of (15.34) for the sample average over the n observations. The derivative

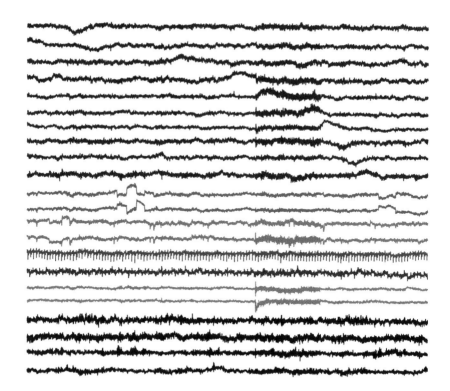

FIGURE 15.6. *Twenty-two independent components (ICs) extracted from the MEG data. Visible in this display are the saccades (IC11, IC12; light-blue curves), blinking (IC13, IC14; green curves), digital watch (IC16; dark-blue curve), and biting (IC17, IC18; orange curves) artifacts, and also cardiac cycle (IC15; red curve) and bump-like overlearning or breathing (IC1–IC10; purple curves) artifacts. The last four ICs are noise components.*

of $\log \mathcal{L}(\mathbf{W})$ with respect to $\mathbf{W}$ is given by the matrix gradient,

$$
\begin{aligned}
n^{-1} \frac{\partial \log \mathcal{L}(\mathbf{W})}{\partial \mathbf{W}} &= (\mathbf{W}^\tau)^{-1} + \mathrm{E}\left\{\frac{\partial}{\partial \mathbf{W}} \sum_{j=1}^{m} \log p_{S_j}(\mathbf{w}_j^\tau \mathbf{X})\right\} \\
&= (\mathbf{W}^\tau)^{-1} + \mathrm{E}\left\{(g_1(\mathbf{w}_1^\tau \mathbf{X}), \cdots, g_m(\mathbf{w}_m^\tau \mathbf{X})) \mathbf{X}^\tau\right\} \\
&= (\mathbf{W}^\tau)^{-1} + \mathrm{E}\{\mathbf{g}(\mathbf{W}\mathbf{X})\mathbf{X}^\tau\}, \quad\quad (15.35)
\end{aligned}
$$

where

$$
\mathbf{g}(\mathbf{W}\mathbf{X}) = (g_1(\mathbf{w}_1^\tau \mathbf{X}), \cdots, g_m(\mathbf{w}_m^\tau \mathbf{X})) \quad\quad (15.36)
$$

and

$$
g_j(\mathbf{w}_j^\tau \mathbf{X}) = \frac{p'_{S_j}(\mathbf{w}_j^\tau \mathbf{X})}{p_{S_j}(\mathbf{w}_j^\tau \mathbf{X})} \qu\quad\quad (15.37)
$$

is the jth score function. The update rule for the kth iterate of $\mathbf{W}$ is

$$\mathbf{W}_k = \mathbf{W}_{k-1} - \alpha \left. \frac{\partial \log \mathcal{L}(\mathbf{W})}{\partial \mathbf{W}} \right|_{\mathbf{W}=\mathbf{W}_{k-1}}, \tag{15.38}$$

where α is the step-size parameter of the optimization rule, depending upon n and possibly k. Setting $\Delta\mathbf{W} = \mathbf{W}_k - \mathbf{W}_{k-1}$ as the difference between successive iterates of $\mathbf{W}$, and using (15.35), we can write (15.38) in the form

$$\Delta\mathbf{W} \propto (\mathbf{W}^\tau)^{-1} + \mathrm{E}\{\mathbf{g}(\mathbf{W}\mathbf{X})\mathbf{X}^\tau\}. \tag{15.39}$$

Postmultiplying the right-hand side of (15.39) by $\mathbf{W}^\tau\mathbf{W}$ gives the fixed-point algorithm,

$$\mathbf{W} \leftarrow \mathbf{W} + \mu[\mathbf{I}_m + \mathrm{E}\{\mathbf{g}(\mathbf{Y})\mathbf{Y}^\tau\}]\mathbf{W}, \tag{15.40}$$

where $\mathbf{Y} = \mathbf{W}\mathbf{X}$ and μ is the *learning rate*, which may be reduced in size until convergence. This modification produces an algorithm that avoids the matrix inversions of (15.39) and speeds up convergence considerably.

Hyvärinen (1999) recognized that (15.40) is really just a special case of the FastICA algorithm. The link between the two algorithms can be seen if we write Step 5 of the parallel FastICA algorithm in Table 15.3 in matrix form as

$$\mathbf{W} \leftarrow \mathbf{W} + \mathbf{D}_\alpha \left[\mathbf{D}_\lambda - \mathrm{E}\{\mathbf{g}(\mathbf{Y})\mathbf{Y}^\tau\} \right] \mathbf{W}, \tag{15.41}$$

where $\mathbf{Y} = (Y_1, \cdots, Y_m)^\tau$, $Y_i = \mathbf{w}_i^\tau\mathbf{X}$, $\lambda_i = \mathrm{E}\{Y_i g(Y_i)\}$, $\alpha_i = 1/(\mathrm{E}\{g'(Y_i) - \lambda_i\}$, $\mathbf{D}_\alpha = \mathrm{diag}\{\alpha_i\}$, and $\mathbf{D}_\lambda = \mathrm{diag}\{\lambda_i\}$. The second term on the right-hand side of (15.41) can be rearranged to give

$$\mathbf{W} \leftarrow \mathbf{W} + \mathbf{D}_{\alpha\lambda}[\mathbf{I}_m - \mathbf{D}_\lambda^{-1}\mathrm{E}\{\mathbf{g}(\mathbf{Y})\mathbf{Y}^\tau\}]\mathbf{W}, \tag{15.42}$$

where $\mathbf{D}_{\alpha\lambda} = \mathrm{diag}\{\alpha_i\lambda_i\}$ and $\mathbf{D}_\lambda^{-1} = \mathrm{diag}\{\lambda_i^{-1}\}$. Thus, the FastICA algorithm as given in Table 15.4 can be interpreted as maximizing the likelihood (15.34), thereby directly obtaining the ML estimate of $\mathbf{W}$. Comparing (15.42) with (15.40), we see that the scalar learning rate μ has now become a more flexible part of the iterative process. Furthermore, simulation studies have demonstrated that careful choice of $\{\alpha_i\}$ and $\{\lambda_i\}$ can speed up convergence of the FastICA algorithm.

15.3.14   Kernel ICA

A radically different approach to ICA was developed by Bach and Jordan (2002). Their approach, which they call Kernel ICA, still involves building an appropriate objective function and then optimizing that objective function using a numerical algorithm. The difference between the Kernel ICA approach and those of the more "traditional" approaches described in

TABLE 15.4. `FastICA` *algorithm for obtaining the maximum likelihood estimate of a square separating matrix* $\mathbf{W}$.

---

1. Center the data, and then sphere the result to give $\mathbf{X}$.

2. Decide on the number, m, of independent components to be extracted.

3. Randomly initialize a separating matrix $\mathbf{W}$.

4. Compute $\mathbf{Y} = \mathbf{W}\mathbf{X}$.

5. Compute $\lambda_i = \mathrm{E}(Y_i g(Y_i))$, $\alpha_i = 1/(\mathrm{E}(g'(Y_i)) - \lambda_i)$, $i = 1, 2, \ldots, m$. The function g is usually taken to be the `tanh` function (see Table 15.1). Set $\mathbf{D}_\alpha = \mathrm{diag}\{\alpha_i\}$ and $\mathbf{D}_\lambda = \mathrm{diag}\{\lambda_i\}$.

6. Update $\mathbf{W}$ by $\mathbf{W} \leftarrow \mathbf{W} + \mathbf{D}_\alpha \left[\mathbf{D}_\lambda - \mathrm{E}(\mathbf{g}(\mathbf{Y})\mathbf{Y}^\tau)\right] \mathbf{W}$. In practice, the expectation is estimated using a sample average.

7. Carry out a symmetric orthogonalization of $\mathbf{W}$ by $\mathbf{W} \leftarrow (\mathbf{W}\mathbf{W}^\tau)^{-1/2}\mathbf{W}$.

8. If convergence has not occurred, return to step 4.

---

this chapter is that the development consists of searching the functions in a reproducing kernel Hilbert space. This approach reduces to finding the eigenvalues and eigenvectors of a certain matrix, which we show is derived from a kernelized version of CVA.

Kernel CVA

The CVA method of Section 7.3 has been generalized to the nonlinear case using similar ideas as were developed for support vector machines. (We will see nonlinear PCA in Chapter 16.) The resulting methodology has been applied to problems as varied as that of extracting correlated gene clusters from multiple genomic data to cross-language latent semantic indexing. In many multivariate applications, the standard CVA method will not be feasible if the dimensionality of the problem is too large or if the data cannot be represented as vectors.

The nonlinear version of CVA that we describe here assumes that we carry out a nonlinear transformation, $\Phi_1 : \Re^r \to \mathcal{H}_1$, of one set of input data, $\mathbf{X}_i \in \Re^r$, $i = 1, 2, \ldots, n$, and another nonlinear transformation, $\Phi_2 : \Re^s \to \mathcal{H}_2$, of a second set of input data, $\mathbf{Y}_i \in \Re^s$, $j = 1, 2, \ldots, n$.

CVA in Feature Space

We now carry out CVA between the two transformed sets of input data, $\{\Phi_1(\mathbf{X}_i), i = 1, 2, \ldots, n\}$ and $\{\Phi_2(\mathbf{Y}_i), i = 1, 2, \ldots, n\}$, where we assume that both sets of transformed data have been centered. We wish to find

$f_1 \in \mathcal{H}_1$ and $f_2 \in \mathcal{H}_2$ such that the features $f_1(\mathbf{X}) = \langle \Phi_1(\mathbf{X}), f_1 \rangle$ and $f_2(\mathbf{Y}) = \langle \Phi_2(\mathbf{Y}), f_2 \rangle$ have maximal correlation.

We search for f_1 and f_2 in the linear spaces, $\mathcal{S}_1$ and $\mathcal{S}_2$, respectively, which are spanned by these Φ-images. These are reproducing kernel Hilbert spaces (rkhs). For a given f_1, f_2, we can write

$$f_1 = \sum_{i=1}^{n} \alpha_{1i} \Phi_1(\mathbf{X}_i) + f_1^{\perp}, \quad f_2 = \sum_{i=1}^{n} \alpha_{2i} \Phi_2(\mathbf{Y}_i) + f_2^{\perp}, \tag{15.43}$$

where $f_1^{\perp}$ and $f_2^{\perp}$ are orthogonal to $\mathcal{S}_1$ and $\mathcal{S}_2$, respectively. Then, we can write $f_1(\mathbf{X}) = \langle \Phi_1(\mathbf{X}), f_1 \rangle = \sum_{i=1}^{n} \alpha_{1i} \langle \Phi_1(\mathbf{X}), \Phi_1(\mathbf{X}_i) \rangle$ and $f_2(\mathbf{Y}) = \langle \Phi_2(\mathbf{Y}), f_2 \rangle = \sum_{i=1}^{n} \alpha_{2i} \langle \Phi_2(\mathbf{Y}), \Phi_2(\mathbf{Y}_i) \rangle$.

We could maximize the covariance of $f_1(\mathbf{x})$ and $f_2(\mathbf{y})$ subject to constraints on the variances as we did previously. However, we consider instead the equivalent problem of maximizing the (canonical) $\mathcal{H}$-correlation ($\mathcal{H} = \mathcal{H}_1 \times \mathcal{H}_2$) between $f_1(\mathbf{X})$ and $f_2(\mathbf{X})$ as defined by

$$\widehat{\rho}_{\mathcal{H}}(\mathbf{X}, \mathbf{Y}) = \max_{(f_1, f_2) \in \mathcal{H}_1 \times \mathcal{H}_2} \frac{\widehat{\mathrm{cov}}\{f_1(\mathbf{X}), f_2(\mathbf{Y})\}}{(\widehat{\mathrm{var}}\{f_1(\mathbf{X})\})^{1/2} (\widehat{\mathrm{var}}\{f_2(\mathbf{Y})\})^{1/2}}, \tag{15.44}$$

where

$$
\begin{aligned}
\widehat{\mathrm{cov}}\{f_1(\mathbf{X}), f_2(\mathbf{Y})\} &= n^{-1} \sum_{i=1}^{n} f_1(\mathbf{X}_i) f_2(\mathbf{Y}_i) \\
&= n^{-1} \sum_{i=1}^{n} \langle \Phi_1(\mathbf{X}_i), f_1 \rangle \langle \Phi_2(\mathbf{Y}_i), f_2 \rangle \\
&= n^{-1} \sum_{i=1}^{n} \sum_{j=1}^{n} \sum_{k=1}^{n} \alpha_{1j} K_1(\mathbf{X}_i, \mathbf{X}_j) K_2(\mathbf{Y}_i, \mathbf{Y}_k) \alpha_{2k} \\
&= n^{-1} \boldsymbol{\alpha}_1^{\tau} \mathbf{K}_1 \mathbf{K}_2 \boldsymbol{\alpha}_2 && (15.45) \\
\widehat{\mathrm{var}}\{f_1(\mathbf{X})\} &= n^{-1} \boldsymbol{\alpha}_1^{\tau} \mathbf{K}_1^2 \boldsymbol{\alpha}_1 && (15.46) \\
\widehat{\mathrm{var}}\{f_2(\mathbf{Y})\} &= n^{-1} \boldsymbol{\alpha}_2^{\tau} \mathbf{K}_2^2 \boldsymbol{\alpha}_2, && (15.47)
\end{aligned}
$$

$\boldsymbol{\alpha}_1 = (\alpha_{11}, \cdots, \alpha_{1n})^{\tau}$, $\boldsymbol{\alpha}_2 = (\alpha_{21}, \cdots, \alpha_{2n})^{\tau}$, and the matrices $\mathbf{K}_1$ and $\mathbf{K}_2$ are the $(n \times n)$ Gram matrices associated with $\{\mathbf{X}_i, i = 1, 2, \ldots, n\}$ and $\{\mathbf{Y}_i, i = 1, 2, \ldots, n\}$, respectively. The kernelized version of the CVA problem is, therefore, given by

$$\widehat{\rho}_{\mathcal{H}}(\mathbf{K}_1, \mathbf{K}_2) = \max_{\boldsymbol{\alpha}_1, \boldsymbol{\alpha}_2 \in \Re^n} \frac{\boldsymbol{\alpha}_1^{\tau} \mathbf{K}_1 \mathbf{K}_2 \boldsymbol{\alpha}_2}{(\boldsymbol{\alpha}_1^{\tau} \mathbf{K}_1^2 \boldsymbol{\alpha}_1)^{1/2} (\boldsymbol{\alpha}_2^{\tau} \mathbf{K}_2^2 \boldsymbol{\alpha}_2)^{1/2}}. \tag{15.48}$$

Differentiating (15.48) with respect to $\boldsymbol{\alpha}_1$ and $\boldsymbol{\alpha}_2$ and then setting the results equal to zero yields the generalized eigenequation,

$$\mathcal{K}\boldsymbol{\alpha} = \lambda \mathcal{D}\boldsymbol{\alpha}, \tag{15.49}$$

where

$$
\mathcal{K} = \begin{pmatrix} \mathbf{0} & \mathbf{K}_1\mathbf{K}_2 \\ \mathbf{K}_2\mathbf{K}_1 & \mathbf{0} \end{pmatrix}, \quad \mathcal{D} = \begin{pmatrix} \mathbf{K}_1^2 & \mathbf{0} \\ \mathbf{0} & \mathbf{K}_2^2 \end{pmatrix}, \quad \boldsymbol{\alpha} = \begin{pmatrix} \boldsymbol{\alpha}_1 \\ \boldsymbol{\alpha}_2 \end{pmatrix}. \quad (15.50)
$$

The problem with this eigenequation is that $\mathcal{D}$ will be singular because centering renders both Gram matrices, $\mathbf{K}_1$ and $\mathbf{K}_2$, singular. It also turns out that all pairs of "kernel canonical variates" in feature space will be perfectly correlated, which will happen even if the non-centered $\mathbf{K}_1$ and $\mathbf{K}_2$ are invertible. As it stands, then, this "naive" kernel method cannot provide us with a useful estimate of the population canonical correlation, $\rho_{\mathcal{H}}(\mathbf{X}, \mathbf{Y})$.

Regularization

One way out of this predicament is to apply *regularization* to the problem. This solution is in the same spirit as ridge regression and smoothing in functional CVA (Leurgans, Moyeed, and Silverman, 1993). In this case, penalizing the $\mathcal{H}_1$-norm of f_1 and the $\mathcal{H}_2$-norm of f_2 each by the same small constant value $\kappa > 0$ means replacing $\mathbf{K}_1^2$ by $(\mathbf{K}_1 + \kappa\mathbf{I}_n)^2$ and $\mathbf{K}_2^2$ by $(\mathbf{K}_2 + \kappa\mathbf{I}_n)^2$ in the definition of $\mathcal{D}$ in (15.50). This can be seen as follows: if θ is a regularization parameter, then,

$$
\begin{aligned}
\widehat{\mathrm{var}}\{f_1(\mathbf{X}_i)\} + \theta\|f_1\|_{\mathcal{H}_1}^2 &= n^{-1}\boldsymbol{\alpha}_1^\tau\mathbf{K}_1^2\boldsymbol{\alpha}_1 + \theta\boldsymbol{\alpha}_1^\tau\mathbf{K}_1\boldsymbol{\alpha}_1 \\
&\approx n^{-1}\boldsymbol{\alpha}_1^\tau(\mathbf{K}_1 + \kappa\mathbf{I}_n)^2\boldsymbol{\alpha}_1 \quad (15.51) \\
\widehat{\mathrm{var}}\{f_2(\mathbf{Y}_j)\} + \theta\|f_2\|_{\mathcal{H}_1}^2 &\approx n^{-1}\boldsymbol{\alpha}_2^\tau(\mathbf{K}_2 + \kappa\mathbf{I}_n)^2\boldsymbol{\alpha}_2, \quad (15.52)
\end{aligned}
$$

where $\kappa = n\theta/2$ (Bach and Jordan, 2002).

The regularized version of (15.48) is given by

$$
\widehat{\rho}_{\mathcal{H}}(\mathbf{K}_1, \mathbf{K}_2) =
$$

$$
\max_{\boldsymbol{\alpha}_1, \boldsymbol{\alpha}_2 \in \Re^n} \frac{\boldsymbol{\alpha}_1^\tau\mathbf{K}_1\mathbf{K}_2\boldsymbol{\alpha}_2}{(\boldsymbol{\alpha}_1^\tau(\mathbf{K}_1 + \kappa\mathbf{I}_n)^2\boldsymbol{\alpha}_1)^{1/2}(\boldsymbol{\alpha}_2^\tau(\mathbf{K}_2 + \kappa\mathbf{I}_n)^2\boldsymbol{\alpha}_2)^{1/2}}. \quad (15.53)
$$

We see in (15.53) that the covariance term in the numerator is to be compared with the variance and the penalty function of each term in the denominator. The value of κ determines the weight to be placed upon the penalty terms compared with the variance terms. As κ gets close to zero, the variance term dominates, whereas as κ gets larger, the variance term becomes more affected by the amount of roughness allowed by the penalty term. Some careful compromise is needed here when deciding upon the value of κ.

Differentiating (15.53) with respect to $\boldsymbol{\alpha}_1$ and $\boldsymbol{\alpha}_2$ and then setting the results equal to zero yields two equations, which can be written in matrix form as

$$
\mathcal{K}\boldsymbol{\alpha} = \lambda\mathcal{D}_\kappa\boldsymbol{\alpha}, \quad (15.54)
$$

where $\mathcal{K}$ is given by (15.50),

$$\mathcal{D}_\kappa = \begin{pmatrix} (\mathbf{K}_1 + \kappa \mathbf{I}_n)^2 & \mathbf{0} \\ \mathbf{0} & (\mathbf{K}_2 + \kappa \mathbf{I}_n)^2 \end{pmatrix}, \qquad (15.55)$$

and $\boldsymbol{\alpha}$ is given by (15.50). This is a generalized eigenequation, which has $2n$ paired eigenvalues

$$\{\lambda_1, -\lambda_1, \ldots, \lambda_n, -\lambda_n\}, \qquad (15.56)$$

each of which lies between -1 and 1. The first eigenvalue, λ_1, is the largest canonical correlation. The equation (15.54) can be written in the alternate form,

$$\mathcal{K}_\kappa \boldsymbol{\alpha} = (1 + \lambda) \mathcal{D}_\kappa \boldsymbol{\alpha}, \qquad (15.57)$$

where

$$\mathcal{K}_\kappa = \begin{pmatrix} (\mathbf{K}_1 + \kappa \mathbf{I}_n)^2 & \mathbf{K}_1 \mathbf{K}_2 \\ \mathbf{K}_2 \mathbf{K}_1 & (\mathbf{K}_2 + \kappa \mathbf{I}_n)^2 \end{pmatrix}, \qquad (15.58)$$

$\mathcal{D}_\kappa$ is given by (15.55) and $\boldsymbol{\alpha}$ is given by (15.50). Equation (15.57) has paired eigenvalues $\{1 + \lambda_1, 1 - \lambda_1, \ldots, 1 + \lambda_n, 1 - \lambda_n\}$.

Note that (15.57) can be expressed as a standard eigenproblem,

$$\widetilde{\mathcal{K}}_\kappa \widetilde{\boldsymbol{\alpha}} = \widetilde{\lambda} \widetilde{\boldsymbol{\alpha}}, \qquad (15.59)$$

where $\widetilde{\mathcal{K}}_\kappa = \mathcal{D}_\kappa^{-1/2} \mathcal{K}_\kappa \mathcal{D}_\kappa^{-1/2}$, $\widetilde{\boldsymbol{\alpha}} = \mathcal{D}_\kappa^{-1/2} \boldsymbol{\alpha}$, and $\widetilde{\lambda} = 1 + \lambda$. We are, therefore, interested in the eigenvalues and eigenvectors of the $(2n \times 2n)$-matrix

$$\widetilde{\mathcal{K}}_\kappa = \begin{pmatrix} \mathbf{I}_n & \mathbf{K}_1^\kappa \mathbf{K}_2^\kappa \\ \mathbf{K}_2^\kappa \mathbf{K}_1^\kappa & \mathbf{I}_n \end{pmatrix}, \qquad (15.60)$$

where $\mathbf{K}_j^\kappa = (\mathbf{K}_j + \kappa \mathbf{I}_n)^{-1} \mathbf{K}_j$, $j = 1, 2$.

The kernel canonical variate scores are then given by $f_1(\mathbf{X}) = \mathbf{K}_1 \boldsymbol{\alpha}_1$ and $f_2(\mathbf{Y}) = \mathbf{K}_2 \boldsymbol{\alpha}_2$.

Choice of Parameter Values

It is important that the parameters in the eigenproblem be chosen carefully. There are two "free" parameters that have to be chosen by the user:

1. Bach and Jordan (2002) recommend that the regularization parameter θ be set to $\theta = 2 \times 10^{-3}$ for $n > 1,000$ and $\theta = 2 \times 10^{-2}$ for $n \le 1,000$. Leurgans, Moyeed, and Silverman (1993), in a slightly different context, consider cross-validation as a method for determining a good choice of θ; they found, however, that cross-validation works much better for the leading canonical variate than it does for subsequent canonical variates.

2. If a Gaussian radial basis kernel is used as the kernel in this method, Bach and Jordan recommend that the scale parameter σ be assigned the value $\sigma = 1/2$ for $n > 1,000$ and $\sigma = 1$ for $n \leq 1,000$.

Kernel ICA

The kernel CVA results can be generalized to $m > 2$ by using an analogue of (15.57). In this case, the equation can be written as

$$\mathcal{K}_\kappa \boldsymbol{\alpha} = (1 + \lambda)\mathcal{D}_\kappa \boldsymbol{\alpha}, \tag{15.61}$$

where

$$\mathcal{K}_\kappa = \begin{pmatrix} (\mathbf{K}_1 + \kappa\mathbf{I}_n)^2 & \mathbf{K}_1\mathbf{K}_2 & \cdots & \mathbf{K}_1\mathbf{K}_m \\ \mathbf{K}_2\mathbf{K}_1 & (\mathbf{K}_2 + \kappa\mathbf{I}_n)^2 & \cdots & \mathbf{K}_2\mathbf{K}_m \\ \vdots & \vdots & & \vdots \\ \mathbf{K}_m\mathbf{K}_1 & \mathbf{K}_m\mathbf{K}_2 & \cdots & (\mathbf{K}_m + \kappa\mathbf{I}_n)^2 \end{pmatrix} \tag{15.62}$$

is the $(mn \times mn)$ covariance matrix of the m vectors $\mathbf{y}_1, \ldots, \mathbf{y}_m$,

$$\mathcal{D}_\kappa = \begin{pmatrix} (\mathbf{K}_1 + \kappa\mathbf{I}_n)^2 & \mathbf{0} & \cdots & \mathbf{0} \\ \mathbf{0} & (\mathbf{K}_2 + \kappa\mathbf{I}_n)^2 & \cdots & \mathbf{0} \\ \vdots & \vdots & & \vdots \\ \mathbf{0} & \mathbf{0} & \cdots & (\mathbf{K}_m + \kappa\mathbf{I}_n)^2 \end{pmatrix} \tag{15.63}$$

is the $(mn \times mn)$ block-diagonal matrix of the individual covariance matrices, and $\boldsymbol{\alpha} = (\boldsymbol{\alpha}_1^\tau, \cdots, \boldsymbol{\alpha}_m^\tau)^\tau$. Note that (15.61) can be expressed as a standard eigenproblem,

$$\widetilde{\mathcal{K}}_\kappa \widetilde{\boldsymbol{\alpha}} = \widetilde{\lambda} \widetilde{\boldsymbol{\alpha}}, \tag{15.64}$$

where $\widetilde{\mathcal{K}}_\kappa = \mathcal{D}_\kappa^{-1/2} \mathcal{K}_\kappa \mathcal{D}_\kappa^{-1/2}$, $\widetilde{\boldsymbol{\alpha}} = \mathcal{D}_\kappa^{1/2}\boldsymbol{\alpha}$, and $\widetilde{\lambda} = 1 + \lambda$. Thus, the eigenvector $\boldsymbol{\alpha}$ of $\mathcal{K}_\kappa$ gets transformed into the eigenvector $\widetilde{\boldsymbol{\alpha}}_j = (\mathbf{K}_j + \kappa\mathbf{I}_n)\boldsymbol{\alpha}_j$ of $\widetilde{\mathcal{K}}_\kappa$, with an identical eigenvalue. We are, therefore, interested in the eigenvalues and eigenvectors of the $(mn \times mn)$-matrix

$$\widetilde{\mathcal{K}}_\kappa = \begin{pmatrix} \mathbf{I}_n & \mathbf{K}_1^\kappa\mathbf{K}_2^\kappa & \cdots & \mathbf{K}_1^\kappa\mathbf{K}_m^\kappa \\ \mathbf{K}_2^\kappa\mathbf{K}_1^\kappa & \mathbf{I}_n & \cdots & \mathbf{K}_2^\kappa\mathbf{K}_m^\kappa \\ \vdots & \vdots & & \vdots \\ \mathbf{K}_m^\kappa\mathbf{K}_1^\kappa & \mathbf{K}_m^\kappa\mathbf{K}_2^\kappa & \cdots & \mathbf{I}_n \end{pmatrix}, \tag{15.65}$$

where $\mathbf{K}_j^\kappa = (\mathbf{K}_j + \kappa\mathbf{I}_n)^{-1}\mathbf{K}_j$, $j = 1, 2, \ldots, m$.

From (15.65), Bach and Jordan suggest two possible objective functions for ICA:

- $\widehat{\mathcal{I}}_{\mathcal{H}_K}(\mathbf{K}_1, \ldots, \mathbf{K}_m) = -\frac{1}{2} \log \lambda_{min}(\widetilde{\mathcal{K}}_\kappa)$,

- $\widehat{\mathcal{I}}_{\mathcal{H}_K}(\mathbf{K}_1,\dots,\mathbf{K}_m) = -\frac{1}{2}\log\det(\widetilde{\mathcal{K}}_\kappa),$

where $\lambda_{min}(\widetilde{\mathcal{K}}_\kappa)$ is the smallest eigenvalue of $\widetilde{\mathcal{K}}_\kappa$, and $\det(\widetilde{\mathcal{K}}_\kappa)$ is the *kernel generalized variance* associated with the eigenproblem. Both objective functions are functions of the Gram matrices $\mathbf{K}_1,\dots,\mathbf{K}_m$ through the separating matrix $\mathbf{W}$ and, hence, can be optimized with respect to that matrix.

As one would expect with such huge $(mn \times mn)$-matrices, computational issues become paramount to the success of this method. The solution implemented by Bach and Jordan reduces the dimensionality of the problem by using low-rank approximations to the m Gram matrices $\{\mathbf{K}_j^\kappa, j = 1, 2, \dots, m\}$. Computations are based upon incomplete Cholesky decompositions and a deflation algorithm similar to that outlined in Table 15.3.

Extensive simulations and comparisons with other ICA algorithms show Kernel ICA to have greater accuracy and to be more robust to outliers and insensitive to asymmetry of the source distributions. Because of its computational complexity, however, running time is somewhat slower than that of the other ICA algorithms.

15.4 Exploratory Factor Analysis

Tukey's distinction between exploratory and confirmatory data analysis has been extended to the techniques of factor analysis. What was once known as "common factor analysis" is now considered as exploratory methodology, and is referred to as *exploratory factor analysis (EFA)*.

The main contributors to the development of EFA as a statistical procedure were Thurstone, Spearman, Harman, Lawley, Guttman, Kaiser, Joreskog, Rao, Harris, and many others. The fact that so many were involved in its growth perhaps reflects the many divergent opinions as to the direction it should ultimately follow. The procedure has been used extensively in its different guises by social and behavioral scientists (especially in education, sociology, and psychology), who have used EFA to study latent characteristics such as mental ability, intellect, personality, and individuality through large batteries of tests. Lately, research workers in marketing, medicine, archaeology, meteorology, and other sciences have noted its usefulness and have applied it to many interesting problems.

However, this is not to say that EFA has been completely accepted. Indeed, it is still regarded by many with a marked degree of skepticism. This may be due, in part, to the type of data commonly used as input to factor analysis programs; in part, to the many subjective judgments involved in using the technique of EFA; and, in part, to the very personal interpretations of what exactly the derived factors represent. Computer packages that include a factor analysis routine now provide enough methodological options to satisfy any factor analyst. The plethora of such available methods, however, can also create a sense of confusion for the researcher. Furthermore, such extensive automation of the subject has also produced its fair share of mindless abuse.

15.4.1   The Factor Analysis Model

The linear mixing version of the noisy ICA model,

$$\mathbf{X} = \mathbf{A}\mathbf{S} + \mathbf{e}, \tag{15.66}$$

where $\mathbf{A} = (a_{ij})$ is a full-rank $(r \times m)$ mixing matrix with unknown coefficients, is usually associated with exploratory factor analysis (Lawley and Maxwell, 1971; Harman, 1976). If we assume that the noise component $\mathbf{e}$ has zero mean and a diagonal $(r \times r)$ covariance matrix, $\text{cov}(\mathbf{e}) = \mathbf{\Psi}$, with positive diagonal entries, and that $\mathbf{S}$ and $\mathbf{e}$ are uncorrelated, $\text{E}(\mathbf{S}\mathbf{e}^\tau) = \mathbf{0}$, then (15.66) reduces to the classical *common factor analysis model* (FA), where the sources are called *factors*. For the model (15.66), $\text{E}\{\mathbf{X}\} = \boldsymbol{\mu} = \mathbf{0}$ and

$$\mathbf{\Sigma}_{XX} = \mathbf{A}\mathbf{A}^\tau + \mathbf{\Psi}. \tag{15.67}$$

The EFA (as well as the BSS and ICA) problem is to estimate $\mathbf{A}$ and recover $\mathbf{S}$.

Assume that each of the r observed input variables $X_1, X_2, \ldots, X_r$ has been standardized to have zero mean and unit variance. We can write the *EFA model* in (15.66) by the following system of linear equations:

$$X_j = a_{1j}S_1 + a_{2j}S_2 + \cdots + a_{mj}S_m + e_j, \quad j = 1, 2, \ldots, r, \tag{15.68}$$

where $S_l, S_2, \ldots, S_m$ are m unobservable random variables (usually called *latent variables* or *common factors*), the $\{a_{ij}\}$ are unknown constants (referred to as *factor loadings*), and the $e_l, e_2, \ldots, e_r$ are unobservable random variables that are called *specific (or unique) factors* because e_j only appears in the equation involving X_j. We can also think of e_j as the unobservable error in fitting the jth equation. We assume that the relationships between the observed input variables, $X_1, \ldots, X_r$, are explained only by the underlying common factors and not by the errors. Thus, we assume that the $\{S_j\}$ are independent of the $\{e_j\}$, and that the $\{e_j\}$ are independent. The common factors, $\{S_j\}$, are called *orthogonal* if they are pairwise uncorrelated, while if they are correlated, they are called *oblique* factors.

From (15.67), we see that the ith diagonal entry of $\mathbf{\Sigma}_{XX}$ is given by $1 = h_i^2 + \psi_{ii}$, where $h_i^2 = \sum_j a_{ij}^2$ is called the *communality* and ψ_{ii} is the *uniqueness* given by the ith diagonal entry of $\mathbf{\Psi}$.

15.4.2 Principal Components FA

Without making any distributional assumption (e.g., Gaussian) for the sources (factors) in (15.66), we can determine $\mathbf{A}$ using a least-squares approach. In fact, premultiplying (15.66) by the Moore–Penrose generalized inverse, $\mathbf{B} = (\mathbf{A}^\tau \mathbf{A})^{-1} \mathbf{A}^\tau$, of $\mathbf{A}$, and then substituting the result in terms of $\mathbf{S}$ back into (15.66), we can re-express the model as

$$\mathbf{X} = \mathbf{CX} + \mathbf{E}, \tag{15.69}$$

where $\mathbf{C} = \mathbf{AB}$ has rank m, $\mathbf{A}$ and $\mathbf{B}$ are full-rank matrices each of rank m, $\mathbf{E} = (\mathbf{I} - \mathbf{C})\mathbf{e}$, and $\mathbf{X}$ and $\mathbf{E}$ both have mean zero. The model (15.69) is the multivariate reduced-rank regression model corresponding to principal component analysis (see Chapters 6 and 7). The least-squares criterion,

$$\mathrm{E}\{(\mathbf{X} - \mathbf{ABX})^\tau (\mathbf{X} - \mathbf{ABX})\} \tag{15.70}$$

is, therefore, minimized by setting

$$\mathbf{A} = (\mathbf{v}_1, \cdots, \mathbf{v}_m) = \mathbf{B}^\tau, \tag{15.71}$$

where $\mathbf{v}_j$ is the eigenvector corresponding to the jth largest eigenvalue of $\mathbf{\Sigma}_{XX}$. The rows of the matrix $\mathbf{B}$ give the coefficients of the m principal components scores, $\mathbf{v}_j^\tau \mathbf{X}$, $j = 1, 2, \ldots, m$, and the eigenvalues of $\mathbf{\Sigma}_{XX}$, which are usually ordered from largest to smallest, measure the variance (or *power*) of the m sources. This approach, which essentially ignores the matrix $\mathbf{\Psi}$, is usually referred to as the *principal components method*.

Typically, $\mathbf{\Sigma}_{XX}$ will be unknown and so we estimate it from the standardized input data by $\widehat{\mathbf{\Sigma}}_{XX}$, the sample correlation matrix. Estimates of $\mathbf{A}$ and $\mathbf{B}$ are given by

$$\widehat{\mathbf{A}} = (\widehat{\mathbf{v}}_1, \cdots, \widehat{\mathbf{v}}_m) = \widehat{\mathbf{B}}^\tau, \tag{15.72}$$

respectively, where $\widehat{\mathbf{v}}_j$ is the eigenvector corresponding to the jth largest eigenvalue of $\widehat{\mathbf{\Sigma}}_{XX}$, $j = 1, 2, \ldots, m$. One of the difficult problems faced by factor analysts is to determine the value of m, the number of common factors. Because the r eigenvalues of $\widehat{\mathbf{\Sigma}}_{XX}$ sum to r (the trace of $\widehat{\mathbf{\Sigma}}_{XX}$), a popular decision rule (Kaiser, 1960) is that m should be taken to be the number of those sample eigenvalues that are greater than unity.

The m-vector of estimated factor scores corresponding to a standardized sample observation $\mathbf{X} = (X_1, \cdots, X_r)^\tau$ is given by

$$\widehat{\mathbf{f}} = \widehat{\mathbf{B}} \mathbf{X} = (\widehat{\mathbf{v}}_1^\tau \mathbf{X}, \cdots, \widehat{\mathbf{v}}_m^\tau \mathbf{X})^\tau. \tag{15.73}$$

For a sample of n sample observations, $\mathbf{X}_1, \mathbf{X}_2, \ldots, \mathbf{X}_n$, it is common to plot the estimated factor scores corresponding to the first two factors on a scatterplot, where possible outliers can be identified.

Because $\mathbf{C} = (\mathbf{AT})(\mathbf{T}^\tau \mathbf{B})$ for any orthogonal $(m \times m)$-matrix $\mathbf{T}$, we can only determine $\mathbf{A}$ (and, hence, also $\mathbf{S}$) up to a rotation. In factor analysis, this is generally referred to as the problem of *factor indeterminancy*. Although this leads to a problem of identifiability, it can be made to work in our favor. We would like to choose a *rotation matrix* $\mathbf{T}$ so that $\widehat{\mathbf{f}}^* = \mathbf{T}^\tau \widehat{\mathbf{f}}$ has some desirable property. For example, can $\mathbf{T}$ be chosen to make $\widehat{\mathbf{A}}\mathbf{T}$ have an interesting interpretive structure? When the elements of $\widehat{\mathbf{A}}\mathbf{T}$ have a particular pattern so that certain elements are zero, that matrix is said to have *simple structure*. The problem of choosing such a $\mathbf{T}$ is known as the problem of *factor rotation*, for which there exist many different approaches. Probably the most popular rotation method is the *varimax* rotation (Kaiser, 1958), which seeks to find an orthogonal transformation $\mathbf{T}$ to maximize the sum, over all factors, of the variance of the squares of the scaled loadings (the estimated loadings divided by h_i, the square-root of the communalities) for each factor.

A modification of the principal components method, which takes account of the diagonal matrix $\mathbf{\Psi}$, is the *principal-factor method*. In this method, the correlation matrix $\mathbf{\Sigma}_{XX}$, with ones along the main diagonal, is replaced in the eigenanalysis by the *reduced correlation matrix* $\mathbf{\Sigma}_{XX} - \mathbf{\Psi}$, which has instead the communalities $\{h_j^2\}$ along the diagonal. In practice, $\mathbf{\Psi}$ is also unknown and, hence, the communalities have to be estimated. The most common estimate of h_j^2 is the squared multiple correlation between X_j and the remaining $r - 1$ input variables, which can be obtained as $\widehat{h}_j^2 = 1 - (1/r^{jj})$, where r^{jj} is the jth diagonal element of the inverse of the sample correlation matrix. The matrix $\widehat{\mathbf{\Sigma}}_{XX} - \widehat{\mathbf{\Psi}}$, with numbers less than unity in the main diagonal, will not necessarily be positive-definite, so that its eigenvalues will be both positive and negative. Because the sum of the positive eigenvalues exceeds the sum of the communalities, the number of factors, m, is usually taken to be at most the maximum number of positive eigenvalues whose sum is less than $\mathrm{tr}(\widehat{\mathbf{\Sigma}}_{XX} - \widehat{\mathbf{\Psi}})$.

Although many analysts have abandoned the principal factor method in favor of the maximum-likelihood (ML) method because of computational issues, this method still occupies a prominent place in many factor analysis programs.

15.4.3 Maximum-Likelihood FA

The ML method (MLFA) assumes a fully parametric model in which the m sources in (15.66) are distributed as multivariate Gaussian, $\mathbf{S} \sim \mathcal{N}_m(\mathbf{0}, \mathbf{I}_m)$, independent of the noise, which is also multivariate Gaussian, $\mathbf{e} \sim \mathcal{N}_r(\mathbf{0}, \mathbf{\Psi})$, where $\mathbf{\Psi}$ is diagonal. In some formulations, $\mathbf{\Psi} = a^2 \mathbf{I}_r$, where a is an unknown constant. These assumptions in turn imply that $\mathbf{X}$ is also multivariate Gaussian, $\mathbf{X} \sim \mathcal{N}_r(\mathbf{0}, \mathbf{\Sigma}_{XX})$, where $\mathbf{\Sigma}_{XX}$ is given by (15.67).

Given n independent observations, $\mathbf{X}_1, \ldots, \mathbf{X}_n$, on $\mathbf{X}$, we compute the sample covariance matrix $\widehat{\boldsymbol{\Sigma}}_{XX}$ as before, which has a Wishart distribution: $n\widehat{\boldsymbol{\Sigma}}_{XX} \sim \mathcal{W}_r(n, \boldsymbol{\Sigma}_{XX})$. ML estimators of $\mathbf{A}$ and $\boldsymbol{\Psi}$ are obtained by maximizing the logarithm of the likelihood function,

$$\log_e \mathcal{L} = -\frac{n}{2} \log_e |\mathbf{A}\mathbf{A}^\tau + \boldsymbol{\Psi}| - \frac{n}{2} \text{tr}\{\widehat{\boldsymbol{\Sigma}}_{XX}(\mathbf{A}\mathbf{A}^\tau + \boldsymbol{\Psi})^{-1}\}, \qquad (15.74)$$

where we have ignored constants and terms that do not involve $\mathbf{A}$ or $\boldsymbol{\Psi}$.

We apply the EM algorithm to maximize $\log_e \mathcal{L}$ with respect to $\mathbf{A}$ and $\boldsymbol{\Psi}$ (Rubin and Thayer, 1982). See Table 15.5. The algorithm treats the unobservable source scores $\{\mathbf{s}_i\}$ as if they were missing data. If the $\{\mathbf{s}_i\}$ were actually observed, the complete-data likelihood would be given by the joint distribution of the $\{\mathbf{s}_i\}$ and the $\{\mathbf{e}_i = \mathbf{x}_i - \mathbf{A}\mathbf{s}_i\}$,

$$\begin{aligned} Lik \;&=\; \prod_{i=1}^{n} \left\{ (2\pi)^{r/2} |\boldsymbol{\Psi}|^{-1/2} e^{-\frac{1}{2}\mathbf{e}_i^\tau \boldsymbol{\Psi}^{-1} \mathbf{e}_i} (2\pi)^{-r/2} e^{-\frac{1}{2}\mathbf{f}_i^\tau \mathbf{s}_i} \right\} \\ &=\; \left\{ (2\pi)^r \prod_{j=1}^{r} \psi_{jj} \right\}^{-n/2} e^{-\frac{1}{2}\sum_{i=1}^{n}\sum_{j=1}^{r} \frac{(x_{ij}-\mathbf{A}_j \mathbf{s}_i)^2}{\psi_{jj}}} \\ &\qquad \times \left\{ (2\pi)^r \right\}^{-n/2} e^{-\frac{1}{2}\sum_{i=1}^{n} \mathbf{s}_i^\tau \mathbf{s}_i}, \end{aligned} \qquad (15.75)$$

where x_{ij} is the jth component of $\mathbf{x}_i$, $\mathbf{A}_j$ is the jth row of $\mathbf{A}$, and ψ_{jj} is the jth diagonal element of the diagonal matrix $\boldsymbol{\Psi}$. Given the observed data $\{x_{ij}\}$ and the current estimated values of the parameters, the conditional expectation of (15.75), taken over the distribution of the missing data $\{\mathbf{S}_i\}$, is equal to $e^{\log_e \mathcal{L}}$.

The logarithm of (15.75) is

$$\log_e(Lik) = -\frac{n}{2} \sum_{j=1}^{r} \log_e(\psi_{jj}) - \frac{1}{2} \sum_{i=1}^{n}\sum_{j=1}^{r} \frac{(x_{ij} - \mathbf{A}_j \mathbf{s}_i)^2}{\psi_{jj}} - \frac{1}{2} \sum_{i=1}^{n} \mathbf{A}_i^\tau \mathbf{s}_i. \qquad (15.76)$$

The E-step of the EM algorithm entails finding the conditional expectation of (15.76), given the observed data $\{\mathbf{x}_i\}$ and the current values of the parameters $\mathbf{A}$ and $\boldsymbol{\Psi}$. Because the joint distribution of $\mathbf{x}_i$ and $\mathbf{s}_i$ given $\mathbf{A}$ and $\boldsymbol{\Psi}$, is $(r + t)$-variate Gaussian, the conditional distribution of $\mathbf{s}_i$ given $\mathbf{x}_i$ is

$$(\mathbf{s}_i | \mathbf{x}_i, \mathbf{A}, \boldsymbol{\Psi}) \sim \mathcal{N}_t(\boldsymbol{\delta}\mathbf{x}_i, \boldsymbol{\Delta}), \qquad (15.77)$$

where

$$\boldsymbol{\delta} = \mathbf{A}^\tau (\mathbf{A}\mathbf{A}^\tau + \boldsymbol{\Psi})^{-1} \qquad (15.78)$$

$$\boldsymbol{\Delta} = \mathbf{I}_t - \mathbf{A}^\tau (\mathbf{A}\mathbf{A}^\tau + \boldsymbol{\Psi})^{-1} \mathbf{A}. \qquad (15.79)$$

TABLE 15.5. *EM algorithm for maximum-likelihood factor analysis.*

---

1. Let $\widehat{\mathbf{A}}_0$ and $\widehat{\mathbf{\Psi}}_0$ be initial guesses for the parameter matrices $\widehat{\mathbf{A}}$ and $\widehat{\mathbf{\Psi}}$, respectively.

2. For $k = 1, 2, \ldots$, iterate between the following two steps:

   - *E-Step:* Compute

$$\mathbf{C}_{XX} = n^{-1} \sum_{i=1}^{n} \mathbf{X}_i \mathbf{X}_i^{\tau}$$

$$\mathbf{C}_{XS}^{(k-1)} = \mathbf{C}_{XX} \boldsymbol{\delta}_{k-1}^{\tau}$$

$$\mathbf{C}_{SS}^{(k-1)} = \boldsymbol{\delta}_{k-1} \mathbf{C}_{XX} \boldsymbol{\delta}_{k-1}^{\tau} + \boldsymbol{\Delta}_{k-1}$$

   where

$$\boldsymbol{\delta}_{k-1} = \widehat{\mathbf{A}}_{k-1}^{\tau} (\widehat{\mathbf{A}}_{k-1} \widehat{\mathbf{A}}_{k-1}^{\tau} + \widehat{\mathbf{\Psi}}_{k-1})^{-1}$$

$$\boldsymbol{\Delta}_{k-1} = \mathbf{I}_t - \boldsymbol{\delta}_{k-1} \widehat{\mathbf{A}}_{k-1}.$$

   - *M-Step:* Update the parameter estimates,

$$\widehat{\mathbf{A}}_k \leftarrow \mathbf{C}_{XS}^{(k-1)} (\mathbf{C}_{SS}^{(k-1)})^{-1}$$

$$\widehat{\mathbf{\Psi}}_k \leftarrow \mathrm{diag}\{\mathbf{C}_{XX} - \mathbf{C}_{XS}^{(k-1)} (\mathbf{C}_{SS}^{(k-1)})^{-1} \mathbf{C}_{XS}^{(k-1)\tau}\}.$$

3. Stop when convergence has been attained.

To find the expectation of (15.77), we need to find the expectations of the following sufficient statistics,

$$\mathbf{C}_{XX} = n^{-1} \sum_{i=1}^{n} \mathbf{X}_i \mathbf{X}_i^{\tau}, \quad \mathbf{C}_{XS} = n^{-1} \sum_{i=1}^{n} \mathbf{X}_i \mathbf{S}_i^{\tau}, \quad \mathbf{C}_{SS} = n^{-1} \sum_{i=1}^{n} \mathbf{S}_i \mathbf{S}_i^{\tau}.$$

Given the data $\{\mathbf{X}_i = \mathbf{x}_i\}$ and parameters $\mathbf{A}$ and $\mathbf{\Psi}$, the expectations are

$$\mathbf{C}_{XX}^* = \mathrm{E}(\mathbf{C}_{XX} | \{\mathbf{x}_i\}, \mathbf{A}, \mathbf{\Psi}) = \mathbf{C}_{XX} \tag{15.80}$$

$$\mathbf{C}_{XS}^* = \mathrm{E}(\mathbf{C}_{XS} | \{\mathbf{x}_i\}, \mathbf{A}, \mathbf{\Psi}) = \mathbf{C}_{XX} \boldsymbol{\delta}^{\tau} \tag{15.81}$$

$$\mathbf{C}_{SS}^* = \mathrm{E}(\mathbf{C}_{SS} | \{\mathbf{x}_i\}, \mathbf{A}, \mathbf{\Psi}) = \boldsymbol{\delta} \mathbf{C}_{XX} \boldsymbol{\delta}^{\tau} + \boldsymbol{\Delta}. \tag{15.82}$$

Equations (15.80) through (15.82) define the *E*-step based upon the observed data $\{\mathbf{x}_i\}$ and the current values of the parameter estimates $\mathbf{A}$ and $\mathbf{\Psi}$.

The *M*-step provides the updated versions of the ML estimates by using the regression estimates,

$$\widehat{\mathbf{A}} = \mathbf{C}_{XS}^* \mathbf{C}_{SS}^{*-1} \tag{15.83}$$

$$\widehat{\mathbf{\Psi}} = \mathrm{diag}\{\mathbf{C}_{XX}^* - \mathbf{C}_{XS}^* \mathbf{C}_{SS}^{*-1} \mathbf{C}_{XS}^{*\tau}\}. \tag{15.84}$$

The current estimates (15.83) and (15.84) are substituted for $\mathbf{A}$ and $\mathbf{\Psi}$, respectively, in (15.78) and (15.79) to get updated estimates of $\boldsymbol{\delta}$ and $\boldsymbol{\Delta}$, which are then used to recompute $\mathbf{C}^*_{XS}$ and $\mathbf{C}^*_{SS}$, and get new values of $\widehat{\mathbf{A}}$ and $\widehat{\mathbf{\Psi}}$. The method is iterated until we arrive at convergence.

15.4.4 Example: Twenty-four Psychological Tests

This classic data set in the factor analysis literature consists of 24 psychological tests administered to 301 seventh and eighth grade students (with ages ranging from 11 to 16) in a suburb of Chicago: a group of 156 students (74 boys, 82 girls) from the Pasteur School and a group of 145 students (72 boys, 73 girls) from the Grant–White School (Holzinger and Swineford, 1939).[3] The 24 psychological tests are as follows:

(1) visual perception, (2) cubes, (3) paper form board, (4) flags, (5) general information, (6) paragraph comprehension, (7) sentence completion, (8) word classification, (9) word meaning, (10) addition, (11) code, (12) counting dots, (13) straight-curved capitals, (14) word recognition, (15) number recognition, (16) figure recognition, (17) object-number, (18) number-figure, (19) figure-word, (20) deduction, (21) numerical puzzles, (22) problem reasoning, (23) series completion, (24) arithmetic problems.

Many of these tests were multiple-choice and all of the tests were timed, ranging from 2 minutes to 24 minutes. Actually, the students from the Grant–White school took 26 tests, where the two additional tests — 25 (paper form board "b") and 26 (flags "b") — were attempts to develop better tests than tests 3 and 4. When analyzing only the 145 Grant-White school students, it is common practice (see, e.g., Harman, 1976, pp. 123–124) to use variables 25 and 26 in place of variables 3 and 4. We note that the means, standard deviations, and correlation matrix of all 24 tests (1, 2, 25, 26, 5–24) obtained in this example are slightly different from those given by Harman.

The estimated loadings, uniquenesses, and sum-of-squares of the loadings for the 5-factor MLFA solution are given in Table 15.6. We see that the first factor, S_1, is a "verbal" factor because it loads heavily on tests 5–9; the second factor, S_2, is a "deduction of relations" factor because it loads heavily on tests 1, 2, 25, 26, and 23; the third factor, S_3, is a "speed" factor because it loads heavily on tests 10–13; the fourth factor, S_4, is a "memory" factor because it loads heavily on tests 14–18; and the fifth factor, S_5, is another "speed" factor because it loads heavily on test 13.

---

[3] The raw data can be downloaded from the book's website. Source: www.psych.yorku.ca/friendly/lab/files/psy6140/data/psych24r.sas. Also available on the website are more detailed descriptions of the 24 tests.

TABLE 15.6. *The Grant–White student data. Estimated loadings for the five-factor MLFA solution with varimax rotation. The ith factor is denoted by S_i. The rightmost column lists the uniquenesses for each test, and the last row gives the sum-of-squares of the loadings for each factor. The largest loadings for each factor are printed in boldface.*

Test	S_1	S_2	S_3	S_4	S_5	Unique
1	0.165	**0.655**	0.124	0.181	0.208	0.453
2	0.108	**0.442**	0.087	0.095	0.003	0.777
25	0.134	**0.559**	−0.048	0.111	0.094	0.646
26	0.230	**0.533**	0.089	0.081	0.014	0.648
5	**0.738**	0.189	0.191	0.149	0.056	0.357
6	**0.772**	0.187	0.031	0.248	0.125	0.291
7	**0.798**	0.214	0.143	0.088	0.051	0.286
8	**0.571**	0.343	0.239	0.127	0.044	0.481
9	**0.808**	0.203	0.033	0.219	−0.007	0.257
10	0.181	−0.108	**0.845**	0.180	0.029	0.208
11	0.195	0.066	**0.422**	0.436	0.419	0.413
12	0.030	0.232	**0.694**	0.102	0.131	0.436
13	0.186	0.432	**0.477**	0.077	**0.540**	0.253
14	0.185	0.061	0.044	**0.552**	0.080	0.649
15	0.104	0.122	0.059	**0.509**	−0.002	0.712
16	0.070	0.406	0.056	**0.509**	0.055	0.565
17	0.154	0.072	0.210	**0.595**	−0.026	0.572
18	0.032	0.300	0.322	**0.458**	0.006	0.596
19	0.156	0.221	0.144	0.378	0.046	0.761
20	0.373	0.462	0.127	0.293	−0.193	0.509
21	0.172	0.398	0.431	0.238	0.002	0.569
22	0.364	0.423	0.114	0.320	−0.068	0.568
23	0.361	**0.542**	0.249	0.231	−0.113	0.447
24	0.368	0.179	**0.495**	0.321	−0.066	0.480
SS	3.639	2.958	2.450	2.386	0.633	

For comparison purposes, an MLFA (with varimax rotation) was conducted separately on the data collected from the Grant-White students and from the Pasteur students, where we used the first 24 variables common to both sets of students. The results are very similar (with certain exceptions). A scatterplot of the first two factor scores from the rotated MLFA solution for each school is given in Figure 15.7; we see that there is little difference in the structure of the individual plots.

15.4.5   Critiques of MLFA

The ML method still has not been universally accepted among factor analysts, and a certain amount of controversy surrounds it. Critics have charged that:

1. MLFA, which is based upon Gaussian assumptions, has been routinely applied to non-Gaussian or discrete data. Whereas deviations

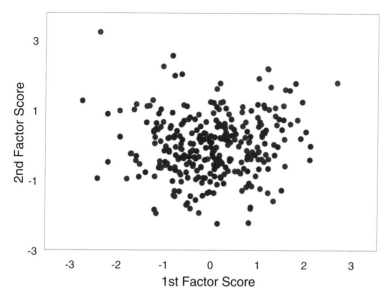

FIGURE 15.7. *MLFA (with varimax rotation) of the 24 psychological tests of Holzinger and Swineford. Superimposed scatterplot of the first two factor scores of the 145 students from the Grant-White school (blue points) and the 156 students from the Pasteur school (red points).*

from normality in social survey data are often short-tailed in nature (due to the discreteness of questions with finite range), we should expect heavy tails to be the more relevant consideration in biometric or geological applications.

2. There are substantial numerical problems that have long plagued the MLFA method, such as the existence of multiple local maxima of the likelihood function. Factor analysts try to obtain a view of the likelihood surface by comparing the solutions obtained from starting the iterative process at several points.

3. MLFA enables approximate standard errors to be obtained in a relatively simple manner using the second-derivative matrix evaluated at a mode; however, in instances where the likelihood function is multimodal, the use of such standard errors can be viewed as being of dubious value (Rubin and Thayer, 1982).

4. *Heywood cases*, which occur when the sample correlation matrix is singular and some squared multiple correlations have values greater than unity, appear in too many (over half) of the MLFA applications.

Furthermore, in these days of high-performance computing, there should be no reason to restrict attention to linear models for FA, especially when subject-matter theory suggests nonlinear relationships between test scores and factors. Indeed, some progress has been made toward formulating nonlinear latent variable models and deriving iterative algorithms for nonlinear MLFA (Yalcin and Amemiya, 2001).

15.4.6 Confirmatory Factor Analysis

During the past 40 years, EFA has been supplemented by the work of Karl Jöreskog and his colleagues, who introduced and developed *confirmatory factor analysis (CFA)* (Jöreskog, 1969). In CFA, the number of common factors is specified, certain elements of the factor loadings matrix are set to zero, and factor variances are specified; then, using Gaussian distribution assumptions, the remaining unknown parameters of the restricted factor model are estimated by maximum likelihood.

The specified factor structure is more likely to be regarded by a researcher as a theory-testing model, and such a restricted model can be evaluated using an appropriate (e.g., chi-squared) goodness-of-fit criterion. If the proposed model is not supported by the data, the model is rejected as a possible representation of the correlation structure of the underlying variables. It is not unusual to find more than one CFA model (i.e., different specifications of zero loadings) that fits the data.

There are a number of additional models that Jöreskog developed to provide more flexiblity in carrying out a confirmatory analysis of factor structures. Such models include the *analysis of covariance structures* (Jöreskog, 1970) and *structural equations modeling* (Jöreskog, 1977).

15.5 Independent Factor Analysis

Although MLFA is a very popular multivariate statistical technique, it cannot solve the BSS problem. For example, Figure 15.8 shows the first four factor scores obtained from an MLFA of the ECG signals recorded from a pregnant woman (see Section 15.2.3). These recovered signals do not separate the mother's ECG signal from the fetus's ECG signal, as did ICA.

This inability of MLFA to solve the BSS problem is due precisely to its use of Gaussian assumptions for the probability distributions of the factors. Gaussian variables that are mutually uncorrelated are also automatically independent, and so MLFA only requires that the sources be uncorrelated. Furthermore, MLFA suffers from a similar ailment as does principal component FA: the likelihood function is rotationally invariant in factor space,

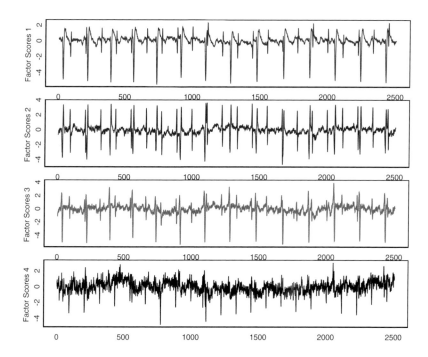

FIGURE 15.8. *The first four sets of factor scores from an EFA of the ECG signals recorded on a pregnant woman. The factor scores do not exhibit any visible separation between the mother's ECG signal and the ECG signal of the fetus.*

and so the sources $\mathbf{S}$ and the mixing matrix $\mathbf{A}$ in the BSS problem can only be defined up to an arbitrary rotation.

Independent factor analysis (IFA) (Attias, 1999) was proposed as an alternative to ICA to deal with the BSS problem and also as an alternative to EFA. IFA essentially adopts the MLFA model but employs arbitrary non-Gaussian densities for the factors. Specifically, the model is still given by

$$\mathbf{X} = \mathbf{AS} + \mathbf{e}, \quad \mathbf{e} \sim \mathcal{N}_r(\mathbf{0}, \mathbf{\Psi}), \tag{15.85}$$

with $\mathbf{\Psi}$ not necessarily diagonal, but now each unobserved source signal S_j is assumed to be independently distributed according to a non-Gaussian density $q_{S_j}(s_j | \boldsymbol{\theta}_j)$ characterized by the parameter vector $\boldsymbol{\theta}_j, j = 1, 2, \ldots, m$. In this set-up, the collection of parameters is given by $(\mathbf{A}, \mathbf{\Psi}, \boldsymbol{\theta})$, where $\boldsymbol{\theta} = (\boldsymbol{\theta}_1, \cdots, \boldsymbol{\theta}_m)$.

In the IFA model, each source density, $q_{S_j}(s_j | \boldsymbol{\theta}_j)$, is modeled parametrically by an arbitrary mixture of univariate Gaussian (MoG) densities,

TABLE 15.7. *Eight-channel ECG recordings of a pregnant woman: esti-*
mated loadings for the four independent sources IFA solution, where the ith
source is denoted by S_i. The rightmost column lists the uniquenesses (i.e.,
the diagonal entries of $\mathbf{\Psi}$) for each channel.

Channel	S_1	S_2	S_3	S_4	Unique
1	0.684	0.447	0.384	0.067	0.101
2	-0.964	0.509	0.176	-0.017	0.015
3	0.967	-0.150	-0.157	0.078	0.072
4	0.112	-0.746	0.099	-0.133	0.445
5	1.010	-0.216	0.054	0.012	0.039
6	0.990	-0.032	-0.009	-0.118	0.012
7	-0.965	-0.093	0.008	-0.111	0.011
8	-0.810	-0.398	0.008	-0.090	0.028

$$q_{S_j}(s_j|\boldsymbol{\theta}_j) = \sum_{i=1}^{I_j} w_{ij}\phi(s_j|\boldsymbol{\eta}_{ij}), \quad j=1,2,\ldots,m, \tag{15.86}$$

where $\phi(s|\boldsymbol{\eta}_{ij})$ is $\mathcal{N}(\mu_{ij},\sigma_{ij}^2)$, $\boldsymbol{\eta}_{ij} = (\mu_{ij},\sigma_{ij}^2)$, and $w_{ij} > 0$ is the *mix-ing proportion* attached to the ith component of the jth source density, $i = 1,2,\ldots,I_j$, with $\sum_{i=1}^{I_j} w_{ij} = 1$, $j = 1,2,\ldots,m$. Note that $\boldsymbol{\theta}_j = \{(w_{ij},\mu_{ij},\sigma_{ij}), i = 1,2,\ldots,I_j\}$. The MoG density (15.86) can mimic both super-Gaussian and sub-Gaussian densities by using a large enough set of component densities. The main disadvantage of working with MoG densi-ties is that the total number of parameters can grow to be very large.

The model parameters $(\mathbf{A},\mathbf{\Psi},\boldsymbol{\theta})$ are estimated by ML using an appro-priate version of the EM algorithm in Table 15.5. Details may be found in Attias (1999). When the model source densities are Gaussian, IFA reduces to EFA. Reconstructing the sources $\mathbf{S}$ can be carried out by least-squares or by Bayesian MAP estimation.

As an illustration of IFA, consider again the example of the 8-channel ECG signals recorded on a pregnant woman. We specified four independent sources, modeled each source distribution as a mixture of three Gaussians, and then used the EM algorithm to find the IFA solution. The resulting estimates are as follows: an estimate of the mixing matrix $\mathbf{A}$ is given in Table 15.7; the estimated distributions of the four independent sources, each a mixture of three Gaussians, with estimated weights, means, and variances, are given by

$$
\begin{aligned}
S_1 \quad &\sim \quad (0.199)\mathcal{N}(0.745,0.360) + (0.755)\mathcal{N}(0.032,0.024) \\
&\quad + (0.046)\mathcal{N}(-3.774,3.344), \\
S_2 \quad &\sim \quad (0.376)\mathcal{N}(-0.170,0.067) + (0.538)\mathcal{N}(0.031,0.009) \\
&\quad + (0.086)\mathcal{N})0.544,10.803),
\end{aligned}
$$

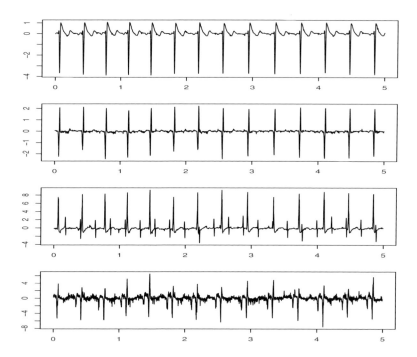

FIGURE 15.9. *Four sets of IFA scores of the ECG signals recorded on a pregnant woman. The horizontal axis is measured in seconds. The source distributions were each taken as a mixture of three Gaussians. We see traces of the mother's ECG signal in all four sets of IFA scores, and hints of traces of the fetal ECG signal in the third and fourth IFA scores, but these plots do not exhibit any visible separation between the mother's ECG signal and the ECG signal of the fetus.*

$$
\begin{aligned}
S_3 \ \sim \ & (0.302)\mathcal{N}(0.294, 0.286) + (0.396)\mathcal{N}(-0.106, 0.106) \\
& + (0.150)\mathcal{N}(0.379, 5.909), \\
S_4 \ \sim \ & (0.361)\mathcal{N}(0.294, 0.286) + (0.396)\mathcal{N}(0.004, 0.131) \\
& + (0.243)\mathcal{N}(-0.430, 3.169);
\end{aligned}
$$

and an estimate of the diagonal matrix $\boldsymbol{\Psi}$ is given by the rightmost column, titled "Unique" in Table 15.7. In Figure 15.9, we display time plots of the four sets of IFA scores. All four plots show traces of the mother's ECG signals, and two of them show hints of the fetus's ECG signals, but no clear separation is visible between the mother's and the fetus's ECG signals as we saw in the ICA solution.

One of the main difficulties with (ML-via-EM-MoG) IFA is that it is an extremely computationally intensive procedure when there are many sources to be separated; this occurs because the MoG model is quite complex, and EM is a slow algorithm that does not necessarily converge to a global maximum of the log-likelihood. Another important aspect of the IFA procedure that has to be resolved is the determination of the number of Gaussians in the mixture for each component and whether such an MoG formulation appears justified. Furthermore, simple toy examples have indicated that IFA does not seem to be appropriate for all BSS situations: in particular, there appears to be identifiability aspects of the method, and it is not yet understood whether an additive noise model such as IFA gains anything over the ICA model with no additive noise component.

15.6   Software Packages

ICA can be carried out in S-PLUS and R using the FASTICA library; FAS-TICA is also available in MATLAB as an ICA TOOLBOX. The KernelICA algorithm is implemented as a MATLAB program, which can be downloaded from the website cmm.ensmp.fr/~bach/kernel-ica/. KernelICA employs two parameters to be set by the user: the regularization parameter κ and the width of the Gaussian kernel σ. See Section 15.6.3 for recommended values of these parameters.

Factor analysis programs are standard in almost every major statistical package. The general acceptance of CFA techniques, especially in the sociometric, psychometric, and even biometric sciences is primarily due to the ready availability of good software (e.g., LISREL, AMOS, EQS, MPLUS) to carry out the extensive computations. IFA models can be fitted using the EM algorithm in the R package ifa (written by Cinzia Viroli).

Bibliographical Notes

Although the concept of ICA was introduced in 1982 in a neurophysiological context, its name was coined by Herault and Jutten (1986). See Jutten (2000) for the early history. Since then, theoretical insights, computational algorithms, and new applications have been developed to enhance and understand the ICA technique. Several books (Stone, 2004; Cichocki and Amari, 2003; Hyvärinen, Karhunen, and Oja, 2001; Lee, 1998) and edited volumes (Roberts and Everson, 2001; Girolami, 2000; Nandi, 1999) have appeared and a huge number of articles have been published on the topic. There is also an international workshop on ICA and related topics held annually in different countries.

Latent variable models and factor analysis models are discussed in the books by Everitt (1984) and Bartholomew (1987). Factor analysis is covered in almost every textbook on multivariate analysis. More specialized books on factor analysis include Harman (1976) and Lawley and Maxwell (1971).

Exercises

15.1 Let a and c be constants. If X is a random variable, show that (i) $\mathcal{H}(X + c) = \mathcal{H}(X)$, (ii) $\mathcal{H}(aX) = \mathcal{H}(X) + \log|a|$.

15.2 Let $\mathbf{X}$ be a random r-vector and let $\mathbf{W}$ be an $(r \times r)$-matrix of constants. Show that $\mathcal{H}(\mathbf{WX}) = \mathcal{H}(\mathbf{X}) + \log|\det(\mathbf{W})|$.

15.3 Suppose $\mathbf{X}$ is a random r-vector with zero mean and covariance matrix $\mathbf{\Sigma}$. Show that $\mathcal{H}(\mathbf{X}) \leq (1/2)[r + \log\{(2\pi)^r\}|\det(\mathbf{\Sigma})|]$.

15.4 Suppose $\mathbf{X} \sim \mathcal{N}_r(\mathbf{0}, \mathbf{\Sigma})$. Show that the differential entropy of $\mathbf{X}$ is given by $\mathcal{H}(\mathbf{X}) = (1/2)[r + \log\{(2\pi)^r\}|\det(\mathbf{\Sigma})|]$. This shows that the multivariate Gaussian distribution maximizes differential entropy among all multivariate distributions having the same covariance matrix $\mathbf{\Sigma}$.

15.5 Show that the differential entropy of the Cauchy distribution, $p(x) = \pi^{-1}(1 + x^2)^{-1}$, $x \in \Re$, is $\log(4\pi) \approx 2.531$.

15.6 Show that the differential entropy of the logistic distribution, $p(x) = e^{-x}(1 + e^{-x})^{-2}$, $x \in \Re$, is 2.

15.7 Generate $n = 500$ values for $X_1(t) = \cos(t)$ and $X_2(t) = e^{-t} - 5e^{-t/5}$. Let $S_1(t) = 0.7X_1(t) + 0.4X_2(t)$ and $S_2(t) = 0.2X_1(t) - 0.5X_2(t)$, $t = 1, 2, \ldots, 500$. Using either the FastICA algorithm or by writing a program to perform ICA, carry out an independent component analysis of the resulting data.

15.8 Define the measure of *kurtosis* as $\kappa_4(X) = \mathrm{E}\{X^4\} - 3[\mathrm{E}\{X^2\}]^2$. Show that for a Gaussian random variable, $\kappa_4 = 0$.

15.9 Let X_1 and X_2 be two independent random variables. Show that, if $\kappa_4(X)$ denotes the kurtosis of the random variable X, then $\kappa_4(X_1 + X_2) = \kappa_4(X_1) + \kappa_4(X_2)$ and, if c is a scalar, $\kappa_4(cX_j) = c^4\kappa(X_j)$, $j = 1, 2$.

15.10 The joint entropy $\mathcal{H}(\mathbf{X}, \mathbf{Y})$ of two random vectors $\mathbf{X}$ and $\mathbf{Y}$ is defined as $\mathcal{H}(\mathbf{X}, \mathbf{Y}) = -\int p(\mathbf{x}, \mathbf{y}) \log p(\mathbf{x}, \mathbf{y}) d\mathbf{x}d\mathbf{y}$, and the conditional entropy of $\mathbf{Y}$ given $\mathbf{X}$ is $\mathcal{H}(\mathbf{Y}|\mathbf{X}) = -\int p(\mathbf{x}, \mathbf{y}) \log p(\mathbf{y}|\mathbf{x}) d\mathbf{x}d\mathbf{y}$. Show that $\mathcal{H}(\mathbf{X}, \mathbf{Y}) = \mathcal{H}(\mathbf{X}) + \mathcal{H}(\mathbf{Y}|\mathbf{X}) = \mathcal{H}(\mathbf{Y}) + \mathcal{H}(\mathbf{X}|\mathbf{Y})$,

15.11 Use the raw data (tests 1–24) to find the MLFA (and varimax) solution to the 24 psychological tests for the combined 301 students from both schools. Give interpretations of the factors you obtain. Compare the solution with the solutions of each school separately.

15.12 Using the combined MLFA solution derived in Exercise 13.11, compare different factor rotation methods. There are two types of rotation methods: orthogonal and oblique rotations, and they attempt to transform the FA solution to simple structure. Read about the orthogonal *quartimax* method and compare it with the *varimax* method by trying it out on these data. Then, read about the oblique rotation methods, *oblimin*, *promax*, and *quartimin*, and try them out on these data. Does it make any difference which rotation method is used?

15.13 Let X and Y be iid random variables with unit variance. Show that $Z = (X + Y)/\sqrt{2}$ has unit variance.

15.14 Let X and Y be iid random variables with unit variance. Let $\mathcal{H}(X)$ denote the entropy of X. Let Z be the normalized version of $X + Y$ as in Exercise 15.13. Show that $\mathcal{H}(Z) = \mathcal{H}(X + Y) - \frac{1}{2}\log_e 2$.

15.15 For X and Y both iid and having unit variance, show that $\mathcal{H}(X + Y) > \max\{\mathcal{H}(X), \mathcal{H}(Y)\}$. Is this relationship still true if $X + Y$ is normalized as in Exercise 15.13? Generalize your results to the sum of n iid random variables, each having unit variances.

16

Nonlinear Dimensionality Reduction and Manifold Learning

16.1 Introduction

We have little visual guidance to help us identify any meaningful low-dimensional structure hidden in high-dimensional data. The linear projection methods of Chapter 7 can be extremely useful in discovering low-dimensional structure when the data actually lie in a linear (or approximately linear) lower-dimensional subspace (called a *manifold*) $\mathcal{M}$ of input space $\Re^r$. But what can we do if we know or suspect that the data actually lie on a low-dimensional *nonlinear* manifold, whose structure and dimensionality are both assumed unknown? Our goal of dimensionality reduction then becomes one of identifying the nonlinear manifold in question. The problem of recovering that manifold is known as *nonlinear manifold learning*.

If we manually search for visual hints of low-dimensional nonlinear structure in high-dimensional data by looking at scatterplot matrices or by spinning three-dimensional scatterplots, we can easily be misled, for such perceived nonlinearity may actually be due to a small group of multivariate outliers present in the data. In other cases, whatever visual guidance we do possess may not help us. Even though we may observe no unusual behavior in 2D or 3D scatterplots, the data may indeed lie close to a low-dimensional

A.J. Izenman, *Modern Multivariate Statistical Techniques*,
doi: 10.1007/978-0-387-78189-1_16,
© Springer Science+Business Media, LLC 2008

curved manifold $\mathcal{M}$, which would be invisible to linear projection methods such as PCA. In such a case, the data would satisfy nonlinear constraints, and it would then be desirable to determine a nonlinear coordinate system that, when suitably reduced, would best explain the data.

When a linear representation of the data is unsatisfactory, we turn to specialized methods designed to recover nonlinear structure. Even so, we may not always be successful in our attempts because extracting nonlinear structure from data is a difficult problem in general. If the data lie on some intrinsically weird, nonlinear manifold of input space (e.g., a one-dimensional helix or a two-dimensional "Swiss roll" embedded in three-dimensional space; see Figure 16.7), then the manifold learning problem becomes even harder, especially when the input dimension is very high.

Nonlinear dimensionality reduction and nonlinear manifold learning have become very active research topics. Some methods were found by generalizing linear multivariate methods. For example, an attractive feature of linear PCA is that it can be derived using a variety of approaches, such as variance maximization and least-squares optimality, and that these approaches yield identical solutions. Unfortunately, these equivalences in the linear case do not transfer to the nonlinear case. Thus, authors usually reformulate one of the defining characteristics of linear PCA so that it fits the nonlinear case. As a result, there can be different nonlinear versions of PCA, depending upon how one defines a nonlinear analogue of linear PCA. Furthermore, there may be technical difficulties inherent in the nonlinear versions of PCA that do not appear in linear PCA.

16.2 Polynomial PCA

How should we generalize PCA to the nonlinear case? One possibility is to transform the set of input variables using a quadratic, cubic, or higher-degree polynomial, and then apply linear PCA (Gnanadesikan and Wilk, 1969). The resulting *polynomial PCA* again boils down to an eigenanalysis, but this time attention is focused on the *smallest* few eigenvalues for nonlinear dimensionality reduction.

In the quadratic PCA case, for example, the r-vector $\mathbf{X}$ is transformed into an extended r'-vector $\mathbf{X}'$, where $r' = 2r + r(r-1)/2$. Here, $\mathbf{X}'$ includes the original r variables plus r quadratic powers and $r(r-1)/2$ cross-products of the elements of $\mathbf{X}$. Thus, for the bivariate case ($r = 2$), quadratic PCA transforms $\mathbf{X} = (X_1, X_2)$ to $\mathbf{X}' = (X_1, X_2, X_1^2, X_2^2, X_1 X_2)$, and a linear PCA is carried out on the five transformed variables of $\mathbf{X}'$. If the bivariate observations follow an exact quadratic curve, the smallest eigenvalue of the covariance matrix of the extended vector will be zero, and

TABLE 16.1. *Quadratic PCA for the bivariate data* (X_1, X_2), *where* $X_1 = -1.5(0.01)0.5$, $X_2 = 4X_1^2 + 4X_1 + 2$, *and* $n = 201$. *Eigenanalysis of the covariance matrix of the variables* $(X_1, X_2, X_1^2, X_2^2, X_1 X_2)$ *for the noiseless and noisy cases. The noisy case is obtained by replacing* X_1 *by* $X_1 + Z$ *and, independently,* X_2 *by* $X_2 + Z$, *where* $Z \sim \mathcal{N}(0, 1)$.

		Noiseless Case			
Eigenvalues	46.722	4.912	0.052	0.050	0.000
		Eigenvectors			
X_1	0.003	−0.253	0.620	0.115	0.696
X_2	−0.173	−0.013	0.337	−0.909	−0.174
X_1^2	−0.046	0.243	−0.578	−0.342	0.696
X_2^2	−0.979	−0.102	−0.063	0.165	0.000
$X_1 X_2$	0.097	−0.929	−0.333	−0.129	0.000
		Noisy Case			
Eigenvalues	74.617	10.229	2.073	0.336	0.247
		Eigenvectors			
X_1	0.012	−0.271	−0.081	−0.380	−0.880
X_2	−0.165	0.000	0.009	−0.906	0.388
X_1^2	−0.019	0.357	−0.934	−0.014	−0.019
X_2^2	−0.980	−0.120	−0.027	0.160	−0.043
$X_1 X_2$	0.121	−0.886	−0.348	0.089	0.268

the scores of the last principal component will be constant with a value of zero.

Consider, for example, the noiseless case in which $n = 201$ bivariate observations, (X_1, X_2), are generated to lie exactly on the quadratic curve $X_2 = 4X_1^2 + 4X_1 + 2$, where $X_1 = -1.5(0.01)0.5$. Suppose we carry out a linear PCA on the extended vector $(X_1^2, X_2^2, X_1, X_2, X_1 X_2)$ and obtain five sets of principal component scores. See the upper panel of Table 16.1 for the eigenanalysis. The scatterplot matrix of the first four pairs of PC scores is given in Figure 16.1 and shows the pretzel-like shapes of the pairwise PCs. The last PC is not displayed because all its values are zero. The hyperplane defined by the zero eigenvalue is $0.696X_1 - 0.0174X_2 + 0.696X_1^2 = 0$ or $X_2 = 4X_1^2 + 4X_1$, which recovers the original quadratic curve (except for the constant). By varying the constant a, we can display a family of possible quadratic curves $X_2 = 4X_1^2 + 4X_1 + a$, and the constant a can be recovered from that curve that passes through each data point. The last PC (actually, $PC5/0.0174 + X_2$) is plotted in Figure 16.2 against X_1, for $a = 0, 1, 2, 3$, where we see that $a = 2$.

Suppose we now add standard Gaussian noise (mean 0, variance 1) independently to the X_1 and X_2-coordinates of each observation and then repeat the linear PCA on the resulting extended vector. How would the eigenanalysis and the PCA scatterplot matrix of the noiseless case be

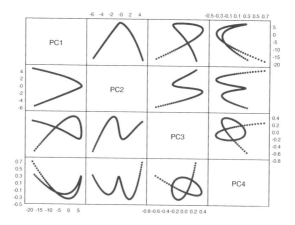

FIGURE 16.1. *Scatterplot matrix of the pairwise scores of the first four principal components from quadratic PCA using the covariance matrix. The last principal component has all its values equal to zero and is not displayed.*

affected? For this noisy case, see the lower panel of Table 16.1. The eigenvalues are each greater than the respective eigenvalues from the noiseless case, with the smallest eigenvalue now 0.247. As we would expect, some of the well-defined patterns in the scatterplot matrix become blurred in the noisy case. Even if we significantly reduce the variance of the added noise component, the results of the quadratic PCA will still be strongly affected by the noisiness of the data.

Some problems inevitably arise when using quadratic PCA. First, the variables in $\mathbf{X}'$ will not be uniformly scaled, especially for large r, and so a standardization of all r' variables may be desirable. Second, the size of the extended vector $\mathbf{X}'$ for quadratic PCA increases quickly with increasing r: when $r = 10$, $r' = 65$, and when $r = 20$, $r' = 230$. For higher-degree polynomials, the size of $\mathbf{X}'$ increases even faster. In practical terms, this introduces a lower bound on the sample size n, which has to be larger than r', the dimensionality of $\mathbf{X}'$.

16.3 Principal Curves and Surfaces

Since the Gnanadesikan and Wilk (1969) article appeared, many attempts have been made to define a more general nonlinear version of PCA. The first such attempt was principal curves and surfaces.

Suppose $\mathbf{X}$ is a continuous random r-vector having density $p_{\mathbf{X}}$, zero mean, and finite second moments. Suppose further that the data observed

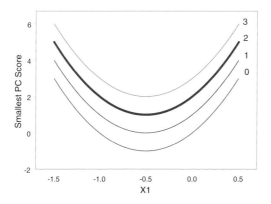

FIGURE 16.2. *Graphs of $X_2 = 4X_1^2 + 4X_1 + a$, where $a = 0, 1, 2, 3$. Superimposed on these graphs are the scores (red points) from the smallest PC (actually, $PC5/0.0174 + X_2$) derived from a quadratic PCA on the generated data.*

on $\mathbf{X}$ lie close to a smooth nonlinear manifold of low dimension. A *principal curve* (Hastie, 1984; Hastie and Stuetzle, 1989) is a smooth one-dimensional parameterized curve $\mathbf{f}$ that passes through the "middle" of the data, regardless of whether the "middle" is a straight line or a nonlinear curve. A *principal surface* is a generalization of principal curve to a smooth two- (or higher-) dimensional curve. Here, we use an analogue of least-squares optimality as the defining characteristic: we determine the principal curve or surface by minimizing the average of the squared distances between the data points and their projections onto that curve.

This idea can be interpreted in terms of the relationship between data points and points on the curve. If every point on the curve is the average of all those data points that project onto it, then the curve can be said to pass through the "middle" of the data set. In this way, it is a nonlinear generalization of the first principal component line.

Before we define principal curves and surfaces, it will be useful, first, to describe the basic ideas behind one-dimensional curves and the notion of curvature.

16.3.1 Curves and Curvature

A one-dimensional curve in an r-dimensional space is an analogue of a straight line in $\Re^r$. To formalize this notion, we define a one-dimensional curve in $\Re^r$ as a function $\mathbf{f} : \Lambda \rightarrow \Re^r$, for $\Lambda \subseteq \Re$, so that

$$\mathbf{f}(\lambda) = (f_1(\lambda), \cdots, f_r(\lambda))^\tau \tag{16.1}$$

is an r-vector parameterized by $\lambda \in \Lambda$. For example, the unit circle in $\Re^2$, $\{(x_1, x_2) \in \Re^2 : x_1^2 + x_2^2 = 1\}$, is a one-dimensional curve that can be parameterized as

$$\mathbf{f}(\lambda) = (f_1(\lambda), f_2(\lambda))^\tau = (\cos\lambda, \sin\lambda)^\tau, \quad \lambda \in [0.2\pi). \tag{16.2}$$

If we take the *coordinate functions* of λ, $\{f_h(\lambda)\}$, to be as smooth as needed (usually, C^∞, functions that have any number of continuous derivatives), then we say that $\mathbf{f}$ is a *smooth* curve. The curve $\mathbf{f}$ is said to be *closed* if it is periodic, i.e., if $\mathbf{f}(\lambda + \alpha) = \mathbf{f}(\lambda)$, for all $\lambda, \lambda + \alpha \in \Lambda$. For example, the unit circle is a closed curve.

We need a notion of how fast something can move along a smooth curve such as $\mathbf{f}$. Accordingly, we define the *velocity* (or *tangent*) *vector* at the point λ as the vector of first derivatives, $\mathbf{f}'(\lambda) = (f_1'(\lambda), \cdots, f_r'(\lambda))^\tau$, where $f_j'(\lambda) = df_j(\lambda)/d\lambda$. For the closed unit circle, $\mathbf{f}'(\lambda) = (-\sin\lambda, \cos\lambda)^\tau$. The length of the velocity vector,

$$\|\mathbf{f}'(\lambda)\| = \left\{ \sum_{j=1}^{r} [f_j'(\lambda)]^2 \right\}^{1/2}, \tag{16.3}$$

is called the *speed* of the curve $\mathbf{f}$ at the point λ. If the speed is never zero, then $\mathbf{f}(\lambda)$ is called a *regular curve*, and if $\|\mathbf{f}'(\lambda)\| = 1$, the curve is said to have "unit speed." The *acceleration vector* of $\mathbf{f}$ is defined as the vector of second derivatives, $\mathbf{f}''(\lambda) = (f_1''(\lambda), \cdots, f_r''(\lambda))^\tau$, where $f_j''(\lambda) = df_j'(\lambda)/d\lambda$. For the unit circle, $\|\mathbf{f}'(\lambda)\| = 1$ and $\mathbf{f}''(\lambda) = (-\cos\lambda, -\sin\lambda)^\tau$.

Distance on a smooth curve $\mathbf{f}$ is given by arc-length, which is measured from a fixed point λ_0 on that curve. Usually, the fixed point is taken as the origin, $\lambda_0 = 0$, defined to be one of the two endpoints of the data. The *arc-length* along the curve $\mathbf{f}$ from λ_0 to λ_1 is defined as the integral of the speed of the curve between those two points,

$$L(\mathbf{f}) = \int_{\lambda_0}^{\lambda_1} \|\mathbf{f}'(\lambda)\| d\lambda. \tag{16.4}$$

We use arc-length as a natural parameterization of the curve $\mathbf{f}$. If two curves have the same arc-length, they are said to be *isometric*. If a curve has unit speed, then its arc-length equals $\lambda_1 - \lambda_0$. We can define a one-dimensional curve uniquely by parameterizing it by arc-length and starting from a given point λ_0 having unit speed. We, henceforth, assume that $\mathbf{f}$ has been scaled to be a unit-speed, (arc-length) parameterized curve.

We next introduce a notion of curvature as a way of distinguishing a curve from a straight line. For a unit-speed curve, the acceleration vector $\mathbf{f}''(\lambda)$ is always orthogonal to the tangent vector $\mathbf{f}'(\lambda)$, so that the two vectors span a plane. The *circle of curvature* of $\mathbf{f}$ at λ is a unique unit-speed circle in the plane with radius $r(\lambda)$, which is tangent to the curve $\mathbf{f}$

at the point λ. An interesting result is that $r(\lambda)$ is a concave function of λ (i.e., $d^2 r(\lambda)/d\lambda^2 \leq 0$).

We say that the curve $\mathbf{f}$ has *radius of curvature* $r(\lambda)$ at λ and that its *curvature* at λ is $K(\lambda) = 1/r(\lambda) = \|\mathbf{f}''(\lambda)\|$; the center of the circle is the *center of curvature* of $\mathbf{f}$ at λ. Knowing the curvature for all values of arc-length λ means that the curve is completely known. If the curvature of a curve is constant and nonzero, it must be a circle (or part of a circle). A straight line is just a curve with everywhere-zero curvature.

16.3.2   Principal Curves

Consider a data point $\mathbf{x} \in \Re^r$ and let $\mathbf{f}(\lambda)$ be a curve. Project $\mathbf{x}$ to a point on $\mathbf{f}(\lambda)$ that is closest (in Euclidean distance) to $\mathbf{x}$. Let

$$\lambda_{\mathbf{f}}(\mathbf{x}) = \sup_{\lambda} \left\{ \lambda : \|\mathbf{x} - \mathbf{f}(\lambda)\| = \inf_{\mu} \|\mathbf{x} - \mathbf{f}(\mu)\| \right\} \qquad (16.5)$$

be the *projection index*, $\lambda_{\mathbf{f}} : \Re^r \to \Re$, which produces a value of λ for which $\mathbf{f}(\lambda)$ is closest to $\mathbf{x}$. In the unlikely event that there are multiple points on the curve closest to $\mathbf{x}$ (called *ambiguity points*), the projection index will pick that point with the largest value of the projection index. Note that $\lambda_{\mathbf{f}}$ can be a discontinuous function.

We define the *reconstruction error* as the expected squared distance between $\mathbf{X}$ (or its associated density) and $\mathbf{f}$,

$$D^2(\mathbf{X}, \mathbf{f}) = \mathrm{E}\left\{ \|\mathbf{X} - \mathbf{f}(\lambda_{\mathbf{f}}(\mathbf{X}))\|^2 \right\}. \qquad (16.6)$$

If $\mathbf{f}(\lambda)$ satisfies

$$\mathbf{f}(\lambda) = \mathrm{E}\{\mathbf{X}|\lambda_{\mathbf{f}}(\mathbf{X}) = \lambda\}, \quad \text{for almost every } \lambda \in \Lambda, \qquad (16.7)$$

then $\mathbf{f}(\lambda)$ is said to be *self-consistent* or a principal curve for $\mathbf{X}$ (or its associated density $p_{\mathbf{X}}$). Thus, for any point on the curve, $\mathbf{f}(\lambda)$ is the average of all those data values that project to that point.

In trying to show that the principal curve $\mathbf{f}$ minimizes the reconstruction error (16.6), Hastie and Stuetzle (1989) proved the important result that, in a variational sense, the principal curve $\mathbf{f}$ is a stationary (or critical) value of the reconstruction error. Specifically, if we perturb $\mathbf{f}$ slightly so that it becomes $\mathbf{f} + \epsilon\mathbf{g}$, where $\mathbf{g}$ is a suitably smooth curve, then

$$\frac{\partial D^2(\mathbf{X}, \mathbf{f} + \epsilon\mathbf{g})}{\partial \epsilon} \Big|_{\epsilon=0} = 0. \qquad (16.8)$$

Furthermore, principal curves are the solutions of a second-order ordinary differential equation, which makes them computable (Duchamp and Stuetzle, 1996). However, *all* principal curves are saddle points and can never be local minima of the reconstruction error. Thus, cross-validation cannot be used for choosing the model complexity when estimating principal curves.

16.3.3 Projection-Expectation Algorithm

The goal is derive an estimate $\widehat{\mathbf{f}}$ of a principal curve $\mathbf{f}$ using n observations, $\{\mathbf{X}_i\}$, on $\mathbf{X}$. To do this, we minimize an estimated reconstruction error,

$$D^2(\{\mathbf{X}_i\}, \widehat{\mathbf{f}}) = \min_{\mathbf{f}} \sum_{i=1}^{n} \|\mathbf{X}_i - \mathbf{f}(\lambda_{\mathbf{f}}(\mathbf{X}_i))\|^2, \qquad (16.9)$$

by using an algorithm that alternates between a projection step (estimating λ assuming a fixed $\mathbf{f}$) and an expectation step (estimating $\mathbf{f}$ assuming a fixed λ).

We start the algorithm by taking the first principal component line as the initial curve $\mathbf{f}^{(0)}$ Next, the n observations $\{\mathbf{X}_i\}$ are each projected onto this line, yielding the n points $\lambda_{\mathbf{f}^{(0)}}(\mathbf{X}_i) = \lambda_i^{(1)}$, $i = 1, 2, \ldots, n$. Then, the updated curve $\mathbf{f}^{(1)}$ is computed by invoking the self-consistency property,

$$\mathbf{f}^{(1)}(\lambda_i^{(1)}) = \mathrm{E}\{\mathbf{X}|\lambda_{\mathbf{f}^{(0)}}(\mathbf{X}_i) = \lambda_i^{(1)}\}, \quad i = 1, 2, \ldots, n. \qquad (16.10)$$

The kth iteration consists of two steps:

Projection step: Given the current iterate, $\mathbf{f}^{(k-1)}$, of the principal curve, we project $\mathbf{x}_i$ onto that curve to get an updated value of λ:

$$\lambda_{\mathbf{f}^{(k-1)}}(\mathbf{x}_i) = \lambda_i^{(k)}, \quad i = 1, 2, \ldots, n. \qquad (16.11)$$

Expectation step: Given the set $\{\lambda_i^{(k)}, \ i = 1, 2, \ldots, n\}$ from the projection step, we compute the next iterate of the principal curve by averaging all those points that project to nearby points on the curve:

$$\mathbf{f}^{(k)}(\lambda_i^{(k)}) = \mathrm{E}\left\{\mathbf{X}|\lambda_{\mathbf{f}^{(k-1)}}(\mathbf{X}) = \lambda_i^{(k)}\right\}, \quad i = 1, 2, \ldots, n. \qquad (16.12)$$

At the kth iteration, let $\lambda_{(i)}^{(k)}$ denote the ith order statistic of the set of projected points, $\{\lambda_1^{(k)}, \ldots, \lambda_n^{(k)}\}$, and let $\mathbf{x}_{(i)}^{(k)}$ denote the data point whose projection is $\lambda_{(i)}^{(k)}$, $i = 1, 2, \ldots, n$. Because the order of the projected points depends upon the particular iterate, then so do the corresponding data points. Let $N_{\mathbf{f}^{(k)}}(\lambda)$ be a neighborhood on the principal curve around λ. Then, let

$$N_{(i)}^{(k)} = \left\{\mathbf{x}_{(\ell)}^{(k)} : \lambda_{(\ell)}^{(k)} \in N_{\mathbf{f}^{(k)}}(\lambda_{(i)}^{(k)})\right\}. \qquad (16.13)$$

The *span* is the fraction of data points that fall into $N_{(i)}^{(k)}$. The conditional expectations (16.12) are estimated by

$$\mathbf{f}^{(k)}(\lambda_{(i)}^{(k)}) = \widehat{\mathrm{E}}\left\{\mathbf{X}|N_{(i)}^{(k)}\right\}, \qquad (16.14)$$

where we use a local averaging procedure for $\widehat{E}$ in which each coordinate function f_h, $h = 1, 2, \ldots, r$, of $\mathbf{f}$ is independently estimated. Local averaging for estimating f_h is accomplished using a *scatterplot smoother* (e.g., kernel, cubic spline, or locally weighted running-line smoother).

We can define a measure of goodness-of-fit of $\mathbf{f}^{(k)}$ by an estimate of the reconstruction error,

$$D^2(\{\mathbf{X}_i\}, \mathbf{f}^{(k)}) = n^{-1} \sum_{i=1}^{n} \|\mathbf{X}_i - \mathbf{f}^{(k)}(\lambda_{\mathbf{f}^{(k-1)}}(\mathbf{X}_i))\|^2, \qquad (16.15)$$

which is the average squared distance of the data values to their projections on the principal curve. The convergence criterion is the relative change in the reconstruction error in going from the $(k-1)$st iteration to the kth iteration,

$$\texttt{thresh}^{(k)} = \frac{|D^2(\{\mathbf{X}_i\}, \mathbf{f}^{(k-1)}) - D^2(\{\mathbf{X}_i\}, \mathbf{f}^{(k)})|}{D^2(\{\mathbf{X}_i\}, \mathbf{f}^{(k-1)})}. \qquad (16.16)$$

We repeat the alternating projection-expectation process until $\texttt{thresh}$ is reduced below some specified threshold, such as 0.001.

The "final" iteration yields a discrete set of n tuples, $(\widehat{\lambda}_i, \widehat{\mathbf{f}}_i), i = 1, 2, \ldots, n$, the elements of which are ordered by increasing $\widehat{\lambda}$-values. The principal curve $\widehat{\mathbf{f}}(\lambda)$ is then the polygon produced by joining up these n tuples. Convergence of this algorithm has not yet been proved; indeed, empirical evidence suggests that, in certain circumstances, the algorithm can converge to a poor "local" solution.

As an example, we generated 100 points in two dimensions, where X_2 is a quadratic function of X_1 plus Gaussian error with mean 0 and standard deviation 0.1. The scatterplot and principal curve are given in Figure 16.3; the left panel shows the first principal component as initial iteration, with $D^2 = 1023.3$, and the right panel shows the fifth (and final) iteration of the principal curve, with $D^2 = 0.54$.

16.3.4 *Bias Reduction*

If segments of $\mathbf{f}$ have high curvature, the projection-expectation algorithm yields a biased estimate of the principal curve. Bias also enters into the estimation procedure because of the smoothing used to estimate the conditional expectations: the bigger the span, the larger the estimation bias. A modification of this algorithm (Banfield and Raftery, 1992) allows principal curves to be estimated in a way which reduces bias.

The Banfield–Raftery enhancement of the original algorithm evolved as a means of charting the outlines of ice floes above a certain size from satellite images of the polar regions. In this particular application, ice floe outlines

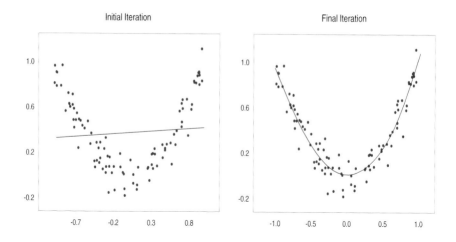

FIGURE 16.3. *Principal curve fitted to 100 randomly generated obser-vations in two dimensions, where X_2 is a quadratic function of X_1 plus Gaussian noise with mean 0 and standard deviation 0.1. Left panel: ini-tial iteration, first principal component, $D^2 = 1023.3$. Right panel: final iteration, principal curve, $D^2 = 0.54$.*

are modeled as closed principal curves. The original algorithm could not do this because of a basic assumption that the curve does not intersect itself. A further modification was added to ignore the effect of outliers on the estimation procedure.

16.3.5   Principal Surfaces

The idea of principal curves has been extended to *principal surfaces* for two (or higher) dimensions (Hastie, 1984; LeBlanc and Tibshirani, 1994).

A continuous two-dimensional surface in $\Re^r$ is a function $\mathbf{f} : \Lambda \to \Re^r$, where $\Lambda \subseteq \Re^2$, so that

$$\mathbf{f}(\boldsymbol{\lambda}) = (f_1(\boldsymbol{\lambda}), \cdots, f_r(\boldsymbol{\lambda}))^\tau = (f_1(\lambda_1, \lambda_2), \cdots, f_r(\lambda_1, \lambda_2))^\tau \qquad (16.17)$$

is an r-vector of smooth continuous coordinate functions parameterized by $\boldsymbol{\lambda} = (\lambda_1, \lambda_2) \in \Lambda$. The projection index for a bivariate surface $\boldsymbol{\Gamma}$ is defined as

$$\boldsymbol{\lambda_f}(\mathbf{x}) = \sup_{\boldsymbol{\lambda}} \left\{ \boldsymbol{\lambda} : \|\mathbf{x} - \mathbf{f}(\boldsymbol{\lambda})\| = \inf_{\boldsymbol{\mu}} \|\mathbf{x} - \mathbf{f}(\boldsymbol{\mu})\| \right\}, \qquad (16.18)$$

which is the value of $\boldsymbol{\lambda}$ corresponding to the point on the surface closest to $\mathbf{x}$. Then, a principal surface satisfies the self-consistency property,

$$\mathbf{f}(\boldsymbol{\lambda}) = \mathrm{E}\{\mathbf{X} | \boldsymbol{\lambda_f}(\mathbf{X}) = \boldsymbol{\lambda}\}, \quad \text{for almost every } \boldsymbol{\lambda} \in \Lambda. \qquad (16.19)$$

Given n observations $\{\mathbf{X}_i\}$ on $\mathbf{X}$, we estimate $\mathbf{f}$ by minimizing the residual sum of squares,

$$RSS(\mathbf{f}) = \sum_{i=1}^{n} \|\mathbf{X}_i - \mathbf{f}(\lambda_{\mathbf{f}}(\mathbf{X}_i))\|^2. \qquad (16.20)$$

Defining a suitable analogue of the "unit-speed" property for parameterizing principal surfaces is a lot more complicated than its definition for principal curves, and so an alternative approach is necessary. Toward this end, LeBlanc and Tibshirani (1994) describe an adaptive formulation and algorithm for the computation of principal surfaces, and they give some examples. Malthouse (1998) gives other possible types of parameterizations.

16.4 Multilayer Autoassociative Neural Networks

Another version of nonlinear PCA has been constructed using a special type of artificial neural network (ANN) architecture: a five-layer autoassociative ANN (Kramer, 1991). An *autoassociative* (or *autoencoder* or *self-supervised*) network is an ANN that is trained to learn its own inputs. The network connects r input nodes to r output nodes in such a way that the output values are trained to approximate the inputs.

16.4.1 Main Features of the Network

The main features of a five-layer autoassociative ANN are

- three hidden layers of nodes (second, third, and fourth layers), where the *mapping* (or *encoding*) *layer* (second) and the *demapping* (or *decoding*) *layer* (fourth) both have nonlinear (sigmoidal) activation functions;

- an internal *bottleneck layer* (third) with fewer (linear or sigmoidal) nodes than either the mapping (second) or demapping (fourth) layers;

- feedforward connections trained by backpropagation.

The number of nodes in the mapping and demapping layers will depend upon how complicated the nonlinearity feature of the network is required to be. In fact, we should not expect the mapping and demapping layers to have the same number of nodes. Too few mapping/demapping nodes sacrifice accuracy whereas too many such nodes encourage overfitting. Furthermore, it may be better in certain circumstances to have more than one mapping or demapping layer.

The bottleneck layer is the most important feature of the network because it reduces the dimensionality of the inputs through data compression.

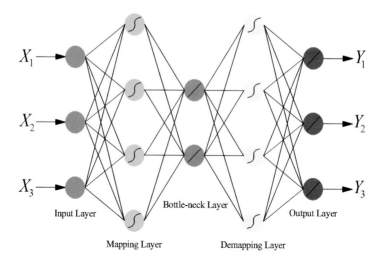

FIGURE 16.4. *Multilayer autoassociative neural network with $r = 3$ input (X) and output (Y) nodes, and three hidden nodes — a mapping layer of four nodes, a demapping layer of four nodes, and a bottleneck layer of $t = 2$ nodes. The outputs are trained to approximate the inputs.*

Without the bottleneck layer, the network is only capable of producing either linear combinations of the inputs (given linear output nodes) or sigmoidally compressed outputs (given nonlinear sigmoidal output nodes).

We saw in Chapter 10 that a three-layer ANN with nonlinear activation functions in the hidden layer can be represented by a function of the form $\sum_j \alpha_j \sigma(\boldsymbol{\beta}_j^\tau \mathbf{x})$, where α_j and the vector $\boldsymbol{\beta}_j$ are weights, and $\sigma(\cdot)$ is a sigmoidal-shaped function. Recall also that such a network with linear output nodes can approximate any continuous function uniformly on compact sets provided that the number of nodes in the hidden layer is sufficiently large (Cybenko, 1989). A five-layer network, such as the one displayed in Figure 16.4, can then be viewed as the composition of two three-layer subnetworks (layers 1, 2, and 3; layers 3, 4, and 5). In order for each of these two subnetworks to represent continuous functions, the second and fourth layers have to consist of nonlinear activation functions. If we remove the mapping and demapping layers and if we set the nodes in the bottleneck layer to be linear, then the resulting network corresponds to linear PCA.

16.4.2 Relationship to Principal Curves

The first part of the autoassociative ANN (layers 1, 2, and 3, with one bottleneck node) can be used to model a continuous one-dimensional function $\lambda_{\mathbf{f}} : \Re^r \to \Re$, which we call a *projection index*. The second part of the ANN (layers 3, 4, and 5) can be used to model the function $\mathbf{f} : \Re \to \Re^r$. The first three layers project the original data onto a curve, and the projected data values are then given by the activation values of the bottleneck node. The weights in the network are found by solving the least-squares problem,

$$\min_{\mathbf{f},\lambda_{\mathbf{f}}} \sum_{i=1}^{n} \|\mathbf{x}_i - \mathbf{f}(\lambda_{\mathbf{f}}(\mathbf{x}_i))\|^2, \tag{16.21}$$

which reduces to a similar minimization problem as we used to find principal curves, but where the same criterion was minimized only over $\mathbf{f}$. For modeling a t-dimensional principal surface, we set the functions $\lambda_{\mathbf{f}} : \Re^r \rightarrow \Re^t$ and $\mathbf{f} : \Re^t \rightarrow \Re^r$, where $t \geq 2$ nodes are set in the bottleneck layer.

A crucial distinction between principal curves and this type of ANN is that the projection index $\lambda_{\mathbf{f}}$ defined for principal curves is allowed to be discontinuous. The fact that the ANN version of $\lambda_{\mathbf{f}}$ is a continuous function causes severe problems with its application as a nonlinear PCA technique (Malthouse, 1998):

1. If $\mathbf{f}$ has any ambiguity points for the data point $\mathbf{x}$, then the ANN must avoid becoming discontinuous at the ambiguity point by projecting $\mathbf{x}$ to the "wrong" point on the curve (i.e., a point that is not closest to $\mathbf{x}$).

2. The ANN cannot model any curves or surfaces that intersect themselves (such as the circle in $\Re^2$). Recall that the original version of principal curves did not allow the curves to intersect themselves, but modifications by Banfield and Raftery (1992) now allow closed curves to be modeled.

For these reasons, we should be very cautious in using this type of ANN to model nonlinear PCA.

16.5   Kernel PCA

An approach that also generalizes polynomial PCA is given by *Kernel PCA* (Scholkopf, Smola, and Muller, 1996). This is an application of so-called *kernel methods*, which we have already seen in studying SVMs (see Chapter 11).

Let $\mathbf{X}_i \in \Re^r$, $i = 1, 2, \ldots, n$, be the input data points. We can think of kernel PCA as a two-step process:

1. Nonlinearly transform the ith input data point $\mathbf{X}_i \in \Re^r$ into a point $\Phi(\mathbf{X}_i)$ in an $N_{\mathcal{H}}$-dimensional *feature space* $\mathcal{H}$, where

$$\Phi(\mathbf{X}_i) = (\phi_1(\mathbf{X}_i), \cdots, \phi_{N_{\mathcal{H}}}(\mathbf{X}_i))^\tau \in \mathcal{H}, \quad i = 1, 2, \ldots, n. \tag{16.22}$$

The map $\Phi : \Re^r \to \mathcal{H}$ is called a *feature map*, and each of the $\{\phi_j\}$ is a nonlinear map.

2. Given $\Phi(\mathbf{X}_1), \ldots, \Phi(\mathbf{X}_n) \in \mathcal{H}$, with $\sum_{i=1}^n \Phi(\mathbf{X}_i) = \mathbf{0}$, solve a linear PCA problem in feature space $\mathcal{H}$, which will have a *higher* dimensionality than that of input space (i.e., $N_{\mathcal{H}} > r$).

The argument is that any low-dimensional structure may be more easily discovered when it becomes embedded in the larger space $\mathcal{H}$, which could be infinite dimensional (i.e., we allow the possibility that $N_{\mathcal{H}} = \infty$). Although we do not need to define Φ explicitly, we have to assume in step 2 that the data have been centered in feature space. We return to this assumption in Section 16.5.2. Unless otherwise stated, we also assume that $N_{\mathcal{H}} < n$.

In the following, we take $\mathcal{H}$ to be an $N_{\mathcal{H}}$-dimensional Hilbert space with inner product $\langle \cdot, \cdot \rangle$ and norm $\| \cdot \|_{\mathcal{H}}$. For example, if $\boldsymbol{\xi}_j = (\xi_{j1}, \cdots, \xi_{jN_{\mathcal{H}}})^\tau \in \mathcal{H}$, $j = 1, 2$, then, $\langle \boldsymbol{\xi}_1, \boldsymbol{\xi}_2 \rangle = \sum_{i=1}^{N_{\mathcal{H}}} \xi_{1i} \xi_{2i}$, and if $\boldsymbol{\xi} = (\xi_1, \cdots, \xi_{N_{\mathcal{H}}})^\tau \in \mathcal{H}$, then $\| \boldsymbol{\xi} \|_{\mathcal{H}}^2 = \langle \boldsymbol{\xi}, \boldsymbol{\xi} \rangle = \sum_{i=1}^{N_{\mathcal{H}}} \xi_i^2$.

16.5.1 PCA in Feature Space

In order to carry out linear PCA in feature space so that it mimics the standard treatment of PCA (as carried out in input space), we have to find eigenvalues $\lambda \geq 0$ and nonzero eigenvectors $\mathbf{v} \in \mathcal{H}$ of the estimated covariance matrix,

$$\mathbf{C} = n^{-1} \sum_{i=1}^n \Phi(\mathbf{X}_i) \Phi(\mathbf{X}_i)^\tau, \qquad (16.23)$$

of the centered and nonlinearly transformed input vectors. The eigenequation $\mathbf{Cv} = \lambda \mathbf{v}$, where $\mathbf{v}$ is the eigenvector corresponding to the eigenvalue $\lambda \geq 0$ of $\mathbf{C}$, can be written in an equivalent form as

$$\langle \Phi(\mathbf{X}_i), \mathbf{Cv} \rangle = \lambda \langle \Phi(\mathbf{X}_i), \mathbf{v} \rangle, \quad i = 1, 2, \ldots, n. \qquad (16.24)$$

Because

$$\mathbf{Cv} = n^{-1} \sum_{i=1}^n \Phi(\mathbf{X}_i) \langle \Phi(\mathbf{X}_i), \mathbf{v} \rangle, \qquad (16.25)$$

all solutions $\mathbf{v}$ with nonzero eigenvalue λ are contained in the span of $\Phi(\mathbf{X}_1), \ldots, \Phi(\mathbf{X}_n)$. So, there exist coefficients, α_i, $i = 1, 2, \ldots, n$, such that

$$\mathbf{v} = \sum_{i=1}^n \alpha_i \Phi(\mathbf{X}_i). \qquad (16.26)$$

Substituting (16.26) for $\mathbf{v}$ in (16.24), we get that

$$n^{-1} \sum_{j=1}^{n} \alpha_j \left\langle \Phi(\mathbf{X}_i), \sum_{k=1}^{n} \Phi(\mathbf{X}_k) \right\rangle \langle \Phi(\mathbf{X}_k), \Phi(\mathbf{X}_j) \rangle = \lambda \sum_{k=1}^{n} \alpha_k \langle \Phi(\mathbf{X}_i), \Phi(\mathbf{X}_k) \rangle,$$

(16.27)

for all $i = 1, 2, \ldots, n$. Define the $(n \times n)$-matrix $\mathbf{K} = (K_{ij})$, where

$$K_{ij} = \langle \Phi(\mathbf{X}_i), \Phi(\mathbf{X}_j) \rangle.$$

(16.28)

Note that $\mathbf{K}$ will generally be a huge matrix. Then, the eigenequation (16.27) can be written as $\mathbf{K}^2 \boldsymbol{\alpha} = n\lambda \mathbf{K} \boldsymbol{\alpha}$, where $\boldsymbol{\alpha} = (\alpha_1, \cdots, \alpha_n)^\tau$, or as

$$\mathbf{K} \boldsymbol{\alpha} = \widetilde{\lambda} \boldsymbol{\alpha},$$

(16.29)

where $\widetilde{\lambda} = n\lambda$. Note that we can express the eigenvalues and vectors, $(\widetilde{\lambda}, \boldsymbol{\alpha})$, of $\mathbf{K}$ in terms of those, $(\lambda, \mathbf{v})$, for $\mathbf{C}$.

Denote the ordered eigenvalues of $\mathbf{K}$ by $\widetilde{\lambda}_1 \geq \widetilde{\lambda}_2 \geq \ldots \geq \widetilde{\lambda}_n \geq 0$, with associated eigenvectors $\boldsymbol{\alpha}_1, \ldots, \boldsymbol{\alpha}_n$, where $\boldsymbol{\alpha}_i = (\alpha_{i1}, \cdots, \alpha_{in})^\tau$. If we require that $\langle \mathbf{v}_i, \mathbf{v}_i \rangle = 1, i = 1, 2, \ldots, n$, then, using the expansion (16.26) for $\mathbf{v}_i$ and the eigenequation (16.27), we have that

$$\begin{aligned} 1 &= \sum_{j=1}^{n} \sum_{k=1}^{n} \alpha_{ij} \alpha_{ik} \langle \Phi(\mathbf{X}_j), \Phi(\mathbf{X}_k) \rangle \\ &= \sum_{j=1}^{n} \sum_{k=1}^{n} \alpha_{ij} \alpha_{ik} K_{jk} \\ &= \langle \boldsymbol{\alpha}_i, \mathbf{K} \boldsymbol{\alpha}_i \rangle = \widetilde{\lambda}_i \langle \boldsymbol{\alpha}_i, \boldsymbol{\alpha}_i \rangle, \end{aligned}$$

(16.30)

which determines the normalization for the vectors $\boldsymbol{\alpha}_1, \ldots, \boldsymbol{\alpha}_n$.

If $\mathbf{X}$ is a test point, then the *nonlinear principal component scores* of $\mathbf{X}$ corresponding to Φ are given by the projection of $\Phi(\mathbf{X}) \in \mathcal{H}$ onto the eigenvectors $\mathbf{v}_k \in \mathcal{H}$,

$$\langle \mathbf{v}_k, \Phi(\mathbf{X}) \rangle = \lambda_k^{-1/2} \sum_{i=1}^{n} \alpha_{ki} \langle \Phi(\mathbf{X}_k).\Phi(\mathbf{X}) \rangle, \quad k = 1, 2, \ldots, n,$$

(16.31)

where the $\lambda_k^{-1/2}$ term is included so that $\langle \mathbf{v}_k, \mathbf{v}_k \rangle = 1$.

Using the kernel trick (see Section 11.3.2), the nonlinear principal component scores of $\mathbf{X}$ can be expressed as

$$\langle \mathbf{v}_k, \Phi(\mathbf{X}) \rangle = \lambda_k^{-1/2} \sum_{i=1}^{n} \alpha_{ki} K(\mathbf{X}_i, \mathbf{X}), \quad k = 1, 2, \ldots, n.$$

(16.32)

If we set $\mathbf{X} = \mathbf{X}_m$ in (16.32), we get that $\langle \mathbf{v}_k, \Phi(\mathbf{X}_m) \rangle = \lambda_k^{-1/2} \sum_i \alpha_{ki} K_{im} = \lambda_k^{-1/2}(\mathbf{K}\boldsymbol{\alpha}_k)_m = \lambda_k^{-1/2}(\lambda_k \boldsymbol{\alpha}_k)_m \propto \alpha_{km}$, where $(\mathbf{A})_m$ stands for the mth row of $\mathbf{A}$.

16.5.2   Centering in Feature Space

So far, we assumed that the Φ-images in feature space have been centered, and that, through (16.28), we can work with the matrix $\mathbf{K}$. Although it may not be possible to do this in feature space, there is a way it can be accomplished back in the original input space. We do not actually need to have a centered Φ to work with, but we do need $\mathbf{K}$.

We can apply the following simple adjustment to the non-centered version of the matrix $\mathbf{K}$,

$$\widetilde{\mathbf{K}} = \mathbf{HKH}, \tag{16.33}$$

where $\mathbf{H} = \mathbf{I}_n - n^{-1}\mathbf{J}_n$ is the centering matrix, $\mathbf{J}_n = \mathbf{1}_n \mathbf{1}_n^\tau$ is an $(n \times n)$-matrix of all ones, and $\mathbf{1}_n$ is an n-vector of all ones. The resulting

$$\widetilde{\mathbf{K}} = \mathbf{K} - \mathbf{K}(n^{-1}\mathbf{J}_n) - (n^{-1}\mathbf{J}_n)\mathbf{K} + (n^{-1}\mathbf{J}_n)\mathbf{K}(n^{-1}\mathbf{J}_n) \tag{16.34}$$

corresponds to starting with a centered Φ as required by the above development (Scholkopf, Smola, and Muller, 1998).

16.5.3   Example: Food Nutrition (Continued)

Consider again the example in Section 7.2.1 on the nutritional value of food. Previously, we had computed the PCA of the data and displayed the scatterplot of the first two principal component scores. Here, we compute the kernel principal components for the data ($n = 961, r = 6$) using a radial basis (Gaussian) function kernel with scale parameter σ. Figure 16.5 displays the scatterplots of the first two kernel PC scores using $\sigma = 0.005, 0.01, 0.1$, and 0.5. The eigenvalues are

σ	λ_1	λ_2
0.005	0.0336	0.0033
0.01	0.0602	0.0066
0.1	0.2200	0.0820
0.5	0.2738	0.1139

Notice that both eigenvalues increase in size as σ increases.

We see that as we increase σ, the shape of the kernel PC plot changes significantly. The scatterplots for $\sigma \geq 0.01$ show an obvious nonlinear configuration of points: for each of these three plots, there is a "head" and a "tail" to the "curve." The head contains those points that have the largest magnitudes for at least one of the six variables, whereas the tail contains only data with negligible values for all variables. In the curve for $\sigma = 0.01$, the head is on the left; in the curve for $\sigma = 0.1$, the head is at the top left; and in the plot for $\sigma = 0.5$, the head is at the center of the plot. There are a small number of stray points falling inside each of the $\sigma = 0.01$ and 0.1 curves; most of these points correspond to foods that are very high in `cholesterol`, and which become the head of the the the curve for $\sigma = 0.5$.

In the scatterplot corresponding to $\sigma = 0.5$ (lower-right panel of Figure 16.5), the points having the largest magnitudes of each of the six variables are annotated with the dominating variable name. We see an ordering of the six variables along the nonlinear curve, starting at `cholesterol`, and continuing with `saturated fat`, `fat/food energy`, `carbohydrates`, and `protein`, in that order. There is very little difference between foods that are high in fat and those that are high in calories (food energy). This display provides a "food-nutrition ordering" similar in spirit to the classic "color wheel." Similar interpretations can be obtained from the other three scatterplots.

16.5.4 Kernel PCA and Metric MDS

We note that kernel PCA with an isotropic kernel function is closely related to metric MDS (Williams, 2001). In feature space, we can compute the distance (i.e., dissimilarity), $\widetilde{\delta}_{ij}^2 = \|\Phi(\mathbf{X}_i) - \Phi(\mathbf{X}_j)\|^2$. Expanding and using the kernel trick, we have that $\widetilde{\delta}_{ij}^2 = 2(1 - K(\mathbf{X}_i, \mathbf{X}_j))$. The matrix $\mathbf{A}$ has ijth entry $a_{ij} = -\frac{1}{2}\widetilde{\delta}_{ij}^2 = K(\mathbf{X}_i, \mathbf{X}_j) - 1$, whence, $\mathbf{A} = \mathbf{K} - \mathbf{J}_n$. Furthermore, $\mathbf{HAH} = \mathbf{HKH}$, because $\mathbf{HJ}_n = \mathbf{0}$.

Thus, carrying out metric MDS on the kernel matrix $\mathbf{K}$ produces an equivalent configuration of points as the distances $\widetilde{\delta}_{ij} = \sqrt{2(1 - K(\mathbf{X}_i, \mathbf{X}_j))}$ computed in feature space. If the kernel $K(\mathbf{X}_i, \mathbf{X}_j)$ is isotropic, it depends only on the distance, $\delta_{ij} = \|\mathbf{X}_i - \mathbf{X}_j\|$, in input space, so that $K(\mathbf{X}_i, \mathbf{X}_j) = k(\delta_{ij})$. It follows that $\widetilde{\delta}_{ij} = \sqrt{2(1 - k(\delta_{ij}))}$, which makes the feature-space distances $\widetilde{\delta}_{ij}^2$ a nonlinear function of the input-space distances δ_{ij}. This shows that this formulation is a special case of metric MDS.

16.6 Nonlinear Manifold Learning

We now discuss some exciting new algorithmic techniques: Isomap, Local Linear Embedding, Laplacian Eigenmap, and Hessian Eigenmap. The goal of each of these algorithms is to recover the full low-dimensional representation of an unknown nonlinear manifold $\mathcal{M}$

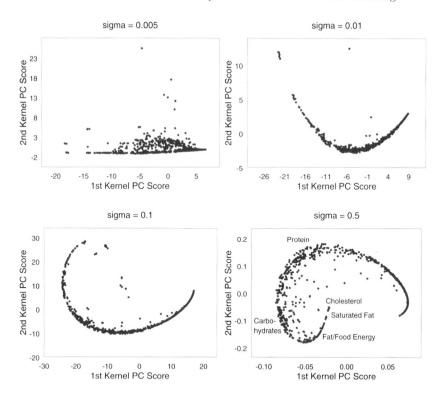

FIGURE 16.5. *The nutritional value of food example. Scatterplots of first and second kernel principal component scores, computed using a radial basis (Gaussian) function kernel with scale parameter σ. Upper-left panel: $\sigma = 0.005$; upper-right panel: $\sigma = 0.01$; lower-left panel: $\sigma = 0.1$; lower-right panel: $\sigma = 0.5$.*

embedded[1] in some high-dimensional space, where it is important to retain the neighborhood structure of $\mathcal{M}$. Although closely related to nonlinear dimensionality reduction, these algorithms are mainly concerned with recovering the manifold $\mathcal{M}$. When $\mathcal{M}$ is highly nonlinear, such as the S-shaped manifold in the left panel of Figure 16.6, these algorithms have outperformed linear techniques. Each algorithm is designed to emphasize simplicity while avoiding optimization problems that could produce local minima.

Although the algorithms use different philosophies for recovering nonlinear manifolds, they each consist of a three-step approach. The first and third steps are common to all algorithms: the first step incorporates

---

[1]A space $\mathcal{A}$ is said to be *embedded* in a bigger space $\mathcal{B}$ if the properties of $\mathcal{B}$ when restricted to $\mathcal{A}$ are identical to the properties of $\mathcal{A}$.

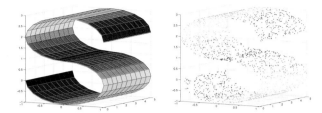

FIGURE 16.6. *Left panel: The S-curve, a two-dimensional S-shaped manifold embedded in three-dimensional space. Right panel: 2,000 data points randomly generated to lie on the surface of the S-shaped manifold.*

neighborhood information from each data point to construct a weighted graph having the data points as vertices, and the third step is an embedding step that involves an $(n \times n)$-eigenequation computation. The second step is specific to the algorithm, taking the weighted neighborhood graph and transforming it into suitable input for the embedding step.

Manifold learning involves concepts from differential geometry. So, before we describe these algorithms, we first discuss what we mean by a manifold and what it means for it to be embedded in a higher-dimensional space.

16.6.1 Manifolds

It is not easy to give a simple description of a "manifold" because of the complex mathematical notions involved in its definition. Even the great mathematician Élie Cartan wrote that "La notion générale de veriété est assez difficile à définir avec précision"[2] (Cartan, 1946, p. 56). However, we will try to give some of the flavor of its definition.

Imagine an ant at a picnic, where there are all sorts of items from cups to doughnuts. The ant crawls all over the picnic items, but because of its diminutive size, the ant sees everything on a very small scale as flat and featureless. A *manifold* (also referred to as a *topological manifold*) can be thought of in similar terms, as a topological space that locally looks flat and featureless and behaves like Euclidean space. To prevent crazy, counterintuitive situations, a manifold also satisfies certain topological conditions. A *submanifold* is just a manifold lying inside another manifold of higher dimension.

In 1854, Georg Friedrich Bernhard Riemann (1826–1866) introduced the idea of a manifold where one could carry out differential and integral calculus. If a topological manifold $\mathcal{M}$ is continuously differentiable to any order

---

[2] "The general notion of manifold is quite difficult to define with precision."

(i.e., $\mathcal{M} \in \mathcal{C}^{\infty}$), we call it a *smooth* (or *differentiable*) *manifold*. All smooth manifolds are topological manifolds, but the reverse is not necessarily true.

If we endow a smooth manifold $\mathcal{M}$ with a metric $d^{\mathcal{M}}$, which calculates distances between points on $\mathcal{M}$, we have a *Riemannian manifold*, $(\mathcal{M}, d^{\mathcal{M}})$. If $\mathcal{M}$ is connected, it is a metric space and $d^{\mathcal{M}}$ determines its structure. More specifically, we take $d^{\mathcal{M}}$ to be a manifold metric defined by

$$d^{\mathcal{M}}(\mathbf{y}, \mathbf{y}') = \inf_{c}\{L(c) | c \text{ is a curve in } \mathcal{M} \text{ which joins } \mathbf{y} \text{ and } \mathbf{y}'\}, \quad (16.35)$$

where $\mathbf{y}, \mathbf{y}' \in \mathcal{M}$ and $L(c)$ is the arc-length of the curve c; see Section 16.3.1. Thus, $d^{\mathcal{M}}$ finds the shortest curve (or *geodesic*) between any two points on $\mathcal{M}$, and the arc-length of that curve is the *geodesic distance* between the points. By Nash's embedding theorem (Nash, 1965), we can embed a smooth manifold $\mathcal{M}$ into a high-dimensional Euclidean space $\mathcal{X}$, which we take to be input space $\Re^{r}$.

16.6.2 Data on Manifolds

The methods we discuss in this section operate under the assumption that finitely many data points, $\{\mathbf{y}_i\}$, are randomly sampled from a smooth t-dimensional manifold $\mathcal{M}$ with metric given by geodesic distance $d^{\mathcal{M}}$; these points are then nonlinearly embedded by a smooth map ψ into high-dimensional input space $\mathcal{X} = \Re^{r}$ ($t \ll r$) with Euclidean metric $\|\cdot\|_{\mathcal{X}}$. This embedding yields the input data $\{\mathbf{x}_i\}$; see, for example, the right panel of Figure 16.7, where 2,000 points in three dimensions are randomly generated to lie on the surface of a two-dimensional S-shaped curve. Thus, $\psi : \mathcal{M} \to \mathcal{X}$ is the embedding map, and a point on the manifold, $\mathbf{y} \in \mathcal{M}$, can be expressed as $\mathbf{y} = \phi(\mathbf{x})$, $\mathbf{x} \in \mathcal{X}$, where $\phi = \psi^{-1}$. The goal is to recover $\mathcal{M}$ and find an implicit representation of the map ψ (and, hence, recover the $\{\mathbf{y}_i\}$), given only the input data points $\{\mathbf{x}_i\}$ in $\mathcal{X}$.

Each of these algorithms computes estimates $\{\widehat{\mathbf{y}}_i\} \subset \Re^{t'}$ of the manifold data $\{\mathbf{y}_i\} \subset \Re^{t}$, for some t'. We consider such a *reconstruction* to be successful if $t' = t$, the true dimensionality of $\mathcal{M}$. Because t' will most likely be too large for practical usage and because we require a low-dimensional display for a visual representation of the solution, we take only the first two or three of the coordinate vectors and plot the corresponding elements of those vectors against each other to yield n points in two- or three-dimensional space.

16.6.3 Isomap

The *isometric feature mapping* (or Isomap) algorithm (Tenenbaum, de Silva, and Langford, 2000) assumes that the smooth manifold $\mathcal{M}$ is a convex

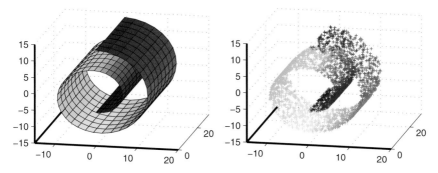

FIGURE 16.7. *Left panel: The Swiss Roll: a two-dimensional manifold embedded in three-dimensional space. Right panel: 20,000 data points lying on the surface of the swiss-roll manifold.*

region of $\Re^t$ ($t \ll r$) and that the embedding $\psi : \mathcal{M} \to \mathcal{X}$ is an isometry. This assumption has two key ingredients:

- *Isometry:* The geodesic distance is invariant under the map ψ. For any pair of points on the manifold, $\mathbf{y}, \mathbf{y}' \in \mathcal{M}$, the geodesic distance between those points equals the Euclidean distance between their corresponding coordinates, $\mathbf{x}, \mathbf{x}' \in \mathcal{X}$; i.e.,

$$d^{\mathcal{M}}(\mathbf{y}, \mathbf{y}') = \|\mathbf{x} - \mathbf{x}'\|_{\mathcal{X}}, \qquad (16.36)$$

  where $\mathbf{y} = \phi(\mathbf{x})$ and $\mathbf{y}' = \phi(\mathbf{x}')$.

- *Convexity:* The manifold $\mathcal{M}$ is a convex subset of $\Re^t$.

Thus, ISOMAP regards $\mathcal{M}$ as a convex region that may have been distorted in any of a number of ways (e.g., by folding or twisting). The so-called *Swiss roll*,[3] which is a flat two-dimensional rectangular submanifold of $\Re^3$, is one such example; see Figure 16.7. ISOMAP appears to work best for intrinsically flat submanifolds of $\mathcal{X} = \Re^r$ that look like rolled-up sheets of paper. In certain situations, the isometry assumption appears to be reasonable, while the convexity assumption may be too restrictive (Donoho and Grimes, 2003).

ISOMAP uses the isometry and convexity assumptions to form a nonlinear generalization of multidimensional scaling (MDS). As we saw in Section 13.3, MDS searches for a low-dimensional subspace in which to embed input data while preserving the Euclidean interpoint distances. ISOMAP extends

---

[3]The Swiss roll is generated as follows: for $y_1 \in [3\pi/2, 9\pi/2]$ and $y_2 \in [0, 15]$, set $x_1 = y_1 \cos y_1$, $x_2 = y_1 \sin y_1$, $x_3 = y_2$.

the MDS paradigm by attempting to preserve the global geometry proper-ties of the underlying nonlinear manifold, and it does this by approximating *all* geodesic distances (i.e., lengths of the shortest paths) on the manifold. In this sense, ISOMAP gives a *global* approach to manifold learning.

The ISOMAP algorithm consists of three steps:

1. Neighborhood graph. Fix either an integer K or an $\epsilon > 0$. Calculate the distances,

$$d_{ij}^{\mathcal{X}} = d^{\mathcal{X}}(\mathbf{x}_i, \mathbf{x}_j) = \|\mathbf{x}_i - \mathbf{x}_j\|_{\mathcal{X}}, \qquad (16.37)$$

between all pairs of data points $\mathbf{x}_i, \mathbf{x}_j \in \mathcal{X}$, $i, j = 1, 2, \ldots, n$. These are generally taken to be Euclidean distances but may be a different distance metric. Determine which data points are "neighbors" on the manifold $\mathcal{M}$ by connecting each point either to its K nearest neighbors or to all points lying within a ball of radius ϵ of that point. Choice of K or ϵ controls neighborhood size and also the success of Isomap.

This gives us a *weighted neighborhood graph* $\mathcal{G} = \mathcal{G}(\mathcal{V}, \mathcal{E})$, where the set of *vertices* $\mathcal{V} = \{\mathbf{x}_1 \ldots, \mathbf{x}_n\}$ are the input data points, and the set of *edges* $\mathcal{E} = \{e_{ij}\}$ indicate neighborhood relationships between the points. The edge e_{ij} that joins the neighboring points $\mathbf{x}_i$ and $\mathbf{x}_j$ has a weight w_{ij} associated with it, and that weight is given by the "distance" $d_{ij}^{\mathcal{X}}$ between those points. If there is no edge present between a pair of points, the corresponding weight is zero.

2. Compute graph distances. Estimate the unknown true *geodesic distances*, $\{d_{ij}^{\mathcal{M}}\}$, between pairs of points in $\mathcal{M}$ by *graph distances*, $\{d_{ij}^{\mathcal{G}}\}$, with respect to the graph $\mathcal{G}$. The graph distances are the shortest path distances between all pairs of points in the graph $\mathcal{G}$. Points that are not neighbors of each other are connected by a sequence of neighbor-to-neighbor links, and the length of this path (sum of the link weights) is taken to approximate the distance between its endpoints on the manifold.

If the data points are sampled from a probability distribution that is supported by the entire manifold, then, asymptotically (as $n \to \infty$), it turns out that the estimate $d^{\mathcal{G}}$ converges to $d^{\mathcal{M}}$ if the manifold is flat (Bernstein, de Silva, Langford, and Tenenbaum, 2001).

An efficient algorithm for computing the shortest path between every pair of vertices in a graph is *Floyd's algorithm* (Floyd, 1962), which works best with dense graphs (graphs with many edges).

3. Embedding via multidimensional scaling. Let $\mathbf{D}^{\mathcal{G}} = (d_{ij}^{\mathcal{G}})$ be the sym-metric $(n \times n)$-matrix of graph distances. Apply "classical" MDS to $\mathbf{D}^{\mathcal{G}}$ to give the reconstructed data points in a t-dimensional feature space $\mathcal{Y}$, so that the geodesic distances on $\mathcal{M}$ between data points is preserved as much as possible:

- Form the "doubly centered," symmetric, $(n \times n)$-matrix of squared graph distances,

$$\mathbf{A}_n^{\mathcal{G}} = -\frac{1}{2}\mathbf{H}\mathbf{S}^{\mathcal{G}}\mathbf{H}, \qquad (16.38)$$

where $\mathbf{S}^{\mathcal{G}} = ([d_{ij}^{\mathcal{G}}]^2)$, $\mathbf{H} = \mathbf{I}_n - n^{-1}\mathbf{J}_n$, and $\mathbf{J}_n = \mathbf{1}_n\mathbf{1}_n^{\tau}$ is an $(n \times n)$-matrix of ones. The matrix $\mathbf{A}_n^{\mathcal{G}}$ will be nonnegative-definite of rank $t < n$.

- The embedding vectors $\{\widehat{\mathbf{y}}_i\}$ are chosen to minimize $\|\mathbf{A}_n^{\mathcal{G}} - \mathbf{A}_n^{\mathcal{Y}}\|$, where $\mathbf{A}_n^{\mathcal{Y}}$ is (16.38) with $\mathbf{S}^{\mathcal{Y}} = ([d_{ij}^{\mathcal{Y}}]^2)$ replacing $\mathbf{S}^{\mathcal{G}}$, and $d_{ij}^{\mathcal{Y}} = \|\mathbf{y}_i - \mathbf{y}_j\|$ is the Euclidean distance between $\mathbf{y}_i$ and $\mathbf{y}_j$. The optimal solution is given by the eigenvectors $\mathbf{v}_1, \ldots, \mathbf{v}_t$ corresponding to the t largest (positive) eigenvalues, $\lambda_1 \geq \cdots \geq \lambda_t$, of $\mathbf{A}_n^{\mathcal{G}}$.

- The graph $\mathcal{G}$ is embedded into $\mathcal{Y}$ by the $(t \times n)$-matrix

$$\widehat{\mathbf{Y}} = (\widehat{\mathbf{y}}_1, \cdots, \widehat{\mathbf{y}}_n) = (\sqrt{\lambda_1}\mathbf{v}_1, \cdots, \sqrt{\lambda_t}\mathbf{v}_t)^{\tau}. \qquad (16.39)$$

The ith column of $\widehat{\mathbf{Y}}$ yields the embedding coordinates in $\mathcal{Y}$ of the ith data point. The Euclidean distances between the n t-dimensional columns of $\widehat{\mathbf{Y}}$ are collected into the $(n \times n)$-matrix $\mathbf{D}_t^{\mathcal{Y}}$.

The ISOMAP algorithm appears to work most efficiently with $n \leq 1{,}000$. Changes to the ISOMAP code (see below) enable us to work with much larger data sets.

As a measure of how closely the ISOMAP t-dimensional solution matrix $\mathbf{D}_t^{\mathcal{Y}}$ approximates the matrix $\mathbf{D}^{\mathcal{G}}$ of graph distances, we plot $1 - R_t^2$ against dimensionality t (i.e., $t = 1, 2, \ldots, t^*$, where t^* is some integer such as 10), where $R_t^2 = [\mathrm{corr}(\mathbf{D}_t^{\mathcal{Y}}, \mathbf{D}^{\mathcal{G}})]^2$ is the squared correlation coefficient of all corresponding pairs of entries in the matrices $\mathbf{D}_t^{\mathcal{Y}}$ and $\mathbf{D}^{\mathcal{G}}$. The intrinsic dimensionality is taken to be that integer t at which an "elbow" appears in the plot.

Consider, for example, the two-dimensional Swiss roll manifold embedded in three-dimensional space. Suppose we are given 20,000 points randomly drawn from the surface of that manifold.[4] The 3D scatterplot of the data is given in the right panel of Figure 16.7. Using all 20,000 points as input to the ISOMAP algorithm proves to be computationally too big, and so we use only the first 1000 points for illustration. Taking $n = 1{,}000$ and $K = 7$ neighborhood points, Figure 16.8 shows a plot of the values of $1 - R_t^2$ against t for $t = 1, 2, \ldots, 10$, where an elbow correctly shows $t = 2$; the 2D ISOMAP neighborhood-graph solution is given in Figure 16.9.

---

[4]These 3D data, stored as a $(3 \times 20{,}000)$-matrix, are available in the data file swiss_roll_data on the ISOMAP website isomap.stanford.edu.

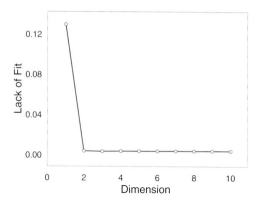

FIGURE 16.8. ISOMAP *dimensionality plot for the* $n = 1,000$ *Swiss roll data points. The number of neighborhood points is* $K = 7$. *The plotted points are* $(t, 1 - R_t^2)$, $t = 1, 2, \ldots, 10$.

The ISOMAP algorithm has difficulty with manifolds that contain holes, have too much curvature, or are not convex. In the case of noisy data, it depends upon how the neighborhood size (either K or ϵ) is chosen; if K or ϵ are chosen neither too large (that it introduces false connections into $\mathcal{G}$) nor too small (that $\mathcal{G}$ becomes too sparse to approximate geodesic paths accurately), then ISOMAP should be able to tolerate moderate amounts of noise in the data.

LANDMARK ISOMAP

When a data set is very large, the performance of the ISOMAP algorithm is significantly degraded by having to store in memory the complete $(n \times n)$-matrix $\mathbf{D}_G$ (step 2) and carry out an eigenanalysis of the $(n \times n)$-matrix $\mathbf{A}_n$ for the MDS reconstruction (step 3). If the data are truly scattered around a low-dimensional manifold, then the vast majority of pairwise distances will be redundant; to speed up the MDS embedding step, we have to eliminate as many of the redundant distance calculations as we can.

In LANDMARK ISOMAP, the researcher tries to eliminate such redundancy by specifying a *landmark* subset of m of the n data points (de Silva and Tenenbaum, 2003). For example, if $\mathbf{x}_i$ is designated as one of the m landmark points, we calculate only those distances between each of the n points and $\mathbf{x}_i$. Input to the LANDMARK ISOMAP algorithm is, therefore, an $(m \times n)$-matrix of distances. The landmark points may be selected by random sampling or by a judicious choice of "representative" points. The number of such landmark points is left to the researcher, but $m = 50$ works

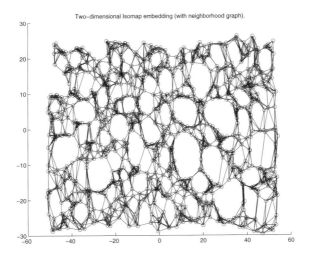

Two–dimensional Isomap embedding (with neighborhood graph).

FIGURE 16.9. *Two-dimensional* ISOMAP *embedding, with neighborhood graph, of the n = 1,000 Swiss roll data points. The number of neighborhood points is K = 7.*

well. In the MDS embedding step, the object is to preserve only those distances between all points and the subset of landmark points. Step 2 in LANDMARK ISOMAP uses *Dijkstra's algorithm* (Dijkstra, 1959), which is faster than Floyd's algorithm for computing graph distances and is generally preferred when the graph is sparse. Dijkstra's algorithm is also recommended as a replacement for Floyd's algorithm in the original ISOMAP algorithm.

Applying LANDMARK ISOMAP to the $n = 1,000$ Swiss roll data points with $K = 7$ and the first $m = 50$ points taken to be landmark points results in an elbow at $t = 2$ in the dimensionality plot; the 2D LANDMARK ISOMAP neighborhood-graph solution is given in Figure 16.10. This is a much faster solution than the one we obtained using the original ISOMAP algorithm. The main differences between Figures 16.9 and 16.10 are roundoff error and a rotation due to sign changes.

Because of the significant increase in computational speed, we can apply LANDMARK ISOMAP to all 20,000 points (using $K = 7$ and $m = 50$); an elbow again correctly appears at $t = 2$ in the dimensionality plot, and the resulting 2D LANDMARK ISOMAP neighborhood-graph solution is given in Figure 16.11.

16.6.4 Local Linear Embedding

The *local linear embedding (LLE)* algorithm (Roweis and Saul, 2000; Saul and Roweis, 2003) for nonlinear dimensionality reduction is similar

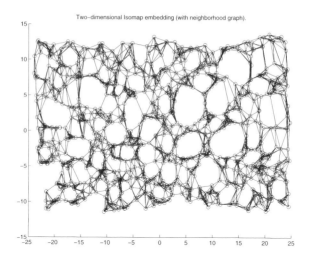

FIGURE 16.10. *Two-dimensional* LANDMARK ISOMAP *embedding, with neighborhood graph, of the n = 1,000 Swiss roll data points. The number of neighborhood points is K = 7 and the number of landmark points is m = 50.*

in spirit to the ISOMAP algorithm, but because it attempts to preserve local neighborhood information on the (Riemannian) manifold (without estimating the true geodesic distances), we view LLE as a *local* approach rather than as the ISOMAP's global approach.

Like ISOMAP, the LLE algorithm also consists of three steps:

1. Nearest neighbor search. Fix $K \ll r$ and let N_i^K denote the "neighborhood" of $\mathbf{x}_i$ that contains only its K nearest points, as measured by Euclidean distance (K could be different for each point $\mathbf{x}_i$).

The success of LLE depends (as does ISOMAP) upon the choice of K: it must be sufficiently large so that the points can be well-reconstructed but also sufficiently small for the manifold to have little curvature.

The LLE algorithm is best served if the graph formed by linking each point to its neighbors is connected. If the graph is not connected, the LLE algorithm can be applied separately to each of the disconnected subgraphs.

2. Constrained least-squares fits. Reconstruct $\mathbf{x}_i$ by a linear function of its K nearest neighbors,

$$\widehat{\mathbf{x}}_i = \sum_{j=1}^{n} w_{ij}\mathbf{x}_j, \tag{16.40}$$

where w_{ij} is a scalar weight for $\mathbf{x}_j$ with unit sum, $\sum_j w_{ij} = 1$, for translation invariance; if $\mathbf{x}_\ell \notin N_i^K$, then set $w_{i\ell} = 0$ in (16.40). Set $\mathbf{W} = (w_{ij})$ to

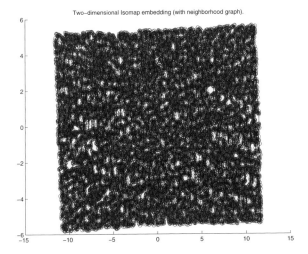

Two–dimensional Isomap embedding (with neighborhood graph).

FIGURE 16.11. *Two-dimensional* LANDMARK ISOMAP *embedding, with neighborhood graph, of the complete set of* $n = 20{,}000$ *Swiss-Roll data points. The number of neighborhood points is* $K = 7$, *and the number of landmark points is* $m = 50$.

be a sparse $(n \times n)$-matrix of weights (there are only nK nonzero elements). Find optimal weights $\{\widehat{w}_{ij}\}$ by solving

$$\widehat{\mathbf{W}} = \arg \min_{\mathbf{W}} \sum_{i=1}^{n} \|\mathbf{x}_i - \sum_{j=1}^{n} w_{ij}\mathbf{x}_j\|^2, \tag{16.41}$$

subject to the *invariance constraint* $\sum_j w_{ij} = 1$, $i = 1, 2, \ldots, n$, and the *sparseness constraint* $w_{i\ell} = 0$ if $\mathbf{x}_\ell \notin N_i^K$.

The matrix $\widehat{\mathbf{W}}$ can be obtained as follows. For a given point $\mathbf{x}_i$, the summand of (16.41) can be written as

$$\|\sum_j w_{ij}(\mathbf{x}_i - \mathbf{x}_j)\|^2 = \mathbf{w}_i^\tau \mathbf{G} \mathbf{w}_i, \tag{16.42}$$

where $\mathbf{w}_i = (w_{i1}, \cdots, w_{in})^\tau$, only K of which are non-zero, and $\mathbf{G} = (G_{jk})$, $G_{jk} = (\mathbf{x}_i - \mathbf{x}_j)^\tau(\mathbf{x}_i - \mathbf{x}_k)$, $j, k \in N_i^K$, is a symmetric, nonnegative-definite, $(n \times n)$-matrix. Using the Lagrangean multiplier μ, we minimize the function

$$f(\mathbf{w}_i) = \mathbf{w}_i^\tau \mathbf{G} \mathbf{w}_i - \mu(\mathbf{1}_n^\tau \mathbf{w}_i - 1).$$

Differentiating $f(\mathbf{w}_i)$ with respect to $\mathbf{w}_i$ and setting the result equal to zero yields $\widehat{\mathbf{w}}_i = \frac{\mu}{2}\mathbf{G}^{-1}\mathbf{1}_n$. Premultiplying this last result by $\mathbf{1}_n^\tau$ gives us the optimal weights

$$\widehat{\mathbf{w}}_i = \frac{\mathbf{G}^{-1}\mathbf{1}_n}{\mathbf{1}_n^\tau \mathbf{G}^{-1}\mathbf{1}_n},$$

where it is understood that for $\mathbf{x}_\ell \notin N_i^K$, the corresponding element, $\widehat{w}_{i\ell}$, of $\widehat{\mathbf{w}}_i$ is zero. Note that we can also write $\mathbf{G}(\frac{2}{\mu}\widehat{\mathbf{w}}_i) = \mathbf{1}_n$; so, the same result can be obtained by solving the linear system of n equations $\mathbf{G}\widehat{\mathbf{w}}_i = \mathbf{1}_n$, where any $\mathbf{x}_\ell \notin N_i^K$ has weight $\widehat{w}_{i\ell} = 0$, and then rescaling the weights to sum to one. Collect the resulting optimal weights for each data point (and all other zero-weights) into a sparse $(n \times n)$-matrix $\widehat{\mathbf{W}} = (\widehat{w}_{ij})$ having only nK nonzero elements.

3. Eigenproblem. Fix the optimal weight matrix $\widehat{\mathbf{W}}$ found at step 2. Find the $(t \times n)$-matrix $\mathbf{Y} = (\mathbf{y}_1. \cdots, \mathbf{y}_n)$, $t \ll r$, of embedding coordinates that solves

$$\widehat{\mathbf{Y}} = \arg\min_{\mathbf{Y}} \sum_{i=1}^{n} \|\mathbf{y}_i - \sum_{j=1}^{n} \widehat{w}_{ij}\mathbf{y}_j\|^2, \qquad (16.43)$$

subject to the constraints $\sum_i \mathbf{y}_i = \mathbf{Y}\mathbf{1}_n = \mathbf{0}$ and $n^{-1} \sum_i \mathbf{y}_i\mathbf{y}_i^\tau = n^{-1}\mathbf{Y}\mathbf{Y}^\tau = \mathbf{I}_t$.

These constraints are imposed to fix the translation, rotation, and scale of the embedding coordinates so that the objective function will be invariant. We can show that (10.60) can be written as

$$\widehat{\mathbf{Y}} = \arg\min_{\mathbf{Y}} \operatorname{tr}\{\mathbf{Y}\mathbf{M}\mathbf{Y}^\tau\} \qquad (16.44)$$

where $\mathbf{M}$ is the sparse, symmetric, and nonnegative-definite $(n \times n)$-matrix $\mathbf{M} = (\mathbf{I}_n - \widehat{\mathbf{W}})^\tau (\mathbf{I}_n - \widehat{\mathbf{W}})$.

The objective function $\operatorname{tr}\{\mathbf{Y}\mathbf{M}\mathbf{Y}^\tau\}$ in (16.56) has a unique global minimum given by the eigenvectors corresponding to the *smallest* $t + 1$ eigenvalues of $\mathbf{M}$. The smallest eigenvalue of $\mathbf{M}$ is zero with corresponding eigenvector $\mathbf{v}_n = n^{-1/2}\mathbf{1}_n$. Because the sum of coefficients of each of the other eigenvectors, which are orthogonal to $n^{-1/2}\mathbf{1}_n$, is zero, if we ignore the smallest eigenvalue (and associated eigenvector), this will constrain the embeddings to have mean zero. The optimal solution then sets the rows of the $(t \times n)$-matrix $\widehat{\mathbf{Y}}$ to be the t remaining n-dimensional eigenvectors of $\mathbf{M}$,

$$\widehat{\mathbf{Y}} = (\widehat{\mathbf{y}}_1, \ldots, \widehat{\mathbf{y}}_n) = (\mathbf{v}_{n-1}, \cdots, \mathbf{v}_{n-t})^\tau, \qquad (16.45)$$

where $\mathbf{v}_{n-j}$ is the eigenvector corresponding to the $(j+1)$st smallest eigenvalue of $\mathbf{M}$. The sparseness of $\mathbf{M}$ enables eigencomputations to be carried out very efficiently.

Because LLE preserves local (rather than global) properties of the underlying manifold, it is less susceptible to introducing false connections in $\mathcal{G}$ and can successfully embed nonconvex manifolds. However, like ISOMAP, it has difficulty with manifolds that contain holes.

16.6.5   Laplacian Eigenmaps

The *Laplacian eigenmap* algorithm (Belkin and Niyogi, 2002) also consists of three steps. The first and third steps of the Laplacian eigenmap algorithm are very similar to the first and third steps, respectively, of the LLE algorithm.

1. Nearest-neighbor search. Fix an integer K or an $\epsilon > 0$. The neighborhoods of each data point are symmetrically defined: for a K-neighborhood N_i^K of the point $\mathbf{x}_i$, let $\mathbf{x}_j \in N_i^K$ iff $\mathbf{x}_i \in N_j^K$; similarly, for an ϵ-neighborhood N_i^ϵ, let $\mathbf{x}_j \in N_i^\epsilon$ iff $\|\mathbf{x}_i - \mathbf{x}_j\| < \epsilon$, where the norm is Euclidean norm. In general, let N_i denote the neighborhood of $\mathbf{x}_i$.

2. Weighted adjacency matrix. Let $\mathbf{W} = (w_{ij})$ be a symmetric $(n \times n)$ *weighted adjacency matrix* defined as follows:

$$
w_{ij} = \begin{cases} \exp\left\{ -\frac{\|\mathbf{x}_i - \mathbf{x}_j\|^2}{2\sigma^2} \right\}, & \text{if } \mathbf{x}_j \in N_i; \\ 0, & \text{otherwise.} \end{cases} \tag{16.46}
$$

These weights are determined by the isotropic Gaussian kernel (also known as the *heat kernel*), with scale parameter σ. A simpler $\mathbf{W}$ is given by $w_{ij} = 1$ if $\mathbf{x}_j \in N_i$, and 0 otherwise. Denote the resulting weighted graph by $\mathcal{G}$. If $\mathcal{G}$ is not connected, apply step 3 to each connected subgraph.

3. Eigenproblem. Embed the graph $\mathcal{G}$ into the low-dimensional space $\Re^t$ by the $(t \times n)$-matrix $\mathbf{Y} = (\mathbf{y}_1, \cdots, \mathbf{y}_n)$, where the ith column of $\mathbf{Y}$ yields the embedding coordinates of the ith point. Let $\mathbf{D} = (d_{ij})$ be an $(n \times n)$ diagonal matrix with diagonal elements $d_{ii} = \sum_{j \in N_i} w_{ij} = (\mathbf{W}\mathbf{1}_n)_i$, $i = 1, 2, \ldots, n$. The $(n \times n)$ symmetric matrix $\mathbf{L} = \mathbf{D} - \mathbf{W}$ is known as the *graph Laplacian* for the graph $\mathcal{G}$. Let $\mathbf{y} = (y_i)$ be an n-vector. Then, $\mathbf{y}^\tau \mathbf{L} \mathbf{y} = \frac{1}{2} \sum_{i=1}^n \sum_{j=1}^n w_{ij}(y_i - y_j)^2$, so that $\mathbf{L}$ is nonnegative definite. The graph Laplacian can be regarded as an approximation to the continuous *Laplace–Beltrami operator* Δ defined on the manifold $\mathcal{M}$.

The matrix $\mathbf{Y}$ is determined by minimizing the following objective function:

$$
\sum_i \sum_j w_{ij} \|\mathbf{y}_i - \mathbf{y}_j\|^2 = \text{tr}\{\mathbf{Y}\mathbf{L}\mathbf{Y}^\tau\}. \tag{16.47}
$$

In other words, we seek the solution,

$$
\widehat{\mathbf{Y}} = \arg \min_{\mathbf{Y}\mathbf{D}\mathbf{Y}^\tau = \mathbf{I}_t} \text{tr}\{\mathbf{Y}\mathbf{L}\mathbf{Y}^\tau\}, \tag{16.48}
$$

where we restrict $\mathbf{Y}$ such that $\mathbf{Y}\mathbf{D}\mathbf{Y}^\tau = \mathbf{I}_t$ to prevent a collapse onto a subspace of fewer than $t - 1$ dimensions. This problem boils down to solving the generalized eigenequation, $\mathbf{L}\mathbf{v} = \lambda \mathbf{D}\mathbf{v}$, or, equivalently, finding the eigenvalues and eigenvectors of the matrix $\widetilde{\mathbf{W}} = \mathbf{D}^{-1/2}\mathbf{W}\mathbf{D}^{-1/2}$. The smallest eigenvalue, λ_n, of $\widetilde{\mathbf{W}}$ is zero. If we ignore the smallest eigenvalue (and its corresponding constant eigenvector $\mathbf{v}_n = \mathbf{1}_n$), then the best embedding in $\Re^t$ is similar to that given by LLE; that is, the rows of $\widehat{\mathbf{Y}}$ are the eigenvectors,

$$\widehat{\mathbf{Y}} = (\widehat{\mathbf{y}}_1, \cdots, \widehat{\mathbf{y}}_n) = (\mathbf{v}_{n-1}, \cdots, \mathbf{v}_{n-t})^\tau, \qquad (16.49)$$

corresponding to the next t smallest eigenvalues, $\lambda_{n-1} \leq \cdots \leq \lambda_{n-t}$, of $\widehat{\mathbf{W}}$.

16.6.6 Hessian Eigenmaps

We noted earlier that, in certain situations, the convexity assumption for ISOMAP may be too restrictive. It may be more realistic in such situations to require instead that the manifold $\mathcal{M}$ be locally isometric to an open, connected subset of $\Re^t$. Examples include families of "articulated" images (i.e., translated or rotated images of the same object, possibly through time) that are selected from a high-dimensional, digitized-image library (e.g., faces, pictures, handwritten numbers or letters). If the pixel elements of each 64-pixel-by-64-pixel digitized image are represented as a 4,096-dimensional vector in "pixel space," it can be very difficult to show that the images really live on a low-dimensional manifold, especially if that *image manifold* is unknown.

Such images can be modeled using a vector of smoothly varying *articulation parameters* $\boldsymbol{\theta} \in \Theta$. For example, digitized images of a person's face that are varied by pose and illumination can be parameterized by two pose parameters (expression [happy, sad, sleepy, surprised, wink] and glasses–no glasses) and a lighting direction (centerlight, leftlight, rightlight, normal); similarly, handwritten "2"s appear to be parameterized essentially by two features, bottom loop and top arch (Tenenbaum, de Silva, and Langford, 2000; Roweis and Saul, 2000). To some extent, learning about an underlying image manifold depends upon data quality: are the images sufficiently scattered around the manifold to enable us to identify the manifold, and how good is the quality of digitization of each image?

Hessian eigenmaps have been proposed for recovering manifolds of high-dimensional libraries of articulated images where the convexity assumption is often violated (Donoho and Grimes, 2003). Assume the parameter space is $\Theta \subset \Re^t$ and suppose that $\phi : \Theta \to \Re^r$, where $t < r$. Assume $\mathcal{M} = \phi(\Theta)$ is a smooth manifold of articulated images. The isometry and convexity requirements of ISOMAP are replaced by the following weaker requirements:

- *Local Isometry:* ϕ is a locally isometric embedding of Θ into $\Re^r$. For any point $\mathbf{x}'$ in a sufficiently small neighborhood around each point $\mathbf{x}$ on the manifold $\mathcal{M}$, the geodesic distance equals the Euclidean distance between their corresponding parameter points $\boldsymbol{\theta}, \boldsymbol{\theta}' \in \Theta$;

i.e.,

$$d^{\mathcal{M}}(\mathbf{x}, \mathbf{x}') = \|\boldsymbol{\theta} - \boldsymbol{\theta}'\|_{\Theta}, \tag{16.50}$$

where $\mathbf{x} = \phi(\boldsymbol{\theta})$ and $\mathbf{x}' = \phi(\boldsymbol{\theta}')$.

- *Connectedness:* The parameter space Θ is an open, connected subset of $\Re^t$.

The goal is to recover the parameter vector $\boldsymbol{\theta}$ (up to a rigid motion).

First, consider the differentiable manifold $\mathcal{M} \subset \Re^r$. Let $\mathcal{T}_{\mathbf{x}}(\mathcal{M})$ be a tangent space of the point $\mathbf{x} \in \mathcal{M}$, where $\mathcal{T}_{\mathbf{x}}(\mathcal{M})$ has the same number of dimensions as $\mathcal{M}$ itself. We endow $\mathcal{T}_{\mathbf{x}}(\mathcal{M})$ with a (non-unique) system of orthonormal coordinates having the same inner product as $\Re^r$. We can view $\mathcal{T}_{\mathbf{x}}(\mathcal{M})$ as an affine subspace of $\Re^r$ that is spanned by vectors tangent to $\mathcal{M}$ and pass through the point $\mathbf{x}$, with the origin $0 \in \mathcal{T}_{\mathbf{x}}(\mathcal{M})$ identified with $\mathbf{x} \in \mathcal{M}$. Let $N_{\mathbf{x}}$ be a neighborhood of $\mathbf{x}$ such that each point $\mathbf{x}' \in N_{\mathbf{x}}$ has a unique closest point $\boldsymbol{\xi}' \in \mathcal{T}_{\mathbf{x}}(\mathcal{M})$; a point in $N_{\mathbf{x}}$ has local coordinates, $\boldsymbol{\xi} = \boldsymbol{\xi}(\mathbf{x}) = (\xi_1(\mathbf{x}), \ldots, \xi_t(\mathbf{x}))^\tau$, say, and these coordinates are referred to as *tangent coordinates*.

Suppose $f : \mathcal{M} \to \Re$ is a C^2-function (i.e., a function with two continuous derivatives) near $\mathbf{x}$. If the point $\mathbf{x}' \in N_{\mathbf{x}}$ has local coordinates $\boldsymbol{\xi} = \boldsymbol{\xi}(\mathbf{x}) \in \Re^t$, then the rule $g(\boldsymbol{\xi}) = f(\mathbf{x}')$ defines a C^2-function $g : U \to \Re$, where U is a neighborhood of $0 \in \Re^r$. The *tangent Hessian* matrix, which measures the "curviness" of f at the point $\mathbf{x} \in \mathcal{M}$, is defined as the ordinary $(t \times t)$ Hessian matrix of g,

$$\mathbf{H}_f^{\mathrm{tan}}(\mathbf{x}) = \left(\frac{\partial^2 g(\boldsymbol{\xi})}{\partial \xi_i \partial \xi_j} \Big|_{\boldsymbol{\xi}=\mathbf{0}} \right). \tag{16.51}$$

The average "curviness" of f over $\mathcal{M}$ is then the quadratic form,

$$\mathcal{H}(f) = \int_{\mathcal{M}} \| \mathbf{H}_f^{\mathrm{tan}}(\mathbf{x}) \|_F^2 \, d\mathbf{x}, \tag{16.52}$$

where $\| \mathbf{H} \|_F^2 = \sum_i \sum_j H_{ij}^2$ is the squared Frobenius norm of a square matrix $\mathbf{H} = (H_{ij})$. Note that even if we define two different orthonormal coordinate systems for $\mathcal{T}_{\mathbf{x}}(\mathcal{M})$, and hence two different tangent Hessian matrices, $\mathbf{H}_f$ and $\mathbf{H}'_f$, at $\mathbf{x}$, they are related by $\mathbf{H}'_f = \mathbf{U}\mathbf{H}_f\mathbf{U}^\tau$, where $\mathbf{U}$ is orthogonal, so that their Frobenius norms are equal and $\mathcal{H}(f)$ is well-defined.

Donoho and Grimes showed that $\mathcal{H}(f)$ has a $(t+1)$-dimensional nullspace consisting of the constant function and a t-dimensional space of functions spanned by the original isometric coordinates, $\theta_1, \ldots, \theta_t$, which can be recovered (up to a rigid motion) from the null space of $\mathcal{H}(f)$.

The *Hessian Locally Linear Embedding (HLLE)* algorithm computes a discrete approximation to the Hessian $\mathcal{H}$ using the data lying on $\mathcal{M}$. There are, again, three steps to this algorithm, which essentially substitutes a quadratic form based upon the Hessian instead of one based upon the Laplacian.

1. Nearest-Neighbor Search. We begin by identifying a neighborhood of each point as in Step 1 of the LLE algorithm. Fix an integer K and let N_i^K denote the K nearest neighbors of the data point $\mathbf{x}_i$ using Euclidean distance.

2. Estimate Tangent Hessian Matrices. Assuming local linearity of the manifold $\mathcal{M}$ in the region of the neighborhood N_i^K, form the $(r \times r)$ covariance matrix $\mathbf{M}_i$ of the K neighborhood-centered points $\mathbf{x}_j - \bar{\mathbf{x}}_i$, $j \in N_i^K$, where $\bar{\mathbf{x}}_i = n^{-1} \sum_{j \in N_i^K} \mathbf{x}_j$, and compute a PCA of the matrix $\mathbf{M}_i$. Assuming $K \geq t$, the first t eigenvectors of $\mathbf{M}_i$ yield the tangent coordinates of the K points in N_i^K and provide the best-fitting t-dimensional linear subspace corresponding to $\mathbf{x}_i$. Next, construct a LS estimate, $\widehat{\mathbf{H}}_i$, of the local Hessian matrix $\mathbf{H}_i$ as follows: build a matrix $\mathbf{Z}_i$ by putting all squares and cross-products of the columns of $\mathbf{M}_i$ up to the tth order in its columns, including a column of 1s; so, $\mathbf{Z}_i$ has $1 + t + t(t+1)/2$ columns and K rows. Then, apply a Gram–Schmidt orthonormalization to $\mathbf{Z}_i$. The estimated $(t(t+1)/2 \times K)$ tangent Hessian matrix $\widehat{\mathbf{H}}_i$ is given by the transpose of the last $t(t+1)/2$ orthonormal columns of $\mathbf{Z}_i$.

3. Eigenanalysis. The estimated local Hessian matrices, $\widehat{\mathbf{H}}_i, i = 1, 2, \ldots, n$, are used to construct a sparse, symmetric, $(r \times r)$-matrix $\widehat{\mathcal{H}} = (\widehat{\mathcal{H}}_{k\ell})$, where

$$\widehat{\mathcal{H}}_{k\ell} = \sum_i \sum_j ((\widehat{\mathbf{H}}_i)_{jk}(\widehat{\mathbf{H}}_i)_{j\ell}. \tag{16.53}$$

$\widehat{\mathcal{H}}$ is a discrete approximation to the functional $\mathcal{H}$. We now follow Step 3 of the LLE algorithm, this time performing an eigenanalysis of $\widehat{\mathcal{H}}$. To obtain the low-dimensional representation that will minimize the curviness of the manifold, find the *smallest* $t + 1$ eigenvectors of $\widehat{\mathcal{H}}$; the smallest eigenvalue will be zero, and its associated eigenvector will consist of constant functions; the remaining t eigenvectors provide the embedding coordinates for $\widehat{\boldsymbol{\theta}}$.

16.6.7 Other Methods

There are several other methods for nonlinear manifold learning, including an algorithm for "charting" manifolds (Brand, 2003), which uses parametric density estimation and a Bayesian approach, and a local tangent space alignment algorithm (Zhang and Zha, 2004).

16.6.8 Relationships to Kernel PCA

The three algorithms of Isomap, LLE, and Laplacian eigenmaps have close connections with kernel PCA (Ham, Lee, Mika, and Scholkopf, 2003).

For each algorithm, the individual elements of the kernel matrix depend upon all the input data, unlike traditional kernel matrices whose entries each depend only upon a pair of input points.

ISOMAP For isotropic kernels, it can be shown that kernel PCA is closely related to metric MDS (Williams, 2001). Thus, ISOMAP is equivalent to kernel PCA if

$$\mathbf{K}_{\mathrm{Isomap}} = \mathbf{A}_n^{\mathcal{G}} \tag{16.54}$$

is used as the appropriate kernel matrix. However, $\mathbf{A}_n$ is not guaranteed to be nonnegative definite for finite n. It is nonnegative definite only in an asymptotic sense (i.e., as $n \to \infty$).

LLE If the largest eigenvalue of $\mathbf{M}$ is λ_1, then the $(n \times n)$ Gram matrix

$$\mathbf{K}_{LLE} = \lambda_1 \mathbf{I}_n - \mathbf{M} \tag{16.55}$$

is nonnegative definite, the eigenvector corresponding to the zero eigenvalue of $\mathbf{K}_{LLE}$ is $n^{-1/2}\mathbf{1}_n$, and eigenvectors 2 through $t+1$ of $\mathbf{K}_{LLE}$ give the LLE embedding. Furthermore, the LLE embedding is equivalent (up to the scaling factors $\sqrt{\lambda_k}$) to the kernel PCA scores (see Section 10.6.2) based upon $\mathbf{K}_{LLE}$. The form of the kernel function K corresponding to $\mathbf{K}_{LLE}$ is also not explicitly known.

An alternative version of LLE uses a kernel representation of the input data. Instead of finding the K nearest neighbors of each point in input space, we can use kernel methods to find the nearest neighbors in feature space (DeCoste, 2001). The Euclidean distance between two points in feature space is given by

$$d_{ij} = \|\Phi(\mathbf{x}_i) - \Phi(\mathbf{x}_j)\| = \sqrt{K_{ii} - 2K_{ij} + K_{jj}}. \tag{16.56}$$

Using this definition of distance in feature space, nearest neighbors of $\Phi(\mathbf{x}_i)$ can be found by using an efficient algorithm that supports such distances (e.g., Yianilos, 1998). Corresponding to the matrix $\mathbf{G}$ in step 2 of the algorithm, we can define the matrix $\widetilde{\mathbf{G}} = (\widetilde{G}_{jk})$ in feature space, where, for all $\mathbf{x}_j, \mathbf{x}_k \in N_i^K$,

$$\begin{aligned} \widetilde{G}_{jk} &= \langle \Phi(\mathbf{x}_i) - \Phi(\mathbf{x}_j), \Phi(\mathbf{x}_i) - \Phi(\mathbf{x}_k) \rangle \\ &= K_{ii} - K_{ij} - K_{ik} + K_{jk}. \end{aligned} \tag{16.57}$$

Replacing $\mathbf{G}$ by $\widetilde{\mathbf{G}}$ in step 2 of the LLE algorithm, we find the matrix of optimal weights $\widetilde{\mathbf{W}}$ (replacing $\mathbf{W}$) and the embedding vectors (corresponding to step 3).

Laplacian Eigenmaps As we saw in step 3 of the algorithm, the embedding is obtained by finding the eigenvectors corresponding to the smallest eigenvalues of the graph Laplacian $\mathbf{L}$. This solution can also be justified in terms of arguments involving heat flow and diffusion on a graph. Without going into details (which involve the notion of commute times of diffusion on a graph), it can be shown that if we take as kernel

$$\mathbf{K}_{LE} = \mathbf{L}^{-}, \tag{16.58}$$

where $\mathbf{L}^{-}$ is a generalized-inverse of $\mathbf{L}$, then the embedding solution is equivalent to performing kernel PCA on the matrix $\mathbf{K}_{LE}$.

16.7   Software Packages

The website `www.iro.umontreal.ca/~kegl/research/pcurves` gives a review of the area of principal curves and gives an introduction to algorithms and software. The S-PLUS/R computer packages `princurve` and `pcurve`, both based on S-code originally written by Hastie, are available for fitting a principal curve to multivariate data. MATLAB code for principal curves is available at `lear.inrialpes.fr/ verbeek/software`.

There are several publicly available computer programs for performing kernel PCA; see, for example, the `kcpa` function included in the R package `kernlab`, which can be downloaded from CRAN.

MATLAB code for implementing ISOMAP, LLE, and HLLE is publicly available at the following websites:

ISOMAP: `isomap.stanford.edu`

LLE: `www.cs.toronto.edu/~roweis/lle/`

Laplacian Eigenmaps:

`people.cs.uchicago.edu/~misha/ManifoldLearning/index.html`

HLLE: `basis.stanford.edu/WWW/HLLE/frontdov.htm`

See Martinez and Martinez (2005, Section 3.2 and Appendix B). There is also a `Matlab_Toolbox_for_Dimensionality_Reduction`, which is downloadable from the website

`www.cs.unimaas.nl/l.vandermaaten/Laurens_van_der_Maaten`

and includes all the methods discussed in this chapter and many data sets. There is, at present, no S-PLUS/R code for ISOMAP, LLE, Laplacian eigenmaps, or HLLE.

Bibliographical Notes

Much of our discussion of nonlinear dimensionality reduction and manifold learning has its roots in differential geometry. A text that gives an excellent panoramic view of the historical development of both Euclidean and Riemannian aspects of differential geometry is Berger (2003). Another useful book is Thorpe (1994).

The website www.iro.umontreal.ca/~kegl/research/pcurves gives a list of references for principal curves. A detailed study of the concept of self-consistency is given by Tarpey and Flury (1996). Linear PCA can also be generalized by considering additive functions $\sum_i \phi_i(X_i)$, where the $\{\phi_i\}$ satisfy normalization and orthogonality conditions. This nonlinear generalization is called *additive principal components* (Donnell, Buja, and Stuetzle, 1994). Another version of nonlinear PCA is given by Salinelli (1998). Our treatment of kernel PCA is based upon the work of Scholkopf, Smola, and Muller (1998). See also Scholkopf and Smola (2002).

Exercises

16.1 Generate $n = 150$ trivariate ($r = 3$) observations on (X_1, X_2, X_3) so that they lie on the surface of the sphere $X_1^2 + X_2^2 + X_3^2 = 36$. Compute the $2r + r(r-1)/2 = 9$ variables $(X_1, X_2, X_3, X_1^2, X_2^2, X_3^2, X_1X_2, X_1X_3, X_2X_3)$ and carry out an error-free quadratic PCA of the extended vector. Then, add an independent $Z \sim \mathcal{N}(0, 0.25)$ variate to the X_2 variable and carry out a noisy quadratic PCA of the extended vector.

16.2 Using the kernels listed in Table 11.1, check whether (or not) they each have the property that $[K(\mathbf{x}, \mathbf{y})]^2 \leq K(\mathbf{x}, \mathbf{x})K(\mathbf{y}, \mathbf{y})$.

16.3 Using the Food Nutrition data, compute the first two kernel principal component scores and plot them for different values of σ for the RBF kernel. In the scatterplot, identify which of the six variables dominates each point. Add another identification for points that have very low values for each variable. Replot the kernel PC scores using different colors for the seven classes. Comment on your findings.

16.4 Using the `pendigits` data (Section 7.2.10), which consist of 10,992 handwritten digits (0, 1, 2, ..., 9), compute the kernel PC scores and plot them for different values of σ for the RBF kernel. Use different colors for the 10 digits. Comment on your findings and compare your results with Figure 7.4.

16.5 Generate $n = 500$ independent data values from the multivariate Gaussian distribution $\mathcal{N}_4(\mathbf{0}, \mathbf{R})$, where $\mathbf{R}$ is the correlation matrix,

$$\mathbf{R} = \begin{pmatrix} 1.0 & 0.5 & 0.7 & -0.6 \\ 0.5 & 1.0 & 0.3 & -0.5 \\ 0.7 & 0.3 & 1.0 & -0.7 \\ -0.6 & -0.5 & -0.7 & 1.0 \end{pmatrix}.$$

Run these simulated data through a kernel PCA program and make a scatterplot of the first two kernel PC scores. Choose a range of values of σ for the RBF kernel and vary the values of the correlations in the matrix $\mathbf{R}$. Comment on your findings.

16.6 In kernel PCA and in MDS, we "double-center" a symmetric $(n \times n)$-matrix $\mathbf{A} = (a_{ij})$ by the transformation,

$$\mathbf{B} = (\mathbf{I}_n - n^{-1}\mathbf{J}_n)\mathbf{A}(\mathbf{I}_n - n^{-1}\mathbf{J}_n),$$

where $\mathbf{J}_n = \mathbf{1}_n\mathbf{1}_n^\tau$ and $\mathbf{1}_n$ is an n-vector of all ones. Show that the ijth entry of $\mathbf{B}$ can be expressed as

$$b_{ij} = a_{ij} - a_{i\cdot} - a_{\cdot j} + a_{\cdot\cdot},$$

where a dot-subscript indicates averaging over that subscript.

16.7 In MDS, the matrix $\mathbf{A} = (a_{ij})$ in Ex. 16.7 is a *dissimilarity matrix* with $a_{ij} = -\frac{1}{2}d_{ij}^2$, where d_{ij} is the interpoint distance (or *dissimilarity*). Note: d_{ij} is a *dissimilarity* if $d_{ii} = 0$, $d_{ij} \geq 0$, and $d_{ij} = d_{ji}$. Show that, for the MDS case, the matrix $\mathbf{B}$ in Ex. 16.7 ($\mathbf{A}_n^{\mathcal{G}}$ in (16.55)) is nonnegative-definite.

16.8 Show that $(\mathbf{I}_n - n^{-1}\mathbf{J}_n)\mathbf{1}_n = \mathbf{0}$ and, hence, that $\mathbf{B}\mathbf{1}_n = \mathbf{0}$, where $\mathbf{B}$ is given in Ex. 16.7. Let $\bar{\mathbf{y}} = n^{-1}\widehat{\mathbf{Y}}^\tau\mathbf{1}_n$, where $\widehat{\mathbf{Y}}$ is the embedding matrix given by (16.56). Use the spectral decomposition of $\mathbf{B}$, assuming $\mathbf{B}$ has rank $t < n$, to show that $n^2\bar{\mathbf{y}}^\tau\bar{\mathbf{y}} = \mathbf{0}$ and, hence, that $\bar{\mathbf{y}} = \mathbf{0}$. Thus, the IsoMAP embeddings have mean zero.

16.9 Download the `helix.mat` data set from the

  `Matlab_Toolbox_for_Dimensionality_Reduction`.

Run PCA, kernel PCA, IsoMAP, LLE, Lapacian Eigenmaps, and HLLE algorithms on the `helix` data, report your results, and compare solutions.

16.10 Download the `COIL20` dataset from the

  `Matlab_Toolbox_for_Dimensionality_Reduction`.

Run PCA, kernel PCA, IsoMAP, LLE, Lapacian Eigenmaps, and HLLE algorithms on the `COIL20` data, report your results, and compare solutions.

17
Correspondence Analysis

17.1 Introduction

Correspondence analysis is an exploratory multivariate technique for simultaneously displaying scores representing the row categories and column categories of a two-way contingency table as the coordinates of points in a low-dimensional (two- or possibly three-dimensional) vector space. The objective is to clarify the relationship between the row and column variates of the table and to discover a low-dimensional explanation for possible deviations from independence of those variates. The methodology has its own nomenclature, and its approach is decidedly geometric, especially for interpreting the resulting graphical displays.

For two-way contingency tables, correspondence analysis is known as *simple* correspondence analysis. For three-way and higher contingency tables, it is known as *multiple* correspondence analysis. Variants of correspondence analysis are *dual* (or *optimal*) *scaling, reciprocal averaging, perceptual mapping*, and *social space analysis*. In general, correspondence analysis is applicable when the variates are discrete with many categories and, hence, is well-suited for analyzing large contingency tables. It can also be used for continuous variates, such as age, which can be segmented into a finite

A.J. Izenman, *Modern Multivariate Statistical Techniques*,
doi: 10.1007/978-0-387-78189-1_17,
© Springer Science+Business Media, LLC 2008

number of ranges, but discretization of a continuous variate usually entails some loss of information.

17.1.1 Example: Shoplifting in The Netherlands

These data[1] were taken from van der Heijden, de Falguerolles, and de Leeuw (1989). It is a three-way contingency table of 33,101 individuals, classified by gender and age, who were suspected of stealing specific goods in The Netherlands in 1978 and 1979. The data were obtained from a survey of about 350 Dutch stores and big retail shops. Cases in which shoplifting consisted of more than a single type of good, or in which more than one person was suspected, were omitted from the study. Age was divided into nine nonoverlapping categories, and shoplifted items were classified into 13 types of goods.

For this example, we arranged the original $2 \times 9 \times 13$ three-way contingency table into a $(2 \times 9) \times 13$ two-way contingency table in which gender has been introduced as separate sets of nine male and nine female rows of ages. The ages were coded by groups: < 12 (1 for boys and 10 for girls), 12–14 (2 and 11), 15–17 (3 and 12), 18–20 (4 and 13), 21–29 (5 and 14), 30–39 (6 and 15), 40–49 (7 and 16), 50–64 (8 and 17), and 65+ (9 and 18). The graphical display from the resulting correspondence analysis is given in Figure 17.1.

We can make the following observations from Figure 17.1. First, points representing males and females are well-separated at each age group, suggesting that their shoplifting profiles are quite different. Second, for both males and females, the age category points are clearly ordered from younger than 12 years old on the left-hand side to older than 65 on the right-hand side, with both sets of points doubling back toward the left after 30 years of age. Third, while there are larger distances between males at the younger age groups than those at older age groups, suggesting that shoplifting behavior changes substantially more for younger than for older males, the distances between female age groups are largest at both the younger and older ages (and, hence, more rapidly changing shoplifting behavior), with smaller distances appearing in the middle age groups (18–49).

The configuration of points in Figure 17.1 also tempts us to identify column points (which types of goods are shoplifted more than average) with nearby row points (age groups), possibly leading to the identification of significant age × goods interactions. Although interrow distances and intercolumn distances can be compared, row-to-column distances are undefined and, therefore, are essentially meaningless (see, e.g., Greenacre and Hastie,

---

[1]The contingency table can be downloaded from the book's website.

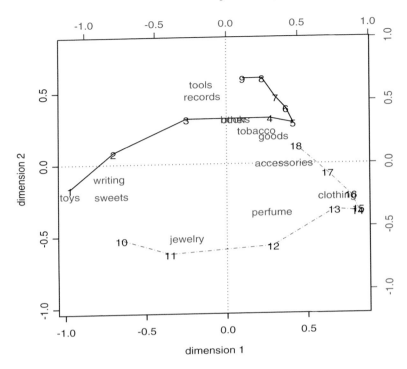

FIGURE 17.1. *Correspondence map for the shoplifting example. The red words are the items shoplifted, the points joined by a solid line represent the progression in male ages (1–9), and the points joined by a dotted line represent the progression in female ages (10–18).*

1987). In other words, row points should not be associated with neighboring column points (and vice versa). Using row percentages obtained from the contingency table, we summarize in Table 17.1 the types of goods most often shoplifted by males and by females at each of the different age groups. In the light of the above comments, it is perhaps instructive for the reader to compare Figure 17.1 with Table 17.1.

17.2 Simple Correspondence Analysis

17.2.1 Two-Way Contingency Tables

Categorical data are count data that are collected in a contingency table **N**. A *two-way* $(r \times s)$ *contingency table* with r rows (labelled $A_1, A_2, \ldots, A_r$) and s columns (labelled $B_1, B_2, \ldots, B_s$) has rs cells. The ijth cell has entry

TABLE 17.1. *Types of goods most often shoplifted by males and by females at each age group, as derived from the two-way contingency table of the example. Superscripts show the percentages of that type of good stolen for that age group and gender. Also listed in parentheses for each age group and gender are those goods that are stolen more than 20% of the time.*

Age	Males	Females
< 12	Toys $^{26.2}$ (writing materials $^{23.5}$)	Writing materials $^{23.8}$
12–14	Writing materials $^{25.1}$	Jewelry $^{26.5}$
15–17	Writing materials $^{14.8}$	Clothing $^{32.3}$ (jewelry $^{20.5}$)
18–20	Clothing $^{22.8}$	Clothing $^{45.4}$
21–29	Clothing $^{27.3}$	Clothing $^{55.8}$
30–39	Clothing $^{25.9}$	Clothing $^{57.2}$
40–49	Clothing $^{21.7}$	Clothing $^{51.7}$
50–64	Hobbies, tools $^{22.6}$	Clothing $^{39.4}$
65+	Provisions, tobacco $^{27.3}$ (hobbies, tools $^{20.9}$)	Provisions, tobacco $^{30.1}$ (clothing $^{24.2}$)

n_{ij}, representing the observed frequency in row category A_i and column category B_j, $i = 1, 2, \ldots, r$, $j = 1, 2, \ldots, s$. The ith *marginal row total* is $n_{i+} = \sum_{j=1}^{s} n_{ij}$, $i = 1, 2, \ldots, r$, and the jth *marginal column total* is $n_{+j} = \sum_{i=1}^{r} n_{ij}$, $j = 1, 2, \ldots, s$. If $n = \sum_{i=1}^{r} \sum_{j=1}^{s} n_{ij}$ individuals are classified by row and column categories, then Table 17.2, which is also called a *correspondence table*, shows the cell frequencies, marginal totals, and total sample size. For interpretation purposes, it is important to distinguish when the n individuals are randomly selected from some very large population or when they actually constitute the entire population of interest.

We denote by π_{ij} the probability that an individual has the properties A_i and B_j, $i = 1, 2, \ldots, r$, $j = 1, 2, \ldots, s$. In the event that the row variable A is independent of the column variable B, we have that $\pi_{ij} = \pi_{i+}\pi_{+j}$, where $\pi_{i+} = \sum_j \pi_{ij}$ and $\pi_{+j} = \sum_i \pi_{ij}$, for all $i = 1, 2, \ldots, r$ and $j = 1, 2, \ldots, s$. We are generally interested in assessing whether A and B are indeed independent variables. Such a question can alternatively be posed in terms of *homogeneity* of the row or column probability distributions; that is, whether all the rows have the same probability distributions across columns, or, equivalently, whether all the columns have the same probability distributions across rows.

17.2.2   Row and Column Dummy Variables

For a two-way contingency table, we are interested in the relationship between the row categories and the column categories. We define two sets of dummy variates, an r-vector $\mathbf{X}_i = (X_{ij})$ to indicate which of the n

TABLE 17.2. *Two-way contingency table, showing observed cell frequencies, row and column marginal totals, and total sample size.*

Row Variable	B_1	B_2	$\cdots$	B_j	$\cdots$	B_s	Row Total
			Column Variable				
A_1	n_{11}	n_{12}	$\cdots$	n_{1j}	$\cdots$	n_{1s}	n_{1+}
A_2	n_{21}	n_{22}	$\cdots$	n_{2j}	$\cdots$	n_{2s}	n_{2+}
$\vdots$	$\vdots$	$\vdots$		$\vdots$		$\vdots$	$\vdots$
A_i	n_{i1}	n_{i2}	$\cdots$	n_{ij}	$\cdots$	n_{is}	n_{i+}
$\vdots$	$\vdots$	$\vdots$		$\vdots$		$\vdots$	$\vdots$
A_r	n_{r1}	n_{r2}	$\cdots$	n_{rj}	$\cdots$	n_{rs}	n_{r+}
Column total	n_{+1}	n_{+2}	$\cdots$	n_{+j}	$\cdots$	n_{+s}	n

observations fall into the ith row, and an s-vector $\mathbf{Y}_j = (Y_{ij})$ to indicate which of the n observations fall into the jth column; that is,

$$X_{ij} = \begin{cases} 1, & \text{if the } j\text{th individual belongs to } A_i \\ 0, & \text{otherwise} \end{cases}$$

$$Y_{ij} = \begin{cases} 1, & \text{if the } i\text{th individual belongs to } B_j \\ 0, & \text{otherwise} \end{cases}$$

$i = 1, 2, \ldots, r$, $j = 1, 2, \ldots, s$. These indicator vectors can be collected into two matrices, an $(r \times n)$-matrix $\mathcal{X}$ and an $(s \times n)$-matrix $\mathcal{Y}$. Note that even though both $\mathcal{X}$ and $\mathcal{Y}$ are defined by the specific distribution of cell frequencies in the contingency table, it turns out that the summary information will be the same as if we assume, for convenience, that $\mathcal{X}$ and $\mathcal{Y}$ are given by

$$\overset{r \times n}{\mathcal{X}} = \begin{pmatrix} 1 & \cdots & 1 & 0 & \cdots & 0 & \cdots & 0 & \cdots & 0 \\ 0 & \cdots & 0 & 1 & \cdots & 1 & \cdots & 0 & \cdots & 0 \\ \vdots & & \vdots & \vdots & & \vdots & & \vdots & & \vdots \\ 0 & \cdots & 0 & 0 & \cdots & 0 & \cdots & 1 & \cdots & 1 \end{pmatrix}, \tag{17.1}$$

$$\overset{s \times n}{\mathcal{Y}} = \begin{pmatrix} 1 & \cdots & 1 & 0 & \cdots & 0 & \cdots & 0 & \cdots & 0 \\ 0 & \cdots & 0 & 1 & \cdots & 1 & \cdots & 0 & \cdots & 0 \\ \vdots & & \vdots & \vdots & & \vdots & & \vdots & & \vdots \\ 0 & \cdots & 0 & 0 & \cdots & 0 & \cdots & 1 & \cdots & 1 \end{pmatrix}, \tag{17.2}$$

respectively.

Matrices derived from $\mathcal{X}$ and $\mathcal{Y}$ reproduce the *observed cell frequencies* and their marginal totals. The $(r \times s)$-matrix $\mathcal{X}\mathcal{Y}^T$ reproduces the observed cell frequencies of the contingency table,

$$\mathcal{X}\mathcal{Y}^\tau = \begin{pmatrix} n_{11} & n_{12} & \cdots & n_{1s} \\ n_{21} & n_{22} & \cdots & n_{2s} \\ \vdots & \vdots & \ddots & \vdots \\ n_{r1} & n_{r2} & \cdots & n_{rs} \end{pmatrix} = \mathbf{N}. \tag{17.3}$$

The $(r \times r)$ matrix $\mathcal{X}\mathcal{X}^\tau$ and the $(s \times s)$ matrix $\mathcal{Y}\mathcal{Y}^\tau$ are both diagonal, $\mathcal{X}\mathcal{X}^\tau$ having as diagonal entries the r marginal row totals and $\mathcal{Y}\mathcal{Y}^\tau$ having as diagonal entries the s marginal column totals,

$$\mathcal{X}\mathcal{X}^\tau = \text{diag}\{n_{1+}, \cdots, n_{r+}\}, \tag{17.4}$$

$$\mathcal{Y}\mathcal{Y}^\tau = \text{diag}\{n_{+1}, \cdots, n_{+s}\}. \tag{17.5}$$

Collecting (17.3), (17.4), and (17.5) together, we can form the $(r+s) \times (r+s)$ block matrix,

$$\begin{pmatrix} \mathcal{X} \\ \mathcal{Y} \end{pmatrix} \begin{pmatrix} \mathcal{X} \\ \mathcal{Y} \end{pmatrix}^\tau = \begin{pmatrix} n\mathbf{D}_r & \mathbf{N} \\ \mathbf{N}^\tau & n\mathbf{D}_c \end{pmatrix}, \tag{17.6}$$

where

$$\mathbf{D}_r = n^{-1}\mathcal{X}\mathcal{X} = \text{diag}\{n_{1+}/n, \ldots, n_{r+}/n\}, \tag{17.7}$$

$$\mathbf{D}_c = n^{-1}\mathcal{Y}\mathcal{Y}^\tau = \text{diag}\{n_{+1}/n, \ldots, n_{+s}/n\}. \tag{17.8}$$

The matrix (17.6) is known as a *Burt matrix* (Burt, 1950) for a two-way contingency table. It is nonnegative definite and symmetric and is the analogue in the discrete case (after dividing through by n) of the sample covariance matrix of two sets of continuous variates.

17.2.3   Example: Hair Color and Eye Color

This classic two-way contingency table $\mathbf{N}$ with $r = 4$ and $s = 5$ (see Table 17.3) was analyzed by R.A. Fisher (1940) and others. It relates to data on hair color and eye color of a sample of 5,387 schoolchildren from Caithness, Scotland. It is given as a (4×5)-matrix by:

$$\mathbf{N} = \mathcal{X}\mathcal{Y}^\tau = \begin{pmatrix} 326 & 38 & 241 & 110 & 3 \\ 688 & 116 & 584 & 188 & 4 \\ 343 & 84 & 909 & 412 & 26 \\ 98 & 48 & 403 & 681 & 85 \end{pmatrix}.$$

The matrices $\mathcal{X}\mathcal{X}^\tau$ and $\mathcal{Y}\mathcal{Y}^\tau$ are given by:

$$\mathcal{X}\mathcal{X}^\tau = \begin{pmatrix} 718 & 0 & 0 & 0 \\ 0 & 1580 & 0 & 0 \\ 0 & 0 & 1774 & 0 \\ 0 & 0 & 0 & 1315 \end{pmatrix}$$

TABLE 17.3. *Relationship of Hair Color to Eye Color of Scottish School-children.*

Eye Color	Fair	Red	Hair Color Medium	Dark	Black	Totals
Blue	326	38	241	110	3	718
Light	688	116	584	188	4	1,580
Medium	343	84	909	412	26	1,774
Dark	98	48	403	681	85	1,315
Totals	1,455	286	2,137	1,391	118	5,387

$$
\mathcal{YY}^T = \begin{pmatrix} 1455 & 0 & 0 & 0 & 0 \\ 0 & 286 & 0 & 0 & 0 \\ 0 & 0 & 2137 & 0 & 0 \\ 0 & 0 & 0 & 1391 & 0 \\ 0 & 0 & 0 & 0 & 118 \end{pmatrix},
$$

respectively. The matrices $\mathbf{D}_r$ and $\mathbf{D}_c$ are obtained by dividing both $\mathcal{XX}^T$ and $\mathcal{YY}^T$ by $n = 5,387$:

$$
\mathbf{D}_r = \begin{pmatrix} 0.1333 & 0 & 0 & 0 \\ 0 & 0.2933 & 0 & 0 \\ 0 & 0 & 0.3293 & 0 \\ 0 & 0 & 0 & 0.2441 \end{pmatrix}
$$

$$
\mathbf{D}_c = \begin{pmatrix} 0.2701 & 0 & 0 & 0 & 0 \\ 0 & 0.0531 & 0 & 0 & 0 \\ 0 & 0 & 0.3967 & 0 & 0 \\ 0 & 0 & 0 & 0.2582 & 0 \\ 0 & 0 & 0 & 0 & 0.0219 \end{pmatrix}.
$$

17.2.4 Profiles, Masses, and Centroids

The $(r \times s)$-matrix

$$
\mathbf{P} = n^{-1}\mathbf{N} \tag{17.9}
$$

converts the contingency table $\mathbf{N}$ into a *correspondence matrix*. See Table 17.4. If the n individuals constitute a random sample, the entry, $p_{ij} = n_{ij}/n$, in the ith row and jth column of $\mathbf{P}$ can be characterized as either the uniformly minimum variance unbiased (UMVU) estimator or the maximum likelihood (ML) estimator of π_{ij}. For the hair-color/eye-color example,

$$
\mathbf{P} = \begin{pmatrix} 0.0605 & 0.0071 & 0.0447 & 0.0204 & 0.0006 \\ 0.1277 & 0.0215 & 0.1084 & 0.0349 & 0.0007 \\ 0.0637 & 0.0156 & 0.1687 & 0.0765 & 0.0048 \\ 0.0182 & 0.0089 & 0.0748 & 0.1264 & 0.0158 \end{pmatrix}.
$$

TABLE 17.4. *Correspondence matrix, showing observed cell relative frequencies* $\mathbf{P}$ $(p_{ij} = n_{ij}/n)$, *row marginal totals* $\mathbf{r}$ $(p_{i+} = n_{i+}/n)$, *and column marginal totals* $\mathbf{c}^\tau$ $(p_{+j} = n_{+j}/n)$

		Column Variable					
Row Variable	B_1	B_2	$\cdots$	B_j	$\cdots$	B_s	Row Total
A_1	p_{11}	p_{12}	$\cdots$	p_{1j}	$\cdots$	p_{1s}	p_{1+}
A_2	p_{21}	p_{22}	$\cdots$	p_{2j}	$\cdots$	p_{2s}	p_{2+}
$\vdots$	$\vdots$	$\vdots$		$\vdots$		$\vdots$	$\vdots$
A_i	p_{i1}	p_{i2}	$\cdots$	p_{ij}	$\cdots$	p_{is}	p_{i+}
$\vdots$	$\vdots$	$\vdots$		$\vdots$		$\vdots$	$\vdots$
A_r	p_{r1}	p_{r2}	$\cdots$	p_{rj}	$\cdots$	p_{rs}	p_{r+}
Column total	p_{+1}	p_{+2}	$\cdots$	p_{+j}	$\cdots$	p_{+s}	1

The row totals and column totals of $\mathbf{P}$ are given by the diagonal elements of $\mathbf{D}_r$ and $\mathbf{D}_c$, respectively.

The $(r \times s)$-matrix $\mathbf{P}_r$ of *row profiles* of $\mathbf{N}$ (or $\mathbf{P}$) consists of the rows of $\mathbf{N}$ divided by their appropriate row totals (e.g., n_{ij}/n_{i+}, which, under random sampling, can be characterized as either the UMVU or ML estimator of π_{ij}/π_{i+}, the conditional probability that an individual has property B_j given that he or she has property A_i), and can be computed as the regression coefficient matrix of $\mathcal{Y}$ on $\mathcal{X}$; that is,

$$\mathbf{P}_r = (\mathcal{X}\mathcal{X}^\tau)^{-1}\mathcal{X}\mathcal{Y}^\tau = \mathbf{D}_r^{-1}\mathbf{P} = \begin{pmatrix} \mathbf{a}_1^\tau \\ \vdots \\ \mathbf{a}_r^\tau \end{pmatrix}, \qquad (17.10)$$

where

$$\mathbf{a}_i^\tau = \left(\frac{n_{i1}}{n_{i+}}, \cdots, \frac{n_{is}}{n_{i+}} \right) \qquad (17.11)$$

is the ith row profile, $i = 1, 2, \ldots, r$. For the hair-color/eye-color example,

$$\mathbf{P}_r = \begin{pmatrix} 0.4540 & 0.0529 & 0.3357 & 0.1532 & 0.0042 \\ 0.4354 & 0.0734 & 0.3696 & 0.1190 & 0.0025 \\ 0.1933 & 0.0474 & 0.5124 & 0.2322 & 0.0147 \\ 0.0745 & 0.0365 & 0.3065 & 0.5179 & 0.0646 \end{pmatrix}.$$

Similarly, the $(s \times r)$-matrix $\mathbf{P}_c$ of *column profiles* of $\mathbf{N}$ (or $\mathbf{P}$) consists of the columns of $\mathbf{N}$ divided by their appropriate column totals (e.g., n_{ij}/n_{+j}, which, under random sampling, can be characterized as the UMVU or ML estimator of π_{ij}/π_{+j}, the conditional probability that an individual has property A_i given that he or she has property B_j), and computed as the regression coefficient matrix of $\mathcal{X}$ on $\mathcal{Y}$; that is,

$$\mathbf{P}_c = (\mathcal{Y}\mathcal{Y}^\tau)^{-1}\mathcal{Y}\mathcal{X}^\tau = \mathbf{D}_c^{-1}\mathbf{P}^\tau = \begin{pmatrix} \mathbf{b}_1^\tau \\ \vdots \\ \mathbf{b}_s^\tau \end{pmatrix}, \tag{17.12}$$

where

$$\mathbf{b}_j^\tau = \left(\frac{n_{1j}}{n_{+j}}, \cdots, \frac{n_{rj}}{n_{+j}} \right) \tag{17.13}$$

is the jth column profile, $j = 1, 2, \ldots, s$. For the hair-color/eye-color example,

$$\mathbf{P}_c = \begin{pmatrix} 0.2241 & 0.4729 & 0.2357 & 0.0674 \\ 0.1329 & 0.4056 & 0.2937 & 0.1678 \\ 0.1128 & 0.2733 & 0.4254 & 0.1886 \\ 0.0791 & 0.1352 & 0.2962 & 0.4896 \\ 0.0254 & 0.0339 & 0.2203 & 0.7203 \end{pmatrix}.$$

The row means of the contingency table $\mathbf{N}$ are the row sums of $\mathbf{P}$,

$$\mathbf{P1}_s = \begin{pmatrix} \bar{X}_1 \\ \vdots \\ \bar{X}_r \end{pmatrix} = \begin{pmatrix} n_{1+}/n \\ \vdots \\ n_{r+}/n \end{pmatrix} = \begin{pmatrix} p_{1+} \\ \vdots \\ p_{r+} \end{pmatrix} = \mathbf{r}, \tag{17.14}$$

and the column means of $\mathbf{N}$ are the column sums of $\mathbf{P}$ (or row sums of $\mathbf{P}^\tau$),

$$\mathbf{P}^\tau \mathbf{1}_r = \begin{pmatrix} \bar{Y}_1 \\ \vdots \\ \bar{Y}_s \end{pmatrix} = \begin{pmatrix} n_{+1}/n \\ \vdots \\ n_{+s}/n \end{pmatrix} = \begin{pmatrix} p_{+1} \\ \vdots \\ p_{+s} \end{pmatrix} = \mathbf{c}, \tag{17.15}$$

where $\mathbf{1}_a$ denotes an a-vector each of whose entries is 1. The vectors $\mathbf{r}$ and $\mathbf{c}$ can be formed from the diagonal elements of $\mathbf{D}_r$ and $\mathbf{D}_c$, respectively; that is, $\mathbf{D}_r = \mathrm{diag}\{\mathbf{r}\}$ and $\mathbf{D}_c = \mathrm{diag}\{\mathbf{c}\}$. For the hair-color/eye-color example,

$$\mathbf{r} = \begin{pmatrix} 0.1333 \\ 0.2933 \\ 0.3293 \\ 0.2441 \end{pmatrix}, \quad \mathbf{c} = \begin{pmatrix} 0.2701 \\ 0.0531 \\ 0.3967 \\ 0.2582 \\ 0.0219 \end{pmatrix}.$$

Powers of these diagonal matrices are given by $\mathbf{D}_r^\alpha = \mathrm{diag}\{\mathbf{r}^\alpha\}$ and $\mathbf{D}_c^\alpha = \mathrm{diag}\{\mathbf{c}^\alpha\}$, where $\mathbf{r}^\alpha$ and $\mathbf{c}^\alpha$ are the column vectors (17.14) and (17.15), respectively, with each entry raised to the αth power. In this chapter, we will be interested in situations where $\alpha = -\frac{1}{2}$ or -1.

The ith element, $p_{i+} = n_{i+}/n$, of the r-vector $\mathbf{r}$ is called the ith *row mass* and, under random sampling, is an estimate of the unconditional probability, π_{i+}, of belonging to A_i. Similarly, the jth element, $p_{+j} = n_{+j}/n$, of the s-vector $\mathbf{c}$ is called the jth *column mass* and is an estimate of the unconditional probability, π_{+j}, of belonging to B_j. In correspondence analysis, $\mathbf{r}$ is called the *average column profile* and $\mathbf{c}$ is called the *average row profile* of the contingency table. The vector $\mathbf{c}$ is also referred to as the *row centroid* because it can be expressed as the weighted average of the row profiles, namely,

$$\mathbf{c} = \sum_{i=1}^{r} p_{i+} \mathbf{a}_i, \tag{17.16}$$

where the weights are the row masses. Similarly, the vector $\mathbf{r}$ is referred to as the *column centroid* because it can be expressed as the weighted average of the column profiles, namely,

$$\mathbf{r} = \sum_{j=1}^{s} p_{+j} \mathbf{b}_j, \tag{17.17}$$

where the weights are the column masses. It is not difficult to show that the relationship between $\mathbf{r}$ and $\mathbf{c}$ is given by $\mathbf{r} = \mathbf{P}^\tau \mathbf{D}_c^{-1} \mathbf{c}$ and $\mathbf{c} = \mathbf{P}^\tau \mathbf{D}_r^{-1} \mathbf{r}$.

17.2.5 Chi-squared Distances

In correspondence analysis, it is important to be able to visualize distances between different row profiles (i.e., rows of $\mathbf{P}_r$) or between different column profiles (i.e., rows of $\mathbf{P}_c$). To do this, we use the *chi-squared metric* as a measure of distance.

Row Distances

Consider the ith and i'th row profiles, $\mathbf{a}_i$ and $\mathbf{a}_{i'}$, respectively. We will need the fact that $\mathbf{a}_i - \mathbf{a}_{i'}$ is an s-vector whose jth entry is $n_{ij}/n_{i+} - n_{i'j}/n_{i'+}$. The squared χ^2-distance between $\mathbf{a}_i$ and $\mathbf{a}_{i'}$ is defined as the quadratic form,

$$d^2(\mathbf{a}_i, \mathbf{a}_{i'}) \equiv (\mathbf{a}_i - \mathbf{a}_{i'})^\tau \mathbf{D}_c^{-1} (\mathbf{a}_i - \mathbf{a}_{i'}) \tag{17.18}$$

$$= \sum_{j=1}^{s} \frac{n}{n_{+j}} \left(\frac{n_{ij}}{n_{i+}} - \frac{n_{i'j}}{n_{i'+}} \right)^2. \tag{17.19}$$

We see from (17.19) that the jth column mass, n_{+j}/n, enters the squared distance between row profiles $\mathbf{a}_i$ and $\mathbf{a}_{i'}$ as an inverse element of the jth term in the sum. It follows that those categories having fewer observations contribute more to the inter-row profile distances.

Recall that $\mathbf{c}$ is the row centroid. The $(r \times s)$-matrix of *centered row profiles* $\mathbf{P}_r - \mathbf{1}_r \mathbf{c}^\tau$, where $\mathbf{P}_r = \mathbf{D}_r^{-1} \mathbf{P}$, has ith row $(\mathbf{a}_i - \mathbf{c})^\tau$, with jth entry $n_{i+}^{-1}(n_{ij} - n_{i+}n_{+j}/n)$, $i = 1, 2, \ldots, r$, $j = 1, 2, \ldots, s$. The squared χ^2-distance between $\mathbf{a}_i$ and $\mathbf{c}$ is, therefore,

$$
\begin{aligned}
d^2(\mathbf{a}_i, \mathbf{c}) &= (\mathbf{a}_i - \mathbf{c})^{\tau} \mathbf{D}_c^{-1}(\mathbf{a}_i - \mathbf{c}) \\
&= \frac{1}{n_{i+}} \sum_{j=1}^{s} \frac{n}{n_{i+}n_{+j}} \left(n_{ij} - \frac{n_{i+}n_{+j}}{n} \right)^2. \quad (17.20)
\end{aligned}
$$

Summing (17.20) over all row profiles yields

$$
n \sum_{i=1}^{r} p_{i+} d^2(\mathbf{a}_i, \mathbf{c}) = \sum_{i=1}^{r}\sum_{j=1}^{s} \left(n_{ij} - \frac{n_{i+}n_{+j}}{n} \right)^2 \bigg/ \left(\frac{n_{i+}n_{+j}}{n} \right), \quad (17.21)
$$

which is the Pearson's chi-squared statistic,

$$
X^2 = \sum_i \sum_j \frac{(O_{ij} - E_{ij})^2}{E_{ij}}, \quad (17.22)
$$

where the *observed cell frequency* O_{ij} and the *expected cell frequency* E_{ij} (assuming independence of row and column variates) are given by

$$
O_{ij} = n_{ij}, \quad E_{ij} = \frac{n_{i+}n_{+j}}{n}, \quad (17.23)
$$

respectively, $i = 1, 2, \ldots, r$, $j = 1, 2, \ldots, s$. Under random sampling, X^2 has approximately (large n) the χ^2 distribution with $(r-1)(s-1)$ degrees of freedom (see, e.g., Rao, 1965, Section 6d.2).

Column Distances

In a similar manner, we define the squared χ^2-distance between the jth and j'th column profiles, $\mathbf{b}_j$ and $\mathbf{b}_{j'}$, respectively, as the quadratic form,

$$
\begin{aligned}
d^2(\mathbf{b}_j, \mathbf{b}_{j'}) &\equiv (\mathbf{b}_j - \mathbf{b}_{j'})^{\tau} \mathbf{D}_r^{-1}(\mathbf{b}_j - \mathbf{b}_{j'}) \quad (17.24) \\
&= \sum_{i=1}^{r} \frac{n}{n_{i+}} \left(\frac{n_{ij}}{n_{+j}} - \frac{n_{ij'}}{n_{+j'}} \right)^2. \quad (17.25)
\end{aligned}
$$

The squared χ^2-distance between the jth column profile and the column centroid is, therefore, given by

$$
\begin{aligned}
d^2(\mathbf{b}_j, \mathbf{r}) &= (\mathbf{b}_j - \mathbf{r})^{\tau} \mathbf{D}_r^{-1}(\mathbf{b}_j - \mathbf{r}) \\
&= \frac{1}{n_{+j}} \sum_{i=1}^{r} \frac{n}{n_{i+}n_{+j}} \left(n_{ij} - \frac{n_{i+}n_{+j}}{n} \right)^2. \quad (17.26)
\end{aligned}
$$

Summing (17.26) over all column profiles yields

$$
n \sum_{j=1}^{s} p_{+j} d^2(\mathbf{b}_j, \mathbf{r}) = X^2, \quad (17.27)
$$

where X^2 is given by (17.22).

Thus, the weighted average of the squared χ^2-distances of all row profiles to the row centroid (or of all column profiles to the column centroid), where the weights are the row masses (column masses), is the quantity X^2/n. If the row and column variates are independent, then X^2/n will be small, in which case every component of X^2/n — either the $\{p_{i+}d^2(\mathbf{a}_i, \mathbf{c})\}$ or the $\{p_{+j}d^2(\mathbf{b}_j, \mathbf{r})\}$ — will be small. On the other hand, if X^2/n is large, that means that at least one of the $\{p_{i+}d^2(\mathbf{a}_i, \mathbf{c})\}$ or at least one of the $\{p_{+j}d^2(\mathbf{b}_j, \mathbf{r})\}$ will be large. This type of information will be important in determining where independence in the table fails.

For the hair-color/eye-color example, the matrix $\mathbf{E} = (E_{ij})$ of expected cell frequencies is given by:

$$\mathbf{E} = \begin{pmatrix} 193.93 & 38.12 & 284.83 & 185.40 & 15.73 \\ 426.75 & 83.88 & 626.78 & 407.98 & 34.61 \\ 479.15 & 94.18 & 703.74 & 458.07 & 38.86 \\ 355.17 & 69.81 & 521.65 & 339.55 & 28.80 \end{pmatrix}.$$

Compare this matrix with $\mathbf{N} = (O_{ij})$ above. The matrix of values of $(O_{ij} - E_{ij})^2/E_{ij}$ is given by:

$$\begin{pmatrix} 89.95 & 0.00 & 6.74 & 30.66 & 10.30 \\ 159.93 & 12.30 & 2.92 & 118.61 & 27.07 \\ 38.69 & 1.10 & 59.87 & 4.63 & 4.26 \\ 186.22 & 6.82 & 26.99 & 343.36 & 109.63 \end{pmatrix}.$$

The sum of all these values is $X^2 = 1240.05$, which should be compared with 21.03, the tabulated 95th-percentile of the χ^2_{12} distribution. Clearly, independence of row and column variates fails for these data.

17.2.6 Total Inertia and Its Decomposition

We see that using dummy variables for representing a two-way contingency table enables us to view the problem as a special case of canonical variate analysis. The situation is, however, different in that instead of extracting the correlation structure between two sets of stochastic data vectors, we are dealing with the correlation structure of two sets of dummy variables.

Let $\mathbf{x} = (x_{ij})$, where $x_{ij} = X_{ij} - \bar{X}_i$ is either $1 - (n_{i+}/n)$ or $-n_{i+}/n$. Similarly, let $\mathbf{y} = (y_{ij})$, where $y_{ij} = Y_{ij} - \bar{Y}_j$ is either $1 - (n_{+j}/n)$ or $-n_{+j}/n$. Then, the covariance matrices are

$$n^{-1}\mathbf{x}\mathbf{x}^\tau = n^{-1}\mathcal{X}(\mathbf{I}_n - n^{-1}\mathbf{J}_n)\mathcal{X}^\tau = \mathbf{D}_r - \mathbf{r}\mathbf{r}^\tau, \tag{17.28}$$

$$n^{-1}\mathbf{y}\mathbf{y}^\tau = n^{-1}\mathcal{Y}(\mathbf{I}_n - n^{-1}\mathbf{J}_n)\mathcal{Y}^\tau = \mathbf{D}_c - \mathbf{c}\mathbf{c}^\tau, \tag{17.29}$$

where $\mathbf{J}_a = \mathbf{1}_a \mathbf{1}_a^T$ is an $(a \times a)$-matrix of 1s. The matrices $\mathbf{xx}^T$ (of rank $r-1$) and $\mathbf{yy}^T$ (of rank $s-1$) are both singular and, hence, their inverses do not exist. We could sidestep this problem by deleting one of the row dummy variables and one of the column dummy variables (see Exercise 17.2), but this would reduce the dimensionality and we would not be able to recover the points from the missing dimensions.

The standard assumption of contingency table analysis is that the row and column totals are considered fixed and the cell frequencies in $\mathbf{N}$ are allowed to vary within those constraints. Accordingly, we center the elements of $\mathbf{N}$ at the values we expect them to have under independence (instead of centering the data $\mathbf{N}$ at the mean). Thus, (17.9) becomes the *relative frequency matrix*,

$$n^{-1}\mathcal{X}(\mathbf{I}_n - n^{-1}\mathbf{J}_n)\mathcal{Y}^T = \mathbf{P} - \mathbf{rc}^T = \widetilde{\mathbf{P}}. \tag{17.30}$$

For the hair-color/eye-color example,

$$\widetilde{\mathbf{P}} = \begin{pmatrix} 0.0245 & -0.0000 & -0.0081 & -0.0140 & -0.0024 \\ 0.0485 & 0.0060 & -0.0079 & -0.0408 & -0.0057 \\ -0.0253 & -0.0019 & 0.0381 & -0.0086 & -0.0024 \\ -0.0477 & -0.0040 & -0.0220 & 0.0634 & 0.0104 \end{pmatrix}.$$

The matrix $\widetilde{\mathbf{N}} = n\widetilde{\mathbf{P}}$ is often called the matrix of *residuals* because its ijth entry, $\tilde{n}_{ij} = O_{ij} - E_{ij}$, shows the difference between the observed cell frequency (O_{ij}) and its expected cell frequency (E_{ij}), assuming independence between row and column variates, $i = 1, 2, \ldots, r, j = 1, 2, \ldots, s$ (see (17.23)). Note that because $\widetilde{\mathbf{N}}\mathbf{1}_s = (\mathbf{N} - n\mathbf{rc}^T)\mathbf{1}_s = \mathbf{N}\mathbf{1}_s - n\mathbf{rc}^T\mathbf{1}_s = n\mathbf{r} - n\mathbf{r} = \mathbf{0}$, the rank of $\widetilde{\mathbf{N}}$ (and, hence, of $\widetilde{\mathbf{P}}$) is at most $s-1$.

The $(s \times s)$-matrix $\mathbf{R}$ in (8.76) plays a central role in canonical variate analysis, and it has an obvious analogue in this development. The correspondences between (8.76) and (17.6) are given by

$$\Sigma_{XX} \leftrightarrow \mathbf{D}_r, \quad \Sigma_{YY} \leftrightarrow \mathbf{D}_c, \quad \Sigma_{XY} \leftrightarrow \widetilde{\mathbf{P}}. \tag{17.31}$$

Accordingly, we use (17.7), (17.8), and (17.30) to compute the $(s \times s)$-matrix,

$$\mathbf{R}_0 = \mathbf{D}_c^{-1/2}\widetilde{\mathbf{P}}^T\mathbf{D}_r^{-1}\widetilde{\mathbf{P}}\mathbf{D}_c^{-1/2}, \tag{17.32}$$

where $\mathbf{D}_r^{-1} = \text{diag}\{\mathbf{r}^{-1}\}$ and $\mathbf{D}_c^{-1/2} = \text{diag}\{\mathbf{c}^{-1/2}\}$. The entry in the jth row and j'th column of $\mathbf{R}_0$ is given by

$$(n_{+j}n_{+j'})^{-1/2}\sum_{i=1}^{r}\frac{1}{n_{i+}}\left(n_{ij} - \frac{n_{i+}n_{+j}}{n}\right)\left(n_{ij'} - \frac{n_{i+}n_{+j'}}{n}\right) \tag{17.33}$$

and the jth diagonal entry of $\mathbf{R}_0$ is obtained by setting $j = j'$,

$$\frac{1}{n_{+j}}\sum_{i=1}^{r}\frac{1}{n_{i+}}\left(n_{ij} - \frac{n_{i+}n_{+j}}{n}\right)^2. \tag{17.34}$$

For the hair-color/eye-color example,

$$
\mathbf{R}_0 = \begin{pmatrix}
0.0881 & 0.0160 & -0.0044 & -0.0798 & -0.0420 \\
0.0160 & 0.0038 & -0.0001 & -0.0156 & -0.0080 \\
-0.0044 & -0.0001 & 0.0179 & -0.0148 & -0.0099 \\
-0.0798 & -0.0156 & -0.0148 & 0.0923 & 0.0507 \\
-0.0420 & -0.0080 & -0.0099 & 0.0507 & 0.0281
\end{pmatrix}.
$$

The trace of $\mathbf{R}_0$, which is also the sum of the eigenvalues of $\mathbf{R}_0$, is

$$
\sum_{j=1}^{s} \lambda_j^2 = \operatorname{tr}\{\mathbf{R}_0\} = \sum_{i=1}^{r}\sum_{j=1}^{s} \frac{1}{n_{i+}n_{+j}}\left(n_{ij} - \frac{n_{i+}n_{+j}}{n}\right)^2 = \frac{X^2}{n}, \qquad (17.35)
$$

where X^2 is given by (17.22).

If the value of X^2 is very large, as it is in the shoplifting example where $X^2 = 19,949.97$ on $17 \times 12 = 204$ degrees of freedom, the hypothesis of independence of the row and column variates in the contingency table has to be rejected. It then becomes of interest to determine where the deviations from independence occur. Understanding which characteristics of the data are important may be useful for further study.

The quantity X^2/n is referred to as the amount of *total inertia* in the contingency table. The eigenvalues (or *principal inertias*) of $\mathbf{R}_0$ form a decomposition of the total inertia. The accumulated contribution of the first t principal inertias is given by

$$
\frac{\lambda_1^2 + \cdots + \lambda_t^2}{\sum_{j=1}^{s} \lambda_j^2}, \qquad (17.36)
$$

which is an analogue of the percentage of total variance explained by the first t principal components, where we usually take t to be 2 or 3.

For the hair-color/eye-color example, the eigenvalues of $\mathbf{R}_0$ (and their individual percentages of the total, $\operatorname{tr}(\mathbf{R}_0) = 0.2302$) are 0.1992 (86.6%), 0.0301 (13.1%), 0.0009 (0.4%), 0, and 0. Clearly, the first two eigenvalues account for almost all of the total inertia.

Table 17.5 lists the 12 principal inertias (eigenvalues of $\mathbf{R}_0$) for the shoplifting example. The total inertia is $X^2/n = 19,949.97/33,101 = 0.6027$. We see that the first three eigenvalues account for about 90% of the total inertia, which suggests that almost all of the deviations from independence can be attributed to the first three dimensions. The two-dimensional plot (see Figure 17.1) accounts for about 78% of the total inertia.

17.2.7 *Principal Coordinates for Row and Column Profiles*

The matrix $\mathbf{R}_0$ in (17.32) can be expressed as

$$
\mathbf{R}_0 = \mathbf{M}^\tau \mathbf{M}, \qquad (17.37)
$$

TABLE 17.5. *Shoplifting example: Principal inertias (eigenvalues λ_j^2), total inertia, the proportions of total inertia explained by each eigenvalue, and the cumulative proportions.*

Axis	Inertia	Percentage	Cumulative
1	0.3504	58.13	58.13
2	0.1192	19.78	77.91
3	0.0700	11.61	89.52
4	0.0382	6.35	95.86
5	0.0112	1.86	97.72
6	0.0086	1.43	99.14
7	0.0031	0.51	99.66
8	0.0009	0.15	99.81
9	0.0006	0.10	99.91
10	0.0003	0.06	99.97
11	0.0001	0.02	99.99
12	0.0001	0.01	100.00
Total	0.6027		

where the $(r \times s)$-matrix

$$\mathbf{M} = \mathbf{D}_r^{-1/2} \widetilde{\mathbf{P}} \mathbf{D}_c^{-1/2} \tag{17.38}$$

has ijth entry given by the *Pearson residual*,

$$m_{ij} = (n_{i+}n_{+j})^{-1/2} \left(n_{ij} - \frac{n_{i+}n_{+j}}{n} \right), \tag{17.39}$$

$i = 1, 2, \ldots, r$, $j = 1, 2, \ldots, s$. For the hair-color/eye-color example,

$$\mathbf{M} = \begin{pmatrix} 0.1292 & -0.0003 & -0.0354 & -0.0754 & -0.0437 \\ 0.1723 & 0.0478 & -0.0233 & -0.1484 & -0.0709 \\ -0.0847 & -0.0143 & 0.1054 & -0.0293 & -0.02811 \\ -0.1859 & -0.0356 & -0.0708 & 0.2525 & 0.1427 \end{pmatrix}.$$

Thus, from (17.35), the sum of squares of all rs Pearson residuals in the contingency table is the total inertia. Note that because $\text{rank}(\widetilde{\mathbf{P}}) \le s - 1$, it follows that $\mathbf{M}$ in (17.38) also has rank at most $s - 1$. The singular value decomposition of $\mathbf{M}$ is, therefore, given by

$$\mathbf{M} = \mathbf{U} \mathbf{D}_\lambda \mathbf{V}^\tau, \tag{17.40}$$

where $\mathbf{U}$ is an $(r \times s)$-matrix, $\mathbf{U}^\tau \mathbf{U} = \mathbf{I}_s$, whose columns are the eigenvectors, $\{\mathbf{u}_j\}$, corresponding to the $s - 1$ nonzero eigenvalues of the $(r \times r)$-matrix

$$\mathbf{MM}^\tau = \mathbf{D}_r^{-1/2} \widetilde{\mathbf{P}} \mathbf{D}_c^{-1} \widetilde{\mathbf{P}}^\tau \mathbf{D}_r^{-1/2} = \mathbf{R}_1, \tag{17.41}$$

$\mathbf{V}$ is an $(s \times s)$-matrix, $\mathbf{V}^T\mathbf{V} = \mathbf{I}_s$, whose columns are the eigenvectors, $\{\mathbf{v}_j\}$, corresponding to the eigenvalues of the $(s \times s)$-matrix $\mathbf{M}^T\mathbf{M} = \mathbf{R}_0$, and $\mathbf{D}_\lambda = \text{diag}\{\lambda_1, \cdots, \lambda_s\}$ is an $(s \times s)$ diagonal matrix with its principal diagonal having entries the *singular values* (the positive square-roots of the nonzero eigenvalues of either $\mathbf{R}_0$ or $\mathbf{R}_1$).

Combining (17.38) and (17.40), we can write

$$\widetilde{\mathbf{P}} = (\mathbf{D}_r^{1/2}\mathbf{U})\mathbf{D}_\lambda(\mathbf{V}^T\mathbf{D}_c^{1/2}) = \mathbf{A}\mathbf{D}_\lambda\mathbf{B}^T, \tag{17.42}$$

where

$$\mathbf{A} = \mathbf{D}_r^{1/2}\mathbf{U}, \quad \mathbf{B} = \mathbf{D}_c^{1/2}\mathbf{V}. \tag{17.43}$$

For the hair-color/eye-color example,

$$\mathbf{A} = \begin{pmatrix} -0.1195 & 0.1271 & -0.2917 & -0.1333 & 0 \\ -0.2896 & 0.1496 & 0.3179 & -0.2933 & 0 \\ 0.0248 & -0.4651 & -0.0624 & -0.3293 & 0 \\ 0.3843 & 0.1885 & 0.0362 & -0.2441 & 0 \end{pmatrix}.$$

$$\mathbf{B} = \begin{pmatrix} -0.3292 & 0.2707 & -0.1154 & 0.2741 & 0 \\ -0.0277 & 0.0148 & 0.2138 & 0.0421 & -0.0680 \\ -0.0373 & -0.4764 & -0.0438 & 0.4071 & 0.0259 \\ 0.3406 & 0.1547 & -0.0891 & 0.2186 & -0.2501 \\ 0.0537 & 0.0362 & 0.0345 & 0.0433 & 0.1210 \end{pmatrix}.$$

Note that

$$\mathbf{A}^T\mathbf{D}_r^{-1}\mathbf{A} = \mathbf{I}_s, \quad \mathbf{B}^T\mathbf{D}_c^{-1}\mathbf{B} = \mathbf{I}_s. \tag{17.44}$$

The expression (17.42) (and (17.44)) is the *generalized singular value decomposition* of $\widetilde{\mathbf{P}}$ in the metrics $\mathbf{D}_r^{-1}$ and $\mathbf{D}_c^{-1}$. The columns of $\mathbf{A}$ and $\mathbf{B}$ are called the *principal axes* of the row and column profiles.

The squared χ^2-distance (in the metric $\mathbf{D}_c^{-1}$) between the $(r \times s)$-matrices of centered row profiles $\mathbf{P}_r - \mathbf{1}_r\mathbf{c}^T$ and $\mathbf{B}$ is given by

$$\begin{aligned} \mathbf{G}_P^T &= (\mathbf{P}_r - \mathbf{1}_r\mathbf{c}^T)\mathbf{D}_c^{-1}\mathbf{B} \\ &= (\mathbf{D}_r^{-1}\widetilde{\mathbf{P}}\mathbf{D}_c^{-1})\mathbf{B} \\ &= \mathbf{D}_r^{-1}(\mathbf{A}\mathbf{D}_\lambda\mathbf{B}^T)\mathbf{D}_c^{-1}\mathbf{B} \\ &= \mathbf{D}_r^{-1}\mathbf{A}\mathbf{D}_\lambda, \end{aligned} \tag{17.45}$$

where we have used (17.10), $\mathbf{1}_r = \mathbf{D}_r^{-1}\mathbf{r}$, (17.41), and (17.43). Similarly, we can show that the squared χ^2-distance (in the metric $\mathbf{D}_r^{-1}$) between the $(s \times r)$-matrices of centered column profiles $\mathbf{P}_c - \mathbf{1}_c\mathbf{r}^T$ and $\mathbf{A}$ is given by

$$\begin{aligned} \mathbf{H}_P^T &= (\mathbf{P}_c - \mathbf{1}_c\mathbf{r}^T)\mathbf{D}_r^{-1}\mathbf{B} \\ &= \mathbf{D}_c^{-1}\mathbf{B}\mathbf{D}_\lambda. \end{aligned} \tag{17.46}$$

Substituting (17.42) for the $\mathbf{A}$ and $\mathbf{B}$ in (17.44) and (17.45), respectively, we have that

$$\mathbf{G}_P^\tau = \mathbf{D}_r^{-1/2}\mathbf{U}\mathbf{D}_\lambda, \quad \mathbf{H}_P^\tau = \mathbf{D}_c^{-1/2}\mathbf{V}\mathbf{D}_\lambda. \tag{17.47}$$

For the hair-color/eye-color example,

$$\mathbf{G}_P^\tau = \begin{pmatrix} -0.4003 & 0.1654 & -0.0642 & 0 \\ -0.4407 & 0.0885 & 0.0318 & 0 \\ 0.0336 & -0.2450 & -0.0056 & 0 \\ 0.7027 & 0.1339 & 0.0043 & 0 \end{pmatrix},$$

$$\mathbf{H}_P^\tau = \begin{pmatrix} -0.5440 & 0.1738 & -0.0125 & 0 \\ -0.0233 & 0.0483 & 0.1181 & 0 \\ -0.0420 & -0.2083 & -0.0032 & 0 \\ 0.5887 & 0.1040 & -0.0101 & 0 \\ 1.0944 & 0.2864 & 0.0461 & 0 \end{pmatrix}.$$

The columns of $\mathbf{G}_P^\tau$ and $\mathbf{H}_P^\tau$ are called the *principal coordinates* of the row and column profiles, respectively (hence the subscript P). The matrices $\mathbf{G}_P^\tau$ and $\mathbf{H}_P^\tau$ are related to each other. It can be shown (see Exercise 17.5) that

$$\mathbf{G}_P^\tau = \mathbf{D}_r^{-1}\mathbf{P}\mathbf{H}_P^\tau\mathbf{D}_\lambda^{-1}, \quad \mathbf{H}_P^\tau = \mathbf{D}_c^{-1}\mathbf{P}^\tau\mathbf{G}_P^\tau\mathbf{D}_\lambda^{-1}. \tag{17.48}$$

Similar results can also be obtained directly from the canonical variate analysis developed in Chapter 8 and the correspondences given in (17.31). From (8.46) and (8.47), we compute the $(s \times r)$-matrix $\mathbf{G}_S$ and the $(s \times s)$-matrix $\mathbf{H}_S$, where

$$\mathbf{G}_S = \mathbf{U}^\tau\mathbf{D}_r^{-1/2}, \quad \mathbf{H}_S = \mathbf{V}^\tau\mathbf{D}_c^{-1/2}. \tag{17.49}$$

Note that $\mathbf{G}_S\mathbf{D}_r\mathbf{G}_S^\tau = \mathbf{I}_r$ and $\mathbf{H}_S\mathbf{D}_c\mathbf{H}_S^\tau = \mathbf{I}_s$. The columns of $\mathbf{G}_S^\tau$ and $\mathbf{H}_S^\tau$ in (17.49) are known as the *standard coordinates* of the row and column profiles, respectively (hence the subscript S). Instead of defining the row and column coordinates as (17.49), however, they are generally scaled as in (17.47).

17.2.8 Graphical Displays

In correspondence analysis, one has the choice between analyzing only the row profiles, or analyzing only the column profiles, or analyzing both the row and column profiles together. The graphical displays formed from plotting the row and column coordinates in Table 17.6 are scatterplots that can be of two types:

Symmetric map: Both row and column coordinates are expressed as principal coordinates.

TABLE 17.6. *The t-dimensional formulas for row and column coordinates are the columns of the first t rows of the following matrices, where t is two or three.*

Problem	Row Coordinates	Column Coordinates
Row Profiles	$\mathbf{G}_P = \mathbf{D}_\lambda \mathbf{U}^\tau \mathbf{D}_r^{-1/2}$	$\mathbf{H}_S = \mathbf{V}^\tau \mathbf{D}_c^{-1/2}$
Column Profiles	$\mathbf{G}_S = \mathbf{U}^\tau \mathbf{D}_r^{-1/2}$	$\mathbf{H}_P = \mathbf{D}_\lambda \mathbf{V}^\tau \mathbf{D}_c^{-1/2}$
Both Profiles	$\mathbf{G}_P = \mathbf{D}_\lambda \mathbf{U}^\tau \mathbf{D}_r^{-1/2}$	$\mathbf{H}_P = \mathbf{D}_\lambda \mathbf{V}^\tau \mathbf{D}_c^{-1/2}$

Asymmetric map: The row (or column) coordinates are expressed as principal coordinates while the other is expressed as standard coordinates.

Most users of correspondence analysis prefer to view a symmetric map of both the row and column principal coordinates (17.47) in a two- (or three-) dimensional scatterplot. First, we make a scatterplot of each of the r rows of the first two (or three) columns of $\mathbf{G}_P^\tau$. Then, on the same scatterplot, we overlay a plot of each of the s rows of the first two (or three) columns of $\mathbf{H}_P^\tau$. In Figure 17.2, we have drawn the symmetric correspondence map for the eye-color/hair-color example. If the three-dimensional points are plotted on a dynamic scatterplot, then the display can be rotated in all three dimensions for better viewing. These merged displays provide interpretable views of different features in the data.

There will be $r + s$ points in these scatterplots, which are called *correspondence maps*. For clearer interpretation, different symbols should be used for the row points and column points. It is also useful (unless the plot would look overly cluttered) to identify each point in the plot by a tag showing its corresponding category name. If the row (or column) categories are ordered in some way, such as time-order by year or successive age ranges (as in the shoplifting example), then it is visually helpful to connect those category points in the plot with each other to indicate such order-dependence.

In general, points in the scatterplot that appear "close" to each other tend to correspond to categories that are closely related. More specifically,

- if row points are close, then those rows have similar conditional distributions across columns;

- if column points are close, then those columns have similar conditional distributions across rows;

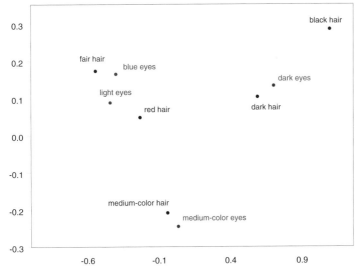

FIGURE 17.2.   *Correspondence map for the hair-color/eye-color example. The points exhibit a U-shaped plot with the first principal coordinate (horizontal axis) displaying gradations along the fair-red-medium-dark-black hair scale and the light-blue-medium-dark eyes scale, and the second principal coordinate (vertical axis) displaying a difference between medium-color hair and eyes and the other hair and eye colors.*

- if a row point is close to a column point, then that configuration suggests a particular deviation from independence.

In general, we should not try to compare the positions of row points with the positions of column points and say, for example, that if a particular row point is very close to a particular column point then the corresponding row and column categories are related to each other. (A dissenting view that supports identifying row points with neighboring column points is given by van der Heijden et al, 1989.)

17.3   Square Asymmetric Contingency Tables

An important special case of two-way contingency tables consists of square tables, where $r = s$ and the rows have the same categories as the columns. Examples of square tables include:

- Individuals who are naturally paired, such as husbands and wives or fathers and sons, are classified by occupational or social status.

- Experiments conducted on naturally paired items, such as vision grades of left eye and right eye.

- Two investigators or event judges independently rate each subject in a study using the same Likert-type scale.

- Individuals in a sample are categorized by region of residence at two distinct points in time.

- To study accuracy of a classification rule, the rows give the classes to which the data were assigned by the rule, the columns define the true classes (possibly determined from reference data), the cell entries show how much the classified data and the reference data agree, and the diagonal cells show the numbers of correct classifications.

If a square table $\mathbf{N}$ is symmetric with respect to the r^2 cell frequencies (i.e., $\mathbf{N}^\tau = \mathbf{N}$), then the correspondence map will display coincident pairs of row and column points. In each of the examples listed above, however, the square tables are asymmetric in the sense that $\mathbf{N}^\tau \neq \mathbf{N}$. Unlike rectangular contingency tables, analyzing asymmetric square tables using correspondence analysis has not been very successful. The reason is similar to that for models that try to analyze square tables for symmetry: the data along the principal diagonal tend to have too great an influence on the results.

An innovative way of analyzing square asymmetric tables was proposed by Gower (1977) and Constantine and Gower (1978). Consider a square asymmetric contingency table $\mathbf{N}$ that yields the correspondence table $\mathbf{P}$, also square and asymmetric. Gower showed that $\mathbf{P}$ can be decomposed, prior to analysis, into two orthogonal component tables,

$$\mathbf{P} = \mathbf{M} + \mathbf{Q}, \tag{17.50}$$

where

$$\mathbf{M} = \frac{1}{2}(\mathbf{P} + \mathbf{P}^\tau), \quad \mathbf{Q} = \frac{1}{2}(\mathbf{P} - \mathbf{P}^\tau). \tag{17.51}$$

In (17.51), $\mathbf{M}$ is a symmetric table ($\mathbf{M}^\tau = \mathbf{M}$) and $\mathbf{Q}$ is a skew-symmetric table ($\mathbf{Q}^\tau = -\mathbf{Q}$). Because of the orthogonality of the decomposition (see Exercise 17.4), separate analyses of $\mathbf{M}$ and $\mathbf{Q}$ can be carried out. See van der Heijden et al. (1989). If r is even, the singular vectors of $\mathbf{Q}$ occur in pairs corresponding to pairs of equal singular values (principal inertias). If r is odd, the last singular value of $\mathbf{Q}$ equals zero.

Greenacre (2000) used the decomposition (17.50) to obtain separate correspondence maps of $\mathbf{M}$ and $\mathbf{Q}$. Greenacre showed that these maps could be obtained from a single application of simple correspondence analysis to the $(2r \times 2r)$ block matrix,

$$\mathbf{N}^* = \begin{pmatrix} \mathbf{N} & \mathbf{N}^\tau \\ \mathbf{N}^\tau & \mathbf{N} \end{pmatrix}, \tag{17.52}$$

with correspondence matrix,

$$\mathbf{P}^* = \frac{1}{4} \begin{pmatrix} \mathbf{P} & \mathbf{P}^\tau \\ \mathbf{P}^\tau & \mathbf{P} \end{pmatrix},$$

(17.53)

and row and column totals,

$$\mathbf{w}^* = \frac{1}{2} \begin{pmatrix} \mathbf{w} \\ \mathbf{w} \end{pmatrix},$$

(17.54)

where $\mathbf{w} = (\mathbf{r} + \mathbf{c})/2$. Whereas the usual correspondence analysis is to analyze $\tilde{\mathbf{P}} = \mathbf{P} - \mathbf{rc}^\tau$ in the metrics $\mathbf{D}_r^{-1}$ and $\mathbf{D}_c^{-1}$, in this case, we analyze $\mathbf{P} - \mathbf{ww}^\tau$ in the metrics $\mathbf{D}_w^{-1}$ and $\mathbf{D}_w^{-1}$. Thus, (17.50) becomes $\mathbf{P} - \mathbf{ww}^\tau = \mathbf{M} - \mathbf{ww}^\tau + \mathbf{Q}$. We should expect the total inertia attributed to $\mathbf{P} - \mathbf{ww}^\tau$ to be larger than the usual total inertia (e.g., (17.35)) because $\mathbf{ww}^\tau$ is not the rank-1 matrix closest to $\mathbf{P}$. The extent of the difference will depend upon how different are $\mathbf{r}$ and $\mathbf{c}$ from each other.

The dimensionality of $\mathbf{N}^*$ is $2r - 1$, of which $r - 1$ dimensions belong to $\mathbf{M}$ and the remaining r dimensions to $\mathbf{Q}$. The correspondence map of $\mathbf{M}$ displays pairs of coincident row and column points (so that it suffices to plot only one set of points). We can, therefore, detect deviations of $\mathbf{N}$ from symmetry by concentrating on the correspondence map of $\mathbf{Q}$.

Thus, there will be two separate correspondence maps for $\mathbf{N}$, one map for the symmetric component $\mathbf{M}$ and the other map for the skew-symmetric component $\mathbf{Q}$. Each map consists of a single set of points. Greenacre recommends that both correspondence maps be scaled equally for comparing the relative sizes of the principal inertias.

17.3.1 Example: Occupational Mobility in England

This 14×14 contingency table (see Table 17.7) of the occupations of a sample of 775 males and their fathers in England was originally studied by Pearson (1904). Figure 17.3 shows the two-dimensional correspondence map of Table 17.7. The total inertia of the contingency table is 1.2974, of which 50.97% is accounted for by the map.

The above decomposition of $\mathbf{P}$ into a symmetric component $\mathbf{M}$ and a skew-symmetric component $\mathbf{Q}$ is accomplished by using (17.52). The resulting total inertia increases by 0.3016 to 1.5990 due to the different type of centering involved. The total symmetric inertia is 1.1484, and the total skew-symmetric inertia is 0.4506. In Table 17.8, we list the 27 principal inertias, of which 13 correspond to the symmetric correspondence analysis and 14 (= 7 pairs) to the skew-symmetric correspondence analysis. Also listed in Table 17.8 are the percentages of the two sets of principal inertias relative to the total symmetric and skew-symmetric inertias. The first pair

TABLE 17.7. *Occupations of fathers and their sons in England (Pearson, 1904). The occupational categories are A army; B art; C teaching, clerical work, civil service; D crafts; E divinity; F agriculture; G landownership; H law; I literature; J commerce; K medicine; L navy; M politics and court; N scholarship and science. Uppercase letters represent occupations of the father and lowercase letters represent occupations of the son. The Pearson chi-squared test for independence gives* $X^2 = 874.9$ *on 169 degrees of freedom, so that an hypothesis of independence is rejected.*

	Sons														
Fathers	*a*	*b*	*c*	*d*	*e*	*f*	*g*	*h*	*i*	*j*	*k*	*l*	*m*	*n*	Totals
A	28	0	4	0	0	0	1	3	3	0	3	1	5	2	50
B	2	51	1	1	2	0	0	1	2	0	0	0	1	1	62
C	6	5	7	0	9	1	3	6	4	2	1	1	2	7	54
D	0	12	0	6	5	0	0	1	7	1	2	0	0	10	44
E	5	5	2	1	54	0	0	6	9	4	12	3	1	13	115
F	0	2	3	0	3	0	0	1	4	1	4	2	1	5	26
G	17	1	4	0	14	0	6	11	4	1	3	3	17	7	88
H	3	5	6	0	6	0	2	18	13	1	1	1	8	5	69
I	0	1	1	0	4	0	0	1	4	0	2	1	1	4	19
J	12	16	4	1	15	0	0	5	13	11	6	1	7	15	106
K	0	4	2	0	1	0	0	0	3	0	20	0	5	6	41
L	1	3	1	0	0	0	1	0	1	1	1	6	2	1	18
M	5	0	2	0	3	0	1	8	1	2	2	3	23	1	51
N	5	3	0	2	6	0	1	3	1	0	0	1	1	9	32
Totals	84	108	37	11	122	1	15	64	69	24	57	23	74	86	775

of symmetric principal inertias (1 and 2) accounts for 33.85% + 20.20% = 54.05% of the total symmetric inertia, suggesting that higher dimensions contain additional significant information. The first pair of skew-symmetric principal inertias (3 and 4) accounts for 35.15% + 35.15% = 70.30% of the total skew-symmetric inertia (compared with only 9.90% + 9.90% = 19.80% of the total inertia). The symmetric dimensions are, therefore, 1, 2, 5–9, 12, 13, 16, 21, 24, and 27, and the remainder, which occur in pairs, are the skew-symmetric dimensions.

Figure 17.4 shows the correspondence maps of dimensions 1 and 2, and 3 and 4, respectively. The top panel of Figure 17.4 shows the symmetric portion of the table. The points representing the arts (B) and crafts (D) occupations are clearly separated from the other points, but these two points are also not close to each other. One can also argue that these two points account for much of the difference in inertias between the symmetric and skew-symmetric analyses because the variation in points is not that different without points B and D. Points that are close together in this map reflect the fact that there is a lot of movement from father to son

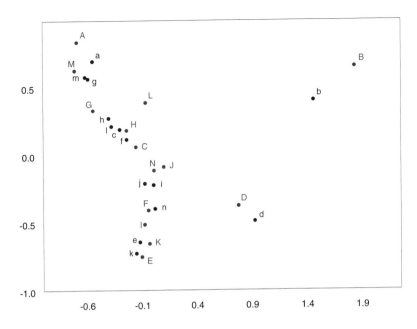

FIGURE 17.3. *Correspondence map for the occupational mobility example. The horizontal axis represents the first principal coordinate and the vertical axis the second principal coordinate. On the left of the map, there is a steady progression in occupations from A to E (and from a to k). The two occupations of B and D (and b and d), representing arts and crafts, stand out from the rest.*

between those occupations, whereas points that are far apart from each other indicate relatively little movement. If we ignore points B and D, there appears to be a progression in the occupations, from the topmost points down through several clusters of points, such as

- army (A), and politics and court (M)

- teaching, clerical work, civil service (C), landownership (G), law (H), and navy (L)

- agriculture (F), literature (I), commerce (J), and scholarship and science (N)

- divinity (E) and medicine (K)

These clusters suggest that occupational mobility from father to son is typically confined to movements within the various clusters only and not between clusters.

TABLE 17.8. *Occupational mobility example: Principal inertias (eigenvalues λ_j^2), total inertia, the percentages and cumulative percentages of total inertia explained by each eigenvalue, and the percentages corresponding to the symmetric (S) and skew-symmetric (SS) correspondence analyses. The total symmetric inertia is 1.1484, and the total skew-symmetric inertia is 0.4506.*

Principal Axis	Principal Inertia	% Inertia	Cumulative	%-S	%-SS
1	0.3887	24.31	24.31	33.85	
2	0.2320	14.51	38.82	20.20	
3	0.1584	9.90	48.72		35.15
4	0.1584	9.90	58.62		35.15
5	0.1439	9.00	67.62	12.53	
6	0.1238	7.74	75.36	10.78	
7	0.0818	5.12	80.48	7.12	
8	0.0707	4.42	84.91	6.16	
9	0.0498	3.12	88.02	4.34	
10	0.0418	2.62	90.64		9.28
11	0.0418	2.62	93.25		9.28
12	0.0229	1.43	94.68	1.99	
13	0.0220	1.38	96.06	1.92	
14	0.0129	0.81	96.87		2.86
15	0.0129	0.81	97.67		2.86
16	0.0104	0.65	98.32	0.91	
17	0.0076	0.47	98.80		1.69
18	0.0076	0.47	99.27		1.69
19	0.0031	0.19	99.46		0.69
20	0.0031	0.19	99.66		0.69
21	0.0017	0.10	99.76	0.15	
22	0.0011	0.07	99.83		0.24
23	0.0011	0.07	99.90		0.24
24	0.0006	0.04	99.94	0.00	
25	0.0004	0.02	99.97		0.00
26	0.0004	0.02	99.99		0.00
27	0.0001	0.01	100.00	0.00	
Total	1.5990				

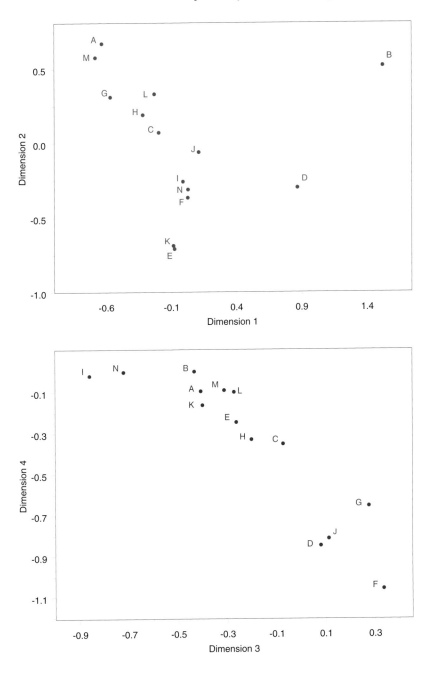

FIGURE 17.4. *Correspondence analysis of the symmetric component (top panel) and skew-symmetric component (bottom panel) for the occupational mobility example.*

The bottom panel of Figure 17.4 shows the deviations from symmetry. Asymmetry between any two points can be envisioned by a triangle constructed with vertices at those two points and the origin; the greater the area of that triangle, the greater the degree of asymmetry between the points. Points that yield triangles with no area (i.e., points on a line through the origin) have no asymmetric relationship. Points that are close to the origin indicate small asymmetries. In this map, there are no points clustered around the origin, suggesting some asymmetry between all occupations. Indeed, all the points in this map lie on one side of a line drawn through the origin, indicating that circular triads are absent in the data. The more drastic asymmetries are those points furthest from the origin, literature (I) and scholarship and science (N) at one extreme and agriculture (F) at the other. The greatest deviation from symmetry is from a father's occupation of literature (I) to a son's occupation in agriculture (F).

17.4   Multiple Correspondence Analysis

Multiple correspondence analysis is intended to be a generalization of simple correspondence analysis, in the sense that it is designed to deal with the graphical representation of contingency tables that have more than two categorical variables. The fact that as currently conceived it is not a true generalization (in the sense that simple correspondence analysis is not a special case) has not, however, detracted from its usefulness. Accordingly, there is much research currently taking place on this topic.

17.4.1   The Multivariate Indicator Matrix

As we did in Section 17.2.2, we can define a dummy (or indicator) variable for each of the Q categorical variables that make up the table. Suppose that the qth variable has J_q categories and that $J = \sum_{q=1}^{Q} J_q$ is the total number of categories over all variables. Suppose further that there are n individuals in the study (who may be some part — a sample — or all of a population). Let $\mathbf{Z} = (Z_{ij})$ be a $(J \times n)$-matrix, where

$$Z_{ij} = \begin{cases} 1, & \text{if the } j\text{th individual belongs to the } i\text{th category} \\ 0, & \text{otherwise,} \end{cases} \qquad (17.55)$$

$i = 1, 2, \ldots, J$, $j = 1, 2, \ldots, n$. We assume that there is no row of $\mathbf{Z}$ that contains all 0s. Each column of $\mathbf{Z}$ sums to Q and all Jn entries sum to nQ. The matrix $\mathbf{Z}$ is often called a *multivariate indicator matrix*. One interpretation of the concept of multiple correspondence analysis is that of carrying out a simple correspondence analysis of the multivariate indicator matrix $\mathbf{Z}$.

We can partition the J rows of $\mathbf{Z}$ into blocks by variable so that

$$\mathbf{Z} = \begin{pmatrix} \mathbf{Z}_1 \\ \vdots \\ \mathbf{Z}_Q \end{pmatrix}, \tag{17.56}$$

where $\mathbf{Z}_q$ is a $(J_q \times n)$-matrix corresponding to the qth categorical variable having J_q categories, $q = 1, 2, \ldots, Q$. The following properties of $\mathbf{Z}$ are given in Greenacre (1984). In $\mathbf{Z}_q$, there are $\mathbf{1}_{J_q}^\tau \mathbf{Z}_q \mathbf{1}_n = n$ 1s, $q = 1, 2, \ldots, Q$. Following (17.15), the row masses of $\mathbf{Z}_q$ are defined by the J_q-vector,

$$\mathbf{c}_q^Z \equiv (nQ)^{-1} \mathbf{Z}_q \mathbf{1}_n. \tag{17.57}$$

Because the row masses of $\mathbf{Z}_q$ sum to $\mathbf{1}_{J_q}^\tau \mathbf{c}_q^Z = (nQ)^{-1} n = Q^{-1}$, each of the Q categorical variables has the same total mass. As a result, the row masses over all Q variables sum to 1. The row centroid is a weighted average of the J_q rows of $\mathbf{Z}_q$, where the weights are the row masses,

$$\frac{(\mathbf{c}_q^Z)^\tau \mathbf{Z}_q}{(\mathbf{c}_q^Z)^\tau \mathbf{1}_{J_q}} = \frac{(nQ)^{-1} \mathbf{1}_n^\tau \mathbf{Z}_q^\tau \mathbf{Z}_q}{Q^{-1}} = n^{-1} \mathbf{1}_n^\tau, \tag{17.58}$$

because $\mathbf{Z}_q^\tau \mathbf{Z}_q = \mathbf{I}_n$. Thus, the qth block of J_q row profiles has a row centroid (17.58) that does not depend upon q. Those J_q row profiles are dispersed within a subspace having at most $J_q - 1$ dimensions. All J row profiles are, therefore, dispersed within a subspace having at most $\sum_q (J_q - 1) = J - Q$ dimensions.

17.4.2   The Burt Matrix

A second interpretation of the idea of multiple correspondence analysis is based upon analyzing the $(J \times J)$-matrix

$$\mathbf{B} = \mathbf{Z}\mathbf{Z}^\tau = \begin{pmatrix} \mathbf{Z}_1\mathbf{Z}_1^\tau & \mathbf{Z}_1\mathbf{Z}_2^\tau & \cdots & \mathbf{Z}_1\mathbf{Z}_Q^\tau \\ \mathbf{Z}_2\mathbf{Z}_1^\tau & \mathbf{Z}_2\mathbf{Z}_2^\tau & \cdots & \mathbf{Z}_2\mathbf{Z}_Q^\tau \\ \vdots & \vdots & & \vdots \\ \mathbf{Z}_Q\mathbf{Z}_1^\tau & \mathbf{Z}_Q\mathbf{Z}_2^\tau & \cdots & \mathbf{Z}_Q\mathbf{Z}_Q^\tau \end{pmatrix}, \tag{17.59}$$

which is called a *Burt matrix*. See (17.6) for a Burt matrix with $Q = 2$. $\mathbf{B}$ is a symmetric matrix with block structure. The qth diagonal block submatrix, $\mathbf{Z}_q\mathbf{Z}_q^\tau = n\mathbf{D}_q$, say, is a diagonal matrix of the row totals of $\mathbf{Z}_q$ $(q = 1, 2, \ldots, Q)$, where $\mathbf{D}_q$ is the diagonal matrix of row or column masses for the qth variable. The off-diagonal (u, v)-block submatrix, $\mathbf{Z}_u\mathbf{Z}_v^\tau = \mathbf{N}_{uv}$, say, $(u \neq v)$, is a two-way contingency table between the uth variable and the vth variable $(u, v = 1, 2, \ldots, Q)$. Because the total of all entries in each submatrix $\mathbf{Z}_i \mathbf{Z}_j^T$ in $\mathbf{B}$ is n, the total of all entries of $\mathbf{B}$ is $b = nQ^2$. The Burt matrix (17.59) is the analogue in the discrete case of the covariance matrix of Q continuous variables.

17.4.3   Equivalence and an Implication

The two primary approaches to multiple correspondence analysis turn out to be equivalent to one another (Greenacre, 1984). From the symmetry of $\mathbf{B}$, a simple correspondence analysis of $\mathbf{B}$ produces the same sets of row and column coordinates, so that one of the two sets can be ignored. Furthermore, the standard coordinates of the rows of $\mathbf{B}$ are identical to the standard coordinates of the rows of $\mathbf{Z}$, and the principal coordinates obtained by analyzing $\mathbf{B}$ are directly related to those obtained by analyzing $\mathbf{Z}$ because the principal inertias of $\mathbf{B}$ are the squares of those of $\mathbf{Z}$.

This equivalence between the two approaches has the following implication. Although the multivariate indicator matrix $\mathbf{Z}$ incorporates information from all Q categorical variables, its multiple correspondence analysis provides no more information than an analysis of all pairs of categorical variables. In other words, multiple correspondence analysis of either $\mathbf{Z}$ or $\mathbf{B}$ offers no insight into three- or higher-way interactions that may be present in the contingency table.

17.4.4   Example: Satisfaction with Housing Conditions

This data set was studied by Madsen (1976) in a study of housing conditions in selected areas of Copenhagen, Denmark. A total of $n = 1,681$ residents living in rented homes built during 1960–1968 were surveyed about their satisfaction (categorized as low (ls), medium (ms), high (hs)), the amount of contact with other residents (low (lc), high (hc)), and their feeling of influence on apartment management (low (li), medium (mi), high (hi)). The rental units were categorized as tower blocks (tb), apartments (ap), atrium houses (ah), and terraced houses (th). The purpose of the study was to assess whether there was any association between degrees of contact, influence, and satisfaction and the type of housing.

The Burt table is given in Table 17.9. The χ^2-statistics for the off-diagonal two-way contingency tables are $X_{12}^2 = 16.660$, $X_{13}^2 = 39.121$, $X_{14}^2 = 60.286$, $X_{23}^2 = 17.586$, $X_{24}^2 = 106.175$, and $X_{34}^2 = 5.140$, where "1" = Housing, "2" = Influence, "3" = Contact, and "4" = Satisfaction. Assuming these two-way tables are independent of each other, we conclude that both housing and influence appear not to be related to either contact or satisfaction. The sum of these χ^2-values is $X^2 = 244.968$.

TABLE 17.9. *Burt table of data on satisfaction with housing conditions in Copenhagen, Denmark (Madsen, 1976). The variables are type of housing (tower blocks:* tb; *apartments:* ap; *atrium houses:* ah; *terraced houses:* th), *influence on apartment management (low:* li; *medium:* mi; *high:* hi), *contact with other residents (low:* lc; *high:* hc), *and satisfaction (low:* ls; *medium:* ms; *high:* hs). For this table, $Q=4$, $J_1 = 4$, $J_2 = 3$, $J_3 = 2$, $J_4 = 3$, $J = 12$, and $n = 1681$.*

	Housing				Influence			Contact		Satisfaction		
	tb	ap	ah	th	li	mi	hi	lc	hc	ls	ms	hs
tb	400	0	0	0	140	172	88	219	181	99	101	200
ap	0	765	0	0	268	297	200	317	448	271	192	302
ah	0	0	239	0	95	84	60	82	157	64	79	96
th	0	0	0	227	124	106	47	95	182	133	74	70
li	140	268	95	124	627	0	0	234	393	282	170	175
mi	172	297	84	106	0	659	0	279	380	206	189	264
hi	88	200	60	47	0	0	395	200	195	79	87	229
lc	219	317	82	95	234	279	200	713	0	262	178	273
hc	181	448	157	182	393	380	195	0	968	305	268	395
ls	99	271	64	133	282	206	79	262	305	567	0	0
ms	101	192	79	74	170	189	87	178	268	0	446	0
hs	200	302	96	70	175	264	229	273	395	0	0	668

The two-dimensional multiple correspondence map is given in Figure 17.5. The first axis orders from right to left the low, medium, and high categories of the influence and satisfaction variables, whereas the reverse ordering occurs for the contact variable. The second axis separates the high levels from the low levels of influence, contact, and satisfaction, and also separates th and tb from ah, and ap is positioned at the center of the map.

Certain points are close to each other and indicate associations. Thus, high influence on management is related to residents being highly satisfied, whereas high contact with other residents produces medium satisfaction. Residents of atrium houses tend to have high contact with other residents and enjoy medium satisfaction, apartment residents have medium influence on management, residents of tower blocks tend to have low contact with other residents, and residents of terraced housing appear to have both low influence and low satisfaction.

17.4.5   A Weighted Least-Squares Approach

There are $Q(Q - 1)/2$ distinct two-way contingency tables above the diagonal of **B**; the tables below the diagonal are transposes of those above. Although we could carry out a simple correspondence analysis for every one of those $Q(Q - 1)/2$ tables, such extensive and exhaustive analyses

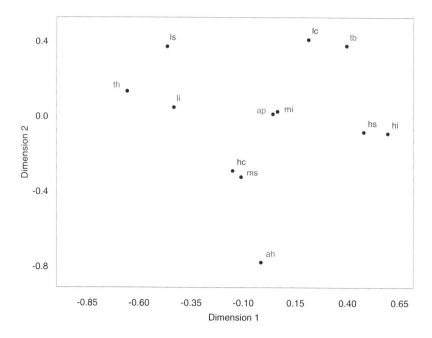

FIGURE 17.5. *Correspondence map for the housing conditions example. The factors in the study were: type of housing (tower blocks,* tb*; apartments,* ap*; atrium houses,* ah*; terraced houses,* th*), influence on apartment management (low,* li*; medium,* mi*; high,* hi*), contact with other residents (low,* lc*; high,* hc*), and satisfaction (low,* ls*; medium,* ms*; high,* hs*).*

would violate the principles of parsimony, efficiency, and dimensionality reduction.

With this in mind, we mention an alternative approach by Greenacre (1988), who proposed a *matrix approximation method* that (*a*) simultaneously fits all the $Q(Q-1)/2$ tables in the upper-triangle of **B**, and (*b*) reduces to simple correspondence analysis of $\mathbf{N} = \mathbf{N}_{12}$ when $Q = 2$. The idea is to approximate **B** by another matrix $\hat{\mathbf{B}}$, say, having reduced rank that minimized the weighted least-squares criterion

$$n^{-1}\mathrm{tr}\{\mathbf{D}^{-1/2}(\mathbf{B} - \hat{\mathbf{B}})\mathbf{D}^{-1}(\mathbf{B} - \hat{\mathbf{B}})^{\tau}\mathbf{D}^{-1/2}\}, \qquad (17.60)$$

where $\mathbf{D} = Q\mathbf{D}_r$ is Q times the diagonal matrix, $\mathbf{D}_r$, of row (or column) masses of **B** and is defined so that all its elements sum to 1 (cf. Exercise 17.3). Greenacre suggested the use of an alternating least-squares algorithm as a means of obtaining $\hat{\mathbf{B}}$ but could not guarantee that the minimum of (17.60) would be achieved by that procedure.

17.5   Software Packages

Many of the popular statistical software packages contain simple and multiple correspondence analysis routines. R has the `ca` package; see Charnomordic and Holmes (2001) and the details in Greenacre (2007, Appendix C). MINITAB has a correspondence analysis routine that appears to be matched to the output in Greenacre (1984). There is also a program `CodonW`, written by John Peden and available at `codonw.sourceforge.net`, which provides correspondence analysis of codon and amino acid usage.

Bibliographical Notes

Correspondence analysis was developed by many individuals. Initial work showing the correlation structure of a two-way contingency table appeared during the mid-1930s by H.O. Hirschfield (later Hartley), P. Horst, and others. At the start of the 1940s, R.A. Fisher and L. Guttman constructed scaling theories for contingency tables for biometric and psychometric contexts, respectively. The methodology found its champion, J.-P. Benzecri, in the early 1960s when Benzecri and a group of French statisticians constructed a theory of associations between rows and columns of a two-way contingency table. This was called *analyse des correspondances* in French, which was later loosely translated as "correspondence analysis." Others who have had major impacts on the subject include M.O. Hill, M.J. Greenacre, and L.A. Goodman.

Much of this chapter has benefitted from the treatment of the topic in books and articles by Greenacre; specifically, Greenacre (1981, 1984, 1988, 2000, 2007) and Greenacre and Hastie (1987). An interesting collection of articles on applications of correspondence analysis (and other related topics) is the book edited by Blasius and Greenacre (1998). See also the articles by Gower and Digby (1981) (who provide a general tour of techniques for graphically representing multivariate data), van der Heijden, de Falguerolles, and de Leeuw (1989) (who studied the correspondence analysis of residuals from fitting a log-linear model to a contingency table), and Pack and Jolliffe (1992) (who proposed measures for detecting influential observations in correspondence analysis).

Exercises

17.1 The 4×4 contingency table in Table 17.10 was originally analyzed by Stuart (1953) and has since been studied by many statisticians. It contains frequency data on eye tests, specifically, the right-eye grade and the corresponding left-eye grade in unaided distance vision for 7,477 women,

TABLE 17.10. *Right-eye grade and left-eye grade of 7,477 women with respect to unaided distance vision (Stuart, 1953). The Pearson chi-squared test for independence gives* $X^2 = 8,096.877$ *on 9 degrees of freedom, so that an hypothesis of independence is rejected.*

		Left-Eye Grade			
Right-Eye Grade	Best	Second	Third	Worst	Totals
Best	1,520	266	124	66	1,976
Second	234	1,512	432	78	2,256
Third	117	362	1,772	205	2,456
Worst	36	82	179	492	789
Totals	1,907	2,222	2,507	841	7,477

aged 30–39, employed in Royal Ordinance factories in Britain, where each eye was graded in one of four categories from best to worst. Carry out a correspondence analysis for this square contingency table and interpret the results.

17.2 Suppose we omit the last row of $\mathcal{X}$ and last row of $\mathcal{Y}$, so that $\mathcal{X}$ has $r - 1$ rows and n columns and $\mathcal{Y}$ has $s - 1$ rows and n columns. Suppose we center $\mathcal{X}$ and $\mathcal{Y}$ at their means.

(a) Show that

$$(\mathcal{X}_c \mathcal{X}_c^T)^{-1} = \text{diag}\left[n_{1+}^{-1}, n_{2+}^{-1}, \ldots, n_{r-1,+}^{-1}\right] + n_{r+}^{-1} \mathbf{J}_{r-1},$$

$$(\mathcal{Y}_c \mathcal{Y}_c^T)^{-1} = \text{diag}\left[n_{+1}^{-1}, n_{+2}^{-1}, \ldots, n_{+,s-1}^{-1}\right] + n_{+s}^{-1} \mathbf{J}_{s-1}.$$

(b) Show that the entry in the jth row and ith column of the full-rank regression coefficient matrix, $\hat{\Theta} = \mathcal{Y}_c \mathcal{X}_c^T (\mathcal{X}_c \mathcal{X}_c^T)^{-1}$, is

$$\theta_{ji} = \frac{n_{ij}}{n_{i+}} - \frac{n_{rj}}{n_{r+}}, \quad i = 1, 2, \ldots, r - 1, \; j = 1, 2, \ldots, s - 1,$$

which is just the difference between the ith and rth row proportions for the jth column of the contingency table. Similarly, show that the entry in the ith row and jth column of $\mathcal{X}_c \mathcal{Y}_c^T (\mathcal{Y}_c \mathcal{Y}_c^T)^{-1}$ is

$$\frac{n_{ij}}{n_{+j}} - \frac{n_{is}}{n_{+s}}, \quad i = 1, 2, \ldots, r - 1, \; j = 1, 2, \ldots, s - 1.$$

(c) From these two matrices, show that the trace of $\hat{\mathbf{R}}$ is given by

$$\sum_{i=1}^{r} \sum_{j=1}^{s} \frac{1}{n_{i+} n_{+j}} \left(n_{ij} - \frac{n_{i+} n_{rj}}{n_{r+}}\right) \left(n_{ij} - \frac{n_{is} n_{+j}}{n_{+s}}\right),$$

and, under independence of A and B, that tr$\{\hat{\mathbf{R}}\}$ reduces to X^2 in (17.22).

TABLE 17.11. *Number of children in a family versus yearly income (in units of 1,000 Kroner) for n = 25263 Swedish families (Cramér, 1946). The Pearson chi-squared test for independence gives $X^2 = 568.57$ on 12 degrees of freedom, so that an independence hypothesis is rejected.*

Number of Children	Yearly Income (1000s Kroner)				
	0–1	1–2	2–3	3+	Total
0	2,161	3,577	2,184	1,636	9,558
1	2,755	5,081	2,222	1,052	11,110
2	936	1,753	640	306	3,635
3	225	419	96	38	778
≥ 4	39	98	31	14	182
Total	6,116	10,928	5,173	3,046	25,263

(d) Show that the $s - 1$ eigenvalues of $\hat{\mathbf{R}}$ are identical to the nonzero eigenvalues of $\mathbf{R}_0$ (or $\mathbf{R}_1$).

17.3 (Greenacre, 2000). Another way of deriving the results of simple correspondence analysis is to find an $(r \times s)$-matrix $\hat{\mathbf{P}}$ having reduced-rank $t < \min(r, s)$ that approximates $\mathbf{P}$ by minimizing the weighted least-squares criterion,

$$\mathrm{tr}\{\mathbf{D}_r^{-1/2}(\mathbf{P} - \hat{\mathbf{P}})\mathbf{D}_c^{-1}(\mathbf{P} - \hat{\mathbf{P}})^\tau \mathbf{D}_r^{-1/2}\}.$$

Using the Eckart–Young Theorem, find the matrix $\hat{\mathbf{P}}$ that yields the best reduced-rank approximation of $\mathbf{P}$ in the above sense. Show that the best "rank-1" approximation to $\mathbf{P}$ is the *trivial solution* $\hat{\mathbf{P}} = \mathbf{rc}^\tau$.

17.4 Let $\mathbf{M} = [m_{ij}]$ and $\mathbf{Q} = [q_{ij}]$ be defined as in (17.51) and let $\mathbf{N} = \mathbf{M} + \mathbf{Q}$. Consider $\mathrm{tr}\{(vec\ \mathbf{N})(vec\ \mathbf{N})^\tau\}$. Show that the cross-product term $\mathrm{tr}\{(vec\ \mathbf{M})(vec\ \mathbf{Q})^\tau\} = 0$, whence, we have the identity,

$$\sum_i \sum_j n_{ij}^2 = \sum_i \sum_j m_{ij}^2 + \sum_i \sum_j q_{ij}^2.$$

17.5 Show that $\mathbf{G}_P^\tau$ and $\mathbf{H}_P^\tau$ are related to each other by proving that $\mathbf{G}_P^\tau = \mathbf{D}_r^{-1}\mathbf{P}\mathbf{H}_P^\tau\mathbf{D}_\lambda^{-1}$ and $\mathbf{H}_P^\tau = \mathbf{D}_c^{-1}\mathbf{P}^\tau\mathbf{G}_P^\tau\mathbf{D}_\lambda^{-1}$.

17.6 The 5×4 contingency table in Table 17.11 is due to Cramér (1946, p. 444); see also Diaconis and Efron (1985). It contains a sample of frequency data from a Swedish census of March 1936 in which 25,263 married couples residing in country districts, who had been married for at most five years, each listed the number of children in their family and their yearly income (in units of 1,000 Kroner). Carry out a correspondence analysis for this table and interpret the results.

17.7 Construct four different contingency tables, each with five rows and three columns, with the restriction that each of the column totals in each table equals 50. Compute the weights in the chi-squared statistic for each table. Compute the inertia for each table and arrange the four tables by increasing inertia. Plot the row profiles for each table as points in a triangular scatterplot. What is the relationship between inertia and these plots?

References

[1] Abbott, A. and Hrycak, A. (1990). Measuring sequence resemblance, *American Journal of Sociology*, **96**, 144–185.

[2] Aizerman, M., Braverman, E., and Rozoener, L. (1964). Theoretical foundations of the potential function method in pattern recognition learning, *Automation and Remote Control*, **25**, 821–837.

[3] Aldrin, M. (1996). Moderate projection pursuit regression for multivariate response data, *Computational Statistics & Data Analysis*, **21**, 501–531.

[4] Alimoglu, F. (1995). Combining multiple classifiers for pen-based handwritten digit recognition, M.Sc. thesis, Institute of Graduate Studies in Science and Engineering, Bogazici University, Istanbul, Turkey.

[5] Alon, U., Barkai, N., Notterman, D., Gish, K., Ybarra, S., Mack, D., and Levine, A. (1999). Broad patterns of gene expression revealed by clustering analysis of tumor and normal colon tissues probed by oligonucleotide arrays, *Proceedings of the National Academy of Sciences*, **96**, 6745–6750.

[6] Altschul, S.F., Gish, W., Miller, W., Myers, E.W., and Lipman, D.J. (1990). Basic local alignment search tool, *Journal of Molecular Biology*, **215**, 403–410.

[7] Akaike, H. (1973). Information theory and an extension of the maximum likelihood principle, In *2nd International Symposium on*

Information Theory, eds. B.N. Petrov and F. Csaki, Budapest: Akademia Kiado, pp. 267–281.

[8] Akaike, H. (1978). A Bayesian analysis of the minimum AIC procedure, *Annals of the Institute of Statistical Mathematics, Series A*, **30**, 9–14.

[9] Amit,Y. and Geman, D. (1996). Shape quantization and recognition with randomized trees. *Neural Computation*, **9**, 1545–1588.

[10] Anderson, J.A. (1982). Logistic discrimination, In: *Handbook of Statistics, Volume 2*, P.R. Krishnaiah and L.N. Kanal (eds.), pp. 169–191, Amsterdam: North-Holland.

[11] Anderson, R.L. and Bancroft, T.A. (1952). *Statistical Theory in Research*, New York: McGraw-Hill.

[12] Anderson, T.W. (1951). Estimating linear restrictions on regression coefficients for multivariate normal distributions, *Annals of Mathematical Statistics* **22**: 327–351.

[13] Anderson, T.W. (1984a). Estimating linear statistical relationships, The 1982 Wald Memorial Lectures, *The Annals of Statistics*, **12**, 1–45.

[14] Anderson, T.W. (1984b). *An Introduction to Multivariate Statistical Analysis, Second Edition*, New York: Wiley.

[15] Anderson, T.W. (1999). Asymptotic distribution of the reduced rank regression estimator under general conditions, *The Annals of Statistics*, **27**, 1141–1154.

[16] Andrews, D.F. and Herzberg, A.M. (1985). *Data*, New York: Springer.

[17] Ashton, K.H., Oxnard, C.E., and Spence, T.F. (1965). Scapular shape and primate classification, *Proceeding of the Zoological Society, London*, **145**, 125–142.

[18] Asimov, D. (1985). The grand tour: a tool for viewing multidimensional data, *SIAM Journal on Scientific and Statistical Computing*, **6**, 128–143.

[19] Attias, H. (1999). Independent factor analysis, *Neural Computation*, **11**, 803–852.

[20] Bach, F.R. and Jordan, M.I. (2002). Kernel independent component analysis, *Journal of Machine Learning Research*, **3**, 1–48.

[21] Baker, H.V. (2003). Comment on "Statistical Challenges in Functional Genomics" by Sebastiani, Gussoni, Kohane, and Ramoni, *Statistical Science*, **18**, 60–64.

[22] Baker, S.G. (2001), Analyzing a randomized cancer prevention trial with a missing binary outcome, an auxiliary variable, and all-or-none compliance. *Journal of the American Statistical Association*, **95**, 43–50.

[23] Banfield, J.D. and Raftery, A.E. (1992). Ice floe identification in satellite images using mathematical morphology and clustering about principal curves, *Journal of the American Statistical Assiciation*, **87**, 7–16.

[24] Barber, D. and Bishop, C.M. (1998). Ensemble learning in Bayesian Neural networks, In: *Neural Networks and Machine Learning* (C.M. Bishop, ed.), Berlin, Germany: Springer.

[25] Barnett, V. and Lewis, T. (1994). *Outliers in Statistical Data, Third Edition*, New York: Wiley.

[26] Bartholomew, D.J. (1987). *Latent Variable Models and Factor Analysis*, London: Charles Griffin & Co. Ltd.

[27] Basalaj, W. and Eilbeck, K. (2003). Straight-line drawings of protein interactions, In *Lecture Notes in Computer Science*, **1731**, New York: Springer.

[28] Belhumeur, P.N., Hespanha, J.P., and Kriegman, D.J. (1997). Eigenfaces vs. Fisherfaces: recognition using class specific linear projection, *IEEE Transactions on Pattern Analysis and Machine Intelligence*, **19**, 711–720.

[29] Bellman, R.E. (1961). *Adaptive Control Processes*, Princeton, NJ: Princeton University Press.

[30] Bellman, R. (1970). *Introduction to Matrix Analysis, Second Edition*, New York: McGraw-Hill.

[31] Belkin, M. and Niyogi, P. (2002). Laplacian eigenmaps and spectral techniques for embedding and clustering, *Advances in Neural Information Processing Systems 14* (T.G. Dietterich, S. Becker, and Z. Ghahramani. eds.), Cambridge, MA: MIT Press, pp. 585–591.

[32] Belsley, D.A., Kuh, E., and Welsch, R.E. (1980). *Regression Diagnostics: Identifying Influential Data and Sources of Collinearity*, New York: Wiley.

[33] Berger, J.O. (1985). *Statistical Decision Theory and Bayesian Analysis, Second Edition*. New York: Springer.

[34] Berger, M. (2003). *A Panoramic View of Riemannian Geometry*, New York: Springer.

[35] Bernardo, J.M. and Smith, A.F.M. (1994). *Bayesian Theory*, New York: Wiley.

[36] Bernstein, M., de Silva, V., Langford, J.C., and Tenenbaum, J.B. (2000). Graph approximations to geodesics on embedded manifolds. Unpublished Technical Report, Stanford University.

[37] Besag, J., Green, P., Higdon, D., and Mengersen, K. (1995). Bayesian computation and stochastic systems (with discussion), *Statistical Science*, **10**, 3–66.

[38] Bickel, P.J. and Ritov, Y. (2004). The golden chain, a discussion of three articles on boosting, *Annals of Statistics*, **32**, 91–96.

[39] Bishop, C.M. (1995). *Neural Networks for Pattern Recognition*, Oxford, U.K.: Clarendon Press.

[40] Bishop, C.M. (2006). *Pattern Recognition and Machine Learning*, New York: Springer.

[41] Blasius, J. and Greenacre, M. (1998). *Visualization of Categorical Data*, New York: Academic Press.

[42] Bolton, R.J. and Krzanowski, W.J. (1999). A characterization of principal components for projection pursuit, *The American Statistician*, **53**, 108–109.

[43] Borg, I. and Groenen, P. (1997). *Modern Multidimensional Scaling: Theory and Applications*, New York: Springer.

[44] Boser, B.E., Guyon, I.M., and Vapnik, V.N. (1992). A training algorithm for optimal margin classifiers, In: *Proceedings of the Fifth Conference on Computational Learning Theory* (ed., D. Haussler), 144–152, New York: Association of Computing Machinery Press.

[45] Box, G.E.P. and Behnken, D. (1960). Some new three level designs for the study of quantitative variables, *Technometrics*, **2**, 455–475.

[46] Boyd, S. and Vandenberghe, L. (2004). *Convex Optimization*, Cambridge, U.K.: Cambridge University Press.

[47] Brand, M. (2003). Charting a manifold, In: *Advances in Neural Information Processing Systems*, **15**, 961–968, Cambridge, MA: MIT Press.

[48] Breiman, L. (1992). The little bootstrap and other methods for dimensionality selection in regression: X-fixed prediction error, *Journal of the American Statistical Association*, **87**, 738–754.

[49] Breiman, L. (1994). The 1991 census adjustment: Undercount or bad data, *Statistical Science*, **9**, 458–475.

[50] Breiman, L. (1995). Better subset regression using the nonnegative garotte, *Technometrics*, **37**, 373–384.

[51] Breiman, L. (1996a). Heuristics of instability and stabilization in model selection, *The Annals of Statistics*, **24**, 2350–2383.

[52] Breiman, L. (1996b). Bagging predictors, *Machine Learning*, **26**, 123–140.

[53] Breiman, L. (1996c). Out-of-bag estimation, Technical Report, Department of Statistics, University of California, Berkeley.

[54] Breiman, L. (1999). Prediction games and arcing algorithms, *Neural Computation*, **11**, 1493–1517.

[55] Breiman, L. (2000). Discussion of Friedman, Hastie, and Tibshirani (2000), *Annals of Statistics*, **26**, 374–377.

[56] Breiman, L. (2001a). Statistical modeling: The two cultures (with discussion), *Statistical Science*, **16**, 199–231.

[57] Breiman, L. (2001b). Random forests, *Machine Learning*, **45**, 5–32.

[58] Breiman, L. (2002). Machine learning. 2002 Wald Memorial Lectures.

[59] Breiman, L. (2004). Population theory for boosting ensembles, *Annals of Statistics*, **32**, 1–11.

[60] Breiman, L. and Cutler, A. (2004). *Random Forests*, Short course given at the 36th Symposium of the Interface, Baltimore, MD, March 25, 2004.

[61] Breiman, L. and Friedman, J.H. (1997). Predicting multivariate responses in multiple linear regression (with discussion), *Journal of the Royal Statistical Society, Series B*, **59**, 3–54.

[62] Breiman, L., Friedman, J., Olshen, R., and Stone, C. (1984). *Classification and Regression Trees*. Boca Raton, FL: Wadsworth.

[63] Breiman, L. and Spector, P. (1992). Submodel selection and evaluation in regression: the X-random case, *International Statistical Review*, **60**, 291–319.

[64] Brillinger, D.R. (1969). The canonical analysis of stationary time series, In *Multivariate Analysis II* (ed P.R. Krishnaiah), pp. 331–350. New York: Academic Press.

[65] Brin, S. and Page, L. (1998). The anatomy of a large-scale hypertextual Web search engine, *Proceedings of the Seventh International World-Wide Web Conference*, Brisbane, Australia, pp. 107–117.

[66] Brown, L.D. (1971). Admissible estimators, recurrent diffusions, and insoluble boundary-value problems, *Annals of Mathematical Statistics*, **42**, 855–903.

[67] Brown, P.J. (1993). *Measurement, Regression, and Calibration.* Oxford: Clarendon Press.

[68] Brown, P.J., Firth, D., and Payne, C.D. (1999). Forecasting on British election night 1997, *Journal of the Royal Statistical Society, Series A*, **162**, 211–226.

[69] Brown, P.J. and Zidek, J.V. (1980). Adaptive multivariate ridge regression, *The Annals of Statistics*, **8**, 64–74.

[70] Brown, P.J. and Zidek, J.V. (1982). Multivariate regression shrinkage with unknown covariance matrix, *Scandanavian Journal of Statistics*, **9**, 209–215.

[71] Bryan, J. (2004). Problems in gene clustering based on gene expression data, *Journal of Multivariate Analysis*, **90**, 44–66.

[72] Buhlmann, P. and Yu, B. (2003). Boosting with the L_2 loss: regression and classification, *Journal of the American Statistical Association*, **98**, 324–339.

[73] Buntine, W.L. and Weigend, A.S. (1991). Bayesian back-propagation, *Complex Systems*, **5**, 603–643.

[74] Burges, C.J.C. (1998). A tutorial on support vector machines for pattern recognition, *Data Mining and Knowledge Discovery*, **2**, 121–167.

[75] Burt, C. (1950). The factorial analysis of qualitative data, *British Journal of Psychology*, Statistics Section **3**, 166-185.

[76] Butler, N.A. and Denham, M.C. (2000). The peculiar shrinkage properties of partial least squares regression, *Journal of the Royal Statistical Society, Series B*, **62**, 585–593.

[77] Buja, A. and Swayne, D.F. (2002). Visualization methodology for multidimensional scaling, *Journal of Computational and Graphical Statistics*,

[78] Buja, A., Swayne, D.F., Littman, M.L., Dean, N., and Hofmann, H. (2002). XGvis: Interactive data visualization with multidimensional scaling, *Journal of Computational and Graphical Statistics*,

[79] Buta, R. (1987). The structure and dynamics of ringed galaxies III: surface photometry and kinematics of the ringed nonbarred spiral NGC7531, *The Astrophysical Journal*, **64**, 1–37.

[80] Cacoullos, T. (1966). Estimation of a multivariate density, *Annals of the Institute of Statistical Mathematics*, **18**, 179–189.

[81] Calvin, W.H. and Ojemann, G.A. (1994). *Conversations with Neil's Brain: The Neural Nature of Thought and Language*, Reading, MA: Addison-Wesley.

[82] Cardoso, J.-F. (1998). Blind signal separation: statistical principles, *Proceedings of the IEEE*, **86**, 2009–2025.

[83] Carlin, B.P. and Louis, T.A. (2000). *Bayes and Empirical Bayes Methods for Data Analysis, Second Edition*, New York: Chapman & Hall/CRC.

[84] Cartan, E. (1946). *Leçons sur la géómetrie des espaces de Riemann, 2e édition, revue et augmentée*, Paris: Gauthier-Villars. (This was reprinted in 1988 by Éditions Jacques Gabay, Sceaux. An English translation by J. Glazebrook is *Geometry of Riemannian Spaces*, Brookline, MA: Math Sci Press, 1983.)

[85] Casella, G. and Berger, R.L. (1990). *Statistical Inference*, Belmont, CA: Wadsworth & Brooks/Cole.

[86] Cattell, R.B. (1966). The scree test for the number of factors, *Multivariate Behavioral Research*, **1**, 245–276.

[87] Chang, W.C. (1983). On using principal components before separating a mixture of two multivariate normal distributions, *Applied Statistics*, **32**, 267–275.

[88] Charnomordic, B. and Holmes, S. (2001). Correspondence analysis with R, *Statistical Computing & Statistical Graphics Newsletter*, **12**, 19–25.

[89] Chen, C., Liaw, A., and Breiman, L. (2004). Using random forest to learn imbalanced data, unpublished technical report.

[90] Cheng, B. and Titterington, D.M. (1994). Neural networks: a review from a statistical perspective (with discussion), *Statistical Science*, **9**, 2–54.

[91] Cheng, Y. and Church, G.M. (2000). Biclustering of expression data, In: *Proceedings of the Eighth International Conference on Intelligent Systems for Molecular Biology*, 93–103.

[92] Chernick, M.R. (1999). *Bootstrap Methods: A Practioner's Guide*, New York: Wiley.

[93] Cherry, E.C. (1953). Some experiments in the recognition of speech, with one and two ears, *Journal of the Acoustical Society of America*, **25**, 975–979.

[94] Chipman, H., Hastie, T.J., and Tibshirani, R. (2003). Clustering microarray data, in: *Statistical Analysis of Gene Expression Microarray Data* (T. Speed, ed.), New York: Chapman & Hall/CRC, pp. 159–200.

[95] Chipman, J.S. (1964). On least squares with insufficient observations, *Journal of the American Statistical Association*, **59**, 1078–1111.

[96] Cristianini, N. and Shawe-Taylor, J. (2000). *Support Vector Machines, And Other Kernel-Based Learning Methods*, Cambridge, U.K.: Cambridge University Press.

[97] Cichocki, A. and Amari, S. (2003). *Adaptive Blind Signal and Image Processing*. New York: Wiley.

[98] Clark, L.A. and Pregibon, D. (1992). Tree-based models, In: *Statistical Models in S* (J.M. Chambers and T.J. Hastie, eds.), Boca Raton, FL: Wadsworth, pp. 377–420.

[99] Cleveland, W.S. and Sun, D.X. (2000). Internet traffic data, *Journal of the American Statistical Association*, **95**, 979985.

[100] Codd, E.F. (1970). A relational model of data for large shared data banks, *Communications of the ACM*, **13**. Reprinted in Milestones of Research – Selected Papers 1958–1982, *Communications of the CACM*, **26** (1983).

[101] Connolly, T.M. and Begg, C.E. (2002). *Database Systems, Third Edition*, New York: Addison Wesley.

[102] Constantine, A.G. (1963). Some noncentral distribution problems in multivariate analysis, *The Annals of Mathematical Statistics*, **34**, 1270–1285.

[103] Constantine, A.G. and Gower, J.C. (1978). Graphical representation of asymmetric matrices, *Applied Statistics*, **27**, 297–304.

[104] Cook, D., Buja, A., and Cabrera, J. (1993). Projection pursuit indexes based on orthogonal function expansions, *Journal of Computational and Graphical Statistics*, 2, 225–250.

[105] Cook, D., Buja, A., Cabrera, J., and Hurley, H. (1995). Grand tour and projection pursuit, *Journal of Computational and Graphical Statistics*, **2**, 225–250.

[106] Cortes, C. and Vapnik, V.N. (1995). Support vector networks, *Machine Learning*, **20**, 273–297.

[107] Cover, T. and Thomas, J. (1991). *Elements of Information Theory, Volume 1*. New York: Wiley.

[108] Cowles, M.K. and Carlin, B.P. (1996). Markov chain Monte Carlo convergence diagnostics: a comparative review, *Journal of the American Statistical Association*, **91**, 883–904.

[109] Cox, T.F. and Cox, M.A.A. (2001). *Multidimensional Scaling, Second Edition*, London: Chapman and Hall.

[110] Cramér, H. (1946). *Mathematical Methods of Statistics*, Princeton, NJ: Princeton University Press.

[111] Craven, P. and Wahba, G. (1979). Smoothing noisy data with spline functions: estimating the correct degree of smoothing by the method of generalized cross-validation, *Numerische Mathematik*, **31**, 377–403.

[112] Crippen, G.M. and Havel, T.F. (1988). *Distance Geometry and Molecular Conformation*, New York: Wiley.

[113] Cristanini, N. and Shawe-Taylor, J. (2000). *An Introduction to Support Vector Machines*, Cambridge, U.K.: Cambridge University Press.

[114] Culp, M., Johnson, K., and Michailidis, G. (2006). ada: an R package for stochastic boosting, *Journal of Statistical Software*, **17**, 2.

[115] Cybenko, G. (1989). Approximation by superpositions of a sigmoidal function, *Mathematical Control Signal & Systems*, **2**, 303–314.

[116] Date, C.J. (2000). *An Introduction to Database Systems, Seventh Edition*, Reading, MA: Addison Wesley.

[117] Davis, R. and Anderson, J. (1989). Exponential survival trees, *Statistics in Medicine*, **8**, 947–962.

[118] Davis, J.B. and McKean, J.W. (1993). Rank-based methods for multivariate linear models, *Journal of the American Statistical Association*, **88**, 245–251.

[119] DeCoste, D. (2001). Visualizing Mercer kernel feature spaces via kernelized locally linear embeddings, In: *Proceedings of the Eighth International Conference on Neural Information Processing*, Shanghai, China.

[120] de Duve, C. (1984). *A Guided Tour of the Living Cell, Volumes 1 and 2*, New York: Scientific American Library.

[121] de Jong, S. (1993). SIMPLS: an alternative approach to partial least squares regression, *Chemometrics and Intelligent Laboratory Systems*, **18**, 251–263.

[122] de Jong, S. (1995). PLS shrinks, *Journal of Chemometrics*, **9**, 323–326.

[123] de Lathauwer, L., de Moor, B., Vandewalle, J. (2000). Fetal electro-cardiogram extraction by blind source subspace separation, *IEEE Transactions on Biomedical Engineering*, **47**, 567–573. *Proceedings of the IEEE SP/Athos Workshop on Higher-Order Statistics*, Girona, Spain, pp. 134–138.

[124] Delicado, P. (2001). Another look at principal curves and surfaces, *Journal of Multivariate Analysis*, **77**, 84–116.

[125] Dempster, A.P., Laird, N.M., and Rubin, D.B. (1977). Maximum likelihood from incomplete data via the EM algorithm (with discussion), *Journal of the Royal Statistical Society, B*, **39**: 1–38.

[126] de Montricher, G.M., Tapia, R.A., and Thompson, J.R. (1975). Nonparametric maximum likelihood estimation of probability densities by penalty function methods, *The Annals of Statistics*, **3**, 1329–1348.

[127] Deonier, R.C., Tavaré, S., and Waterman, M.S. (2005). *Computational Genome Analysis*, New York: Springer.

[128] de Silva, V. and Tenenbaum, J.B. (2003). Unsupervised learning of curved manifolds, In: *Nonlinear Estimation and Classification* (D.D Denison, M.H. Hansen, C.C. Holmes, B. Mallick, B. Yu, eds.), *Lecture Notes in Statistics*, **171**, New York: Springer, pp. 453–466.

[129] Devroye, L. (1983). The equivalence of weak, strong, and complete convergence in L_1 for kernel density estimates, *The Annals of Statistics*, **11**, 896–904.

[130] Devroye, L. and Gyorfi, L. (1985). *Nonparametric Density Estimation: The L1 View*, New York: Wiley.

[131] Devroye, L. and Penrod, C.S. (1984). The consistency of automatic kernel density estimates, *The Annals of Statistics*, **12**, 1231–1249.

[132] Diaconis, P. and Efron, B. (1985). Testing for independence in a two-way table: new interpretations of the chi-square statistics (with discussion), *The Annals of Statistics*, **13**, 845–913.

[133] Diaconis, P. and Freedman, D. (1984). Asymptotics of graphical projection pursuit, *Annals of Statistics*, **12**, 793–815.

[134] Diaconis, P. and Shahshahani, M. (1984). On nonlinear functions of linear combinations, *SIAM Journal of Scientific and Statistical Computing*, **5**, 175–191.

[135] Di Battista, G., Eades, P., Tamassia, R., and Tollis, I. (1994). Algorithms for drawing graphs: An annotated bibliography, *Computational Geometry*, **4**, 235–282.

[136] Dietterich, T.G. (2000). An experimental comparison of three methods for constructing ensembles of decision trees: bagging, boosting, and randomization, *Machine Learning*, **40**, 139–158.

[137] Dijkstra, E.W. (1959). A note on two problems in connection with graphs, *Numerische Mathematik*, **1**, 269–271.

[138] Do, K.-A,, Broom, B., and Wen, S. (2003). GENECLUST, In: *The Analysis of Gene Expression Data: Methods and Software* (G. Parmigiani, E.S. Garrett, R.A. Irizarry, and S.L. Zeger, eds.), Chapter 15, New York: Springer.

[139] Donnell, D.J., Buja, A., and Stuetzle, W. (1994). Analysis of additive dependencies and concurvities using smallest additive principal components (with discussion), *The Annals of Statistics*, **22**, 1635–1673.

[140] Donoho, D. (1981). On minimum entropy deconvolution, In *Applied Time Series Analysis II*, D.A. Finley (ed.), New York: Academic Press, pp. 565–608.

[141] Donoho, D. and Grimes, C. (2003). Hessian eigenmaps: locally linear embedding techniques for high-dimensional data, *Proceedings of the National Academy of Sciences*, **100**, 5591–5596.

[142] Draper, N. and Smith, H. (1981). *Applied Regression Analysis, Second Edition*. New York: Wiley.

[143] Drăghici, S. (2003). *Data Analysis Tools for DNA Microarrays*, Boca Raton, FL: Chapman & Hall/CRC.

[144] Duane, S., Kennedy, A.D., Pendleton, B.J., and Roweth, D. (1987). Hybrid Monte Carlo, *Physics Letters B*, **195**, 216–222.

[145] Duchamp, T. and Stuetzle, W. (1996). Extremal properties of principal curves in the plane, *The Annals of Statistics*, **24**, 1511–1520.

[146] Duda, R.O., Hart, P.E., and Stork, D.G. (2001). *Pattern Classification, Second Edition*, New York: Wiley.

[147] Dudoit, S., Fridlyand, J., and Speed, T.P. (2002). Comparison of discrimination methods for the classification of tumors using gene expression data, *Journal of the American Statistical Association*, **97**, 77–87.

[148] Durbin, R., Eddy, S., Krogh, A., and Mitchison, G. (1998). *Biological Sequence Analysis: Probabilistic Models of Proteins and Nucleic Acids*, Cambridge, U.K.: Cambridge University Press.

[149] Eckart, C. and Young, G. (1936). Approximation of one matrix by another of lower rank, *Psychometrika*, **1**, 211–218.

[150] Efron, B. (1979). Bootstrap methods: another look at the jackknife, *Annals of Statistics*, **7**, 1–26.

[151] Efron, B. (1982). *The Jackknife, the Bootstrap, and Other Resampling Plans*. Philadelphia: Society for Industrial and Applied Mathematics.

[152] Efron, B. (1983). Estimating the error rate of a prediction rule: some improvements on cross-validation, *Journal of the American Statistical Association*, **78**, 316–331.

[153] Efron, B. (1986). How biased is the apparent error rate of a prediction rule? *Journal of the American Statistical Association*, **81**, 461–470.

[154] Efron, B. and Morris, C. (1973). Stein's empirical rule and its competitors — an empirical Bayes approach, *Journal of the American Statistical Association*, **68**, 117–130.

[155] Efron, B. and Morris, C. (1975). Data analysis using Stein's estimator and its generalizations, *Journal of the American Statistical Association*, **70**, 311–319.

[156] Efron, B. and Morris, C. (1977). Stein's paradox in statistics, *Scientific American*, **236**(5), 119–127.

[157] Efron, B. and Tibshirani, R.J. (1993). *An Introduction to the Bootstrap*. London: Chapman and Hall.

[158] Efron, B. Tibshirani, R. (1997). Improvements on cross-validation: The .632+ bootstrap method, *Journal of the American Statistical Association*, **92**, 548–560.

[159] Efron, B., Hastie, T., Johnstone, I., and Tibshirani, R. (2004). Least angle regression (with discussion), *Annals of Statistics*, **32**, 407–499.

[160] Ekman, G. (1954). Dimensions of color vision, *Journal of Psychology*, **38**, 467–474.

[161] Epanechnikov, V.K. (1969). Nonparametric estimation of a multivariate probability density, *Theory of Probability and its Applications*, **14**, 153–158.

[162] Everitt, B.S. (1984). *An Introduction to Latent Variable Models*, London, U.K.: Chapman and Hall.

[163] Everitt, B.S. and Hand, D.J. (1981). *Finite Mixture Distributions*. London: Chapman and Hall.

[164] Fairley, W.B., Izenman, A.J., and Crunk, S.M. (2001). Combining incomplete information from independent assessment surveys for estimating masonry deterioration, *Journal of the American Statistical Association*, **96**, 488–499.

[165] Farid, H. (2001). Detecting steganographic messages in digital images, Department of Computer Science, Technical Report 412, Dartmouth College, Hanover, NH.

[166] Fayyad, U.M., Piatetsky-Shapiro, G., Smyth, P., and Uthurusamy, R. (1996). *Advances in Knowledge Discovery and Data Mining*, Menlo Park, CA: AAAI Press/MIT Press.

[167] Findley, D.F. and Parzen, E. (1995). A conversation with Hirotugu Akaike, *Statistical Science*, **10**, 104–117.

[168] Fine, T.L. (1999). *Feedforward Neural Network Methodology*, New York: Springer.

[169] Fisher, R.A. (1915). Frequency distribution of the values of the correlation coefficient in samples from an indefinitely large population, *Biometrika*, **10**, 507–521.

[170] Fisher, R.A. (1936). The use of multiple measurements in taxonomic problems, *Annals of Eugenics*, **7**, 179–188.

[171] Fisher, R.A. (1940). The precision of discriminant functions, *Annals of Eugenics*, **10**, 422–429.

[172] Fisherkeller, M.A., Friedman, J.H., and Tukey, J.W. (1974). PRIM-9, an interactive multidimensional data display and analysis system, In: *Proceedings of the Pacific ACM Regional Conference*. [Also in the *Collected Works of John W. Tukey*, **V** (1988), 307–327].

[173] Floyd, R.W. (1962). Algorithm 97, *Communications of the ACM*, **5**, 345.

[174] Fraley, C. and Raftery, A.E. (1998). MCLUST: Software for model-based cluster analysis, Technical Report No. 342, Department of Statistics, University of Washington.

[175] Fraley, C. and Raftery, A.E. (2002). Model-based clustering, discriminant analysis, and density estimation, *Journal of the American Statistical Association*, **97**, 611–631.

[176] Frank, I.E. and Friedman, J.H. (1993). A statistical view of some chemometrics regression tools (with discussion), *Technometrics*, **35**, 109–148.

[177] Freedman, D.A. (1981). Bootstrapping regression models, *The Annals of Statistics*, **9**, 1218–1228.

[178] Freund, Y. (1995). Boosting a weak learning algorithm by majority, *Information and Computation*, **121**, 256–285.

[179] Freund, Y. and Schapire, R.E. (1997). A decision-theoretic general-ization of on-line learning and an application to boosting, *Journal of Computer and Systems Sciences*, **55**, 119–139.

[180] Freund, Y. and Schapire, R.E. (1998). Discussion of Breiman, 1998, *Annals of Statistics*, **26**, 824–832.

[181] Friedman, J.H. (1984). *SMART User's Guide*, Technical Report LCM001, Department of Statistics, Stanford University.

[182] Friedman, J. (1987). Exploratory projection pursuit, *Journal of the American Statistical Association*, **82**, 249–266.

[183] Friedman, J.H. (1989). Regularized discriminant analysis, *Journal of the American Statistical Association*, **84**, 165–175.

[184] Friedman, J.H. (1991). Multivariate adaptive regression splines (with discussion), *The Annals of Statistics*, **19**, 1–141.

[185] Friedman, J.H. (2001). Greedy function approximation: A gradient boosting machine, *Annals of Statistics*, **29**, 1189–1232.

[186] Friedman, J.H. and Stuetzle, W. (1981). Projection pursuit regres-sion, *Journal of the American Statistical Association*, **76**, 817–823.

[187] Friedman, J.H. and Stuetzle, W. (1982). Projection pursuit meth-ods for data analysis, In: *Modern Data Analysis*, R.L. Launer and A.F. Siegel (eds.), pp. 123–147, New York: Academic Press.

[188] Friedman, J.H. and Stuetzle, W. (2002). John Tukey's work on in-teractive graphics, *The Annals of Statistics*, **30**, 1629–1639.

[189] Friedman, J. and Tukey, J. (1974). A projection pursuit algorithm for exploratory data analysis, *IEEE Transactions on Computers, Series C*, **23**, 881–889.

[190] Friedman, J., Hastie, T., and Tibshirani, R. (2000). Additive logistic regression: a statistical view of boosting (with discussion), *Annals of Statistics*, **28**, 337–407.

[191] Friedman, J.H., Stuetzle, W., and Schroeder, A. (1984). Projection pursuit density estimation, *Journal of the American Statistical As-sociation*, **79**, 599–608.

[192] Funahashi, K. (1989). On the approximate realization of continuous mappings by neural networks, *Neural Networks*, **2**, 183–192.

[193] Furnival, G. and Wilson, R. (1974). Regression by leaps and bounds, *Technometrics*, **16**, 499–511. Reprinted in *Technometrics*, **42** (2000), 69–79.

[194] Geisser, S. (1974). A predictive approach to the random effects model, *Biometrika*, **61**, 101–107.

[195] Geisser, S. (1975). The predictive sample reuse method with applications, *Journal of the American Statistical Association*, **70**, 320–328.

[196] Gelfand, A.E. and Smith, A.F.M. (1990). Sampling-based approaches to calculating marginal densities, *Journal of the American Statistical Association*, **85**, 398–409.

[197] Gelman, A., Carlin, J.B., Stern, H.S., and Rubin, D.B. (1995). *Bayesian Data Analysis*, London: Chapman & Hall.

[198] Geman, S. and Geman, D. (1984). Stochastic relaxation, Gibbs distributions, and the Bayesian restoration of images, *IEEE Transactions on Pattern Analysis and Machine Intelligence*, **6**, 721–741.

[199] Gilks, W.R., Richardson, S., and Spiegelhalter, D.J. (1996). Markov Chain Monte Carlo in Practice, New York: Chapman & Hall.

[200] Girolami, M. (ed.) (2000). *Advances in Independent Component Analysis*. New York: Springer-Verlag.

[201] Glunt, W., Hayden, T.L., and Raydan, M. (1993). Molecular conformations from distance matrices, *Journal of Computational Chemistry*, **14**, 114–120.

[202] Gnanadesikan, R. (1977). *Methods for Statistical Data Analysis of Multivariate Observations*, New York: Wiley.

[203] Gnanadesikan, R. and Kettenring, J.R. (1972). Robust estimates, residuals, and outlier detection with multiresponse data, *Biometrics*, 28, 81–124.

[204] Gnanadesikan, R. and Wilk, M.B. (1969). Data analytic methods in multivariate statistical analysis, In *Multivariate Analysis II* (P.R. Krishnaiah, ed.), New York: Academic Press.

[205] Goldstein, M. and Smith, A.F.M. (1974). Ridge-type estimatrs for regression analysis, *Journal of the Royal Statistical Society, Series B*, **36**, 284–319.

[206] Golub, T.R., Slonim, D.K., Tamayo, P., Huard, M., Gaasenbeek, J.P,, Mesirov, J.P., Coller, H., Loh, M.L., Downing, J.R., Caligiuri, M.A., Bloomfield, C.D., and Lander, E.S. (1999). Molecular classification of cancer: class discovery and class prediction by gene expression monitoring, *Science*, **286**, 531–537.

[207] Good, I.J. and Gaskins, R.A. (1971). Nonparametric roughness penalties for probability densities, *Biometrika*, **58**, 255–277.

[208] Good, I.J. and Gaskins, R.A. (1980). Density estimation and bump-hunting by the penalized likelihood method exemplified by scattering and meteorite data (with discussion), *Journal of the American Statistical Association*, **75**, 42–73.

[209] Gordon, L. and Olshen, R.A, (1985). Tree-structured survival analysis, *Cancer Treatment Reports*, **69**, 1065–1069.

[210] Goutis, C. (1996). Partial least squares algorithm yields shrinkage estimators, *The Annals of Statistics*, **24**, 816–824.

[211] Gower, J.C. (1966). Some distance properties of latent root and vector methods used in multivariate analysis, *Biometrika*, **53**, 325–338.

[212] Gower, J.C. (1977). The analysis of asymmetry and orthogonality, In: *Recent Developments in Statistics*, (eds. J. Barra et al), Amsterdam: North-Holland.

[213] Gower, J.C. and Digby, P.G.N. (1981). Expressing complex relationships in two dimensions, In: *Interpreting Multivariate Data*, (ed. V. Barnett), 83–118. New York: Wiley.

[214] Green, P.J. (1984). Iteratively reweighted least squares for maximum likelihood estimation, and some robust and resistant alternatives (with discussion), *Journal of the Royal Statistical Society, Series B*, **46**, 149–192.

[215] Green, P.J. (1990). Bayesian reconstructions from emission tomography data using a modified EM algorithm, *IEEE Transactions on Medical Imaging*, **16**, 516–526.

[216] Greenacre, M.J. (1981). Practical correspondence analysis, In: *Interpreting Multivariate Data*, (ed. V. Barnett), 119–146, New York: Wiley.

[217] Greenacre, M.J. (1984). *Theory and Applications of Correspondence Analysis*, New York: Academic Press.

[218] Greenacre, M.J. (1988). Correspondence analysis of multivariate categorical data by weighted least-squares, *Biometrika*, **75**, 457–467.

[219] Greenacre, M. (2000). Correspondence analysis of square asymmetric matrices, *Applied Statistics*, **49**, 297–310.

[220] Greenacre, M. (2007). *Correspondence Analysis in Practice, Second Edition*, New York: Chapman & Hall/CRC.

[221] Greenacre, M. and Hastie, T. (1987). The geometric interpretation of correspondence analysis, *Journal of the American Statistical Association*, **82**, 437–447.

[222] Hadi, A.S. and Ling, R.F. (1998). Some cautionary notes on the use oif principal components regression, *The American Statistician*, **52**, 15–19.

[223] Haitovsky, Y. (1987). On multivariate ridge regression, *Biometrika*, **74**, 563–570.

[224] Hall, P. (1989). On polynomial-based projection indices for exploratory projection pursuit, *Annals of Statistics*, **17**, 589–605.

[225] Hall, P. (1992). *The Bootstrap and Edgeworth Expansion.* New York: Springer.

[226] Hall, P. and Marron, J.S. (1987). Extent to which least-squares cross-validation minimises integrated square error in nonparametric density estimation, *Probability Theory and Related Fields*, **74**, 567–581.

[227] Hall, P. and Wand, M.P. (1988). Minimizing L_1 distance in nonparametric density estimation, *Journal of Multivariate Analysis*, **26**, 59–88.

[228] Ham, J., Lee, D.D., Mika, S., and Schölkopf, B. (2003). A kernel view of the dimensionality reduction of manifolds, Technical Report TR-110, Max Planck Institut fur biologische Kybernetik, Germany.

[229] Hampel, F. (2002). Some thoughts about classification, Research Report No. 102, Seminar für Statistik, Eidegenössische Technische Hochschule, CH–8092 Zürich, Switzerland.

[230] Hand, D.J. (1982). *Kernel Discriminant Analysis*, Chichester, U.K.: Research Studies Press.

[231] Hand, D., Mannila, H., and Smyth, P. (2001). *Principles of Data Mining*, Cambridge, MA: MIT Press.

[232] Harman, H.H. (1976). *Modern Factor Analysis, Third Edition Revised.* Chicago: The University of Chicago Press.

[233] Hartigan, J.A. (1972). Direct clustering of a data matrix, *Journal of the American Statistical Association*, **6**, 123–129.

[234] Hartigan, J.A. (1975). *Clustering Algorithms*, New York: Wiley.

[235] Hartigan, J.A. (1974). Block voting in the United Nations, In: *Exploring Data Analysis: The Computer Revolution in Statistics* (W.J. Dixon and W.L. Nicholson, eds.), Berkeley and Los Angeles, CA: University of California Press, Chapter 2, pp. 79–112.

[236] Hartigan, J.A. and Hartigan, P.M. (1985). The dip test of unimodality, *The Annals of Statistics*, **13**, 70–84.

[237] Hassibi, B., Stork, D.G., Wolff, G., and Wanatabe, T. (1994). Optimal brain surgery: expensions and performance comparisons, *Advances in Neural Information Processing Systems*, **6**, 263–270.

[238] Hastie, T. (1984). Principal curves and surfaces, Technical Report, Department of Statistics, Stanford University.

[239] Hastie, T. and Stuetzle, W. (1989). Principal curves, *Journal of the American Statistical Association*, **84**, 502–516.

[240] Hastie, T. and Tibshirani, R. (1986). Generalized additive models, *Statistical Science*, **1**, 295–318.

[241] Hastie, T. and Tibshirani, R. (1990). *Generalized Additive Models*, London: Chapman and Hall.

[242] Hastie, T. and Tibshirani, R. (2000). Bayesian backfitting (with discussion), *Statistical Science*, **15**, 196–223.

[243] Hastie, T., Tibshirani, R., and Friedman, J. (2001). *The Elements of Statistical Learning: Data Mining, Inference, and Prediction*, New York: Springer-Verlag.

[244] Hastie, T. and Tibshirani, R. (2000). Bayesian backfitting (with discussion), *Statistical Science*, **15**, 196–223.

[245] Hastie. T., Tibshirani, R., Eisen, M.B., Alzadeh, A., Levy, R., Staudt, L., Chan, W.C., Botstein, D., and Brown, P.O. (2000). 'Gene shaving' as a method for identifying distinct sets of genes with similar expression patterns, *Genome Biology*, **1**(2), research0003.1–research0003.21.

[246] Havel, T.F. (1991). An evaluation of computational strategies for use in the determination of protein structure from distance constraints obtained by nuclear magnetic resonance, *Progress in Biophysics and Molecular Biology*, **56**, 43–78.

[247] Havel, T.M., Kuntz, I.D., and Crippen, G.M. (1983). The combinatorial distance geometry approach to the calculation of molecular conformation. I. A new approach to an old problem, *Journal of Theoretical Biology*, **104**, 359–381; II. Sample problems and computational statistics, *Journal of Theoretical Biology*, **104**, 383–400.

[248] Haykin, S. (1999). *Neural Networks: A Comprehensive Foundation, Second Edition*, New York: Macmillan.

[249] Hebb, D.O. (1949). *The Organization of Behavior: A Neuropsychological Theory*, New York: Wiley.

[250] Helland, I.S. (1988). On the structure of partial least squares regression, *Communications in Statistics: Simulation and Computation*, **17**, 581–607.

[251] Helland, I.S. (2000). Some theoretical aspects of partial least squares regression, *Chemometrics and Intelligent Laboratory Systems* (to appear).

[252] Henikoff, J.G. and Henikoff, S. (1996). Using substitution probabilities to improve position-specific scoring matrices, *Computer Applications in the Biosciences*, **12**, 135–143.

[253] Herault, J. and Jutten, C. (1986). Space or Time Processing by Neural Network Models, in *Proceedings of the AIP Conference: Neural Networks for Computing* (ed.: J.S. Denker), 151, New York: American Institute for Physics.

[254] Hersh, W., Buckley, C., Leone, T.J., and Hickam, D. (1994). OSHUMED: an interactive retrieval evaluation and new large test collection for research, *Proceedings of the 17th Annual International ACM SIGIR Conference on Research and Development in Information Retrieval*, pp. 192–201.

[255] Hinton, G.E. (1987). Connectionist learning procedures, In: *Machine Learning: Paradigms and Methods* (J. Carbonell, ed.), pp. 185–234, Cambridge, MA: MIT Press.

[256] Hinton, G.E., Plaut, D.C., and Shallice, T. (1993). Simulating brain damage, *Scientific American*, 76–82.

[257] Ho, T.K. (1998). The random subspace method for constructing decision forests, *IEEE Transactions on Pattern Analysis and Machine Intelligence.* **20**, 832–844.

[258] Hoerl, A.E. and Kennard, R. (1970a). Ridge regression: Biased estimation for non-orthogonal problems, *Technometrics* **12**: 55–67. Reprinted in *Technometrics*, **42** (2000), 80–86.

[259] Hoerl, A.E. and Kennard, R. (1970b). Ridge regression: applications to non-orthogonal problems, *Technometrics*, **12**, 69–82 [correction, **12**, 733].

[260] Hoeting, J.A., Madigan, D., Raftery, A.E., and Volinsky, C.T. (1999). Bayesian model averaging, *Statistical Science*, **14**, 382–417.

[261] Hoffman, A.J. and Wielandt, H.W. (1953). The variation of the spectrum of a normal matrix, *Duke Mathematical Journal*, **20**, 37–39.

[262] Holm, L. and Sander, C. (1996). Mapping the protein universe, *Science*, **273**, 595–603.

[263] Holzinger, K. and Swineford, F. (1939). A study in factor analysis: the stability of a bifactor solution, *Supplementary Educational Monograph*, **48**, Chicago, IL: University of Chicago Press.

[264] Hornick, K., Stinchcombe, M., and White, H. (1989). Multilayer feedforward network are universal approximators, *Neural Networks*, **2**, 359–366.

[265] Horton, I.F., Russell, J.S., and Moore, A.W. (1968). Multivariate-covariance and canonical analysis: a method for selecting the most effective discriminators in a multivariate situation, *Biometrics*, **24**, 845–858.

[266] Hotelling, H. (1933). Analysis of a complex of statistical variables into principal components, *Journal of Educational Psychology*, **24**, 417–441, 498–520.

[267] Hotelling, H. (1936). Relations between two sets of variates, *Biometrika*, **28**, 321–377.

[268] Hou, J., Sims, G.E., Zhang, C., and Kim, S.-H. (2003). A global representation of the protein fold space, *Proceedings of the National Academy of Sciences*, **100**, 2386–2390.

[269] Huang, X. and Miller, W. (1991). A time-efficient, linear-space local similarity algorithm, *Advances in Applied Mathematics*, 12, 337–357.

[270] Huber, P.J. (1964). Robust estimation of a location parameter, *The Annals of Mathematical Statistics*, **53**, 73–101.

[271] Huber, P.J. (1985). Projection pursuit, *Annals of Statistics*, **53**, 73–101.

[272] Hurri, Gävert, Särelä, and Hyvärinen, A. (1998). The `FastICA` package for MATLAB, `isp.imm.dtu.dk/toolbox/`

[273] Hwang, J.-N., Li, H., Maechler, M., Martin, D., and Schimert, J. (1992). A comparison of projection pursuit and neural network regression modeling, *NIPS*, **4**, 1159–1166.

[274] Hwang, J.T.G. and Ding, A.A. (1997). Prediction intervals for artificial neural networks, *Journal of the American Statistical Association*, **92**, 748–757.

[275] Hwang, J.T.G. and Nettleton, D. (2003). Principal components regression with data-chosen components and related methods, *Technometrics*, **45**, 70–79.

[276] Hyvärinen, A. (1999). The fixed-point algorithm and maximum likelihood estimation for independent component analysis, *Neural Processing Letters*, **10**, 1–5.

[277] Hyvärinen, A., Karhunen, J. and Oja, E. (2001). *Independent Component Analysis*. New York: Wiley.

[278] Intrator, O. and Kooperberg, C. (1995). Trees and splines in survival analysis, *Statistical Methods in Medical Research*, **4**, 237–261.

[279] Izenman, A.J. (1972). *Reduced-Rank Regression for the Multivariate Linear Model, Its Relationship to Certain Multivariate Techniques, and Its Application to the Analysis of Multivariate Data*, Ph.D. dissertation, University of California, Berkeley.

[280] Izenman, A.J. (1975). Reduced-rank regression for the multivariate linear model, *Journal of Multivariate Analysis* **5**, 248–264.

[281] Izenman, A.J. (1980). Assessing dimensionality in multivariate regression, *in* Krishnaiah (ed), *Handbook of Statistics* **1**, Amsterdam: North-Holland, pp. 571–591.

[282] Izenman, A.J. (1991). Recent developments in nonparametric density estimation, *Journal of the American Statistical Association*, **86**, 205–224.

[283] Izenman, A.J. and Sommer, C.J. (1988). Philatelic mixtures and multimodal densities, *Journal of the American Statistical Association*, **83**, 941–953.

[284] Jackson, J.E. (2003). *A User's Guide to Principal Components*, New York: Wiley.

[285] James, A.T. (1960). Distribution of the latent roots of the covariance matrix, *Annals of Mathematical Statistics*, **31**, 151–158.

[286] James, A.T. (1964). Distribution of matrix variates and latent roots derived from normal samples, *The Annals of Mathematical Statistics*, **35**, 475–501.

[287] James, W. and Stein, C. (1961). Estimation with quadratic loss, *Proceedings of the Fourth Berkeley Symposium on Mathematical Statistics and Probability*, **1**, 361–380. Berkeley, CA: University of California Press.

[288] Jee, J.R. (1987). Exploratory projection pursuit using nonparametric density estimation, *Proceedings of the Statistical Computing Section of the American Statistical Association*, 335–339.

[289] Jiang, W. (2004). Process consistency for AdaBoost, *Annals of Statistics*, **32**, 13–29.

[290] Joachims, T. (2002). *Learning to Classify Text Using Support Vector Machines: Methods, Theory, and Algorithms*, New York: Kluwer Academic Publishers.

[291] Johnson, R.A. and Wichern, D.W. (1998). *Applied Multivariate Statistical Analysis, Fourth Edition*, Upper Saddle River, NJ: Prentice Hall.

[292] Johnstone, I.M. (2001). On the distribution of the largest eigenvalue in principal components analysis, *The Annals of Statistics*, **29**, 295–327.

[293] Johnstone, I.M. (2006). High-dimensional statistical inference and random matrices, *Proceedings of the International Congress of Mathematicians*. eprint ARXIV: math/0611589.

[294] Johnstone, I.M. and Silverman, B.W. (1990). Speed of estimation in positron emission tomography and related inverse problems, *The Annals of Statistics*, **18**, 251–280.

[295] Jolliffe, I.T. (1982). A note on the use of principal components in regression, *Applied Statistics*, **31**, 300–303.

[296] Jolliffe, I.T. (1986). *Principal Component Analysis*. New York: Springer.

[297] Jones, M.C. and Sibson, R. (1987). What is projection pursuit? *Journal of the Royal Statistical Society, Series A*, **150**, 1–36.

[298] Jones, M.C., Marron, J.S., and Sheather, S.J. (1996). A brief survey of bandwidth selection for density estimation, *Journal of the American Statistical Association*, **91**, 401–407.

[299] Jöreskog, K.G. (1969). A general approach to confirmatory maximum likelihood factor analysis, *Psychometrika*, **34**, 183–202.

[300] Jöreskog, K.G. (1970). A general method for analysis of covariance structures, *Biometrika*, **57**, 239–251.

[301] Jöreskog, K.G. (1977). Structural equation models in the social sciences: specification, estimation, and testing, in: *Applications of Statistics* (P.R. Krishnaiah, ed.), pp. 265–287, Amsterdam: North-Holland.

[302] Jusczyk, P.W. and Klein, R.M. (1980). *The Nature of Thought: Essays in Honor of D.O. Hebb*, Hillsdale, NJ: Lawrence Erlbaum Associates.

[303] Jutten, C. (2000). Source separation: from dusk till dawn, in *Proceedings of the 2nd International Workshop on Independent Component Analysis and Blind Source Separation (ICA 2000)*, 15–26, Helsinki, Finland.

[304] Kahn, D. (1996). The history of steganography, *Proceedings of Information Hiding, First International Workshop*, Cambridge, U.K.

[305] Kaiser, H.F. (1958). The varimax criterion for analytic rotation in factor analysis, *Psychometrika*, **23**, 187–200.

[306] Kaiser, H.F. (1960). The application of electronic computers to factor analysis, *Educational and Psychological Measurement*, **20**, 141–151.

[307] Karlin, S. and Altschul, S.F. (1990). Methods for assessing the statistical significance of molecular sequence features by using general scoring schemes, *Proceedings of the National Academy of Sciences, USA*, **87**, 2264–2268.

[308] Kass, G.V. (1980). An exploratory technique for investigating large quantities of categorical data, *Applied Statistics*, **29**, 119–127.

[309] Kass, R.E. and Raftery, A.E. (1995). Bayes factors, *Journal of the American Statistical Association*, **90**, 773–795.

[310] Kass, R.E., Tierney, L., and Kadane, J.B. (1988). Approximate methods for assessing influence and sensitivity in Bayesian analysis, *Biometrika*, **76**, 663–674.

[311] Kasser, I.S. and Bruce, R.A. (1969). Comparative effects of aging and coronary heart disease on submaximal and maximal exercise, *Circulation*, **39**, 759–774.

[312] Kaufman, L. and Rousseeuw, P.J. (1990). *Finding Groups in Data: An Introduction to Cluster Analysis*, New York: Wiley.

[313] Kendall, M.G. and Stuart, A. (1969). *The Advanced Theory of Statistics*, **1**, *Distribution Theory*, London, U.K.: Charles Griffin & Co. Ltd.

[314] Kernighan, B.W. and Pike, R. (1984). *The Unix Programming Environment*. Englewood Cliffs, NJ: Prentice-Hall.

[315] Khan, J., Wei, J.S., Ringner, M., Saal, L.H., Ladanyi, M., Westermann, F., Berthold, F., Schwab, M., Antonescu, C.R., Peterson, C., and Meltzer, P.S. (2001). Classification and diagnostic prediction of cancers using gene expression profiling and artificial neural networks, *Nature Medicine*, **7**, 673–679.

[316] Khattree, R. and Naik, D.N. (1999). *Applied Multivariate Statistics With SAS Software, Second Edition*, Cary, NC: SAS Institute, Inc.

[317] Kimeldorf, G. and Wahba, G. (1971). Some results on Tchebychef-fian spline functions, *Journal of Mathematical Analysis and its Applications*, **33**, 82–95.

[318] Kirkendall, N. (1997). Massive data sets: a view from the Federal Statistical System, *Proceedings of the Statistical Computing Section of the American Statistical Association*, 14–17.

[319] Kohonen, T. (1982). Self-organized formation of topologically-correct feature maps, *Biological Cybernetics*, **43**, 59–69.

[320] Kohonen, T. (2001). *Self-Organizing Maps, Third Edition*, New York: Springer.

[321] Kolmogorov, A.N. (1957). On the representation of continuous functions by superposition of continuous functions of one variable and addition, *Doklady Akademiia Nauk SSSR*, **114**, 953–956.

[322] Korf, I., Yandell, M., and Bedell, J. (2003). *BLAST: An Essential Guide to the Basic Local Alignment Search Tool*. Sebastopol, CA: O'Reilly & Associates, Inc.

[323] Kramer, M.A. (1991). Nonlinear principal component analysis using autoassociative neural networks, *AIChE Journal*, **37**, 233–243.

[324] Krieger, A., Long, C., and Wyner, A. (2001). Boosting noisy data, *Proceedings of the Eighteenth International Conference on Machine Learning*, San Francisco: Morgan Kauffmann Publishers, Inc., pp. 274–281.

[325] Kruskal, J.B. (1964a). Multidimensional scaling by optimizing goodness of fit to a nonmetric hypothesis, *Psychometrika*, **29**, 1–27.

[326] Kruskal, J.B. (1964b). Nonmetric multidimensional scaling: A numerical method, *Psychometrika*, **29**, 115–129.

[327] Kruskal, J.B. (1969). Toward a practical method which helps uncover the structure of a set of multivariate observations by finding the linear transformation which optimizes a new 'index of condensation', In *Statistical Computation* (R.C. Milton and J.A. Nelder, eds.), pp. 427–440, New York: Academic Press.

[328] Kruskal, J.B. (1972). Linear transformation of multivariate data to reveal clustering, In *Multidimensional Scaling: Theory and Applications in the Behavioural Sciences, Volume 1* (R.N. Shepard, A.K. Romney, and S.B. Nerlove, eds.), pp. 179–191, London: Seminar Press.

[329] Kruskal, J.B. and Seery, J.B. (1980). Designing network diagrams, *Proceedings of the First General Conference on Social Graphics*, Washington, D.C.: U.S. Bureau of the Census, pp. 22–50.

[330] Lampinen, J. and Vehtari, A. (2001). Bayesian approach for neural networks — review and case studies, *Neural Networks*, **14**, 257–274.

[331] Lander, E.S. and Waterman, M.S. (1995). *Calculating the Secrets of Life*, Washington, D.C.: National Academy Press.

[332] Laplace, P.S. (1774/1986). Memoire sur la probabilite des causes par les evenements, *Memoirs Academie Science Paris*, **6**, 621–656. Memoir on the probability of the causes of events (English translation by S.M. Stigler), *Statistical Science*, **1**, 364–378.

[333] Lattin, J., Carroll, J.D., and Green, P.E. (2003). *Analyzing Multivariate Data*, Pacific Grove, CA: Thomson-Brooks/Cole.

[334] Lawley, D.N. and Maxwell, A.E. (1971). *Factor Analysis as a Statistical Method*, London: Butterworth.

[335] Lazzeroni, L. and Owen, A. (2002). Plaid models for gene expression data, *Statistica Sinica*, **12**, 61–86.

[336] LeBlanc, M. and Tibshirani, R. (1994). Adaptive principal surfaces, *Journal of the American Statistical Association*, **89**, 53–64.

[337] LeCun, Y. (1985). Une procedure d'apprentissage pour reseau a seuil assymetrique, In: *In Cognitiva 85: A la Frontiere de l'Intelligence Artificielle des Sciences de la Connaissance des Neurosciences*, pp. 599–604, CESTA, Paris.

[338] Lee, T.-W. (1998). *Independent Component Analysis — Theory and Applications*. Boston: Kluwer.

[339] Lee, Y., Lin, Y., and Wahba, G. (2004). Multicategory support vector machines: theory and application to the classification of microarray data and satellite radiance data, *Journal of the American Statistical Association*, **99**, 67–81.

[340] Lehmann, E.L. (1983). *Theory of Estimation*, New York: Wiley.

[341] Lesteven, S., Poinçot, P., and Murtagh, F. (2001). Visual exploration of astronomical documents, In: *Astronomical Data Analysis Software and Systems X* (F.R. Harnden Jr., F.A. Primini, and H.E. Payne, eds.), 78, *ASP Conference Series*, **238**, San Francisco, CA: American Society of the Pacific.

[342] Leurgans, S.E., Moyeed, R.A., and Silverman, B.W. (1993). Canonical correlation when the data are curves, *Hournal of the Royal Statistical Society, Series B*, **55**, 725–740.

[343] Lewis, D.D., Yang, Y., Rose, T.G., and Li, F. (2004). RCV1: a new benchmark collection for text categorization research, *The Journal of Machine Learning Research*, **5**, 361–397.

[344] Lin, Y. (2002). Support vector machines and the Bayes rule in classification, *Data Mining and Knowledge Discovery*, **6**, 259–275.

[345] Lindley, D.V. and Smith, A.F.M. (1972). Bayes estimates for the linear model (with discussion), *Journal of the Royal Statistical Society, Series B*, **34**, 1–41.

[346] Lingjaerde, O.C. and Christophersen, N. (2000). Shrinkage structure of partial least squares, *Scandanavian Journal of Statistics*, **27**, 459–473.

[347] Little, R.J.A. and Rubin, D.B. (1987). *Statistical Analysis With Missing Data*, New York: Wiley.

[348] Lloyd, C.J. (1999). *Statistical Analysis of Categorical Data*, New York: Wiley.

[349] Loader, C.R. (1999). Bandwidth selection: classical or plug-in? *The Annals of Statistics*, **27**, 415–438.

[350] Lodhi, H., Saunders, C., Shawe-Taylor, J., Cristianini, N., and Watkins, C. (2002). Text classification using string kernels, *Journal of Machine Learning*, **2**, 419–444.

[351] Lugosi, G. and Vayatis, N. (2004). On the Bayes-risk consistency of regularized boosting methods, *Annals of Statistics*, **32**, 30–55.

[352] Macrae, E.C. (1974). Matrix derivatives with an application to an adaptive linear decision problem, *The Annals of Statistics*, **2**, 337–346.

[353] MacKay, D.J.C. (1991). *Bayesian Methods for Adaptive Models*, doctoral dissertation, California Institute of Technology.

[354] MacKay, D.J.C. (1992a). A practical Bayesian framework for back-propagation networks, *Neural Computation*, **4**, 448–472.

[355] MacKay, D.J.C. (1992b). The evidence framework applied to classification networks, *Neural Computation*, **4**, 720–736.

[356] MacKay, D.J.C. (1994). Hyperparameters: optimize or integrate out? In: *Maximum Entropy and Bayesian Methods*, Santa Barbara, Dordrecht: Kluwer.

[357] MacKay, D.J.C. (2003). *Information Theory, Inference, and Learning Algorithms*, Cambridge, U.K.: Cambridge University Press.

[358] MacNaughton-Smith, P., Williams, W.T., Dale, N.B., and Mockett, L.G. (1964). Dissimilarity analysis: a new technique of hierarchical subdivision, *Nature*, **202**, 1034–1035.

[359] MacQueen, J.B. (1967). Some methods for classification and analysis of multivariate observations, *Proceedings of the Fifth Berkeley Symposium on Mathematical Statistics and Probability*, **1**, Berkeley, CA: University of California Press.

[360] Madsen, M. (1976). Statistical analysis of multiple contingency tables, *Scandanavian Journal of Statistics*, **3**, 97–106.

[361] Mallows, C.L. (1973). Some comments on C_p, *Technometrics*, **15**, 661–667. Reprinted in *Technometrics*, **42** (2000), 87–94.

[362] Mallows, C.L. (1995). More comments on C_p, *Technometrics*, **37**, 362–372.

[363] Malthouse, E.C. (1998). Limitations of nonlinear PCA as performed with generic neural networks, *IEEE Transactions on Neural Networks*, **9**, 165–173.

[364] Mangasarian, O.L., Street, W.N., and Wolberg, W.H. (1995). Breast cancer diagnosis and prognosis via linear programming, *Operations Research*, **43**, 570–577.

[365] Marčenko, V.A. and Pastur, L.A. (1967). Distributions of eigenvalues of some sets of random matrices, *Math. USSR-Sb.*, **1**, 507–536.

[366] Marchini, J.L., Heaton, C., and Ripley, B.D. (2003). The fastICA package, version 1.1-3, http://www.stats.ox.ac.uk/~marchini/software.html

[367] Mardia, K.V. (1978). Some properties of classical multidimensional scaling, *Communications in Statistical Theory and Methods, Series A*, **7**, 1233–1241.

[368] Mardia, K., Kent, J., and Bibby, J. (1979). *Multivariate Analysis*, New York: Academic Press.

[369] Marquardt, D.W. (1970). Generalized inverses, ridge regression, biased linear estimation, and nonlinear estimation, *Technometrics*, **12**, 591–612.

[370] Martens, H. and Naes, T. (1989). *Multivariate Calibration*, New York: Wiley.

[371] Martinez, W.L. and Martinez, A.R. (2005). *Exploratory Data Analysis with MATLAB*, Boca Raton, FL: Chapman & Hall/CRC.

[372] Massy, W.F. (1965). Principal component regression in exploratory statistical research, *Journal of the American Statistical Association*, **60**, 234–256.

694   References

[373] McCullagh, P. and Nelder, J.A. (1989). *Generalized Linear Models*, New York: Chapman and Hall.

[374] McCullogh, W.S. and Pitts, W. (1943). A logical calculus of the ideas immanent in the nervous activity, *Bulletin of Mathematical Biophysics*, **5**, 115–133.

[375] McLachlan, G.J. and Basford, K.E. (1988). *Mixture Models: Inference and Applications to Clustering*. New York: Dekker.

[376] McLachlan, G.J., Peel, D., Basford, K.E., and Abrams, P. (1999). The EMMIX software for the fitting of mixtures of normal or *t*-components, *Journal of Statistical Software*, **4**.

[377] McLachlan, G.J. and Krishnan, T. (1997). *The EM Algorithm and Extensions*, New York: Wiley.

[378] Mease, A. and Wyner, A.J. (2007). Evidence contrary to the statistical view of boosting (with discussion), *Journal of Machine Learning Research*, **9**, 131–201.

[379] Mease, D., Wyner, A.J., and Buja, A. (2007). Boosted classification trees and class probability/quantile estimation, *Journal of Machine Learning Research*, **8**, 409–439.

[380] Mercer, J. (1909). Functions of positive and negative type and their connection with the theory of integral equations, *Philosophical Transactions of the Royal Society, London, Series A*, **209**, 415–446.

[381] Metropolis, N., Rosenbluth, A.W., Rosenbluth, M.N., Teller, A.H., and Teller, E. (1953). Equations of state calculations by fast computing machines, *Journal of Chemical Physics*, **21**, 1087–1091.

[382] Miller, A.J. (2002). *Subset Selection in Regression, Second Edition*. London: Chapman and Hall.

[383] Milner, P.M. (1957). The cell assembly: Mk. II. *Psychological Review*, **64**, 242–252.

[384] Minsky, M.L. and Papert, S.A. (1969). *Perceptrons*, Cambridge, MA: MIT Press.

[385] Moguerza, J.M. and Munoz, A. (2006). Support vector machines with applications (with discussion), *Statistical Science*, **21**, 322–336.

[386] Monk, H. (2000). An Intel Solution for: The U.S. Census Bureau. *DM Review*, April 2000. (p.34)

[387] Morgan, J.N. and Sondquist, J.A. (1963). Problems in the analysis of survey data, and a proposal, *Journal of the American Statistical Association*, **58**, 415–434.

[388] Mosteller, F. and Tukey, J.W. (1977). *Data Analysis and Regression.* Reading, MA: Addison-Wesley.

[389] Nandi, A.K. (ed.) (1999). *Blind Estimation Using Higher-Order Statistics.* Boston: Kluwer.

[390] Nash, J. (1965). The embedding problem for Riemannian manifolds, *Annals of Mathematics*, **63**, 20–63.

[391] Neal, R.M. (1996). *Bayesian Learning for Neural Networks*, Lecture Notes in Statistics, **118**, New York: Springer.

[392] Needleman, S.B. and Wunsch, C.D. (1970). A general method applicable to the search for similarities in the amino-acid sequence of two proteins, *Journal of Molecular Biology*, **48**, 443–453.

[393] Nerbonne, J., Heeringa, W., and Kleiweg, P. (1999). Edit distance and dialect proximity, In: *Time Warps, String Edits, and Macromolecules: The Theory and Practice of Sequence Comparison, 2nd edition* (D. Sankoff and J. Kruskal, eds.), pp. v–xv, Stanford, CA: CSLI.

[394] Oh, M.-S. and Raftery, A.E. (2001). Bayesian multidimensional scaling and choice of dimension, *Journal of the American Statistical Association*, **96**, 1031–1044.

[395] Oja, E. and Kaski, S. (2003). *Kohonen Maps*, Amsterdam: Elsevier.

[396] Ojelund, H., Brown, P.J., Madsen, H., and Thyregod, P. (2002). Prediction based on mean subset, *Technometrics*, **44**, 369–378.

[397] O'Sullivan, F. (1986). A statistical perspective on ill-posed inverse problems, *Statistical Science*, **1**, 502–527.

[398] Owen, A.B. (2000). *Plaid Users Guide*, Stanford, CA: Stanford University.

[399] Pack, P. and Jolliffe, I.T. (1992). Influence in correspondence analysis. *Applied Statistics*, **41**, 365–380.

[400] Parker, D.B. (1985). Learning logic, Technical Report TR–47, Center for Computational Research in Economics and Management Sciences, MIT.

[401] Parmigiani, G., Garrett, E.S., Irizarry, R.A., and Zeger, S.L. (2003). *The Analysis of Gene Expression Data*, New York: Springer.

[402] Parzen, E. (1962). On estimation of a probability density function and mode, *The Annals of Mathematical Statistics*, **33**, 1065–1076.

[403] Pearson, K. (1904). On the theory of contingency and its relation to association and normal correlation, *Draper's Company Research Memoirs, Biometric Series 1*. Reprinted in 1948 in *Karl Pearson's Early Papers*. Cambridge, U.K.: Cambridge University Press.

[404] Pearson, W.R. and Lipman, D.J. (1988). Improved tools for biological sequence comparison, *Proceedings of the National Academy of Science USA*, **85**, 2444–2448.

[405] Pechura, M. and Martin, J.B. (1991) (eds.). *Mapping the Brain and Its Functions: Integrating Enabling Technologies into Neuroscience Research*, Washington, D.C.: National Academy Press.

[406] Peña, D. and Prieto, F.J. (2001), Multivariate outlier detection and robust covariance matrix estimation (with discussion), *Technometrics*, 43, 286–310.

[407] Petitcolas, F.A.P., Anderson, R., and Kuhn, M.G. (1999). Information hiding: a survey. *Proceedings of the IEEE*, **87**, 1062–1078.

[408] Phatak, A. and de Hoog, F. (2002). Exploiting the connection betwen PLSR, Lanczos, and conjugate gradients: alternative proofs of some properties of PLSR, *Journal of Chemometrics*, **16**, 371–367.

[409] Pinsker, S. and Prince, A. (1988). On language and connectionism: analysis of a parallel distributed processing model of language acquisition, *Cognition*, **23**, 73–193.

[410] Platt, J. (1999). Fast training of support vector machines using sequential minimal optimization, In: B. Schölkopf, C.J.C. Burges, and A.J. Smola, eds., *Advances in Kernel Methods — Support Vector Learning*, pp. 185–208, Cambridge, MA: MIT Press.

[411] Pregibon, D. (1991). Incorporating statistical expertise into data analysis software, In: *The Future of Statistical Software: Proceedings of a Forum*, National Research Council, Washington, D.C.: National Academies Press, 51–62.

[412] Pregibon, D. and Gale, W. (1984). REX: an expert system for regression analysis, In: *Proceedings of COMPSTAT 84* (T. Havranek, Z. Sidak, and M. Novak, eds.), Heidelberg, Germany: Physica-Verlag, 242–248.

[413] Press, S.J. (1989). *Bayesian Statistics: Principles, Models, and Applications*, New York: Wiley.

[414] Quinlan, J.R. (1993). *C4.5: Programs for Machine Learning*, San Mateo, CA: Morgan Kaufmann.

[415] Ramsay, J.O. (1982). Some statistical approaches to multidimensional scaling (with discussion), *Journal of the Royal Statistical Society, Series A*, **145**, 285–312.

[416] Ramsay, J.O. (1988). Monotone regression splines (with discussion), *Statistical Science*, **3**, 425–461.

[417] Ramsay, J.O. and Silverman, B.W. (1997). *Functional Data Analysis*, New York: Springer.

[418] Rao, C.R. (1965). The use and interpretation of principal components in applied research, *Sankhya (A)*, **26**, 329–358.

[419] Rao, C.R. (1965). *Linear Statistical Inference and Its Applications*, New York: Wiley.

[420] Rao, C.R. (1979). Separation theorems for singular values of matrices and their applications in multivariate analysis, *Journal of Multivariate Analysis*, **9**, 362–377.

[421] Rätsch, G. and Warmuth, M.K. (2005). Efficient margin maximizing with boosting, *Journal of Machine Learning Research*. **6**, 2131–2152.

[422] Redner, R.A, and Walker, H.F. (1984). Mixture densities, maximum likelihood, and the EM algorithm, *SIAM Review*, **26**, 195–239.

[423] Reinsel, G.C. and Velu, R.P. (1998). *Multivariate Reduced-Rank Regression*, Lecture Notes in Statistics **136**, New York: Springer.

[424] Rencher, A.C. (2002). *Methods of Multivariate Analysis, Second Edition*, New York: Wiley.

[425] Ripley, B.D. (1994a). Neural networks and related methods for classification (with discussion), *Journal of the Royal Statistical Society, Series B*, **56**, 409–456.

[426] Ripley, B.D. (1994b). Flexible non-linear approaches to classification, In: *From Statistics to Neural Networks*, (V. Cherkassky, J.H. Friedman, and H. Wechsler, eds.), New York: Springer.

[427] Ripley, B.D. (1996). *Pattern Recognition and Neural Networks*, Cambridge, U.K.: Cambridge University Press.

[428] Robert, C.P. and Casella, G. (1999). *Monte Carlo Statistical Methods*. New York: Springer.

[429] Roberts, S. and Everson, R. (eds.) (2001). *Independent Component Analysis: Principles and Practice*, Cambridge, U.K.: Cambridge University Press.

[430] Rojas, R. (1996). *Neural Networks — A Systematic Introduction*, New York: Springer.

[431] Roosen, C.B. and Hastie, T.J. (1994). Automatic smoothing spline projection pursuit, *Journal of Computational and Graphical Statistics*, **3**, 235–248.

[432] Rosenblatt, F. (1958). The perceptron: a probabilistic model for information storage and organization in the brain, *Psychological Review*, **65**, 386–408.

[433] Rosenblatt, F. (1962). *Principles of Neurodynamics: Perceptrons and the Theory of Brain Mechanisms*, Washington, D.C.: Spartan Books.

[434] Rosenblatt, M. (1956). Remarks on some nonparametric estimates of a density function, *The Annals of Mathematical Statistics*, **27**, 832–837.

[435] Rothkopf, E.Z. (1957). A measure of stimulus similarity and errors in some paired-associate learning, *Journal of Experimental Psychology*, **53**, 94–101.

[436] Rousseeuw, P.J. (1987). Silhouettes: a graphical aid to the interpretation and validation of cluster analysis, *Journal of Computational and Applied Mathematics*, **20**, 53–65.

[437] Rousseeuw, P.J. and Leroy, A.M. (1987). *Robust Regression and Outlier Detection*, New York: Wiley.

[438] Roweis, S.T. and Saul, L.K. (2000). Nonlinear dimensionality reduction by locally linear embedding, *Science*, **290**, 2323–2326.

[439] Rubin, D.B. (1987). *Multiple Imputation for Nonresponse in Surveys*, New York: Wiley.

[440] Rubin, D.B. and Thayer, D.T. (1982). EM algorithms for ML factor analysis, *Psychometrika*, 47, 69–76.

[441] Rudin, C., Daubechies, I., and Schapire, R.E. (2004). The dynamics of ADABOOST: Cyclic behavior and convergence of margins, *Journal of Machine Learning*, **5**, 1557–1295.

[442] Rudin, C., Schapire, R.E., and Daubechies, I. (2007). Analysis of boosting algorithms using the smooth margin function, *The Annals of Statistics*, **35**, 2723–2768.

[443] Rumelhart, D.E. and McClelland, J. (1986a). *Parallel Distributed Processing: Explorations in the Microstructure of Cognition*, **1**, Cambridge, MA: MIT Press.

[444] Rumelhart, D. and McClelland, J. (1986b). On learning the past tenses of English verbs, In: *Parallel Distributed Processing*, **1** D. Rumelhart, J. McClelland, and the PDP Research Group (eds.), Chapter 18, Cambridge, MA: MIT Press.

[445] Salinelli, E. (1998). Nonlinear principal components I: Absolutely continuous random variables With positive bounded densities, *Annals of Statistics*, **26**, 596–616.

[446] Sammon, J.W. (1969). A non-linear mapping for data structure analysis, *IEEE Transactions on Computers*, **C–18**, 401–409.

[447] Saul, L.K. and Roweis, S.T. (2003). Think globally, fit locally: unsupervised learning of low dimensional manifolds, *Journal of Machine Learning Research*, **4**, 119–155.

[448] Schafer, J.L. (1997). *Analysis of Incomplete Multivariate Data*, London, U.K.: Chapman and Hall/CRC Press.

[449] Schapire, R.E. (1990). The strength of weak learnability, *Machine Learning*, **5**, 197–227.

[450] Schapire, R.E., Freund, Y., Bartlett, P., and Lee, W.S. (1998). Boosting the margin: A new explanation for the effectiveness of voting methods, *Annals of Statistics*, **26**, 1651–1686.

[451] Schapire, R.E. and Singer, Y. (1999). Improved boosting algorithms using confidence-rated predictions, The Eleventh Annual Conference on Computational Learning Theory, *Machine Learning*, **37**, 297–336.

[452] Schölkopf, B. and Smola, A.J. (2002). *Learning With Kernels*, Cambridge, MA: MIT Press.

[453] Schölkopf, B. and Smola, A.J. (2003). A short introduction to learning with kernels, In *Advanced Lectures on Machine Learning* (S. Mendelson and A.J. Smola, eds.), New York: Springer.

[454] Schölkopf, B., Smola, A.J., and Muller, K.-R. (1998). Nonlinear component analysis as a kernel eigenvalue problem, *Neural Computation*, **10**, 1299–1319.

[455] Schwarz, G. (1978). Estimating the dimension of a model, *The Annals of Statistics*, **6**, 461–464.

[456] Scott, D.W. (1985a). Average shifted histograms: effective nonparametric density estimators in several dimensions, *The Annals of Statistics*, **13**, 1024–1040.

[457] Scott, D.W. (1985b). Frequency polygons, *Journal of the American Statistical Association*, **80**, 348–354.

[458] Scott, D.W. (1992). *Multivariate Density Estimation: Theory, Practice, and Visualization*, New York: Wiley.

[459] Scott, D.W. and Terrell, G.R. (1987). Biased and unbiased cross-validation in density estimation, *Journal of the American Statistical Association*, **82**, 1131–1146.

[460] Scott, D.W., Tapia, R.A., and Thompson, J.R. (1980). Nonparametric probability density function by discrete maximum penalized-likelihood criteria, *The Annals of Statistics*, **8**, 820–832.

[461] Sebastiani, P., Gussoni, E., Kohane, I.S., and Ramoni, M.F. (2003). Statistical challenges in functional genomics, *Statistical Science*, **18**, 33–70.

[462] Seber, G.A.F. (1984). *Multivariate Observations*, New York: Wiley.

[463] Segal, M.R. (1988). Regression trees for censored data, *Biometrics*, **44**, 35–48.

[464] Sejnowski, T. and Rosenberg, C. (1987). Parallel networks that learn to pronounce English text, *Complex Systems*, **1**, 145–168.

[465] Shawe-Taylor, J. and Cristianini, N. (2004). *Kernel Methods for Pattern Analysis*, Cambridge, U.K.: Cambridge University Press.

[466] Sheather, S.J. and Jones, M.C. (1991). A reliable data-based bandwidth selection method for kernel density estimation, *Journal of the Royal Statistical Society, Series B*, **53**, 683–690.

[467] Shepard, R.N. (1962). The analysis of proximities: multidimensional scaling with an unknown distance function I. *Psychometrika*, **27**, 125–140; II *Psychometrika*, **27**, 219–246.

[468] Shepard, R.N. (1963). Analysis of proximities as a technique for the study of information processing in man, *Human Factors*, **5**, 33–48.

[469] Shiffrin, R.M. and Börner, K. (2004). Mapping knowledge domains, *Proceeding of the National Academy of Sciences of the USA*, **101** (Supplement 1), 5183–5185.

[470] Silverman, B.W. (1981). Using kernel density estimates to investigate multimodality, *Journal of the Royal Statistical Society, Series B*, **43**, 97–99.

[471] Silverman, B.W. (1983). Some properties of a test for multimodality based on kernel density estimates, In: *Probability, Statistics, and Analysis* (J.F.C. Kingman and G.E.H. Reuter, eds.), Cambridge, U.K.: Cambridge University Press, pp. 248–259.

[472] Silverman, B.W. (1986). *Density Estimation for Statistics and Data Analysis*, New York: Chapman and Hall.

[473] Simon, R.M., Korn, E.L., McShane, L.M., Radmacher, M.D., Wright, G.W., and Zhao, Y. (2004). *Design and Analysis of DNA Microarray Investigations*, New York: Springer.

[474] Simonoff, J. (1996). *Smoothing Methods in Statistics*, New York: Springer.

[475] Smith, T.F. and Waterman, M.S. (1981). Identification of common molecular subsequences, *Journal of Molecular Biology*, **147**, 195–197.

[476] Speed, T. (ed.) (2003). *Statistical Analysis of Gene Expression Data*, Boca Raton, FL: Chapman & Hall/CRC.

[477] Spence, I. and Graef, J. (1974). The determination of the underlying dimensionality of an empirically obtained matrix of proximities, *Multivariate Behavioral Research*, **9**, 331–341.

[478] Sprecher, D.A. (1965). On the structure of continuous functions of several variables, *Transactions of the American Mathematical Society*, **115**, 340–355.

[479] Stein, C. (1955). Inadmissibility of the usual estimator for the mean of a multivariate normal distribution, *Proceedings of the Third Berkeley Symposium on Mathematical Statistics and Probability*, **1**, 197–206. Berkeley, CA: University of California Press.

[480] Stern, H.S. (1996). Neural networks in applied statistics (with discussion), *Technometrics*, **38**, 205–220.

[481] Stone, M. (1974). Cross-validatory choice and assessment of statistical predictions (with discussion), *Journal of the Royal Statistical Society, Series B*, **36**, 111–147.

[482] Stone, J.V. (2004). *Independent Component Analysis: A Tutorial Introduction*, Cambridge, MA: MIT Press.

[483] Stovel, K., Savage, M., and Bearman, P. (1996). Ascription into Achievement: Models of career systems at Lloyds Bank, 1890–1970. *American Journal of Sociology*, **102**, 358–399.

[484] Street, W.N., Wolberg, W.H., and Mangasarian, O.L. (1993). Nuclear feature extraction for breast tumor diagnosis, *IS&T/SPIE International Symposium on Electronic Imaging: Science and Technology* (San Jose, CA), **1905**, 861–870.

[485] Stuart, A. (1953). The estimation and comparison of strengths of association in contingency tables. *Biometrika*, **40**, 105–110.

[486] Swayne, D.F., Cook, D., and Buja, A. (1998). XGobi: interactive dynamic data visualization in the X-window system, *Journal of Computational and Graphical Statistics*, **7**, 113–130.

[487] Swierenga, H., de Weijer, A.P., van Wijk, R.J., and Buydens, L.M.C. (1999). Stategy for constructing robust multivariate calibration models, *Chemometrics and Intelligent Laboratory Systems*, **49**, 1–17.

[488] Takane, Y., Young, F.W., and de Leeuw, J. (1977). Nonmetric individual differences multidimensional scaling: An alternating least-squares method with optimal scaling features, *Psychometrika*, **42**, 7–67.

[489] Tarpey, T. and Flury, B. (1996). Self-consistency: a fundamental concept in statistics, *Statistical Science*, **11**, 229–243.

[490] Tenenbaum, J.B., de Silva, V., and Langford, J.C. (2000). A global geometric framework for nonlinear dimensionality reduction, *Science*, **290**, 2319–2323.

[491] Terrell, G.R. and Scott, D.W. (1980). On improving convergence rates for nonnegative kernel density estimators, *The Annals of Statistics*, **8**, 1160–1163.

[492] Therneau, T.M. and Atkinson, E.J. (1997). An introduction to recursive partitioning using the RPART routine, Technical Report 61, Section of Statistics, Mayo Foundation.

[493] Theus, M. and Schonlau, M. (1998). Intrusion detection based on structural zeroes, *Statistical Computing & Graphics Newsletter*, **9**, 12–17.

[494] Thisted, R.A. (1976). Ridge regression, minimax estimation, and empirical Bayes methods, *Technical Report No. 28*, Division of Biostatistics, Stanford University, Stanford, CA.

[495] Thisted, R.A. (1980). Comment on "A critique of some ridge regression methods" by G. Smith and F. Campbell, *Journal of the American Statistical Association*, **75**, 81–86.

[496] Thisted, R.A. (1988). *Elements of Statistical Computing: Numerical Computations*, New York: Chapman and Hall.

[497] Thorpe, J.A. (1994). *Elementary Topics in Differential Geometry*, New York: Springer.

[498] Tibshirani, R. (1996b). Regression shrinkage and selection via the lasso, *Journal of the Royal Statistical Society, Series B*, **58**, 267–288.

[499] Tibshirani, R., Walther, G., and Hastie, T. (2001). Estimating the number of clusters in a data set via the gap statistic, *Journal of the Royal Statistical Society, Series B*, **63**, 411–423.

[500] Tierney, L. and Kadane, J.B. (1986). Accurate approximations for posterior moments and marginal densities, *Journal of the American Statistical Association*, **81**, 82–86.

[501] Titterington, D.M. (2004). Bayesian methods for neural networks and related models, *Statistical Science*, **19**, 128–139.

[502] Titterington, D.M., Smith, A.F.M., and Makov, U.E. (1985). *Statistical Analysis of Finite Mixture Distributions*. New York: Wiley.

[503] Tomayo, P., Slonim, D., Mesirov, J., Zhu, Q., Kitareewan, S., Dmitrovsky, E., Lander, E.S., and Golub, T.R. (1999). Interpreting patterns of gene expression with self-organizing maps, *Proceedings of the National Academy of Sciences USA*, **96**, 2907–2912.

[504] Torgerson, W.S. (1952). Multidimensional scaling: I. Theory and method, *Psychometrika*, **17**, 401–419.

[505] Torgerson, W.S. (1958). *Theory and Methods of Scaling*, New York: Wiley.

[506] Trosset, M.W. (1998). Applications of multidimensional scaling to molecular conformation, *Computing Science and Statistics*, **29**, 148–152.

[507] Tukey, J.W. (1960). A survey of sampling from contaminated distributions, In: *Contributions to Probability and Statistics: Essays in Honor of Harold Hotelling*, (I. Olkin, S.G. Ghurye, W. Hoeffding, W.G. Madow, and H.B. Mann, eds.), Stanford, CA: Stanford University Press, pp. 448–485.

[508] Tukey, J.W. (1977). *Exploratory Data Analysis*, Reading, MA: Addison-Wesley.

[509] Tukey, P.A. and Tukey, J.W. (1981). Graphical display of data sets in 3 or more dimensions, In *Interpreting Multivariate Data* (V. Barnett, ed.), pp. 187–275, New York: Wiley.

[510] Turk, M.A. and Pentland, A.P. (1991). Face recognition using eigen-faces, *Proceedings of the IEEE Conference on Computer Vision and Pattern Recognition*, Maui, Hawaii, 586–591.

[511] Ultsch, A. and Siemon, H.P. (1990). Kohonen's self-organizing feature maps for exploratory data analysis, In: *Proceedings of the International Neural Network Conference*, Dortrecht, The Netherlands: Kluwer, pp. 305–308.

[512] Valdivia-Granda, W.A. and Dwan, C. (2006). Microarray data management, an enterprise information approach: implementation and challenges, Chapter 6, Orion Integrated Biosciences Inc., New York. ARXIV e-print, q-bio.GN/0605005.

[513] Uitdenbogerd, A. and Zobel, J. (1999). Melodic matching techniques for large music databases, *ACM Multimedia*, **1**, 57–66.

[514] Valiant, L.G. (1984). A theory of the learnable, In: *Proceedings of the 16th Annual ACM Symposium on Theory of Computing*, New York: ACM Press, pp. 436–445.

[515] van der Heijden, P.G.M., de Falguerolles, A., and de Leeuw, J. (1989). A combined approach to contingency table analysis using correspondence analysis and log-linear analysis, *Applied Statistics*, **38**, 249–292.

[516] van der Leeden, R. (1990). *Reduced-Rank Regression With Structured Residuals*, Leiden: DSWO Press.

[517] Vapnik, V.N. (1998). *Statistical Learning Theory*, New York: Wiley.

[518] Vapnik, V.N. (2000). *The Nature of Statistical Learning Theory, Second Edition*, New York: Springer.

[519] Vapnik, V. and Chervonenkis, A. (1971). On the uniform convergence of relative frequencies of events to their probabilities, *Theory of Probability and its Applications*, **16**, 264–280.

[520] Venables, W.N. and Ripley, B.D. (2002). *Modern Applied Statistics with S, Fourth Edition*, New York: Springer-Verlag.

[521] Vigário, R,, Jousmäki, V., Hämäläinen, M., Hari, R., and Oja, E. (1998). Independent component analysis for identification of artifacts in magnetoencephalographic recordings, In: *Advances in Neural Information Processing Systems*, **10**, pp. 229–235, Cambridge, MA: MIT Press.

[522] Vinod, H. (1969). Integer programming and the theory of grouping, *Journal of the American Statistical Association*, **64**, 506–517.

[523] Wald, A. (1944). On cumulative sums of random variables, *The Annals of Mathematical Statistics*, **15**, 283–296.

[524] Walker, A.M. (1969). On the asymptotic behaviour of posterior distributions, *Journal of the Royal Statistical Society, B*, **31**, 80–88.

[525] Wanatabe, M. and Yamaguchi, K. (eds.) (2004). *The EM Algorithm and Related Statistical Models*, New York: Marcel Dekker.

[526] Watnik, M.R. (1998). Pay for play: are baseball salaries based on performance? *Journal of Statistics Education*, **6**.

[527] Weigend, A.S., Rumelhart, D.E., and Huberman, B.A. (1991). Generalization by weight-elimination with application to forecasting, In: *Advances in Neural Information Processing Systems*, **3** (R.P. Lippmann, J.E. Moody, and D.S. Touretzky, eds.), pp. 875–882, San Mateo, CA: Morgan Kaufmann.

[528] Weir, I.S. (1997). Fully Bayesian SPECT reconstructions, *Journal of the American Statistical Association*, **92**, 49–60.

[529] Weir, I.S. and Green, P.J. (1994). Modelling Data From Single Photon Emission Computed Tomography, In K.V. Mardia (ed.), *Statistics and Images*, **2**, 313–338. Abingdon: Carfax.

[530] Weisberg, S. (1985). *Applied Linear Regression, Second Edition*, New York: Wiley.

[531] Wenk, C. (1999). Applying an edit distance to the matching of tree ring sequences in dendrochronology, Technical Report B 99–01, Institute fur Informatik, Freie Universitat, Berlin.

[532] Werbos, P.J. (1974). *Beyond regression: new tools for prediction and analysis in the behavioral sciences*, Ph.D. dissertation, Harvard University.

[533] Wilkinson, L. (2005). *The Grammar of Graphics*, New York: Springer.

[534] Williams, C.K.I. (2001). On a connection between kernel PCA and metric multidimensional scaling, In *Advances in Neural Information Processing Systems 13* (T.K. Leen, T.G. Dietterich, V. Tresp, eds.), Cambridge, MA: MIT Press.

[535] Wishart, J. (1928). The generalized product moment distribution in samples from a normal population, *Biometrika*, **20A**, 32–52.

[536] Witten, I.H. and Frank, E. (2005). *Data Mining: Practical Machine Learning Tools and Techniques, 2nd Edition*, San Francisco, CA: Morgan Kaufmann.

[537] Wold, H. (1975). Soft modelling by latent variables: The nonlinear iterative partial least squares (NIPALS) approach, *Perspectives in Probability and Statistics: Papers in Honor of M.S. Bartlett* (J. Gani, ed.), London, U.K.: Academic Press, pp. 117–144.

[538] Wold, S. (1978). Cross-validatory estimation of the number of components in factor and principal component models, *Technometrics*, **20**, 397–405.

[539] Wold, S., Martens, H., and Wold, H. (1983). The multivariate calibration method in chemistry solved by the PLS method, In: Proceedings of a Conference on Matrix Pensils, Pitea, Sweden (A. Ruhe and B. Kagstrom, eds.), *Lecture Notes in Mathematics*, Heidelberg: Springer, pp. 286–293.

[540] Wolfe, J.H. (1970). Pattern clustering by multivariate mixture analysis, *Multivariate Behavioral Research*, **5**, 329–350.

[541] Wolf. C., Meisenheimer, M., Kleinheinrich, M., Borch, A., Dye, S., Gray, M., Wisotski, L., Bell, E.F., Rix, H.-W., Cimatti, A., Hasinger, G., and Szokoly, G. (2004). A catalogue of the Chandra Deep Field South with multi-colour classification and photometric redshifts from COMBO-17, *Astronomy & Astrophysics*, `arXiv:astro-ph/0403666v1`.

[542] Wolpert, D.H. (1993). On the use of evidence in neural networks, *Advances in Neural Information Processing Systems*, **5** (S.J. Hanson, J.D. Cowan, and C.L. Giles, eds.), pp. 539–546, San Mateo, CA: Morgan Kauffman.

[543] Woodroofe, M. (1970). On choosing a delta-sequence, *Annals of Mathematical Statistics*, **41**, 1665–1671.

[544] Wu, C.F.J. (1983). On the convergence properties of the EM algorithm, *The Annals of Statistics*, **11**, 95–103.

[545] Wu, C.F.J. (1986). Jacknife, bootstrap and other resampling methods in regression analysis, *The Annals of Statistics*, **14**, 1261–1350.

[546] Yalcin, I. and Amemiya, Y. (2001). Nonlinear factor analysis as a statistical method, *Statistical Science*, **16**, 275–294.

[547] Yianilos, P.N. (1998). Excluded middle vantage point forests for nearest neighbor search, Technical Report, NEC Research Institute, Princeton, NJ.

[548] Young, G. and Householder, A.S. (1938). Discussion of a set of points in terms of their mutual distances, *Psychometrika*, **3**, 19–22.

[549] Zhang, H. (1998). Classification trees for multiple binary responses, *Journal of the American Statistical Association*, **93**, 180–193.

[550] Zhang, H. and Singer, B. (1999). *Recursive Partitioning in the Health Sciences*, New York: Springer.

[551] Zhang, Z. and Zha, H. (2004). Principal manifolds and nonlinear dimension reduction via local tangent space alignment, *SIAM Journal of Scientific Computing*, **26**, 313–338.

[552] Zou, H. and Hastie, T. (2005). Regularization and variable selection via the elastic net, *Journal of the Royal Statistical Society, Series B*, **67**, 301–320.

Index of Examples

Author Index

Subject Index

Springer

the language of science

springer.com

Statistical Design

George Casella

The goal of this book is to describe the principles that drive good design, paying attention to both the theoretical background and the problems arising from real experimental situations. Designs are motivated through actual experiments, ranging from the timeless agricultural randomized complete block, to microarray experiments, which naturally lead to split plot designs and balanced incomplete blocks.

2008. 310 pp. (Springer Texts in Statistics) Hardcover
ISBN 0-387-75964-7

Time Series Analysis with Applications in R

Johnathon D. Cryer and Kung-Sik Chan

Time Series Analysis With Applications in R, Second Edition, presents an accessible approach to understanding time series models and their applications. Although the emphasis is on time domain ARIMA models and their analysis, the new edition devotes two chapters to the frequency domain and three to time series regression models, models for heteroscedasticty, and threshold models. All of the ideas and methods are illustrated with both real and simulated data sets. A unique feature of this edition is its integration with the R computing environment.

2008. 2nd Ed., 494 pp. (Springer Texts in Statistics) Hardcover
ISBN 0-387-75958-6

Pattern Recognition and Machine Learning

Christopher M. Bishop

This is the first textbook on pattern recognition to present the Bayesian viewpoint. The book presents approximate inference algorithms that permit fast approximate answers in situations where exact answers are not feasible. It uses graphical models to describe probability distributions when no other books apply graphical models to machine learning. No previous knowledge of pattern recognition or machine learning concepts is assumed. Familiarity with multivariate calculus and basic linear algebra is required, and some experience in the use of probabilities would be helpful though not essential. The book includes an introduction to basic probability theory.

2006, 740 pp. (Information Science and Statistics) Hardcover
ISBN 978-0-387-31073-2

Easy Ways to Order▶ Call: Toll-Free 1-800-SPRINGER • E-mail: orders-ny@springer.com • Write: Springer, Dept. S8113, PO Box 2485, Secaucus, NJ 07096-2485 • Visit: Your local scientific bookstore or urge your librarian to order.

Printed in the United States of America